Biomedical Science Practice

fundamentals of biomedical science
OXFORD
TRANSFUSION &
TRANSPLANTATION SCIENCE
Robin Knight

fundamentals of biomedical science
OXFORD
CLINICAL
IMMUNOLOGY
Angela Hall, Chris Scott & Matthew Buckland

fundamentals of biomedical science
OXFORD
CYTOPATHOLOGY

fundamentals of biomedical science
OXFORD
CLINICAL BIOCHEMISTRY
Nessar Ahmed
SECOND EDITION

fundamentals of biomedical science
OXFORD
DATA HANDLING
AND ANALYSIS
Andrew Blann

fundamentals of biomedical science
OXFORD
HAEMATOLOGY
Gary Moore, Gavin Knight & Andrew Blann

fundamentals of biomedical science
OXFORD
MEDICAL
MICROBIOLOGY
Michael Ford

fundamentals of biomedical science
OXFORD
CELL STRUCTURE
& FUNCTION
Guy Orchard & Brian Nation

fundamentals of biomedical science
OXFORD
HISTOPATHOLOGY
Guy Orchard & Brian Nation

fundamentals OF
biomedical science

Fundamentals of Biomedical Science

Biomedical Science Practice

Experimental & professional skills

Second edition

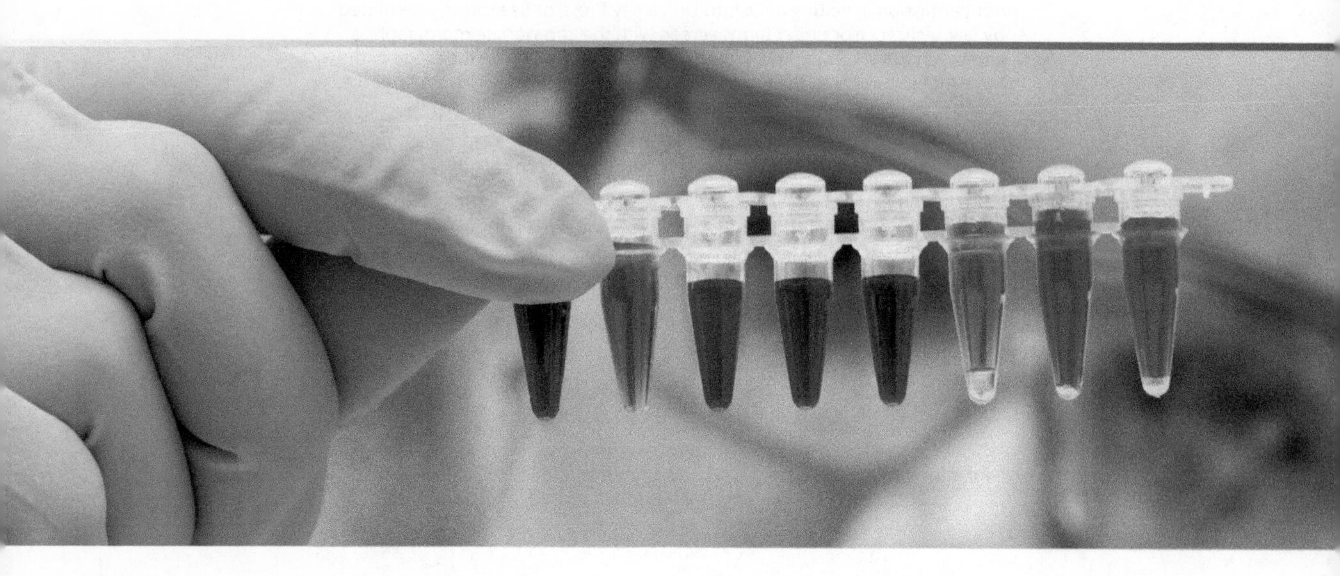

Edited by

Nessar Ahmed
Manchester Metropolitan University

Hedley Glencross
Queen Alexandra Hospital, Portsmouth

Qiuyu Wang
Manchester Metropolitan University

OXFORD UNIVERSITY PRESS

OXFORD
UNIVERSITY PRESS

Great Clarendon Street, Oxford, OX2 6DP,
United Kingdom

Oxford University Press is a department of the University of Oxford.
It furthers the University's objective of excellence in research, scholarship,
and education by publishing worldwide. Oxford is a registered trade mark of
Oxford University Press in the UK and in certain other countries

First edition published 2011
Impression: 7

Published in the United States of America by Oxford University Press
198 Madison Avenue, New York, NY 10016, United States of America

British Library Cataloguing in Publication Data
Data available

Library of Congress Control Number: 2015960998

ISBN 978-0-19-871731-7

Printed in Great Britain by
CPI Group (UK) Ltd, Croydon, CR0 4YY

Contents

For my children

Dr Nessar Ahmed

For my wife and son

Hedley Glencross

For my family

Dr Qiuyu Wang

An Introduction to the Fundamentals of Biomedical Science series

Biomedical scientists form the foundation of modern healthcare, from cancer screening to diagnosing HIV, from blood transfusion for surgery to infection control. Without biomedical scientists, the diagnosis of disease, the evaluation of the effectiveness of treatments, and research into the causes and cures of disease would not be possible. However, the path to becoming a biomedical scientist is a challenging one: trainees must not only assimilate knowledge from a range of disciplines, but must understand and demonstrate how to apply this knowledge in a practical, hands-on environment.

The *Fundamentals of Biomedical Science* series has been written to reflect the challenges of biomedical science education and training today. It blends essential basic science with insights into laboratory practice to show you how an understanding of the biology of disease is coupled to the analytical approaches that lead to diagnosis. Produced in collaboration with the Institute of Biomedical Science, the series provides coverage of the full range of disciplines to which a biomedical scientist might be exposed. Alongside volumes exploring specific biomedical themes and related laboratory diagnosis, the overarching *Biomedical Science Practice* volume provides a grounding in the general professional skills and laboratory techniques with which every biomedical scientist should be familiar.

Learning from the series

The *Fundamentals of Biomedical Science* series draws on a range of learning features to help readers master both biomedical science theory, and biomedical science practice.

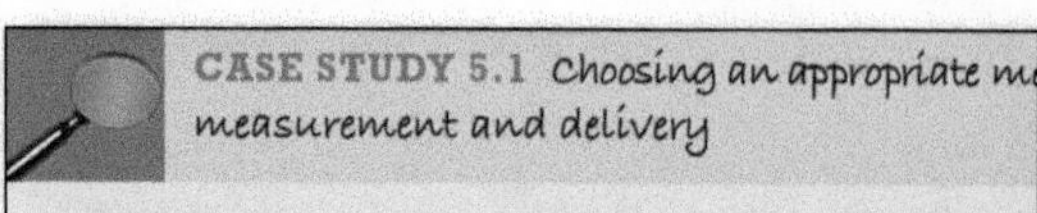
CASE STUDY 5.1 Choosing an appropriate m
measurement and delivery

The purpose of this case study is to help you to think about the importance of choosing an appropriate volume measurement method for a particular situation. Details about alternatives will be given later in this section.
Consider the following scenarios, A and B.

curve is required to estimat
To produce an appropriate
centrations for this standar
volumes (in the range 0-1 c
will be required. The volum

Case studies illustrate how the biomedical science theory and practice presented throughout the series relate to situations and experiences that are likely to be encountered routinely in the biomedical science laboratory. Answers to questions posed by case studies are available at the back of the book.

BOX 6.1 Procedure for the collection of high-

- Samples must be collected by the clinician requesting the test or by an experienced and fully trained phlebotomist.
- A disposable plastic apron and disposable gloves must be worn. It is recommended that suitable eye protection such as goggles or safety spectacles be worn, but a face mask is
- Any spillages must be cl pochlorite solution.
- Used swabs must be disc cally designated for sharp
- Gloves and apron must b incineration (colour-code

Additional information to augment the main text appears in **boxes**.

Method boxes walk through the key protocols that the reader is likely to encounter in the laboratory on a regular basis.

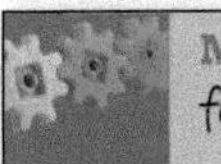

METHOD Setting up of a microscope for optimal illumination

- Whatever type of microscope is being used, the basic principles of setting it up for critical or Köhler illumination are essentially the same. First, adjust the condenser to as high a position as possible and then focus on a sample using a low-power objective lens (normally 10× or 4×). Second, close the field diaphragm in the condenser and sharply focus the diaphragm onto the image of the sample using the condenser focus controls

Key points reinforce the key concepts that the reader should master from having read the material presented, while **Summary** points act as end-of-chapter checklists for readers to verify that they have remembered correctly the principal themes and ideas presented within each chapter.

suboptimal way.

Key points

Samples for histology or cytology are often unique and it is not possible to take a repeat sample if something goes wrong.

Each title in the series features a **glossary**, which provides explanations of key terms with which the reader may not be familiar.

Glossary

Absorption spectrum graph illustrating the fractions of incident electromagnetic radiation absorbed by a material over a range of different wavelengths.

Acceleration is the rate of speed change.

Acceleration voltage the potential difference in volts neces-

Anode the negatively charged electrode.

Anode plate disc in the electron gun of an electro scope that is maintained at earth potential and c cloud of electrons produced by the filament of the accelerated into the microscope column.

Self-check questions throughout each chapter provide the reader with a ready means of checking that they have understood the material they have just encountered; answers to self-check questions are available at the back of the book.

blue bands against a light background. The inclusion of positively charged resins speeds up the destaining time considerably.

SELF-CHECK 12.5

What is the structural difference between Coommassie R-250 and G-250?

Discussion questions are provided at the end of each chapter to encourage the reader to analyse and reflect on the material they have just read. Answers to the end-of-chapter questions can be found at the back of the book.

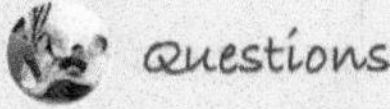

Questions

1. You normally separate fragments of DNA by electrophoresis using a potential difference of 100 mV. The distance between the electrodes in your apparatus is 150 mm. However, this apparatus is unavailable and the larger system available does not have a protocol or manufacturer's instructions. The distance between its electrodes is 200 mm. What potential difference is necessary to produce the electric field you would normally use?

Cross references help the reader to see biomedical science as a unified discipline, making the connections between topics presented within each volume, and across all volumes in the series.

fying the site where the modification has occurred. are molecules of small M_r, present in a biological s, signalling molecules, and the chemical intermedi- it 3000 metabolites are essential for normal human hese metabolites could be helpful in determining if ug therapy, or responding to another form of stress. n optimizing therapy and management of a disease. analysing metabolomes. However, in this role MS aration method such as GLC or high performance

Cross reference

You can read about the various forms of chromatography in Chapter 11.

Online learning materials

Each title in the *Fundamentals of Biomedical Science* series is supported by an Online Resource Centre, which features additional materials for students, trainees, and lecturers.

www.oxfordtextbooks.co.uk/orc/fbs

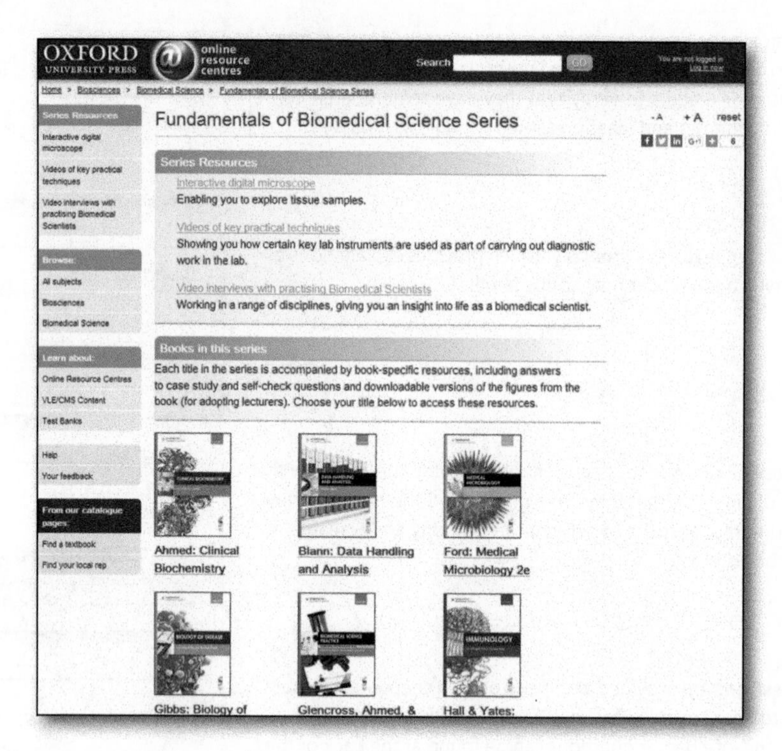

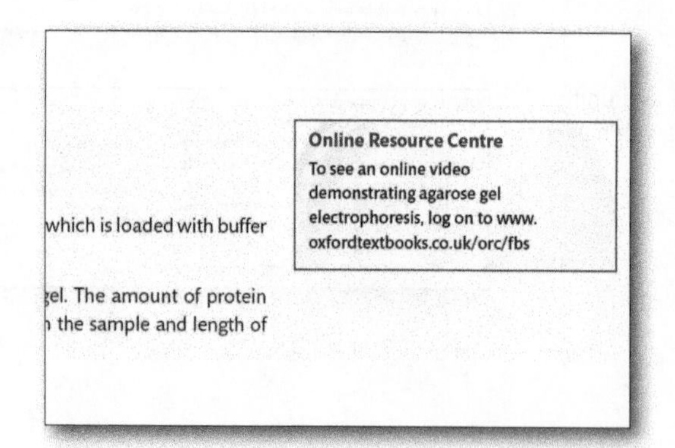
which is loaded with buffer

gel. The amount of protein
the sample and length of

Online Resource Centre
To see an online video demonstrating agarose gel electrophoresis, log on to www.oxfordtextbooks.co.uk/orc/fbs

Guides to key experimental skills and methods

Video walk-throughs of key experimental skills are provided to help you master the essential skills that are the foundation of biomedical science practice. Margin notes in the text direct you to key videos relating to that particular section.

Biomedical science in practice

Interviews with practising biomedical scientists working in a range of disciplines give a valuable insight into the reality of work in a biomedical science laboratory.

Virtual microscope

Visit the library of microscopic images and investigate them with the powerful online microscope to help gain a deeper appreciation of cell and tissue morphology.

Lecturer support materials

The Online Resource Centre for each title in the series also features figures from the book in electronic format, for registered adopters to download for use in lecture presentations, and other educational resources.

To register as an adopter visit **www.oxfordtextbooks.co.uk/orc/ahmed_practice2e/** and follow the on-screen instructions.

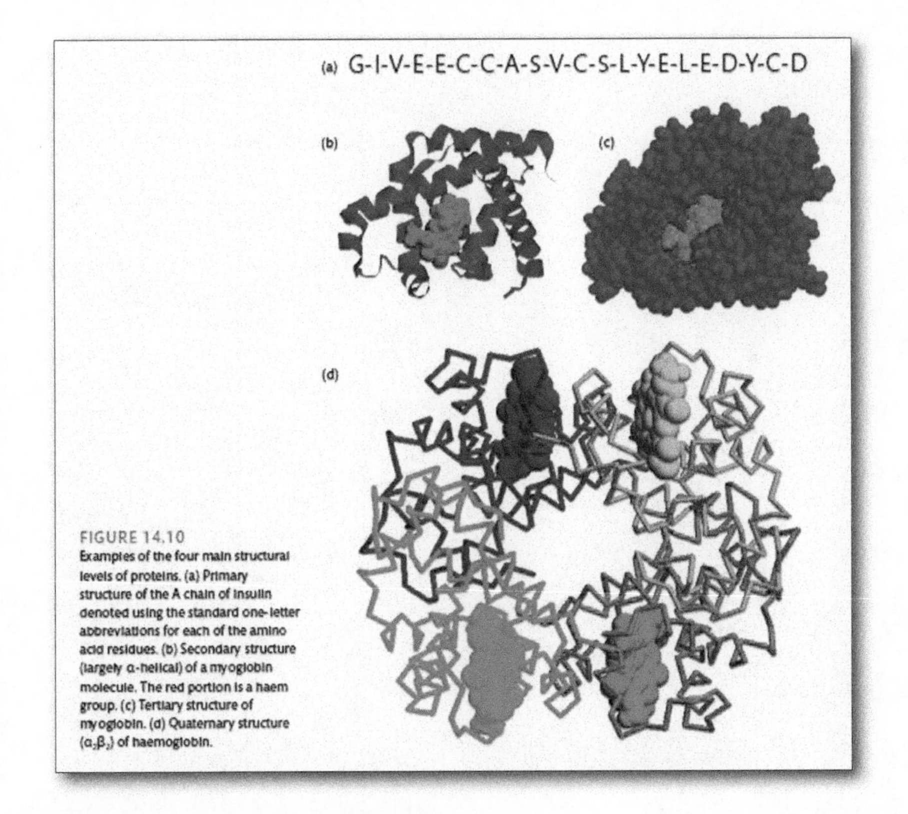

FIGURE 14.10 Examples of the four main structural levels of proteins. (a) Primary structure of the A chain of insulin denoted using the standard one-letter abbreviations for each of the amino acid residues. (b) Secondary structure (largely α-helical) of a myoglobin molecule. The red portion is a haem group. (c) Tertiary structure of myoglobin. (d) Quaternary structure ($\alpha_2\beta_2$) of haemoglobin.

Any comments?

We welcome comments and feedback about any aspect of this series.
visit **www.oxfordtextbooks.co.uk/orc/feedback** and share your views.

Preface

This second edition of *Biomedical Science Practice: Experimental and Professional Skills* covers the practical and professional aspects of biomedical science. It is written to meet the needs of undergraduate students studying BSc (Hons) degree programmes in Biomedical and Healthcare Sciences. This textbook should also be suitable for graduates entering the hospital pathology service as trainee biomedical or healthcare scientists and, indeed, those studying relevant modules on MSc programmes in Biomedical or Healthcare Sciences.

This second edition has the same features as our highly successful first edition but any obsolete material has been removed, including the chapter on radioactivity and radiation. The chapter on statistics and data handling has also been removed as it is now part of a separate new textbook on *Data Handling and Analysis*, as part of the *Fundamentals of Biomedical Science* series. All chapters have been updated and the number of case studies increased.

We are grateful to Dr Chris Smith for his valuable advice and input during the preparation of this and the previous edition of this textbook. We are also grateful to Mick Hoult for his assistance with the illustrations. We would like to thank Jonathan Crowe (Editor in Chief) and Alice Roberts (Publishing Editor), both from Oxford University Press, for their constant support and encouragement. We take full responsibility for any errors or omissions. Please do write and tell us if you detect any, and let us know how we can improve this textbook for future editions.

Dr Nessar Ahmed
Hedley Glencross
Dr Qiuyu Wang

Contributors

Nessar Ahmed
School of Healthcare Science, Manchester Metropolitan University

Hedley Glencross
Cytopathology Department, Queen Alexandra Hospital, Portsmouth

Ian Graham
Formerly of School of Biology, Chemistry and Health Science, Manchester Metropolitan University

Tim James
Department of Clinical Biochemistry, Oxford Radcliffe Hospitals NHS Trust

Georgina Lavender
Laboratory Training Officer and freelance tutor

Garry McDowell
School of Healthcare Science, Manchester Metropolitan University

Helen Montgomery
Kratos Analytical, Manchester

Elaine Moore
Pathology Department, St Peter's Hospital, Surrey

Joyce Overfield
School of Healthcare Science, Manchester Metropolitan University

Lynda Petley
Partnership Pathology Services, Frimley Park Hospital Foundation Trust

Peter Robinson
College of Science and Technology, University of Central Lancashire, Preston

Tony Sims
Formerly of Histopathology Department, University Hospitals of Leicester NHS Trust

Chris Smith
Formerly of School of Biology, Chemistry and Health Science, Manchester Metropolitan University

Jan Still
Formerly of Department of Chemical Pathology, Watford General Hospital

Alison Taylor
Histology Department, Basingstoke and North Hampshire Hospital

Qiuyu Wang
School of Healthcare Science, Manchester Metropolitan University

Christine Yates
Formerly of Immunology Department, Hull and East Yorkshire Hospitals NHS Trust

Online materials developed by

Sheelagh Heugh, Principal Lecturer in Biomedical Science, Faculty of Human Sciences, London Metropolitan University

Dr Ken Hudson, Lecturer in Biomedical Science, Faculty of Human Sciences, London Metropolitan University

William Armour, Lecturer in Biomedical Science, Faculty of Human Sciences, London Metropolitan University

Abbreviations

% v/v	volume-volume percentage
% w/v	mass-volume or weight-volume
% w/w	mass percentage
μ	electrophoretic mobility
A&E	Accident and Emergency department
a	absorptivity
A	absorbance
ABTS	azino-*bis*-ethylbenzthiazoline sulfonic acid
ACDP	Advisory Committee on Dangerous Pathogens
ACOP	Approved Codes of Practice
ADR	accord européen relative au transport international des marchandises dangereuses parroute
AGE	agarose gel electrophoresis
AHG	antihuman globulin
AIDS	acquired immune deficiency syndrome
ALP	alkaline phosphatase
AMP	adenosine 5´-monophosphate
Amp^R	ampicillin resistance gene
ANCA	anti-neutrophil cytoplasmic antibodies
ATP	adenosine 5´-triphosphate
BAC	bacterial artificial chromosome
BJBS	*British Journal of Biomedical Science*
bp	base pairs
BSA	bovine serum albumin
BSE	bovine spongiform encephalopathy
BSHI	British Society of Histocompatibility and Immunogenetics
BSI	British Standards Institute
BSQR	Blood Safety and Quality Regulations
c	concentration
c	velocity of light
ccc	supercoiled covalently closed circular
CD	cluster of differentiation
cDNA	complementary DNA
CE	capillary electrophoresis
CF	cystic fibrosis
CHART	continuous hyperfractionated radiotherapy
CI	chemical ionization
CID	collisional dissociation cell
CJD	Creutzfeldt-Jakob disease
COSHH	Control of Substances Hazardous to Health
CPA	Clinical Pathology Accreditation
CPD	continuing professional development
CPSM	Council for Professions Supplementary to Medicine
CSF	cerebrospinal fluid
CT	computed tomography
CTP	cytidine 5´-monophosphate
d	distance
dAMP	deoxyadenosine 5´-monophosphate
dATP	deoxythymidine 5´-monophosphate
dCTP	deoxycytidine 5´-monophosphate
ddNTP	dideoxynucleoside triphosphate
DGA	DiGeorge anomaly
dGMP	deoxyguanosine 5´-monophosphate
DGSA	Dangerous Goods Safety Advisor
DIG	digoxygenin
DNA	deoxyribonucleic acid
DPA	Data Protection Act
dRib	deoxyribose
DSE	display screen equipment
DSEAR	Dangerous Substances and Explosive Atmosphere Regulations
DSS	3-trimethylsilyl 1-propane sulfonate
dTDP	deoxythymidine 5´-diphosphate
dTMP	deoxythymidine 5´-monophosphate
E	electric field or potential gradient; energy of electromagnetic radiation
E	electronic component of electromagnetic radiation
ECD	electron capture detector
EDTA	ethylenediamintetraacetic acid
EI	electron ionization
EIm	electron impact
ELISA	enzyme-linked immunosorbent assay
ELISPOT	enzyme-linked immunospot
EQA	external quality assessment
EQF	European Qualification Framework

ESI	electrospray ionization
EU	European Union
eV	electron volt (1.6×10^{-19} J)
F	Faraday constant
f	frictional coefficient
FAB	fast atom bombardment
FACS	fluorescence-activated cell sorter
FID	flame ionization detector
FISH	fluorescent *in situ* hybridization
FOI	Freedom of Information
FTP	Fitness to Practise
g	earth's gravitational field
G	centrifugal field or acceleration
GHS	globally harmonized system
GLC	gas–liquid chromatography
GMP	guanosine 5′-monophosphate
h	Planck's constant
H	height of a theoretical plate in a chromatography column; magnetic component of electromagnetic radiation
H_0	applied magnetic field
HASAWA	Health and Safety at Work Act
Hb	haemoglobin
Hb^A	adult haemoglobin
Hb^S	haemoglobin S
HCPC	Health and Care Professions Council
HETP	height equivalent to a theoretical plate
HIS	Hospital Information System
HIV	human immunodeficiency virus
HPLC	high-performance liquid chromatography
HRP	horseradish peroxidase
HSA	human serum albumin
HSC	Health and Safety Commission
HSD	Higher Specialist Diploma
HSE	Health and Safety Executive
HSG	Health and safety guidance publications
HTA	Human Tissue Act (also used for Human Tissue Authority)
I	intensity of transmitted light
I_0	intensity of incident light
IBMS	Institute of Biomedical Science
IEF	isoelectric focusing
Ig	immunoglobulin
IIF	indirect immunofluorescence
IOSH	Institute of Occupational Safety and Health
IPTG	isopropylthio-β-galactoside
IQC	internal quality control
IR	infrared
IRT	immunoreactive trypsin
IS	internal standard
ISE	ion-selective electrode
ISFET	ion-selective field effect transistor
ISO	International Standards Organization
IT	information technology
IUPAC	International Union of Pure and Applied Chemistry
IVD	*in vitro* diagnostic device
JBL	journal-based learning
Kbp	kilobase pairs
K_d	distribution coefficient
K_{ow}	octanol–water partition coefficient
KSF	Knowledge and Skills Framework
L or l	path length
L	length
LAS	lithium (Li), aluminium (Al), silicon (Si) glass used in ion-selective electrodes
LDH	lactate dehydrogenase
LEV	local exhaust ventilation
LIMS	laboratory information management system
LIS	laboratory information system
LOLER	Lifting Operations and Lifting Equipment Regulations
m	mass
mAB	monoclonal antibody
MALDI	matrix-assisted laser desorption ionization
MALDI-MS	matrix-assisted laser desorption ionization-mass spectrometry
MBL	mannose-binding lectin
MHO	Manual Handling Operations
MHRA	Medicines and Healthcare products Regulatory Agency
MM	multiple myeloma
mol	mole - the SI unit denoting the amount of a substance
MOSFET	metal oxide semiconductor field effect transistors
M_r	relative molecular mass
MRI	magnetic resonance imaging
mRNA	messenger RNA

MRSA	methicillin-resistant *Staphylococcus aureus*
MS	mass spectrometry
MSDS	Material Safety Data Sheet
MSU	midstream urine
N	Avogadro's number
N	the number of theoretical plates in a chromatography column
NAS	sodium (Na), aluminium (Al), silicon (Si) glass used in ion-selective electrodes
NEQAS	National External Quality Assessment Service
NHS	National Health Service
NHSLA	NHS Litigation Authority
NMR	nuclear magnetic resonance
nt	number of nucleotides
oc	open circular or nicked conformation
OH	occupational health
PAGE	polyacrylamide gel electrophoresis
PAS	periodic acid–Schiff
PAT	portable appliance testing
PCR	polymerase chain reaction
PDP	personal development plan
PFGE	pulsed field gel electrophoresis
Phage	bacteriophage
pI	isoelectric point
PKU	phenylketonuria
PMF	peptide mass fingerprinting
POCT	point-of-care testing
PPE	personal protective equipment
PUWER	Provision and Use of Work Equipment Regulations
PVA	polyvinyl alcohol
PVDF	polyvinylidene difluoride
*P*X	partial pressure (concentration) of the gas X
Q	quadrupole
QA	quality assurance
QAA	Quality Assurance Agency for Higher Education
QC	quality control
QIT	quadrupole ion-trap
QMS	quality management system
qPCR	quantitative real-time polymerase chain reaction (PCR); also called real-time PCR
R	electrical resistance
R	universal gas constant
RCA	root cause analysis
RCF	relative centrifugal field
RE	restriction endonuclease
Real-time PCR	real-time polymerase chain reaction; also called quantitative real-time or qPCR
rev min^{-1}	revolutions per minute
RF	radiofrequency (also used for rheumatoid factor)
R_f	relative to front
RFLP	restriction fragment length polymorphisms
Rh	Rhesus blood group
Rib	ribose
RIDDOR	Reporting of Injuries, Diseases, and Dangerous Occurrences Regulations
RNA	ribonucleic acid
rRNA	ribosomal RNA
RSI	repetitive strain injury
RT-PCR	reverse transcriptase-PCR
s	sedimentation coefficient
S	Nernst factor
SABRE	serious adverse blood reactions and events
SCOT	support coated open tubular
SDS	sodium dodecyl sulfate
SDS-PAGE	sodium dodecyl sulfate polyacrylamide gel electrophoresis
SETS	Standards of Education and Training
SHOT	serious hazard of transfusion
SI	Le Système International d'Unités
SLE	systemic lupus erythematosus
SOP	standard operating procedures
SRID	single radial immunodiffusion
t	retention time
T	transmitted light
t_o	dead time
TAG	triacylglycerol
TAT	turnaround time
Tb	turbidity
TB	tubercle bacilli/tuberculosis
TCA	trichloroacetic acid
TEMED	N,N,N′,N′-tetramethylethylenediamine
TLC	thin-layer chromatography
TMB	tetramethylbenzidine
TMS	tetramethylsilane ($Si(CH_3)_4$)
TOF	time of flight
TREM	transport emergence card
Tris	2-amino-2-hydroxymethylpropane-1,3-diol

tRNA	transfer RNA
UKAS	United Kingdom Accreditation Service
UMP	uridine 5′-monophosphate
UV	ultraviolet
ν	frequency of electromagnetic radiation
v	(sedimentation) velocity
v′	wave number
$\bar{v}$	partial specific volume
V_0	void volume
V_e	elution volume
Vis	visible
VLCFA	very long-chain fatty acids
V_t	total volume
W	width at the base of a peak
WEL	workplace exposure limit
W_h	width at half the height of a peak
WRULD	work-related upper limb disorders
X-gal	artificial substrate, 5-bromo-4-chloro-3-indolyl-β-galactoside, of β-galactosidase
z	charge
α	selectivity
ε	molar extinction coefficient
λ	wavelength of electromagnetic radiation
ρ	density of the solvent

Units

Ω	Ohm
A	Ampere – the SI unit of electrical current
Å	angstrom (10^{-10} m)
Da	dalton
Hz	hertz
nm	nanometre
ppm	parts per million
s	second
S	Svedberg unit
T	temperature in kelvin
V	volt

1

Biomedical science and biomedical scientists

Hedley Glencross

Learning objectives

After studying this chapter, you should be able to:

- Define biomedical science as a scientific discipline
- Outline the academic and scientific contents of biomedical science degree programmes
- Discuss the training required to become a biomedical scientist
- Discuss how the education and training of a biomedical scientist are interrelated

Introduction

Biomedical science is often described as the application of the basic sciences, but especially the biological sciences, to the study of medicine, in particular, the causes, consequences, diagnosis, and treatment of human diseases. Biomedical scientists are registered practitioners of the subject, who normally work in departments of clinical pathology. Thus, they are vital healthcare, scientifically qualified professionals who, for example, perform diagnostic tests on clinical samples of blood and urine. Given their role(s), biomedical scientists are typically active in more laboratory-based work and tend to have a relatively limited contact with patients compared with other healthcare professionals.

Approximately 70% of medical decisions or interventions, from diagnoses to monitoring medical treatments to screening populations at risk of a particular clinical condition, are based on the activities of biomedical scientists or require the knowledge and skills of biomedical science. This may be as simple as a preoperative blood test or something more complex, such as using an image-guided needle biopsy to remove tissue samples to help diagnose breast or prostate cancers. Unfortunately, while everyone knows their doctor and, indeed, doctors and nurses have a high degree of recognition in society, in many ways the same is not true of biomedical scientists. Perhaps considering the following

FIGURE 1.1
A typical, modern clinical laboratory.

scenario may reinforce the point. A 52-year-old, overweight male smoker presents to his local hospital's Emergency Department with chest pains, shortage of breath, and complaining of a generalized weakness and tiredness. It is apparent that he may be experiencing any one of a number of problems. These include, for example, a panic attack, a chest infection, an allergic reaction, or he may be at the early stage of a heart attack. How can an appropriate clinical decision be reached? Biomedical scientists earn their living by helping in these decision-making processes!

Routine tests, such as measuring the patient's height, weight, temperature, and blood pressure, as well as taking X-rays and performing detailed laboratory investigations on samples from a patient may all help provide answers. The laboratory investigations may include blood tests to look for increased activities of enzymes or substances that act as markers for a heart attack, microbiology tests to identify infection-causing microorganisms, while immunological tests may help eliminate the possibility of an allergic reaction. Hence, clinical laboratories are often busy places (Figure 1.1).

Online Resource Centre
To see video interviews with practising Biomedical Scientists, log on to www.oxfordtextbooks.co.uk/orc/fbs

The education and training needed to become a biomedical scientist are detailed and lengthy. However, if this is your career choice, then the knowledge you learn and the skills you acquire have the potential to impact greatly upon many individuals and their lives. They will also give you an interesting and worthwhile career.

1.1 What is biomedical science?

Biomedical science is more than a broad collection of scientific topics. Instead, it refers to a defined area of study related to the clinical investigation of disease within a degree programme. Degree programmes of these types are designed to allow the student to meet the educational standards and, in some cases, the professional skills required of a registered **biomedical scientist** by the **Health and Care Professions Council** (HCPC). Only those individuals who have met the appropriate standards of proficiency of the HCPC will be legally allowed to call themselves a biomedical scientist in the UK. The HCPC is a statutory regulatory body for a variety of therapeutic and scientific health professions.

Cross reference
The HCPC will be discussed further in Chapter 2, 'Fitness to practise'.

The definition of biomedical science that this book and the accompanying volumes in the series will be using is that of the **Quality Assurance Agency for Higher Education** (QAA) and is described in Section 1.2 below.

1.2 Biomedical science degree programmes

Biomedical science degrees are concerned with the study of a variety of subjects, including:

- biology
- chemistry
- pathobiology of human disease
- analytical methods.

Although biology and chemistry are key elements of the academic programme, **pathobiology** is a fundamental component of a biomedical science degree. Pathobiology involves studying anatomy, physiology, biochemistry, genetics, immunology, microbiology, cell biology, and molecular biology. It is also concerned with knowledge of disease processes, which are investigated in the clinical laboratory specialities of histology, cytology, biochemistry, immunology, haematology, transfusion science, and microbiology. Any student successfully completing such a course of study will have a graduate-level proficiency in the science of the causes, consequences, diagnosis, treatment, and management of human diseases.

Biomedical science requires multidisciplinary approaches to the study of human diseases and, as such, you must know why and how the development of diseases affects the normal structures and functions of the human body. You will also need to know the current methods used for investigating, diagnosing, and monitoring diseases, as well as research methodologies, which will include an appreciation of how new methods to combat diseases are developed, evaluated, and integrated into clinical practice. Biomedical science also requires knowledge of several scientific and clinical subject areas, which have been summarized by the QAA. You can see a list of these subjects in Table 1.1.

TABLE 1.1 **Quality Assurance Agency for Higher Education subject areas in biomedical science**

Scientific subject areas	Clinical subject areas
Human anatomy and physiology	Histology
Biochemistry	Clinical biochemistry
Cell biology	Cytology
Immunology	Clinical immunology
Molecular biology	Haematology
Genetics	Transfusion science
Microbiology	Medical microbiology

1.2.1 Practical and transferrable skills

Biomedical science is very much a *hands-on* science and acquiring some of the discipline or subject-specific skills associated with its practice is an essential part of a biomedical science degree programme. These skills include:

- learning how to handle patient samples in a safe and responsible manner when preparing these samples for clinical investigation
- using aseptic procedures
- using laboratory instruments safely and appropriately to ensure accuracy, precision, and reproducibility
- interpreting patient-related data in an appropriate manner.

Biomedical science, as described by the QAA, is an *honours* degree programme of study and so this will incorporate a **project** or **dissertation**, which themselves are definitive components of honours

TABLE 1.2 Summary of organizations and work areas for biomedical science graduates

Organizations
NHS Blood and Transplantation
Public Health England (and equivalents in Scotland, Wales and Northern Ireland)
Independent pathology laboratories
Veterinary laboratories
Agricultural laboratories
The World Health Organization
Universities
Areas of practice
Clinical genetics
Forensic science
Research
Pharmaceutical diagnostics
Medical devices
Clinical trials
Commerce
Education
Food and food safety
Biotechnology

degree programmes. They are individually undertaken by a student although, of course, they are under the guidance of and studied with the support of professional staff. An honours degree will help in the development of transferrable skills, such as literature searching and appraisal of literature, experimental design, scientific communication, research methodology related to ethics, governance, audit, and statistical analysis. As part of the assessment of your dissertation or project you may be required to present this work to a group of peers (fellow students).

You have now seen the extensive and complex nature of biomedical science degrees in terms of their scientific and professional contents, but these degree programmes are not only just about science. They will enable you to develop a wide range of generic skills related to team-working, communication, negotiation, numeracy, data analysis, and information technology. Developing these generic skills is as essential as the gaining of underlying scientific knowledge upon which biomedical science is based. Although the knowledge and skills acquired from biomedical science degrees are essential for work in National Health Service (NHS) laboratories of the UK, they also provide a sound academic and professional base and the potential to work in any of the organizations and areas you can see in Table 1.2.

SELF-CHECK 1.1

What generic skills may be acquired during the study for a biomedical science degree?

1.3 What is a biomedical scientist?

This section title may seem a slightly odd question to ask as you may naturally infer that graduates with a biomedical science degree course are all biomedical scientists. However, as we have already explained in Section 1.1, the title *biomedical scientist* is legally protected and only those persons who are registered as such with the HCPC are permitted to use this title within the UK. Thus, the title describes

FIGURE 1.2
Institute of Biomedical Science, 12 Coldbath Square, London.

a person who has met the appropriate minimum or threshold standards of proficiency required by the HCPC (and has demonstrated that their character and health are satisfactory, too). These standards are met by the successful completion of an HCPC-approved degree programme.

Although all of the standards are important, standard 13 requires a registrant biomedical scientist to understand the key concepts of the knowledge base relevant to their profession. To meet this standard you will need to demonstrate an understanding of the knowledge and practice of all of the major areas of biomedical science. The underlying scientific knowledge was outlined in Section 1.2 and the associated professional skills will be discussed in Section 1.4.

Cross reference

The HCPC, their Standards of Proficiency, and the degree approval process will be discussed further in Chapter 2, 'Fitness to practise'.

Amongst the HCPC-approved programmes are a number of honours degrees that are also accredited by the **Institute of Biomedical Science** (IBMS), which is the professional body that biomedical scientists may join; its offices are shown in Figure 1.2.

The IBMS offers professional qualifications and sets standards of professional behaviour for its members. As part of its work, the IBMS has also developed a training programme using a portfolio of professional evidence (the *registration portfolio*), which when successfully completed will allow an individual to apply to become registered with HCPC. A **Certificate of Competence** is awarded upon successful completion and verification of the registration portfolio. The IBMS also provides this information direct to HCPC following receipt of a laboratory feedback form after a successful verification visit. The concept of a portfolio should not be new to most readers. However, a portfolio worthy of the award needs to focus on the evidence necessary to demonstrate to an external verifier that you have acquired the knowledge and skills to meet the threshold standards of proficiency necessary to join the HCPC.

Cross reference

The IBMS, their professional guidelines, and the degree accreditation process will be discussed further in Chapter 2, 'Fitness to practise'.

Another newly formed organisation, *The Academy for Healthcare Science*, also undertakes degree accreditation, and is, like the IBMS, an accredited education provider for the HCPC. The academy also has developed a specific healthcare science honours degree programme, leading to the award of *Healthcare Science Practitioner*. Practitioners are also able, if graduating from an HCPC-approved degree, to apply to register as biomedical scientiststs. While the degree programme itself and the professional training are similar to that of biomedical science, the degree and its title are specifically tailored to professional careers within the NHS. As such, the programmes of study and training are outside the scope of this textbook.

1.4 Professional skills

The professional training of a biomedical scientist can only be delivered and the relevant skills acquired by physically working in a *clinical laboratory* to gain the necessary hands-on experience.

Key points

Throughout this chapter and the rest of the accompanying volumes in the series, *clinical laboratory* will be used to describe a *collection* of departments. *Department* refers to an individual discipline-specific area of work.

Professional skills may be acquired in a variety of ways. One way may be through the completion of a full-time HCPC-approved degree. However, these skills may also be gained during employment, while simultaneously completing an accredited degree programme and the portfolio of professional evidence described in Section 1.3. Professional skills may also be acquired after graduation, during employment in a relevant department by completing the portfolio of professional evidence while also undertaking some discipline-specific training. This is the route followed if you have gained an unaccredited degree whose content needs to be supplemented with further study. Case study 1.1 illustrates how this was achieved by a now-successful biomedical scientist.

Case study 1.2 shows a typical reflective essay written at the end of a laboratory placement. A Certificate of Competence is then awarded upon graduation.

SELF-CHECK 1.2

How are professional skills acquired?

1.5 Initial employment as a biomedical scientist

Following registration with the HCPC, it is likely that newly qualified biomedical scientists will be employed in a discipline-specific department, such as histology, biochemistry, cytology, haematology, immunology, microbiology, blood transfusion, or virology. However, it is becoming apparent that the divisions between these traditional disciplines are becoming blurred and new departments are beginning to emerge through advances in science and technology or through changes in clinical practice. For example, some laboratories combine the routine diagnostic investigations performed in haematology or biochemistry into a department often called *blood science* or *blood sciences*. Further, the use of **point-of-care testing** (POCT) by bench-top or hand-held devices in settings outside of the laboratory, including a patient's own home, is increasing (Box 1.1). Often POCT devices allow simple microbiological, biochemical, or haematological tests to be performed in combination.

It is only after registration and employment that you will acquire the full range of more detailed discipline-specific skills. This may involve completing an IBMS Specialist Portfolio, which is explained in Chapter 2, 'Fitness to practise'.

BOX 1.1 Point-of-care testing

POCT involves using hand-held or bench-top analysers to test patient samples in settings other than a clinical laboratory. These relatively simple analysers often use capillary blood sampling and may be encountered in hospital areas such as intensive care or within a primary care setting, for example, an anticoagulation clinic. Increasingly, POCT devices and tests are being offered directly to patients for part of self-testing or as part of the self-management of chronic diseases such as diabetes.

1.5.1 Training

Properly structured and delivered training programmes are essential to the development of a biomedical scientist. Whether the training is part of a work placement or as a new employee in a laboratory, all newly employed biomedical scientists will be assigned a training lead who will facilitate their training. All of the training will be guided by the contents of an individual training programme and the HCPC requirements for minimum standards of proficiency, the front cover of which is illustrated in Figure 1.3.

Cross reference

POCT is explored in Chapter 16.

Cross reference

You can read more about the HCPC Standards of Proficiency in Chapter 2.

FIGURE 1.3
Front cover of the Health and Care Professions Council Standards of Proficiency for Biomedical Scientists. Copyright © Health and Care Professions Council.

Training programmes are, by their nature, individual to the person concerned because of a variety of issues, such as prior knowledge and expertise, and the individual's own learning style. However, any one programme is likely to contain the basic underlying knowledge needed, together with an indication of the level of skill, a definition of competency, and the assessment processes. As training is a continuous process, any training programme will need to be negotiated and agreed by a trainee and the training lead. Training is also an interactive process that requires equal commitment from the trainee and the employer and must take place without affecting the service provided by the laboratory. So its delivery must be in a manner that minimizes its effects on the normal working of a department, but maximizes the benefits to the trainee.

During training it may be necessary to work alongside support staff or undertake some of the basic work of the department, such as loading machines, making solutions, and filing. Hence, professional training may have to fit around the needs of an individual department. Such tasks should not affect the delivery of training. In fact, these types of tasks are likely to benefit a trainee, as they will increase his or her understanding of all the functions of a laboratory and the associated benefits of collaborative work. All individuals are members of a team, and need to help colleagues and support all the different areas and activities of a department. This will also provide time for more experienced colleagues to deliver professional training to others. After successful completion of such a training programme, the trainee should have become a competent, confident, and independent practitioner known as a *registered biomedical scientist.*

1.6 What do biomedical scientists do?

Until now we have considered the knowledge and skills that are necessary to be a biomedical scientist, but what will you do as a professional biomedical scientist in your daily working life? Much of the work undertaken by a laboratory is determined by a combination of the population served by the hospital, the associated range of illnesses that this population may develop, and the clinical specialities that are practised within the hospital. For example, in a *teaching* hospital, you would probably undertake more specialized work or research-related activities, along with the routine work associated with investigating, diagnosing, and treating disease. In a *general* hospital, the work may be less specialized, but there may also be opportunities for involvement in other activities. Wherever you work, it is likely that you will be busy, and it will be expected that you and the other biomedical scientists perform your duties and fulfil your roles in a helpful and professional manner, remembering that all samples originated from a *patient* and, in the end, all test results affect him or her.

1.6.1 Laboratory tests

As mentioned in the Introduction, it is estimated that over 70% of clinical or medical decisions or interventions are based wholly or partly on the results of laboratory tests and by inference the work of biomedical scientists. The significance of these tests cannot be underestimated.

Laboratory tests can be divided into three general categories, diagnostic, screening, and monitoring, which you can see outlined in Table 1.3.

TABLE 1.3 Types and role of laboratory tests

Type of test	Role of test
Diagnostic	Investigating an individual showing signs or symptoms of an illness
Screening	Identification of 'at risk' individuals within a defined population who do not show any symptoms, but may have an underlying undiagnosed clinical condition
Monitoring	Checking the health status of an individual before undertaking a medical procedure Assessing an individual following diagnosis and treatment Monitoring the progress of long-term or chronic conditions

Diagnostic tests are used to investigate a patient who has signs or symptoms of illness. For example, someone who experiences pain on urinating may have an infection of the urinary tract. Samples of urine from the patient must be investigated for evidence of disease-causing microorganisms. Screening tests are used to identify an 'at risk' individual from a defined population, who does not demonstrate symptoms, but may have an undiagnosed illness or condition. Thus, for example, in the UK all females of reproductive age (and up to the age of 65) are invited to have a sample taken from their cervix to investigate for any potential cellular or viral changes that may indicate the development of cancer. Monitoring tests are used to check the status of an individual before a medical procedure, following diagnosis and treatment, or to monitor his or her progress during a long-term condition. For example, someone who goes into hospital for an operation will be subjected to blood and urine tests designed to give information about their overall health and physiological functions. Increasingly, microbiological investigations form part of this test profile to ascertain if an individual has one of the so-called 'super bugs', such as methicillin-resistant *Staphylococcus aureus* (MRSA). It may be that the blood and urine tests show the patient's liver, kidneys, and heart are healthy, and are not a bar to the planned operation. However, the microbiology tests may show he or she is suffering an infection that needs to be treated before an operation can take place. This type of situation shows why it is essential for biomedical scientists to understand the work of all the main departments and why training programmes (discussed above) emphasize these aspects of biomedical science.

We will now, briefly, consider the work carried out in each of the main discipline-specific departments.

1.6.2 Histopathology

Histology involves the microscopic study of tissues to identify disease that may be present. Tissues may be removed from living individuals following an operation or from deceased individuals at post-mortem. The tissue samples range from small biopsies to whole organs. Slides are prepared from these samples, which are stained and examined using a microscope. Diseases such as inflammation, infections, and benign and malignant tumours can all be diagnosed using histology, as you can see in Figure 1.4. Biomedical scientists are involved at all stages of the preparation processes and work closely with medical histopathologists in this respect.

Cross reference

You can read more about the roles of biomedical scientists in histology in the companion text, *Histopathology*.

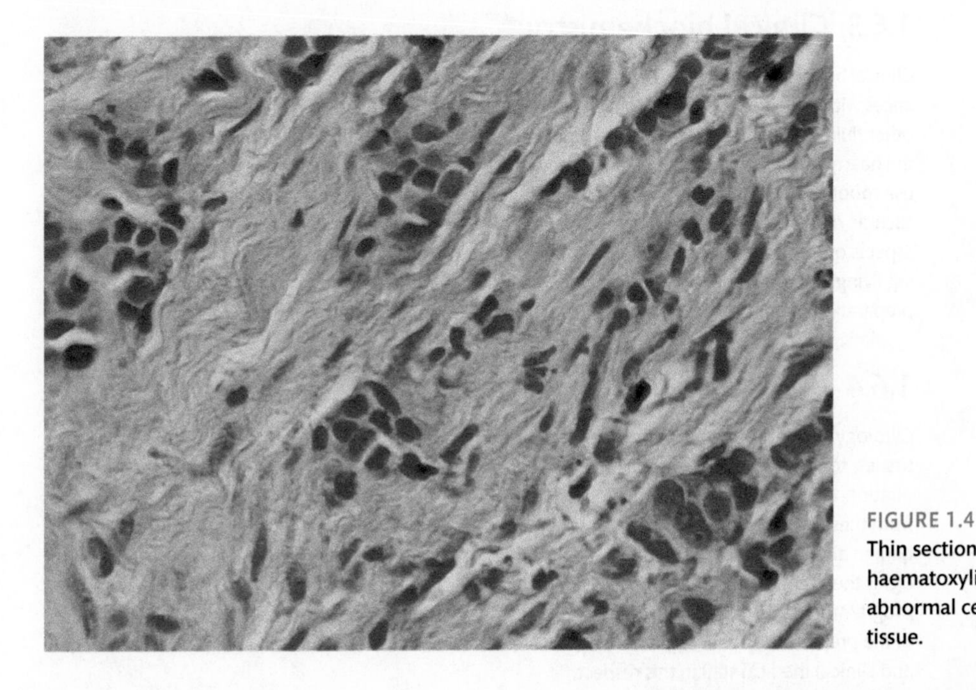

FIGURE 1.4

Thin section of breast stained with haematoxylin and eosin showing abnormal cells infiltrating connective tissue.

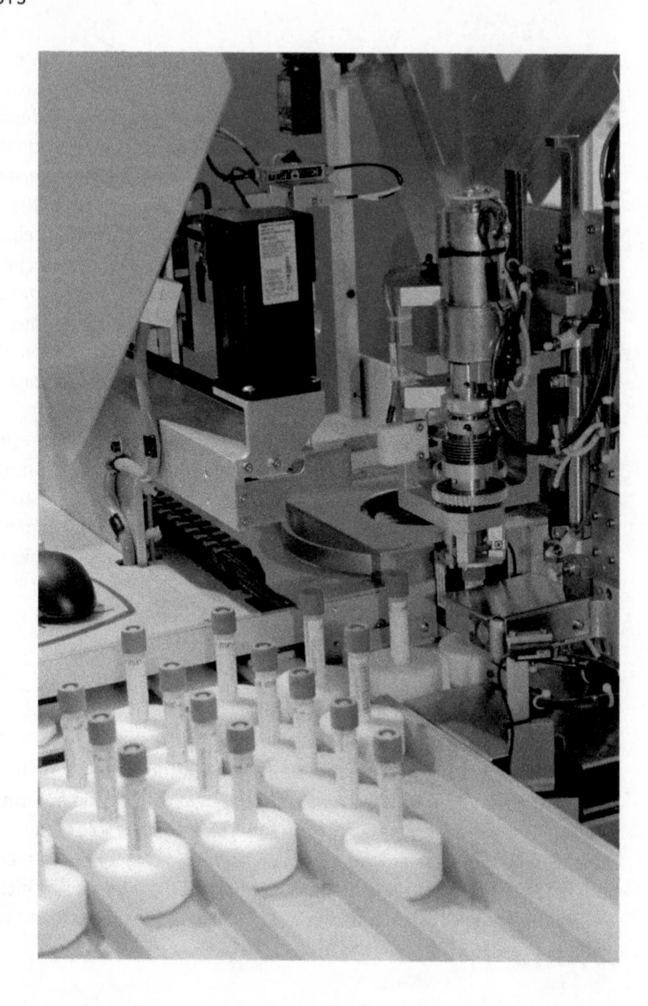

FIGURE 1.5
Samples waiting to be loaded onto a chemistry analyser.

1.6.3 Clinical biochemistry

Clinical biochemistry involves the study of biochemicals and biochemical mechanisms, and their imbalances within the body. Clinical biochemistry tests are usually performed on blood or urine samples, but other fluids may also be tested, for example, cerebrospinal fluid or fluids that abnormally collect around the heart, lungs, or abdominal organs. Many of the tests are performed automatically on machines that use robotics to manage many samples simultaneously (Figure 1.5). Malfunctions of organs or diseases, such as diabetes, heart attacks, and kidney failure, may be diagnosed using biochemical tests. Other aspects of biochemistry are related to screening for drug use and abuse, which may involve investigating living patients or deceased individuals. Again, biomedical scientists are involved at all stages of the preparation process, and work closely with clinical scientists or chemical pathologists in this respect.

Cross reference

You can read more about the work of clinical biochemists in *Clinical Biochemistry*.

1.6.4 Cytopathology

Cytology (like histology) involves microscopic studies, but in this case cells are studied, rather than tissues, to identify disease that may be present. The cells are obtained through natural shedding, aspiration, or mechanical abrasion, and often involve the analysis of body fluids. Slides are prepared from these samples and examined using a microscope. Diseases such as inflammation, infections, and benign and malignant tumours are diagnosed using cytology. Probably the best-known use of cytology is the examination of samples obtained from the uterine cervix as part of the UK cervical screening programme, as can be seen in Figure 1.6. Once again, biomedical scientists are involved at all stages of the preparation and interpretation processes, and work very closely with medical cytopathologists and clinical medical staff in this respect.

Cross reference

You can learn more about clinical cytology in *Cytopathology*.

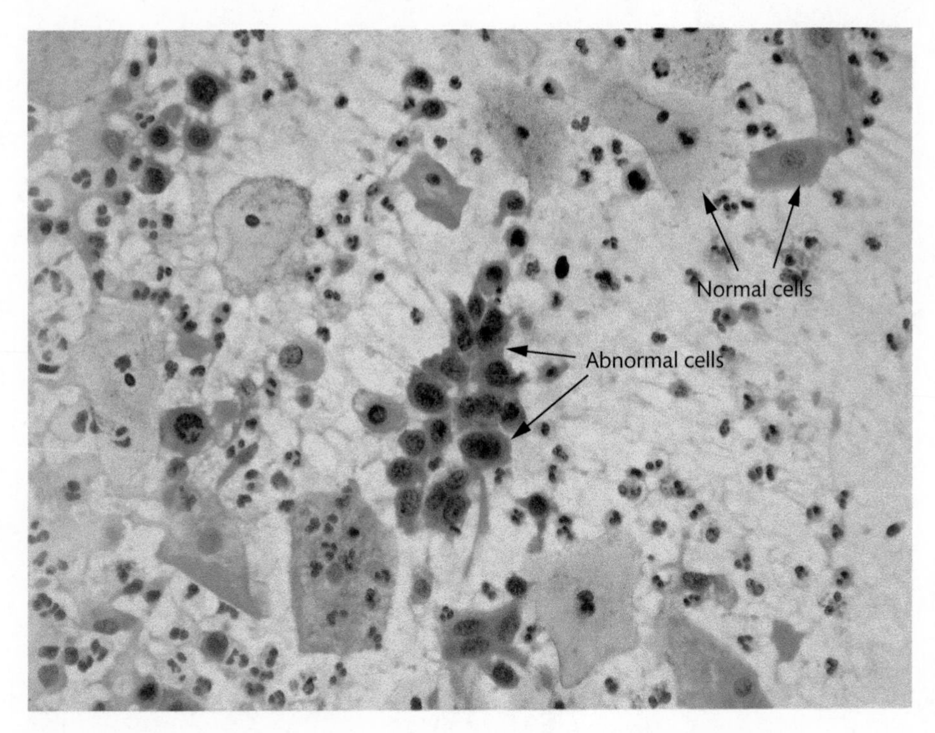

FIGURE 1.6
Normal and abnormal cells in a cervical smear stained by Papanicolaou's method.

1.6.5 Immunology

Immunology involves the study of the immune functions of the body, as a system or its responses to a stimulus. Immunological specimens include blood and other fluid or tissue samples that are subjected to a variety of manual or automated clinical tests (Figure 1.7). Diseases of the immune system include immunodeficiency where it fails to respond to appropriate stimuli, autoimmune disease when the immune system 'attacks' the tissues it is meant to protect, or hypersensitivity when it responds inappropriately to normally harmless stimuli. Immunology is also involved in tissue and

Cross reference
Read more about clinical aspects of immunology in the companion volume, *Immunology*.

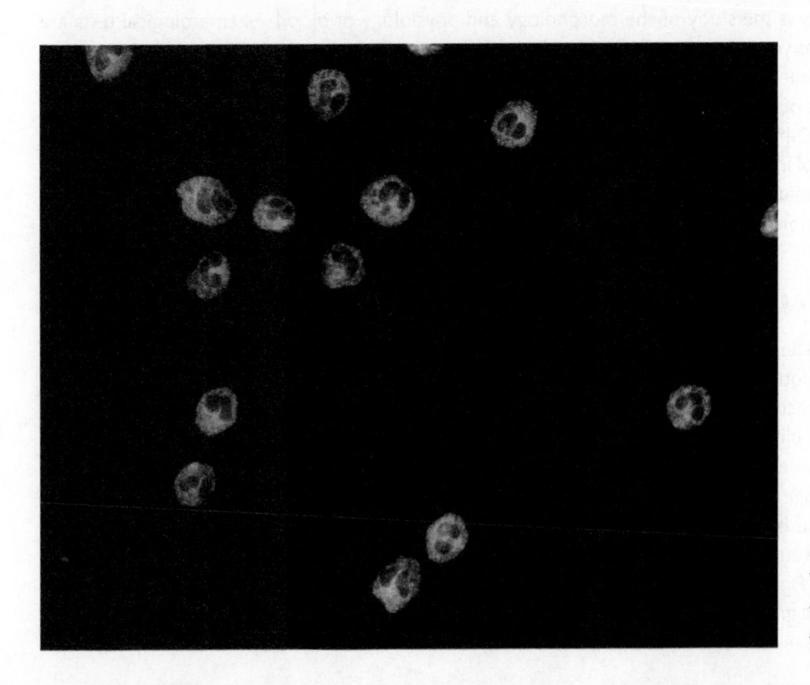

FIGURE 1.7
White blood cells showing green fluorescence after being labelled with fluorescein.

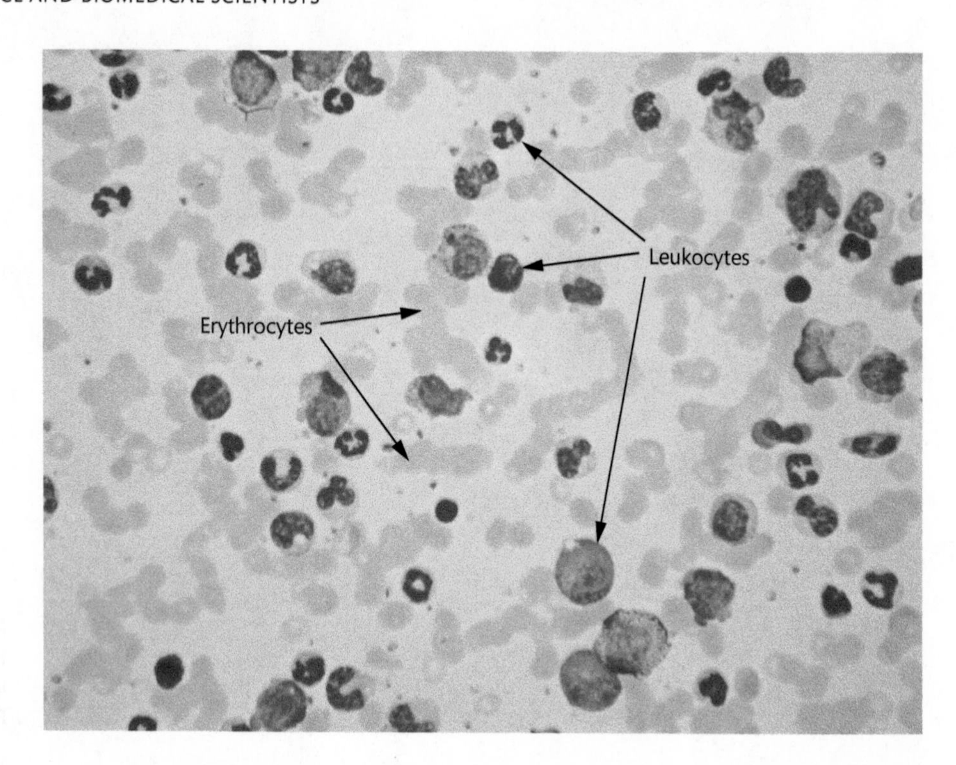

FIGURE 1.8
Normal blood cells in a blood film stained by May-Grünwald and Giemsa's method.

organ transplantation through the matching of donor and recipient tissues, or the determination of rejection. Biomedical scientists, clinical scientists, and medical immunologists are all concerned with such studies.

1.6.6 Haematology

Haematology is the study of the morphology and physiology of blood. Haematological tests are performed, as you might expect, on blood samples; such tests are increasingly carried out by automated instruments. Diseases such as anaemia or leukaemia can be diagnosed following analyses of the numbers of erythrocytes or the haemoglobin content, or the numbers of leucocytes present in the sample. Other aspects of haematology are related to abnormalities of blood coagulation or the monitoring of individuals on long-term medication using anticoagulants. Biomedical scientists and medical haematologists collaborate during these studies. If you examine Figure 1.8 you can see the presence of normal blood cells on a microscope slide of a stained blood film.

Cross reference

You can read more about haematological investigations in *Haematology*.

1.6.7 Transfusion science

Transfusion science involves the provision of blood and blood products to patients who have anaemia, blood clotting problems, and those who are suffering from acute blood loss or have been involved in an accident. Donated blood is the primary source of blood for these purposes. Acquiring donations of blood for transfusion requires its appropriate collection, storage, preparation, and the issue of suitable blood and blood products. Other work in transfusion science is related to the collection, storage, and preparation of donated cells and tissues. Work in blood transfusion is highly regulated and is subject to European Union Directives that are now part of UK law. Owing to the nature of the supply aspects of these products, biomedical scientists tend to work very closely with clinical staff. Figure 1.9 shows blood that has been cross-matched in order to establish its ABO and Rhesus blood group.

Cross reference

The companion volume *Transfusion and Transplantation Science* will give you more detailed information on this topic.

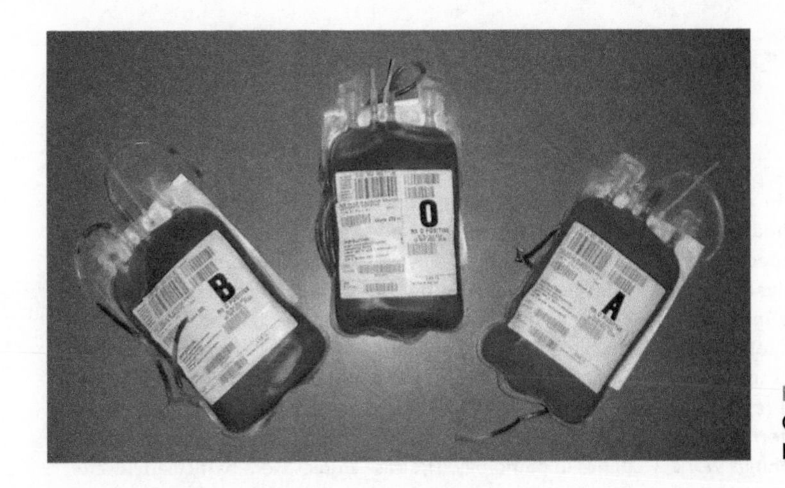
B
O
A

FIGURE 1.9
Cross-matched blood showing blood group types.

1.6.8 Microbiology and virology

Microbiology and virology are involved in the detection, isolation, and identification of microorganisms, including bacteria, fungi, parasites, and viruses that cause disease. Samples may be taken from any part of a body, which may be of a living or deceased individual. Traditionally, microorganisms were grown on nutrient agar plates, as shown in Figure 1.10, in broths, or in cell cultures, but the use of molecular (Chapter 14) or spectroscopic detection methods such as matrix-assisted laser desorption ionization-time-of-flight (MALDI-TOF) and full automation are now becoming the norm in many laboratories. Tests that used to take weeks to confirm the identity of a microorganism can now be done in hours or days. Often when examining a bacterium, its sensitivity to antibiotics is also determined, to ensure the optimal treatment for the patient is chosen and to minimize the misuse of these drugs. Medical microbiologists and virologists work closely with biomedical scientists in microbiology and virology investigations.

Cross reference

You can read more about the roles of biomedical scientists in medical microbiology in *Medical Microbiology*.

SELF-CHECK 1.3

Using histopathology as the discipline, briefly explain the uses of diagnostic tests.

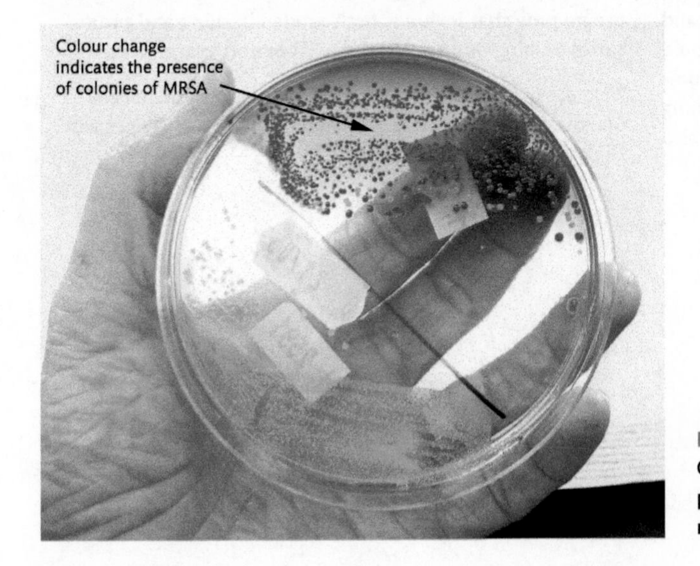
Colour change indicates the presence of colonies of MRSA

FIGURE 1.10
Chromogenic (colour-identifying) agar plate showing the presence of colonies of methicillin-resistant *Staphylococcus aureus* (MRSA).

CASE STUDY 1.1 Supplementary study and professional skills

I graduated with a first-class honours degree in anatomy. Biology had always been my strength throughout school, and I enjoyed every minute of my degree, especially the pathology elements. My final year project involved a lot of practical histology. Consequently, I decided to investigate histology as a potential career option and I was told that to do histology for a living, I would have to either undertake a medical degree, or 'do' a PhD and become a researcher. I chose the latter option, as I have always preferred to be 'behind the scenes', finding out how and why things work, rather than dealing with the public directly. I embarked on a PhD studying the morphology and formation of the mineral phase of bone, and how it alters in osteoporosis and osteoarthritis.

Part-way through my PhD, I found information about biomedical science and biomedical scientists. I had never previously given much thought to who tested patient samples in hospitals, assuming those people to be doctors. I realized that this was what I had been looking for in a career, investigating anatomy, histology, and pathology 'behind the scenes'. Deciding that the PhD would still be very useful to me, and given that I was enjoying it and wanted to complete it, I continued with my research, while finding out what I would need to do to become a biomedical scientist. I contacted the IBMS, who assessed my BSc, and I was told that since my degree was not accredited by the IBMS, I would have to take a course of supplementary education.

Once I had completed my PhD laboratory work, and had begun the daunting task of writing up my thesis while studying as an undergraduate again, I undertook six 10-credit undergraduate modules in clinical microbiology, haematology and immunohaematology, transfusion science, immunology, clinical biochemistry, and cytopathology. While studying and writing up, I contacted a local laboratory to ask if they would be willing to have me visit for work experience. The laboratory agreed, and I spent one day each week there learning how a diagnostic histology department operates while also refreshing my practical skills.

Shortly after completing my undergraduate modules, I was offered a full-time, paid, 14-week placement in the histology department, which allowed me to make a proper start on my IBMS Registration Portfolio. I made good progress with it, and during that time I applied for and was offered a permanent trainee biomedical scientist position in the same laboratory. For the pieces of work that made up my portfolio, I tried to pull in as many aspects of biomedical science as I could, to show that I was aware of how histology related to the other disciplines in pathology. The case studies were particularly useful in demonstrating this, and by researching how other disciplines applied to particular areas of pathology, I was able to deepen my understanding of histopathology at the same time.

Following success in my PhD, soon afterwards I completed my registration portfolio and verification visit. The laboratory tour went well, although I was understandably nervous. I passed, but my assessor did say that I should have made my portfolio less impersonal—it was all neatly typed, but the evidence should have been annotated with my comments or reflections.

Currently, I am working in another department and I am about to be assessed for my Specialist Diploma in Cellular Pathology. I contacted the IBMS regarding my MSc and was told that because my PhD was relevant to my field, an MSc* was not needed in my case. This means I am able to move directly on to the Higher Specialist Diploma, and I plan to spend the next 3 years gaining experience and gathering evidence for this, before sitting the examinations. Eventually, I hope to train in histological dissection, as to me this is the most exciting and appropriate use of the varied skills I have developed over the years of my training.

Finally, these skills can be gained through full-time study and completion of an accredited biomedical science degree programme that includes both academic study and professional training through a series of laboratory placements.

* Note: since the time of writing, the absolute requirement to undertake an MSc prior to entry into the IBMS Higher Specialist Diploma has been removed.

CASE STUDY 1.2 Example of a placement reflective essay

My placement took place at a District General Hospital in Scotland, in a multidisciplinary laboratory consisting of the histopathology, biochemistry, haematology and blood transfusion, and microbiology departments. The placement was scheduled for the second semester of the third year of the course and the outcome was to gather a training portfolio over a 12-week period. Being trained in a multidisciplinary laboratory was particularly advantageous for me in that I was able to experience work in different fields, so this gave me an insight on what were my disciplines of preference.

It all started with an induction week, which involved tutorials on important laboratory requirements such as laboratory accreditation, the roles of regulatory and professional organizations (HCPC and IBMS), quality management and quality assurance, health and safety, and data protection, and this list is not exhaustive. My view of that first week was a little bit of apprehension because not only was the '9 to 5' routine difficult to cope with, most of the concepts were new, very prescriptive, and appeared to have nothing to do with what was being taught at university. However, it didn't take long for me to realize that the life of patients is what is at the end of the line, explaining why the biomedical profession is so tightly regulated.

Two weeks of discipline-specific training were then received in each of the departments and it was interesting to see how the knowledge and skills gained at university were applied to work in the laboratory, this in terms of test principles, test procedure and techniques, interpretation, and statistical analysis of results. My first 2 weeks of training were in the biochemistry department, where I was introduced to the modular analysers. Despite the fact that the system was highly automated, standard operating procedures (SOP) indicated that different test procedures and detection principles were exactly the same as those performed manually in practical classes at university. For instance, most analytes (e.g. glucose, lipids) were measured using colour reactions and spectrophotometer readings; hormones, enzymes, and tumour markers were measured using immunoassay systems, and depending on their molecular size, either sandwich methods (measurement of troponin T) or competitive assays (as for free thyroxin) were used. Although these methods were also automated and the detection principles varied (fluorescence, chemiluminescence), the basic principles behind the test procedures were the same as those taught in the immunology module. I made similar observations when I moved on to other departments.

It was a relief not to have to perform manual cell counting in haematology or in microbiology as at university; meanwhile, it was good to see that the university was in line with new advances in biomedical science, such as molecular methods, which are increasingly used in clinical laboratories. It was also interesting to see how most of the techniques learned at university could be used interchangeably across disciplines, namely molecular techniques, microscopic techniques, and immunological methods. My only reservation was about the content of a particular university practical class where no practical work was actually done (not even the haematoxylin and eosin staining routinely done in histopathology laboratories) except for microscopic identification on ready-made slides.

The importance of maintenance, calibration of equipment, and quality control (QC) was much more emphasized than at university. Other important notions, such as prioritization, effective communication, time management, and cost-effectiveness are only some of the other topics with which students are not really familiarized; it was a privilege to be introduced to these at an early stage and this gave me a feel of what is expected in the working place.

To sum it up, I would say that going on to a placement was an excellent opportunity for me to appreciate the practical relevance of every single module taught in the biomedical science course, from the first year of study. It is evident that without the basic knowledge and skills gained from clinical modules it would be very difficult to understand work in the laboratory. Working in a NHS laboratory is not just about applying a SOP to patient samples to obtain some results. I've learned that a biomedical scientist should be able to know why a test is being requested, when and how the sample should be properly collected, what sort of additives (if indicated) are needed and what will be their impact on the tests, what test methods are more appropriate for a particular specimen, and what precautions should be taken during processing. He or she must also be able to analyse the results obtained and make informed comments about their significance.

The rewards of the placement were, in fact, greater than I initially thought, as it wasn't until I started working on my honours project that I realized how much I had learned in just a few weeks. I became conscious as soon as I started the project that my planning, organizational, and practical skills were much more enhanced. I also had a totally independent and more confident approach to it compared with the execution of my previous laboratory assignments.

In addition, the placement turned out to be an integral work experience on my CV, which helped earn me a job in a NHS laboratory, so I'm happy today to say that it was worth doing.

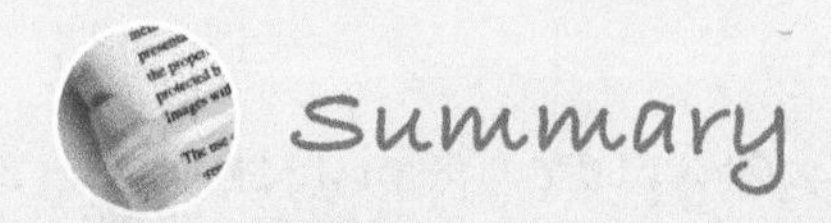

Summary

- Biomedical science is a term used to describe the study of the causes, consequences, diagnosis, and treatment of human diseases.
- Biomedical science is also an academic subject and can be studied as part of an honours degree programme, the content of which has been defined by the QAA.
- To become a registered biomedical scientist, an individual must meet defined minimum standards of proficiency by successfully completing an approved degree programme that includes a period of professional training.
- The study of biomedical science will include acquiring knowledge and professional skills in all of the major discipline-specific areas.
- Once registered, biomedical scientists normally pursue their career by professional practice and the extended study of one of the major discipline-specific areas.

Further reading

- **Kumar V, Abbas A, Aster J. *Pathologic Basis of Disease*. 9th ed. Philadelphia, PA: Elsevier, 2015.**
- **Lakhani S, Dilly S, Finlayson C. *Basic Pathology: An Introduction to the Mechanisms of Disease*. 4th ed. London: Hodder Arnold, 2009.**
- **Lodish H, Berk A, Kaiser C, et al. *Molecular Cell Biology*. 27th ed. New York: WH Freeman, 2012.**
- **Pocock G, Richards D, Richards C. *Human Physiology: The Basis of Medicine*. 34th ed. Oxford: OUP, 2013.**

Useful Websites

- www.hcpc-uk.org.uk
- www.ibms.org
- www.nhsbt.nhs.uk

Questions

1. Which of the following statements is/are true?
   - **(a)** Screening tests are used to investigate signs and symptoms
   - **(b)** Populations are subjected to screening tests
   - **(c)** Screening tests can identify illnesses
   - **(d)** Screening tests are used to identify cancers only
   - **(e)** Screening tests are 100% accurate.
2. Discuss why it is necessary for prospective HCPC registrants to study all of the major areas of biomedical science.

3. If the outcomes are similar, why do training programmes differ between organizations and individuals?

4. Which of the following are role(s) of a test to monitor health?

   (a) Checking the general health status of an individual prior to performing a medical procedure

   (b) Identification of at-risk individuals who have no symptoms

   (c) Investigating an individual who is showing symptoms

   (d) Assessing the progress of long-term clinical conditions

   (e) Follow-up assessments of treated individuals.

Answers to self-check and end-of-chapter questions are available at the end of the book.

2

Fitness to practise

Hedley Glencross

Learning objectives

After studying this chapter, you should be able to:

- Define the functions of a regulatory authority
- Discuss the roles of a professional body
- Explain how a regulatory body determines and maintains the fitness to practise of its members
- Demonstrate how a professional body may help you meet the requirements imposed by regulatory authorities
- Explain how a professional body may help you develop your career

Introduction

Fitness to practise is the ability of an appropriately trained and registered individual to perform his or her professional tasks in a safe manner, and to make appropriate judgements with a degree of autonomy. In Chapter 1 you were introduced to the Health and Care Professions Council (HCPC) and the Institute of Biomedical Science (IBMS). In this chapter we describe these organizations (Figure 2.1a,b), discuss how their respective, but complementary, approaches to fitness to practise differ, and explore their contributions to the training and career development of biomedical scientists.

Cross reference
Chapter 1, 'Biomedical science and biomedical scientists'.

2.1 Healthcare regulation

Biomedical scientists have been regulated by legal statutes since 1960 under the Professions Supplementary to Medicine Act, when biomedical scientists, or rather medical laboratory technicians as they were then known, were registered and directed by the **Council for Professions Supplementary to Medicine** (CPSM). The CPSM regulated nine professions and maintained a list of individuals who had met their (threshold) standards. However, the CPSM did not set other professional standards, so that once a biomedical scientist was admitted to the then CPSM *state register* there was no requirement to demonstrate any continuing competence or further education to remain on the register. As a consequence, it was difficult for the CPSM to remove someone from the register, other than for a criminal offence. However, following a number of then well-publicized incidents, such as the murder of four children by the nurse Beverley Allitt in 1991 and the murders of many elderly

(a)

(b)

FIGURE 2.1
Logos of (a) Health and Care Professions Council (HCPC) and (b) Institute of Biomedical Science (IBMS).
Both logos reproduced with permission from the respective organizations.

patients by the general practitioner Harold Shipman between 1995 and 1998, the issue of regulation of all healthcare professionals (including medical staff) was brought into focus during the 1990s. Consequently, in 2001 a Parliamentary Order for the Health Professions, which included provision for the formation of the Health Professions Council (now known as the Health and Care Professions Council (HCPC)), was approved. The HCPC was given (and maintains) a UK-wide remit for the registration of competent healthcare professionals.

Cross reference
Chapter 4, 'Health and safety', discusses some of the legal framework relevant to healthcare professionals.

2.2 Health and Care Professions Council

The HCPC is principally concerned with patients and their safety. Indeed, it was established as a regulator of healthcare standards with a duty to protect the public. Thus, the work of the HCPC is designed to safeguard the public from being harmed by the actions or inactions of healthcare professionals. Parliament has granted legal powers for the HCPC to carry out its statutory roles by establishing and maintaining the **HCPC register** of **competent** individuals who meet its standards of training, professional skills, behaviour, and health. Competence is defined as the ability of a registered individual to perform repeatedly any action related to his or her profession to a defined standard, and in a safe and effective manner. Any individual who wishes to be admitted to the HCPC register must demonstrate minimum or threshold *standards of proficiency* and also provide evidence they are of good character and that their health is such that 'their ability to practise is not impaired'. Additionally from the 2015 renewal cycle, all biomedical scientists (and all other registered professions except Social Workers in England) will need to declare (upon application to HCPC or at renewal of registration) that they have appropriate professional indemnity insurance in place. This latter requirement may be satisfied through employment, trade union, or professional body membership. For its members, the IBMS provides a group malpractice insurance policy that satisfies the HCPC requirements.

Cross reference

Chapter 1, 'Biomedical science and biomedical scientists'.

Appropriate evidence that these professional proficiency standards have been satisfied would be by successfully completing an HCPC-approved degree course, as discussed in Chapter 1. Students who have not studied in the UK, and who also have overseas laboratory experience, may apply to the HCPC for registration by its international route, where they will need to have their qualifications, experience, and competence individually assessed.

The HCPC currently registers 16 professions, including therapeutic professionals, such as physiotherapists and radiographers, community-based professionals, such as social workers (in England), as well as scientific professionals, for example clinical scientists and biomedical scientists (Table 2.1). The HCPC fulfils its principal role of regulating the activities of these professionals by restricting the use of protected titles, such as physiotherapist or biomedical scientist to those individuals it deems competent. The HCPC also has the legal rights to investigate and remove from its register any individuals who it finds to be incompetent. Individuals may be removed from the register because of unprofessional conduct, inappropriate personal actions, or because of criminal activities.

The HCPC sets both generic and profession-specific standards (as appropriate to the regulated professions and their respective roles) in the following areas:

- proficiency
- education and training
- conduct, performance, and ethics
- continuing professional development
- character
- health
- prescribing.

The standards of proficiency are the relevant knowledge and skills that a registrant must demonstrate before he or she can be admitted to the register. To remain registered, all registrants must maintain their knowledge and skills as appropriate to their developing role during the progress of their careers. The standards of education and training (SETS) are those that providers of education and training must meet to ensure all students who complete an HCPC-approved educational programme meet their standards of proficiency. Successful students are then eligible to apply to HCPC for registration. Standards of conduct, performance, and ethics apply to registrants once admitted to the register. These are the codes of behaviour, attitude, and aptitude to which registrants must adhere if they are

TABLE 2.1 Professionals regulated by the HCPC

Therapeutic professionals	Scientific professionals	Community-based professionals
Arts therapists	Biomedical scientists	Social workers in England
Chiropodists/podiatrists	Clinical scientists	
Dieticians		
Hearing aid dispensers		
Occupational therapists		
Operating department practitioners		
Orthoptists		
Paramedics		
Physiotherapists		
Practitioner psychologists		
Prosthetists and orthotists		
Radiographers		
Speech and language therapists		

BOX 2.1 Continuing professional development

Upon joining the IBMS, new members are automatically enrolled on the IBMS CPD Scheme (although there is an option to opt out). The variety of activities associated with the scheme and their independent validation will help its members in fulfilling the HCPC CPD requirements, too.

to remain a registered professional. It is the continuing enforcement of these standards that allows the public, patients, and the registrants' colleagues to be confident of their suitability to practise and remain on the register. HCPC statements regarding character and health are fairly self-explanatory in that registrants must be of good character and their health must not impair their ability to perform a professional role. Indeed, registrants are required to inform the HCPC if their health is affected such that they can no longer carry out their day-to-day duties. Standards of continuing professional development (CPD) also apply to registrants. Individual registrants are required to demonstrate that they are keeping up to date in their developing role. They must also show, through a range of CPD activities, how these activities benefit themselves professionally, the wider health service, and service users or patients alike. The IBMS also contributes to CPD, as outlined in Box 2.1.

The SETS and CPD will be discussed in more detail in Chapter 18. The standards of proficiency and the standards of conduct, performance, and ethics are explored in more detail in Section 2.3.

Cross reference
Chapter 18 'Personal development'.

2.2.1 Health and Care Professions Council approval of degrees

The HCPC has developed standards of education and training that education providers must satisfy if their degree courses are to achieve HCPC approval. These generic standards ensure that any individual who successfully completes an approved programme will also reach the minimum standard of proficiency. As a consequence, he or she will then be eligible to apply to be admitted to the HCPC register. For biomedical scientists, both a number of biomedical science and healthcare science degrees are HCPC approved.

Key points

The major advantage of successfully graduating from an HCPC-approved degree is that you will be able to apply immediately for admittance to the HCPC register.

The key areas considered when approval for a biomedical science degree is sought are:

- level of entrance qualifications
- programme admissions
- programme management and resources
- curriculum
- practice placements
- assessments.

The HCPC approval process is designed to be robust but not too onerous for the education providers. The approval of a new degree programme by the HCPC will require that they make an initial visit to the education provider. The HCPC delegation may visit as part of a wider assessment group or in isolation. Once approval has been granted by the HCPC, it is 'open ended' and subject to satisfactory monitoring. For example, there is an annual review of the degree programme to ensure its structure and content have not been subject to major revisions and that it continues to meet the HCPC SETS. If there have been major changes in an approved programme, the HCPC will consider their effects on the standards and act accordingly. Thus, an annual review may be the trigger for further visits by the HCPC for the re-approval of significantly altered biomedical science degree programmes.

Although degree approval is normally done on behalf of individuals who wish to gain HCPC registration, a number of post-registration degree programmes are also approved by the HCPC. These include, for example, prescribing or local anaesthetic programmes, and do not apply to biomedical scientists.

2.3 Health and Care Professions Council Standards of Conduct, Performance and Ethics

The HCPC has developed 10 'duties' that define their Standards of Conduct, Performance and Ethics (Figure 2.2), and to which all registrants from the 16 regulated professions must conform. These duties are listed in Table 2.2. The duties describe in general terms what is expected of registrants as an autonomous

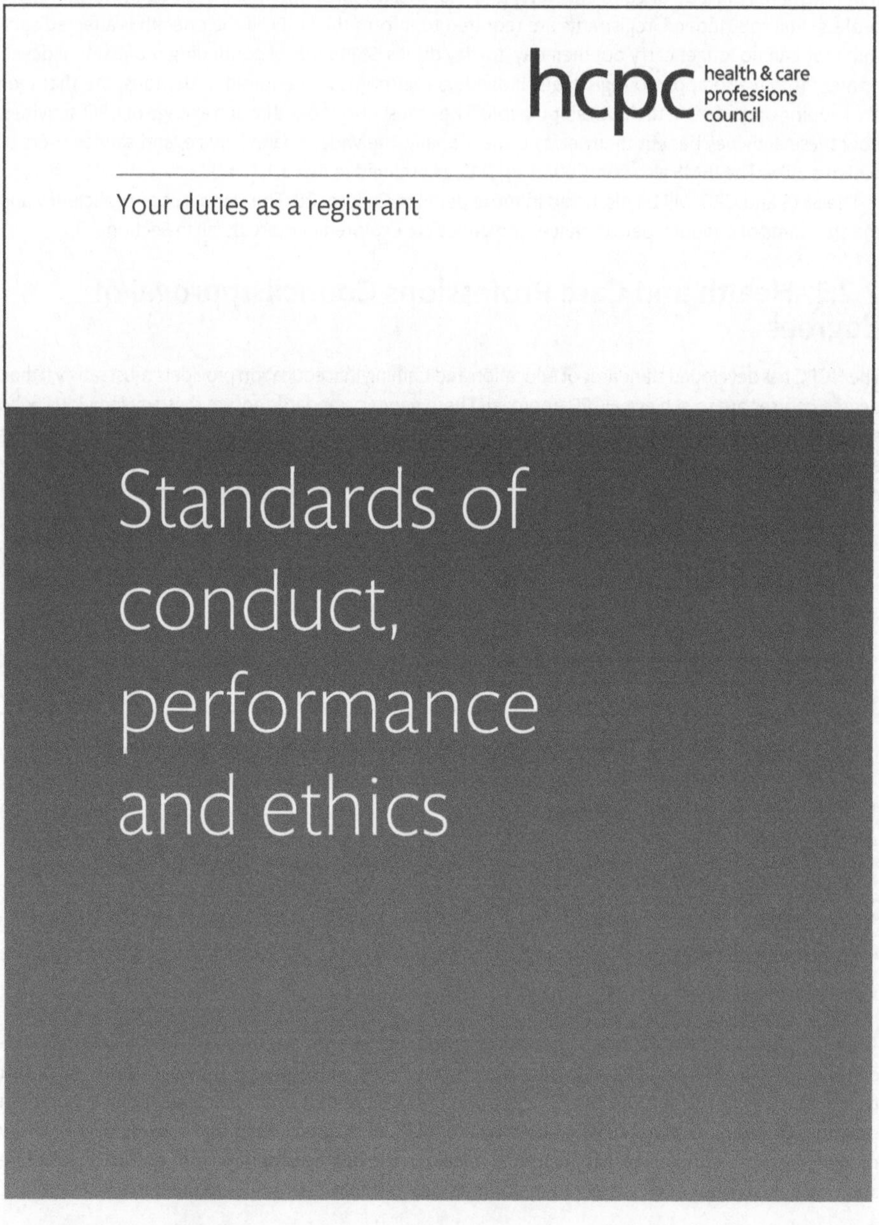

hcpc health & care professions council

Your duties as a registrant

Standards of conduct, performance and ethics

FIGURE 2.2
Front cover of *HCPC Standards of Conduct, Performance and Ethics*. Copyright © Health and Care Professions Council.

TABLE 2.2 Health and Care Professions Council standards of conduct, performance and ethics

Duty	You must:
1	Promote and protect the interests of service users and carers
2	Communicate appropriately and effectively
3	Work within the limits of your knowledge and skills
4	Delegate appropriately
5	Respect confidentiality
6	Manage risk
7	Report concerns about safety
8	Be open when things go wrong
9	Be honest and trustworthy
10	Keep accurate records of your work

and accountable professional. A rather simpler summary is that the standards describe the levels of honesty, integrity, and probity required of a registered individual at all times in his or her personal and professional life. These standards have, necessarily, been written in a style that allows the general public to understand them and also to allow their use (in conjunction with the standards of proficiency) to decide on a course of action if any concerns are raised about the practice of a specific individual.

These standards may be met in a number of ways. For example, a registrant may seek professional advice and support from professional bodies as part of their roles in representing and promoting the interests of their members. This may be as simple as joining and maintaining membership of a professional body or it may be by seeking expert advice from that professional body. As will be discussed later (Section 2.6), the IBMS also defines good practice for its members and other individuals by providing guidance to help clarify professional or scientific matters.

Once an individual has joined the register, he or she is only required to sign (every 2 years at renewal of registration) a declaration that their ability to practise has not been affected by any changes in his or her health status. This declaration covers all aspects of health and abilities, in particular the decision-making skills and awareness required by registered individuals. It is expected that the HCPC will be informed immediately if the health status of a registrant changes in a way that affects their role(s) as a registered individual. From 2007 this declaration has also included a statement regarding CPD, as will be discussed in Section 2.4.

2.3.1 Health and Care Professions Council Standards of Proficiency

The HCPC has also developed standards of proficiency for each of the 16 professions (see Chapter 1, Figure 1.3). These standards were updated again in late 2014 following widespread consultation with appropriate stakeholders. This new update revised the standards into 15 generic statements shared by all of the regulated professions, and by sharing gives continuity of proficiency standards across all of the 16 regulated professions. The individual profession-specific standards then sit underneath each of these 15 statements. The generic statements are reproduced in Table 2.3.

In summary, these statements (or standards) describe the minimum level of knowledge and skills required by a registered biomedical scientist to be able to practise. It is essential that these standards are met as they are the threshold standards needed for an individual to be admitted to the HCPC register and describe the **scope of practice** for a newly registered biomedical scientist (or a registrant from one of the other 15 regulated professions). Scope of practice defines the range of activities in which a registered individual has been trained, assessed, and deemed competent to perform. The initial scope of practice defines the threshold standards for admittance to the HCPC register, which are met on successfully completing an HCPC-approved degree or obtaining an IBMS Certificate of Competence. Although the statements (or standards) are too numerous to explore fully within this book, the education and training programmes undertaken by a prospective biomedical scientist (when HCPC approved) will provide the knowledge, skills, and evidence to ensure that these minimum standards of proficiency are met.

Cross reference

We described the Certificate of Competence in Chapter 1, 'Biomedical science and biomedical scientists'.

TABLE 2.3 Health and Care Professions Council (HCPC) Standards of Proficiency

All HCPC registrants must:
1. be able to practise safely and effectively within their scope of practice
2. be able to practise within the legal and ethical boundaries of their profession
3. be able to maintain fitness to practise
4. be able to practise as an autonomous professional, exercising their own professional judgement
5. be aware of the impact of culture, equality and diversity on practice
6. be able to practise in a non-discriminatory manner
7. understand the importance of and be able to maintain confidentiality
8. be able to communicate effectively
9. be able to work appropriately with others
10. be able to maintain records appropriately
11. be able to reflect on and review practice
12. be able to assure the quality of their practice
13. understand the key concepts of the knowledge base relevant to their profession
14. be able to draw on appropriate knowledge and skills to inform practice
15. understand the need to establish and maintain a safe practice environment

SELF-CHECK 2.1

Why is it considered necessary that the HCPC apply standards of proficiency for prospective registrants?

A scope of practice is not static and changes during a registrant's career. Thus, as a career progresses the associated scope of practice also evolves as new skills are acquired or new roles are undertaken. However, any new scope of practice also provides a different context for the HCPC Standards of Proficiency and their Standards of Conduct, Performance and Ethics. For instance, if a registrant no longer practises in the laboratory but works as, for example, a researcher, lecturer, or manager, and can show that he or she continues to meet the standards of the HCPC, they can remain on the register. It is only when these standards cannot be met that a registrant may wish to remove him or herself from the register voluntarily or be removed by the HCPC itself.

If a registrant has voluntarily come off the register or been removed from it, the HCPC does allow a suitable individual to be readmitted and has developed processes, which include remedial training, supervised practice, or a combination of both, for this purpose.

SELF-CHECK 2.2

How may registrants who no longer practise in a laboratory continue to meet the HCPC standards?

2.3.2 Health and Care Professions Council sanctions on practice

The HCPC has statutory powers to apply sanctions or restrictions that limit a registered individual's ability to practise as a biomedical scientist (or as one of the other 15 appropriately registered professionals). Information regarding the competency of a registered individual will result in an investigation by the HCPC. If this raises sufficient concern, it will be followed by a *Fitness to Practise* (FTP) hearing on either competence or health grounds. You can see an outline of the FTP hearing in Figure 2.3.

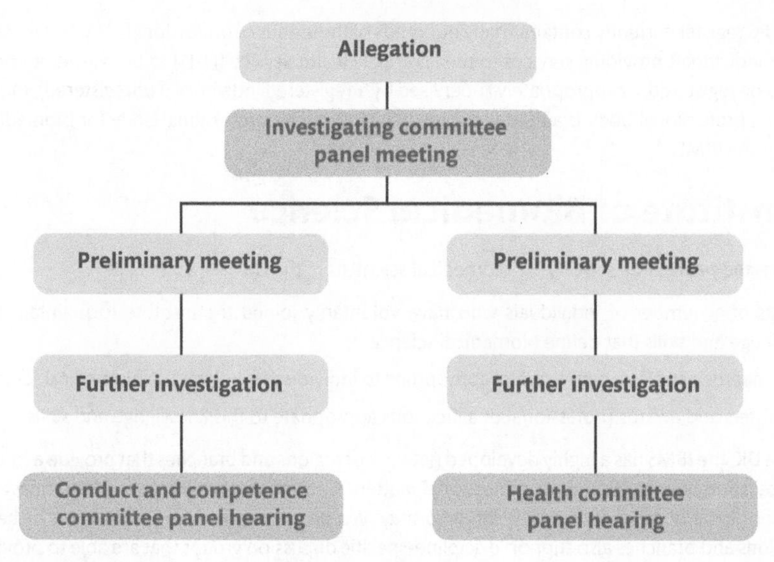

FIGURE 2.3
Schematic illustrating the Health and Care Professions Council's Fitness to Practise process.

Depending on the outcome, the hearing can lead to sanctions against the individual concerned, which include restrictions on the right to practise, supervision of practice, suspension from the register, or, ultimately, removal from the register. FTP hearings are detailed and rigorous but apply the lesser civil, not criminal, court burden of proof, as these hearings relate only to professional practice and public safety.

Look at the summaries of three FTP hearings in Case studies 2.1, 2.2, and 2.3, which show the detail in which the HCPC investigates all allegations made against registrants from all of its 16 registered professions.

All registered professionals who are found guilty of a criminal offence will have the details of the offence, trial, and any sentence passed to the HCPC for investigation, hearing, and possible sanctions as described above.

Full transcripts of all FTP hearings, including those for biomedical scientists, are made public and may be found on the HCPC website.

SELF-CHECK 2.3

List five reasons for which a biomedical scientist may be suspended or removed from the HCPC register.

2.4 Renewal of Health and Care Professions Council registration

To remain on the HCPC register, an individual registrant is required to sign a personal declaration every 2 years confirming that he or she continues to meet the HCPC Standards of Conduct, Performance and Ethics, and their Standards of Proficiency. Since November 2007 this declaration by biomedical scientists also included a statement that they also continue to meet the HCPC standards of health and CPD. From 2015, this declaration will additionally include a statement that they have appropriate professional indemnity insurance in place. To check the validity of the CPD statement, 2.5% of the total registrants from each of the regulated professions are required to provide evidence of their CPD activities during the preceding 2 years. Failure to provide sufficient evidence within a given time period results in action by the HCPC against those individuals, such as suspension or even removal from their register, although further time to submit evidence may also be given. All new registrants are automatically exempt from this requirement for the first 2 years of their registration period. The first CPD audits for biomedical scientists occurred in 2009.

The HCPC register currently contains many hundreds of thousands of professionals. While it is mandatory for individuals providing services to the National Health Service (NHS) to be registered (or in training to be registered, or appropriately supervised by a registered individual if unregistered), membership of a professional body is a matter of personal choice. The professional body for biomedical scientists is the IBMS.

2.5 Institute of Biomedical Science

The IBMS is the **professional body** for biomedical scientists in the UK and, as such, it:

- consists of a number of individuals who have voluntarily joined the Institute to maintain the knowledge and skills that define biomedical science
- sets standards of professional conduct appropriate to individuals who practise biomedical science
- administers and awards professional qualifications appropriate to this knowledge and skills.

Within the UK, the IBMS has a highly developed network of regions and branches that provide a forum for the discussion of scientific and/or professional matters. Commonly, there is also a CPD officer and a network of local hospital representatives, who may also be approached for information or help. Some regions and branches also support discipline-specific discussion groups that are able to provide another forum where scientific or professional matters may be discussed. Upon joining the IBMS, an individual is assigned to a specific branch, which is determined by the member by choosing their home or work address as their principal point of contact. It is possible to change branch if he or she wishes to join a branch closer to their home address.

The vast majority of IBMS members reside and practise in the UK; however, it also has members in many countries and supports branches in Cyprus, Gibraltar, Hong Kong, and (most recently) Sudan.

The IBMS promotes and develops biomedical science and those who practise biomedical science, and is therefore principally concerned with professional matters and issues related to its members;

TABLE 2.4 **Grades of the Institute of Biomedical Science (IBMS) membership effective from 1 January 2014**

Membership grade	Comments and progression
Licentiate	Licentiate is the grade of corporate membership for someone who holds a BSc Honours degree in biomedical science or a related subject Licentiates are entitled to use the letters LIBMS after their name Licentiates of the IBMS have the opportunity to work towards an IBMS Specialist Diploma or Diploma in Biomedical Science and are eligible to take on a variety of professional roles, such as a portfolio verifier The qualifications above will allow individuals to apply for the next class of IBMS membership: *Member*
Member	Member is the next grade of corporate membership for someone who has an IBMS Specialist Diploma, Diploma in Biomedical Science (or an equivalent level professional qualification), or who holds an EQF level 7 (Masters) qualification *and* also has at least 2 years of professional experience in biomedical science Members are entitled to use the letters MIBMS after their name Members of the IBMS have the opportunity to work towards an IBMS Higher Specialist Diploma and are able to take on a wider variety of professional roles, such as an Institute examiner The qualifications above will allow individuals to apply for the highest class of IBMS membership: *Fellow*
Fellow	Fellow is the highest corporate grade of IBMS membership for someone holds an IBMS Higher Specialist Diploma (or equivalent level professional qualification) or who holds an EQF level 8 (Doctoral) qualification *and* also has at least 5 years of professional experience in biomedical science Fellows are entitled to use the letters FIBMS after their name
e-Student	e-Student is a non-corporate grade of membership for someone who is studying biomedical science at undergraduate or postgraduate level. This membership is available at a marginal rate but gives access to the majority of membership benefits, predominately in an online environment
Associate	Associate is a non-corporate grade of IBMS membership for someone who is working in the field of, or related to, biomedical science, who holds a qualification between EQF level 2 (GCSE) and level 5 (ordinary/foundation degree)

you should not, however, infer that it is not concerned about patients or with patient safety. The activities of the IBMS are supportive and will allow appropriate (HCPC-registered) members to meet the HCPC standards described in Section 2.2.

2.5.1 Grades of Institute of Biomedical Science membership

The IBMS has five grades of membership: *e-Student, Associate, Licentiate, Member, and Fellow.* You can see a summary of the grades of membership and how to progress through them in Table 2.4. As an undergraduate student, an individual is admitted to the grade of e-Student upon joining the IBMS. Following graduation he or she will be eligible to apply for Licentiate membership. Once admitted as a Licentiate, further upgrade of membership grade to Member and ultimately to Fellow is encouraged. Equivalent routes for entry and membership grade upgrade for all corporate grades of IBMS membership exist for those individuals whose education and training are different from what might be described as 'standard'.

Whatever their membership grade, members of the IBMS are expected to apply the same principles of good professional conduct and practice in biomedical science.

SELF-CHECK 2.4

How may the structure of the IBMS membership help biomedical scientists develop their careers?

2.6 Institute of Biomedical Science Code of Conduct and Good Professional Practice

The *IBMS Code of Conduct* is embedded within the IBMS publication, *Code of Good Professional Practice for Biomedical Scientists.* Like the HCPC Standards of Conduct, Performance and Ethics (Section 2.3), this publication states in general terms what is expected of a member of the IBMS in both their personal and professional lives. The IBMS Code of Conduct is reproduced in Table 2.5.

TABLE 2.5 Institute of Biomedical Science Code of Conduct

A member of the Institute of Biomedical Science will:
1. Professionalism
1.1 Uphold the name and reputation of the Institute of Biomedical Science and the biomedical science profession and practice according to its responsibilities, standards, ethics and laws
1.2 Maintain the highest standards of professional practice and act in the best interests of the service, patients and other professionals
1.3 Respect the confidentiality of patients, employer and service users unless disclosure is permitted by law and justified in the patient's interest
1.4 Not practise, nor impose upon others to practise in conditions where professional integrity, standards and laws would be compromised
2. Competence
2.1 Understand and work within the limits of their professional knowledge, skills and experience
2.2 Never delegate a task or duty to anyone who is not trained, qualified or experienced sufficiently to undertake it without supervision
2.3 Ensure that colleagues under their management are fully supervised and supported
2.4 Exercise and continually develop their professional knowledge and skill throughout their professional life
2.5 Communicate effectively and meet all applicable reporting standards
3. Behaviour
3.1 Not allow bias, conflict of interest, or the undue influence of others, override their professional judgement
3.2 Take action without delay if patient safety or service delivery is at risk according to local and national 'whistleblowing' guidelines
3.3 Treat all patients, service users and colleagues respectfully and equally without any discrimination or prejudice that could compromise their professional roles or duty of care
3.4 Co-operate with employer and professional colleagues in the interests of providing a safe and high-quality service

Compare the 13 major points of the code with the 15 statements in the Standards of Conduct, Performance and Ethics of the HCPC (see Table 2.2); you can see that both are essentially similar. However, the ways in which they are written are rather different. For example, statement 1 of the HCPC requires biomedical scientists to act safely and effectively within their scope of practice. This requirement is incorporated into a number of points (including 1.2, 2.1, 2.4, and 3.4) of the IBMS code.

SELF-CHECK 2.5

Which statements in the IBMS Code of Conduct would allow you to meet the HCPC standard 1 that requires a biomedical scientist to act in the best interests of service users?

The IBMS *Code of Good Professional Practice for Biomedical Scientists* is only one of a number of its publications (Box 2.2).

BOX 2.2 IBMS publications and website

The IBMS produces a variety of publications and leaflets; however, its two principal journals, which have been published for over 50 years are:

1. *The Biomedical Scientist*
2. *British Journal of Biomedical Science.*

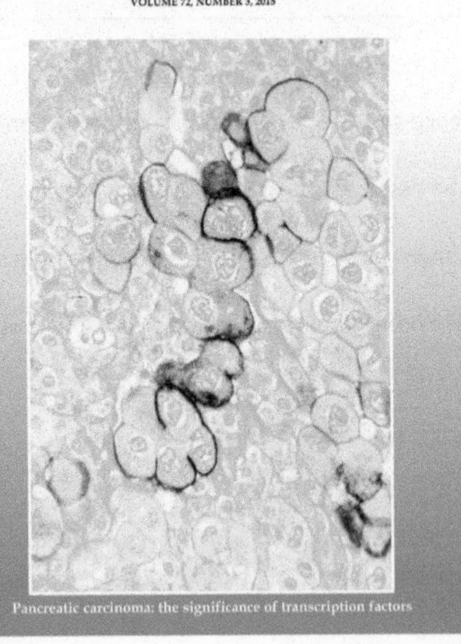

FIGURE 2.4
Front cover of an issue of the *British Journal of Biomedical Science.* Reproduced with kind permission from the Institute of Biomedical Science.

The Biomedical Scientist is published and issued monthly to all members (either as hard copy or electronically). It contains news, articles of scientific interest, CPD activities, IBMS activities, and a classified advertising section. *The British Journal of Biomedical Science (BJBS)* is published quarterly (by default electronically, but hard copy may be requested) and is the largest worldwide circulation peer-reviewed journal dedicated to biomedical science (Figure 2.4). It includes a wide range of papers covering basic science, scientific advances, and the scientific and clinical practice of biomedical science.

The website of the IBMS provides access to the latest biomedical science news, information, online CPD, and other learning resources. You can see the welcome page in Figure 2.5.

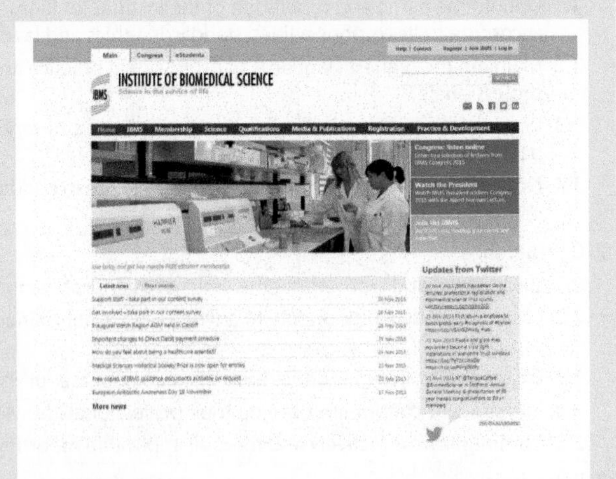

FIGURE 2.5
Welcome page of the Institute of Biomedical Science (IBMS) website (www.ibms.org). Reproduced with kind permission from the IBMS.

FIGURE 2.6
Front cover of *IBMS Code of Good Professional Practice*. Copyright © Institute of Biomedical Science.

2.6.1 Good professional practice

The IBMS has defined what it regards as good professional practice in the context of biomedical science and the behaviour of biomedical scientists. Its code of conduct describes the attributes and the professional behaviour expected of IBMS members and complements the HCPC Standards of Conduct, Performance and Ethics, which we described in Section 2.3. The IBMS code is explained in more detail in the IBMS publication, *Code of Good Professional Practice for Biomedical Scientists* (Figure 2.6). However, in many ways the IBMS code does not necessarily describe the full range of good professional practice. This would be difficult anyway, as observations as to what is good or bad practice can be subjective, and it may be difficult to always give an objective viewpoint on such matters. In this volume, we will discuss in detail health and safety, quality assurance, and professional development in the relevant chapters. However, the education and training programme required of a biomedical scientist in the UK will ensure an individual works within the context of the IBMS Code of Good Professional Practice and meets the various standards of the HCPC.

Cross reference
Health and safety, quality assurance and management, and personal development are discussed in Chapters 4, 17, and 18, respectively.

2.7 Institute of Biomedical Science and education

The IBMS has been involved in formal educational activities since its founding in 1912; indeed, the IBMS motto is 'Learn that you may improve'. Its major educational activities in higher education institutions are the accreditation of appropriate degree programmes and its involvement in the provision of professional qualifications.

2.7.1 Degree accreditation

The programme of biomedical science degrees is defined for registration purposes by the Quality Assurance Agency for Higher Education, while their standards are subject to approval by the HCPC (Section 2.2). However, the IBMS, in its role as a professional body also accredits the contents of degrees in biomedical science and as such has been recognized by the HCPC as an approved education provider. **Accreditation** of a professional degree is a process of peer review set against a number of

FIGURE 2.7
Logo for use on, for example, stationery by an Institute of Biomedical Science (IBMS)-accredited university.

defined learning outcomes. The IBMS accredits both undergraduate and postgraduate degree programmes. Accreditation of any particular degree in biomedical science by the IBMS is not designed to replace or replicate the HCPC approval process or, indeed, the internal validation exercises undertaken by the universities concerned. The IBMS considers the content of the units or modules constituting the curriculum of the degree and uses the definitions and subject areas already discussed in Chapter 1 as the basis for their undergraduate accreditation process.

Accreditation by the IBMS requires the formation of a local liaison group consisting of academic staff of the university, NHS employees (largely hospital laboratory staff), and any independent part-time teachers who may be involved. This group is required to meet on a regular basis to oversee matters relevant to the course. Written evidence regarding the formation of the group and its meetings must be provided for scrutiny by the IBMS. It is generally expected that senior and experienced laboratory staff from local hospitals will contribute to the delivery of both undergraduate and postgraduate courses through part-time lecturing as 'experts' in clinical subjects, particularly where expertise in those areas cannot be provided by university staff.

Successfully completing an IBMS degree at an accredited university offers a number of advantages to the graduate (Figure 2.7). Thus, it eases the entry of a graduate into the IBMS and contributes towards registration with the HCPC.

The accreditation of a specific degree in biomedical science provided by a particular university may only apply to the academic component. This allows successful students to work in a variety of occupations or areas, which we listed in Chapter 1. It also allows the IBMS to tailor its accreditation of postgraduate education and degrees to the needs of biomedical scientists during their professional careers. A number of universities have developed co-terminus or applied biomedical science degrees. These degrees consist of an accredited programme of academic study together with laboratory-based teaching and training during a series of work placements. The advantage of completing this type of degree is that professional training is integrated into the degree programme. A Certificate of Competence is awarded upon graduation, allowing an individual to apply immediately for inclusion on the HCPC register.

FIGURE 2.8
Logo for use on, for example, stationery by an Institute of Biomedical Science (IBMS)-approved training laboratory.

Graduates who have completed a degree course that is not accredited by the IBMS may be required to undertake supplementary or extra undergraduate studies to make their non-accredited degree equivalent to an accredited honours degree in biomedical science. Normally, the major areas of extra study involve the clinical disciplines, which are often not fully covered in non-accredited degrees. If a degree programme of study does not include any professional training in a working laboratory (work placements), the Certificate of Competence must be obtained by a different route. For example, it can be obtained by working in a laboratory following graduation. Alternatively, it can be gained while working in a laboratory and studying for an appropriate degree part-time. Such study pathways require the registration portfolio to be verified by IBMS appointed examiners, who must visit the place of work. The verifiers must also assess the role of the laboratory in the training and have the remit to award training approval to appropriate laboratories (Figure 2.8).

2.7.2 Professional qualifications awarded by the Institute of Biomedical Science

The IBMS has developed the following professional qualifications:

- Specialist Diploma
- Diploma in Biomedical Science
- Diploma of Specialist Practice
- Certificate of Expert Practice
- Higher Specialist Diploma
- Diploma of Higher Specialist Practice
- Diploma of Expert Practice
- Advanced Specialist Diploma.

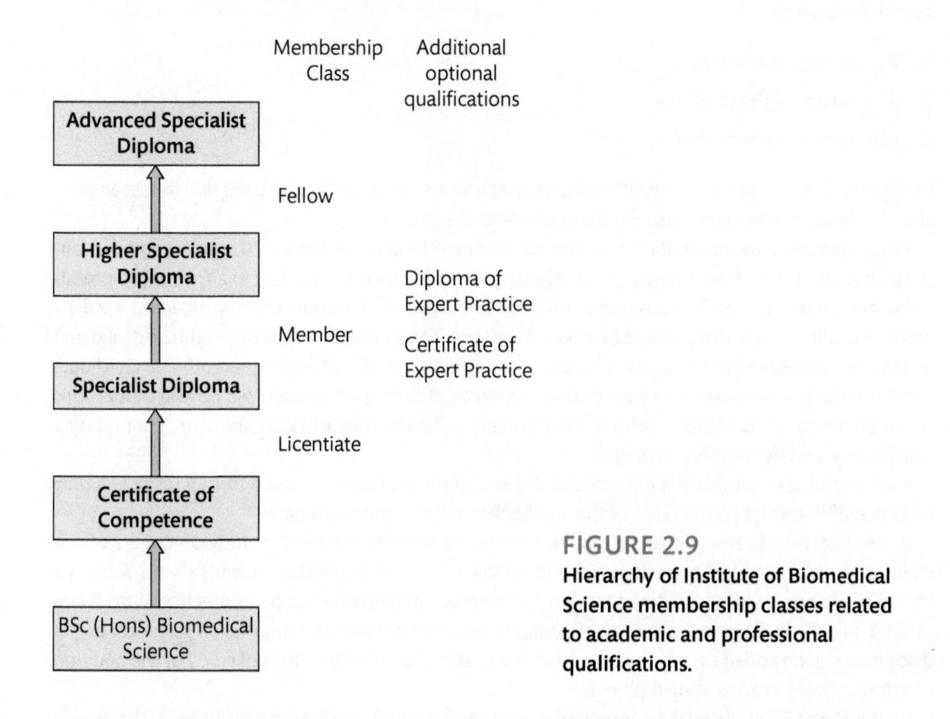

FIGURE 2.9
Hierarchy of Institute of Biomedical Science membership classes related to academic and professional qualifications.

These qualifications are linked to grades of IBMS membership (Section 2.5) and also to a career in the NHS. The IBMS qualification structure is outlined in Figure 2.9.

The Specialist Diploma or Diploma of Biomedical Science is likely to be the first IBMS qualification encountered following registration at the beginning of a career in biomedical science. Qualification for a Specialist Diploma involves the professional study of a single discipline and diplomas are awarded in the principal, discipline-specific areas of biomedical science:

- clinical biochemistry
- cytopathology
- cellular pathology
- haematology with hospital transfusion practice
- histocompatibility and immunogenetics
- clinical immunology
- medical microbiology
- transfusion science
- virology.

Some of these programmes have been produced in partnership with other organizations, such as the British Society of Histocompatibility and Immunogenetics (BSHI). Specialist Diplomas consist of discipline-specific elements that naturally differ from subject to subject but which are standardized across all of the diplomas. One recently introduced, but important, change to the format of specialist diplomas is to give 'ownership' of the evidence collection to the candidate, with the emphasis on justification and reflection.

Although it is recommended that 12–18 months of study be undertaken to collect the evidence required for the award of a specialist diploma, the IBMS has also placed a time limit on completion of 3 years.

Its award requires an assessment by an IBMS-appointed examiner. The examination process consists of three parts:

1. Candidate presentation
2. Assessment of the portfolio
3. Laboratory tour with *viva*.

Each of these processes will assess knowledge, practice, and skills, and is also designed to test an individual's confidence in demonstrating these points to the examiner.

The presentation by the candidates to the examiner will be expected to last 20 minutes and demonstrate that a transition from a newly qualified and generically trained individual to a specialist biomedical scientist has occurred. The candidate will also need to explain how their training relates to the daily practice within a single discipline of biomedical science. The examiner will examine the portfolio and evidence of achievement to ensure all sections have been completed and signed off. The candidate will then take the examiner on a tour of their department during which they will be asked questions based on the contents of their portfolio. Finally, there will be feedback to the candidate, their training coordinator, and department manager.

If successful, the candidate is informed at the end of the examination, but a formal result will only be issued after receipt by the IBMS of the feedback form from the examiners.

In contrast, qualification for a Diploma of Biomedical Science involves the study of a range of disciplines. This diploma has been developed to enable registered biomedical scientists who work in a multidisciplinary environment to be trained and assessed in this wider scope of practice. Introduced in 2015, initially the diploma covers haematology, clinical chemistry, and transfusion science. Further disciplines will be added as they become available to allow for different mixes of study or the addition of extra modules to an existing diploma.

Certificates of Expert Practice are intended to recognize scientific practice in specific areas that would only be performed by a few individuals who have chosen to specialize, and for whom alternative qualifications do not exist or do not warrant the development of a full portfolio. In this way, they complement the broader discipline-specific post-registration awards already undertaken.

Examples of such qualifications include the *Certificate of Expert Practice in Electron Microscopy* and the online qualifications of *Certificate of Expert Practice in Quality Management*, *Certificate of Expert Practice in Leadership and Management*, and *Certificate of Expert Practice in Training*.

The *Higher Specialist Diploma* (HSD) is also a discipline-specific qualification, although an HSD in *Leadership and Management* has also been developed. The HSD is designed to test and demonstrate a deeper professional knowledge of a discipline than that at diploma level.

The HSD is at the centre of the IBMS qualification process in that although the HSD is the normal requirement for the upgrade from Member to Fellow, it represents only an intermediate step in terms of professional practice. The HSD assessment runs annually over 2 days and includes case studies or scenarios as a major component of the examination. In this way, the ability to draw strands of information or practice together and demonstrate a deeper understanding is assessed.

Diplomas of Expert Practice and *Advanced Specialist Diplomas* of the IBMS recognize knowledge and skills at the highest level of scientific or clinical practice within areas of biomedical science or in the application of biomedical science. They are designed for those who wish to practise at the level of a consultant biomedical scientist.

Diplomas of Expert Practice may be undertaken either before or after an HSD, very much dependent on the career path chosen by the individual themselves. Examples of Diplomas of Expert Practice qualifications include the *Diploma of Expert Practice in Clinical Transfusion*, *Diploma of Expert Practice in Histological Dissection*, and *Diploma of Expert Practice in Non-gynaecological Cytology*.

Advanced Specialist Diplomas are only open to Fellows of the Institute, as holders often have a professional role that contains a high level of autonomous practice, too. Examples of Advanced Specialist Diplomas include the *Advanced Specialist Diploma in Cervical Cytology*, *Advanced Specialist Diploma in Specimen Dissection*, and *Advanced Specialist Diploma in Ophthalmic Pathology*. The cervical cytology advanced diploma was introduced in 2001 and holders have seen their role expand in the intervening

CASE STUDY 2.1 Example of FTP hearing relating to an occupational therapist

An occupational therapist was struck off the register following allegations that he or she failed to maintain adequate records, provided inappropriate treatment to patients, wrote up case notes retrospectively, falsely wrote up case notes, and incorrectly closed cases that required further assessment.

The panel determined that the misconduct found to be proven was wide ranging, covered a period of time, and concerned basic competencies. The allegations relating to note keeping, which included the falsification of records, demonstrated a marked lack of honesty and integrity. Furthermore, the panel concluded that the registrant had not shown insight into his or her failings, or the consequences arising from them.

CASE STUDY 2.2 Example of FTP hearing relating to an operating department practitioner

An operating department practitioner was struck off the register for self-administration of the drug propofol, having accessed the employer's drug store without authorization. The registrant had also received a police caution for this offence.

The panel took into account the fact that for the police caution to have been given, a full admission of the allegation had to have been made. Accordingly, the panel were satisfied that the theft of the drugs had occurred. The panel were also satisfied that the registrant had self-administered the drug.

The panel considered that a caution order would not reflect the severity of the matter and that a 'conditions of practice' order would not be appropriate, given that the registrant was not present at the hearing. It was not known if the registrant was working, with the result that conditions could not be considered. The panel gave careful consideration to imposing a suspension order but concluded that there had been a serious breach of trust on the registrant's part, which had the effect of putting patients and colleagues at risk.

CASE STUDY 2.3 Example of FTP hearing relating to a biomedical scientist

A biomedical scientist was struck off the register following allegations that the registrant had failed to maintain adequate records and not ensured staff were adequately trained, had behaved in an intimidating and aggressive manner towards staff, and discussed confidential personal details of staff with others. These behaviours had continued over a significant period but at no stage had the registrant shown any insight into their inappropriate nature, or remorse for the consequences of these actions on patients and colleagues. The panel concluded that these behaviours were misconduct and showed a clear breach of several of the Standards of Conduct, Performance and Ethics.

period from a starting point of reporting of abnormal cervical cytology results to a number of other areas, including case presentations at multidisciplinary team meetings, audit of invasive cancers, and commissioning. In 2013 the first candidates were successfully examined for the *Advanced Specialist Diploma in Ophthalmic Pathology* and now have a professional role that includes independent reporting of histological samples.

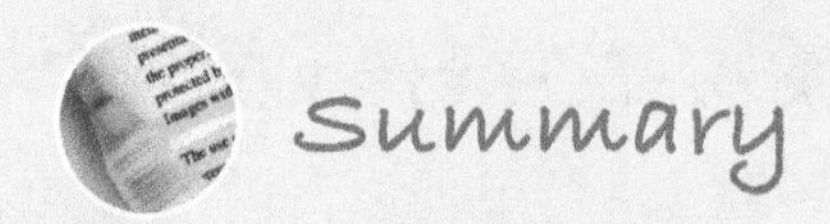

Summary

- The principal duty of the HCPC is to protect the public.
- Biomedical scientists are regulated by the HCPC, which has the legal powers to admit individuals to its register.
- The HCPC possesses the legal powers to restrict practice and remove from its register those registrants who it considers not competent to practise.
- Biomedical scientists may join their professional body, the IBMS.
- The IBMS acts for its members by not only helping them meet the requirements of the HCPC, but also by providing other support and benefits to members.
- The functions and activities of the HCPC and IBMS are complementary, and they work closely together.

Further reading

- **Institute of Biomedical Science. *Good Professional Practice for Biomedical Scientists*. Version 5. London: Institute of Biomedical Science, 2015.**
- **Wood J. The roles, duties and responsibilities of technologists in the clinical laboratory. *Clinica Chimica Acta* 2002; 319: 127–32.**

Useful Websites

- HCPC. www.hcpc-uk.org.uk/
- HCPC international route. www.hcpc-uk.org.uk/apply/international/
- HCPC degree approval, annual monitoring, and major change. www.hcpc-uk.org/education/processes/
- HCPC re-admittance process. www.hcpc-uk.org/apply/readmission/
- HCPC Fitness to Practise. www.hcpc-uk.org/complaints/fitnesstopractise/
- IBMS. www.ibms.org/
- IBMS Code of Good Professional Practice. www.ibms.org/go/practice-development/good-professional-practice
- IBMS degree accreditation process. www.ibms.org/go/qualifications/ibms-courses/accreditation
- IBMS portfolio verification process and laboratory training approval. www.ibms.org/go/registration/become-hcpc-registered/lab-training-and-ibms-reg-training-portfolio
- IBMS regions and branches. www.ibms.org/estudents/go/membership/connectlocal
- IBMS CPD Scheme. www.ibms.org/go/practice-development/cpd
- Department of Health England. www.dh.gov.uk/en/index.htm
- Health of Wales Information Service. www.wales.nhs.uk/
- Scotland's Health on the Web. www.show.scot.nhs.uk/
- Department of Health, Social Services and Public Safety Northern Ireland. www.dhsspsni.gov.uk/

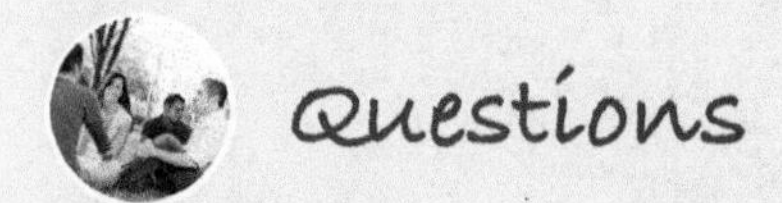

Questions

1. Which of the following statements are applicable to biomedical scientists?
   - **(a)** Must be able to practise in a non-discriminatory manner
   - **(b)** Are not required to use problem-solving skills
   - **(c)** Be able to critically evaluate collected information
   - **(d)** Be able to review their individual knowledge and skills
   - **(e)** Be able to maintain appropriate records
   - **(f)** Have to always practise under direct supervision
   - **(g)** Always work in isolation.
2. Which of the following statements from the HCPC are applicable to biomedical scientists?
   - **(a)** You must keep your professional knowledge and skills up to date
   - **(b)** You must get informed consent to provide care or services (so far as possible)
   - **(c)** You must keep accurate records
   - **(d)** You must limit your work or stop practising if your performance or judgement is affected by your health
   - **(e)** You must make sure any advertising you do is accurate
   - **(f)** You must provide any important information (to us or any other relevant regulator) about your conduct and competence.
3. Why is it considered necessary for an individual to be registered?
4. Why should it be appropriate for registrants to join a professional body, if this is not mandatory?
5. Discuss the value of continuing professional development.

Answers to self-check and end-of-chapter questions are available at the end of the book.

3

Communications in laboratory medicine

Hedley Glencross and Georgina Lavender

Learning objectives

After studying this chapter, you should be able to:

- Explain the need for clear policy for communication
- Explain why there is a need for a generic telephone policy
- Discuss the attributes required of a biomedical scientist when communicating results
- Discuss the regulations that apply to the storage and transmission of electronic data with reference to the current guidelines from the Institute of Biomedical Science
- Explain the roles of regulatory bodies that have interests in clinical laboratory record keeping

Introduction

Communication skills, including written, verbal, and electronic communications, are an essential part of being a biomedical scientist. Biomedical scientists are required to communicate clearly and precisely on all aspects of their work. They must communicate within their local workplace with their colleagues, within the organization that employs them, and they must be able to communicate with other healthcare professionals and any of the public with whom they may have direct or indirect contact.

Communication from the clinical laboratory includes the results of laboratory tests, specimen requirements, changes to methodology, and their effects on reference ranges, changes to service provision, significant changes in laboratory personnel, the introduction of new policies and procedures, or any other information that may affect the relationships between the laboratory and its service users.

One of the principal roles of biomedical scientists is to provide information to appropriate healthcare professionals that may affect the diagnosis, treatment, and prognosis of patients. They provide information regarding prescription drugs, what drugs are suitable, and in what amounts for a variety of diseases. They also contribute to screening programmes for large numbers of 'well' individuals to detect disease in the early stages before symptoms are noticed. The whole process of receiving specimens, performing analytical work, and reporting results from start to finish relies on adequate communications between all the groups involved. This communication may not immediately be obvious to less experienced members of a laboratory team; hence, the purpose of this chapter is to highlight areas of both verbal and **non-verbal communication** that may impact on the relationships between laboratory staff, and other professionals and users. As biomedical scientists, you must be aware of your limitations and the consequences of your actions.

This chapter highlights the importance of adequate communication skills within the clinical laboratory and how communications by laboratory personnel can affect the perceptions of users of the service they provide.

3.1 Communication and the clinical laboratory

Laboratory medicine has a reputation for being one of the more under-rated services within a hospital environment. However, studies and audits have shown how essential the clinical laboratory is in diagnosis and in the treatment expected in the demanding healthcare organizations of today. It has been estimated that the clinical laboratory will have an input into 70% of patients receiving healthcare. Laboratory medicine plays a vital role in the overall function of the healthcare system, but its role may be overlooked, particularly by the general public. This situation has arisen because, traditionally, clinical laboratory personnel appeared to have little contact with other healthcare professionals and patients. The public perception of the work carried out behind the laboratory doors, is that the 'path lab' is a somewhat unpleasant entity where doctors dig around with bits of dead people. However, the reality could not be further from these preconceived ideas. The World Wide Web has raised public awareness of healthcare to new levels, and evidence-based medicine and audit have increased the importance of clinical laboratory investigations with the other healthcare professional groups within the patient care pathways. Laboratory medicine is now recognized by a wider audience within the healthcare community as an essential discipline and given due appreciation. Staff in clinical laboratories have a much higher profile in the provision of patient care. The corollary is that clinical laboratory staff must be aware of the impression that they give to patients, healthcare professionals, and users of the laboratory (those who either request tests or receive results as part of their duties) during the course of their work. Just because the laboratory staff are not necessarily seen in person does not detract from the importance of being aware of the often latent signals that are picked up by other groups with which they have dealings.

3.1.1 Laboratory staff in public areas of hospitals

Clinical laboratory personnel rarely wear a uniform, and are often not immediately recognizable as other hospital staff when in public areas, such as shops, dining rooms, and hospital corridors. As healthcare professionals, individuals have a responsibility to dress and act in a manner that will not cause offence to others, breach their hospital trust policy, or bring the profession into question by any aspect of their behaviour.

Some organizations have a **dress code** to which all staff are expected to adhere. Dress codes are often viewed as draconian by laboratory staff who (a) are not often seen outside the laboratory and (b) are usually largely covered with a white laboratory coat. However, there are attempts by employers to bring standardization and professionalism to their workforces. Clothes are perceived, rightly or wrongly, as a form of non-verbal communication. Staff outside the laboratory, patients, and their relatives and visitors are quickly judgemental of the standard of dress worn throughout the hospital. Biomedical scientists must remember that the general public may not share their personal views regarding what is or is not an acceptable standard of dress at work, and that a dress code has been written by an organization with its *total* staff in mind. When considering standards of dress within a hospital, it is important that aspects of infection control and health and safety are considered. The reason(s) behind certain aspects of dress codes may not immediately be apparent and while biomedical scientists may not normally attend clinics or wards, they are still members of the wider hospital community and cannot be given preferential treatment. Where biomedical scientists are free to dress as they please, they should remember to be clean, tidy, and not wear clothing that may cause offence to others, or that may constitute a breach of the Health and Care Professions Council (HCPC) standards of conduct, performance, and ethics. These were introduced in Chapters 1 and 2. Likewise, if they are provided with a full uniform, as opposed to a laboratory coat and other personal protective equipment, then it would be seen that the dress code of the department is to wear that uniform as and when appropriate.

Cross reference
Chapter 4, 'Health and safety'.

The HCPC has clear standards of behaviour that are expected from its registrants at all times in their working life, and also in their private and personal life. Standards of behaviour are also viewed as non-verbal means of communication, and any unacceptable private behaviour may spill over and be extended to the type of behaviour that should be shown to a member of the public. The biomedical scientist involved in a street brawl may possibly be deemed as having a short temper and be capable of assaulting a patient under certain circumstances. Such behaviour will be investigated by the HCPC where a criminal offence has been proven to have taken place and thus could jeopardize the registrant's fitness to practice.

Cross reference
Chapter 1, 'Biomedical science and biomedical scientists', and Chapter 2, 'Fitness to practise'.

3.1.2 Communications within the clinical laboratory

Clinical laboratories are increasingly subject to extensive legislation. This has led to descriptive procedures and documents being made available to all personnel within the laboratory. The advantage of having such documentation is that it ensures all staff follow procedures to high professional and scientific standards. The advantage of strict written instructions or **standard operating procedures** (SOP) is that all staff who are required to carry out a particular task or procedure do so in the same way. Not only are such SOPs necessary for accreditation purposes, but there is also the guarantee that if a member of staff is taken ill, another member of staff will know exactly what to do to complete the day's work, even though there has been no direct communication between the two individuals!

Cross reference
You can read more about SOPs in Chapter 17, 'Quality assurance and management'.

Communications *within* a clinical laboratory are varied. They include the transfer of instructions and information between staff who work together to provide the end products of the laboratory. These products are the final results of the various measurements and tests performed on clinical specimens. **Teamwork** is the way in which a group of individuals must interact together to achieve a common goal. Thus, maintaining a team structure and working together in an appropriate fashion is crucial. Most working days within a clinical laboratory follow a given pattern in that the work is generated and must be carried out to specific deadlines. It is essential that every member of the team is aware of their responsibilities and the responsibilities of others around them. Each person working in a laboratory, regardless of his or her grade, has their own, sometimes unique, place in that team. Team members have a responsibility to meet deadlines.

Communication within the team is necessary so that the team leader is able to maintain appropriate control and the rest of the team recognize his or her leadership. It is essential that any information that may affect the processing of specimens is communicated throughout the team, both upwards to the most senior member with the responsibility for the service, and downwards to the support staff. Figure 3.1 (a, b) illustrates the ends of the spectrum when information needs to be communicated effectively between members of staff in the same department.

For example, the major failure of a large piece of equipment in the clinical chemistry laboratory may only seem important to those involved. Such events, while unlikely, require a number of immediate remedial actions in addition to repairs to the equipment; all staff must know what is required of them. Any major delays in the processing of samples must be reported to staff in specimen reception, so that they are aware of the situation and can inform other departments or service users as necessary that clinical samples may not be processed with the usual speed and efficiency. Staff involved in the issuing of results and dealing with telephone calls asking for results must also be informed so that they are able to explain to service users why the results they require may not be ready. There is also the need for clear written communication in the laboratory to invoke contingency plans that may involve staff who do not directly work in the laboratory, such as those involved in point-of-care testing using equipment in remote locations or using another laboratory on a different hospital site, or even in another hospital or organization.

Cross reference
Chapter 16, 'Point-of-care testing'.

(a)

"May I start the meeting by thanking you all for your exceptional effort over the last two weeks in dealing with an unprecedented number of clinical samples."

(b)

YOU'RE
FIRED

FIGURE 3.1
(a) Appropriate and (b) inappropriate ways of informing staff of departmental or procedural changes.

Other examples where effective communication is essential include the following:

- occasions when staff are temporarily absent and their duties must be covered by others
- the reporting of failures of **quality control procedures**
- clinical samples that must be processed immediately because test results are required urgently
- stock control where there is a danger of a particular reagent running out before a new delivery is expected.

All these factors may critically affect the routine functioning of the laboratory and will almost certainly require attention from more than one member of staff.

It is also essential to communicate any updated or changed information within the laboratory team, no matter how trivial it may appear. Thus, the introduction of a new method may require a change in the stated reference range of a particular analyte, and this, of course, has implications for the final result appearing to be within normal limits. A change in the batch of a particular reagent on a large analyser may require its recalibration and a re-evaluation of the performance of the method. Even something as apparently simple as a broken door may result in a fire exit becoming unusable and require the selection of an alternative route.

SELF-CHECK 3.1

Why is it necessary to write down any changes in the consignment lot numbers of reagents used in laboratory procedures?

Procedures are often changed by senior members of staff who are not actually involved in the performance of the daily tasks. In practice, new or inexperienced members of staff are often more able to point out a flaw in a change because they are the ones who can most easily see possible consequences of any changes. However, it is also possible that junior members of staff may feel that particular tasks have no real benefit and are irrelevant, and so stop doing them. This too may have adverse effects further along the laboratory chain. It is essential that any changes to a SOP are discussed fully by the whole team and only implemented when there is a written instruction to do so.

Communication within the laboratory also includes completing critical paperwork, such as **quality control charts**, **reagent books**, and **troubleshooting sheets** in the case of an instrument failure. If a staff member carries out maintenance or runs successful quality control, and does not record it in the correct manner in accordance with recognized laboratory procedure, other members of staff will not know that the tasks have already been carried out.

Cross reference

Chapter 17, 'Quality assurance and management'.

3.1.3 Communications between clinical laboratory staff and service users

Service users may be described as any person or group of people who interact with the laboratory. This, of course, includes patients and their relatives. It also includes doctors or other healthcare professionals (both within and outside of a hospital trust), and visitors to clinical laboratories. Table 3.1 gives you an indication of possible service users; it is not exhaustive and should only be used as a guide.

Patients and their relatives form a group that are generally described as *direct* service users, given the manner in which results and information generated by the laboratory affect them. They will be given information about their own health, or that of a relative or friend, that may have a profound impact on their lives. They are rarely medically qualified and communication with this group must be appropriate to their level of understanding but without being in any way patronizing. The communication must also comply with local policy, professional codes of conduct, and current legislation.

Hospital staff groups are varied and information given to them must also be appropriate to their qualifications. It also depends on their *need to know* as part of their own professional duties; not all

TABLE 3.1 List of possible service users

Hospital staff	Community medicine groups	Visitors to the laboratory
Doctors	General practitioners	Company representatives
Nurses	Community nurses	Research scientists and doctors
Radiographers	Midwives	Instrument engineers
Physiotherapists	Surgeries	Formal inspectors
Occupational therapists	Administrative staff	Patients
Cardiographers		
Pharmacists		
Dieticians		
Midwives		
Phlebotomists		
Porters		
Domestic staff		
Maintenance staff		
Ward support staff		
Secretarial staff		

hospital employees need to have access to confidential information about patients. Different departments will have their own rules regarding who needs to know certain pieces of information, and if there is any doubt regarding the level of information that can be passed between staff groups, it is better to be cautious and give too little in the first instance. More detailed information can always be given at a later date, but it is far more difficult to retract something inappropriate.

Community medicine groups are healthcare professionals who do not necessarily have direct connections to the clinical laboratory but are *external* users of the laboratory services. They need to communicate with clinical laboratories regarding patient samples and clinical test results as part of their daily activities. Levels of information and communications between laboratories with these groups depend on local policies.

Visitors to the clinical laboratory are wide ranging, from other healthcare professionals to patients. The common link in this group is that they are entering an environment with which they are unlikely to be familiar. Their levels of knowledge are equally widespread and may be very specialized. Each visitor must be treated appropriately to the needs of the visit.

It is often difficult for biomedical scientists to build relationships with user groups that are not based on preconceived ideas. Biomedical scientists are often disadvantaged in a hospital structure because there are largely unseen, and so is the environment in which they work. Often they will be judged by an unseen audience responding to telephone conversations and **reports**. However, as biomedical scientists, we must also realize that the majority of the work done during the course of a normal day passes in and out of the laboratory without incident, and is processed in a timely manner in accordance with the expectations of the service users. There is no need for a service user to have any direct contact with laboratory staff. It is only when there are deviations from the normal process, such as when a patient's condition becomes critical and a result is needed urgently, a sample is lost and a result not available as expected, or there is a major failure of a piece of laboratory equipment and test results are therefore not available in the usual timeframe. Such occurrences may prompt a user to contact the laboratory staff. It can be that service users who need to communicate with the laboratory are under more pressure or stress than usual, and may be abrupt or rude. However, biomedical scientists should remain professional and act appropriately at all times.

> **Key points**
>
> **It is in everyone's interest that good working relationships are fostered with all service users.**

3.2 Communication and confidentiality

In direct face-to-face contact or over the telephone, laboratory staff must quickly establish exactly who they are communicating with. It is fundamental for a biomedical scientist to know to whom they are speaking before releasing information or patient details that could be considered confidential or sensitive. Confidential information must only be given to certain groups of users. For example, it is usually inappropriate for a biomedical scientist to discuss the results of clinical tests or give such results to a patient, a patient's relative, or members of the public unless there are specific circumstances within written guidelines. Such behaviour, even inadvertent, would be in breach of the standards laid down by the HCPC, and could lead to both disciplinary action by the employer and investigation of the action by the HCPC, and that could then result in the biomedical scientist being removed from the professional register.

> **Cross reference**
>
> **You can read more about the HCPC in Chapter 2, 'Fitness to practise'.**

> **Key points**
>
> **In some cases patients may need to know the result of their test, including pregnancy tests, glucose measurements in diabetes, or when monitoring anticoagulation.**

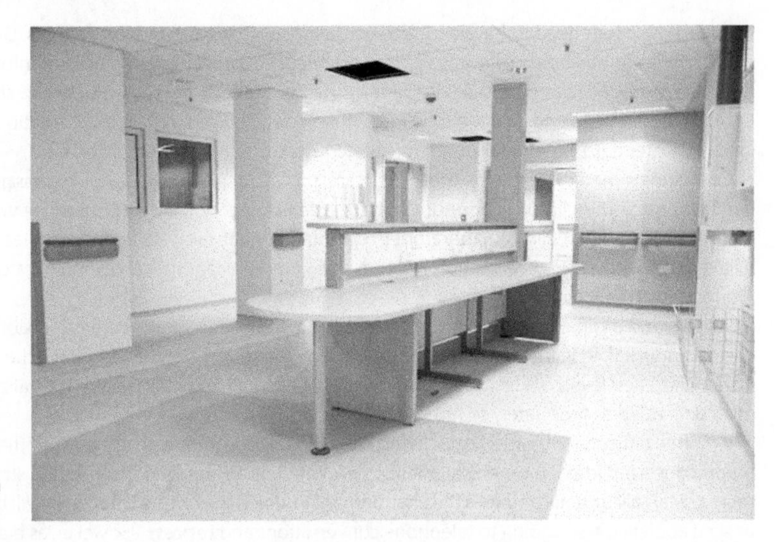

FIGURE 3.2
A typical nursing station on a children's ward. This normally extremely busy area is the major interface between patients and the clinical laboratory. Photograph courtesy of Central Manchester University Hospitals NHS Foundation Trust, UK.

Cross reference

You can read more about point-of-care testing in Chapter 16, and *Haematology* in the series companion book.

These considerations may also apply in some circumstances to commercially sensitive information, for example discussions about pricing structures or costing regarding laboratory equipment with competitor companies as this may be sensitive information if a major purchase is about to go out to tender.

3.3 Methods of communication

The principal form of communication between laboratory staff and most of the user groups listed in Table 3.1 is verbal, but this is not exclusive, as written material is still exchanged. Verbal communication may be one-to-one, an example being telephoning a set of test results concerning a particular patient. It may also be used to inform a large group of a change in policy or procedure. The introduction of a new method may be presented to a large group of hospital staff in the form of a case study to demonstrate why the new method is preferable over the old. Written communications are used to transfer both clinical test results and messages, and may be in paper or electronic forms. Whether meant for a single recipient or for a wide audience, the message must be clear and unambiguous, and presented at a level suitable for the intended recipient. The most important interface for the transfer of results through a hospital is the nurses' station on a ward or clinic (Figure 3.2).

3.3.1 Telephone communications

Telephone communications are often the means by which a laboratory is judged by its users. While few people have face-to-face contact with laboratory staff, a correspondingly significant number use the telephone as the primary means for contacting laboratory staff. For this reason it is important that laboratory staff are helpful, courteous, and confident when using the telephone. These skills are not necessarily innate in laboratory staff and tend to be acquired over time as an individual's knowledge and confidence increases with their developing practice.

All laboratories will each have their own telephone policies and although such policies should conform to certain principles, all individuals must be familiar with the local telephone policy of their own department and strictly adhere to it. Figure 3.3 (a, b) illustrate the correct and incorrect manner in which to deal appropriately with a telephone enquiry.

The following basic rules normally apply to telephone conversations:

- the professional norms applied to face-to-face conversations are exactly the same for telephone conversations
- the recipient or originator of telephone conversations must be established at the start of a telephone call

(a)

"Certainly, I will e-mail those test results to you immediately"

(b)

"Your call is important to us and I'll deal with your emergency tomorrow"

FIGURE 3.3
(a) Appropriate and (b) inappropriate ways of answering telephone enquiries.

- only appropriate discussions must take place
- information of a clinical nature or test results must never be communicated to patients (except in certain defined circumstances, e.g. where patients themselves are responsible for self-monitoring of chronic conditions), their relatives, or those not authorized to receive such information
- records of telephone conversations that involve the dissemination of clinical information or test results, especially when interim or non-validated results are being issued, must be kept for audit purposes
- patient-sensitive or confidential information must never be left on an answerphone or voicemail.

Cross reference
Chapter 17, 'Quality assurance and management'.

General enquiries, such as *what colour tube is used for a particular test?* do not usually require documenting, but it is important for the receiver of the call to realize that there is always a laboratory user at the other end of the telephone. When talking to users, a biomedical scientist must remain professional at all times, no matter how irrelevant the enquiry may seem. However, it is also important that the identity of the user be established and any information given by telephone is appropriate to the level of understanding of the user.

Incoming calls that request specific information may sometimes be beyond the professional expertise of the person taking the call. In such circumstances, these requests should promptly be passed to a more senior staff member, with as much detail as possible, so that the requestor does not have to repeat themselves unnecessarily. This minimizes the need for a service user to keep giving the same information when handed from person to person on the telephone, which reflects poorly on the communication strategies within the department. Ideally, systems should be in place to help ensure all enquiries are managed with as few referrals as possible.

It is also helpful to have a locally generated message pad near each telephone, not only to write down messages, but also to act as a prompt to ensure that the appropriate contact details are transcribed, as well as any messages or records of conversations.

The telephone policy may cover giving results or this may be guided by a separate policy. As a minimum, when telephoning a result you must identify the person taking the call. This individual should write down the result, the time, and date of the call, and a record of the person giving the result. The details should then be read back to the caller in the laboratory to ensure the correct information has been communicated and understood. Once communicated, a record should also be made in the laboratory that results have been communicated verbally, by whom, and to whom together with the date and time of the conversation. These records form part of an audit trail and, while at first glance may seem somewhat lengthy, it is necessary to keep such records in case something goes wrong or there has been misunderstanding of the conversation. During such exchanges, if there are any doubts about the abilities of the receiver to transcribe and communicate the results correctly to the requestor, then the biomedical scientist should request that someone more suitable be asked to take over the call. This is essential for the safety of the patient and biomedical scientists in such situations must be able to demonstrate independent and confident practice, and not feel intimidated in challenging the competence of the receiver of the results. A cautionary example is outlined in Case study 3.1.

SELF-CHECK 3.2

What is the simplest check one can make to ensure information given out over the telephone has been correctly received?

3.3.2 Paper reports

In our increasingly paperless society, the number of hard-copy laboratory reports is in decline. It is felt by many (and there have been many studies to show this) that the majority of paper reports never reach their final destination, usually the patient's written medical record, and that it is largely a waste of resources and the time needed to produce them in quantity. However, because the majority of reports are available electronically does not mean that the presentation of laboratory data as a written document, as a means of visual communication, can be disregarded or dismissed as unimportant. The written report is often the professional standard that the outside world uses to judge the status of the laboratory. Poor reports imply poor mechanisms to produce the data. The production of pathology reports of appropriate quality may not be considered as a means of communication in the first instance but, for many users, the request form and the final report are the only direct interface they have with the clinical laboratory service. The report must be clear and unambiguous, and written with the correct use of English grammar and language. This is particularly important for added comments or clinical codes. Many laboratories now only offer their locally developed coded comments or only allow certain groups of staff to add comments. While an individual may understand the comment they are writing, it is essential to ask how that comment may be viewed by the reader. Is the comment of relevance and does it add value to the report? Is it written in good English? Will the service user understand the comment? Has the correct medical terminology been used?

When information is presented as a paper report, all laboratory data should be accompanied by appropriate units, reference ranges, useful clinical information, and details of the issuing laboratory, as well as the obvious information, such as complete patient demographics and specimen collection

BOX 3.1 Details are required

As examples, a patient with sepsis is likely to have a variety of microbiological investigations. The results of a swab with no details of the site may render the clinical information available useless. Likewise, serial blood samples are often taken (a poorly controlled diabetic may have several blood glucose samples taken over a 24-hour period) and the date on a report with no indication of the time the blood sample was taken may make the serial results of little value.

details. When such information is not complete, time and effort will be wasted by the recipient of the report in trying to find the missing data, as indicated in Box 3.1.

3.3.3 Electronic data

As the generation of paper communication is decreasing, so the transfer of electronic data is increasing. The transfer of such data is the primary method of communication in many laboratories. This allows **requests** to enter the laboratory using ward-based requesting programs linked to pathology networks. Results can also be returned automatically to the patient electronic record and downloaded directly to general practitioner surgeries and outreach clinics. The use of websites and email to 'cascade' information is reducing the need for writing letters, telephone conversations, and the need to print information on paper.

Electronic data must be viewed as being the same as written communication. The rules applied to paper reports about, for example, including units, reference ranges, useful clinical information, details of the issuing laboratory, complete patient demographics, and specimen collection details, also apply to electronic reports. Again, only standard, recognized abbreviations should be used, and any possibly ambiguous language such as 'text speak' must be avoided. The National Health Service (NHS) and the Department of Health have standardized acronyms for medical and other frequently applied terminology. Previous references to use of a **Laboratory Information Management System** (LIMS) in respect to coded comments and standardization apply, as in the same way a paper report may be the only user–laboratory interface, for some groups of users, the electronic copy of the report may be the only interface. The use of coded comments through a LIMS in the laboratory makes for much easier standardization.

Email is a popular method of communication. It is fast, easy to use, and an excellent mechanism for cascading information, which, for the most part, has replaced the use of facsimile (fax) transmissions, by allowing the secure delivery of patient sensitive data through systems such as the NHS net. However, there is no excuse for not writing emails in correct English, especially when conveying complex interpretive reports, with complete spelling and no shorthand. Emails should be proofread before sending, as the standard of the English used may be the only basis a service user has to make a judgement on the overall educational standards of a member of the laboratory staff. While electronic data is a boon in today's busy society, it also has serious consequences when things go awry and, like all methods of communication, it is not 100% reliable. All systems can fail and data can be lost or fall into the wrong hands. Thus, steps must always be taken to ensure that procedures are in place to support the protection and recovery of electronic data.

All electronic data must be held on a back-up system. Anyone who has lost personal work due to the failure of a personal computer has experienced the frustration and extra work that could so easily have been prevented by saving the data on a secondary data storage system! Whereas CD-ROMs and memory sticks may be useful ways for students to protect their personal educational material, this method of storage is unsuitable for patient information. Indeed, government data was lost in two high-profile cases involving the Ministry of Defence, who admitted losing personal details of service employees, and the Child Benefit offices, which lost bank details of some claimants. Such cases involving the loss of personal information demonstrate the need for the secure storage of sensitive data pertaining to the general public. Imagine the scale of such a system failure if the data, such as that stored on a LIMS, or on a hospital-wide patient information system, was lost as a result of insufficient back-up mechanisms or stolen owing to poor security. All large systems within the NHS have a back-up mechanism that implements some type of housekeeping and back-up of all accessed files at regular time intervals. This frequently goes unnoticed by the majority of staff, but the usual practice is that a named person has overall responsibility for the security and protection of laboratory data and sets protocols that must be followed to prevent loss of data or even to prevent data falling into the wrong hands. The NHS technology systems have carefully controlled access, with all data being password protected and tracked with audit trails. It is essential that all staff follow all local and Trust-wide policies to ensure the protection and security of electronic data.

The storage and transfer of electronic data is covered by the Data Protection and the Freedom of Information (FOI) Acts, and the consequences of both pieces of legislation require their examination in detail.

3.4 Data protection

The Data Protection Act (DPA) 1998 governs the use of personal information by businesses and other organizations. It came into law in March 2000, and regulates how a business or organization may store and use personal details about individuals, including their names and addresses, bank details, medical records, and opinions they may have expressed. A specific problem arising from the Act is that it only applies to living individuals. Following their death, any data stored about them need no longer be protected in the same way as when they were living. This is seen as a loophole in the existing legislation that is particularly pertinent to health records.

The DPA 1998 applies eight principles for the purpose of protection of personal information. You can see these listed in Table 3.2. These principles ensure that personal information may only be held and used for the purpose for which it was originally intended and cannot be used for anything else subsequently. It also gives individuals the right to view the data held about themselves. Thus, for example, the DPA gave individuals the following legal rights:

- the right to a copy of the data
- the right to prevent processing likely to cause damage or distress
- the right to prevent processing for the purpose of direct marketing
- the right to have inaccurate data rectified, blocked, erased, or destroyed
- rights in relation to automated decision making.

In addition, the Act classifies some personal information as *sensitive*, about which there are even stricter rules. Sensitive information includes:

- racial or ethnic origin
- political opinions
- religious or similar beliefs

TABLE 3.2 Summary of principles listed in the Data Protection Act

Processed fairly and lawfully
Processed for one or more specified and lawful purpose, and not further processed in any way that is incompatible with the original purpose
Adequate, relevant, and not excessive
Accurate and, where necessary, kept up to date
Kept for no longer than is necessary for the purpose for which it is being used
Processed in line with the rights of individuals
Kept secure with the appropriate technical and organizational measures taken to protect the information
Not transferred outside the European Economic Area (The European Union (EU) member states, plus Norway, Iceland, and Liechtenstein), unless there is adequate protection for the personal information being transferred

- trade union membership
- physical or mental health or conditions
- sexual orientation
- offences or alleged offences committed
- proceedings related to these offences or alleged offences.

An organization is only allowed to use sensitive information where they are able to conform with a narrow set of conditions to ensure that such information is only used where the organization has an essential need to use it, or that the individual has given explicit consent for its use.

It is immediately clear that records held in a clinical laboratory, either in paper or electronic forms, and also any paper reports issued by the laboratory, are considered sensitive personal information for the purpose of the DPA and must be treated accordingly. Considering the scale of sensitive information held within a hospital, and indeed the NHS, it is a significant task, and to that end, in 1997 a report was commissioned by the Chief Medical Officer of England, and a committee established under the chair of Dame Fiona Caldicott. A report followed, referred to as the Caldicott Report, highlighting six key principles and making 16 specific recommendations to be applied to all data held by the NHS.

The six key principles as set out in the Caldicott Report are:

1. Justify the purpose
2. Do not use patient-identifiable information unless it is absolutely necessary
3. Use the minimum necessary patient-identifiable information
4. Access to patient-identifiable information should be on a strict need-to-know basis
5. Everyone with access to patient-identifiable information must be aware of their responsibilities
6. Understand and comply with the law.

These principles mean that every single use or transfer of patient-identifiable information that moves within or outside of an organization, such as the NHS, should be necessary and its purpose defined. All transfers of such information must be regularly scrutinized and reviewed by an appropriate person to establish that the continued transfer really is necessary. All NHS Trusts have a named Caldicott Guardian, who is usually a senior member of the medical staff with the overall responsibility for ensuring compliance within the trust.

Patient-identifiable information that is not strictly necessary should be excluded unless it would be dangerous to remove that data. Patients should only be identified where it is essential to the successful movement of that data and, even then, the minimum amount of identification should be used. Current legislation often dictates the amount and type of data needed. Once data have been collected and stored, that data must only be accessed by those people to whom the data are absolutely necessary and those with access must be sure of their own responsibilities to protect the data and how they fit into the legal requirements of the Caldicott principles.

So, having described the legislation, we will now consider how the information held in pathology departments is held in relation to the Caldicott principles. First, *justify the purpose*. Clinical requests must be made on the basis of providing information for diagnosis and treatment and not for research purposes unless consent has been specifically sought. It is clearly paramount that the correct reports be identified with the correct patient, so in terms of using little or no identifying information, this must always be balanced against patient safety. It is generally accepted that patients be identified by surname, forename, date of birth, and a unique number (hospital number, NHS or Community Health Index number; Box 3.2) as minimum information for the purpose of pathology records. Problems do arise when using data transfer mechanisms, such as fax machines or electronic links, where there is a danger that personalized data may not end up in the hands of the person for whom it was intended. There are circumstances where it is necessary to code or encrypt information so that the true identity can only be recognized by the intended recipient. An exception to this general rule, where sets of results need not have a true identity, are data arising from clinical trials or research programmes; in these cases, anonymous reports are acceptable. Access to information on a need-to-know basis has already been discussed in this chapter in relation to use of the telephone as a method of communication (Section 3.3). All healthcare professionals and laboratory support workers have policies and protocols that outline their specific responsibilities with regard to patient information.

BOX 3.2 Use of the NHS and Community Health Index number

The NHS number is now the identifier of choice across different NHS organizations in England and Wales. It is a 10-digit number that is unique to every individual who is registered with the NHS in England and Wales. The NHS is now required to use this number on all formal patient documentation. The NHS number is issued to all babies born in England and Wales, and will be given to anyone not born in those countries who is eligible to register with a general practitioner in England and Wales, and use NHS facilities. In Scotland a similar 10-digit system operates, using an identifier called the Community Health Index number.

SELF-CHECK 3.3

Why is the NHS number now the recognized means of identifying an individual?

3.4.1 Freedom of Information Act

Where the text refers to 'The Act' it covers both the English and Scottish Acts.

The Freedom of Information (FOI) Act 2000 came into force in January 2005. A separate act was also enabled in Scotland in 2002 that covers public bodies that are under direct control of Holyrood, rather than Westminster. For such institutions this act fulfills the same purpose as the 2000 act.

The Act gives a general right of access to all types of recorded information held by public authorities and places obligations on them to disclose information (subject to a range of exemptions); indeed, the information applied for may be retrospective in nature. The FOI Act gives applicants two related rights:

1. The right to be told whether the information exists
2. The right to receive the information within 20 working days, where possible, in the manner requested.

Thus, a public sector organization has a statutory duty to implement the Act and regulates how an organization, such as the NHS, must respond to a request for information that it may have stored. The Act applies to everyone within the organization of a Public Authority (inclusive of all aspects of the NHS) and has a huge impact on the public sector. The purpose of the FOI Act is to contribute to breaking down barriers within the NHS and help organizations learn from each other, as well as improving on their own performance. There are specific exceptions to the data that can be released that need to be discussed.

It does appear at first glance that the DPA and the FOI Act are at odds with one another, but really they are complementary in nature. Under the DPA, individuals have a right to access information about themselves, but third parties (unless appropriate permission has been sought or given) do not have rights to personal information. Under the FOI Act, individuals are not able to access to any personal information, either about themselves or others, but are able to access most other types of information. The main strategic objectives of the combined Acts are listed, and are not specific to the NHS or the hospital environment:

- to improve people's knowledge and understanding of their rights and responsibilities
- seek to encourage an increase in openness in the public sector
- monitor the Code of Practice on Access to Government Information
- develop a data protection policy, which properly balances personal information privacy with the need for public and private organizations to process personal information.

Both the DPA and FOI Act are the responsibility of the Lord Chancellor's Department at government level, namely the Information Commissioner.

3.4.2 Informed consent

The 2000 NHS Plan pledged that proper consent would be sought from all NHS patients and research subjects. The plan promised a review of consent procedures and the establishment of good practice in seeking consent throughout the NHS. The recommendations are that before a doctor or other healthcare professional treats a patient, they need to obtain their consent. The consent may be verbal (but it is better in writing), and outlines the key points and decisions of the procedure. A patient has the right to withdraw his or her consent at any time. The healthcare professional seeking consent has an obligation to ensure that the patient knows enough to be able to make an informed decision, and has had the opportunity to ask questions and consider alternative treatments and procedures. Patients must be told in advance if their treatment is likely to involve the removal of part of their body, and also if invasive sampling, such as taking blood samples, is likely to be part of their care plan. Sample taking during operations may routinely be used for teaching, research, or public health monitoring and, if this is the case, the patient must be made aware of the possible use of such samples and their consent obtained. If prior consent has been provided in writing, particularly in the case of the retention and storage of histological samples, there is no longer ambiguity on the outcome of a given series of events. Likewise, regarding consent to taking blood samples. It is readily assumed that if a patient has a request form and has allowed the phlebotomist to take a blood sample, they have given consent for those tests marked on the form but to no others. It is no longer considered appropriate to use excess samples for research purposes or for additional tests to be added without discussion with the requesting clinician and, in some circumstances, with the patient.

3.4.3 Human Tissue Acts

The Human Tissue Act (HTA) 2004 (England, Wales, and Northern Ireland) and the Human Tissue (Scotland) Act 2006 have replaced a variety of earlier legislation, including the HTA 1961, the Anatomy Act 1984, and the Human Organ Transplant Act 1989. All consent for the use of pathological samples for prolonged storage and/or research activities must be obtained in writing, and the paperwork be sufficiently detailed to establish that the patient is fully aware of the nature of the consent sought. The consent must be given voluntarily, be accompanied by sufficient information, be made by a legally competent individual, and be given with no form of coercion. Research must have the dignity, rights, safety, and well-being of the participant as the primary considerations. All studies must have appropriate written, informed consent, with clear arrangements about how that consent should be obtained. It is for the protection of the researcher that the consent record shows that the process has been two way, with the participant having the opportunity to ask questions and receive answers and, even better, if a friend or relative of the participant was also present at the interview and allowed to participate in the questioning. It needs to be understood that informed consent is a process and not simply a signature on a form, and that the informed consent form indicates that there has been adequate communication between the healthcare professional and the participant or patient. It is increasingly common that, rather than signing consent to a list of inclusive events, patients are asked to sign a general disclaimer; thus, specific consent is not required at a later date.

The HTA was passed by Parliament in the wake of public response to the activities at the Bristol Royal Infirmary and Alder Hey Hospital, which are discussed further in Chapter 17, where specific consent to remove, store, and carry out research on tissues retrieved at post-mortem from children having undergone unsuccessful major surgery had not been requested. It was clear at that time that much of the public had little grasp of post-mortem procedures, nor the fairly common practices of storing tissues for research and teaching purposes. The HTA has made the processes of post-mortem and retrieval of tissue much more transparent. Although Scotland has its own HTA, the contents of both acts are expressions of the same principles, such that an act:

Cross reference
Chapter 17, 'Quality assurance and management'.

- regulates removal, storage, and use of human tissues, which are defined as material that has come from a human body and consists of, or includes, human cells—now the HTA is in force, it is unlawful to carry out these licensable activities without an appropriate licence
- created a new offence of DNA 'theft', which is having human tissue with the intention of analysing its DNA without the consent of the person from whom the tissue came, from 1 September 2006
- made it lawful to take minimum steps to preserve the organs of a deceased person for use in transplantation, while steps are taken to determine the wishes of the deceased or, in the absence of their known wishes, obtain consent from someone in an appropriate relationship

- give specified museums in England discretionary power to move human remains out of their collections if the remains are reasonably believed to be those of a person who died less than 1000 years ago.

These acts also have a list of enforceable offences, with penalties ranging from fines to imprisonment or, exceptionally, both. These offences include:

- removing, storing, or using human tissue for scheduled purposes without appropriate consent
- storing or using human tissue donated for a scheduled purpose for something else
- trafficking human tissue for transplantation purposes
- carrying out licensable activities without holding a licence from HTA (with lower penalties for related lesser offences such as failing to produce records or obstructing the HTA in carrying out its legal responsibilities)
- storing or using human tissue, including hair, nail, and gametes, with the intention of its DNA being analysed without the consent of the person from whom the tissue came or of those close to them if they have died.

The scheduled purposes mentioned in the above list is a general term used to cover the description given to the patient or relative of the procedures and purposes for which that tissue was collected. Scheduled purposes differ between tissue samples and also between patients; they include whether the tissue is to be used only for diagnostic purposes or also for research, whether the tissue can be used by a single institution or be shared, and if it can be stored after use. If it is to be stored, then the scheduled purpose will include details of the storage, including its site and duration.

Clearly, the HTA and the need for informed consent go hand-in-hand. The written communication between healthcare professional and patient/participant acts as a legal document to provide the necessary proof that informed consent has been properly sought and agreed with regard to blood, fluid, and tissue samples requiring investigation, processing, and storage within laboratory medicine. This is particularly important in histopathology when acquired tissue from surgery or post-mortem may be further used for teaching or research purposes (Figure 3.4). Tissue must only be used and stored in the manner for which prior consent has been obtained.

FIGURE 3.4

Interior of a modern mortuary. The removal of any tissues must comply with the Human Tissue Act. Photograph courtesy of the London Borough of Haringey, UK.

3.5 Record keeping

While record keeping has been mentioned briefly as a method of communication within the laboratory (Section 3.1) and is also an HCPC standard, it deserves consideration in its own right within the remit of the Data Protection and the FOI Acts. It has already been mentioned that communication within the laboratory also includes completion of documents such as quality control charts, reagent log books, and troubleshooting sheets in the case of an instrument failure. The primary purpose of this type of documentation is so all laboratory staff can assess situations at a glance, whether certain basic procedures have been followed, whether a particular instrument is functioning correctly, or whether the availability of certain stock is below a certain level without having to find and question a specific member of staff. Record keeping in the laboratory is not only limited to practical procedures carried out on benches, but also covers complete patient and staff records.

Key points

When completing patient records, clearly it is essential that a report correctly identifies the patient, but it is also necessary that an historical record be kept of all the laboratory investigations pertaining to a patient.

When accessing a new set of results on a LIMS for authorization, there is frequently a so-called **delta check** performed, where the new result is compared with the previous existing record. If there is a significant change, the result may not be automatically passed for authorization and release directly by the LIMS but may be held in a queue and require manual validation or authorization by a biomedical scientist who can investigate further (see Case study 3.2).

Cross reference

You can read more about immunology in Chapter 13, 'Immunological techniques'.

3.5.1 Staff records

Staff records must also be kept by departmental managers. Records must exist for human resource purposes, such as sickness and absence rates, the availability of staff on different parts of shift systems, names of staff on professional registers, and details of mandatory training, competence assessment, and continuous personal development, that may be needed to fulfil professional responsibilities.

3.5.2 External bodies

Record keeping in the clinical laboratory also has to satisfy the many external regulatory bodies that have a bearing on current laboratory practices. There are many legal requirements to be met, as well as guidelines for good practice. Such bodies include the United Kingdom Accreditation Services (UKAS)/Clinical Pathology Accreditation (UK) Ltd (CPA), Medicines and Healthcare products Regulatory Agency (MHRA), the Health and Safety Executive (HSE), and the NHS Cervical Screening Programme. We will examine briefly the role of some external bodies in this respect.

United Kingdom Accreditation Services/Clinical Pathology Accreditation (UK) Ltd

CPA was a peer-reviewed organization that provided an external review of clinical pathology services and external quality assessment schemes. CPA itself has now become a wholly owned subsidiary of UKAS and as an ongoing development laboratories will, in future, be inspected and accredited according to International Organization for Standardization (ISO) standards (ISO 15189:2012). Although currently accreditation is desirable rather than mandatory (except for a few named exceptions such as laboratories providing services to the NHS Cervical Screening Programme), this situation could change in the future. Assessors visit laboratories and examine their services to confirm that defined standards of practice are being met in all areas of laboratory medicine. Thus, all laboratory records must comply with the most up-to-date and relevant ISO standards to meet the requirements of the inspectors.

Medicines and Healthcare products Regulatory Agency

The MHRA is an agency of the Department of Health and thus a government organization that was established to protect the health and safety of the public in relation to the dispensation of medicines and use of medical devices. The activities of the MHRA are particularly relevant in the field of blood transfusion, as blood products are classed as pharmaceuticals and administered to a patient following suitable prescription by a qualified person. The MHRA assesses against several EU directives, the majority of which have been incorporated into UK law as the Blood Quality and Safety Regulations, that have set standards of practice in blood transfusion, and which apply to blood establishments and hospital blood banks. Inspectors for the MHRA require evidence that these directives are being met for quality systems of working, the use of equipment and consumables, and that there are suitable traceability and record-keeping requirements for blood banks and facilities that may receive blood products for transfusion.

Health and Safety Executive

The HSE protects people against risks to health or safety arising out of work activities. Its role is discussed more fully in Chapter 4. The HSE is a government body that regulates all employers to ensure that all aspects of health and safety law are met in full. The HSE inspectors have a right to visit workplaces, and inspect practices and records to ensure that all relevant UK and EU legislation is being adhered to in full. The HSE has the power, in extreme circumstances, to close down and prosecute any employer found to be in breach of these regulations, particularly if previous warnings have been issued. All records concerning health and safety, which constitute UK and EU law, must be up to date and available for health and safety inspectors to access at any time.

Cross reference
Chapter 4, 'Health and safety'.

CASE STUDY 3.1 Transposing results

A postoperative patient had a short stay on the intensive care unit following major heart surgery before being deemed well enough to be transferred to a surgical ward. Prior to transfer, the ward clerk in intensive care telephoned the laboratory to ask for the most recent test results on the urea and electrolyte concentrations of the patient. Results were given by the support worker in accordance with the laboratory protocol and recorded by the ward clerk in accordance with the intensive care protocol. The electrolytes and creatinine were normal and the urea was 7.0 mmol/L (reference range: 3.0–8.0 mmol/L). Although this value is near the top of the reference range it is not an unexpected finding after major surgery. In the afternoon, a telephoned request came into the clinical biochemistry laboratory urgently requiring repeat determinations of urea and electrolytes on the patient, following treatment of the hyperkalaemia (high concentration of potassium (K^+) in the plasma) together with measurements of blood insulin and glucose concentrations. The laboratory noticed that there had been no previous hyperkalaemia and quickly assayed the sample, and found the K^+ to be 1.2 mmol/L (reference range: 3.8–5.2 mmol/L). So, what had happened and where had communications broken down?

A check of the record held in the LIMS revealed that the correct procedures had been carried out by the support worker, and a check of the results book in intensive care revealed that the ward clerk had also followed the correct procedures, and the results matched. The nurse transferring the patient had taken the results from the record book on intensive care and written them onto a scrap of paper, which she had taken with the patient to the surgical ward. She then verbally gave the results to the nurse receiving the patient, who also wrote the results on a scrap of paper. Later this nurse transferred the results into the patient's notes. Both nurses agreed that they had written results on paper, but no record of these handwritten notes remained. The patient notes revealed that there had been a transposition of the values of urea and K^+ results when compared with the electronic record held in the LIMS. The consequence was that when the surgical junior doctor came to review the patient, he noticed an apparent hyperkalaemia, which he immediately treated, reducing the concentration of K^+ in the patient's blood.

This course of events ended with the patient receiving an infusion of K+ and suffering no ill effects. However, this could easily have resulted in the tragic death of the patient and highlights the importance of following SOP when communicating results verbally, either by telephone or face to face.

CASE STUDY 3.2 Identifying samples is essential

A patient whose test reveals a blood haemoglobin result of 140 g/L (reference range: 140–180 g/L), 24 hours later has a haemoglobin concentration of only 75 g/L, and therefore requires a blood transfusion. If the patient has undergone major surgery, such as requiring a prolonged period of time on a cardiac bypass machine, or has just had a massive gastrointestinal bleed, the discrepancy between the two results is both explained and appropriate. If there has been no such clinical event to explain the change, then it is unlikely that the samples have originated from the same patient and there has been an error in correctly identifying the sample.

Another area of laboratory medicine where it is essential that full historical records be maintained is within the discipline of blood transfusion. If a patient develops an antibody to a particular red cell antigen, there is a possibility that the titre of that antibody may change with time. Examples of these are the antibodies of the Kidd system, such as anti-Jka. Not all antibodies to erythrocyte antigens remain detectable in the serum at a consistent titre and some may decrease with time. A patient may have presented several years previously with a positive antibody screen and such an antibody identified. When returning in the present day for major surgery and possible transfusion, the antibody screen may be negative, as titres are so low as to be undetected by routine screening methods. If that patient required a blood transfusion, it would still be necessary that Jka-negative blood be used as the introduction of the antigen would promote a massive response in terms of antibody production, and the patient would be likely to suffer an incompatible transfusion reaction that could possibly be fatal.

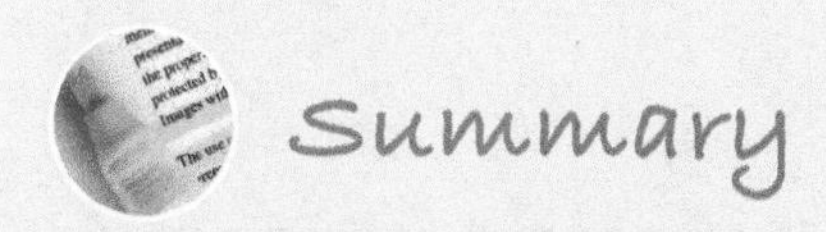

Summary

- Effective communication within a clinical laboratory, between its staff and a variety of external people and organizations, is necessary. This is because biomedical scientists work as part of a laboratory team, must follow local policy and procedures, and comply with legislation, and are involved in a process that provides essential information contributing to patient care.
- Biomedical scientists are healthcare professionals, and must never act in a manner that causes offence to others or to bring the biomedical science profession into disrepute by inappropriate communications of any means.
- Biomedical scientists and other laboratory staff are required to work in structured teams, following set procedures and protocols, and must cooperate and communicate appropriately with others within the laboratory environment.
- Biomedical scientists must communicate in a professional manner and at an appropriate level when dealing with the public, patients, and the wider healthcare groups of professionals and support workers.
- Due to their normally remote working circumstances, biomedical scientists often communicate indirectly, whether by verbal or non-verbal means. This type of indirect communication may be the only way by which biomedical scientists are judged by the wider healthcare community.
- Biomedical scientists must conform to current legislation, such as the Data Protection, Freedom of Information, and Human Tissue Acts, and be aware of the consequences of their actions if inappropriate records are kept and audit trails are not maintained.
- Biomedical scientists must ensure that all work carried out within laboratory medicine has appropriate consent and that confidentiality is always maintained.

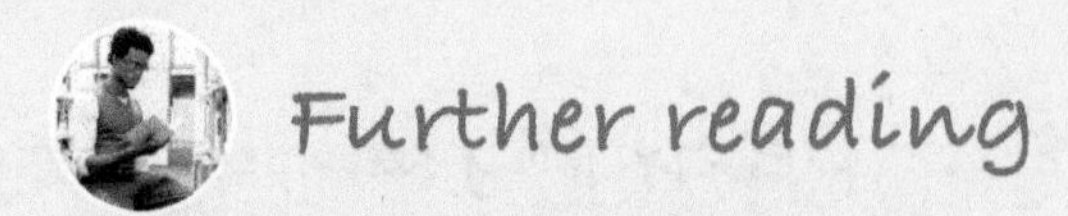

Further reading

- Divan A. ***Communication Skills for the Biosciences***. Oxford: OUP, 2009.
- Pitt SJ, Cunningham JM. ***An Introduction to Biomedical Science in Professional and Clinical Practice***. Chichester: Wiley-Blackwell, 2009.

  Chapter 3, 'Communication for biomedical scientists', is worth browsing through.
- ***Standards of Conduct, Performance and Ethics***. HCPC, London, 2016.
- ***Standards of Proficiency–Biomedical Scientists***. HCPC, London, 2014.

Useful Websites

- www.ibms.org
- www.hcpc-uk.org.uk
- www.ukas.com/services/accreditation-services/clinical-pathology-accreditation/
- www.mhra.gov.uk
- www.hse.gov.uk
- www.hta.gov.uk
- www.nhsprofessionals.nhs.uk

Questions

1. Guidelines issued by the Institute of Biomedical Science concerning the communication of results by telephone:
   - **(a)** Are a legal requirement
   - **(b)** Can be adapted for individual situations
   - **(c)** Are an example of best practice
   - **(d)** Must be adhered to by all members of the IBMS
   - **(e)** Are only of relevance to NHS employees.
2. Under the DPA, sensitive information is defined as:
   - **(a)** Information that relates to the age of an individual
   - **(b)** Data that cannot be sold on to another party
   - **(c)** Information that can be released to a third party under the FOI Act
   - **(d)** Information that can only be held with explicit consent
   - **(e)** Information that must be removed from a database after 5 years.
3. The key principles of the Caldicott Report do not include:
   - **(a)** The appointment of a Caldicott Guardian
   - **(b)** Information held outside of the NHS
   - **(c)** Instructions stating which personnel should have access to medical records held by the NHS
   - **(d)** Instructions informing NHS staff of their legal obligations under the DPA
   - **(e)** Instructions that all staff must comply with the law.

4. Indicate which of the following statements are **TRUE**:
   (a) The NHS number is given to all babies born in the UK
   (b) Telephone calls to the laboratory requesting tests to be added to a previous sample must be documented
   (c) Acronyms for laboratory tests may be used freely, providing the laboratory responds with a published list that explains what they mean
   (d) The DPA applies only to those members of NHS staff who are named on a professional register
   (e) Electronic data on a screen must be viewed as a written method of communication.

5. The text refers to point-of-care blood glucose concentrations at a diabetic clinic and international normalized ratio (INR) results at a coagulation clinic as examples where it may be appropriate to give results to patients. List other occasions where results may be given directly to the patient.

6. With reference to stored electronic data, what are the potential consequences for an individual if medical data are lost?

Answers to self-check and end-of-chapter questions are available at the end of the book.

4

Health and safety

Hedley Glencross and Alison Taylor

Learning objectives

After studying this chapter, you should be able to:

- Identify hazards that represent a risk to yourself or to the safety of others
- List your statutory duties under the Health and Safety at Work Act
- Discuss the regulations that drive health and safety practices in the workplace
- Conduct a basic risk assessment and be able to outline more specialized assessments
- Use personal protective equipment safely and effectively, and be aware of their limitations and maintenance requirements

Introduction

Occupational health and safety are any **system** or procedure established to ensure the health, safety, and welfare of every person coming into a work place. **Health** is preventing illness occurring to the body or mind, and **safety** is the prevention of physical injury. The term **welfare** describes facilities, such as sanitation, heating, lighting, and changing rooms, that are not direct causes of ill health or injury, although deficiencies in these areas could, and almost certainly would, contribute to poor health and safety.

The law defines standards of health and safety (Health and Safety at Work Act, Section 4.3) and, once identified, a **hazard** or deficiency must be corrected. In health and safety practice, a key way to identify deficiencies is by **workplace inspection**, which is a specialized audit, and by performing an assessment of the risk (Section 4.4). Risks are the combined likelihood of a hazardous event occurring, and the severity of injury or ill health that it could cause. Any deficiencies or hazards identified by either route must have corrective actions or control measures implemented to ensure they do not cause ill health or injury. The effectiveness of these controls needs to be assessed when the inspections and/or risk assessments are repeated or reviewed. The regulatory body for health and safety legislation is the **Health and Safety Executive** (HSE), which carries out these types of inspections and publishes guidance to help employers and employees comply with the law (Figure 4.1). The HSE also produces reference materials and institutes advertising campaigns to raise awareness of health and safety risks. Within the UK, the Institute of Occupational Safety and Health (IOSH) is the recognized body for health and safety professionals, providing training, guidance, and information, with members in over 120 countries.

This chapter will examine the legal requirements and regulations governing occupational health and safety, including the methods used to assess and control risks. In combination, these measures attempt to ensure the health and safety of everybody in a workplace. Naturally, however, we will focus on the risks and controls particular to biomedical science laboratories.

FIGURE 4.1
Logo of the Health and Safety Executive (HSE).

4.1 Hazards

A hazard is any substance, activity, or process that may cause harm. On a daily basis, you may encounter many potential hazards (e.g. crossing a busy road). However, by using subconscious risk assessments and applying actions to minimize the dangers of the hazard, you will (we hope!) avoid injury and ill heath. Unfortunately, in working environments, hazards, especially those found in clinical laboratories, may not be easily identifiable. Thus, it is difficult to take actions to minimize their effects. Training, labelling of substances, using standard operating procedures (SOP), and acting on all the relevant information, such as risk assessments, will assist in identifying hazards related to each task undertaken.

In this section we will identify common laboratory hazards and discuss the process of conducting formal risk assessments to assess these hazards, the risks they pose, and the controls required to reduce or eliminate them.

Hazards may be classified into six main types:

1. Biological
2. Physical, mechanical, or equipment
3. Chemical
4. Electrical
5. Fire
6. Environmental.

4.1.1 Biological hazards

Biological hazards take five main forms:

1. Bacteria
2. Viruses
3. Fungi
4. Moulds
5. Prions.

Bacteria are single-celled prokaryotic microorganisms that can live independently or dependent on host organisms. Examples of bacteria include the causative organisms of tuberculosis and legionella. Viruses are nucleoprotein complexes, which are about 200 times smaller than bacterial cells. All viruses are obligate parasites in that they can only grow or replicate within a host cell. Examples of viruses include agents that cause influenza and AIDS. Fungi are single-celled or multicellular eukaryotic organisms, which have cell walls rich in chitin. They reproduce by producing spores. Fungi cause skin conditions such as athlete's foot, some forms of eczema, or they may cause allergic reactions. Moulds are microscopic fungi, which can have similar effects on health to the above, and can also lead to respiratory sensitization because their small size allows them to evade filtration by the respiratory system. Thus, they can cause or contribute to conditions such as asthma. Prions are protein molecules that adopt **toxic** conformations and cause infectious diseases such as scrapie, bovine spongiform encephalopathy (BSE), and Creutzfeldt-Jakob disease (CJD).

> **Cross reference**
> You can read more about the roles of biomedical scientists in Chapter 1, 'Biomedical science and biomedical scientists'.

The Advisory Committee on Dangerous Pathogens (ACDP) classifies all biological hazards into one of four categories or groups (groups 1-4). These are published by the HSE. The group assigned to an agent defines the level of containment required to control the risks associated with it, which are defined in **Control of Substances Hazardous to Health** (COSHH, Section 4.5) regulations.

Group 1 hazards are biological agents that are unlikely to cause human disease. In contrast, group 2 hazards are those agents that can cause human diseases. However, effective treatments and/or prophylaxes are available to combat these agents and so this group is unlikely to spread in the community. Biological agents of hazard group 3 can cause severe human disease and present a high risk to health. There is a risk of the disease spreading in the community, but there is usually effective prophylaxis or treatment(s) available. Hazard group 4 agents can cause severe human disease and present a serious risk. There is also a likelihood of the disease spreading in the community because usually effective prophylaxis or treatments are not available.

4.1.2 Physical, mechanical, and equipment hazards

Physical hazards or those associated with mechanical devices or other equipment include entanglement, entrapment, being struck by or striking an object (impact injuries), abrasion, cutting, and shearing. These hazards most often cause injury but in some cases lead to ill health. Slips/trips/falls also come into this category and are common causes of occupational injuries. Health problems associated with extensive computer or display screen equipment (DSE) work or repetitive muscular actions are included in this group.

4.1.3 Chemical hazards

Chemical hazards can be present in a number of physical forms; these include solids, dusts, gases, **vapours**, mists, fumes, and liquids. The severity of ill health effects they cause can vary considerably.

Chemicals agents can enter the body in a variety of ways, dependent on their physical form. For example, an **irritant** chemical in *solid* pellet form is unlikely to enter the body but on contact with the skin can lead to relatively minor health effects. Hence, its effects are easily controlled by wearing safety gloves (Section 4.7). However, in powder form the same chemical could enter the body by inhalation and have significant effects, such as shortness of breath and asthma. The control of this greater hazard may possibly require the use of a ventilation hood. **Dusts** are categorized separately as they have significant specific risks dependent on the size of the particle. A flammable substance in the form of a fine dust will mix easily with air and can become explosive. Fine dusts, otherwise known as respirable dusts, present a significant hazard because of the ease with which they are inhaled.

Liquids exist in different forms depending on the ambient temperature and their boiling points. Liquids are less predictable and more difficult to handle than solids. Splashes and spills increase the chance of contact with the skin or entry into the body by ingestion or through the eyes. Gases and vapours are produced by heating liquids or when working in higher temperatures. They can enter the body through the skin, gastrointestinal tract, and eyes, or by inhalation. Spraying processes can form mists, even at low temperatures, and this increases the chances of the liquid causing ill health. **Fumes** are mists or vapours containing very small metallic particles. In a laboratory, a fume could be formed when the bulb in a mercury microscope 'blows'.

The different types of chemical hazards and their classification are discussed under the COSHH in Section 4.5. Some chemicals are also radiation hazards and so are classified differently; their use is governed by the Radiation Protection Act.

4.1.4 Electrical hazards

Any electrical equipment, either fixed position or portable, has the potential to cause an electric shock. This can be particularly so in a clinical laboratory when electricity and water are often used in close proximity. They can pose electrical hazards, with consequences ranging from minor injury to serious injury or even death. Portable appliance testing must be carried out regularly to check the safety of the instrument, but there must also be a more frequent inspection of plugs, cables, and casing. Any potential hazards arising from the use of electrical equipment in close proximity to chemicals need assessing and controlling.

4.1.5 Fire hazards

Fires can only start if all parts of the fire triangle—oxygen, fuel, and a heat source—are present (Section 4.6). Therefore, flammable substances, including liquids and gases, or substances or equipment giving off heat can all fall into this group of hazards.

4.1.6 Environmental hazards

Environmental hazards include loud noises, vibrations, light, excessive humidity, and/or temperatures, and a lack of ventilation. Sometimes these are the most difficult hazards to identify, as they are often experienced daily and so become accepted or tolerated. These hazards often cause welfare-related issues. In extreme situations, or when particularly vulnerable people are affected, they can result in ill health.

SELF-CHECK 4.1

List at least two examples of each of the major types of hazard discussed above.

4.2 Routes of entry into the body by chemical and biological agents

Chemical and biological hazards can only cause harm if they enter the body. Controlling these hazards therefore requires knowledge of their possible routes of entry.

Airborne biological and chemical agents can gain entry into the lungs following inhalation through the mouth or nose. Here, they can be absorbed and then transported around the body in the bloodstream. Thus, they can potentially affect every organ in the body. The skin prevents the absorption of many agents into the bloodstream. However, some chemicals can enter the body this way through pores on the surface of the skin. More commonly, absorption occurs through wounds, burns, or breaks in the skin in, for example, conditions like eczema. A number of biological and chemical agents can enter the body through the gastrointestinal tract following their ingestion, and potentially gain access to the bloodstream. This is not a common route of entry into the body, but may occur if good hand hygiene and basic laboratory health and safety practices are not followed. Finally, some agents can enter the body by injection due to accidents with hypodermic needles and compressed air lines, which can puncture the skin giving direct access into the bloodstream.

SELF-CHECK 4.2

To what types of hazards would a person collecting a venous blood sample be potentially exposed? What are the possible routes of entry of the hazard(s) into the body?

There are many regulations governing health and safety; however, in this chapter we will focus on the principal ones most likely to affect you while working in a biomedical laboratory within the UK. Health and safety across Europe is governed by the European Union (EU) through directives, which are incorporated into the laws of the member states.

4.3 Statutory framework for health and safety

Prior to 1974, health and safety legislation in the UK was only targeted at specific industries, leaving many workers unprotected. Other people, such as contractors and visitors, entering a workplace were simply not protected in law. The introduction of new regulations was too slow to reflect changing practices, leading to a system that was reactive rather than proactive. However, the legislative framework changed with the Health and Safety at Work Act 1974 (HASAWA).

4.3.1 Health and Safety at Work Act 1974

The HASAWA resulted from a review of occupational health and safety in 1970 by Lord Robens. The HASAWA is an enabling act, which means the Secretary of State can pass new regulations without the lengthy process of passing them into law by Acts of Parliament. Thus, the process of developing health and safety legislation became more proactive.

The HASAWA established the Health and Safety Commission (HSC) whose duties were to draw up new regulations and to enforce them through the actions of the HSE. In April 2008, the two bodies merged under the name, HSE. The HSE is responsible for issuing **Approved Codes of Practice** (ACOP) for most regulations. The role of ACOPs is to explain the regulations and the detailed requirements necessary to comply with them. The HSE produces two types of **health and safety guidance** (HSG) **documents**, legal and best practice. The legal guidance series of booklets covers the technical aspects of the regulations and usually includes the regulations and ACOP. Best practice guidance is published in the HSG series for specific areas; an example being the Decontamination of Equipment Prior to Inspection, Service and Repair HSG (93) 26. Both ACOPs and HSGs are good sources of information and are readily available on the HSE website, which is listed at the end of the chapter.

The HSE has teams of inspectors to cover all types of workplaces, including hospitals and laboratories, assisted by local authority inspectors covering premises such as hotels, shops, and restaurants. They can inspect at any time and have similar powers to the police in being able to collect evidence, and take and retain samples and photographs. One of two levels of enforcement notification is issued if deficiencies or dangerous practices are identified. An **improvement notice** follows contraventions of a regulation, together with a set date for its resolution. Alternatively, a **prohibition notice** halts the activity until a satisfactory resolution of the problem is established.

Should the HSE prosecute a health and safety offence successfully, the penalties imposed range from fines of up to £5000 for employees and £20,000 for employers, if the case is tried in a magistrates' court. Failure to comply with an enforcement notice can result in imprisonment for up to 6 months. In the crown courts, fines are unlimited with imprisonment for up to 2 years for the same offence.

SELF-CHECK 4.3

List the role(s) of the HSE.

Duties of employers

The HASAWA places a duty on employers to ensure, so far as reasonably practicable, the health, safety and welfare of all employees. It also details what that may entail, such as providing a safe place in which to work and ensuring all procedures associated with work are safe to perform. In addition, the act sets out the need for set policies with associated processes for organizing arrangements such as risk assessments and consultation with employees. The act protects visitors to a workplace, outside contractors, and people living near it or passers-by, by placing a duty of care on the employers to protect all those affected by its activities. Duties are also placed on the employer to ensure all materials, including equipment and substances, are without risk to health. The Act also affects the practice of suppliers by placing on them a duty to design, manufacture, import, and supply equipment or substances that are, again, as far as is reasonably practicable, without risk to health.

Duties of employees

The HASAWA places two main duties on employees:

1. To take reasonable care for the health and safety of yourself and others affected by your acts or omissions
2. As an employee, you must cooperate with your employer to enable them to fulfil their legal obligations.

A further section, often known as the 'horseplay section', impresses a duty on everyone not to interfere or misuse safety equipment. The act also states that employers must provide staff with all **personal protective equipment** (PPE; Section 4.7) required to fulfil their work duties free of charge.

4.3.2 The 'six pack'

The UK is part of the EU and European law, passed by the European Commission, is therefore an integral part of our legal system. The resulting uniform approach to health and safety in the workplace

FIGURE 4.2
Documents contained in the 'six pack', the six sets of regulations that originated in the European Union and which became part of UK health and safety legislation in 1992. © Crown copyright.

across the EU is designed, in part, to remove barriers to trade between its member states. The first effect of the EU on UK health and safety law occurred in 1992, when six sets of regulations were introduced:

1. Management of Health and Safety at Work Regulations
2. Workplace (Health, Safety and Welfare) Regulations
3. Health and Safety (Display Screen Equipment) Regulations
4. Personal Protective Equipment at Work Regulations
5. Manual Handling Operations
6. Provision and Use of Work Equipment Regulations.

These regulations are contained in a set of documents familiarly called the 'six pack', which you can see in Figure 4.2.

Management of Health and Safety at Work Regulations

The Management of Health and Safety at Work Regulations (known colloquially as the 'management regs') were updated in 1999. They provide the framework for the risk assessment process that we will discuss in the next section. None of these regulations contradicts those of the HASAWA but, in general, expands on them and redefines some of the duties of both employer and employee. Thus, the employer must undertake suitable and sufficient risk assessment, and establish health and safety arrangements that include planning, control, monitoring, and review of all work-associated procedures. To achieve effective health and safety management, the employer should use competent specialists to advise them on how to comply with their legal duties.

Employers must ensure that all health and safety information, particularly findings of risk assessments are communicated to all employees. They must also provide specific health and safety training. Employers must put in place emergency procedures and ensure all workers are aware of them. The employer must also co-operate on health and safety matters with other employers sharing a workplace.

With regard to employees, Management of Health and Safety at Work Regulations require that they follow their training and instructions whenever they use equipment or handle substances. They must also report anything that may be dangerous, and any problems with health and safety equipment or arrangements to their manager or health and safety officer.

SELF-CHECK 4.4

Give an appropriate example of something that has to be reported to a manager under the duties imposed on employees by the Management of Health and Safety at Work Regulations.

Workplace (Health, Safety and Welfare) Regulations

The Workplace (Health, Safety and Welfare) Regulations, known as the 'welfare regs', give the minimum requirements for a workplace to ensure the welfare of employees. These are generalized regulations, and cover elements of the working environment such as its ventilation, temperature, lighting, floors, furniture, wastes, and accommodation space (room dimensions). The associated ACOPs give information that is more detailed. The regulations also cover the use of corridors and staircases within premises that give access and egress to and from places of work.

The *safety* aspects covered by the welfare regs include maintenance of equipment, windows, doors, and traffic routes. Maintenance also includes general cleaning and the emphasis should be on planned maintenance, which is preventative, rather than repair after breakdown. The *welfare* section covers the provision of sanitary and washing facilities, drinking water, facilities where workers can rest and eat meals, and secure space to store items such as outdoor coats.

Health and Safety (Display Screen Equipment) Regulations

The Health and Safety (Display Screen Equipment) Regulations, usually known as the 'DSE regs' are designed to establish systems that prevent ill health occurring when using a computer or any instrument that incorporates a display screen or visual display unit. The regulations define a user as someone who works continuously with DSE in excess of 1 hour each working day. They entitle a user to an eye test to be paid for by the employer. If glasses are required specifically for DSE work, only the employer must pay for them. The regulations prescribe suitable and sufficient risk assessment of the workstation, its surrounding area, the software being used, and the user.

The regulations also set minimum specifications for the workstation, giving details on the type of chair, dimensions of the desk or workspace, and how adjustable the DSE must be. The DSE regulations have been written to help prevent the occurrence of ill health, such as visual, musculoskeletal problems and stress. Therefore, taking regular short breaks, away from the DSE when possible, is a requirement. Regular short breaks, to answer a phone or to talk to someone in a different work area are good examples of this, but if these opportunities do not occur, *formal* breaks must be taken.

There is also a duty on the employer to provide information on these aspects of the work, and to train staff to be aware of the risks and how to avoid them.

Cross reference

You can read more information on formal breaks within the European Working Time Directive (WTD) documentation.

The study of how well a person is suited to their work in combination with their workstation is known as **ergonomics**. Ergonomic studies assess the physical and mental capabilities of the individual, including their perceptions, interactions, movements that occur while doing the job, and the surrounding work area. This includes elements of the task such as the positioning of equipment, how easily the individual can reach it without excessive stretching that over time might cause discomfort or injury, and avoiding stress to the individual by allowing them, rather than the process, to control the speed of the work. Workstations must be adjustable, as individuals have different physical and mental capabilities. The study of variations in physical characteristics, such as height, arm, and leg length, is known as **anthropometry**, and is a key component to ergonomic assessments. Poor ergonomics can lead to work-related upper limb disorders (WRULDs) such as repetitive strain injury (RSI), tenosynovitis (inflammation of the tendons), carpal tunnel syndrome (pain in the wrist and hand), and frozen shoulder.

SELF-CHECK 4.5

Name the elements that should be considered in a DSE risk assessment.

Personal Protective Equipment at Work Regulations

The Personal Protective Equipment at Work Regulations place a duty on the employer to provide and maintain PPE when its requirement is identified by risk assessment. PPE, by definition, only protects an individual and so potentially leaves others in the workplace at risk. Thus, it should only be used as a last resort, although it can be combined with other control measures such as working in a fume cupboard or other local exhaust ventilation (LEV) system. As with the other six-pack regulations, there is a duty on the employer to provide information, instruction, and training to anyone who is required to use PPE. These must include the reasons why PPE should be used, including the risks it is designed to control, and how the equipment should be used and maintained. The regulations place a duty on employees to use PPE as they have been instructed and trained to do so, to maintain and store it after use, and to report any obvious defects.

Any item of PPE that is provided to the employee must be to control the risk associated with the process, and be appropriate for the conditions of use and the characteristics of the wearer. For example, face masks provided to prevent inhalation of a dust may not adequately protect someone with facial hair, which prevents them fitting correctly. The regulations place a duty for these considerations to be formally assessed for risk and, when multiple PPE is required, that all of the items must be compatible and not introduce additional hazards. The risk assessment must be reviewed if any part of the process changes, to ensure that PPE is still required and that what is provided is adequate for the purpose.

All PPE must be regularly maintained; the level of maintenance should be proportional to the potential risk and type of PPE. It may also be necessary to keep a written record of all maintenance. For example, a respirator requires regular inspections and planned preventative maintenance that must be recorded, whereas safety gloves may only require a visual inspection by the user. The use of disposable PPE can simplify this process, but all users must be aware of the necessity to discard and replace disposable PPE items. All PPE must be stored appropriately.

SELF-CHECK 4.6

What factors should be considered when choosing PPE for a task?

Manual Handling Operations

Manual Handling Operations (MHO) cover those actions required to move any load by human effort. These include pushing, pulling, lifting, putting down, and carrying loads. The regulations place a duty on the employer to put in place processes that avoid the need for MHO. When this is not possible, the MHO must have been subject to a suitable and sufficient risk assessment. Risk assessments for MHOs must cover five areas:

1. Environment
2. Task
3. Individual
4. Load
5. Equipment used.

These can be remembered using the acronym ETILE. To make an assessment of the possible load, manufacturers are required to indicate its total weight and the heavier side of uneven loads. Once a risk assessment has taken place, there is a duty on the employer to reduce the risk of injury by introducing appropriate controls such as task layout, work rotation, PPE, teamwork, and training. As with all risk assessments (Section 4.4), they must be reviewed if any of the processes change.

These MHO regulations also place duties on employees to follow the safe systems of work and use any equipment provided in a proper manner. In practice, in a laboratory setting, MHOs are not usually a significant part of the work, but MHO should be considered when ordering stock to ensure materials arrive in units that can be moved, stored, and used without risk of causing injury.

SELF-CHECK 4.7

What does the acronym, ETILE stand for?

Provision and Use of Work Equipment Regulations

The Provision and Use of Work Equipment Regulations (PUWER) are designed to ensure that work equipment is safe to use no matter which country it originated in or how old it is. The Lifting Operations and Lifting Equipment Regulations (LOLER) 1998 apply to equipment used such as hoists and trolleys, for lifting or lowering loads, but the duties placed upon employers and employees are similar to those in PUWER.

The PUWER regulations place a duty on the employer to ensure all work equipment is suitable for its purpose and, when deciding what equipment to purchase, the health and safety of its users is taken into consideration. Furthermore, all work equipment must be maintained in an efficient state and working order, and kept in good repair with planned preventative maintenance as required. All maintenance activities and inspections must be recorded. The regulations are prescriptive about numerous features including systems such as emergency stop controls, guarding of dangerous parts, protection against high temperatures, and procedures to isolate equipment from power supplies. As with all the six-pack regulations, the employer has a duty to provide information, instruction, and training to anyone who uses the equipment so that they are fully aware of its operation, risks, and emergency procedures.

4.4 Risk assessment

There is a legal requirement to carry out specific risk assessments for certain types of work such as DSE and MHOs (Section 4.3). The general duty for an employer to conduct risk assessments is imposed by the Management of Health and Safety at Work Regulations. Risk assessments and other arrangements must be fully described in writing ('documented') when there are more than four employees at a workplace.

To clarify the processes involved in risk assessment we should first consider some of the definitions used:

- hazards are any substance, equipment, or process with the potential to cause harm
- risk is the likelihood that the substance, equipment, or process will cause harm.

Using the terms in these specific ways makes it possible to define clearly an activity; for example a process could present a high hazard but a low risk because the hazard is well controlled. The level of risk is governed by the severity of the harm that could be caused, often referred to as impact, and the likelihood that harm will occur.

The phrase used to describe the level of risk assessment required in many of the regulations is *suitable and sufficient.* All risk assessments must identify the major risks and ignore trivial ones, identify controls, and list them in order of importance. Assessment should remain valid for a reasonable period of time. In practice, this means that the level of detail in a risk assessment must reflect the degree of risk, so that high-risk activities have more detailed assessments that focus on the most dangerous hazard. Risk assessments must be regularly reviewed. The time scale for review should reflect the level of risk, with high-risk activity assessments being reviewed more frequently than those of low risk. However, if any part of the process changes, which may include a change in the person performing the activity, the current risk assessment must be reviewed immediately. This cycle is shown in Figure 4.3.

4.4.1 Types of risk assessment

Risk assessment can either be *qualitative*, using terms such as high, medium, and low, or *quantitative* giving a numerical value. Quantitative risk assessments are usually done for high-risk activities (Figure 4.4), and also in large organizations when a guidance tool of numerical values to be used for different levels of risk is provided to give consistency.

FIGURE 4.3
Outline of the risk assessment cycle.

Increasingly, risk assessments are carried out at an organizational level. These assessments look at the same hazards, and also consider the risks to the service or business, and service users or patients. For example, if a fire hazard in a laboratory that stored flammable liquids was poorly controlled, a fire might occur. If a fire does start, it is unlikely that anyone would be seriously hurt provided that the employees followed the correct evacuation procedures. However, a fire that is not rapidly extinguished may cause considerable damage to the laboratory, which may mean it could not operate for some time. The cost of this in terms of repair, effect upon patient care, and the need for a contingency plan would form a part of a business risk assessment.

4.4.2 Risk assessment process

The HSE publishes guidance that outlines the five steps on how to carry out a risk assessment:

1. Identify the hazards
2. Decide who might be harmed and how
3. Evaluate the risks and decide on precautions
4. Record your findings and implement them
5. Review your assessment and update if necessary.

Some systems regard step 3 as two separate steps, because once evaluation of the risks has occurred, they should be arranged in order of severity ('prioritized') before precautions are considered.

Generally, at least two people should perform a risk assessment: one who is familiar with the activity and the workplace, and another who is less familiar with both, as it is easy to overlook potential, but well-controlled, problems if one is used to the activity and environment. A person less familiar with the workplace will see it with 'fresh eyes'. They may be able to identify hazards and suggest alternative controls that those working in the area had not considered. The use of two contrasting assessors therefore increases the likelihood of identifying hazards, which must be recorded together with details of the persons carrying out the assessment, and details of the activity and work area being assessed.

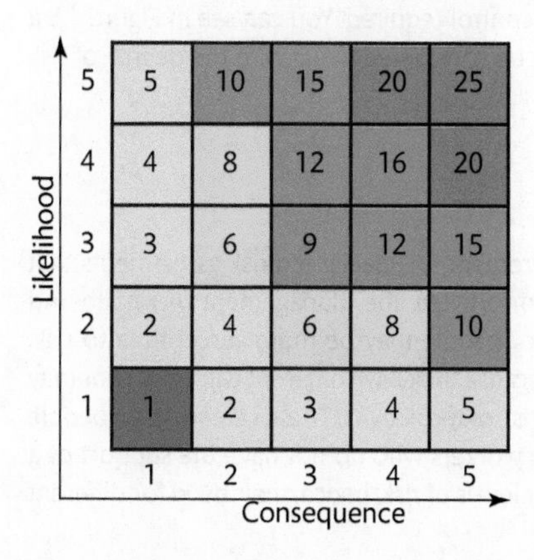

FIGURE 4.4
Example of a 5 × 5 risk matrix used to perform quantitative risk assessments.

When deciding who might be at risk, it is necessary to consider vulnerable persons, such as young workers or expectant mothers, in addition to anyone else who comes into the workplace including visitors and contractors. The HSE provides a sample form in their written guidance, although many organizations have their own format.

Hazards should be evaluated in terms of their potential danger, and then arranged in order of decreasing danger. In most systems, controls already in place are also considered when evaluating risks, but these processes must be checked as part of the assessment. Any actions and controls must have completion dates or priorities set to them. The degree of risk associated with the process (high, medium, or low), and the action completion dates will indicate the timescale for reviewing the assessment. A specific individual must take responsibility for any required action and this individual must be identified in writing.

Involving all staff in risk assessment is essential and, indeed, a legal requirement. It also identifies and introduces controls that will be workable and have the support of the team needing to implement them.

4.4.3 Hierarchy of risk control

The following options must be considered when deciding what are appropriate controls for an identified hazard. If possible, the risk should be removed, but if this is impossible, something less hazardous should be used as a substitute, for example, by replacing a dangerous chemical with one that is non-hazardous or less harmful. When this is not possible, access to the hazard needs to be prevented, such as by placing hazardous liquids in drip bottles, rather than in open baths. Alternatively, exposure should be reduced, for example, by enclosing the process within a fume cabinet or minimizing the time spent doing the activity. As always, defining and following safe systems of work, using PPE, providing adequate information and training, all help to eliminate risks. In reality, a combination of all these approaches is usually required to control the hazards associated with any one particular activity, and provision of training and information is always essential.

4.4.4 Legal interpretation

The HASAWA impresses a duty on the employer to ensure, so far as reasonably practicable, the health, safety and welfare of all employees. This is the most common level of duty in health and safety law, and was first described by Judge Asquith in 1949 in the case of Edwards versus the National Coal Board. That case decided that 'reasonably practicable' was different to 'physically possible'. This means that the level of risk has to *balance* against the sacrifice involved in implementing its control. If the sacrifice grossly outweighs the risk, the employer would have discharged his duty if he did not implement the control, as it was not reasonably practicable. Put in another way, if the risk is relatively small compared with the cost of the measures required to control it no action is necessary. The level of risk must consider both likelihood and severity and the cost of implementing the control should include time, trouble, and expense.

During any risk assessment process, this balance needs consideration for each of the types of controls set out in the hierarchy of control. At some point, the cost of the control will be at an acceptable balance to the risk; this gives the minimum level of control required. You can see in Figure 4.5 a diagrammatic representation of how time and cost must be considered along with the degree of risk when deciding what controls to put in place.

4.4.5 Specialist risk assessments

Some of the regulations we have discussed have a requirement for specialized risk assessments that consider specific risks and control hierarchy. In addition to these, the Management of Health and Safety at Work Regulations recognize that some types of people may be more susceptible to risk. Specific risk assessments are necessary for such people because they have different capabilities or may be more susceptible to hazards, therefore increasing the risk of the activity. These categories of people include the young, expectant or new mothers, and lone workers who do not have the support of a team, and so are subject to additional hazards and higher levels of risk; hence, their need for different systems of work and emergency procedures.

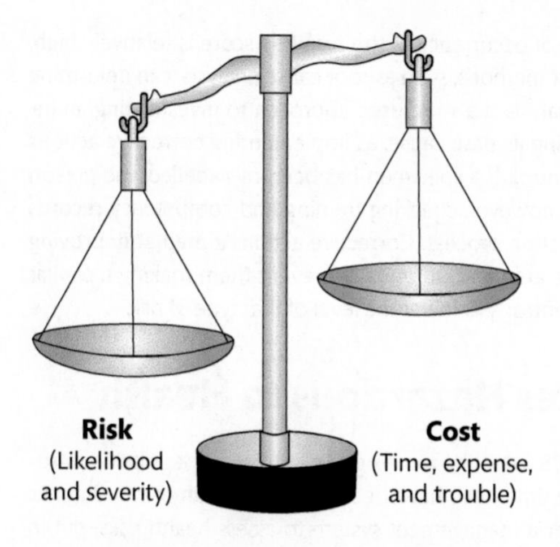

FIGURE 4.5
Schematic indicating how costs should be considered against the degree of risk when deciding which controls should be established.

4.4.6 Reporting of incidents

An **incident** is an unplanned event that causes an adverse outcome, such as injury, ill health, and/or damage to property, equipment, or the environment. **Near misses** are events that could have had an adverse outcome. Incident reporting and an appropriate investigation of any incident are essential parts of a health and safety management system. This can turn into a reactive approach, but the system can be proactive, contributing to improving health and safety, when used to report near misses and hazards. Reporting and investigating near misses are essential as generally 10 near misses occur for every minor injury and the number of minor injuries relates to the occurrences of serious injuries. The Birds Accident Triangle of 1969 shows how many minor incidents occur for every major event (Figure 4.6). Thus, near misses are 'free lessons' that help prevent more serious incidents.

There is a legal requirement, under the Social Security (Claims and Payments) Regulations for employers to keep a written record of any accidents that occur in the workplace. Stationery offices produce an accident book, but the records can be in any form as long as sufficient detail about the accident and persons affected is recorded. The employer is also required to investigate the cause of the accident and enter this into the record.

All organizations have their own systems for reporting incidents and near misses. In general, information about the event, the people involved, and contact details are collated and recorded. In some cases, damaged equipment is quarantined because it may be necessary to examine it as part of the investigation. Often, immediate actions are necessary to make the area and/or activity safe, and these must be noted. If a medical practitioner is consulted, he or she must record details of any injuries.

The severity of an incident is often quantified by scoring its impact against the likelihood of reoccurrence in the same way that risk levels are assessed. Near misses are assessed in the same way, by

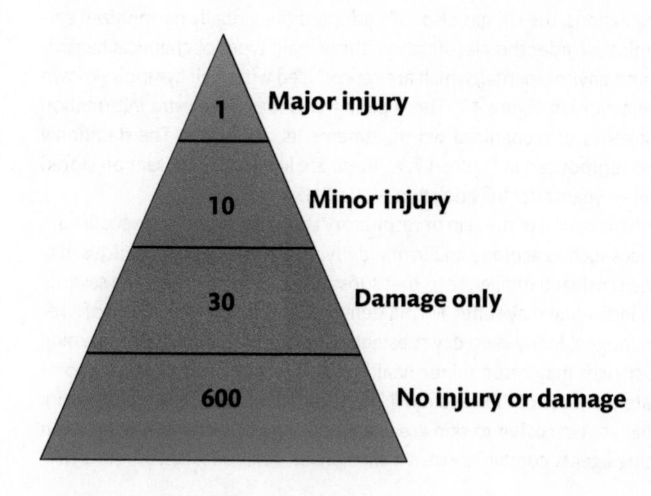

FIGURE 4.6
Bird's accident triangle (1969) showing how many minor occurrences underpin every major accident. See also Figure 17.7.

scoring the likely impact against the likelihood of occurrence. If the resulting score is relatively high, a more formal investigation, including specialist methods, such as root cause analysis, can determine the cause and suitable controls. Root cause analysis is a structured approach to investigating an incident, with the aim of identifying and examining its base cause, as implementing corrective actions at this level will produce the most effective control. If a specimen has been mislabelled, the person responsible may have simply made a mistake; however, checking training and competency records may identify deficiency in the training or induction process. Corrective actions aimed at improving training and induction will affect all staff in the area and potentially prevent them making a similar error. Hopefully, this will lead to much better control and lower the level of that type of risk.

4.5 Control of Substances Hazardous to Health

The COSHH Regulations, first introduced in 1988, were the most significant piece of legislation introduced in the UK after the HASAWA; since that time, they have been subject to many amendments. COSHH regulations set out the requirement for a management system to assess health risks, put in place suitable control procedures, and to monitor these controls. The COSHH regulations pre-dated the 'management regs', and were the first regulations to require employers to perform risk assessments. Thus, they brought about significant changes to health and safety practices. These regulations also established the need to monitor the health of employees. Two potential types of ill health can occur, i.e. acute and chronic. Acute conditions include asthma or fainting. They are usually of short duration with a rapid onset of symptoms following a single or short-term exposure to the causative agent. Although acute ill health can be severe, it is usually reversible. **Chronic ill health** includes cancers caused by exposure to industrial and environmental carcinogens. Chronic ill health normally develops over a long period following repeated exposures to a harmful agent. The onset of symptoms is often so gradual that the effects on health can be undetected and the condition remains undiagnosed for years. Unlike acute conditions, the effects of chronic diseases tend to be irreversible.

The COSHH regulations cover almost all hazardous substances; however, some agents are subject to separate and more specific regulations. These agents include lead, asbestos, and sources of radiation.

4.5.1 Classification of hazards

A chemical substance is considered hazardous if it is classified as dangerous to health under the Classification, Labelling and Packaging (CLP) Regulations (2009). The CLP regulations came into full use in the UK on 1 June 2015 and placed a duty of care on all suppliers when they package and transport hazardous substances. All packages must be clearly labelled as being hazardous and must also have hazard information contained in the form of a Material Safety Data Sheet (MSDS). Both hazard warning labels and MSDS are essential sources of the information required to perform a COSHH assessment. The packaging and labelling of biological substances is also legislated by a number of transport regulations, and is discussed in Section 4.9.

By implementing the CLP Regulations, the UK has also fully adopted the globally harmonized system (GHS) classification of chemicals. Under this classification, three main types of chemical hazards are identified: physical, health, and environmental, which are reproduced with their symbols (known as pictograms in the GHS classification) in Figure 4.7. These pictograms may have extra information added to them as well, using a series of recognized words, statements or phrases. The traditional hazard warning symbols are also reproduced in Figure 4.7, as these are likely to be present on stored chemicals for some months or even years after full adoption of the GHS.

Those chemicals that cause inflammation of the skin or respiratory tract after repeated exposure are normally non-corrosive substances such as acetone and formaldehyde. These types of chemicals may also cause individuals to become sensitized or allergic to that substance, or may increase the severity of other allergies from which an individual may suffer. Formaldehyde, itself, is also the subject of a re-classification as a (potential) carcinogen. Many everyday substances are harmful, for example cleaning products and paints. These substances may cause minor health effects if swallowed or inhaled, or if they penetrate the skin. Fortunately, the use of PPE (Section 4.7) is usually sufficient to be able to work safely with them. Substances that are **corrosive** to skin are usually strong acids or alkalis, which can also attack metals. Many cleaning agents contain corrosive substances. Toxic substances, otherwise

(a)

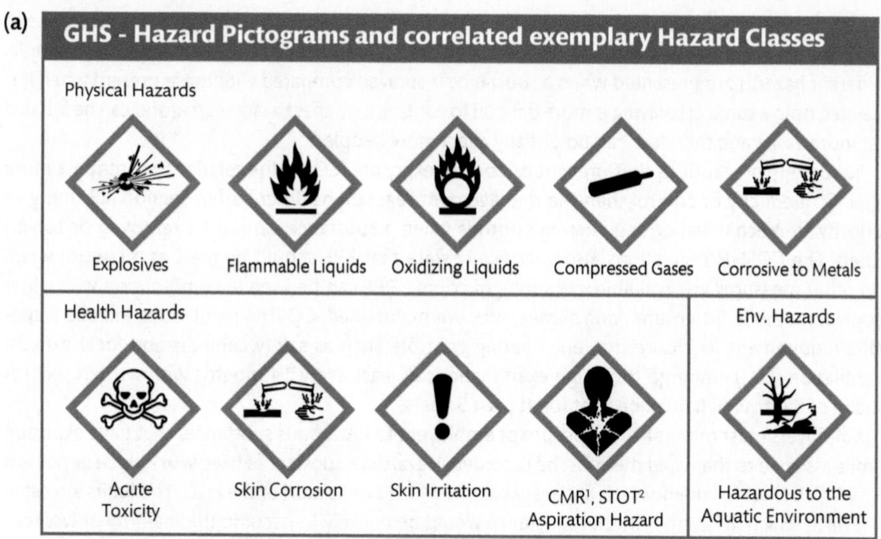

[1]carcinogenic, germ cell mutagenic, toxic to reproduction
[2]specific target organ toxicity

(b)

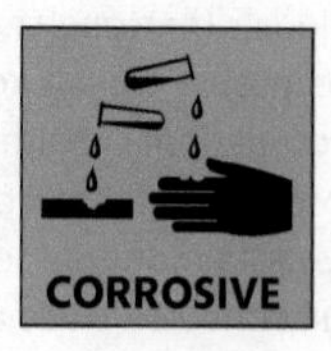

FIGURE 4.7
(a) Pictograms of the globally harmonized system (GHS) of classification and labelling of chemicals, (b) Traditional hazard symbols.

known as poisons, reduce or prevent the function of major organs such as the heart, liver, and kidneys. These may cause either acute or chronic effects. Mercury and carbon monoxide are examples of toxic substances.

Even pure chemicals may have more than one hazard classification, for example some cleaning agents are both harmful and corrosive. Some laboratory reagents contain a number of substances of these types. Under the COSHH regulations, the controls identified and put in place must be adequate to control the most serious hazard to health. Risks associated with **flammable** substances are not covered by COSHH, but are covered under the Dangerous Substances and Explosive Atmosphere Regulations (DSEAR). Further information on DSEAR is supplied in Section 4.5.3. If an inflammable substance has other classifications, such as being harmful or an irritant, these risks must also be considered under COSHH.

Cross reference

You can see examples of acceptable extra information phrases that may be added to the GHS pictograms on the HSE website.

SELF-CHECK 4.8

What are the main types of chemical hazards identified under the GHS classification system?

4.5.2 Complying with COSHH

Seven basic steps must be satisfied if an organization is to comply with the COSHH regulations:

1. Risk assessment
2. Identify precautions
3. Prevent or adequately control exposure
4. Ensure control measures are used and maintained
5. Monitor exposure of employees to hazardous substances
6. Carry out appropriate health surveillance where necessary
7. Ensure employees are informed, trained, and supervised.

A COSHH risk assessment must be detailed and focus on the hazards to health from the substance or mixture of substances, and the way it is used and the environment it is used in. Thus, for example, different hazards are presented when a substance is sprayed compared with those present when it is painted onto a surface; spraying is more difficult to control, may affect a larger area, and can be inhaled far more easily, and therefore can potentially affect more people.

To identify precautions that prevent or control exposure, COSHH regulations prescribe a more specific hierarchy of control than the management regulations described in Section 4.3 that give priority to mechanical or engineering controls when a substance cannot be removed or substituted. The COSHH regulations also specifically state that PPE should be used as a control when no other measures are available, although, of course, PPE can be used in combination with other control methods. To ensure control measures are maintained, COSHH regulations include a specific requirement to ensure that engineering controls, such as safety cabinets and local exhaust ventilation (LEV), undergo thorough examinations at least every 14 months with written records made of them, which must be kept for at least 5 years.

Employers must monitor the exposure of employees to hazardous substances that have exposure limits assigned to them and this must be recorded. Hazardous substances have **workplace exposure limits** (WEL) set for them, which are available in EH40, a publication of the HSC. The limits are set at a level at which ill health from the substance would be unlikely to occur in the majority of workers. The limits may be short term (15 minutes) or long term (8 hours) depending on whether the agent usually causes acute or chronic ill health effects. General records must be kept for at least 5 years, but a record of an individual's (personal) exposure must be available for 40 years. Substances that require the monitoring of exposure usually require the employer to carry out health surveillance. In large organizations, an occupational health department provides and reviews health screening questionnaires for staff exposed to such hazards. The department may also conduct tests, such as skin checks and lung function tests, to monitor the health of staff and detect any possible ill health effects.

The COSHH regulations also place a duty on employers to inform employees of any risks and the relevant control measures in use. They must also train staff in how to use and maintain all control measures, and perform activities safely. There is also a duty on the employer to supervise employees, and ensure the control measures are used correctly and are well maintained.

4.5.3 Dangerous Substances and Explosive Atmosphere Regulations

The DSEAR are designed to protect employees from the risks of substances likely to cause fire, create dangerous atmospheres, and/or explosions. The use, handling, and storage of flammable substances, such as industrial methylated spirits, often used in clinical laboratories, are controlled by these regulations. Flammable substances also fall under these regulations, although like any other hazards to health, they are also covered by COSHH regulations. Like all regulations, DSEAR prescribe systems to be established that allow appropriate risk assessments to be performed and which implement controls that must be monitored and reviewed.

4.6 Fire regulations

In 2005, a major change to fire legislation in the UK was made with the introduction of the Regulatory Reform (Fire Safety) Order. The need for fire certification of premises by the fire brigade was removed and replaced by a requirement by employers to carry out a fire risk assessment. Employers should appoint a person responsible for fire safety and this person should:

- carry out a fire risk assessment to identify the risks and hazards
- consider who may be especially at risk
- eliminate or reduce the risk from fire as far as is reasonably practical
- provide general fire precautions to deal with any residual risk
- take additional measures to ensure fire safety where flammable or explosive materials are used or stored

- plan procedures to deal with any emergency and document these plans
- review the findings of risk assessments as necessary.

The regulations also contain minimum standards for the provision of fire fighting equipment, and fire-related safety signs to show the location of this equipment, alarm call points, and fire exits. There is also a requirement that all employees receive annual training on fire procedures, which should include training in the use of the fire-fighting equipment available in the workplace.

The control of fire risks requires knowledge of how a fire starts. The initiation of a fire has three requirements (or elements): oxygen or air, fuel, and a source of heat, which are known as and often represented as a fire triangle, which you can see in Figure 4.8. The absence or removal of any one of the three will prevent a fire starting. All methods of fire fighting attempt to remove one or more of the three elements.

Heat
Fuel
Oxygen

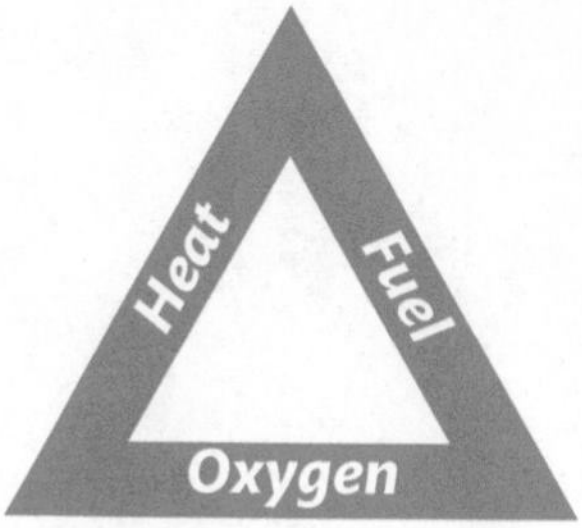

FIGURE 4.8
Fire triangle showing the three elements required to start a fire.

SELF-CHECK 4.9

List two sources of heat and two types of fuel that are routinely found in clinical laboratories.

Fire-fighting equipment, such as fire blankets and fire extinguishers, dampen fires by removing one or more of the three elements required for a fire to burn.

Fire extinguishers are predominately red but have colour-coded labels that easily identify the different types of extinguisher. The commonest types found in laboratories are, for example, red-and black-labelled extinguishers, which are water- and carbon dioxide-based, respectively. Water extinguishers function by removing heat, while carbon dioxide extinguishers reduce oxygen levels. Each type of extinguisher must be used correctly: water extinguishers must not be used on electrical fires as this carries the risk of an electric shock. Figure 4.9 shows some of the different types of extinguishers and the types of fires they are able to extinguish.

Fire Extinguisher Chart

Extinguisher		Type of fire				
Colour	Type	Solids (wood, paper, cloth, etc.)	Flammable Liquids	Flammable Gases	Electrical Equipment	Cooking Oils & Fats
	Water	✓ Yes	✗ No	✗ No	✗ No	✗ No
	Foam	✓ Yes	✓ Yes	✗ No	✗ No	✓ Yes
	Dry Powder	✓ Yes	✓ Yes	✓ Yes	✓ Yes	✗ No
	Carbon Dioxide (CO_2)	✗ No	✓ Yes	✗ No	✓ Yes	✓ Yes

FIGURE 4.9
Fire extinguisher selection chart showing the type of extinguishers and their colour codes, together with their recommended uses.

4.7 Personal protective equipment

We looked at the regulations designed to ensure the provision and use of PPE to control risks that cannot be controlled by other means in Section 4.3. We will now look at some of the common types of PPE that you might use in a biomedical laboratory. Figure 4.10 shows a selection of the types of PPE that are used in biomedical laboratories.

4.7.1 Body protection

Laboratory coats are worn to protect everyday clothing and skin from exposure to hazardous substances. Depending on the nature of the hazards, laboratory coats may need to be supplemented by aprons, or the use of a disposable coat or overalls. For example, disposable coats are used in Category 3 containment laboratories. The fastenings, cuffs, and sleeves of laboratory coats should be inspected before first use and the fit must enable the person to close the coat fully. Overly long coats or sleeves are additional hazards, as parts of the coat could get trapped, or caught on equipment or furniture, causing an accident.

It may be necessary to have transport coats for staff, for use when moving between risk areas of the laboratory to relatively clean ones, such as corridors. A clean coat must always be worn when leaving a laboratory to go to a clean site or to visit clinical areas for laboratory-related purposes. This ensures that the clean area is not contaminated with substances from the laboratory transported on a used laboratory coat.

4.7.2 Hand and foot protection

Many different types of safety gloves are routinely used in clinical laboratories. As with all equipment, the type of glove selected must effectively control the risk; the circumstances of use must also be considered. Liquid-resistant gloves can control the risks associated with handling substances where contact between them and the skin must be avoided. However, many types of glove perish upon contact with commonly handled laboratory reagents, such as xylene; therefore, careful consideration must be given to the material of the glove. Vinyl, latex, nitrile, and rubber gloves are

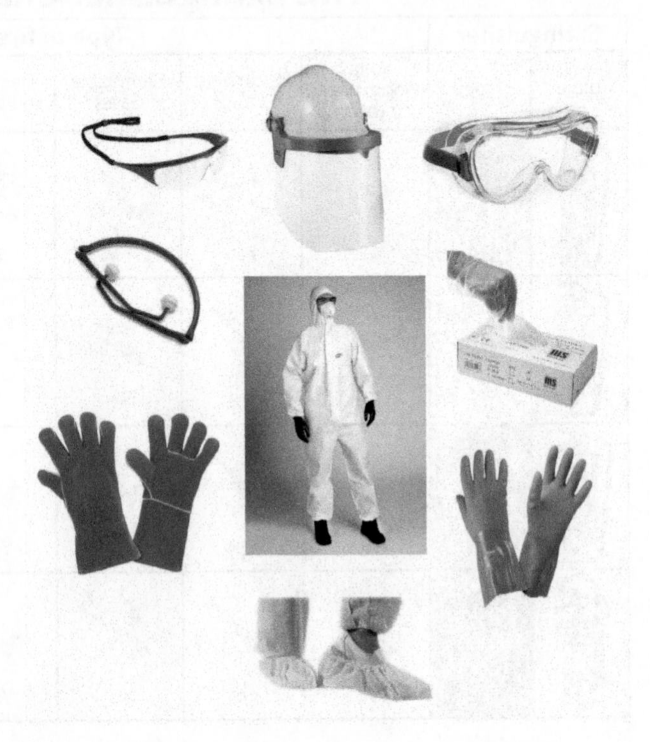

FIGURE 4.10

Selection of personal protective equipment that might be used in a biomedical laboratory. Courtesy of Sperian Protection, UK.

TABLE 4.1 Example of a glove selection chart showing the type of gloves to be used with some common types of chemical and biological hazards

Glove material	Uses	Recommended for:	Not recommended for:	Comments
Latex (natural rubber)	Incidental contact	Weak acids and bases, alcohols, aqueous solutions	Oils, greases, and organic solvents	Good for biological and water-based materials. Poor for organic solvents. Little chemical protection. Can puncture easily. Can cause or trigger latex allergies
Nitrile (synthetic rubber)	Incidental contact	Oils, greases, acids, caustic solutions, aliphatic solvents	Aromatic solvents, many ketones, esters, and chlorinated solvents	Good for solvents, oils, greases, some acids and bases. Clear indication of tears and breaks. Good alternative for those with latex allergies
Polyvinyl alcohol (PVA)	Specific uses	Wide range of aliphatic, aromatic ketones (except acetone), esters, ethers, and chlorinated solvents	Acids, alcohols, bases, water	Good for aromatic and chlorinated solvents. Poor for water-based solutions
Polyvinyl chloride (PVC)	Specific uses	Strong acids and bases, salts and other aqueous solutions, alcohols, glycol ethers	Aliphatic, aromatic, chlorinated solvents, nitro compounds	Good for acids, bases, oils, fats, peroxides, and amines. Good resistance to abrasions. Poor for most organic solvents

This table is for general reference only; for specific recommendations contact the glove manufacturer or the Material Safety Data Sheet.

all available. An additional consideration when selecting safety gloves is that of latex skin allergies or sensitivities. Most workplaces try to remove all latex material from use; however, some non-latex gloves do not, in some people's opinion, give the same dexterity and sensitivity of touch so they may still be used in a limited number of areas. A detailed risk assessment justifying their use should be carried out and anyone at risk clearly identified.

Safety gloves made of a suitable material may also be used to protect against sharp or rough surfaces in MHOs and when handling substances at, or working in, extremes of heat.

Gloves must also be of the correct size to ensure safe handling of the substance: loose gloves can lead to reduced grip and tight gloves to skin irritations. You can see an example of a glove selection chart, showing the type of gloves used when handling some common chemical and biological hazards in Table 4.1.

Foot protection is sometimes required to control the risk of hazardous substances falling onto feet. In most circumstances, rigid, fully enclosed shoes offer sufficient protection. In others, reinforced footwear might be necessary to protect against falling sharp objects or chemicals. Specialist footwear is sometimes used to control other risks, for example, using non-slip soles in hazardous floor conditions, or electrically insulated shoes to protect against electric shock, and shoes incorporating antifatigue properties for people that spend long periods of time standing.

4.7.3 Eye, head, and ear protection

Numerous types of spectacles, goggles, and face shields are available to protect the eyes and/or face from contact with hazardous substances. Here, again, the material used to construct them must be sufficient to control the risks involved. The comfort of the person required to wear the PPE must be considered, as otherwise they are unlikely to wear them consistently.

Head and ear protectors are not commonly used in laboratories. Head protection involves wearing hard hats or bump caps to protect against objects falling from above or when working in confined spaces. Ear plugs or defenders can be used to protect against noise that at some levels and frequencies damages hearing.

4.7.4 Respiratory protection

Filtration masks, respirators, and breathing apparatus can be used to control the hazards of airborne agents when engineering solutions, such as ventilation systems and safety cabinets, cannot be used.

Often these are used in non-routine procedures like conducting maintenance of the normal engineering controls or after a spillage. All masks, respirators, and breathing apparatus must be tested to ensure they fit the individual who will use them. Factors such as beards can compromise the fit and therefore the level of protection given will be reduced. Fume hoods do not suffer this disadvantage; however, they must be fitted with the correct air filters and checked as part of a detailed maintenance schedule.

4.8 Reporting of Injuries, Diseases and Dangerous Occurrences Regulations

The Reporting of Injuries, Diseases and Dangerous Occurrences Regulations (RIDDOR) set out the processes of reporting injuries, ill health, or other dangerous events to the HSE or, in some cases, the local authority. The aim of the regulations is to provide data on those significant events that can be analysed to reveal any national patterns and so lead to new health and safety advice and/or regulations. When serious events are reported, the HSE reviews the information it has received about the incident and may decide to inspect the workplace. The data are also used to collect information about occupational illnesses, which may lead to a better understanding of how they develop and how to control the hazards that cause them.

There are several levels of reporting that depend on the severity of the situation:

- major injuries and dangerous occurrences
- 3-day-plus accidents
- reportable diseases.

Major injuries and dangerous occurrences include deaths, major injuries, or those requiring hospital treatment, fires, electrical shocks, and the release of biological agents. Such events must be notified *immediately*. If an injury sustained at work prevents a person from doing all of their normal work for more than 3 days, it must be reported to the HSE on a specific form within 10 days of the incident becoming reportable. Should an employee be diagnosed with a reportable disease by a medical practitioner, the employer must report this to the HSE on the specified form. Examples of such diseases include occupational dermatitis, occupational asthma, mesothelioma, tuberculosis, and some musculoskeletal disorders.

The Health and Safety (First Aid) Regulations 1981 require an employer to provide **first aid** equipment, facilities, and personnel. The specific nature of the provision required is related to the findings of both general and COSHH risk assessments.

SELF-CHECK 4.10

How should a dangerous occurrence be reported to the HSE?

4.9 Transport regulations

Transporting dangerous goods is regulated by several pieces of legislation. These include, for example, the CLP regulations we discussed in Section 4.5, the Carriage of Dangerous Goods and Use of Transportable Pressure Equipment Regulations (2004), and a European agreement known as *accord européen relative au transport international des marchandises dangereuses parroute* (ADR) of 2009. The Department for Transport publishes guidance on these regulations, and, with the HSE and Vehicle Certification Agency, is also responsible for enforcing them.

Dangerous goods include pathology samples, gas cylinders, chemicals, explosives, flammable liquids, and radioactive substances. Each specific class has defined controls for their packaging and labelling, and what must be used when they are transported.

As with all other modern regulations, risk assessment, training, and information about safe systems of work, risks, and controls is an essential part of the duties imposed. Every employer that transports dangerous goods as part of their business must have a Dangerous Goods Safety Advisor (DGSA), who

FIGURE 4.11
Labelling required for transporting category B specimens.

must have attended a training course and passed the associated examination. The role of the DGSA is to ensure all staff involved in the processes are aware of the requirements needed to meet the regulations. Drivers who transport dangerous goods by road must carry instructions in writing, known as a **transport emergence (TREM) card**, detailing the substance, its category name, United Nations (UN) number and class, details of what to do and who to contact in an emergency, and first aid information in case of spillage of the substance being carried.

4.9.1 Pathology samples

Blood, urine, faeces, swabs, tissue samples, and clinical waste are infectious substances under the regulations, and assigned to class 6.2, which is further subdivided into A and B groups. The majority of samples fall into category B and may contain hazard group 1, 2, or 3 pathogens. They must be labelled *UN3373–Biological substance, category B.* Category A samples are defined as cultures of hazard group 3 pathogens or samples suspected of containing hazard group 4 pathogens. Packages with category A samples must be labelled *UN2814–infectious substance affecting humans.* Only specified carriers may transport category A samples.

4.9.2 Packaging and labelling of biological materials

Pathological specimens may be transported by road, rail, or air if the packaging and labelling instructions given in Packaging Instructions P 650 (found in part 4 of the ADR) are followed. These instructions may be summarized as follows. The primary layer (the sample container or tube) must be watertight and leak proof. There must be an absorbent material, sufficient to soak up the volume of sample, between the primary and secondary layers. The secondary layer (sample bag) must be watertight. A rigid outer packaging that has passed a 1.2 m drop test and has minimum dimensions (on one face) of 100 × 100 mm is required. Cushioning material must be inserted between the secondary layer and the outer packaging. Either the primary or the secondary packaging must be able to withstand a pressure differential of 95 kPa or more. Usually, the request form is attached to the outside of the bag to protect it from damage by any leakage. The package must be labelled with the UN3373 diamond (Figure 4.11) and a biohazard label.

Samples that do not have an associated risk of infection, such as processed tissue blocks, can be transported as 'exempt human specimen', but they must be packed according to the instructions contained in P 650.

4.10 Personal health and safety

Every workplace will have its own specific health and safety instructions that you must follow at all times. The initial training you receive when you first take a job in a laboratory must always cover the hazards, risks, and controls of each process you undertake. However, you will later receive supplementary specific and mandatory training related to the regulations we have already discussed. If you believe that you have received inadequate information, you should consult a senior member of staff. However, some rules and practices are common to all laboratories, and include the following:

- A waterproof dressing must cover all cuts, abrasions, or other types of wounds to the skin, especially of the hands, before you start work. If you puncture your skin in an accident at work, gently encourage it to bleed under cold running water. Report the incident to a senior member of staff and seek medical advice as necessary.
- You must never take personal items, such as pens, pencils, cosmetics, toiletries, and hairbrushes, into the laboratory. Cosmetics are not to be applied in the laboratory; indeed, hands must never be in contact with the mouth, face, eyes, and nose. Always wear a correctly fastened coat when in the laboratory. It must be cleaned regularly and immediately changed if it becomes contaminated.
- Never take food or drink (including sweets and medication) into a laboratory nor eat or drink there. Always follow the correct laboratory procedures. Never deviate from them unless you have received alternative instructions from a senior member of staff for a specific reason. Should circumstances arise that prevent you from following the procedure, always inform a senior member of staff before continuing. Wear gloves when necessary; if this is a requirement, it will be specified in the SOP. Replace the gloves if they become contaminated or perforated. Avoid all practices that can cause splashing or the release of airborne substances unless they are performed in an appropriate containment cabinet. Report all accidents, breakages, spillages, or damaged equipment to a senior member of staff immediately.
- Always wear a clean coat if visiting a clinical area. Clean coats must be stored in a separate, clearly marked area, separate from that for other laboratory coats. Remove your coat and wash your hands before leaving the laboratory.
- If you become ill and consult a doctor always inform him or her where you work, and ask them to talk to your manager if they require further details.
- Lastly, *never* take any unnecessary risks in a laboratory.

4.10.1 Occupational health

Most employers will have their own occupational health (OH) service, or access to one. The role of OH is to assess whether individuals are fit to perform the role in question *before* an offer of employment is given, and to give advice on health issues that relate to the specific area of an employee's work. OH may also provide services such as vaccinations and health monitoring. Staff can also discuss concerns related to their health and welfare with them, in confidence.

4.10.2 Risk awareness

Some people are natural risk takers; others are naturally more cautious. It is essential that in the workplace you take a responsible attitude to risk and avoid it at all times. You should also remember that the severity of a hazard is rarely the same as the level of risk; laboratory procedures contain a wide range of controls that, if followed, greatly reduce the level of risk.

4.10.3 Infection control

Hand hygiene is an essential control to minimize the risk of infection. It is essential that you take great care every time you wash your hands. The seven-step technique, which is usually displayed above sinks designated for hand washing (Figure 4.12), should be followed to ensure every area of the hands, fingers, and wrists is cleaned. Great care should also be taken to dry hands thoroughly as leaving them damp can cause dermatitis. False or long nails and jewellery can harbour dirt, or infectious and hazardous material: never wear them in areas where they pose a risk.

4.10.4 Clothing and personal appearance

Many workplaces have a clothing and personal appearance policy. Such policies are partly intended to encourage health and safety elements, such as wearing enclosed footwear, having hair tied up, and

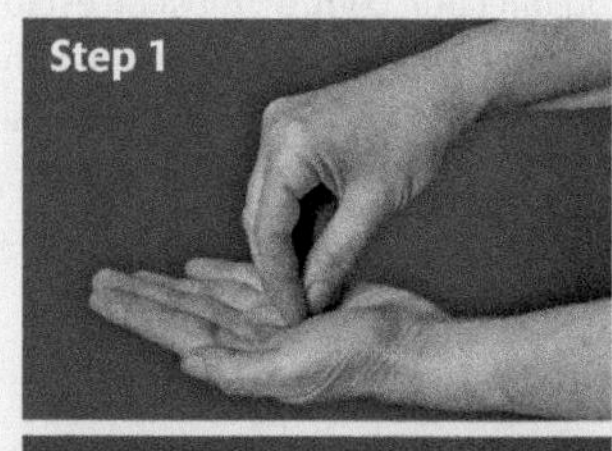

Squeeze a small amount of sanitizer gel or soap over left palm. Dip all the fingers of the right hand into the left palm and vice versa

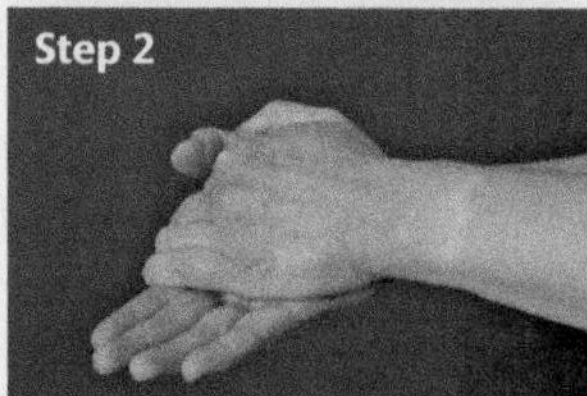

Join hands palm to palm

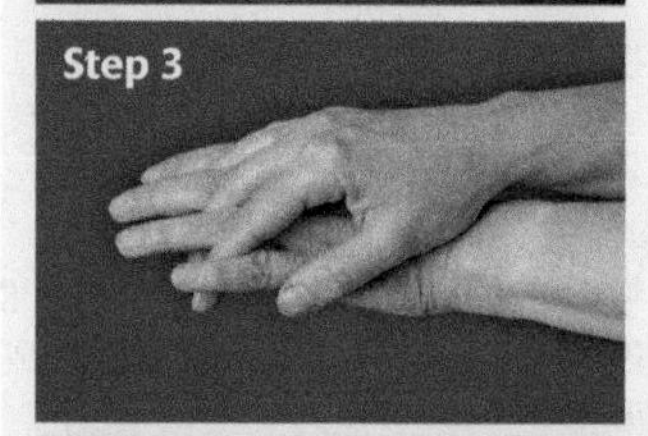

Rub right palm over left hand as shown and then left palm over right hand

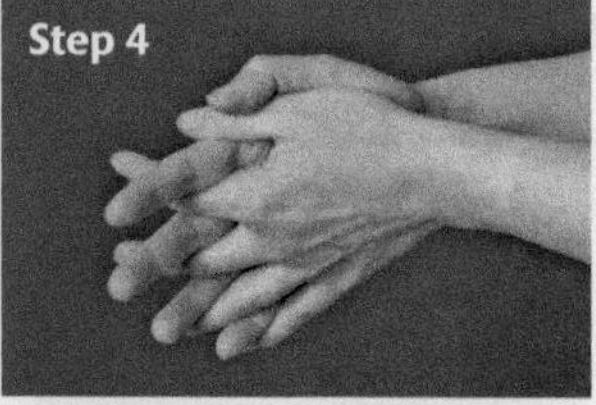

Interlace fingers with hands joined palm to palm

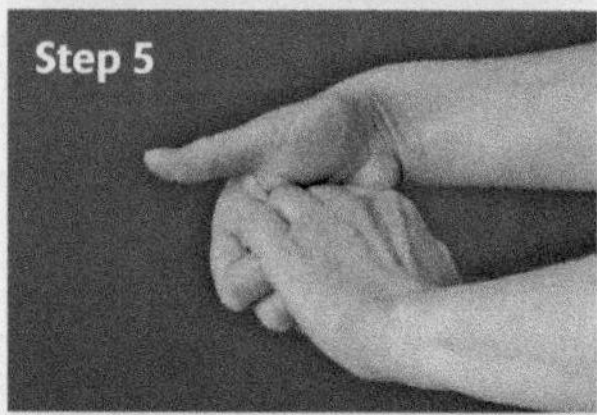

With fingers still interlaced, rub the backs of fingers against opposing palms

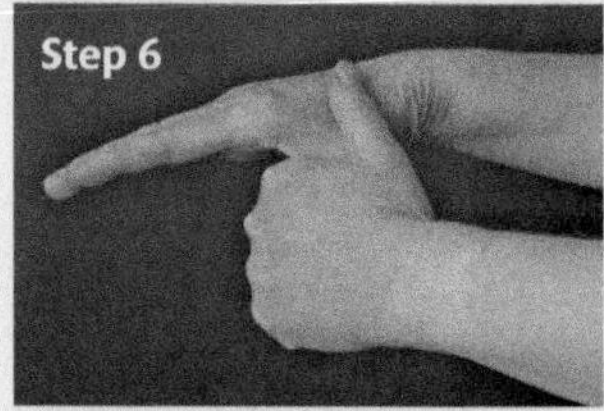

Clasp thumbs in opposing palms and wash with rotational movements

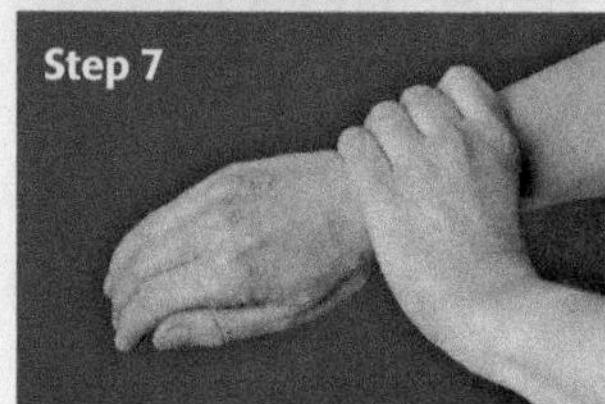

Wash wrists with rotational movements of opposing hand

Rinse and dry thoroughly

FIGURE 4.12
Seven-step technique of hand washing. Once clean, the hands must be thoroughly dried.

not wearing large earrings or necklaces, but they are partly intended to ensure that staff present a professional appearance. Standards of dress are subjective; hence, these policies can set clear guidelines for staff to follow. Dress standards are closely related to infection control policy as long-sleeved tops or shirts and ties can transport infectious material around the workplace. Thus, a balance between presenting a professional appearance, perhaps by wearing a tie, and infection control, must be established. A professional appearance by staff is especially important in service industries such as healthcare, and is an essential part in developing trust and positive working relationships with professionals in other departments, as well as with patients.

4.11 Universal or standard precautions

The Centers for Disease Control in the USA first recommended universal or standard precautions in 1985, in response to the risk of transmitting HIV. It is not always possible to tell if a clinical sample poses an infection risk; hence, universal precautions is a system of work and controls that assume all patient-related contact carries a risk of infection. The key points of universal precaution systems are:

- good hand hygiene
- the wearing of PPE at all times
- the correct use and disposal of sharps, such as needles and blades
- disinfection of body fluid spills
- careful disposal of waste.

Using universal precautions in the wider healthcare setting has been the driver for decontaminating hands after every patient interaction, irrespective of the level of infection risk from the examination or procedure. This has lead to the increased use of hand gel cleaning agents.

In a clinical laboratory, implementing universal precautions means treating all specimens as if they are extremely hazardous, and wearing gloves and possibly eye protection at all times. Wearing of gloves would even be necessary when using equipment such as telephones and personal computers. This approach would certainly simplify the use of precautions and, therefore, make it easier to follow and monitor the use of controls. However, the long-term use of PPE can lead to ill health issues, for example, wearing gloves for extended periods can lead to dermatitis.

CASE STUDY 4.1

Before the introduction of liquid-based cytology, it was not unusual for badly packed slides of sent cervical smears to arrive at the laboratory broken. Where these were repairable, rather than subject the woman to another procedure, the majority of departments would attempt to stick the pieces back together before the slides were stained and examined. This so-called 'splinting' of slides involved the use of a new plain slide as a template upon which the broken pieces were reassembled and 'glued' together using DPX (or a similar mounting medium). The introduction of liquid-based cytology into the system has largely removed the need for slide repair, as slides are no longer sent in, as samples are submitted now, and it is possible to prepare a new slide if one is broken (or lost, or required for educational purposes). Slides do still get broken occasionally and this may be some time after the original sample has been discarded when new preparations can no longer be made.

Although termed 'permanent preparations', slides also get broken in other departments such as histology, as glass is inherently fragile and these, too, may need repair from time to time.

In the department concerned, slide repair involved the use of old hypodermic needles (pushed into a cork board) to ensure that the broken pieces did not move or become displaced while the DPX was setting. Over time, these needles became blunted and bent, and were increasingly coated with a film of DPX, too, making them less easy to handle.

Unfortunately, one of the staff, when undertaking a repair, managed to suffer a needle-stick injury, so much so that it required appropriate first-aid to be administered.

Following this incident, a number of actions were taken.

The immediate response was to remove the needles from the department and to ban their use subsequently by finding a less hazardous way of stabilising slides under repair. The SOP concerned was also updated to reflect this change and all staff were reminded of the need to be vigilant when undertaking potentially hazardous procedures; glass fragments also tend to have sharp edges in their own right, too. Staff were

also reminded that slide repair should only be undertaken as a last resort.

Officially, this was deemed a reportable incident (owing to the skin puncture and consequent bleeding nature of the wound) and was recorded on the Trust's incident reporting system (in this case Datix). Under Datix, this incident had to be reported as a 'needle-stick' injury, which has wider ramifications, owing to the infection potential through transfer of blood-borne disease. In this case, this was not an issue but had to be fully addressed and explained in the report.

While a small incident in itself, this case does show that health and safety issues can occur at almost any point in the laboratory, often in the least expected places or times. This shows that continual vigilance is needed, as laboratories are inherently hazardous places within which to work, and familiarity or complacency are often components recognised in analysis of incidents.

Happily, no ill effects were encountered by the individual concerned and the department now has a much safer, but still effective, slide repair procedure.

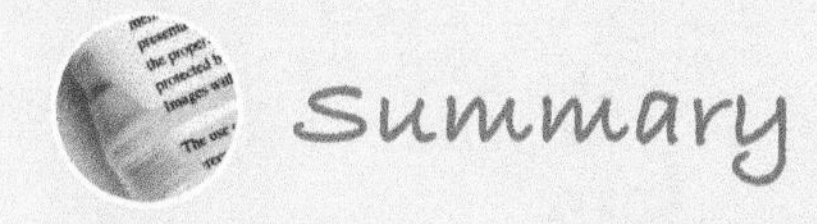

Summary

- The HASAWA and the Management of Health and Safety at Work Regulations form the legal framework that ensures the safety of everyone in the workplace. They place numerous duties not only upon employers, but also some on employees.
- All employees have a duty to ensure their actions or omissions do not cause a risk to themselves or to the safety of others working with them. They also have a duty to use and to care for equipment as instructed, and to report defects in equipment or hazards to the health and safety officer.
- Risk assessment is the primary tool prescribed in legislation to ensure high standards of health and safety. Risk assessments identify hazards and those who are potentially in danger; they evaluate the risks associated with those hazards and suggest how the risks can be reduced to acceptable levels. All risk assessments need reviewing on a regular basis to ensure they are still adequate. Other regulations that require a more detailed form of risk assessment include COSHH, DSE, MHO, and fire safety.
- The ethical duty of everyone to ensure the safety of everybody affected by the workplace is upheld in law. Ensuring a high standard of health and safety in the workplace is also desirable because the overt costs associated with incidents and injuries are considerable; hidden costs, such as insurance, time lost in investigations, and negative publicity, are also substantial.

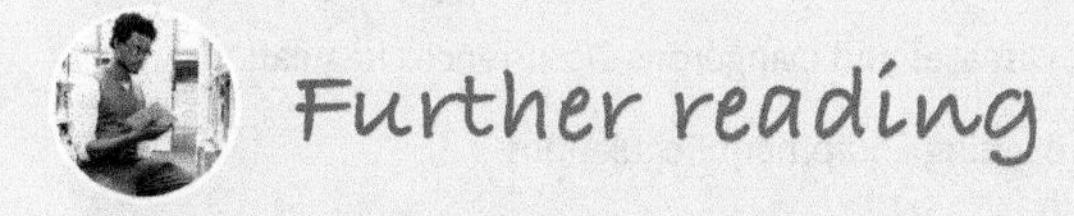

Further reading

- **HSE Books. *Safe Working and the Prevention of Infection in Clinical Laboratories and Similar Facilities*. Bootle: HSE Books, 2003.**
- **HSE Books. *5 Steps to Risk Assessment*. Bootle: HSE Books, 1998. Available at: www.hse.gov.uk/risk/fivesteps.htm (accessed 13 January 2016).**

  This is an essential guide to performing risk assessments.
- **Hughes P, Ferrett E. *Introduction to Health and Safety at Work*. 4th ed. Oxford: Butterworth-Heinemann, 2009.**

  This is an excellent book, covering all areas of health and safety, including the legal aspects and case law.

Useful Websites

- www.hse.gov.uk
- www.hsebooks.co.uk
- www.dft.gov.uk/vca/
- www.coshh-essentials.org.uk

These websites are excellent sources of information about general health and safety issues and COSHH information. Many of the publications are available to download.

Questions

1. The abbreviation **Xn** is used in traditional chemical hazard classification to represent which of the following?
   - **(a)** Flammable
   - **(b)** Irritant
   - **(c)** Harmful
   - **(d)** Corrosive
   - **(e)** Toxic.
2. Which of the following statements are **TRUE**?
   - **(a)** The penalties for failing to comply with HSE enforcement notices include up to 6 months imprisonment
   - **(b)** The act introduced to ensure workplaces were safe was The Safety Act
   - **(c)** Under DSE rules you are entitled to free glasses if you require them to do your job
   - **(d)** RSI stands for recent strain injury
   - **(e)** You must wear gloves when opening specimen containers to get a better grip on the lid.
3. Arrange the following two lists into their most appropriate pairings.

ACDP	Display screen equipment
ACOP	Health and Safety at Work Act
DSE	Personal protective equipment
HSE	Dangerous Substances and Explosive Atmosphere Regulations
WRULD	Reporting of Injuries, Diseases and Dangerous Occurrences Regulations
DSEAR	Lifting Operations and Lifting Equipment Regulations
HASAWA	Health and Safety Executive
HSC	Provision and Use of Work Equipment Regulations
IOSH	Repetitive strain injury
LOLER	Advisory Committee on Dangerous Pathogens
MHO	Work-related upper limb disorder
WEL	Approved codes of practice
PUWER	Manual handling operations
RSI	Workplace exposure limits

RIDDOR	Institute of Occupational Safety and Health
PPE	Health and Safety Commission

4. What are the five steps required to perform a general workplace risk assessment?
5. **(a)** What are the possible routes of entry to the body for a chemical mist? **(b)** Suggest possible control measures to prevent its entry.
6. What are your duties as an employee under the HASAWA?
7. List your duties as an employee under the Management of Health and Safety Regulations.
8. **(a)** What are the three elements of the fire triangle? **(b)** Which of these is eliminated when using a carbon dioxide fire extinguisher?
9. **(a)** Give an example of a mechanical control, as described by the COSHH regulations.

   (b) What are the requirements under COSHH to maintain such equipment and associated records?
10. What are the differences between a laboratory using universal precautions compared with one applying control measures based on risk assessment?

Answers to self-check and end-of-chapter questions are available at the end of the book.

5

Preparing and measuring reagents

Ian Graham

Learning objectives

After studying this chapter, you should be able to:

- Recognize the key features of top pan and analytical balances, and be able to use them to weigh materials
- Demonstrate the need for correct installation, calibration, and maintenance of balances
- Distinguish between ways of measuring and delivering volumes of liquids, and demonstrate an appropriate selection and application of volume measurement
- Recognize the key features of pipettors and demonstrate their correct use in measuring and delivering volumes
- Know why there is a need for calibration and maintenance of pipettors
- Distinguish between different ways of expressing concentrations and show how dilution affects concentration
- Demonstrate the ability to calculate concentrations, know how to prepare solutions of a particular concentration, and make dilutions
- Show an awareness of correct practical procedures and health and safety requirements

Introduction

As a biomedical scientist, you will undertake much practical work where quantities of materials and solutions must be measured and delivered. You may have to dilute a blood sample with a solution such as saline, mix volumes of different solutions in a colorimetric assay, or, perhaps, dilute a microbial suspension. Solutions may be bought in ready prepared or they may be prepared by yourself or others within the laboratory. In large laboratories, many routine tests will be automated and autoanalysers will make volume measurements and mix together different solutions. Although you may not have to prepare many solutions or undertake significant amounts of volume measurement and delivery, it is essential

BOX 5.1 The SI system and units of volume

Europe and much of the rest of the world uses a set of units of measurement called SI units (SI stands for *Le Système International d'Unités*). Within the SI system, volume measurements are based upon the m^3, which is equivalent to 10^3 L. However, the USA and some other countries have not fully adopted SI units, and volumetric glassware and pipettes are usually calibrated in L, mL, and µL. For these reasons, volumes will be quoted in both sets of units as appropriate. Table 5.1 shows the relationship between SI (based on the m^3) and non-SI (based on the L) volume measurements.

TABLE 5.1 A comparison of SI (m^3) and non-SI (L) volume measurements

SI based volume	m^3 equivalent	dm^3/L based equivalent
m^3		10^3 dm^3/L
dm^3	10^{-3} m^3	1 dm^3/L
cm^3	10^{-6} m^3	1 cm^3/mL
mm^3	10^{-9} m^3	1 µL

you understand how to undertake these activities as some assays may be performed manually. You need to be able to use accurately balances and pipettes, or other volume measurement devices, and know how to make solutions, perform dilutions, and understand the effects of dilution upon concentration (see Box 5.1 for the SI system and units of volume). These are essential and fundamental skills for all biomedical scientists and the quality of your work depends upon their correct application.

Cross reference

You can read more about automation in Chapter 15, 'Laboratory automation'.

5.1 Balances and weighing

This section will consider:

- key features of balances and their use
- care and maintenance
- safe working practices.

5.1.1 Types of balance

Several types of balance are available. The simplest are beam balances, but more sophisticated top pan and analytical balances are normally used in clinical laboratories. You should be able to recognize Figure 5.1 as balances. Top pan and analytical balances are mainly used to measure a particular amount of a solid for preparing a solution, but they can also be used to weigh liquids. Simple beam balances have limited applications in laboratories. For example, they can be used in **balancing** a pair of centrifuge tubes that contain a suspension prior to centrifugation.

Cross reference

Centrifugation techniques are described in Chapter 10, 'Centrifugation'.

The type of balance used determines whether **mass** or **weight** is being measured. An item is 'weighed' on a balance, but there is an important difference between mass and weight. Weight is *a force* (measured in, e.g., Newtons, grams, or pounds), reflecting the effect of gravitational force at a particular location upon a mass (measured in, e.g., grams or kilograms). Mass does *not* vary and is a measure of the amount of matter present. Although the two are related, the weight of a given mass will change if the gravitational force varies. A beam balance measures mass, while spring balances and top pan balances measure a force, the weight of the item. Why this difference?

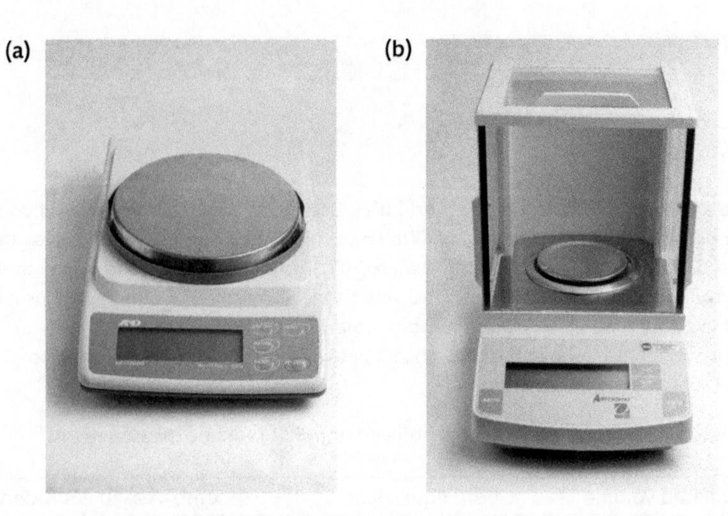

FIGURE 5.1
Photographs showing (a) a top pan and (b) an analytical balance.

A beam balance consists of a beam supported at the midpoint by a near friction-free fulcrum. The sample is placed in a weighing or measuring pan at one end of the beam and one or more standard masses are placed in a scale pan at the other end to achieve equilibrium. The sum of the standard masses in the scale pan when balance is achieved gives the mass of the sample because gravity applies equally to each side of the fulcrum point. Thus, beam balances measure mass not weight. Variants of this arrangement have been introduced over the years. An asymmetric fulcrum arrangement reduces the requirement for large reference masses. For larger-scale applications, a floating platform may be provided by using a cantilever system. By comparison, spring and top pan balances are force-measuring devices, and have different forms of construction and operation that give weight measurements. Although they measure weight, the requirement for top pan and analytical balances to be calibrated means they can be used to measure mass. In practice, the terms *weight* and *weighing* have such common usage, they tend to be applied whether mass or weight is actually being measured.

SELF-CHECK 5.1

If you were a scientist on a manned lunar mission with a top pan balance as part of your equipment, would the mass and weight of an object be the same on earth and the moon or different?

Top pan and analytical balances differ in appearance from a beam balance; they include electronics, have digital displays, and are easier and more convenient to use. High levels of measuring precision are achieved by combining the application of mass changes, and a spring force in achieving the final balance.

Top pan and analytical balances

Top pan and analytical balances work in a different way from the simple mechanical beam balance. They are compact, robust, and measure to high levels of accuracy and precision. The routine top pan balance may weigh to within 0.1 g or 0.01 g and cover a weighing range or **capacity** up to 200 g, 1000 g, or more. They have a broad range of weighing applications, are simple to use, relatively cheap, and robust if not misused. Analytical balances are a variant of the general top pan balance, measuring to within 1 mg (10^{-3} g) or less, and fully enclosed within a draught and dust shield that incorporates several sliding doors. Inevitably, these are more expensive and used for more demanding purposes where only small quantities are to be measured. Catalogues for electronic balances refer to **readability**, which is simply the number of decimal places to which that particular balance can be used. A balance with a readability of 0.1 g would be used in general applications that differ from the more demanding applications of one with a readability of 0.0001 g.

SELF-CHECK 5.2

Which one of the following sets of features would you associate with an analytical balance?

(a) Cheap to buy, high levels of readability, includes a draught and dust shield, generally used for weighing smaller quantities

(b) Relatively expensive to buy, low levels of readability, includes a draught and dust shield, generally used for weighing larger quantities

(c) Relatively expensive to buy, high levels of readability, includes a draught and dust shield, generally used for weighing smaller quantities

(d) Relatively expensive to buy, high levels of readability, lacks a draught and dust shield, used for weighing larger quantities.

Whatever the type, modern electronic balances are designed for a long service life with minimal maintenance if not abused. However, they do require regular recalibration to ensure accurate and consistent measurements.

5.1.2 Uses of laboratory balances

Normally, a balance would be used to weigh dry or partially hydrated solid materials to be used in the production of a solution. Section 5.4 considers matters related to the production of solutions and the effects of dilutions. The amount of material needed depends upon both the concentration and the total volume of the solution to be prepared. Although it is relatively easy to calculate the required amount to be weighed, in a clinical laboratory a standard protocol is usually followed, with amounts already determined. Balances can also be used to measure volumes, as shown in Box 5.2.

5.1.3 Correct installation of a balance

Top pan and analytical balances are portable. It is essential that they are correctly located and installed, as several factors can affect their use, including:

- draughts, which are caused by open windows and doors, rising warm air from radiators, other hot air sources, and air conditioning ducts
- sources of vibration or sudden movements

BOX 5.2 Measuring liquid volumes using a balance

Balances are usually associated with the weighing of solid materials. There are occasions where it is more convenient to weigh a liquid as opposed to measuring and delivering a particular volume of that liquid. To do this a weighing container suitable for the liquid will be required and the *density* of the liquid at the ambient temperature (density is temperature dependent) needs to be known.

$$\text{density} = \text{mass}/\text{volume}$$

Therefore

$$\text{mass} = \text{density} \times \text{volume}$$

The required mass can be calculated using the density of the liquid. For example, if a volume of 25.0 cm^3 is required where the liquid density is 1.06 g cm^{-3} at laboratory temperature, the calculation is:

$$1.06 \text{ g cm}^{-3} \text{ (density)} \times 25.0 \text{cm}^3 \text{ (volume)} = 26.5 \text{ g (mass required)}$$

- magnetic fields from some types of adjacent equipment
- a non-level, unstable support surface.

SELF-CHECK 5.3

Glycerol is a viscous liquid sugar alcohol with a density of 1.261 g cm^{-3}. How could you use a top pan balance to measure a 15 cm^3 volume of this liquid?

These are relatively easy to identify and rectify. For example, to level correctly, balances have a level indicator and several adjustable feet.

5.1.4 Choice of balance

The choice of an appropriate balance to use is dictated by two key factors, the weight to be measured (capacity of the balance) and the degree of accuracy of the measurement (readability of the balance). For example, if 15.624 g of glucose were required to prepare a solution for use in an assay, a 1000 g capacity balance with a readability of 0.1 g would be inappropriate. However, a 65 g capacity balance with a readability of 0.1 mg (10^{-4} g) would be a sensible choice.

5.1.5 Balance calibration

Some anticipation of use is required as balances may take some time to warm up and self-check. In a general purpose laboratory, it is likely that balances will be kept switched on. Once on, the balance needs to be set for the weighing units and mode of use that is required. These are generally gram units and **tare** operation, respectively. The process of taring sets the balance readout to 0 g when an empty container is placed on the pan. When material is added to the empty container the readout is therefore an indication of the total weight of material in the container. This makes the weighing simple. Prior to use, any spilt material left by a previous user must be carefully removed to avoid contamination of future samples and damage to the balance.

Key points

Balances are often operated in tare mode so that the displayed weight only measures the material added to the weighing container, excluding the weight of the container itself.

There must be an established schedule of balance **calibration** in the laboratory, and it is useful to understand how it is achieved. When supplied new, a balance will have been calibrated at the factory, but regular rechecking is required to ensure that the calibration and, hence, accuracy of the instrument is not lost. This is particularly necessary when a balance has been moved owing to the influences of location, temperature, and possibly risk of physical knocks. The most common balances are top pan types, which measure weight. Their calibration involves checking against known standard calibration weights, which may be either built in, or externally supplied and placed on the weighing pan. Two common approaches to calibration are the simpler *span* calibration and more sophisticated *linearity* calibration methods.

Span calibration uses two weight values, zero and a near to or at capacity value to check the balance. The *linearity calibration* approach uses an additional value near the middle of the weighing range. This latter approach reduces the likely extent of deviation between the true and displayed weight. The procedure used will depend upon the balance make and model. Calibration checks on a regular basis are clearly essential. The Method box 'How to use a balance' summarizes these points.

METHOD How to use a balance

- Before commencing the weighing process, you must be aware of any issues that may affect how you proceed. Relevant health and safety information needs to be read and potential problems with the material understood. This will determine how the material can be safely handled, how solid and solution containers should be labelled, and how the material should be disposed of if spilt. Where handling of the material may release a dust you will need to proceed with caution.
- To avoid condensation onto the material and to reduce air currents around the weighing pan, samples should reach ambient temperature before being weighed. Material that is hygroscopic (absorbs or adsorbs water from the surroundings) or that loses water rapidly should be placed in a sealed container before weighing. The same precaution will be required where volatile liquids are weighed.
- Some form of container for the material to be weighed will be required, a plastic or foil weighing boat for small amounts, or possibly some form of glassware for larger amounts. To avoid contamination, a clean weighing container and spatula are essential.
- The most common mode of using a balance is with tare selected, but the procedure will vary with the model of balance. First, the empty weighing container should be placed centrally on the pan, followed by a pause until there is an indication that the balance has stabilized (the response time varies with model). The balance is then tared so that the readout is zero. Material is gradually added to the weighing container using a spatula, remembering to pause until the balance is stabilized before adjusting the amount of added material in order to achieve the desired weight. Many balances have a communications port that allows the attachment of a printer or computer, and the recording of data if this is required.
- The weight of the container into which the material is placed needs to be as low as possible compared with that of material to be added so as to maximize the weight change. For instance, the addition of 0.50 g of material to a 0.50 g weighing boat would result in a 0.50–1.00 g weight change. The addition of the same amount to a 100.00 g glass container would result in a change to 100.50 g, a proportionally smaller weight change.
- Electronic balances are precision instruments and the self-checking procedures occasionally result in error codes on the digital display. Should an error code appear, its meaning can be found in the manufacturer's operation manual.
- A balance must be cleaned of spilt material after each period of use for which a brush can be left near the balance for this purpose.
- Balances require minimal cleaning and a wipe with a soft cloth moistened with a dilute mild detergent solution is usually sufficient.
- There may be circumstances where a balance is located away from the open laboratory because of some potentially harmful feature of the material handled. The reasons for this must be clearly identified near the temporary location, so that other laboratory users are not inadvertently exposed to an unnecessary risk.

SELF-CHECK 5.4

Why is it necessary that top pan and analytical balances are correctly located and calibrated?

Cross reference

Health and safety issues are described in Chapter 4, 'Health and safety'.

5.2 Volume measurements and delivery

This section will consider:

- the selection of appropriate procedures for measuring volumes
- advantages and disadvantages of different approaches.

When preparing solutions by dissolving a solute in a solvent, diluting a more concentrated solution, or delivering specific sample volumes, an appropriate means of volume measurement and delivery must be chosen. Consider the two scenarios given in Case study 5.1 and think about the differing requirements. This will make you aware of the need for an appropriate level of accuracy in different situations.

Cross reference

Electrophoretic methods are described in Chapter 12, 'Electrophoresis'.

5.2.1 Alternatives for measuring and delivering volumes of liquid

Figure 5.2 shows a selection of equipment that is generally available for measuring and delivering volumes of liquids. Such equipment can be grouped in various ways, for example convenience of use or volume range that can be delivered. The most useful grouping is probably according to level of accuracy of the delivered volume. **Accuracy** relates to how close the measured volume is to the actual volume, whereas the **precision** of volume measurements refers to the closeness of repeat deliveries of (supposedly) the same measured volume. At the low end of the accuracy scale lie Pasteur pipettes (droppers) together with beakers and conical flasks that contain a volume mark. Measuring cylinders with a graduated scale provide an intermediate level of accuracy. Higher levels of accuracy are provided by various types of pipettes, burettes, volumetric flasks, and certain syringes. You may be familiar with many if not all of these but may not have thought much about their relative strengths and weaknesses. Certain precautions need to be taken when measuring and handling certain liquids (Box 5.3). As mentioned in the Introduction, volumes can also be delivered by weighing a liquid. This approach can deliver volumes to a high level of accuracy but lacks some convenience compared with the use of a mechanical pipettor.

The range of volume measurement procedures commonly available will now be considered. Most emphasis will be upon mechanical pipettors (such as Gilson pipettes) as they are the most sophisticated and accurate volume measurement devices available in the routine laboratory.

5.2.2 Pasteur pipettes, beakers, and conical flasks

These provide a relatively simple means of measuring and delivering specific volumes but at a low level of accuracy. The original Pasteur pipette was made of glass with a rubber teat to create a partial vacuum to draw up liquid. It is possible to make an approximate calibration of a fixed volume by drawing that volume of sample into the Pasteur pipette, marking the level on the glass. Disposable one-piece plastic Pasteur pipettes are available with several approximate volume markings on the body of the pipette.

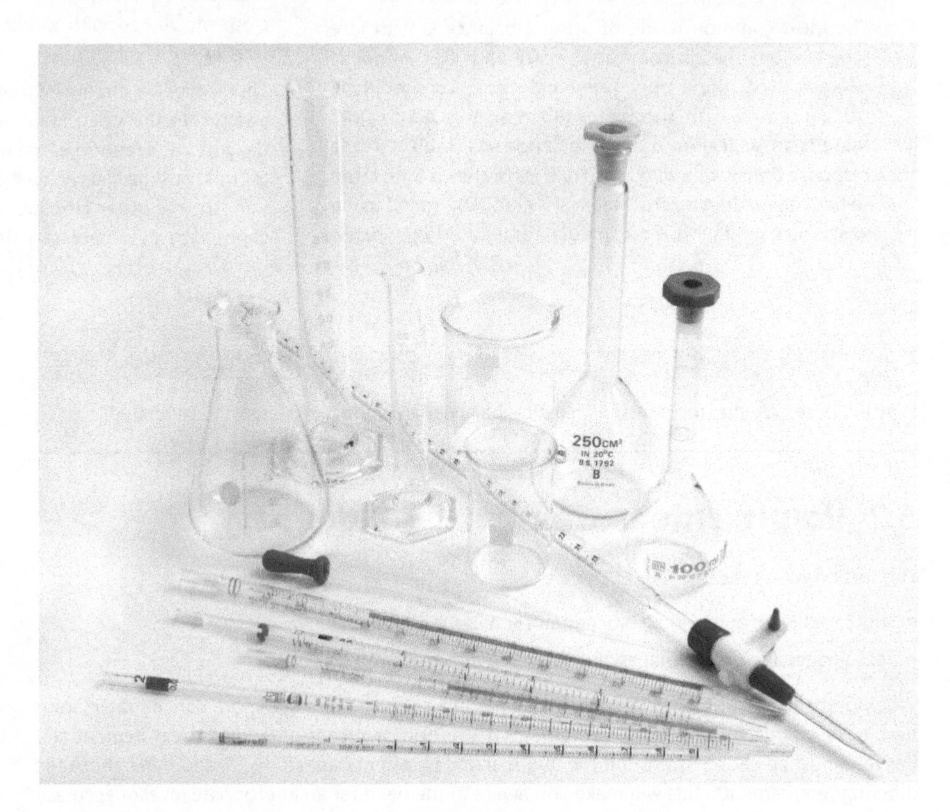

FIGURE 5.2

A selection of equipment to measure volumes of liquids.

BOX 5.3 Precautions to be taken when measuring and handling certain liquids

In addition to the practical considerations related to accuracy and precision, convenience, flexibility, reliability, and the particular nature of a liquid may all influence how it is measured and handled.

- **Some solutions are viscous and care must be taken to ensure complete measurement and delivery of the desired volume. This will include allowing sufficient time for the liquid to be drawn into and to drain from the measuring device.**
- **Sampling from a cell suspension, such as blood or micro-organisms, requires that the suspension is first gently swirled to ensure an even distribution of cells.**
- **Protein solutions can denature at air-water interfaces if excessively shaken, stirred, or poured, and it is necessary to keep frothing to a minimum.**
- **Some solutions such as strong acids are highly toxic or corrosive and care has to be taken to avoid spillage and contamination. It may be appropriate to use disposable 'plasticware' where suitable.**
- **Some organic solvents may evaporate and so exposure to air needs to be minimized by keeping them in appropriate sealed containers. Care has to be taken if using plasticware as some organic solvents can cause deterioration of the plastic.**

Pasteur pipettes are cheap and easy to use where approximate volumes are required, such as dispensing a few drops of a cell suspension or for completely filling wells on an enzyme-linked immunosorbent assay (ELISA) plate when a washing step is required. Where their limitations are acceptable, Pasteur pipettes are useful: they are simple to use, quick, and the problems of cross-contamination are easily avoided with the use of disposable plastic ones.

Beakers and conical flasks are available in a range of volumes, with a horizontal line indicating the fill level for the quoted volume at the stated temperature. They are useful for measuring approximate volumes but must be kept horizontal when checking the volume measurement, and are best used as holding receptacles in solution preparation and mixing.

5.2.3 Burettes, measuring cylinders, and volumetric flasks

Burettes and volumetric flasks are calibrated for volume measurement with high levels of accuracy at or close to room temperature. The temperature at which the calibration has been made is stated on the glassware. Measuring cylinders provide a lower level of accuracy (but higher than that for beakers and conical flasks). Whereas a volumetric flask is used for measuring a fixed volume (although a range of flask sizes are available), measuring cylinders and burettes have calibration subdivisions over a range of volume and, again, can be obtained in various sizes. For making a solution, the use of a volumetric flask containing solute would be the preferred approach, solvent being added up to the defined volume measurement line on the flask. This is both simple and convenient.

Burettes are glass cylinders with a range of volume calibrations along their length and a tap that can be opened and closed near one end. Burettes are useful for measuring and delivering small volumes (aliquots), such as in titrations where, for instance, an alkaline solution is added in small incremental volumes to a solution of acid that is stirred in a conical flask beneath the burette. Care must be taken to clean a burette both before use, by washing with the solution to be used, and also after use, usually by washing with water. The latter is necessary when using solutions such as sodium hydroxide, which can cause the burette tap to seize in its mounting.

The use of measuring cylinders, volumetric flasks, burettes, and pipettes requires *care in observing the volume measurement mark at the base of the liquid meniscus.* A meniscus is the curved surface of liquid when viewed in a narrow cylindrical space, such as the neck of a volumetric flask or the body of a pipette. The meniscus forms as a result of capillary action; aqueous solutions and most other liquids produce a concave meniscus in which the centre is lower than the edges. For consistent measurement, the centre point at the base of the meniscus should be observed at eye level against the appropriate volume calibration mark, as you can see in Figure 5.3. Observing the level of the meniscus is a potential source of measuring error if a consistent and appropriate approach is not adopted. When completing the filling of a volumetric flask, or, indeed, a measuring cylinder, with solvent the final small volume needed to take the solution to the correct meniscus position on the calibration mark has to be made

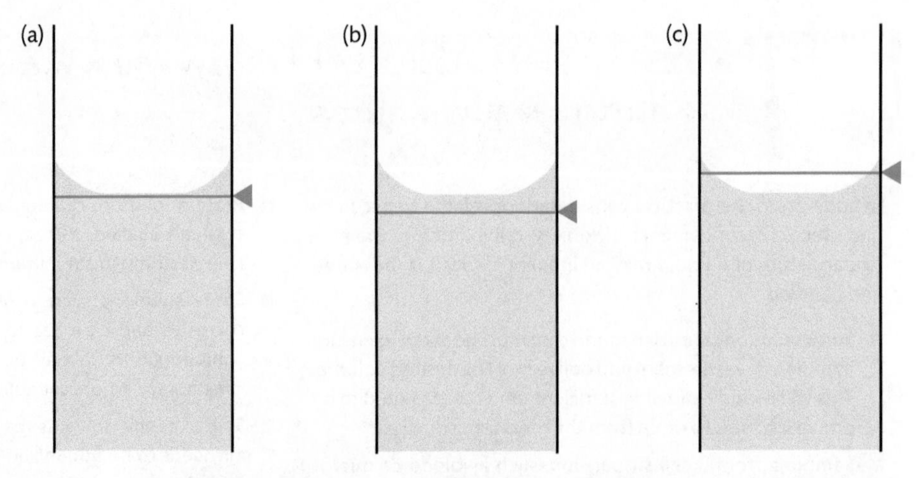

FIGURE 5.3
(a) Correct alignment of the meniscus with the indicated calibration line. (b, c) Volume incorrectly indicated.

by small, dropwise additions using a Pasteur pipette. All solutions need to be appropriately labelled, stating what they are, their concentration, and any associated hazards.

Cross reference
You can read about ELISA in Chapter 13, 'Immunological techniques'.

SELF-CHECK 5.5

What aspects of the uses of a burette and volumetric flask could lead to inconsistencies of measurement?

Although the items mentioned above have traditionally been made of glass, plastic alternatives are becoming more common. Plasticware is safer to use as it is less likely to be damaged than glass and can also be disposable. Glassware needs to be checked before use to identify any cracked or broken items. In solution measurements and transfers between containers, care must be taken to avoid spillages and cross-contamination. Spillages must be dealt with according to local health and safety requirements.

Cross reference
Chapter 4, 'Health and safety'.

5.2.4 Pipettes

Together with certain types of syringes, pipettes measure small volumes in a reproducible and accurate manner. They are commonly used to deliver volumes in the 0–20 cm^3 range, while pipettors can deliver much smaller volumes, as low as 1 μL. Pipettors are complex, precision measuring devices whose construction and operation need to be fully understood to operate correctly and reliably. They will be considered in Section 5.3.

Online Resource Centre
To see an online video showing some different types of pipette, log on to www.oxfordtextbooks.co.uk/orc/fbs

Pipettes were originally made from glass, although they have increasingly been replaced by plastic alternatives. Two types are available. *Bulb* pipettes deliver a fixed volume and have a large bulbous region part way along their length with a single, fixed-volume mark near the top. Although they only measure a single fixed volume, different sizes are available. There is only one calibration mark and so volume measurement mistakes are unlikely. Bulb pipettes are simple to use and potentially more accurate than a graduated pipette. *Graduated* pipettes are straight-walled and calibrated across a particular volume range such as 0–10 cm^3 in 0.1-cm^3 increments. Care must be taken to check the way the scale is presented, for example, 10–0 cm^3 or 0–10 cm^3, and whether the pipette is of the 'blowout' type or not. Blowout pipettes require that the last drop of liquid is expelled from the tip to deliver the intended volume. Non-blowout pipettes rely upon gravity to deliver the volume and allow for a small volume of liquid to remain in the tip of the pipette. This difference is not an issue if the volume to be measured and delivered is less than the total volume of the pipette. For example, a 5 cm^3 pipette could be used to measure and deliver 4.0 cm^3 with the pipette filled to the top mark (5.0 cm^3) and liquid delivered until the level has dropped to the 1.0 cm^3 mark. In this situation, it would not matter whether you used a blowout or non-blowout pipette. With both types of pipette, care has to be taken to be consistent in reading the meniscus against the graduated scale.

SELF-CHECK 5.6

What do the terms *accuracy* and *precision* mean in the context of measuring volumes with a pipette?

From a health and safety perspective, *pipettes must always be operated with a pipette filler*, which creates a partial vacuum to draw the liquid into the pipette. This avoids any chance of liquid entering the mouth as occasionally used to happen before pipette fillers became available. It is good practice to draw the liquid to just above the required mark, slowly releasing the vacuum to allow the meniscus to fall to the mark. The tip of the pipette is placed above the receiving receptacle and the pipette filler operated to release the liquid as required. As with all laboratory procedures, a correct and consistent approach to the operation of a pipette will deliver accurate and consistent results.

5.3 Pipettors

This section will consider:

- the design features and use of mechanical pipettors
- optimal performance and calibration of pipettors.

Pipettors are also known as *autopipettes* and sometimes *micropipettes*. They are usually adjustable volume mechanical pipettes and have become the most common type of pipette in many laboratories. Despite their cost, they are the volume-measuring tool of choice because they are highly accurate, precise, and reliable. They are sophisticated mechanical instruments and, to avoid damage, have to be operated and handled correctly. Electronic versions of pipettors are also available.

In the early 1970s Warren Gilson developed an adjustable-volume, high-precision pipette with a volume readout display. Although many manufacturers produce pipettors, they are still often called 'Gilsons' for this reason. Pipettors are of particular value in biomedical science because they can deliver small volumes, as small as 1 μL or less, which are required in applications such as in molecular biology. They use a *disposable* plastic tip that can be autoclaved, and this adds to their versatility. Why might this be an advantage? There are a number of situations where the use of a 'clean' tip may be essential, such as for work with nucleic acids, where a tip contaminated with nucleases could degrade nucleic acids in the sample. Plastic tips are cheap and so avoidance of cross-contamination is easy to achieve by using different removable tips. Pipettors can be obtained in both single and multichannel forms, different models covering various volume ranges, and to different levels of accuracy. Multichannel pipettors have a special manifold to which can be attached multiple tips (such as 8- or 12-channel types). This permits simultaneous deliveries of a particular volume of solution such as to the wells of an ELISA plate (a multiwell plate) in immunology. You can see examples of single channel and multichannel pipettors in Figure 5.4.

Cross reference

Molecular biological techniques are outlined in Chapter 14, 'Molecular biology techniques'.

Online Resource Centre

To see an online video demonstrating how to use a micropipette, log on to www.oxfordtextbooks.co.uk/orc/fbs

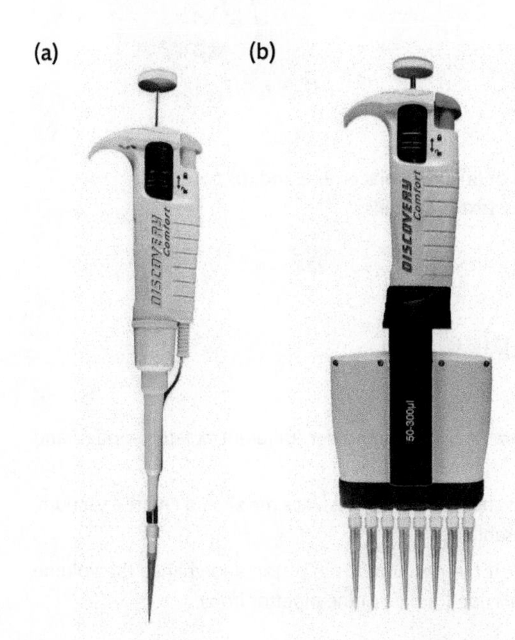

FIGURE 5.4
(a) Single and (b) multichannel pipettors.

BOX 5.4 Air-displacement and positive-displacement pipettors

Air-displacement pipettors have general-purpose uses for dispensing aqueous solutions and are the type you are most likely to use. You can see the general features of their construction in Figure 5.5. These include a piston that moves within the chamber of the pipettor. The piston moves down to release air, and the volume of air remaining in the chamber is inversely proportional to the volume of liquid drawn from the sample when the piston is allowed to rise. The smaller the cushion of air left within the pipettor body, the larger the sample volume. The cushion of air between the liquid being sampled and the piston prevents the piston being contaminated with liquid. In contrast, positive-displacement pipettors work more like a syringe, as you can see in Figure 5.6, which contrasts the workings of the two types. Thus, they lack an air cushion between liquid and piston, which are in direct contact with one another. Hence, the pistons must be disposable because of contamination with liquid. Positive-displacement pipettors have the advantages of being more accurate than air-displacement types because they are less subject to variations in temperature and because they do not form aerosols, which can be produced by the air-displacement types. They are mainly used to measure and dispense viscous, volatile, corrosive, or radioactive liquids.

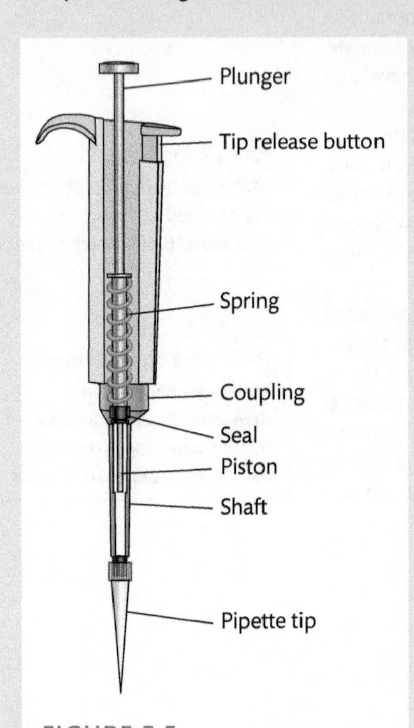

FIGURE 5.5
Schematic showing the construction of an air-displacement pipettor.

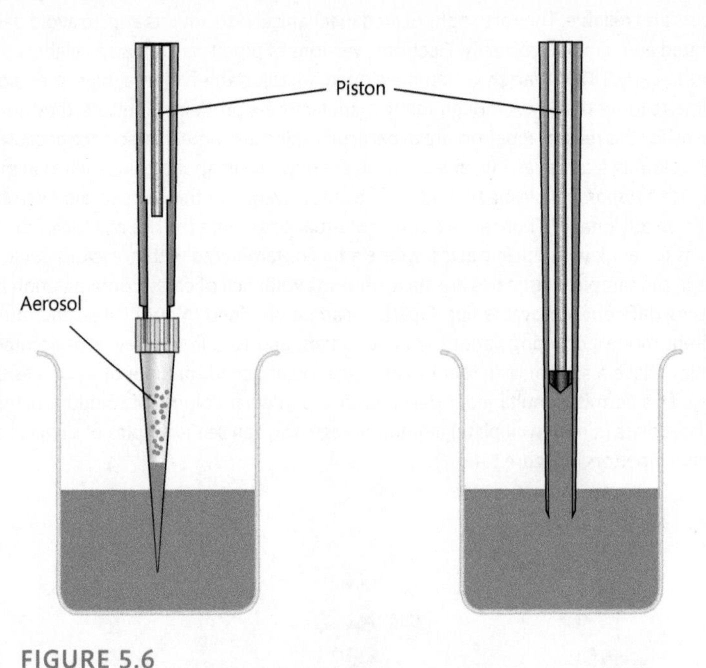

FIGURE 5.6
Comparison of the actions of (a) air-displacement and (b) positive-displacement pipettors. See text for details.

5.3.1 Features and types of pipettors

The features of a pipettor are as follows:

- a disposable plastic tip placed by the operator on the tip holder (different pipettor models and capacities require different types of tips)
- a ceramic or metal piston within an airtight channel that can displace the air and create a vacuum, allowing liquid to be drawn into the disposable tip
- a thumbwheel that can be rotated to adjust the position of the piston and change the volume setting, which is indicated on a digital display on the side of the pipettor body

METHOD How to set up and use a pipettor

- Choose a pipettor appropriate for the task in hand. For a given model, it is essential to use the pipettor only within the volume range indicated. Use beyond this range can damage the pipettor mechanism.
- Organize the work area so that all materials and items of equipment to be used are conveniently placed. The work surface needs to be at a comfortable height. If necessary, this can be achieved by using a height-adjustable seat. Ideally, while holding the pipettor you should be able to rest your elbow of the same arm comfortably on the work surface. This ensures the arm can be rested for some of the time in lengthy procedures while holding the pipettor in a near-vertical position. When not in use, the pipettor must be placed vertically in an appropriate rack.
- Fit an appropriate tip onto the pipettor, applying a little force to get a good, leak-proof fit. This is most readily achieved with the tips stored vertically in a tip rack. A dry tip should be wetted with solution before use. This reduces the surface tension of the inside surface of the tip and will help to minimize evaporation when the sample is drawn.
- Set the volume to be measured, by rotating the adjuster. A Gilson Pipetman® P20 can deliver volumes between 2 and 20 µL. The digital display on the P20 shows three digits. The third, which is in red, represents tenths of a microlitre. Therefore, 020 represents 2.0 µL, 130 is 13.0 µL and 200 is 20.0 µL. For a P200, 095 is 95 µL and for a P1000, 085 represents 850 µL. The meaning of the three digits on the display varies with the model so instructions must be checked otherwise an incorrect volume will be delivered.
- To measure and deliver a volume of sample, the plunger is first depressed as far as the first stop (resistance will be felt) with the pipettor held away from the liquid. Next, the tip should be placed below the liquid surface with the pipettor body held in a near-vertical position. The plunger should be released slowly, to allow the liquid to be drawn into the tip. Care must be taken to ensure that the tip remains within the liquid and that air bubbles are not drawn into the tip. After a few seconds, the pipettor should be withdrawn, keeping it near to vertical.
- To dispense the liquid, the tip is allowed to touch the side of the receiving container by slightly tilting the pipettor away from the vertical, avoiding any other liquid that may have been added already. The plunger is slowly depressed through the first stop and as far as the second stop (the final resistance). After a few seconds, keeping the plunger fully depressed, the pipettor is removed. Only when the pipettor has been removed should the plunger be released.
- The used tip can be removed by pressing the ejector button which moves a collar onto the top of the tip, loosening and releasing it. The used tip must be ejected into an appropriately labelled and suitable waste container.

- a moveable sleeve above the tip that allows the tip to be removed without direct contact with the operator, hence avoiding physical contact with the tip following use and possible cross-contamination with new tips.

There are two different types of pipettor, *air-displacement* and *positive-displacement* (see Box 5.4). Pipettors are available in various volume ranges. For many applications, a range such as the Gilson Pipetman® P is suitable. Table 5.2 shows the volume ranges for this family of pipettors.

TABLE 5.2 The Gilson Pipetman® P range of pipettors

Model	Volume range/µL
P2	0.2–2
P10	1–10
P20	2–20
P100	20–100
P200	50–200
P1000	200–1000
P5000	1000–5000
P10 mL	1000–10,000

SELF-CHECK 5.7

Which of the following features would you *not* associate with a pipettor?

(a) Disposable tip

(b) Permanent or disposable piston

(c) Moveable sleeve for ejecting the tip

(d) Thumbwheel for adjusting the position of the tip.

5.3.2 Using a pipettor

If the steps for use of a pipettor given in the Method box 'How to set up and use a pipettor' are followed in a systematic manner, accurate and reproducible measurements will be achieved. It is worth practising these steps a number of times to achieve consistency.

SELF-CHECK 5.8

In what ways does a pipettor differ from a pipette?

5.3.3 Obtaining optimal performance from a pipettor

Pipettors are designed to deliver high levels of performance, but inconsistencies in results may occur if certain precautions are not taken. Thus, the mode of operation must be followed in a consistent manner. The metal components of low-volume pipettors can expand if heat is transferred from the operator to the plastic handle part of the pipettor. Appropriate pipettor design can partly reduce such heating but the practice of using several pipettors on a rotational basis during prolonged use can help to overcome the potential effect of temperature changes upon the volume determining mechanism. Extended use, particularly if the pipettor is not held in the most comfortable position, may lead to repetitive strain injury. This type of discomfort can lead to the operator holding and using the pipettor in a way that is not consistent with high levels of performance. These problems can largely be overcome by using an electronic pipettor, which removes the need for the operator to exert a significant force to fill and empty the pipettor, and eject the tip. As with all experimental devices, there is a limit to the period of continuous use before operator fatigue begins to compromise performance. The work regime needs to address this issue where concentrated periods of pipettor use occur.

Inexperienced users of pipettors tend to make one or more of the following mistakes which can be overcome with practice:

- the tip may leak if it is not fitted securely to the pipettor; the fitting of the tip must be airtight
- the pipettor can be damaged if used outside the designated volume range
- incorrect volumes may be drawn if the plunger is not depressed to the first stop or if the pipettor is raised during filling so that the tip is above the liquid level
- an incorrect volume may be delivered if the plunger is not fully depressed to the second stop or if the pipettor tip is still in contact with the delivered sample while the plunger is released
- liquid may enter the internal mechanism and damage it, if the pipettor is held near to or horizontal when liquid is in the tip. To avoid this, pipettors should be held as close as possible to the vertical when in use.

It is possible to check for consistency of approach by delivering a measured volume into a small, tared container on a balance. For water, it can be assumed that 1 μL weighs 1 mg. Even if the pipettor is miscalibrated (accuracy is compromised), successive deliveries of the same volume should still give the same measured weight (precision is maintained).

5.3.4 Calibrating a pipettor

In addition to routine inspections of pipettors, calibration checks are also required. An understanding of why pipettors can deteriorate can help minimize deterioration in performance, although this can be compromised in several ways. For example, misuse by the operator, including incorrect volume adjustment, inappropriate handling that results in physical knocks, and dropping of the pipettor can all lead to internal damage. Clearly, the more skilled the operator, the less likely that such damage will occur. It helps if pipettors are stored in a vertical rack when not in use. The use of certain liquids results in volatile, corrosive vapours entering the internal mechanism and can damage the piston, spring, or seals that will affect the airtight seal of the pipettor and the volume that is delivered.

Regular calibration checks are necessary where high levels of accuracy and precision are required. Local laboratory procedures and practices will indicate the required frequency of pipettor servicing and calibration.

How is pipettor calibration checked? This is quite straightforward: the easiest way is to deliver a measured volume of water into a tared receptacle on an analytical balance. Ambient conditions, such as temperature and pressure together with the density of water are taken into account in converting the weight of the sample into the corresponding volume. This volume can then be compared with the volume originally selected on the pipettor. Calibration checks, as well as routine physical inspections, will inform the need for maintenance, which will either be undertaken locally in the laboratory or by the manufacturer.

SELF-CHECK 5.9

Look at Figure 5.7, which shows the digital display of a pipettor. What volume would the display represent if the pipettor was from the Gilson Pipetman® P range and was (a) a P20, (b) a P100, and (c) a P200?

0
6
5

FIGURE 5.7
Digital volume display of a pipettor.

5.4 Preparing reagents: concentrations and dilutions

This section will consider:

- a solution preparation checklist
- the mole (mol) and molar concentrations
- alternative ways of expressing concentration
- different types of dilution and their impact upon concentration.

A **solution**, for instance of glucose in water, consists of a quantity of **solute** (such as glucose) dissolved in a volume of **solvent** (water in this case). The use of balances and volume measurements are the two essential techniques for the preparation of a solution and its dilution. If a set method is to be followed, where all quantities are documented, calculations need not be performed. However, do you know how to weigh an amount of glucose equivalent to 0.1 mol, can you prepare a 0.2 mol dm^{-3} sodium acetate solution, or are you able to calculate the final concentration of a 0.0005 mol dm^{-3} protein solution when 1 cm^3 is diluted to 25 cm^3? If not, then this final part of the chapter will be of help. The precautions to be observed before preparing solutions are explained in Box 5.5.

Online Resource Centre
To see an online tutorial demonstrating how to complete a dilution series, log on to www.oxfordtextbooks.co.uk/orc/fbs

5.4.1 Solutions and concentrations

The *concentration* of a solution is determined by the *quantity* of solute and the *volume* of solvent. as described in Box 5.6.

Key points

Solutions consist of a solute dissolved in a solvent; the concentration of the solution is determined by the relative proportions of solute and solvent.

BOX 5.5 A precautions checklist when preparing solutions

Several matters need to be considered and understood before proceeding to prepare and dilute solutions.

- The health and safety assessments for all materials being used, any precautions that must be observed, procedures for appropriate labelling of solutions in storage containers, dealing with spillages, and safely disposing of waste must be fully adhered with.
- Appropriate protective equipment must be used, including a fully fastened laboratory coat, together with safety glasses and gloves, as determined by the nature of the materials to be handled and procedures to be undertaken.
- Solutions must never be pipetted by mouth.
- The location of all relevant safety equipment in the laboratory, such as an eye wash, safety shower, first aid kit, fire extinguisher, and fire blanket must be known.
- Additional precautions must be observed in some situations, such as the addition of a strong acid to water (and *not* water to the strong acid) because of the heat generated in the process. This is minimized by keeping the concentration of acid low initially and by adding it in small quantities to a much larger volume of water.
- Where a generic health and safety assessment for a particular procedure is documented and followed, a check must be made to ensure that the details of what is being undertaken do not significantly differ from it. For instance, a risk assessment for preparing and using a 1×10^{-5} mol dm^{-3} solution of a slightly hazardous material would need to be modified where a more concentrated, 0.1 mol dm^{-3} solution is being prepared and used, as concentration is a key part of the risk assessment of a potentially harmful solution

Cross reference
Chapter 4, 'Health and safety'.

The different ways in which concentration may be expressed will now be considered. The most common form is **molarity** and this along with the concept of the **mole** will be considered in some detail, followed by other common forms of concentration, **molality**, % (per cent) concentration, and parts per million (ppm). Finally, we will discuss the effects of dilution upon concentration.

5.4.2 Moles, molar concentrations, and molarity

The mole is one of the base units of the SI system of units (Section 5.1). It is a measure of the *amount* of a substance. The mole is that amount of substance that contains as many elementary entities, such as photons, atoms, or molecules, as there are atoms in 0.012 kg (12 g) of carbon (^{12}C). This value is called **Avogadro's number** and is equal to 6.022×10^{23}, and is a constant for a mole of any substance. Although a molecule of water (M_r, relative molecular mass, 18) is small compared with a molecule

BOX 5.6 Concentrations of molecules and ions

The numerical nature of the mole means that a solution contains a particularly large number of individual solute molecules. Aqueous sodium chloride solutions contain sodium and chloride ions (1 NaCl $\rightarrow$ 1 Na^+ + 1 Cl^-). A 0.04 mol dm^{-3} aqueous solution of sodium chloride would have a concentration of 0.04 mol dm^{-3} for each of the individual Na^+ and Cl^- ions but 0.08 mol dm^{-3} for the *combined* concentrations of both.

BOX 5.7 Molecular size and the dalton

The unit called the dalton (Da) is sometimes used to express a mass equal to one-twelfth of that for ^{12}C. It is often adopted in biochemistry and molecular biology texts when referring to the sizes of macromolecules. A protein with M_r of 45,000 could be described as being 45,000 Da or 45 kDa in size.

of haemoglobin (M_r 64,500) (see the next paragraph for an explanation of M_r), a mole of each would contain this same number of individual molecules. Of course, the mass of a mole of these two different molecules would be vastly different.

This numerical nature of the mole is of little help when measuring a particular amount of a substance as counting individual molecules is not a practical proposition! Instead, it is much more convenient to use **molar mass**, which is the mass of *all* the elementary entities in one mole of a substance and has units of g mol^{-1}. For glucose the value is 180.16 g mol^{-1}. A numerically equivalent alternative to molar mass is M_r, which makes reference to the mass of ^{12}C. For an atom the corresponding term is relative atomic mass (A_r):

$$M_r = (\text{mass of one molecule of substance})/(1/12 \text{ mass of one molecule of } ^{12}C)$$

SELF-CHECK 5.10

Distinguish between a mole, molar mass, and relative molecular mass.

As M_r is a ratio of masses, it has no units, but may be expressed in daltons as described in Box 5.7. M_r is equivalent to the now outdated term *molecular weight*, although the latter term is still found on many reagent bottle labels, in many textbooks, and is still used by many biomedical scientists. To obtain one mole of glucose, you need to weigh 180.16 g of it; that is, obtain a mass equivalent to the value of its M_r.

SELF-CHECK 5.11

The amino acid glycine has the formula $C_2H_5NO_2$. How much glycine would you need to weigh to obtain 0.8 moles?

Method box 'Preparation of solutions' gives practical hints on how to prepare, in this example, a solution of glucose.

5.4.3 How are concentrations expressed using the mole?

A one-molar solution contains one mole of solute in a total solution volume of one *dm³* or *litre*. This forms the basis of the molarity form of concentration. Technically, this is not in accordance with the SI system of units as the SI unit of volume is the m^3. However, as the m^3 is 1000 times larger than a dm^3 (which is equivalent to a litre), the use of molarity is numerically more convenient.

The number of moles in a particular volume of a solution can be calculated by applying the following equation:

$$\text{Number of moles (mol)} = \text{molarity (mol dm}^{-3}) \times \text{volume (dm}^3)$$

Key points

Molarity is the most common representation of concentration and is the number of moles of solute in 1 L of solution.

METHOD Preparation of solutions

How could you prepare 1 dm^3 of a 1 mol dm^{-3} aqueous solution of glucose?

You would:

- find out the M_r value for glucose from the reagent bottle label or from data sheets (M_r = 180.16)
- weigh this amount using an appropriate balance then place the glucose in a 1 dm^3 volumetric flask
- add distilled water to the mark on the volumetric flask
- mix well.

However, you do not always prepare 1 dm^3 volumes of solutions, most likely much less than this. How could you prepare 100 cm^3 of a 0.2 mol dm^{-3} solution of glucose? It is easiest to split the calculation into several simple, logical steps:

- 1 dm^3 of a 1 mol dm^{-3} solution will contain 180.16 g of glucose
- 1 dm^3 of a 0.2 mol dm^{-3} solution will contain 180.16 × 0.2 g glucose, which = 36.032 g
- if a volume of 100 cm^3 is needed, the amount of glucose to be weighed is:

$$100\ cm^3 \times (36.032\ g/100\ cm^3) = 3.603\ g$$

Although 3.603 g is equivalent to 0.02 mol glucose (mol = mass/M_r), the concentration is 0.2 mol dm^{-3} because the 0.02 mol glucose is dissolved in a 100 cm^3 (0.1 dm^3) volume. Molarity expresses concentration as though the volume is 1 dm^3 (1 L).

$$\text{Molarity (mol dm}^{-3}) = (\text{mass}/M_r)\ \text{dm}^{-3}$$

SELF-CHECK 5.12

What is the molar concentration of a 6 mg cm^{-3} solution of lactose (M_r = 360.31)? How many moles would be in 50 cm^3 of this solution?

5.4.4 Other common representations of concentration

Molarity is the most common way of expressing concentrations and is therefore the one you are most likely to be familiar with. However, there are several other ways in which concentration may be expressed, three of which we will now consider.

Molality is a form of concentration where the mass of the solvent is used, rather than the volume of solution, as used in molarity. For example, a 0.1 molal solution would contain 0.1 mol solute/kg solvent. For a dilute aqueous solution where the mass of the solute is relatively low compared with the volume of solvent and where water has a density of 1 g cm^{-3}, molal and molar concentrations will be similar in value. With greater amounts of solute, the values will differ as the addition of solute will affect the final volume of solution compared with the volume of solvent used. However, using molality can be convenient because the only measurements are those of mass, unlike molarity, which involves both mass and volume measurements.

SELF-CHECK 5.13

Would you expect any differences between the molal and molar concentrations of 1 dm^3-volumes of aqueous solution containing (a) 0.1 mg of solute and (b) 50 g of solute?

Per cent concentrations are expressed as parts of solute per 100 parts solution. For example, % w/v (*mass-volume or weight-volume percentage*) means mass parts of solute per 100 volume parts of the resulting solution. Thus, a 10% w/v solution of glucose contains 10 g glucose per 100 cm^3 of the glucose solution. Mass *percentage* (% w/w) expresses the concentration as solute mass parts per 100 total mass parts. It is also known as weight percentage or weight-weight percentage. This could be used in the preparation of an aqueous solution of an alcohol. For instance, 25 g alcohol plus 75 g water would make a 25 % w/w solution. Finally, % v/v (*volume-volume* percentage) expresses the concentration as volume parts of solute per 100 volume parts of solution.

SELF-CHECK 5.14

Convert a 0.1 mol dm^{-3} concentration of glucose into a mass-volume percentage (% w/v).

Parts per million (ppm), like per cent concentrations, is a non-SI expression of concentration. Similar forms of expression include parts per thousand. You will have come across ppm if you have carried out flame photometry or atomic absorption spectroscopy work, as these techniques detect relatively low concentrations of analytes. A concentration of 1 ppm means 1 g solute per 10^6 g of solution or the aqueous solution volume equivalent, per 10^6 cm^3. This is the same as 1 μg per g or 1 μg per cm^3.

Per cent concentrations and parts per million are useful ways of expressing concentrations where the M_r of the analyte in question is unknown.

5.4.5 Dilutions

This final part of the chapter will consider how to dilute a concentrated solution (often called a *stock solution*) to make other less concentrated solutions of known value. Where assays and other procedures that require a range of solutions are undertaken on a regular basis, it is convenient to store pre-prepared concentrated stock solutions for subsequent dilution as and when required. This is, of course, only possible where the solution is stable and unaffected by long-term storage, and the dilutions are performed in an accurate and precise manner.

Preparing diluted samples from a stock solution

We will consider the preparation of diluted samples from a stock solution using the practical example given in Case study 5.2.

Using the approach in Case study 5.2 will allow you to handle any situation that involves dilutions. The equation $c_1v_1 = c_2v_2$ also allows you to prepare a particular dilution, where the required final concentration is known because v_1, the volume of stock solution needed for the preparation of the final volume of solution, v_2, can be calculated.

Key points

The equation $c_1v_1 = c_2v_2$ facilitates the calculation of the effect of a dilution and how to produce a specific dilution from a stock solution.

It is helpful to remember that for any dilution the following points apply:

- the final concentration will be less than that of the stock solution because the number of moles of solute taken in the measured volume of stock solution is now dispersed in a larger volume
- the extent of the dilution will determine the concentration change, so the greater the dilution the lower the concentration of the final solution
- the effect of a dilution can be calculated using the expression $c_1v_1 = c_2v_2$ (Case study 5.2)
- the accuracy of a dilution will be influenced by the choice of methods for measuring and delivering the required volumes

- information about a dilution needs to reference both the initial volume to be used and the final volume; for example, a 20-fold dilution would involve adding 19 volumes of diluent to 1 volume of the stock solution to give 20 volumes of diluted solution (this could be called a 1 in 20 dilution).

SELF-CHECK 5.15

What is the concentration of the coenzyme nicotinamide adenine dinucleotide (NADH) in a dehydrogenase enzyme assay in which 0.20 cm^3 of stock 150 mmol dm^{-3} NADH solution is mixed with several other solutions to give a final volume of 3.00 cm^3?

Specific types of dilution

There are certain situations where a particular pattern of dilution (*a dilution series*) is required. These may be a linear or a non-linear series. Where a standard curve is being constructed, equal or near to equal, increments in the dilution series will ensure that there is an even spread of data points. The range of concentrations in your completed version of Table 5.5 is of this type. Sometimes adjustments to the pattern of dilutions are necessary so as to spread evenly the reciprocals of substrate concentrations, as in the Lineweaver–Burk enzyme kinetics plot. There are, however, circumstances in which a wider range of dilutions using non-linear incremental changes are required. This may, for example, reflect an uncertainty of knowledge about the density of cells in a cell suspension or concerning the level of biological activity in a sample, such as may be encountered with ELISA. In these situations, some form of serial dilution is undertaken.

Cross reference

See Chapter 13, 'Immunological techniques'.

A *twofold* or *log*$_2$ dilution is where successive dilutions halve the concentration. This is achieved by measuring a volume of sample and mixing it with an equal volume of diluent and repeating this '1 + 1' approach for each successive dilution. Table 5.3 illustrates the effect of this type of dilution. You can see that after only a few cycles of dilution, the concentration falls rapidly.

A *tenfold* or *log*$_{10}$ dilution occurs when each successive dilution reduces the concentration tenfold. This is accomplished by measuring one volume of solution and adding nine volumes of diluent to it: a '1 + 9' approach for each successive dilution. After several cycles of serial dilutions, the low concentration can become difficult to comprehend easily in a table of data. Tenfold dilutions make graphical representation of concentration data on a linear scale rather meaningless. Table 5.4 shows how logarithmic values can be used to represent such data. The log values could be used on a linear graph scale as an alternative to using logarithmic graph paper.

TABLE 5.3 Cumulative effects of a twofold serial dilution

Dilution step	Dilution
1	2 ×
2	4 ×
3	8 ×
4	16 ×
5	32 ×
6	64 ×
7	128 ×
8	256 ×
9	512 ×
10	1024 ×

SELF-CHECK 5.16

How many cycles of dilution will achieve a minimum of a 1000-fold dilution using (a) a twofold and (b) a tenfold dilution series?

TABLE 5.4 The effect of tenfold dilution steps upon antigen (Ag) concentration in an enzyme-linked immunosorbent assay. The stock [Ag] is 1 mg cm^{-3}. A $\log_{10}$ value change of 1 represents a 10× change in the concentration

Dilution step	[Ag]/mg cm^{-3}	$\log^{10}$ [Ag]
1	0.1	−1
2	0.01	−2
3	0.001	−3
4	0.0001	−4
5	0.00001	−5

BOX 5.8 Experimental kits and automated systems

Biomedical science laboratories to a greater or lesser extent use commercially available kits (see Box 14.1) and/or automated analysers to test clinical samples. Given the advantages conferred by the uses of these systems, particularly the vastly reduced turnaround time, such automation will only increase with future advances in robotics and its underlying information technology.

Reagents in experimental kit are pre-prepared and require minimal dilution. Those used in automated systems are purchased commercially. Thus, there is little need to prepare solutions of reagents. While the solutions of kits do require some dispensing, usually using pipettors, the probes of automated analysers dispense solutions with commendable accuracy and precision. In may seem, superficially, that the general principles involved in weighing solids and liquids, measuring volumes and making and diluting solutions, and the general practices described in this chapter in performing these operations are redundant. However, all scientists should have a firm grasp of the principles and practices that underlie whatever technology they are using.

Cross reference

See Chapter 15, 'Laboratory automation', and Chapter 2, 'Automation', in the accompanying book, *Clinical Biochemistry*.

CASE STUDY 5.1 Choosing an appropriate method for volume measurement and delivery

The purpose of this case study is to help you to think about the importance of choosing an appropriate volume measurement method for a particular situation. Details about alternatives will be given later in this section.

Consider the following scenarios, A and B.

Scenario A

You need to prepare 200 cm^3 of electrophoresis buffer for gel electrophoresis to separate serum proteins. The laboratory stores a stock solution of the buffer that is 20 times more concentrated than required.

What approach might you use to produce sufficient electrophoresis buffer of the required dilution?

Scenario B

You are setting up an assay to measure the concentration of glucose in an unknown sample. A standard (calibration) curve is required to estimate the unknown concentration. To produce an appropriate range of known glucose concentrations for this standard curve, a series of different volumes (in the range 0–1 cm^3) of a stock glucose solution will be required. The volume of each glucose sample will then be made up to a total volume of 1 cm^3 with buffer prior to the addition of a colour generating reagent. This will react with the glucose to generate a colour that will be measured colorimetrically and used in the production of the standard curve.

How might you approach the measurement and delivery of these volumes?

Are the specific volume measuring requirements for B similar to or different from A? You do not need to consider dilutions in detail at this point, only how you would proceed in terms of measuring and delivering the required volumes. The effects of dilutions and the approach to working out how to dilute will be considered in the last section of this chapter.

CASE STUDY 5.2 Quantitatively diluting a stock solution

Cross reference

You can read more about the Beer–Lambert law in Chapter 9, 'Spectroscopy'.

A common form of assay is one where a range of known concentrations of a solution is prepared and a standard curve (calibration curve) produced. The curve can then be used to estimate the concentration of an unknown sample. A property of the solute (such as absorbance due to a coloured compound) that varies with concentration is measured. The Beer–Lambert law provides the basis for a linear relationship between absorbance and concentration.

Usually solutions are colourless and some form of colorimetric assay will be needed to form a coloured derivative of the solute, the absorbance of which can then be measured using a colorimeter or spectrophotometer. For instance, there are several reducing sugar-based colorimetric assays for the detection of glucose. The Somogyi–Nelson method utilizes an alkaline copper solution that reacts with glucose resulting in the conversion of oxidized $Cu(II)^{2+}$ to reduced $Cu(I)^{+}$. The latter reacts with arsenomolybdate to form a blue-coloured complex. The absorbance of the blue solution can be measured.

A simple assay to assess the concentration of haemoglobin in a sample will be used to illustrate the effects of dilution of a stock solution. Look at Table 5.5, which has a blank row and requires your input of the concentration values. The table shows a series of dilutions from a stock solution of haemoglobin which is, of course, coloured. You should be able to see that the dilutions have been achieved by aliquoting a range of volumes of the haemoglobin solution between 0 and 3.00 cm^3 and adding an appropriate volume of saline to each to give a total volume of 3.00 cm^3 in each case. The stock haemoglobin solution has a concentration of 2.00 mg cm^{-3}. You need to work through the following information so as to be able to complete the table.

How might you proceed to fill in the concentration values? Tubes 1, 5, and 7 should pose little difficulty. Tube 1 lacks haemoglobin and so the concentration of its contents will be zero. At the other end of the range, tube 7 contains undiluted haemoglobin and the concentration will therefore be the same as the stock solution (2.00 mg cm^{-3}). Tube 5 has a mixture of equal volumes of haemoglobin and saline, so the concentration in this tube has halved. How can the other concentrations be determined? The extent of dilution will need to be calculated by comparing the initial volume of haemoglobin with the final volume of haemoglobin plus saline. This is readily achieved by using the equation:

$$c_1v_1 = c_2v_2$$

where c_1 = concentration of stock solution; v_1 = volume of stock solution used; c_2 = concentration of the diluted solution; and v_2 = total volume of solution after dilution.

It is essential that both c_1 and c_2 have the same units of concentration. Similarly, v_1 and v_2 must have the same units of volume.

Applying this equation to tube 6:

$$c_1 = 2.00 \text{ mg cm}^{-3}$$

$$v_1 = 2.25 \text{ cm}^3$$

$$c_2 = \text{unknown concentration}$$

$$v_2 = 3.00 \text{ cm}^3 \ (2.25 = 0.75)$$

TABLE 5.5 Preparation of a range of haemoglobin (Hb) concentrations for the construction of a standard curve. Stock [Hb] = 2.00 mg cm^{-3}. The final row in the table is for you to complete

Tubes	1	2	3	4	5	6	7
vol Hb/cm^3	0.00	0.25	0.50	1.00	1.50	2.25	3.00
vol saline/cm^3	3.00	2.75	2.50	2.00	1.50	0.75	0.00
[Hb]/mg cm^{-3}							

N.B. v_2 will be the same value for all tubes.

By rearranging the above equation, the unknown concentration c_2 can be obtained:

$$c_2 = (c_1 \times v_1)/v_2$$
$$= (2.00\ \text{mg cm}^{-3} \times 2.25\ \text{cm}^3)/3.00\ \text{cm}^3$$
$$= 1.50\ \text{mg cm}^{-3}$$

By repeating this approach you can work out the remaining concentrations and complete the last row of Table 5.5.

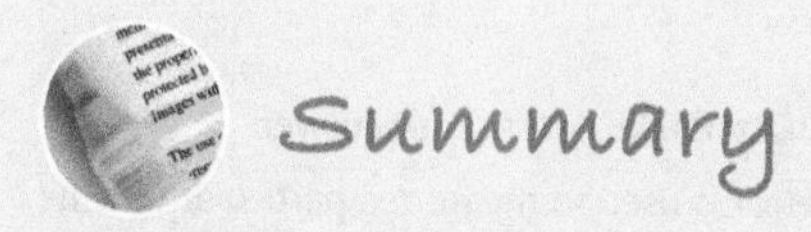

Summary

- The use of balances for weighing and pipettors and other volume measurement methods for volume delivery are key techniques in the production of solutions and their dilution.
- Top pan and analytical balances measure weight; with calibration they can be used to measure a required mass.
- For correct operation, balances must be appropriately sited and calibrated.
- The choice of method for measuring and delivering a volume of liquid depends upon the level of required accuracy and precision.
- Burettes, pipettes, and volumetric flasks provide high levels of accuracy if used correctly.
- Pipettors are the volume measurement tool of choice for highly accurate and precise work. They use disposable tips for convenience, require calibration, and, being precision instruments, must be used with care.
- Pipettors must be used in a correct and consistent manner to achieve accurate and precise performance; to avoid operator fatigue and discomfort in extended use, a comfortable posture must be adopted.
- Solutions are prepared at particular concentrations of solute in solvent, most frequently expressed as molarity, but also in other forms such as molality.
- The production of solutions of a particular molarity requires an understanding of the mole and how the relative proportions of moles of solute and volume of solvent affect the concentration of the solution.
- Dilutions increase the proportion of solvent compared with solute and, therefore, reduce concentration; the application of the equation $c_1v_1 = c_2v_2$ simplifies the calculation of dilutions and concentrations.
- Dilution series can be linear, such as in standard curves or non-linear such as successive two- and ten-fold dilutions.

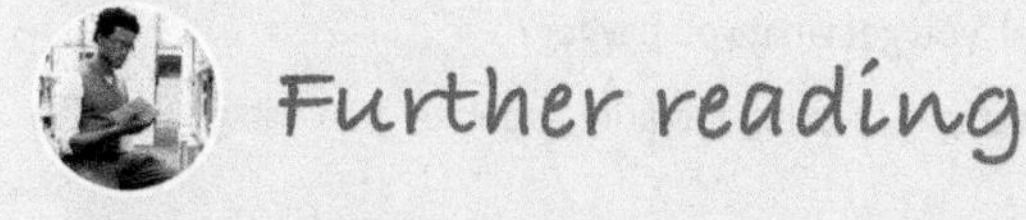

Further reading

- **Fishel LA. Dilution confusion: Conventions for defining a dilution. *Journal of Chemical Education* 2010; 87: 1183–5.**

  This article addresses two conventions for preparing dilutions used in clinical laboratories.

- **Graham I. Difficulties encountered by biology students in understanding and applying the mole concept.** ***Journal of Biological Education*** **1983; 17: 339–42.**

  This article addresses some of the common areas of confusion and misunderstanding in applying the mole concept and considering concentrations and dilutions.

- **Reed R, Holmes D, Weyers J, Jones A.** ***Practical Skills in Biomolecular Sciences.*** **3rd ed. Harlow: Pearson Education, 2007.**

  Chapters 22–24 of this comprehensive basic laboratory skills textbook complement much of this chapter.

Questions

1. Indicate which of the following statements are **TRUE** or **FALSE**. Justify your answer.

   (a) It does not matter what size of pipette or pipettor is used to measure a particular volume of solution.

   (b) When purchasing a balance for a laboratory, it is necessary to obtain the most sophisticated model that funds permit.

   (c) Dilutions for the production of a standard curve should be prepared using equipment with high levels of accuracy and precision.

2. An acid has a M_r of 192 and is used in the preparation of various aqueous solutions. What are the concentrations in mol dm^{-3} of the following solutions?

   (a) 100 cm^3 contains 2 g of acid

   (b) 200 cm^3 contains 4 g

   (c) 5 cm^3 contains 0.05 g

   (c) 2.5 dm^3 contains 45 g

   (d) 10 g of the acid is diluted in 1 dm^3, which is then diluted 25-fold.

3. **(a)** Which of the following represents a 10 ppm concentration of an aqueous solution?

   (i) 10 g per dm^3

   (ii) 1 mg per 100 g

   (iii) 100 μg per kg

   (iv) 1 μg per mL.

   (b) Assuming that (a) (i) is a solution of glucose, calculate (i) the molar and (ii) mass–volume percentage concentrations.

4. There has been an electrical fault in your section of the laboratory and you have to relocate temporarily the analytical balance that you are using. List the key issues that you would need to address before being able to operate the balance in the new location.

5. Construct a table with the column headings cost, simplicity of construction, convenience of use, and level of accuracy. Under each heading, use a simple scoring system ranging from low (--) to high (++) for a Pasteur pipette, beaker, measuring cylinder, pipette, and pipettor. This exercise will help you to summarize the key features of different items of volume measuring equipment. Looking at your table, do you think that 'you get what you pay for'?

6. What are the advantages and disadvantages of using molarity and mass percentage (% w/w) forms of concentration?

7. Thinking about how it is achieved, what is the main precaution that must be taken when preparing a two-fold dilution series?

Answers to self-check and end-of-chapter questions are available at the end of the book.

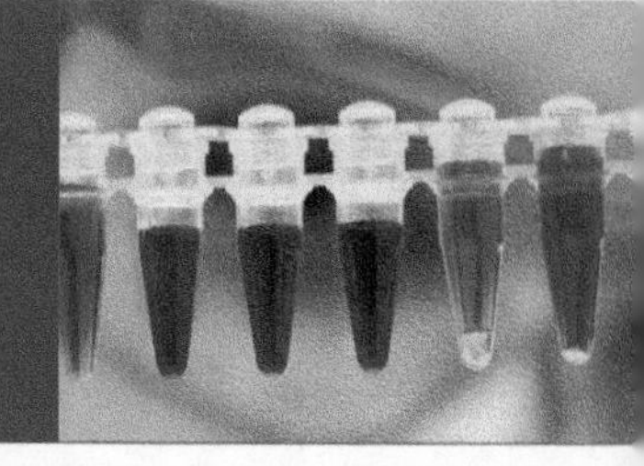

6 Samples and sample collection

Joyce Overfield

Learning objectives

After studying this chapter, you should be able to:

- Identify major factors that contribute to 'sample quality'
- Outline guidelines and procedures for the collection of samples
- Describe the procedures for processing a range of blood samples
- Discuss sampling of histological specimens
- Describe the key steps in obtaining a sample for microbiology testing
- Describe how samples should be handled during the pre-analytical stages

Introduction

Blood and other tissues, and body fluids can be tested in clinical laboratories to aid in preventing, diagnosing, and treating diseases and disorders. A variety of analytical techniques are used, which are being continually developed and improved in accuracy and ease of use by biomedical scientists. Samples for clinical testing include whole blood, plasma, serum, urine, and faeces for chemical and cellular analyses; tissues and **curettings**, which are scrapings of tissues, for histological examination to identify pathological changes; and swabs may be taken from various parts of the body for microbiological culture. Examples of body fluids that may be tested for infection include blood, urine, cerebrospinal, bronchial, pleural, synovial, and bone marrow.

Samples obtained from patients must be placed in suitable containers, transported to the clinical laboratory, and stored correctly, if clinical test results from them are to be accurate. Most clinical laboratories, including those within the National Health Service (NHS) in the UK, have standard operating procedures (SOP), which specify the manner in which samples must be treated. These procedures must always be adhered to.

This chapter will describe the range of samples and the procedures for their collection for analyses. It is not possible here to include sample preparation for every clinical test available and if you have particular interests, the further reading at the end of the chapter is wider ranging.

SELF-CHECK 6.1

Describe the key factors that contribute to pre-analytical quality assurance of clinical samples.

6.1 Blood

Online Resource Centre

To see an online video demonstrating how to process a blood sample, log on to www.oxfordtextbooks.co.uk/orc/fbs

Blood is the most commonly tested type of clinical sample because of its accessibility and the wealth of information that can be obtained by analysing its many different constituents. A clinician or general medical practitioner can request that a range of different blood tests be performed in the clinical diagnostic laboratories by simply ticking or writing on a request form, an example of which you can see in Figure 6.1.

Only a limited number of tests are performed on whole blood; most are done using plasma or serum. Blood tests are performed for a number of clinical reasons, some of which you can see listed in Table 6.1. These include, for example, to assist in the diagnosis of diseases or conditions, or in screening tests, such as testing a cohort of individuals for the presence of HIV. Appropriate testing can also eliminate the possibility of a particular condition even if the patient's symptoms have suggested its

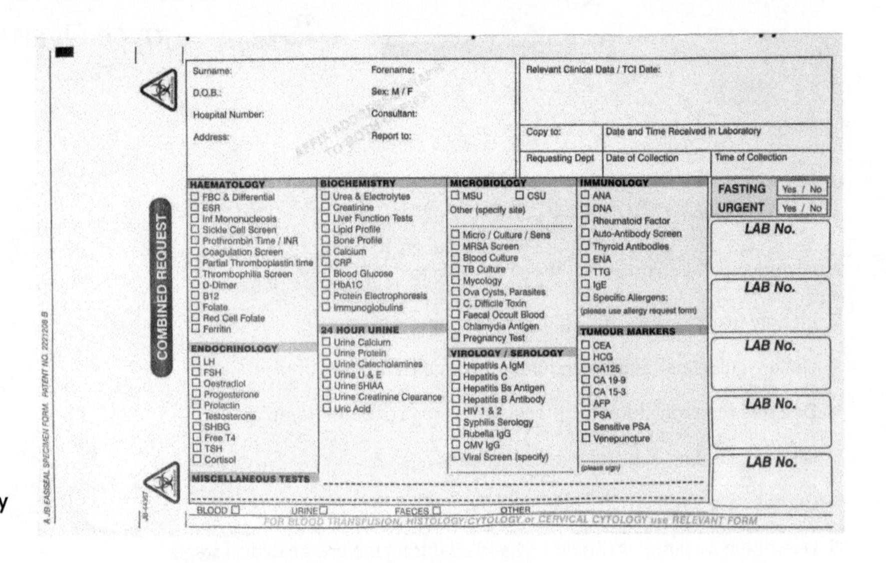

A JB EASISEAL SPECIMEN FORM. PATENT NO. 2221208 B

JB-44387

COMBINED REQUEST

Surname:
Forename:
D.O.B.:
Sex: M / F
Hospital Number:
Consultant:
Address:
Report to:

AFFIX ADDRESSOGRAPH TO BOTH PARTS

Relevant Clinical Data / TCI Date:

Copy to:
Date and Time Received in Laboratory
Requesting Dept
Date of Collection
Time of Collection

HAEMATOLOGY
☐ FBC & Differential
☐ ESR
☐ Inf Mononucleosis
☐ Sickle Cell Screen
☐ Prothrombin Time / INR
☐ Coagulation Screen
☐ Partial Thromboplastin time
☐ Thrombophilia Screen
☐ D-Dimer
☐ B12
☐ Folate
☐ Red Cell Folate
☐ Ferritin

ENDOCRINOLOGY
☐ LH
☐ FSH
☐ Oestradiol
☐ Progesterone
☐ Prolactin
☐ Testosterone
☐ SHBG
☐ Free T4
☐ TSH
☐ Cortisol

BIOCHEMISTRY
☐ Urea & Electrolytes
☐ Creatinine
☐ Liver Function Tests
☐ Lipid Profile
☐ Bone Profile
☐ Calcium
☐ CRP
☐ Blood Glucose
☐ HbA1C
☐ Protein Electrophoresis
☐ Immunoglobulins

24 HOUR URINE
☐ Urine Calcium
☐ Urine Protein
☐ Urine Catecholamines
☐ Urine U & E
☐ Urine 5HIAA
☐ Urine Creatinine Clearance
☐ Uric Acid

MICROBIOLOGY
☐ MSU ☐ CSU
Other (specify site)
☐ Micro / Culture / Sens
☐ MRSA Screen
☐ Blood Culture
☐ TB Culture
☐ Mycology
☐ Ova Cysts, Parasites
☐ C. Difficile Toxin
☐ Faecal Occult Blood
☐ Chlamydia Antigen
☐ Pregnancy Test

VIROLOGY / SEROLOGY
☐ Hepatitis A IgM
☐ Hepatitis C
☐ Hepatitis Bs Antigen
☐ Hepatitis B Antibody
☐ HIV 1 & 2
☐ Syphilis Serology
☐ Rubella IgG
☐ CMV IgG
☐ Viral Screen (specify)

IMMUNOLOGY
☐ ANA
☐ DNA
☐ Rheumatoid Factor
☐ Auto-Antibody Screen
☐ Thyroid Antibodies
☐ ENA
☐ TTG
☐ IgE
☐ Specific Allergens:
(please use allergy request form)

TUMOUR MARKERS
☐ CEA
☐ HCG
☐ CA125
☐ CA 19-9
☐ CA 15-3
☐ AFP
☐ PSA
☐ Sensitive PSA
☐ Venepuncture

(please sign)

FASTING Yes / No
URGENT Yes / No

LAB No.
LAB No.
LAB No.
LAB No.
LAB No.

MISCELLANEOUS TESTS

BLOOD ☐ URINE ☐ FAECES ☐ OTHER

FOR BLOOD TRANSFUSION, HISTOLOGY/CYTOLOGY or CERVICAL CYTOLOGY use RELEVANT FORM

FIGURE 6.1
Example of a request form for laboratory tests.

Online Resource Centre

To see an online video demonstrating blood grouping and cross-matching, log on to www.oxfordtextbooks.co.uk/orc/fbs

TABLE 6.1 Examples of tests performed on blood samples

A full blood count (UK) or complete blood count (US). This repertoire of tests checks for anaemia, and other conditions affecting erythrocytes, leucocytes, and platelets[1]
Blood chemistry profiles[2]
Kidney function (urea, Na^+, K^+, creatinine)[2]
Liver function tests[2]
Hormone (thyroxine, thyroid stimulating hormone) concentrations[2]
Blood glucose (sugar) concentrations[2]
Blood clotting tests[3]
Tests for inflammation[4]
Blood lipids and cholesterol concentrations[2]
Monitoring the doses of specific medications[2]
Immunology or serology tests such as testing for antibodies to specific viruses and bacteria[5]
Blood grouping and screening for blood group antibodies[3]

Cross reference

[1]See the *Haematology* book of this series; [2]see the *Clinical Biochemistry* book of this series; [3]see the *Transfusion and Transplantation Science* book of this series; [4]see the *Histopathology* book of this series; [5]see the *Clinical Immunology* book of this series.

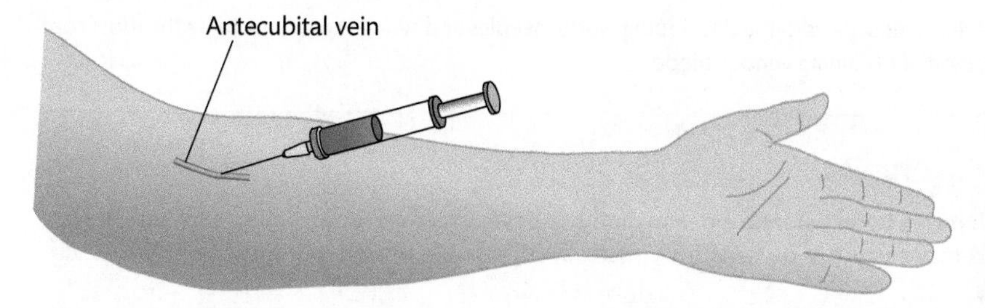

FIGURE 6.2
Schematic illustration of a venepuncture, indicating the position of the patient's arm and antecubital vein.

possibility. Testing of samples of blood from patients is also used to monitor the activity and severity of some clinical conditions. A blood test may help to see if the patient is responding to treatment. For example, a case of iron-deficiency anaemia treated with oral iron supplements should begin to show an increase in the concentration of haemoglobin in the blood. Samples of blood are also required to test the effectiveness of specific organs, such as liver and kidney function, and of the effects of medical or surgical treatment on these organs.

Naturally, a blood sample from a patient is also needed to establish his or her blood group in cases where a blood transfusion is required.

6.1.1 Blood sampling procedures

The healthcare worker responsible for drawing blood samples may be a specialist **phlebotomist**, a clinician, or other healthcare worker trained in blood sampling techniques.

Samples of blood may be peripheral or venous, depending on whether the sample is obtained by pricking a finger or by venepuncture. Finger-prick blood is used mostly for point-of-care tests and is only acceptable providing a free flow of blood can be obtained. However, for laboratory analyses, it is more common to obtain blood by venepuncture. Blood can be drawn from a vein in one of several ways, depending on the age of the patient and clinical tests required. Usually, the antecubital vein, which is located in the crook of the forearm, is used as a source (Figure 6.2). Occasionally, when veins are difficult to locate or have been used many times, blood samples may be taken from the veins on the back of the hand.

In the UK, USA, and Australia most blood samples are collected using commercial vacuum tube systems called **vacutainers**. Under certain circumstances, a syringe may be used, often with a butterfly needle, which is a short needle with an attached plastic **cannula**; the name is derived from the characteristically shaped plastic flanges that aid handling, as you can see in Figure 6.3(b).

(a)

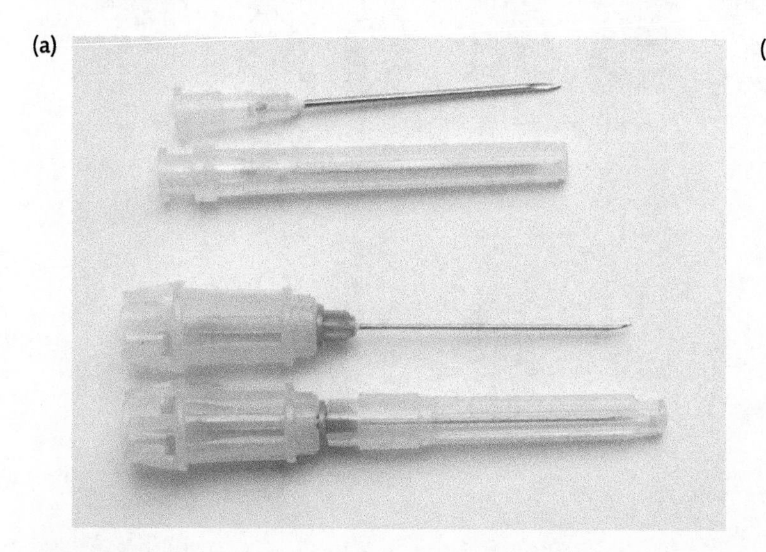

(b)

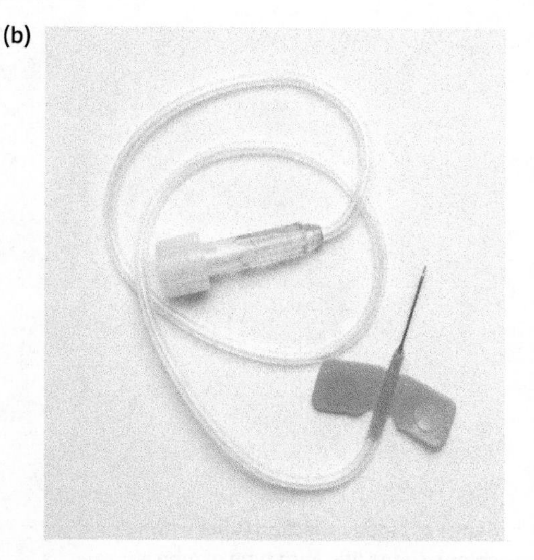

FIGURE 6.3
(a) Selection of conventional syringe needles and (b) a butterfly needle. Note the characteristically shaped plastic handling vanes that give the needle its name.

Cross reference
Chapter 4, 'Health and safety'.

However, in most parts of the developing world, needles and syringes continue to be the most common method of obtaining venous blood.

Key points

All blood is a potential source of infection and poses a risk to the phlebotomist; in particular, blood may carry HIV or hepatitis viruses. Thus, suitable precautions must be taken at all times.

Irrespective of whether blood is collected by vacuum venepuncture or using a syringe and needle, a number of general procedural rules are the same. Any dietary requirements and other appropriate preparatory instructions should be noted before performing investigations. Some blood samples must be taken without a tourniquet (Figure 6.2), so it may be necessary to check with the laboratory for the appropriate procedure. Blood samples must not be contaminated by patient infusions such as a saline drip or blood transfusion.

The correct type of sample tube (see Section 6.1.2) as specified by the **pathology** laboratory must always be used. If a blood sample is added to the wrong type of tube by mistake, it should never be decanted into a second (correct) tube. Blood collection tubes must not be shaken as this can cause **haemolysis**, which is the breakdown of the membrane surrounding the erythrocytes and leakage of their contents. If this does occur, it is likely to invalidate subsequent test results; hence, tubes should be mixed gently by inversion. Finally, it is essential to avoid over- or underfilling tubes that contain an anticoagulant.

Venous blood sampling using vacutainers

Vacuum tubes were first marketed by the US company Becton Dickinson Limited under the trade name Vacutainer® tubes. You can see a range of vacutainer tubes in Figure 6.4. Many companies now sell vacuum tubes, as the patent for this device is now in the public domain.

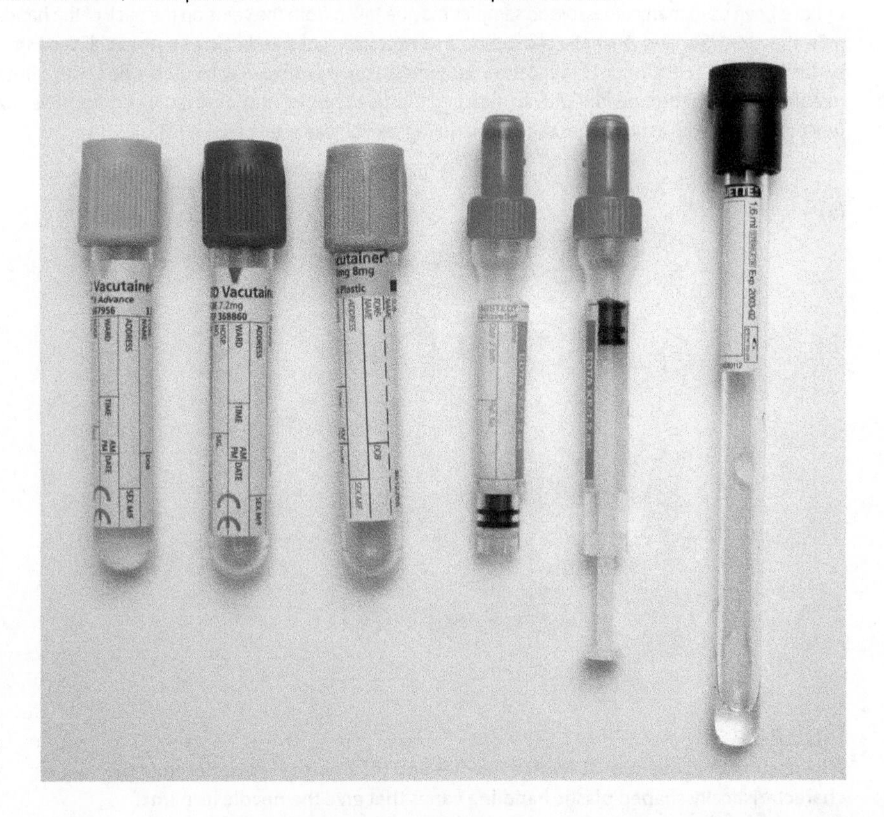

FIGURE 6.4
Range of blood collection tubes with colour-coded lids. Each type of tube is used for different clinical tests. See also Tables 6.2 and 15.2, and Figure 10.14.

A vacutainer consists of a plastic hub, a hypodermic needle, and a vacuum tube. It is based on the use of *evacuated* tubes so that venous blood automatically aspirates through the needle inserted into the tube. Vacutainer systems incorporate a plastic cap known as a multiple sample sleeve, which fits over the posterior end of the needle. This mechanism allows multiple vacuum tubes to be attached and removed in turn from a single needle, allowing numerous blood samples to be obtained from a single venepuncture. This multiple sample sleeve also prevents blood from leaking onto both the phlebotomist and the patient. When using the vacuum tube system, the needle pierces the top of the sample tube and will potentially come into contact with the additives in the tube (see Section 6.1.2). As it is hollow, the needle can carry some additive into the next tube and contaminate it. The additive most likely to cause trouble is ethylenediaminetetraacetic acid (EDTA). This will affect coagulation time assays and may interfere with some clinical biochemistry tests by chelating and removing divalent metal ions, for example Ca^{2+}, and can affect the determination of serum K^+ by reducing erythrocyte integrity. Thus, blood samples for plain tubes should be drawn first and those to be added to EDTA-containing tubes last.

SOP for venepuncture samples must state that the healthcare worker accurately identifies a patient by verbal confirmation, explain the reason(s) for the procedure, and obtain the patient's consent. The patient's arm is examined to identify the most easily accessible vein for the venepuncture. It is advisable not to take blood from an arm containing a drip or transfusion, as this will contaminate the sample. A tourniquet is placed in position on the upper arm without applying pressure (to allow for patient comfort), and the arm positioned comfortably on a support. Alcohol gel or hand washing must be performed between each patient and latex-free safety gloves may be worn if necessary. The tourniquet is tightened to facilitate the distension of the vein. The vein may also be gently palpated.

The vacutainer needle is now attached to the holder and its protective shield removed. The needle is inserted into the distended vein. The vacutainer tubes are gently but firmly pushed in the appropriate order into the needle holder while keeping the holder steady with the other hand. Once filled, all tubes are gently inverted as they are removed from the holder, ensuring that the blood and anticoagulant are mixed to prevent clotting.

When all tubes have been filled, the tourniquet is released and removed. The final tube should be removed from the vacutainer holder before removing the needle from the vein as this practice reduces the risk of a **haematoma**.

To prevent haemorrhage from the puncture wound, the needle is withdrawn and pressure applied through a piece of gauze. The patient should be asked to maintain pressure on the needle site for approximately 5 minutes, to aid wound sealing.

Venous blood sampling using a needle and syringe

A sterile, plastic syringe fitted with a suitable needle may be used to extract venous blood from a patient. Needles for venepuncture are available in a variety of sizes, which are described by their gauge (g) number. The most commonly used needles are:

- 21 g (green top)
- 22 g (black top)
- 21 g (green label) butterfly
- 23 g (blue label) butterfly
- 25 g (orange label) butterfly.

The last is only used in paediatric blood sampling or other extreme cases where the vein is particularly small. A 19 g (brown top) needle has a larger bore and should be used in cases where a large volume of blood is required, for example up to 100 cm^3.

A needle and syringe may be more suitable for obtaining samples from elderly patients, or those with difficult or inaccessible veins. Given that syringes are manually operated and the amount of suction applied can easily be controlled, their use is particularly suitable in patients who have small veins, which collapse under the suction of an evacuated tube. In children or other circumstances where the quantity of blood gained may be limited, it is useful to know how much blood has been obtained before distributing it among the various tubes. Needles and syringes are also used for collecting blood for microbiological examination (Section 6.5). In all cases, the tube contents must be mixed by inverting gently a number of times immediately upon adding the blood to prevent clotting occurring in the tube.

BOX 6.1 Procedure for the collection of high-risk blood samples

- **Samples must be collected by the clinician requesting the test or by an experienced and fully trained phlebotomist.**
- **A disposable plastic apron and disposable gloves must be worn. It is recommended that suitable eye protection such as goggles or safety spectacles be worn, but a face mask is unnecessary.**
- **A waterproof disposable sheet must be placed under the patient's arm.**
- **The patient's skin must be cleaned and the needle inserted in the usual way using a vacutainer or syringe.**
- **Any spillages must be cleaned with freshly made 3% hypochlorite solution.**
- **Used swabs must be discarded into a disposal bin specifically designated for sharp items.**
- **Gloves and apron must be discarded into a plastic sack for incineration (colour-coded yellow in the UK).**
- **Hands must be washed with soap and water from elbow taps and dried with disposable paper towels.**
- **'Danger of infection' labels need to be attached to request forms and samples, which must be placed in plastic sample bags and sealed.**

Once used, the syringe and needle should be disposed of as a whole unit, that is, the needle must not be disconnected, into a bin dedicated to sharp items. Nor must the protective sheath be threaded back onto the needle before discarding it. Needle-stick injuries pose a risk of infection to all healthcare workers because of the exposure to the patient's blood. It is therefore essential to follow good practice. Patients may be identified as 'high risk', that is already diagnosed with infections, for example with HIV or hepatitis, or they may be carriers and unaware of their status. Box 6.1 describes the precautions necessary when obtaining blood from a patient known to be of high risk.

Cross reference

You can read more about the features on health and safety in Chapter 4.

Key points

Irrespective of whether blood is collected by vacuum venepuncture or using a syringe and needle, it must be added to the type of tube appropriate to the clinical tests required.

6.1.2 Blood containers

Once collected, the sample of blood can be added to one of several different containers, depending on the tests required. Some tests must be performed using serum and thus require the prior clotting of the blood. In contrast, tests on plasma require the addition of anticoagulants to the blood sample to prevent its clotting. If you are unclear, Box 6.2 describes the differences between plasma and serum. If the concentration of glucose in the blood is being measured, then the blood is added to a container containing a preservative. Sample bottles with specifically coloured lids (see, e.g., Table 6.2) indicating the additive present and the test profiles for which they can be used have been developed to help phlebotomists in adding blood samples to the appropriate container.

Anticoagulants and other additives

The tubes in which venous or capillary blood is transported to the clinical laboratory may contain a variety of additives or none at all. It is essential to know which anticoagulant or additive is required by the laboratory for particular tests. Table 6.2 describes anticoagulants, additives, and their use.

For most haematology assays, particularly full blood counts, blood group determinations, antibody screening, and transfusion compatibility tests, whole blood needs to be mixed with the anticoagulant EDTA, which acts by chelating and removing Ca^{2+} from solution. This prevents clotting, as Ca^{2+} is essential to the clotting cascade that forms fibrin. However, if blood-clotting tests are required,

BOX 6.2 What is the difference between serum and plasma?

Plasma is formed from blood samples that have been prevented from forming a clot by adding anticoagulants; the erythrocytes, leucocytes, and platelets are then allowed to settle or are separated by centrifugation, leaving the plasma as a separate upper layer in the tube. Portions of plasma can then be removed for analytical tests. Serum, however, is obtained from whole blood that has been allowed to form a clot. The clot is formed by the conversion of soluble fibrinogen to a fibrin gel. The gel consists of a mesh of protein fibres that traps erythrocytes, leucocytes, and platelets. If the clot is allowed to settle or the tube is centrifuged, the serum forms a layer above it. Samples of serum can then be easily removed with a pipette or in a clinical analyser for testing or can be stored in a refrigerator or freezer. Whole-blood samples cannot be stored in this way as the erythrocytes would haemolyse. In addition, many of the analytes estimated in clinical laboratories are present in the serum and using whole blood is not satisfactory for their analyses. Thus, plasma contains fibrinogen, which is absent in serum.

then trisodium citrate is traditionally used as the anticoagulant of choice. It is essential to add appropriate volumes of blood samples to the tubes as under- or overfilling means the proportion of blood to anticoagulant will be incorrect, which will decrease the accuracy of test results. This is especially the case when coagulation tests are to be performed. Some assays require whole blood and in those cases, tubes containing lithium heparin are used.

The majority of clinical biochemistry tests are performed on serum and so a plain tube or a tube containing a clotting accelerator (known as a gel tube; see Figure 10.14) are used. These are usually a plastic or glass tube with kaolin-coated plastic granulate or a polymerized acrylamide resin or gel. The

TABLE 6.2 Commonly used coloured containers and anticoagulants. See also Tables 15.2 and 15.3

Additive	Optimum concentration	Mode of action	Lid colour code[1]	Examples of clinical tests
EDTA[2]	1.2 mg anhydrous salt per cm^3 of blood	Chelates and removes Ca^{2+}	Lilac	Full blood count Sickle cell tests Glandular fever tests
EDTA[2]	As above	As above	Pink	Blood group and compatibility tests Antibody screening
Trisodium citrate[2]	1 volume 0.109 mol dm^{-3} trisodium citrate per 9 volumes of blood	As above	Blue	Coagulation (clotting) tests, e.g. prothrombin time, activated partial thromboplastin time
Trisodium citrate[2]	1 volume 0.109 mol dm^{-3} trisodium citrate per 4 volumes of blood	As above	Black	Erythrocyte sedimentation rate
Lithium heparin, heparin[2]	15 (± 2.5) international units per cm^3 blood	Neutralizes thrombin by inhibiting several clotting factors	Green	Blood glucose, Ca^{2+}
Fluoride oxalate[3]		Fl^- prevents pre-analytical breakdown of glucose (glycolysis). Oxalate binds Ca^{2+}	Grey	Blood glucose
Gel clot activator[3]	Gel substance	Accelerates clotting of blood	Red/brown/ yellow	Tests requiring serum

EDTA, ethylenediaminetetraacetic acid.

[1]Paediatric sample bottles have different coloured tops; the anticoagulant is stated on the label.

[2]Anticoagulant.

[3]Not an anticoagulant.

TABLE 6.3 **Labelling of sample request form**

Labelling of specimen containers	Labelling of request form
Full name of patient	Name of patient–forename and surname
Date of birth and hospital number	Date of birth and hospital number Clinical details–including any information likely to aid the clinician or other healthcare workers
Name of consultant	Name of consultant
Ward or clinic	Ward or clinic

clotting accelerator can interfere with some clinical assays and so a plain tube is recommended for certain tests in which the blood must be allowed to clot and the clot to retract naturally, leaving the serum freely available. If serum is removed before the clot has retracted, it may continue to clot and form a fibrin gel. The advantage of adding a clotting accelerator is that the rapid formation of the clot means tests can be performed immediately after centrifugation.

6.1.3 Labelling of tubes

It is essential that all tubes are legibly and accurately labelled with information that allows subsequent matching of the patient and sample by the laboratory. This information may include the patient's family and given names, date of birth, NHS/Community Health Index number, hospital number, sex, source, date, and time of sample collection. Quality assurance in laboratory professional practice begins with sample collection. Thus, it is essential that healthcare professionals have an understanding of the sampling procedures and the correct labelling of these samples. Each discipline in biomedical science will have particular sample labelling requirements which are fully explained in the companion volumes in this book series, but the bare minimum of information is outlined in Table 6.3.

Once appropriately labelled, the sample tubes are placed in a plastic bag, which is then sealed. The patient should be checked for bleeding from the venepuncture and pressure continued until the bleeding stops if necessary. The needle site may be covered using a fresh piece of folded gauze and micropore tape. Finally, hands must be washed with soap and water after removing gloves, or washed immediately if they have been contaminated with blood.

SELF-CHECK 6.2

What are the minimum identification details to be included on the label of a sample?

6.1.4 Factors in blood sampling that may affect accuracy of results

To perform tests on serum or plasma obtained from a sample of blood, the blood must first be centrifuged. Centrifugation must be performed at the appropriate speed and for the correct length of time to isolate the serum or plasma from the blood. It is essential that cellular components do not leak their contents into the serum or plasma because this contamination will affect the concentrations of some analytes.

Blood must be centrifuged within a suitable time of sampling, normally 4 hours post-venepuncture. It is essential to prevent haemolysis as this will affect many biochemical and haematological tests. High concentrations of blood lipids also affect the accuracy of many clinical tests. This is known as **lipaemia**, and is due to a high fat diet or a lipid abnormality. Figure 6.5 shows the levels of haemolysis and lipaemia in centrifuged blood samples.

Serum must be removed from erythrocytes if the sample is to be stored overnight, to ensure accurate test results, particularly for concentrations of K^+ and glucose. Samples for coagulation tests

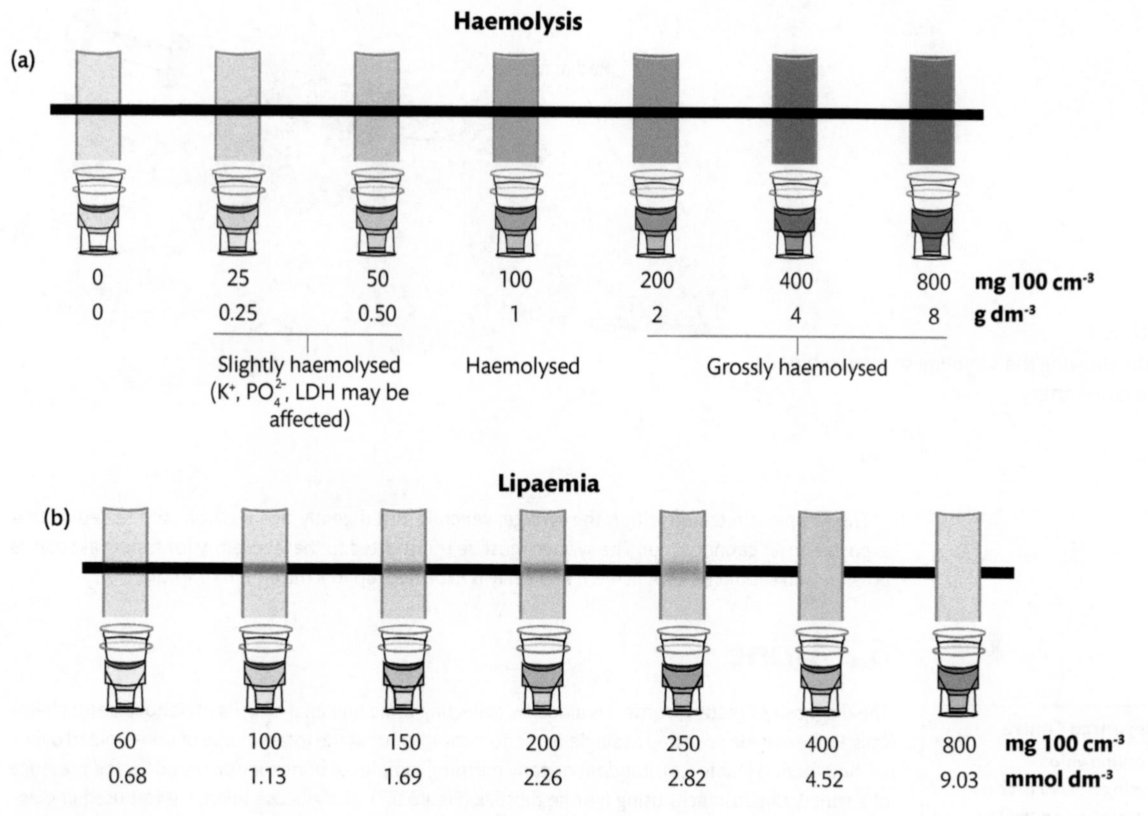

FIGURE 6.5

Levels of (a) haemolysis and (b) lipaemia in centrifuged blood samples. The black line gives an indication of the opacities of the supernatants. LDH: lactate dehydrogenase.

need to be centrifuged and tested as soon as possible as coagulation factors are relatively labile. Alternatively, the plasma may be removed and frozen prior to analysis. When freezing samples, the permitted length of storage may depend on the temperature or the substance being analysed. The normal recommended temperature is −20°C.

Cross reference

You can read about the principles of centrifugation in Chapter 10, 'Centrifugation'.

6.1.5 Arterial blood sampling for blood gas analysis

Arterial blood samples are required when investigating conditions such as diabetes and lung diseases. These can affect the functioning of kidneys or lungs, and result in an imbalance in blood pH, carbon dioxide, and oxygen contents, and/or electrolyte concentrations. Arterial blood transports oxygen to the tissues and venous blood takes waste to the lungs and kidneys; therefore, the gas concentrations and pH differ between these two types of blood vessels. To estimate blood gases, blood is sampled from the radial artery, although in the case of new-born infants it may be taken from the umbilical cord. The radial artery is located on the inside of the wrist, just beneath the thumb, as shown in Figure 6.6, where a pulse is found. In difficult cases an artery in the elbow (brachial) or the groin (femoral) may be used. An arterial blood sample must be obtained by a clinician who is trained in this specific procedure. Blood is drawn into a syringe, which contains the anticoagulant heparin. A small amount of heparin is expelled from the syringe before taking the sample; adequate heparin will remain in the 0.2 cm^3 dead space of the syringe barrel and needle. A minimum of 3 cm^3 of blood is required to avoid excessive dilution by the heparin. Samples containing air bubbles must be discarded because this will cause gas equilibration between the air and the blood, thus lowering the pressure of arterial CO_2 and increasing the pressure of arterial O_2. After sampling, to ensure bleeding has stopped, pressure must continue to be applied to the artery for at least 5 minutes.

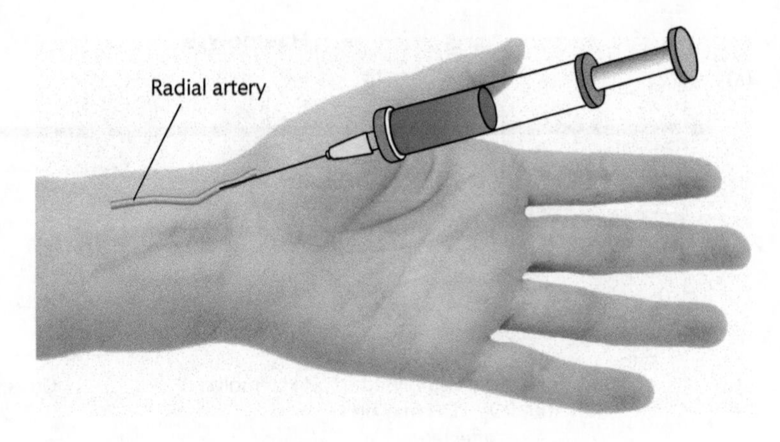

FIGURE 6.6
Schematic showing the sampling of arterial blood from the radial artery.

The sample is retained within the syringe, which is mixed gently by inversion, and sealed so that exposure to air cannot occur. The syringe must be transported to the laboratory for testing as soon as possible and should be kept at 4°C if any delay is expected, up to a maximum of 1 hour.

6.2 Urine

Online Resource Centre
To see an online video demonstrating how to process a urine sample, log on to www.oxfordtextbooks.co.uk/orc/fbs

The diagnosis of many disorders is aided by collecting urine and analysing its metabolites and chemicals. Urine may be collected as single or random samples, or as the total volume of urine voided over a 24-hour period (Table 6.4). Random or early morning samples of urine may be tested for the presence of a variety of substances using a urine dipstick (Figure 6.7). The glucose tolerance test used in diagnosing diabetes mellitus requires three samples of urine, which are collected at specified intervals and analysed for their glucose concentrations.

TABLE 6.4 **Types of urine samples and their corresponding analytes or tests**

Type of urine collection	Analyte/test in urine
Random urine collection	Bence Jones protein Reducing substances Indicans Osmolarity Porphyrins (protect from light) Pregnancy testing
24-hour collection *without* preservative	Creatinine Electrolytes Protein Urate Citrate
24-hour collection *with* preservative	Ca^{2+} Oxalate Phosphate 5-Hydroxyindole acetic acid Catecholamines

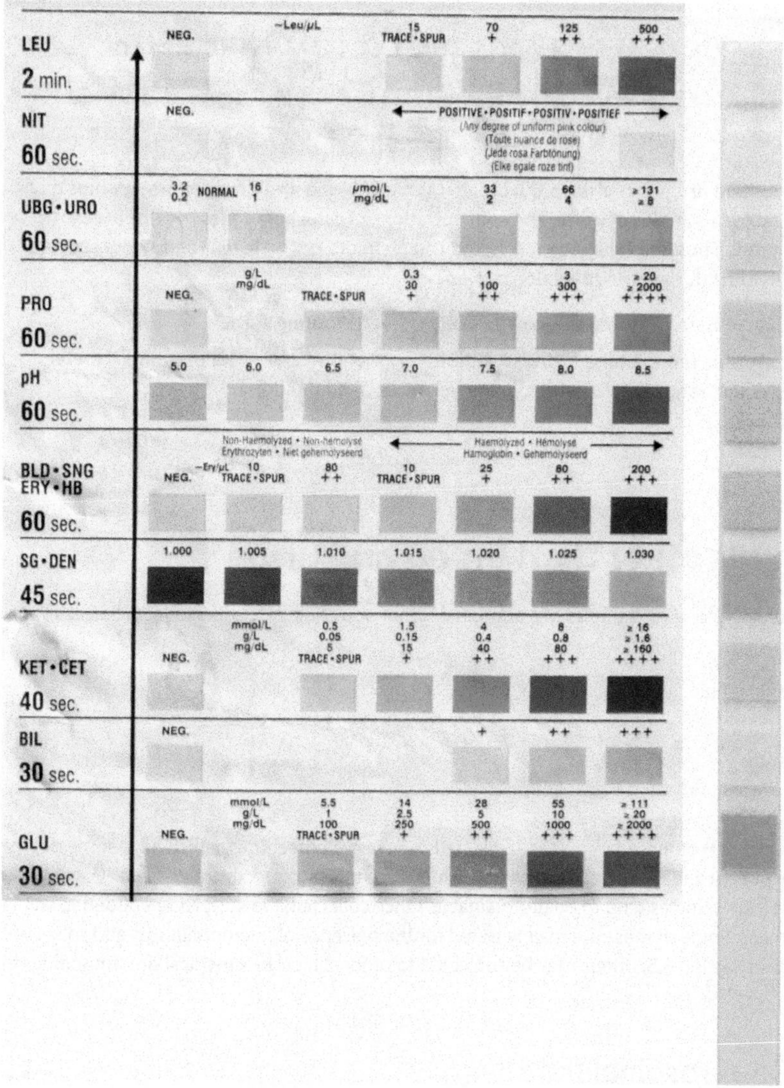

FIGURE 6.7
Renal dipstick used for clinical testing of urine.

6.2.1 Collection of 24-hour urine samples

Twenty-four-hour urine samples give the most accurate estimates of the chosen analyte. However, it is essential to obtain a collection of the total urine output over an *accurately* timed 24 hours. Thus, careful attention to detail is essential. The patient is supplied with a special plastic container, with capacity of up to 2 dm^3. If urine output in the 24 hours exceeds this, a second container is provided. The container should not be emptied or rinsed as it is chemically cleaned, and in some cases may contain a preservative such as hydrochloric or acetic acid. Care must be taken to avoid spillages. It is usual to start collecting urine on rising in the morning. The patient should initially empty their bladder and discard the urine, noting the exact time. In the next 24 hours, all urine voided should be included in the collection and added to the container. On the following day, at the same time as the collection started on the previous day, the bladder is emptied and all of this urine is collected into the container. Urine collection is then complete.

The container must be clearly labelled with the patient's name and the dates and times of collection. Additional instructions may be required for patients whose urine is to be tested for catecholamine levels (see Box 6.3).

Like blood, urine may also be collected for microbiological analyses (Section 6.5).

BOX 6.3 Foods and drugs that affect urinary catecholamine concentrations

Many foods and drugs can disturb the biochemistry of blood and produce metabolites that are detectable in samples of urine. The concentrations of catecholamines in urine are particularly affected. Thus, the following foods and drugs should *not* be taken for 48 hours before and for 24 hours during, the collection of a urine sample:

- *foods:* bananas, fruit, coffee, chocolate, vanilla, and flavouring
- *drugs:* aspirin, monoamine oxidase inhibitors, phenothiazines, imipramine, labetalol, guanethidine, reserpine, levedopa, tetracycline, or α-methyldopa.

6.3 Fluids other than blood and urine

Apart from blood and urine, samples of a number of other body fluids are collected for clinical testing. These include:

- cerebrospinal fluid
- bronchial fluid
- pleural fluid
- bone marrow.

A minimum of 1 cm^3 of these samples are collected by invasive procedures, which have some degree of risk to the patient and must therefore be performed carefully by correctly trained healthcare workers. Often, second samples tend to be unobtainable. Once collected, the samples are added to sterile universal bottles. Some of these are mainly tested for the presence of microorganisms and so we will discuss them in Section 6.5; others may be subjected to other clinical biochemical or immunological analyses.

6.3.1 Cerebrospinal fluid

Samples of cerebrospinal fluid (CSF) are required to investigate infections of the central nervous system (Section 6.5), or cases of suspected subarachnoid haemorrhage. The fluid is removed using a needle from the spinal canal in the lower back, in a procedure called a lumbar puncture (Figure 6.8). Three samples of CSF are needed from patients with suspected subarachnoid haemorrhages.

The samples should be investigated by the microbiology laboratory to determine the numbers of cells and types of microorganisms present. In addition, the clinical biochemistry laboratory may, for example, determine the concentration of protein in the sample.

6.3.2 Bronchial fluid

A fibreoptic bronchoscope is used to obtain fluid from the bronchioles of the lungs, as you can see in Figure 6.9. The tip of the bronchoscope is introduced into the main stem bronchus. The bronchus is usually rinsed with a saline solution. Suction is then applied to the main stem bronchus and other segments of the bronchioles to remove fluid. The bronchial fluid may be filtered through sterile gauze before being added to a sterile container for transport to the laboratory. The sample may be centrifuged to obtain cells for cytological investigations (Section 6.4), or cultured to detect and identify any microorganisms that it may contain.

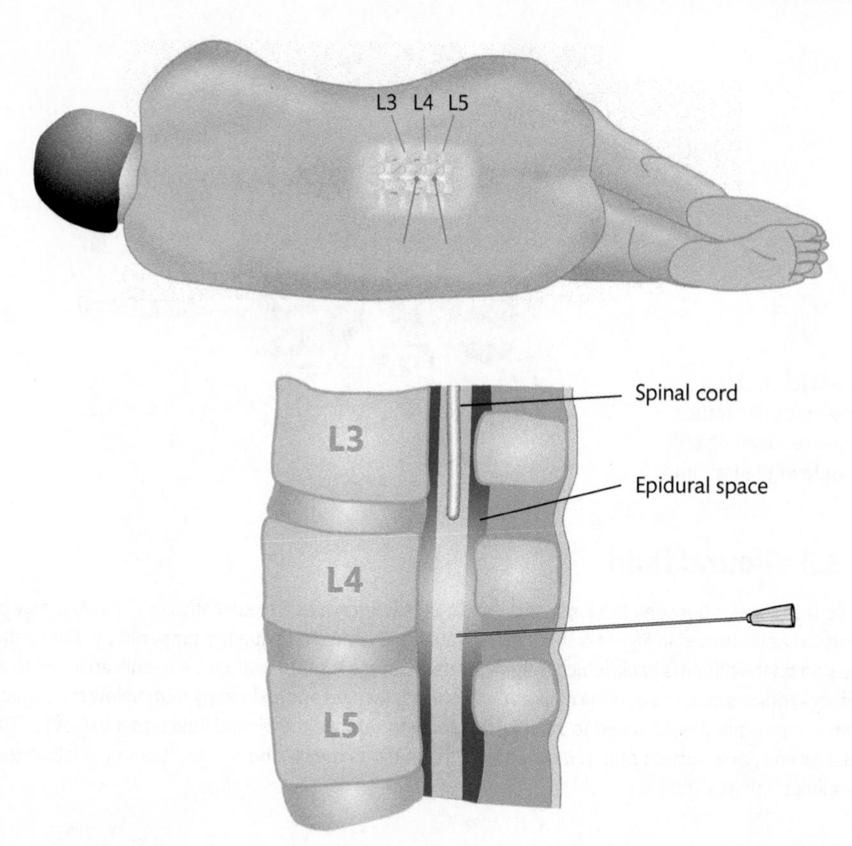

FIGURE 6.8

Simplified illustration of the sites and technique used to perform a lumbar puncture. Note how the needle is inserted between the vertebrae allowing cerebrospinal fluid to be collected.

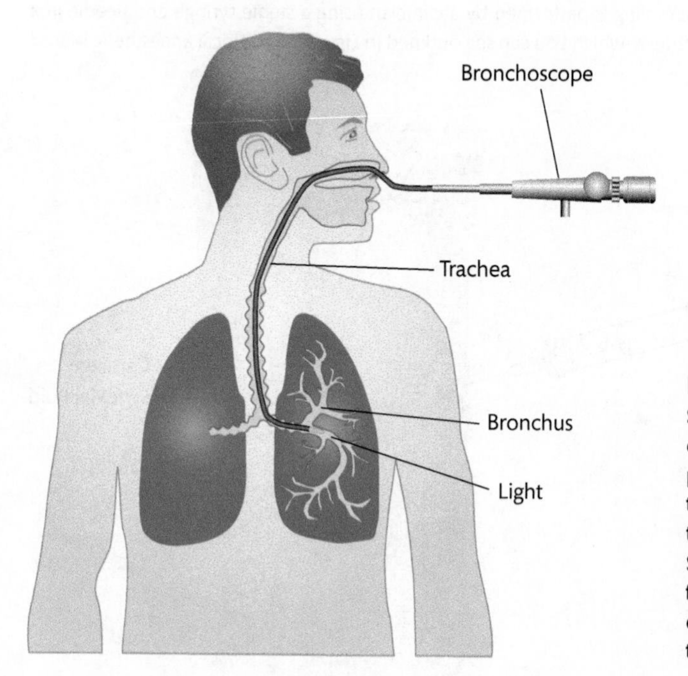

FIGURE 6.9

Schematic illustration of a bronchoscopy, a procedure that allows the respiratory system to be directly viewed. Samples of bronchial fluid can be removed as described in the main text.

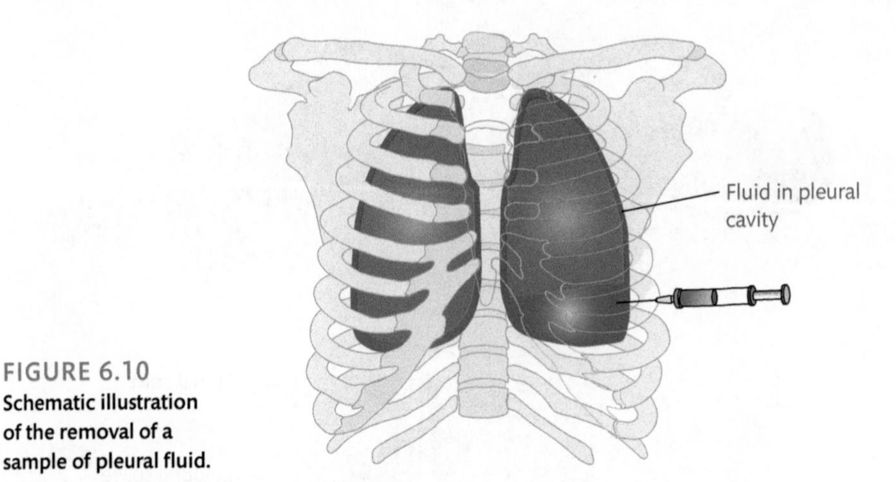

FIGURE 6.10
Schematic illustration of the removal of a sample of pleural fluid.

6.3.3 Pleural fluid

Fluid that collects between the lungs and the rib cage is known as a pleural effusion. The fluid may be sampled as illustrated in Figure 6.10, to investigate for disease. The patient is prepared by disinfecting the skin on the patient's back, which is then numbed using a local anaesthetic. A needle attached to an empty syringe is inserted into the fluid pocket, which is located approximately 3 cm below the surface. The syringe is filled or attached to soft plastic tubing to remove the pleural fluid into a bag or jar. The total sample, or a portion of it, is labelled with the patient's details and transported to the laboratory for clinical investigation.

6.3.4 Synovial fluid

Synovial fluid cushions and lubricates the mobile joints of the body where it is found in small quantities in the spaces contained within the synovial membranes. Analysis of synovial fluid may identify diseases that affect the structure and function of the joints, including osteoarthritis, rheumatoid arthritis, and gout. Synovial fluid sampling is performed by a clinician using a sterile syringe and needle in a procedure called arthrocentesis, which you can see outlined in Figure 6.11. A local anaesthetic is used

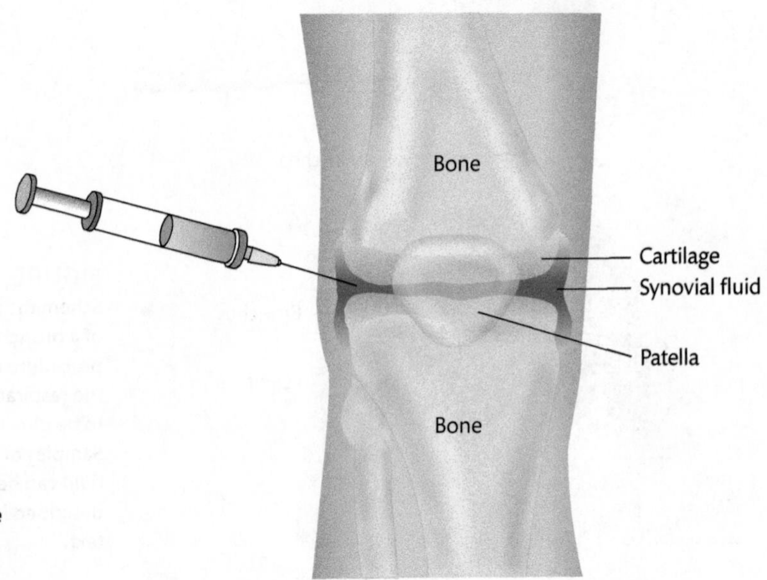

FIGURE 6.11
Schematic illustration of the removal of a sample of synovial fluid, in this case from the knee joint, using a needle and syringe.

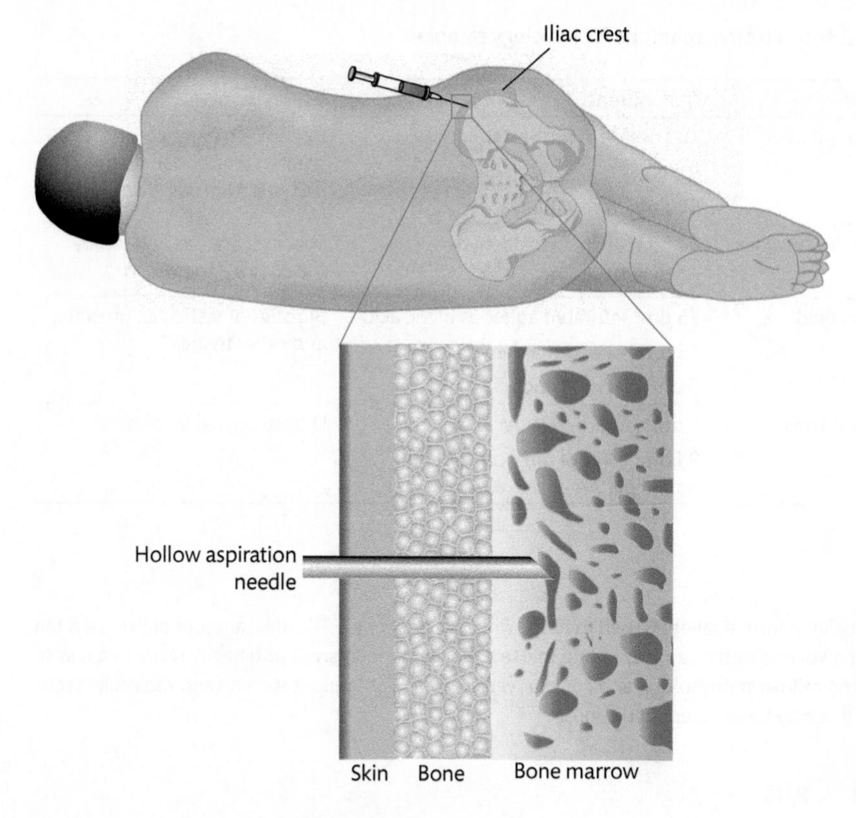

FIGURE 6.12
Schematic illustration of the sites and technique used to obtain a sample of bone marrow from an iliac bone using a robust intraosseous needle and syringe.

to numb the area surrounding the joint and a needle is then gently inserted into the space between the bones, and a sample of fluid sucked out into a syringe. The fluid is placed in a sterile container and transported to the laboratory for analysis. Microscopical (Section 6.4), microbiological culturing (Section 6.5), and biochemical tests, such as those for uric acid and glucose may be performed on the sample.

Cross reference

Gout is described in Chapter 4, 'Hyperuricaemia and gout', in *Clinical Biochemistry*.

6.3.5 Bone marrow

Samples of bone marrow are used diagnostically to observe changes to cellular morphology, and for clinical and immunological investigations. Samples are obtained by aspiration or by trephine biopsy. Bone marrow may be sampled from a variety of sites but the most common are the posterior iliac crest and sternum. As you can see in Figure 6.12, aspiration involves inserting a hollow needle, to which a syringe is attached, into the marrow. The marrow is then sucked into the syringe and spread directly onto glass microscopic slides for analysis or into a sample pot to be used for cytogenetic investigations, culturing the cells present, biochemical analyses, and a variety of other specific investigations.

A trephine biopsy is performed using a trephine needle, known as a Jamshidi bone trephine needle. The sample comprises a core of solid bone, which includes the marrow. This is placed into a solution of formalin for fixation and examined by the histology laboratory.

6.4 Cytopathology and histopathology samples

Cytology is the study of cells and histology that of tissues. Cytopathology and histopathology are the branches of pathology that study and diagnose diseases at the level of the cell and tissue, respectively. The study of cytological and histological abnormalities requires samples of tissues and cells.

In general, the majority of histopathology samples are transported to the laboratory in plastic pots of varying sizes, or glass 'universal' bottles containing a fixative (Table 6.5). In the UK, 4% formaldehyde

TABLE 6.5 Fixative solutions for histology samples

Fixative	Constituents	Use
Formal saline	100 dm^3 40% formaldehyde 9 g NaCl 900 dm^3 tap water	General preparation of paraffin embedded tissue sections from, e.g., liver, kidney
Alcoholic formalin	100 dm^3 40% formaldehyde 900 dm^3 95% ethanol	Specific cases, as advised by the clinical laboratory
Bouin's fluid	75 dm^3 saturated aqueous picric acid 25 dm^3 40% formaldehyde 5 dm^3 glacial acetic acid	Biopsies of testicular, prostate, and other tissues
Glutaraldehyde	5 dm^3 glutaraldehyde 95 dm^3 0.1 mol dm^{-3} cacodylate buffer, pH 7.4	Ultrastructural studies

Online Resource Centre

To see an online video demonstrating how to process a cytology sample, log on to www.oxfordtextbooks.co.uk/orc/fbs

(as formalin or formal saline) is used to fix the majority of samples. The ideal amount of fixative is ten times the volume of the sample. Failure to use adequate amounts will potentially result in a loss of tissue and cellular morphology, and could adversely affect subsequent staining and molecular techniques that may be required for diagnosis.

6.4.1 Cells

Cytological samples are normally provided unfixed, as cell suspensions in transport media or as smears of cells on glass slides. The latter are fixed using alcoholic spray fixatives. The most common cytological samples are taken from the cervix by swabbing (Section 6.5) and are used in smear testing to detect cervical cancer at an early, treatable stage (see Figure 7.2).

Fine-needle aspirates

A clinical sample can be taken from superficial sites such as breast, lymph nodes, thyroid, and salivary glands, and subcutaneous lesions in general by aspiration using a fine needle. Aspirations can usually be performed in an outpatient department immediately by a pathologist or on inpatients in their ward. Aspirations of deep-seated sites may be undertaken using needles of the correct length and using some form of image guidance, such as ultrasound or radiographs. If a pathologist is not available or the clinician wishes to perform the aspiration then a member of the cytopathology laboratory may assist in the spreading, fixation, and transport of the sample. Aspirations are generally considered to be cytopathological samples, unless larger aggregations of cells, akin to tissues, are removed. A diagnosis can normally be reached the same day, allowing for more rapid determination of a treatment programme for the patient.

6.4.2 Tissues

Post-mortem and post-operative samples may be major parts of bodies, portions of organs, or small pieces of tissue taken as biopsies. Sections cut from pieces of tissue are processed and stained on glass slides in a variety of ways to allow them to be visualized using a microscope. Samples of tissue obtained during surgery may be rapidly frozen using liquid nitrogen, which prevents ice crystals forming and disrupting the morphology of the tissues. Thus, intraoperative samples should be sent immediately to the laboratory, without fixation. Sections of the frozen sample can then be cut and stained for microscopic observation. Sections prepared in this way can aid a rapid diagnosis and guide the surgeon during an operation. Samples may also be obtained from patients by curettage, which are essentially scrapings of tissues.

Biopsies such as skin or renal tissue samples, which require testing by immunofluorescence, must be sent to the clinical laboratory as soon as possible in normal or physiological saline solution (9g dm^{-3}). These are then frozen and sections of tissue prepared. Removing biopsies of muscle tissues may require specific techniques that can be obtained by direct discussion with the laboratory. Testicular and prostatic biopsies are recommended to be placed in Bouin's fixative (see Table 6.5). Lymph nodes and other samples on which microbiological or biochemical examination is required must be divided and a portion placed in a sterile container. Lymph nodes removed from patients with suspected lymphoma should be fresh and intact and sent immediately to the histopathology laboratory.

Tissues may also be sampled to investigate specific enzyme activities, such as that of ATPase; activities of this change in different types of muscle fibres in some muscle disorders. In such circumstances, the sections are fixed *following* the investigative technique, to ensure that enzyme activity is not compromised. Muscle biopsies may have specific requirements, which can be obtained after direct discussion with the laboratory.

Cross reference
Chapter 7, 'Microscopy'.

Cross reference
You can read more about immunofluorescence in Chapter 13, 'Immunological techniques'.

6.4.3 Transport of cytological and histological samples

Samples of cells and tissues for cytological and histological investigations are placed in pots with tightly closing lids to prevent leakage. The pots are enclosed in plastic bags for transport to the laboratory. Samples must be accompanied with an appropriate request form in a manner that keeps the form and sample together, but avoids contamination of the form. As with all samples, it is essential to complete the request form, as incorrect or incomplete forms may result in a delay in results. Any fresh samples must be sent to the laboratory as quickly as possible to avoid deterioration of these samples.

When the sample is received in the laboratory, it must be matched to its request form, ensuring all details are legible and correct. Owing to the diagnostic nature of many histological or cytological samples, it is important that full clinical details are provided on any previous medical history, biopsies, or treatments. Failure to provide this information may result in the sample being processed in a suboptimal way.

Key points

Samples for histology or cytology are often unique and it is not possible to take a repeat sample if something goes wrong.

SELF-CHECK 6.3

List three ways in which a sample of tissue can be obtained for histological examination.

6.4.4 Storing cytological and histological samples

Cytological and histological samples are often stored for longer periods than samples for other disciplines, being kept (as a minimum) until a definite diagnosis has been established. The resulting prepared specimens are also stored for long periods of time, which presents its own problems.

Storage of samples and specimens is now governed by the requirements of the various human tissue acts and is the subject of professional guidance from the Institute of Biomedical Science and the Royal College of Pathologists.

Cross reference
More material about the human tissue acts and blood regulations is given in Chapter 17, 'Quality assurance and management'.

6.5 Samples for microbiology testing

Microbiology laboratories of hospitals investigate samples for viruses, bacteria, and fungal infections. Virology laboratories offer routine and specialist tests for the presence of viruses, including the detection of those causing hepatitis, AIDS (HIV), German measles (rubella), and, despite being caused by a bacterium, often syphilis. The bacteriology section of the microbiology laboratory deals with the isolation, identification, and antibiotic sensitivity testing of clinically significant bacterial isolates obtained from a range of patient samples. These include blood, urine, sputum, and other body fluids, faeces, swabs from, for example, wounds, eyes, nose, throat, vagina, and groin. In addition, scrapings and nail clippings are also cultured for microbiological growths. Fungal infections are investigated by the mycology section of the microbiological laboratory.

In all cases, the healthcare worker must ensure that the sample is correctly labelled with the patient's identifying details, and that the appropriate details, including medical and treatment history, are provided on the request form.

6.5.1 Blood

Samples of blood for microbiological culture in aerobic and anaerobic conditions are obtained using a needle and syringe to keep the environment sterile (Section 6.1). Bottles used for the blood cultures do not fit in the standard vacuum tube holders, although it is possible to obtain blood culture bottles with long narrow necks which do fit, or to obtain adaptors that allow the culture bottles to be used as a vacuum tube. When blood cultures are collected using a winged blood collection set, air in its tubing means the phlebotomist must collect the blood sample for aerobic culture first. A blood sample for anaerobic blood culture can then be collected once the air has been flushed out of the tube.

It is essential that the collecting procedure is performed in as sterile a manner as possible to prevent skin **commensals** contaminating the samples. To reduce this danger, an abrasive swab is used to prepare the site by removing upper layers of dead skin cells along with their contaminating bacteria. Povidone-iodine has traditionally been used to clean the skin, prior to sampling, but in the UK a 2% chlorhexidine in 70% ethanol or isopropyl alcohol solution is preferred and time must be allowed for it to dry. The tops of all blood containers should also be disinfected using a similar solution. The bottles are then incubated in specialized incubators for 24 hours. This step allows the very small numbers of bacteria (potentially only one or two organisms) to multiply sufficiently for identification and antibiotic resistance testing. Modern blood culture bottles have an indicator in their base that changes colour in the presence of bacterial growth and can be read automatically by machine. The bar-coded stickers found on these bottles must not be removed as they are used for identification by the automated systems.

6.5.2 Urine

Urinary tract infections are common in certain groups, such as the elderly, immunocompromised patients, or pregnant women. The aim is to collect urine from the patient with minimal contamination so that any pathogenic infection can be detected by culturing and identifying bacteria present in the sample. Patients need to be given clear instructions as to how to collect the urine sample. Obviously, if they require assistance this must be given. Samples of urine should be collected midstream (MSU) or from a urinary catheter (see next section, 'Collecting urine with a catheter'). Early morning urine samples contain the highest numbers of bacteria, as the urine has been 'incubating' in the bladder overnight. The genitals should be thoroughly washed with soap and water, and the first part of the urine stream discarded into the toilet or bedpan. The urine sample is then collected into a sterile utensil such as a foil bowl or sample pot, by intercepting a continuous stream of urine. Sterile plastic screw-capped bottles are used for all samples. Sample pots for MSU contain boric acid as this keeps bacterial numbers constant for at least 24 hours.

If the urine sample is to be used to investigate the possible presence of mycobacteria that cause tuberculosis (TB) then complete (that is *not* midstream) early morning urine samples need to be collected over 3 days. If testing is to be for *Chlamydia trachomatis*, at least 15–20 cm^3 of the **first voided urine** of the day is required.

In all cases, a request form should accompany the urine sample, indicating the type of sample and any patient history of antibiotic treatment. If there is to be a delay in transporting the sample to the laboratory, the sample should be stored refrigerated but never in a food refrigerator!

Collecting urine with a catheter

Sometimes, for example, following an operation or when the patient cannot urinate for themselves, the urine is drained directly from the bladder through a catheter (catheter sample urine) into a drainage bag. Urine samples should not be collected directly from the drainage bag but taken from the sampling port by clamping the drainage tube immediately below the sampling port and waiting for 10 minutes until enough urine has accumulated. Hands must be washed prior to cleansing the sampling port with a medi-swab. A needle and syringe is then used to collect approximately 10 cm^3 of urine from the sample port and transfer it to a sterile pot. The sample pot should be labelled with the patient's details and sent to the microbiological laboratory for appropriate testing.

6.5.3 Collection of sputum

The aim when collecting sputum is to obtain samples of deep respiratory secretions that are not contaminated with upper respiratory tract bacteria. Sample collection begins by explaining the procedure to the patient, encouraging them to breathe deeply, and on exhalation to cough directly into a sterile screw-capped container. Microbiological culturing of sputum samples should use freshly collected specimens. If sputum is being sampled to investigate for the possible presence of TB-causing mycobacteria, at least three separately collected samples should be tested over three consecutive days.

All samples of sputum should be labelled and sent to the microbiological laboratory in a sample bag separate from those for other types of samples, for example blood.

6.5.4 Other fluids

Other fluids collected for microbial analysis include CSF, bronchial, pleura, and ascitic. These fluids are obtained by invasive techniques and are from normally sterile sites, as described in Section 6.3. Samples must be added to sterile containers. Care must be taken by healthcare workers to avoid contaminating the sample with normal microbial flora. If the patient has started antimicrobial therapy prior to sampling, this must be noted. In many cases, the results of culturing these fluids to detect microorganisms present are particularly useful clinically. Therefore, samples should be tested immediately by the laboratory.

The skin must be thoroughly disinfected prior to taking samples of CSF by lumbar puncture (Figure 6.8). In addition to the patient identifier number on the sample bottles, the bottles need to be labelled in the order the samples were collected. Three samples are collected; the first two are used for biochemical analyses. The third should be sent to the microbiological laboratory for examination by Gram staining, appropriate culturing of the bacteria present, which depends on the findings of the Gram stain, and counting any cells that may be present in the sample.

Ascites is the accumulation of fluid in the abdominal cavity in a number of pathological conditions, such as cirrhosis of the liver, heart failure, certain cancers, and infections, for example pancreatitis, chronic hepatitis, and TB. Fluid may be removed by draining or during surgery (Figure 6.13), and subjected to relevant clinical testing.

Sweat testing involves the collection of a sample of body sweat to determine its concentrations of Na^+ and Cl^-. This has been the most widely used diagnostic test for cystic fibrosis. Sweat collection requires specific training as a 'good' sample depends on care and skill in avoiding evaporation and contamination, both of which lead to falsely increased concentrations of Na^+ and Cl^-.

Collecting a sample for a sweat test uses a small, painless electric current to help draw sweat to the surface of the skin, from where it can be collected for analysis. The sweat may be drawn from the thigh or the forearm, depending on the age and size of the patient. The area is prepared by washing and drying, after which two metal electrodes are attached and fastened with straps. Two gauze pads—one soaked in saline or hydrogen carbonate, and the other in pilocarpine, a drug that stimulates sweating, are placed under the electrodes. The electric current is applied to the skin for 5–10 minutes to carry the pilocarpine into the skin where it then stimulates the sweat glands. Once the sweat is being produced the electrodes are removed, and the skin is washed with distilled water and re-dried. A dry piece of filter paper is taped to the area where the pilocarpine was applied. The filter paper, known as a sweat patch, is covered with wax or a sheet of plastic to prevent evaporation. After 20–40 minutes, the plastic

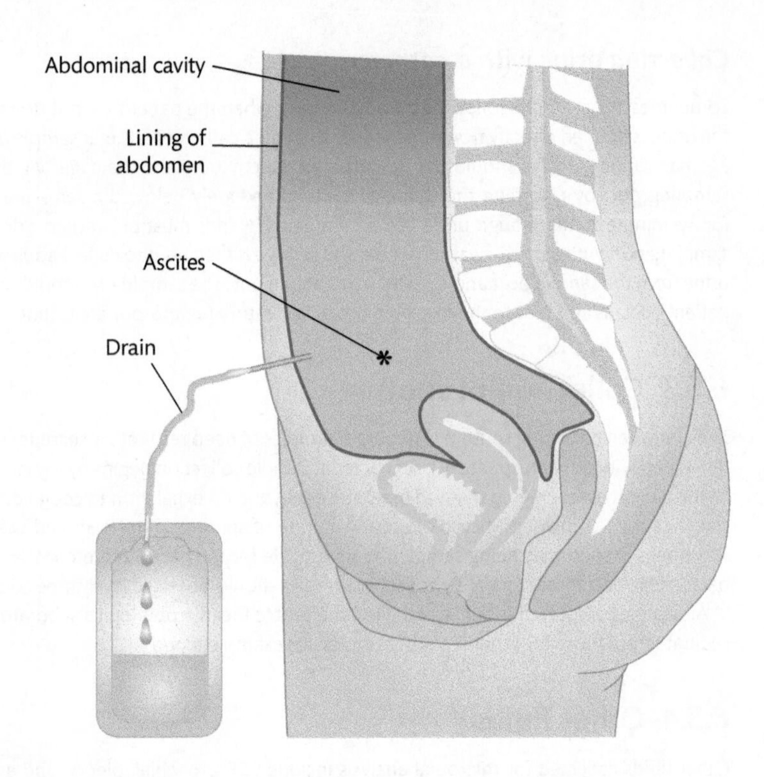

FIGURE 6.13
Schematic illustration of the collection of ascites fluid by drainage.

is removed and the filter paper is placed in a container, which is sealed and labelled with the patient's details for transport to the laboratory.

6.5.5 Collection of faeces

The microbiological examination of faeces is complex. A full clinical history must be taken from the patient, including foods eaten (because of the possibility of food poisoning), any travel to foreign countries, any medication being taken, and the more usual basic information.

A sample of faeces is collected using a spatula. A piece the size of a large pea is adequate. The sample, usually called stool, is transferred to a plastic screw-cap faeces container. If the faecal sample is liquid, the pot is filled to about one-third. Hand-washing and the wearing of safety gloves are essential. The presence of blood in faecal sample, known as **occult blood,** can be detected using specialized cards, as indicated in Figure 6.14. The faecal material is smeared on the relevant portion of the card, which is then sent to the laboratory for analysis. When testing for occult blood, patients must not eat any of the following for 3 days prior to the test: red meat, black pudding, liver, kidneys, fish with dark meat (salmon, tuna, sardines, mackerel), cauliflower, horseradish, tomatoes, radishes, melon, bananas, soya beans, alcohol, iron tablets, aspirin, and ascorbic acid.

6.5.6 Collection of samples using sterile swabs

Simple sterile swabs consist of a plastic stick ending with a pad made from cotton, Dacron, or alginate, which is usually contained in a plastic sleeve containing a transport material to preserve the sample or swabbing (Figure 6.15). Swabbings usually consist of body tissues and fluids for the detection of microbiological **pathogens**. They are transported from the patient to the microbiological laboratory in media contained in the corresponding sleeve. The most common areas sampled by swabbing are:

- high vagina
- nose and ears
- throat

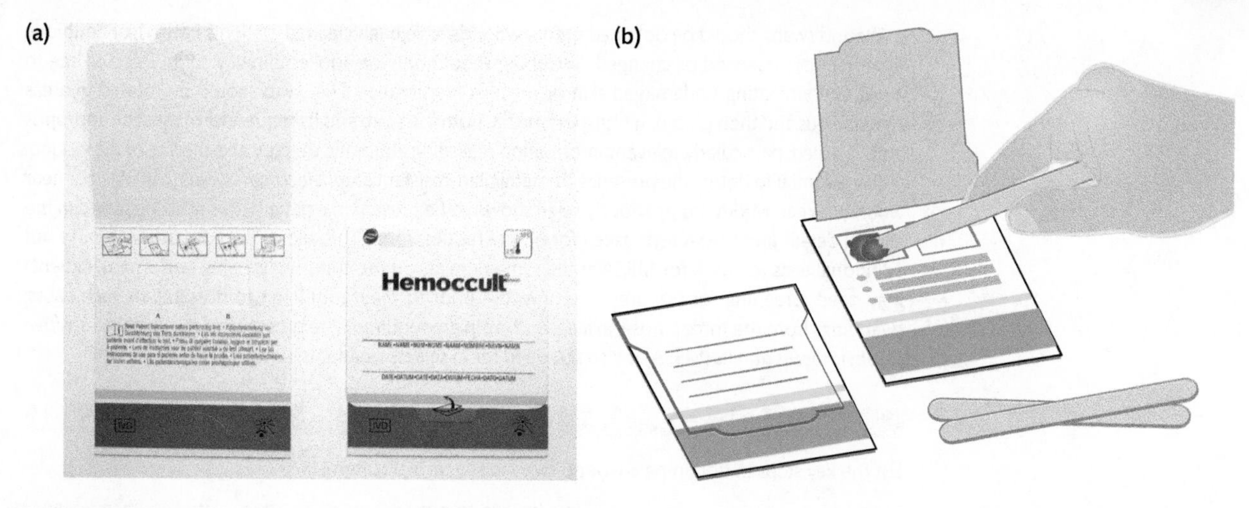

FIGURE 6.14

(a) Photograph of an open and closed occult blood card. (b) Schematic illustration of a faecal sample being added to an occult blood card. Note three cards are provided for collecting samples on different occasions.

- conjunctiva (eye)
- general wounds.

Obtaining a high vaginal swab requires a speculum being placed into the vagina to dilate its walls. The sterile plastic envelope containing the swab is discarded. The swab is removed from its sleeve and inserted as high as possible into the vagina and gently rotated. Any visible discharge is particularly sampled. The swab is returned to its plastic sleeve, which contains the transport medium, and sent to the laboratory for suitable analyses. Plain swabs are suitable for the detection of all vaginal pathogens except *Chlamydia* or herpes, which require the use of specialized types of swabs.

Nasal swabs are obtained by moistening the swab in sterile saline, and then rubbing and rotating it against the anterior nasal hairs. The swab is then placed into its sleeve, where it is moistened in its transport medium. Ear swabs are collected by placing a sterile swab into the outer ear cavity, and rotating it to collect any pus or discharge. The swab is then placed into the sterile sleeve, labelled with the patient's details, including whether the sample is from the left or right ear, and transported to the microbiological laboratory.

Swabbing the throat involves depressing the tongue with a wooden spatula. The swab is then moved to the back of the throat and the area around both tonsils rubbed using gentle rotary movements. The swab is then placed in its plastic sleeve containing transport medium.

Microbial infections of the eyes can lead to a discharge of pus or other fluid. Conjunctival (eye) swabbing involves thoroughly soaking the swab in the exudate by rubbing it across the lower eyelid from the inner to outer corners. It is necessary that the patient remains still during the procedure. The swab is then added to its sleeve with transport medium.

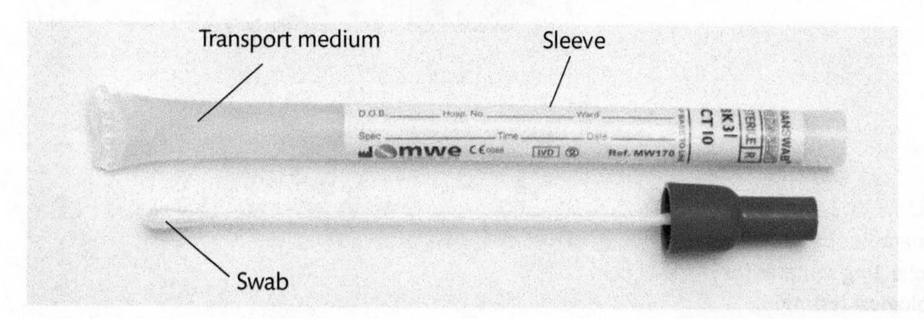

FIGURE 6.15

Photograph of a sterile swab and its container, which contains transport medium.

Wound swabs should be obtained before wounds have been cleaned, patients bathed, or antibiotic treatment commenced or changed. Swabbing should only be applied directly to an infected site to avoid contaminating undamaged skin or mucous membranes. The swab should be rotated in areas showing pus and then placed in transport media. Both the swab and its request form must be appropriately labelled, particularly relevant information regarding antibiotic therapy and the site of the wound.

It is essential to detect the presence of methicillin-resistant *Staphyloccocus aureus* (MRSA) in or near wounds. When MRSA has previously been shown to be present in a patient, this must be stated in the clinical details and MRSA tests asked for on the request form. This allows the laboratory to carry out additional tests to check for MRSA at the same time as routine culturing is being performed. Patients may need screening for the presence of MRSA prior to their admission to hospital. In such cases, swabbings from the throat, nose, axillae, and groin are required. The presence of methicillin-sensitive *Staphylococcus aureus* (MSSA) is also screened for in selected patients.

SELF-CHECK 6.4

List the key steps in obtaining a wound swab for microbial culture.

6.5.7 Samples for investigating fungal pathogens

Samples for mycology testing should be collected into special collection envelopes supplied by the laboratory (Figure 6.16). As with other samples, they are sent with a request form, including details of the patient's clinical symptoms and history.

Cross reference

You can read more about medical microbiology in the companion book of that name.

Material from skin lesions is collected by scraping material from their outer edges, usually with the edge of a glass microscope slide or scalpel blade. The edge of the lesion is the part most likely to contain viable fungal material. Scrapings from the scalp are also obtained in this way but should include

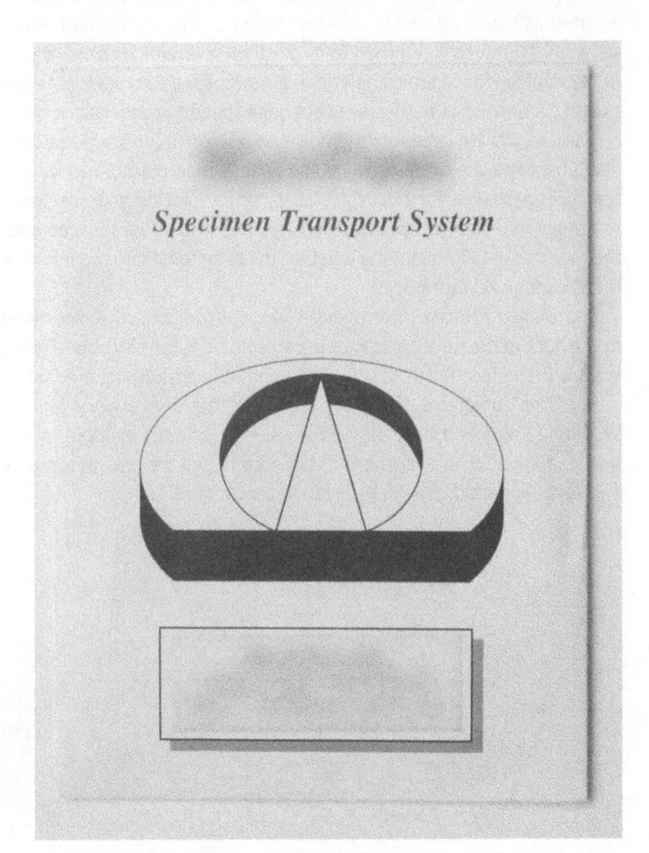

FIGURE 6.16
Photograph of an envelope for collecting samples for mycological testing.

BOX 6.4 Categories of pathogenic microorganisms

The Advisory Committee for Dangerous Pathogens has divided pathogenic organisms into four groups, according to their level of danger. Group I pathogens are the least dangerous, and group IV the most pathogenic. *Mycobacterium tuberculosis*, one of the two pathogenic organisms that cause tuberculosis, is a group III pathogen.

hair stubs. Hairs may be plucked from the scalp, but cut hairs are unsuitable for testing as fungal infections are usually below the surface of the scalp.

Nail clippings for mycological investigation should be taken from the discoloured or brittle parts of the nail, and cut back as far as possible from the free edge as the growth of some fungi are restricted to the lower parts. Scrapings can also be taken from under the nails to supplement clippings; however, even if fungi are present in the nail clippings, they often fail to grow on culturing.

6.5.8 High-risk, hazardous microbiology samples

If it is thought that there is a likelihood of **group III** pathogens (Box 6.4), such as tubercle bacilli and hepatitis B and C viruses, being present in clinical samples, then the samples must be placed in a plastic sample bag, which is then enclosed in a second sample bag attached to the request form. The request form and sample must be marked with a *Danger of Infection* label.

Cross reference
Chapter 4, 'Health and safety'.

CASE STUDY 6.1 Hypocalcaemia—true or false?

The laboratory received a blood sample from a patient suffering from viral hepatitis. The patient's blood result for calcium was 0.1 mmol dm^{-3}, whereas that for albumin was normal. The reference range for ionized calcium is 1.20–1.37 mmol dm^{-3}. A blood calcium concentration below the reference range is referred to as hypocalcaemia. The latter can have serious consequences in affected patients, such as seizures and congestive heart failure, so a value as low as 0.1 mmol dm^{-3} was of concern to the biomedical scientists.

Can you think of a possible explanation for this result?

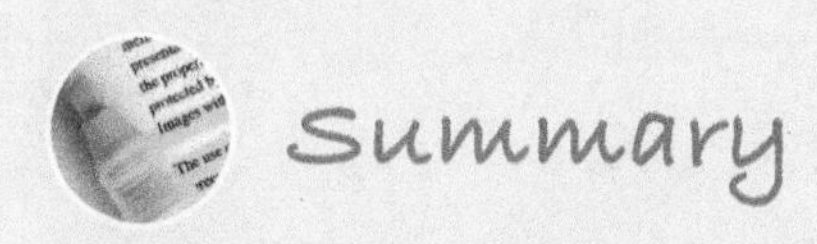

Summary

- The quality assurance in obtaining samples for testing by clinical laboratories includes correctly identifying patients, obtaining samples using correct procedures, meticulous labelling of samples and request forms, adhering to health and safety procedures, and being aware of SOP for sample processing.
- Procedures for obtaining and processing blood samples require selection of the appropriate container. An anticoagulant or additive is necessary for some laboratory analyses. Venepuncture procedures may employ a needle and syringe or needle and vacutainer. All blood samples should be free from haemolysis. Plasma or serum samples are obtained by centrifuging blood samples.

- Samples for histological examination may be obtained as tissue biopsies. Sections cut from pieces of tissue are processed and stained on glass slides, and visualized using a microscope. Samples of tissue obtained during surgery may be rapidly frozen using liquid nitrogen. In some situations, samples may be obtained by aspiration using a fine needle.
- Key steps in obtaining a range of samples for microbiology testing include the relevance of sterility and appropriate use of transport media in which to transport samples to the laboratory.
- Urine samples for biochemical analysis may be collected randomly or over a 24-hour period. Containers for the latter may require the inclusion of a preservative.

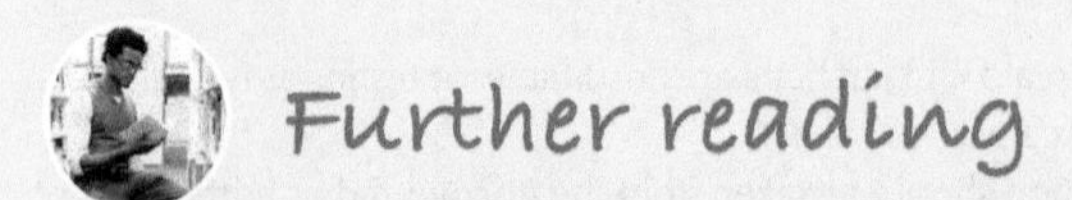

Further reading

- **Bowen RA, Hortin GL, Csako G, Otañez OH, Remaley AT. Impact of blood collection devices on clinical chemistry assays. *Clinical Biochemisty*, 2010; 43: 1-2, 4-25.**
- **Ford M (ed.). *Medical Microbiology*. Oxford: OUP, 2010.**
- **Haverstick DM, Brill LB, Scott MG, Bruns DE. Preanalytical variables in measurement of free (ionized) calcium in lithium heparin-containing blood collection tubes. *Clinica Chimica Acta* 2009; 403: 1-2, 102-104.**

  Two articles and a book that provide details of the types of blood collection tubes, and discuss their effects on blood chemistry analytes.
- **Jacobs DS, DeMott WR, Oxley DK. *Laboratory Test Handbook*. 2nd ed. Cleveland, OH: Lexi-Comp, 2002.**

  An excellent pocket-sized reference book (1348 pages.) containing sample details for all tests imaginable, clinical reference ranges, and much more.
- **Miller WG, Tate JR, Barth JH, Jones GR. Harmonization: the sample, the measurement, and the report. *Annals of Laboratory Medicine* 2014; 34: 187-97.**

  A summary of the harmonization of each area, including sample collection, in the biomedical laboratory.
- **Pitt SJ, Cunningham JM. *An Introduction to Biomedical Science in Professional and Clinical Practice*. Chichester: Wiley-Blackwell, 2009.**

  This textbook covers many aspects of laboratory practice, including an introduction to the organization of sample analyses by clinical pathology laboratories.
- **Polack B. Evaluation of blood collection tubes. *Thrombosis Research*, 2007; 119: 133-4.**

  A journal article focusing on types of tubes for blood samples required for coagulation and thrombosis investigations.

Useful Websites

- Human Tissue Authority. The Human Tissue Act. Available at: www.hta.gov.uk/legislationpoliciesandcodesofpractice/legislation/humantissueact.cfm
- Epsom and St Helier NHS Services. Available at: www.epsom-sthelier.nhs.uk/chemical-pathology. A detailed NHS (UK) website that describes specimen reception for pathology samples.
- Lab Tests Online-UK. Available at: www.labtestsonline.org.uk/. A peer-reviewed public resource on clinical laboratory testing with a patient-centred approach.
- The retention and storage of pathological records and specimens. Available at: www.rcpath.org/resourceLibrary/the-retention-and-storage-of-pathological-records-and-specimens—5th-edition-.html

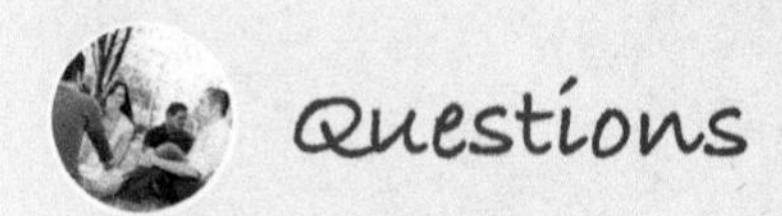

Questions

1. Trisodium citrate is used as an anticoagulant for which of the following?
   - **(a)** Catecholamine tests
   - **(b)** Coagulation tests
   - **(c)** Investigating suspected cholera cases
   - **(d)** Investigating suspected Chlamydia cases
   - **(e)** Clinical chemistry testing.

2. Which of the following statements are **TRUE**?
   - **(a)** The commonest type of clinical sample analysed is blood
   - **(b)** A sample of serum does not contain fibrinogen
   - **(c)** Using an insufficient amount of fixative in preparing histopathology samples results in a loss of cell morphology
   - **(d)** A bone marrow biopsy is called a trephine biopsy
   - **(e)** Patients providing faecal samples for occult blood tests must avoid eating bananas.

3. List the common additives used in blood collection tubes, their modes of action, and the clinical tests with which they are compatible.

4. Explain why clinical samples must be appropriately centrifuged and stored prior to clinical analysis.

5. What instructions would you give to a patient whose urine is to be tested for its concentrations of catecholamines?

Answers to self-check and end-of-chapter questions are available at the end of the book.

7

Microscopy

Tony Sims and Qiuyu Wang

Learning objectives

After studying this chapter, you should be able to:

- Define magnification, contrast, and resolution
- Explain the principles of image formation
- Identify the various components of a microscope and describe their functions
- Describe the major types of microscopes used in clinical laboratories
- Discuss the applications of the various forms of light and electron microscopy used in clinical laboratories
- Set up a microscope with optimum illumination

Introduction

A **microscope** is an instrument that forms an enlarged image of an object that would normally be too small to be seen with the naked eye, as shown in Figure 7.1 (a, b). **Microscopy** is the use of a microscope to examine and analyse such objects. Microscopes that use a single lens are called simple microscopes; those with more than one are compound microscopes.

Microscopes are perhaps the most widely used instruments in biomedical science. They have contributed greatly to our knowledge and understanding of pathological processes, and are used in all branches of biomedical science. There are several different types of microscope, each of which has its own specific applications. *Light* microscopes are used to look at cells and tissues. They are used in many areas of biomedical science and form the work horses of histological examinations or in screening programmes, such as that for cervical cancer (Figure 7.2). The *electron* microscope, with its vastly increased **magnification** and **resolution**, is used to visualize virus particles, explore the structures of bacteria, and observe more fully the subcellular components seen in both normal and diseased cells and tissues.

In this chapter we will describe how the various types of microscope are constructed and work, and how they can be applied to diagnostic biomedical practice.

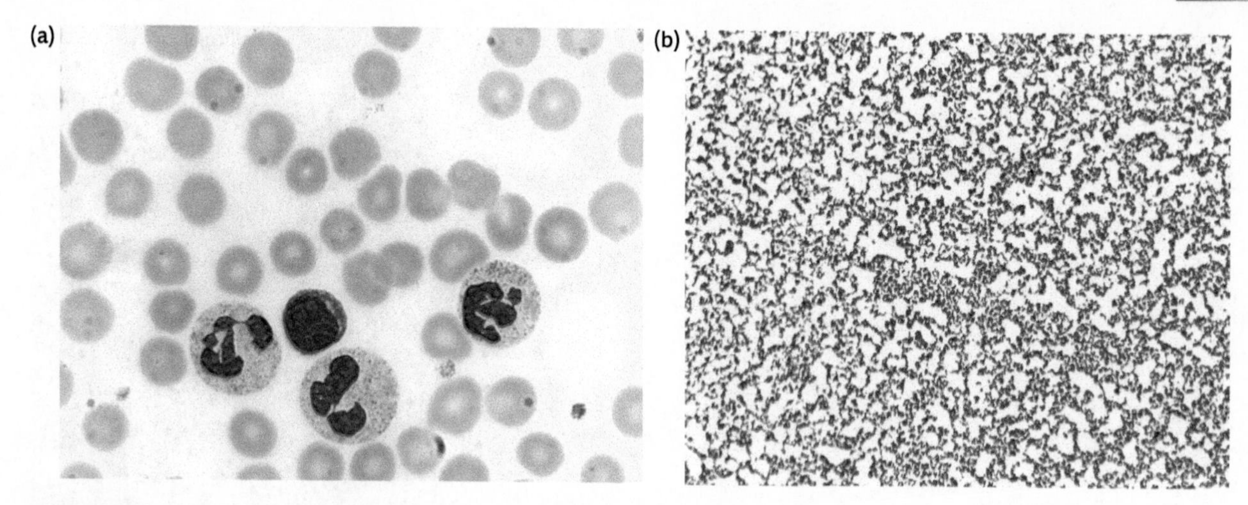

FIGURE 7.1

Microscopical examinations of (a) blood smear and (b) the bacterium *Staphylococcus aureus*. Courtesy of M. Hoult, School of Healthcare Science, Manchester Metropolitan University, UK.

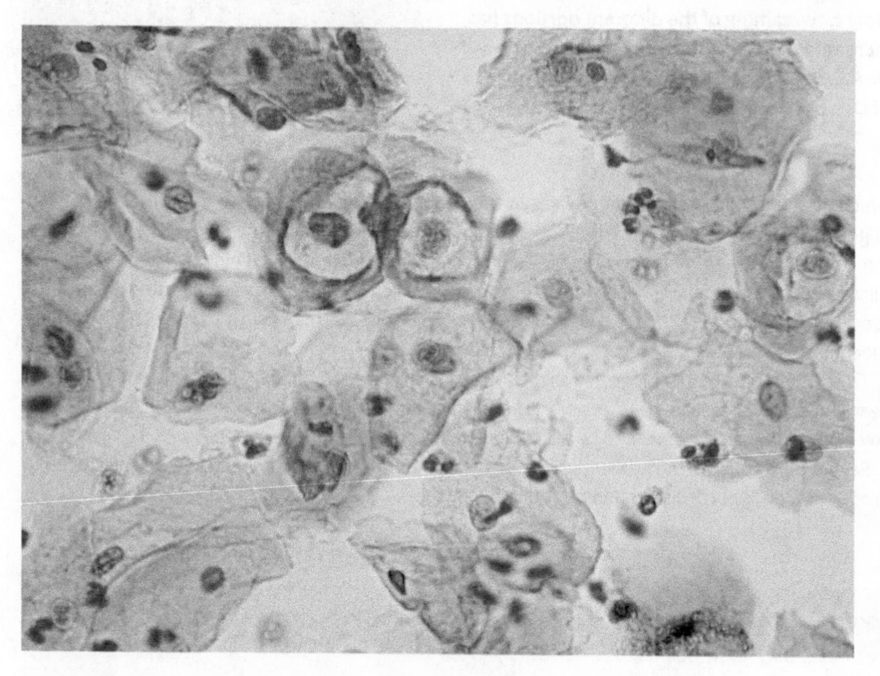

FIGURE 7.2

Light micrograph of a cervical smear showing pink-stained normal squamous cells from the superficial layer of the cervix. The blue stained cells with comparatively large nuclei indicate that they are abnormal and the patient is at risk of developing cancer of the cervix. Courtesy of Manchester Cytology Centre, Manchester Royal Infirmary, UK.

7.1 Microscopy and image formation

Three main features limit the use of any microscope:

1. Magnification
2. Contrast
3. Resolution.

The magnification of a specimen given by any microscope is simply the apparent size of the image formed divided by its real size. We will discuss in Section 7.3 how a microscope produces a magnified image. The *contrast* of an image is the difference between the darkest and lightest coloured areas,

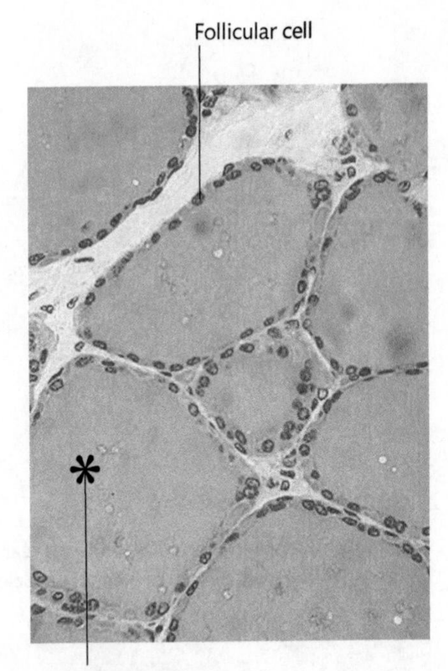

FIGURE 7.3
Histological structure of the thyroid gland. Note how staining of the different portions has increased the contrast of the image. Courtesy of Dr A.L. Bell, University of New England, College of Osteopathic Medicine, USA.

Cross reference
You can read more about health and safety in Chapter 4.

Cross reference
Chapter 6, 'Samples and sample collection', describes how swabs can be used to obtain clinical samples.

which, in microscopy, reflects differences in the densities of different parts of the structure viewed. In general, a biological specimen will show poor contrast and will therefore be difficult to see unless the contrast between different parts of it are increased by differential staining. Specimens for light microscopy are generally stained with coloured dyes, which are only absorbed by some areas of the specimen so boosting the contrast, as you can see in Figure 7.3. You can read an outline of how clinical specimens are prepared for light microscopy in Box 7.1.

The contrast of specimens for electron microscopy is increased by staining specimens with the salts of heavy metals, such as lead, silver, or gold. Even for specimens with excellent contrast, the magnification any microscope can give is limited by the resolution, which we will discuss in Section 7.1.1.

Sections are cut from the wax-embedded tissue using a microtome; these can also be automated to produce sections of consistent and reproducible thickness.

BOX 7.1 Preparing samples for light microscopy

In general, tissue samples are transported to the laboratory in containers holding a fixative to preserve structure (Table 7.1). In the UK, 4% formaldehyde (known as formalin) is the most commonly used fixative. Ideally, about a tenfold volume of fixative to sample is required. Failure to use adequate amounts will potentially result in a loss of tissue and cellular morphology, and could adversely affect subsequent staining and molecular techniques that may be required for diagnosis. Large tissue samples may be sliced open to facilitate optimal fixation.

Once fixed, samples of tissue for histopathological examination must be processed to allow thin (2–3 µm thick) sections to be produced for microscopical study. For routine processing into paraffin wax, tissues are exposed to a number of reagents, for example, transferring the samples through a series of solutions with increasing concentrations of alcohol, to remove fully their water content. The alcohol is then replaced by an intermediate (or link) reagent that is not miscible with water but allows molten wax to penetrate the sample block. In the UK, USA, and Australia, this is routinely undertaken on automated,

TABLE 7.1 **Fixative solutions for histology samples**

Fixative	Constituents	Use
Buffered formalin	100 dm^3 40% formaldehyde 9g NaCl 900 dm^3 tap water	General preparation of paraffin-embedded tissue sections from, e.g., liver, kidney
Alcoholic formalin	100 dm^3 40% formaldehyde 900 dm^3 95% ethanol	Specific cases, as advised by the clinical laboratory
Bouin's fluid	75 dm^3 saturated aqueous picric acid 25 dm^3 40% formaldehyde 5 dm^3 glacial acetic acid	Biopsies of testicular, prostate, and other tissues
Glutaraldehyde	100% glutaraldehyde	Ultrastructural studies

fully enclosed tissue processors. In developing countries, this can still be undertaken manually; however, the nature of the solutions used to process the tissue does pose a health and safety risk to the laboratory worker.

Situations may arise that require the production of sections for rapid diagnosis or investigation of materials. For example, samples of tissue obtained during surgery may be rapidly frozen using liquid nitrogen (–196°C), which prevents ice crystals forming that would disrupt the architecture of the tissue. Sections of the sample are cut and stained for microscopic observation to provide a rapid diagnosis and guide to the surgeon during the operation.

In some instances, for example, renal biopsies, it may be necessary to produce sections of less than 2 µm thickness. Often, this requires processing the tissue sample into resin rather than wax. Sections are cut on an automated microtome using a knife blade strengthened, for example, by coating its cutting edge with tungsten carbide or diamond. Using the more solid medium of resin allows thinner sections to be cut, but it does suffer the disadvantage of absorbing many of the stains used to heighten contrast. Consequently, interpreting these sections can be more difficult. Resin is also used to embed small pieces of tissues that require examination using an electron microscope (see Box 7.3).

Once the tissue has been processed and sectioned, it must be stained to provide contrast between the different tissue and cellular components. There is a multitude of different methods that may be utilized to highlight different tissue components, but the routine stain used is haematoxylin and eosin. Such stains highlight the general architecture of tissues and determine if a pathological process is occurring, or if any further techniques are required to reach a definitive diagnosis.

Samples received in histopathology laboratories are varied: it is not unusual to receive samples of bone or other mineralized materials. These cannot be processed straight to paraffin wax and sectioned using a microtome because the mineralized areas are too tough for normal microtome knives to cut. However, the mineral content of the fixed tissue can be reduced by exposing it to **decalcifiying agents**. For samples with relatively small levels of mineralization, gentle chelating agents such as ethylaminediaminetetraacetic acid may be used. For heavily mineralized tissues, such as bone, harsher agents such as formic or hydrochloric acid must be applied. Over-exposure to the decalcification agents can be detrimental to tissue architecture and have a negative impact on the staining properties of the tissue. In particular, antigenic sites may be destroyed, leading to false-negative results in subsequent investigations. Thus, samples should be examined regularly using, for example, X-rays, to determine the optimal extent of decalcification. The tissue should then be thoroughly rinsed and returned to fixative solution. In samples where the degree of mineralization is crucial to diagnosis, as, for example, when investigating osteoporosis, decalcifying agents cannot be used and tissues should be processed into resin.

In cytopathology, the cells of the sample must be made available on a slide for viewing. This can be achieved in different ways. The sample may be smeared onto the slide, for example, from a swab or a spatula. A specialized centrifuge called a cytospin may be used, to force the cells out of suspension (be it urine or transport medium) and onto the slide. If a sample has very few cells in relation to its volume, it is first subjected to conventional centrifugation to concentrate the cells. The pellet of cells is then resuspended in a relatively small volume, which is subjected to the **cytocentrifuge**.

In the UK, many cervical samples are received as part of the national cervical screening programme (Figure 7.2). These samples are delivered to the cytology laboratory in transport fluid or fixative and the cells are separated by using membrane filtration or density gradient systems that deposit the cells onto glass slides. As with histopathological sections, once the cells are fixed on the slide, they must be stained to increase contrast for visualization by light microscopy.

Key points

All microscopes used in biomedical science are likely to be compound in nature. That is, they consist of two main elements, called objective and eyepiece lenses. By using lenses in this way, vastly improved magnifications can be achieved over the use of a single lens.

7.1.1 Resolution

Resolution is a measure of the clarity or acuteness of an image. It is the smallest distance that can be distinguished between two points and objects smaller than the resolution cannot be observed. The resolution of the unaided eye is approximately 200 μm.

SELF-CHECK 7.1

Define what is meant by the term *resolution*.

Resolution is a limiting feature of all microscopes, as simply magnifying an image will not always provide more information. As magnification increases, resolution or the resolving power of a microscope must also improve. Consider, for example, a digital image that is enlarged to the point of pixellation: clearly, the clarity of the image has been decreased despite the increase in magnification and is known as *empty magnification*.

In mathematical terms, resolution can be represented by the formula:

$$\mathbf{R} = 0.61\lambda/NA$$

Where **R** is resolution, λ is the **wavelength** of the radiation or light, and NA is the **numerical aperture** of the lens.

Numerical aperture is a quantitative representation of the light (or radiation) gathering capacity of a lens and is represented by the formula:

$$NA = n \sin \alpha$$

Where n is the **refractive index** of the medium between the glass coverslip (or the prepared specimen) and the front of the objective lens, and α is the angle between the outermost ray of light (or radiation) that enters the front of the objective lens and the optical axis of the objective lens, as shown in Figure 7.4. If the medium between the coverslip and objective is air then the maximum theoretical numerical aperture is 1.0. However, it is not possible to achieve this, as the lens would need to lie directly on top of the coverslip. If, however, we substitute oil or water for air then the theoretical numerical aperture is increased to 1.3–1.5. In practice, the theoretical values cannot be attained and in air the best possible value of numerical aperture is approximately 0.95 and in oil approximately 1.4. Thus, the resolution of a light microscope is approximately half (0.61/1.4) of the wavelength of visible light.

SELF-CHECK 7.2

Calculate the approximate resolution obtained with a microscope fitted with an oil immersion objective lens and using light of wavelength 600 nm.

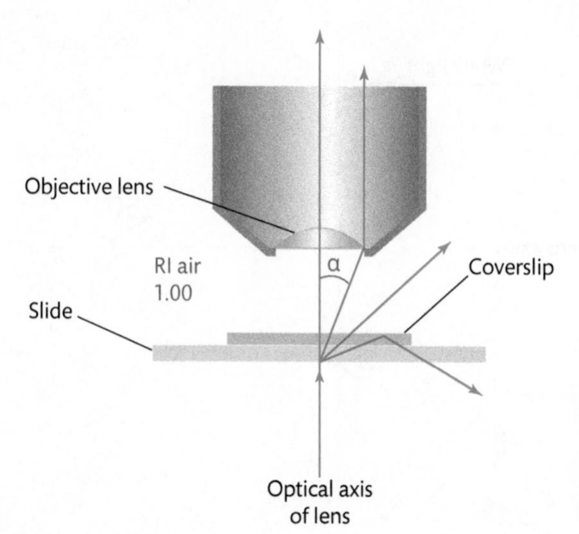

FIGURE 7.4
Schematic illustrating the angle of the light path when calculating the numerical aperture of a lens.
RI: refractive index.

7.1.2 Light and waveforms

Light emitted from a single source can be regarded as a series of waves travelling in straight lines in all directions. The length or periodicity of these waves is termed the 'wavelength' and gives the light its colour. White light consists of a spectrum of light of different wavelengths (and different colours) ranging from short ultraviolet (UV) through to long infrared light.

When light passes through air it travels in a straight line. If, however, it strikes a different medium such as glass, the light path is both retarded and refracted, or deviated at the air/glass interface. Light of different wavelengths is refracted to variable amounts. Refraction is best seen when light passes from air into a glass prism and out again into air, and is shown in Figure 9.7(c).

Cross reference
Chapter 9, 'Spectroscopy', gives a fuller description of electromagnetic radiation.

Monochromatic radiation consists of light of a single wavelength whose amplitude determines the brightness of light. The higher the amplitude of the wave, the brighter the light will be. It is necessary to understand these two properties of the waveform of light to appreciate the effects they have on any image seen in a microscope (Figure 7.5).

7.1.3 Image formation and lens defects

When parallel light rays pass through a **lens**, they are refracted by the lens and brought to a common focus by it forming an image. This focus is called the **focal point**. The distance between the centre of the lens and the focal point is called the **focal length**. However, lenses can suffer from a number of defects or aberrations, which may prevent light being brought to this common focus. This is because at its edge, a lens will also act as a prism, owing to its shape and the greater curvature of the lens, which will exacerbate any aberrations. These defects include:

- **chromatic aberration**
- spherical aberration
- **astigmatism**.

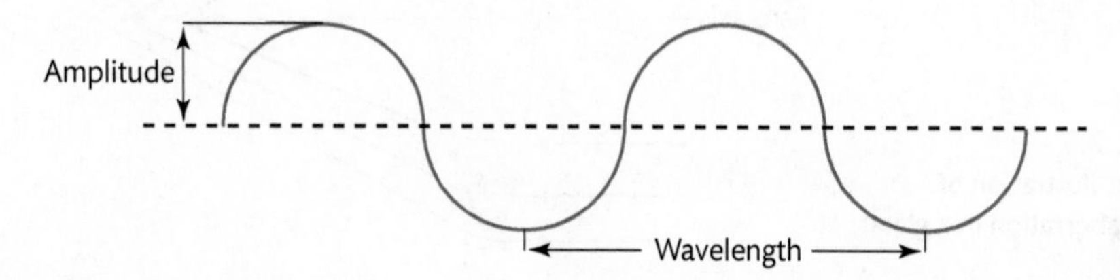

FIGURE 7.5
Schematic illustrating the definitions of amplitude and wavelength.

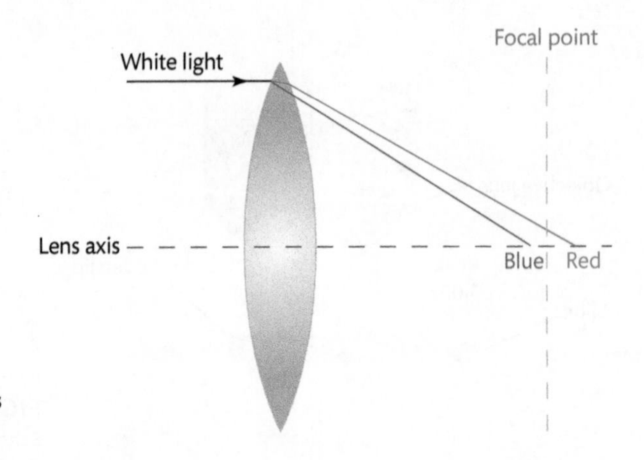

FIGURE 7.6
Schematic illustration of chromatic aberration in a glass lens.

Chromatic aberration

Chromatic aberration is the inability of a lens to bring light of different wavelengths to a common focal point resulting in a blurred image with multiple coloured fringes. It is caused by different wavelengths being refracted to different degrees, as seen in a simple prism, as lenses also act as prisms. Thus, light rays of shorter wavelengths are brought to different focal points than ones of longer wavelengths, as you can see in Figure 7.6.

Key points

Chromatic aberration is the inability of a lens to bring electromagnetic rays of different wavelengths to a common focal point.

Spherical aberration

Spherical aberration is the inability of a lens to bring light passing through the periphery of the lens to the same point as light passing through the central part of the lens (Figure 7.7). This is caused by the lens refracting the light to different degrees, which depends upon the curvature of the lens and the angle at which light enters the lens.

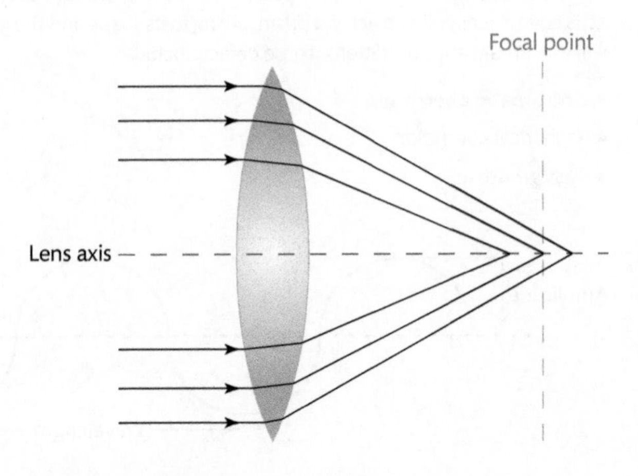

FIGURE 7.7
Schematic illustration of spherical aberration in a glass lens.

Key points

Spherical aberration is the inability of a lens to bring electromagnetic rays of different wavelengths to a common focus.

Both spherical and chromatic aberration result in images that are significantly reduced in quality. These aberrations can both be corrected by using different glass additives and lenses in combination that have differing amounts of curvature and different shapes.

SELF-CHECK 7.3

How can chromatic aberration be corrected?

Astigmatism

Astigmatism results when a lens is unable to bring light passing through one part of a lens to the same focal point as light passing through another part. This results in a distorted blurred image and is best illustrated when viewing a lattice. Astigmatism is perhaps the least important lens aberration as correction of either chromatic or spherical aberration usually results in the elimination of astigmatism.

Key points

Astigmatism is the inability of a lens to bring electromagnetic rays passing through different parts of the lens to a common focal point.

7.2 Components of a microscope

The basic construction of a compound light microscope has changed little since the mid-nineteenth century. All consist of a system of lenses and their associated components, which are used to form a magnified image of the specimen. However, since the 1800s, significant improvements have been made in the quality of lenses, resulting in greatly improved images (Figure 7.8). Much emphasis has also been placed on the ergonomic design of microscopes to provide the user with the most comfortable working position.

7.2.1 Light source

All microscopes require a reliable light source. In early designs, this source came from the sun and a mirror was used to reflect its rays into the microscope condenser. As optics improved, this form of illumination became inadequate, as it was not possible to adjust the brightness. Tungsten filament lamps fitted with a variable rheostat that allows the brightness to be adjusted have proven to be a reliable mechanism. However, tungsten filament bulbs generate considerable heat and may need cooling to be used safely. Consequently, most modern microscopes now use low-voltage halogen light sources. These provide a more intense light and a brighter image without the heat associated with tungsten filament light sources.

Fluorescent microscopes require a light source capable of emitting a high level of UV light, which neither tungsten nor halogen lamps are capable of producing. For this purpose the best light source is a high-pressure mercury vapour lamp. This lamp, however, requires a specialist lamp holder and

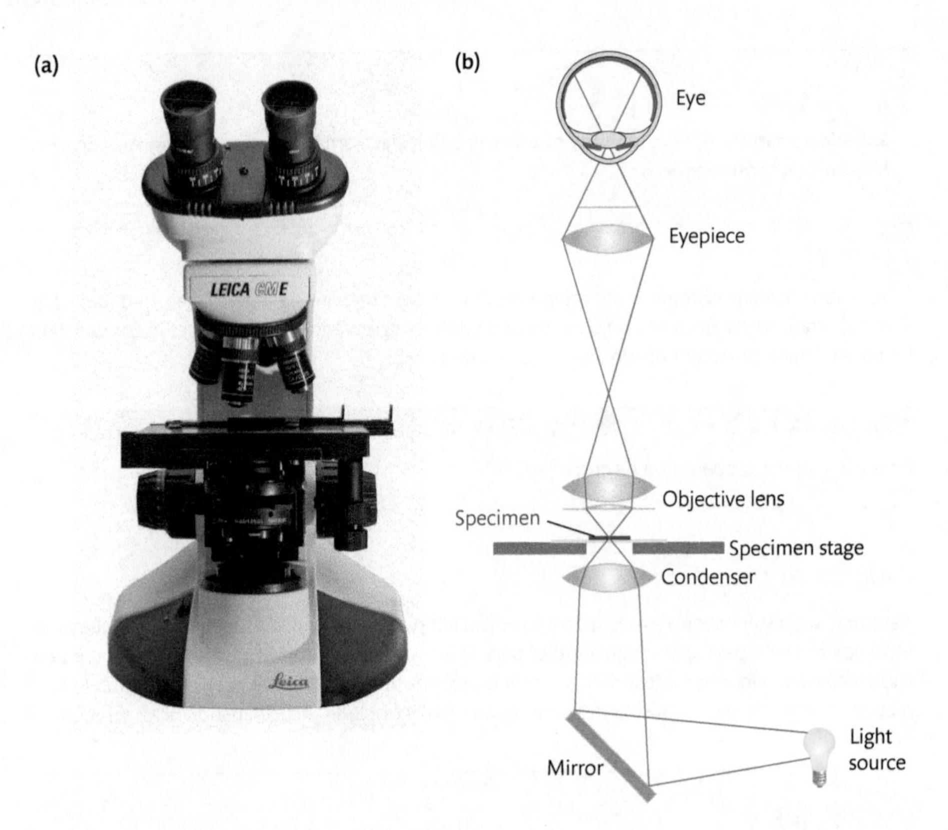

FIGURE 7.8
(a) Modern compound light microscope. (b) Schematic illustrating the light path of a compound microscope.

power supply, and generates a significant amount of heat that needs to be dissipated. Care must also be exercised regarding health and safety issues (see Health & Safety Box, 'The use and disposal of high-pressure mercury vapour lamps').

HEALTH & SAFETY

The use and disposal of high-pressure mercury vapour lamps

The mercury vapour lamp contains mercury gas at high pressure within a glass housing. The intensity of light emission from these lamps diminishes with use and, in addition, the glass becomes increasingly unstable and prone to explode. To ensure that the lamp operates optimally, manufacturers incorporate an elapsed timer to record how long the lamp has been in use and recommend its replacement and correct disposal after a known length of use. Disposal must be controlled to ensure mercury does not enter the environment.

It is also essential that after the lamp has been switched off, it is not re-used until it has cooled down. Failure to observe this cooling period may result in the lamp exploding. Manufacturers now incorporate switching timers into the lamp power supply that prevent the lamp being re-used until an adequate cooling period has elapsed.

Cross reference

Health and safety issues are discussed in Chapter 4.

7.2.2 Illumination of specimens

Effective microscopy relies upon the correct illumination of specimens. This is done by using the condenser or *substage condenser*, as it is sometimes called, to focus the light produced by the lamp into a cone onto the specimen giving it a maximum and even illumination. In its simplest form, the substage condenser consists of one or two lenses with an adjustment that allows vertical focusing of the light to compensate for the varying thickness of microscope slides.

METHOD Setting up of a microscope for optimal illumination

- **Whatever type of microscope is being used, the basic principles of setting it up for critical or Köhler illumination are essentially the same. First, adjust the condenser to as high a position as possible and then focus on a sample using a low-power objective lens (normally 10× or 4×). Second, close the field diaphragm in the condenser and sharply focus the diaphragm onto the image of the sample using the condenser focus controls.**
- **The field diaphragm can be considered to represent the illumination axis and should be adjusted using the x and y condenser adjustment screws. The condenser lens system focuses as much light as possible onto the sample and the lamp is focused onto the plane of the field diaphragm of the condenser. This is termed Köhler illumination. Now open the field diaphragm until it is just outside the field of view.**
- **If the microscope does not have a condenser fitted with a field diaphragm, it is necessary to place a sharply pointed object, for example a pin, onto the glass of the coverslip and focus the illumination onto this point by raising or lowering the condenser. This is termed critical illumination.**

Online Resource Centre

To see an online video demonstrating how to set up a light microscope, log on to www.oxfordtextbooks.co.uk/orc/fbs

For best illumination, it is necessary to be able to align accurately the axis of the condenser with the objective lens, and to be able to adjust the size and brightness of the cone of light hitting the specimen. The condenser is fitted with adjustable screws to move the condenser in the x–y axes to allow the light cone to be centred, and with an iris diaphragm to adjust the diameter of the light cone.

It is possible to set the condenser in one of two ways to illuminate the specimen optimally:

- Köhler illumination
- critical illumination.

Each method has its advantages and applications within biomedical science.

Köhler illumination requires the condenser to be set in such a way that light reaching the specimen evenly illuminates a wide area. This form of illumination is particularly useful when recording photographic images to ensure even illumination across the whole field. **Critical illumination** is the focusing of the light source at a point directly onto the specimen using the substage condenser. It is an effective way of ensuring that the maximum level of illumination reaches the specimen. This method is particularly necessary when using high-magnification objective lenses that have short working distances but when combined with low-magnification objective lenses result in uneven illumination.

You can read an outline of how to set up a microscope for optimal illumination in the Method box 'Setting up of a microscope for optical illumination'.

7.2.3 Specimen stage

The specimen stage is the rigid flat surface immediately above the condenser on which the specimen sits. Most stages incorporate sample clamps attached to mechanical manipulators that allow accurate movement of the specimen. **Vernier scales** are incorporated into the mechanism, which means that when a point of interest in the specimen is observed, its Vernier values can be recorded and therefore the position is known and returned to later with ease.

7.2.4 Stand and tube

The stand of the microscope is the sturdy, rigid, and vibration-free base to which all the other components are attached. It incorporates the coarse and fine focus adjustment mechanisms of the microscope that allow for fine vertical movement of either the stage or the objective lens, depending on the manufacturer.

Above the stage the stand has a *nosepiece* for holding between three and five *objective lenses* of different magnifications. The nosepiece rotates between set points allowing the appropriate objective lens to be moved into the light path. Above the nosepiece is the observation tube that houses the *eyepiece* lens. There are several types of observation tube: the simplest consists of a single eyepiece lens but more usually consists of a binocular or trinocular attachment. A trinocular tube allows a camera to be attached for simultaneous viewing and recording of the image. The distance from the front of the objective lens and the eyepiece is termed the tube length. The length of the tube has a significant impact on the final magnification, as described in Section 7.3.

Recent developments in lens construction have resulted in the production of 'infinity corrected' lenses. Here, the focused image leaving the lens is not brought to a focal point but rather remains parallel at the same focal plane to infinity with no change in magnification. This development has allowed manufacturers to vary the tube length without altering the magnification allowing them to develop and design more ergonomic stands and multiheaded microscopes.

Key points

Microscopes are able to have multiple viewing heads (up to 20) that allow all observers to see exactly the same image. This facility is particularly useful for teaching or demonstration purposes.

7.2.5 Objective lens

The objective lens collects the light from the specimen and forms a **real image** of the specimen. It is perhaps the most important component of the microscope in that it contributes most to the ultimate quality of the image. There are numerous types of objective lenses; each has specific uses and qualities. The selection of an objective lens is determined by the function to which the microscope is to be put.

There are three main types of objective lenses:

1. Acromatic
2. Semiapochromatic or fluorite
3. Apochromatic.

All lenses have other features that contribute to their effective use, and objective lenses are no exception. Two features of note are *depth of field* and *working distance*. Depth of field is the amount of a specimen in focus at any one time. Generally, lower power lenses have a greater depth of field and higher power lenses have a smaller depth of field so require constant refocusing. Working distance is the distance between the coverslip and the bottom of the objective lens and, again, lower power lenses have a longer working distance than higher power ones.

Achromatic

Achromatic lenses are the simplest types of corrected lens available. They correct for chromatic aberration in two wavelengths and for spherical aberration for only one wavelength. Achromatic lenses have a relatively low numerical aperture, which allows a good working distance and depth of field. When using this type of lens, the quality of the image is best in the lens axis. Blurring occurs at the periphery of the image, which makes achromatic lenses unsuitable for colour photomicrography.

Semiapochromatic

Semiapochromatic or fluorite lenses incorporate the mineral fluorite into the glass of the lens. This allows lenses to be constructed with improved numerical apertures and increased image resolution. These lenses still only compensate for chromatic aberration in two colours and spherical aberration is still present; however, they provide superior performance over simple achromatic lenses.

Apochromatic

Apochromatic lenses are the highest quality lenses available. They correct for three colours, and do not exhibit spherical aberration or astigmatism. These lenses provide the highest numerical aperture and, consequently, the highest resolution. The working distances of these lenses are usually short but this can be used to advantage by the addition of high refractive index immersion oil or water between the specimen and the objective lens. As described in Section 7.1, this has the effect of increasing numerical aperture and thus increasing the potential resolving power of the lens. Using apochromatic lenses means it is possible to achieve numerical apertures in excess of 1.4 with full colour correction.

7.2.6 Eyepiece

The eyepiece is the last lens system through which the image passes. It magnifies the image formed by the objective lens and tube stand and allows a **virtual image** to fall on the eye of the observer. Some eyepieces are capable of accommodating a range-measuring graticule. Graticules are glass discs onto which are etched micrometer scales, grids, or points, and which can be used for quantitative and accurate measurements of images. Modern microscopes are fitted with wide field, flat field, and possibly high focal point eyepieces especially designed for spectacle users. In choosing the eyepiece, it is necessary to match its quality to that of the objective lens. It is inappropriate to use the best quality apochromatic objective lens with the simplest of eyepiece lenses.

7.3 Magnification

The magnification produced by a *lens* can be defined by using the distance from the object to the lens and that from the lens to the image, using the formula:

Magnification = Real image distance/Specimen focal distance

The real image distance or optical tube length has been standardized in microscopes to 160 mm. The focal distance is the length from the optical centre of the lens to its focal point.

You can see from the formula that the smaller the focal distance, the larger the magnification obtained. The surface curvature of the lens can be increased to shorten the focal distance, but maximum curvature is limited to that of a spherical lens. By adding multiple lenses it is possible to increase magnification significantly: in contemporary microscopes both objective and eyepiece lenses are constructed from numerous lenses of differing curvatures and diameters.

The total magnification obtained using a compound microscope when the microscope has a standard optical tube length of 160 mm can be obtained by simply multiplying the magnification values of the objective and eyepiece lenses.

For non-standard optical tube lengths the formula in Box 7.2 must be used.

7.4 Dark field microscopy

Conventional bright field illumination will in itself provide only minimal contrast (Section 7.1), and unstained biological samples are virtually invisible as they have insufficient variations in optical densities. To overcome this problem, it is conventional to introduce contrast into specimens by staining them with coloured dyes. However, when it is necessary to examine samples in the living state, this 'normal' preparation may result in death and disruption of the sample so alternative methods of illuminating the specimen are needed to prevent this.

The simplest method of introducing contrast into unstained samples is to use oblique light. In doing so, the sample is illuminated by light from an angle that has the effect of increasing refraction while creating a decrease in direct illumination. Unstained samples thus appear bright against a dark background. The dark field condenser was developed to take advantage of this effect. This consists of a parabolic mirror placed in the substage condenser that only allows light to reach the specimen stage at an acute angle, as illustrated in Figure 7.9. When a specimen is not present on the microscope stage, light does not pass through to the objective lens. When a sample is placed on the stage, the small variations in refraction created by the specimen are amplified to make the sample appear bright against the dark background.

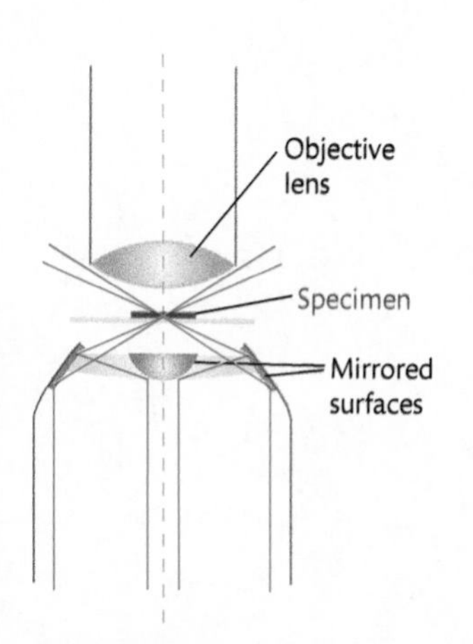

FIGURE 7.9
Schematic outline of light path in dark field illumination.

BOX 7.2 Magnification of a compound microscope

The final magnification produced by a compound microscope can be expressed as objective magnification × tube lens factor × eyepiece magnification:

Tube lens factor = Optical tube length/Object focal length

Example: magnification using a 10× objective lens with a tube lens factor of 1.25 and an 8× eyepiece:

Magnification = 10 x 1.25 x 8 = 100 x

Dark field condensers are occasionally used in conjunction with the transmission fluorescence microscope and this application will be described more fully in Section 7.7.

7.5 Phase contrast microscopy

As you learned in Section 7.1, light is a waveform. When two rays of light originate from the same source they are said to be **coherent**, because they are in phase with each other and their amplitudes coincide. As a consequence, they are able to combine to form a ray with an amplitude twice the height of the original rays resulting in a brighter one. If, however, the two rays become out of phase then they are able to interfere with one another. If the rays were out of phase with one another by half of the wavelength, then this would result in the extinction of the ray. This phenomenon of interference is utilized in the phase contrast microscope. You can see an outline of its operation in Figure 7.10.

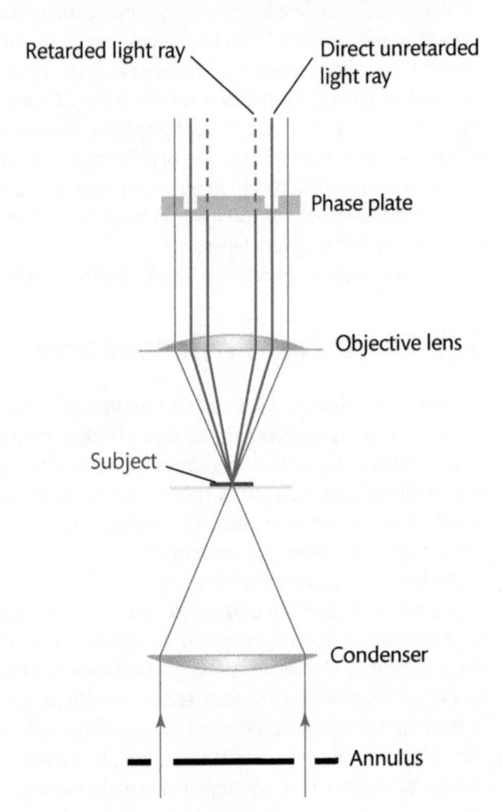

FIGURE 7.10
Schematic illustration of the light path in a phase contrast microscope.

The sample is illuminated with a cone of light produced by an annular stop, which consists of a ring positioned in the bright field substage condenser. The cone of light passes through the sample, which is positioned at its focal point. Light leaving the sample enters the objective lens of the microscope.

Objective lenses in phase contrast microscopes have a **phase plate**, which consists of a disc of glass with a ring etched into it that corresponds exactly to the shape and size of the annulus. The depth of the etching is critical: it must be of such a depth that it retards the light rays passing through the full thickness of the phase plate by ¼ wavelength when compared with the light passing through the etched ring. This retardation, while resulting in interference, does not in itself produce the desired increase in contrast. The sample is responsible for producing this effect. Tissues will also retard light by approximately ¼ of a wavelength; hence, the combined retardation from the phase plate and the specimen produces the total interference.

When light passes through a sample, some of it is scattered and passes into the objective lens in the normal way and passes to the eye of the observer through the non-etched part of the phase plate. When the unaltered rays are focused with the retarded rays they combine to form the real image of the specimen. Subtle changes in refractive index within the sample are seen as varying degrees of brightness in the image against a dark background.

When setting a microscope for phase contrast it is essential to ensure that the annulus and the phase plate are exactly aligned. It is usual to use an auxiliary microscope to view the back focal plane of the phase contrast microscope, which allows the operator to view the phase plate and use adjusting screws fitted in the condenser to align it with the annulus with great accuracy. In practice, each objective lens and phase plate has a matching annulus fitted into a rotating disc set into the base of the substage condenser. Each annulus can be adjusted and set individually to match its objective. If properly adjusted, a good phase contrast microscope should be able to distinguish differences in refractive index of less than 5% easily, producing images of high contrast without the use of staining.

Phase contrast microscopy is used extensively in biomedical science when it is necessary to examine unstained tissue samples, such as examining cell deposits from urine samples from patients suffering from urinary tract infections. Phase contrast microscopy is used to examine and identify the cellular component of such samples.

Cross reference

You can read more about urinary infections in Chapter 3, 'Kidney disease', in the companion text, *Clinical Biochemistry*, and in *Medical Microbiology*.

SELF-CHECK 7.4

In a phase contrast microscope, by how much does the phase plate retard the rays of light?

7.6 Polarization microscopy

As described in Section 7.1, light emitted from a single source can be regarded as electromagnetic waves whose amplitude represents brightness and the wavelength its colour. This waveform is emitted and vibrates in all directions. **Polarized light**, however, only vibrates in a single direction. Certain crystals are capable of producing plane polarized light by splitting incident rays entering the crystal into two components, which are refracted by the crystal in two different planes. This phenomenon is termed **birefringence.**

If we place two similar polarizing crystals so that their planes of polarization are parallel to each other, then any light entering the first crystal would be capable of passing through the second. However, rotating the second crystal through an angle of 90° would result in the extinction of light and none would exit from the second crystal. This phenomenon is exploited in the polarization microscope, which is used to examine and measure substances that are capable of polarizing light.

Cross reference

You can read about the deposition of urate in gout patients in Chapter 4, 'Hyperuricaemia and gout', in the companion volume, *Clinical Biochemistry*.

The polarizing microscope has a **polarizer** filter situated in the substage condenser. This illuminates the specimen with plane polarized light. A second polarizing filter or **analyser** is situated below the eyepiece and is capable of being rotated through 180°. When using the polarizing microscope, the two filters are orientated so that their planes of polarization are at a right angle resulting in light extinction. However, any specimen that contains a birefringent substance will rotate the illuminating plane polarized light so that it is able to pass through the analyser and appear bright against a dark background.

This form of microscopy is particularly useful in the examination of fluids that may contain crystals, such as urine and joint fluids.

7.7 Fluorescence microscopy

The use of shorter wavelength light in the UV range is outside the visible spectrum so its use to improve resolution is limited. However, certain substances when illuminated with high-energy light of a short wavelength (i.e. UV) will re-emit light of a lower energy of longer wavelength that is within the visible spectrum. This is called fluorescence. The use of fluorescent microscopes allows us to visualize small quantities of biological components within tissue samples as they appear bright against a dark background.

Cross reference

Fluorescence is also discussed in Chapter 9, 'Spectroscopy'.

Some substances are auto-fluorescent or exhibit **primary fluorescence** (i.e. they fluoresce spontaneously under UV light). Other substances do not fluoresce but can have fluorescent dyes called fluorochromes attached to them, which will produce a **secondary fluorescence** when stimulated with UV light. Examples of substances that exhibit auto-fluorescence are the vitamin As and some porphyrins. A number of dyes that exhibit fluorescence can be used to demonstrate the presence of specific cellular components in specimens.

The light source in all fluorescent microscopes must be capable of producing a high proportion of light rays in the UV range. Halogen lamps produce significant amounts, but almost all of it is in the UV blue range and so is of limited value. The most common light sources in use are high-pressure mercury vapour or xenon arc lamps. Both are highly efficient but require specific power supplies and appropriate lamp housings.

There are two types of fluorescence microscope, which differ in their manner of illuminating the specimen. In *transmission* fluorescence microscopes, the light source is situated beneath the specimen as in a conventional light microscope. In *incident* light fluorescence microscopes, however, the specimen is illuminated from above and the viewer sees visible light transmitted from the specimen against a black background.

7.7.1 Transmitted light fluorescence

To gain the highest performance from the fluorescence microscope it is necessary to use filters that are able to transmit the specific wavelength of the excited light, while removing unnecessary and unwanted wavelengths that may reduce specificity. A transmission fluorescence microscope has two types of filters (Figure 7.11): an **exciter filter** placed in the light path between the lamp and the condenser and a **barrier filter** situated in the body tube before the eyepiece. The filters must be specifically manufactured to correspond to each type of fluorochrome in use.

The exciter filter allows the maximum amount of UV light of an appropriate wavelength to excite the specimen while filtering out the levels of visible and inappropriate UV light from the illumination. This ensures that the maximum level of fluorescence is emitted from the sample. The light entering the objective lens is a combination of unscattered UV light and the longer wavelength visible light produced by the fluorochrome(s). It would be dangerous for the observer if this light were to fall directly onto the eye as it may cause blindness. This is prevented by the barrier filter, which removes all of the unused UV light and allows only the fluorescent visible light to be transmitted to the eye, thus protecting the observer. Unfortunately, to produce a filter capable of both blocking UV light and allowing transmission of specific wavelength visible light means the overall level of transmission is lowered and significantly reduces the intensity and brightness of any fluorescence produced.

An alternative to the use of a barrier filter is to use a dark ground condenser. This ensures that the only light reaching the eye of the observer is that emitted by the specimen as unscattered, indirect UV light does not reach the eye. This has the added advantage of forming a dark contrasting background that gives an apparent increase in the brightness of the fluorescent light.

SELF-CHECK 7.5

Which light has the shorter wavelength: UV or blue? (You might find it helpful to consult Figure 9.2.)

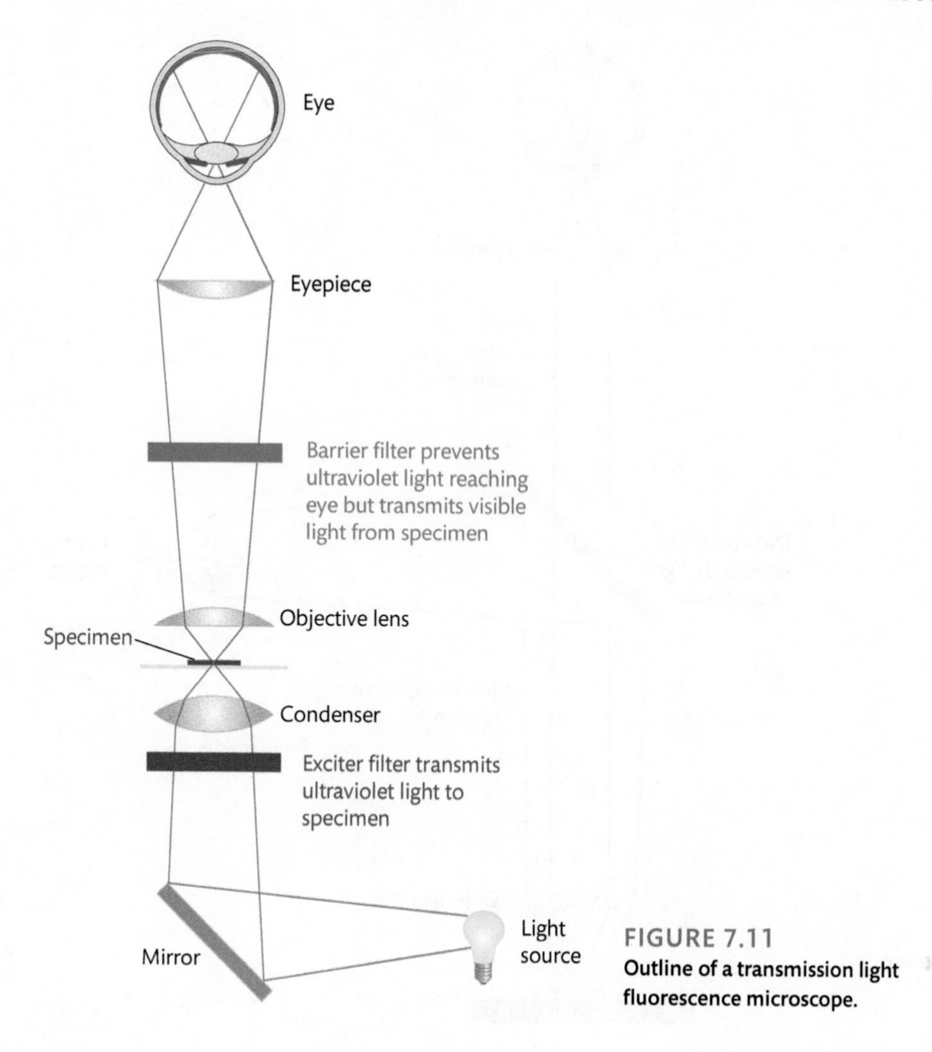

FIGURE 7.11
Outline of a transmission light fluorescence microscope.

7.7.2 Incident light fluorescence

As the fluorescent light from a fluorochrome is emitted in all directions it is possible to illuminate the specimen in a fluorescence microscope from above. This is called *epi-illumination*. Some of the excitation light from the fluorochrome is directed back into the objective for observation by the user.

In operation, the exciting light passes from the lamp and through the excitation filter to a dichroic mirror or prism, as you can see in Figure 7.12, which filters out all the unwanted light, only allowing UV light of a specific wavelength to pass through. The resultant ultraviolet light passes to the specimen where the fluorochrome is excited to generate visible light. The visible light passes into the objective lens and then to the observer through the same dichroic prism, which is able to transmit visible light in the opposite direction to the exciter light, and to the eye of the observer. The image therefore appears brightly coloured against a dark background.

Incident light fluorescence microscopy has significant advantages over transmitted light fluorescence. The reduced intensity of light resulting from the use of barrier filters is eliminated, a substage condenser is not required and the level of light reaching the specimen is considerably increased giving a much brighter image. In addition, as the fluorescence microscope uses light that is produced from sources above the specimen it is possible to use other forms of transmission microscopy simultaneously. This allows both morphological and fluorescent images of the specimen to be seen at the same

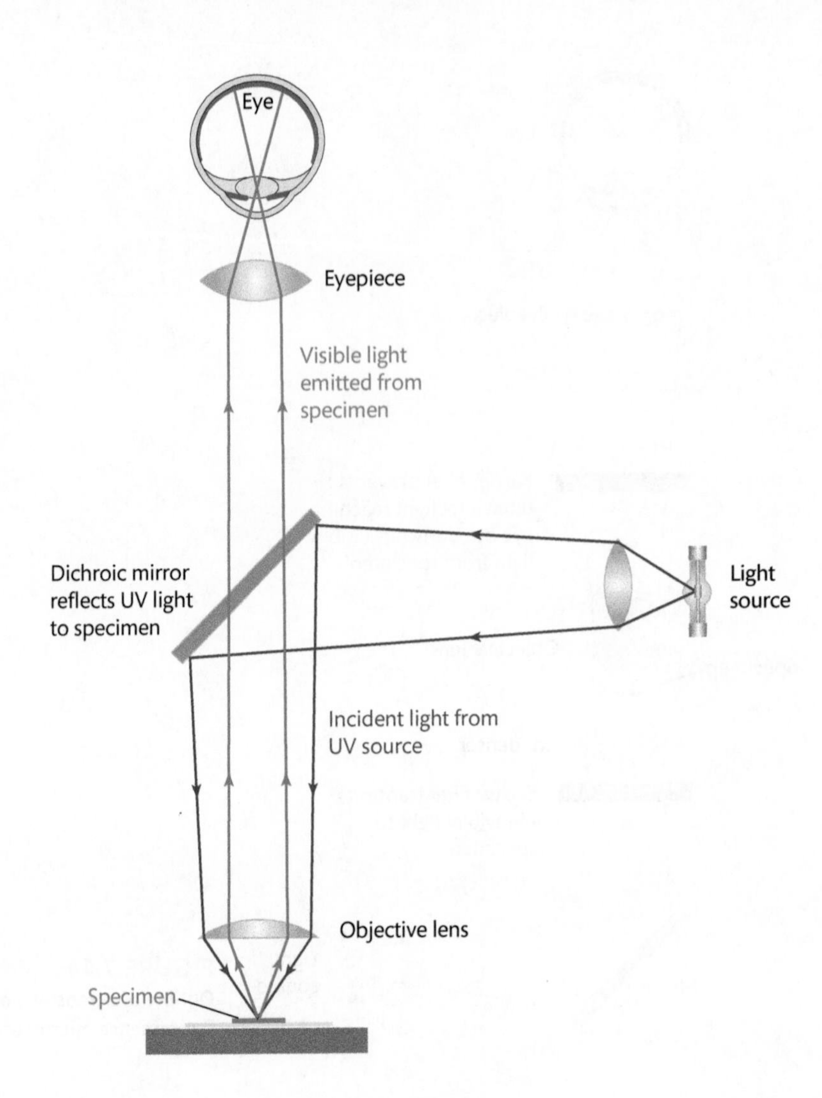

FIGURE 7.12
Outline of an incident light fluorescence microscope. UV: ultraviolet.

time. This is of particular value when additional staining of specimens may mask the fluorescence produced. In this case a phase contrast or dark ground condenser can be used to introduce additional contrast.

7.7.3 Uses of fluorescence microscopy

> **Cross reference**
> You can read more about immunofluorescence in Chapter 13, 'Immunological techniques', and the accompanying volume, *Clinical Immunology.*

Both transmitted and incident light fluorescence microscopy are used extensively in the study of immunological disorders. The technique of **immunofluorescence** is used to identify specific immune responses in disease. These methods rely on specific interactions and binding between antibodies labelled with fluorescent dyes (fluorochromes) and antigens present on cell surfaces. Once bound, the visualization of the fluorescence label confirms the presence of the antigen and also identifies its specific location within the tissue (Figure 7.13).

SELF-CHECK 7.6

How does an incident light fluorescence microscope differ from a transmitted light fluorescence one?

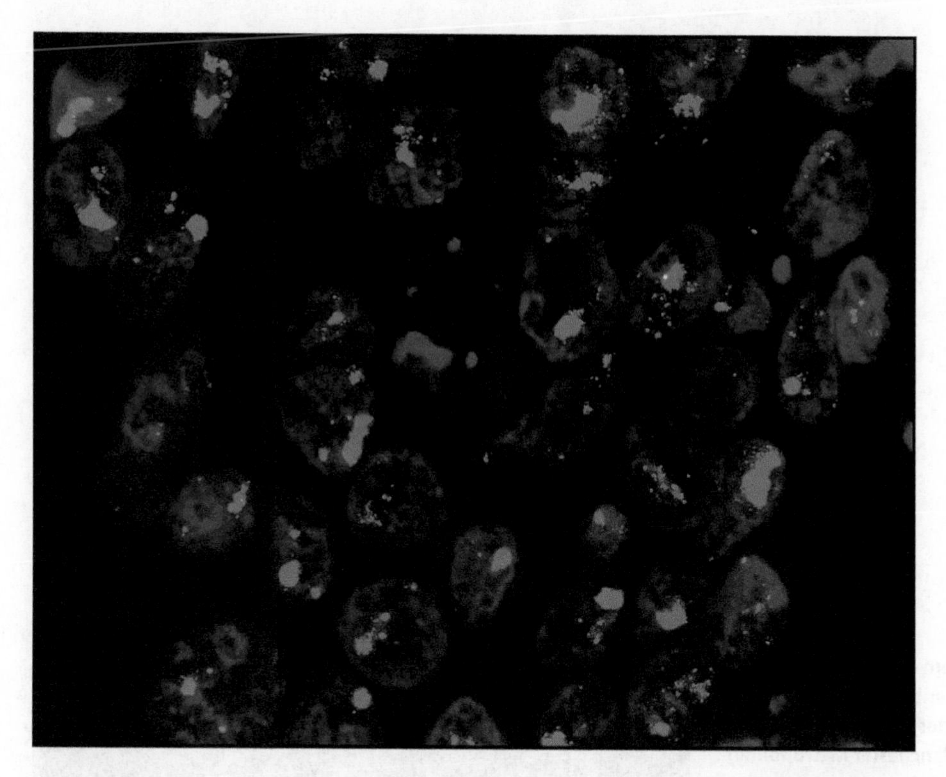

FIGURE 7.13

Incident light fluorescence micrograph of a section of breast tumour stained using fluorescent *in situ* hybridization (FISH) to demonstrate amplification of the *HER2* gene on chromosome 17. This method is used as a prognostic marker for treatment with the drug herceptin. See also Figure 14.26 and its associated text.

7.8 Confocal microscopy

In all forms of microscope, the final image is affected by the thickness of the specimen being examined because of the relatively small depth of focus. This is a particular problem in fluorescence microscopy where the fluorescent light emitted from the specimen is derived from the full thickness of the specimen resulting in blurring of the image caused by those portions of the specimen that are above and below the focal plane. This blurring is minimized in the confocal microscope.

Confocal microscopes use a narrow beam of light from a laser and thus the specimen is illuminated with coherent light of high intensity. The beam of laser light is scanned through the specimen using a system of galvanometer mirrors that oscillate at adjustable speeds to divert the laser beam to produce a scan. Reflected light produced is re-focused by a confocal aperture or pinhole, which only allows light from a specific focal plane to fall onto a photomultiplier and produce a digital signal. A computer then processes this signal to produce an image that has a great depth of field giving the specimen a three-dimensional appearance. If the specimen is successively scanned through a range of focal planes, the images at each level can be electronically stored and used by the computer to produce a full three-dimensional image of it.

To date, confocal microscopy has found limited application in diagnostic biomedical science. It is, however, used extensively in scientific research as biological material can be examined in three dimensions without the need for extensive sample preparation (Figure 7.14).

SELF-CHECK 7.7

How is the quality of an image improved using confocal microscopy?

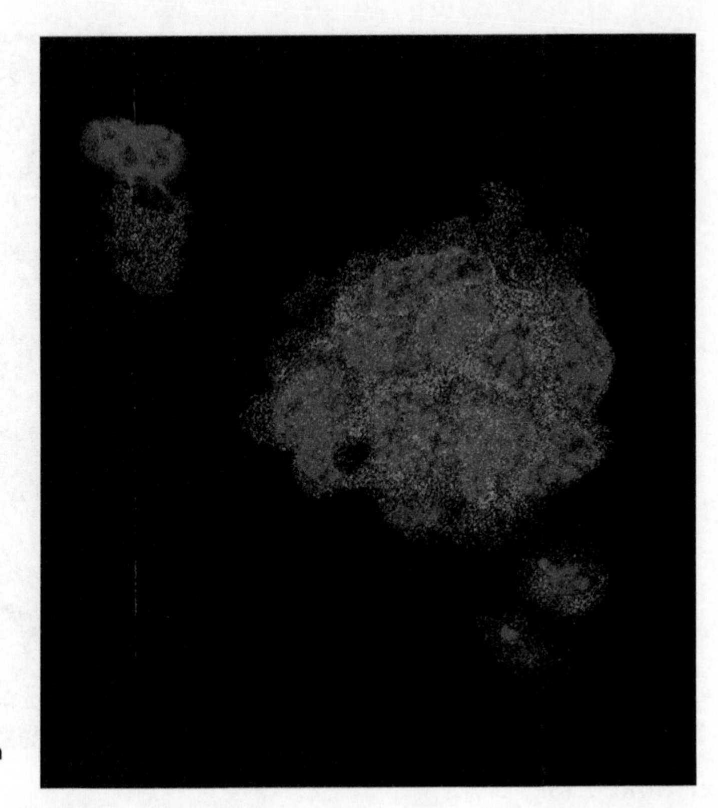

FIGURE 7.14
Confocal image of a group of stem cells. The green areas are parts of the plasma membranes and the purple portions are nuclei. Courtesy of Dr Q. Wang, School of Healthcare Science, Manchester Metropolitan University, UK.

7.9 Inverted microscopes

When examining living material, for example certain types of cells grown in tissue culture, it is necessary to use an *inverted* microscope. This allows the bottom surface of a culture flask, where the cells grow and adhere, to be examined with minimal disturbance to the specimen. The inverted microscope is constructed so that the specimen stage has a large flat surface capable of accommodating tissue culture flasks or Petri dishes (Figure 7.15). The objective lens and condenser are designed to work at much longer working distances than those of conventional microscopes to allow for the thickness of the culture vessel. Inverted microscopes also have the facility to utilize phase contrast or dark field illumination and fluorescence, giving the instrument maximum flexibility.

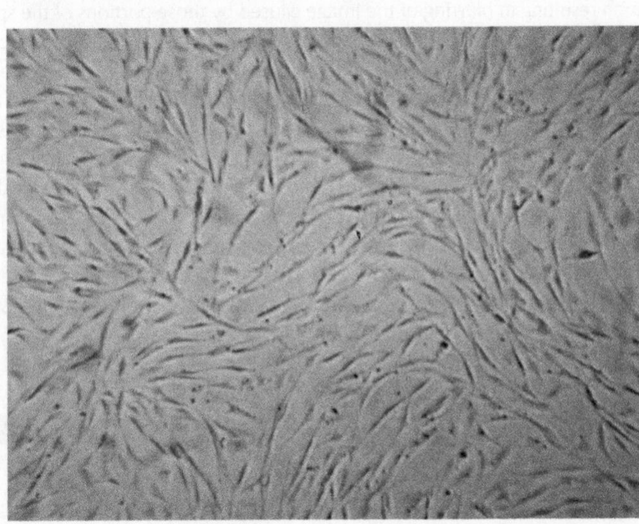

FIGURE 7.15
Myoblasts growing in culture viewed using an inverted microscope. Courtesy of Dr Q. Wang, School of Healthcare Science, Manchester Metropolitan University, UK.

7.10 Electron microscopy

In Section 7.1 we noted that the resolution of a microscope depends largely upon the wavelength of the illuminating light; using visible light limits the resolution to about 200 nm. Using radiation of short wavelengths is a way of increasing the resolution of a microscope. This feature is exploited in the electron microscope. Like light, moving electrons are part of the electromagnetic spectrum; the faster they move the shorter their wavelength.

Modern electron microscopes have resolutions that are less than 0.2 nm when used with biological specimens, which is 1000× the resolution of a light microscope.

Cross reference

The electromagnetic spectrum is described in more detail in Chapter 9, 'Spectroscopy'.

7.10.1 Components of the electron microscope

Electron microscopes were first developed in the 1920s. Their components are analogous to those of the conventional light microscope, as you can see if you compare Figure 7.16 with Figure 7.8.

Electron microscopes have a source of electrons, commonly called an *electron gun*, a condenser system that collects the electrons and focuses them on the specimen, a system of objective lenses to magnify the image formed, and a projector lens system to focus the image onto a phosphorescent screen, thus making it visible. Glass cannot focus beams of electrons, but they can be focused using electromagnetic lenses.

(a)

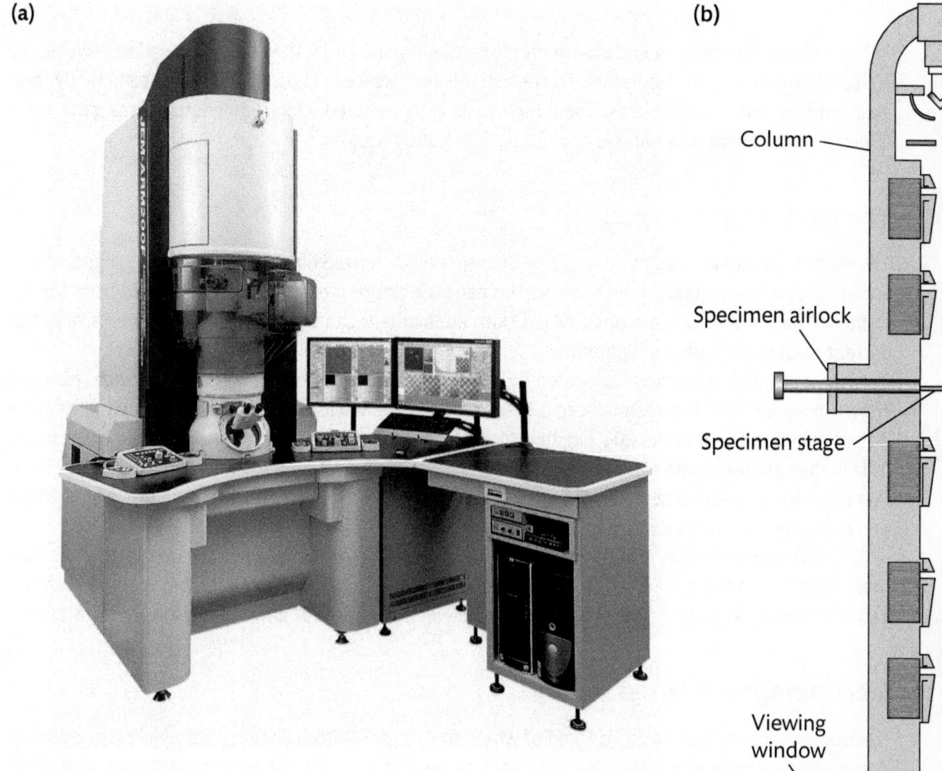

(b)

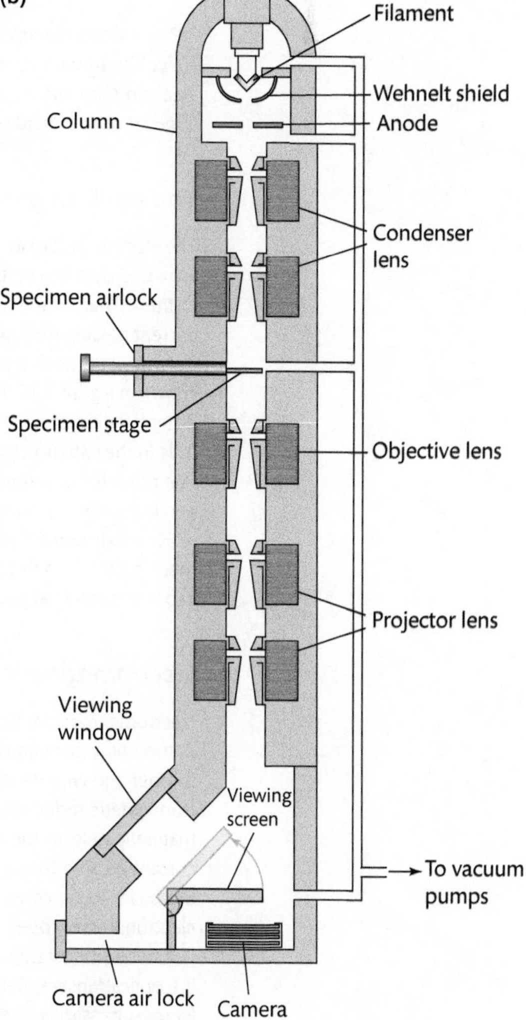

FIGURE 7.16

(a) Modern transmission electron microscope (JEOL JEM-ARM200F). Courtesy of JEOL Ltd. (b) Cross-section diagram of a transmission electron microscope. See text for details.

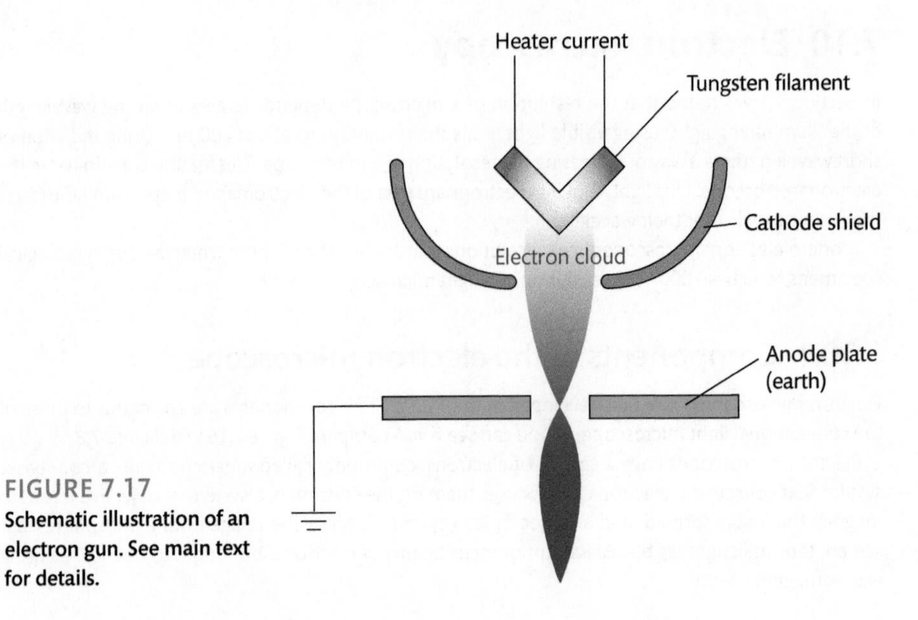

FIGURE 7.17
Schematic illustration of an electron gun. See main text for details.

The electron beam in an electron microscope is energetic and if they were allowed to travel far in air, collisions with its molecules would soon dissipate the beam. Thus, to produce a sufficiently long electron path within the microscope a high vacuum is required. Electromagnetic lenses generate a large amount of heat so a suitable cooling system is also required.

The electron gun

The electron gun of an electron microscope consists of a tungsten filament bent into a V-shape, which is heated by a low voltage direct current. This causes a cloud of electrons to be released from the tip of the V. The number of electrons emitted from the filament can be varied by altering the size of the current passing through the filament.

A *beam* of moving electrons is formed by a Wehnelt cap and a positively charged (anode) plate, as shown in Figure 7.17. The Wehnelt cap is made of metal and encases the filament. It is maintained at a potential difference of about 20 V negative to the filament resulting in an attraction to the electrons. A hole in the cap directs the electrons towards the **anode plate**, which is maintained at earth potential. The potential difference between the filament and the anode determines the **acceleration voltage** and in most electron microscopes can be varied between 20,000 and 100,000 V.

For maximum efficiency and the production of a suitable electron beam, it is essential that the filament is heated to the correct temperature, often referred to as the saturation point, that the Wehnelt cap is accurately aligned, and that the distance between the filament and cap is critically adjustable.

Electromagnetic lenses

Electromagnetic lenses consist of a coil of wire encased in soft iron. They have a pole-piece opening at the point through which the electron beam passes (Figure 7.18). When a direct current is applied through the wire it induces a strong magnetic field through the axis of the coil. As electrons are electromagnetic radiation, an electron beam passing through the pole-piece opening is affected by the magnetic field in the metal coil, which allows the beam to be focused. Varying the strength of the current passing through the coil we can vary the degree by which the electrons are focused allowing the beam to be convergent or divergent. A number of such lenses are built into the column of the electron microscope.

As with all lenses, electromagnetic ones suffer from several aberrations. As you learned earlier in Section 7.1, optical lenses suffer from chromatic and spherical aberrations, as well as astigmatism. Electromagnetic lenses suffer similar problems; however, the ways in which these defects are overcome differ.

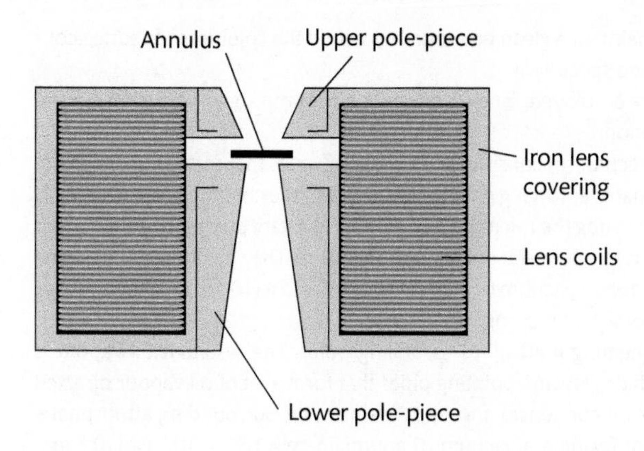

FIGURE 7.18
Schematic showing a cross-section of an electromagnetic lens. See main text for details.

To recap: chromatic aberration is the inability of a lens to bring electromagnetic rays of different wavelengths to a common focal point. Chromatic aberration in electromagnetic lenses is overcome by using a beam of monochromatic electrons. Manufacturers go to great lengths to ensure the high tension voltage between the filament and the anode plate within the electron gun is stable, which ensures a constant accelerating voltage and thus a monochromatic beam of electrons. A further potential source of chromatic aberration in the electron microscope is the energetic interactions between the electron beam and the specimen under examination. This reduces the energy of the electrons and so changes their wavelength to produce a potential polychromatic beam. This form of chromatic aberration can be minimized by using higher acceleration voltages and preparing extremely thin sections of specimens for examination.

Spherical aberration is caused by the inability of a lens to bring electromagnetic waves to a common focal point. It is not possible to correct this type of aberration in electromagnetic lenses because the magnetic field they produce varies across their diameter, and the physical dimensions of the lenses prevents the production of a compound lens. However, spherical aberration can be overcome by using only very small apertures in the centres of the lenses, which only allow electrons at the central axis of the lens to pass through and illuminate the specimen. Electrons at the periphery of the lenses are prevented from doing this. Given electrons travelling at the central axis of the lens exhibit the least spherical aberration, this approach minimizes the defect.

Astigmatism, the inability of a lens to bring electromagnetic rays from a single source to a common focus, is caused by the asymmetry of the electromagnetic field within the lens. It is usually caused by defects in pole-piece manufacture or, more commonly, contamination on the pole-piece. This results in electrons passing through different parts of the lens being focused to different degrees. Naturally, contamination on the pole-pieces must be minimized to reduce astigmatism effects. This can be achieved by ensuring that the microscope column is kept scrupulously clean and by using an anti-contamination trap or cold finger comprising a small metal plate cooled by liquid nitrogen situated close to the specimen. Thus, any contaminating vapour liberated from the specimen during irradiation condenses on this cold plate rather than the lens pole-pieces.

It is not possible to remove the asymmetries within lenses that lead to astigmatism, but it is possible to compensate for them by surrounding the lens with a number of small secondary electromagnets. Individual adjustments of these so-called astigmatism coils make it possible to adjust the degree of astigmatism within the lens and to compensate for the defect.

Vacuum systems

Electrons do not travel far in air as they collide and react with the gas molecules. The distance that an electron can travel without collision is called the mean free path; in modern electron microscopes this needs to be in excess of 2 m. To achieve this distance, it is necessary to create a very high vacuum in the region of 1.33×10^{-2}–1.33×10^{-3} Pa (10^{-4}–10^{-5} mm of mercury) within the column of the microscope. This high vacuum also has the effect of reducing oxidation of the tungsten filament and thus extending its life. In addition, it also helps to maintain a clean environment within the column and reduce contamination of the lens pole-pieces and specimens

Several types of vacuum pumps are employed, for example oil rotary pump and oil diffusion pumps, which are typically used in combination.

Oil rotary pumps have a simple mechanism consisting of a series of vanes that rotate in an oil bath. This mechanism traps gas within a chamber forcing gas molecules out of the column. They are usually employed in the initial stages of evacuating the microscope column. Oil rotary pumps are reliable and have the advantage that they can start to operate at atmospheric pressure. They do, however, have the disadvantage that they can only produce a vacuum of approximately 1.33 Pa (10^{-2} mm of mercury) so they must be used with other types of vacuum pump.

Oil diffusion pumps operate by heating inert oil to its boiling point. The resultant oil vapour is trapped within a series of cones with downward-pointing pipes that form a jet of oil vapour directed towards a cooled surface where the oil condenses and traps gas from the surrounding atmosphere. They are efficient pumps, capable of forming a vacuum of approximately 1.33×10^{-3} Pa (10^{-5} mm mercury) and are silent in operation. However, they do not operate from atmospheric pressure and require an initial vacuum of 1.33–0.133 Pa (10^{-2}–10^{-3} mm mercury) to operate effectively. The assistance of a rotary pump, which is mounted in series, is therefore useful to obtain this degree of vacuum before the diffusion pump begins to operate. It is essential that diffusion pumps are not exposed to vacuums of less than 1.33 Pa (10^{-2} mm mercury) when the oil is at operating temperature, as this may cause back streaming of the oil vapour passing the cooling plates allowing it to enter the microscope column, causing considerable contamination. Manufacturers prevent this from happening by placing a pressure-sensitive baffle plate at the top of the pump, which automatically closes if the vacuum fails or is insufficient.

SELF-CHECK 7.8

Why is it not possible to operate an oil diffusion pump from atmospheric pressure?

Cooling system

Electromagnetic lenses generate considerable heat as a by-product of their action. This heat can produce instabilities in the action of the lenses and can also lead to the thermal expansion of metal components around the lenses that can adversely affect the formation of a well-focused image. To prevent these undesired effects, lenses are water cooled. In addition, oil diffusion pumps also require cooling for them to operate. This is simply achieved by passing water pipes around and through the electromagnetic lenses and the pump, through which chilled water is circulated maintaining the temperature at 18–20 °C.

Viewing screen and recording images

Owing to their very short wavelengths, electrons are invisible to the human eye. Thus, to see the image of a specimen produced in an electron microscope it is necessary to allow the focused electron beam to hit a phosphorescent screen placed in the beam path. The electrons react with the **phosphor**, which emits light. The intensity of this light is proportional to the number of electrons striking the phosphor and thus a visible image of the specimen is formed.

The image formed on the phosphorescent screen is transient, being only present when illuminating the specimen. The traditional method of making permanent reproducible records of the image was to form a photographic record. Fortunately, photographic emulsions are extremely sensitive to electrons and are capable of producing high-resolution copies of an image. Microscope cameras have to be mounted within the column and the vacuum system, as it is necessary for the electron to fall directly onto the photographic emulsion. They can be situated beneath the phosphorescent viewing screen, which is moved out of the path of the beam to allow exposure of the film or placed above the screen immediately beneath the projector lens and moved into the beam path when a photograph is required.

Photographic film contains a large quantity of moisture, which if exposed to the high vacuum environment of the electron microscope column would result in considerable contamination and

BOX 7.3 Preparing samples for electron microscopy

Sample preparation for transmission electron microscopy (TEM) is designed to take advantage of the increased magnification and resolution that TEM offers over light microscopy (Box 7.1). Unfortunately, the majority of the reagents and solutions used in preparing samples for TEM present some form of risk to the user. Hence, sample preparation must only be undertaken when wearing appropriate personal protective equipment and within a fume cupboard.

Electron beams are destructive to tissues, so the preparation of samples must result in specimens that can withstand being bombarded with electrons in a vacuum. This is normally achieved by using extremely potent fixatives, such as glutaraldehyde and embedding the sample in hard epoxy resin, which is capable of being thinly sectioned.

Preparation of slices of tissue or suspensions of cells for electron microscopy begins by fixing them in a strong cross-linking reagent, such as glutaraldehyde or osmium tetroxide, to minimize the effects of autolysis and microbial degradation. This type of fixation also produces samples that are able to withstand the subsequent processing steps. The samples are then dehydrated using a graded series of concentrations of ethanol, ending with 100% strength, before being impregnated in liquid epoxy resin. The samples are then baked to set the resin. Once embedded, sections approximately 1 μm thick are metochromatically stained with, for example, toluidine blue and examined using light microscopy, to identify initially any areas of particular interest. Once such areas are exposed, ultrathin sections of only 50–90 nm are produced using an ultramicrotome. The sections are mounted on support grids, which typically are about 3 mm in diameter and made of copper. The holes in the grid allow the beam of electrons to pass though the sections, while the lattice supports the section. Prior to viewing, the sections are stained using a variety of heavy metal salts including those of lead, uranium, or less often gold, silver, or platinum. Once stained, the grids may then be placed in the electron microscope for visualization of the areas of interest, such as those shown in Figures 7.19–7.22.

deterioration of the vacuum. It is necessary to dry the film prior to placing it into the column and most manufacturers incorporate a desiccation chamber within the microscope that can be maintained under vacuum by the rotary oil pump. Most manufacturers also incorporate an air lock into the camera, which facilitates film change without the need to bring the whole column back to atmospheric pressure.

With the rapid development of digital imaging systems, manufacturers of electron microscopes have been quick to realize the advantages of these types of imaging systems. Digital images are readily accessible, easy to store electronically, and can be transferred to multiple viewers simultaneously by email. They are also of high quality and are the medium of choice for scientific publications.

Digital cameras can be incorporated into the electron microscope in the same way as conventional film cameras: either situated beneath the phosphorescent screen or placed in the column beneath the projector lens. The image electrons are allowed to fall directly onto a small phosphorescent screen placed in the beam and the resultant image is directed to the camera.

To take advantage of increased magnifications and resolution possibilities of the electron microscope requires specimens to be processed in a particular way, as described in Box 7.3.

Cross reference

Chapter 4, 'Health and safety'.
You can read more about electron microscopy and sample preparation in the accompanying volume *Histopathology* in this series.

7.10.2 Application of transmission electron microscopy in pathology

There are several pathological processes that can only be examined using electron microscopy because the cellular changes are so small that they are beyond the resolution of light microscopes (Figure 7.19). This is especially true in the case of some renal diseases. There are changes to the glomerular basement membrane and to the epithelial cell foot processes that can only be seen using an electron microscope, which you can see in Figure 7.20 (a, b).

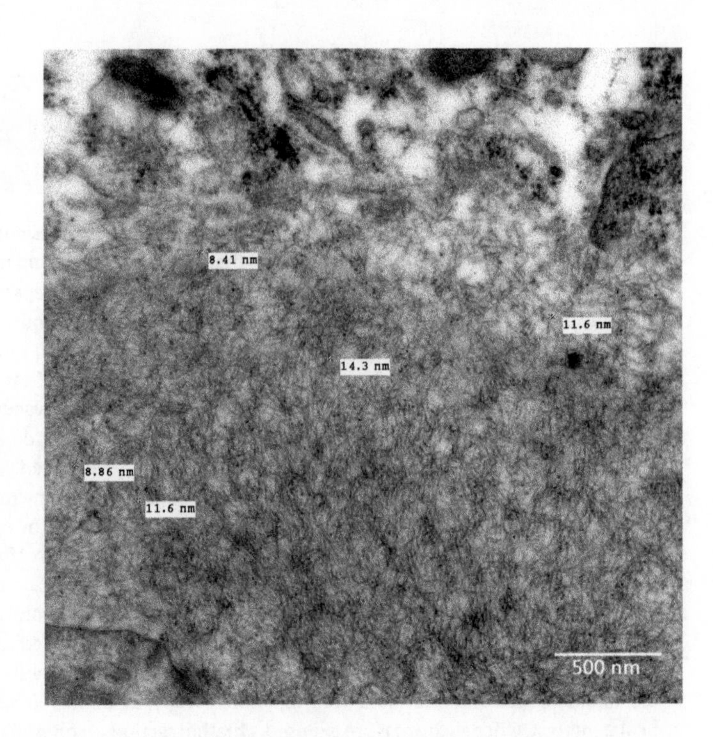

FIGURE 7.19

Electron micrograph of amyloid fibres within the wall of a small blood vessel from a patient with chronic inflammatory disease. Note the characteristic fibrillary nature of this abnormal protein. Electron microscopy is the definitive method of choice for the positive identification of amyloid.

(a)

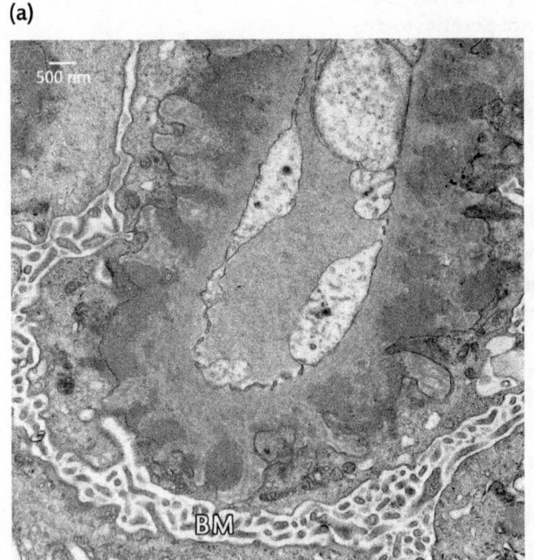

(b)

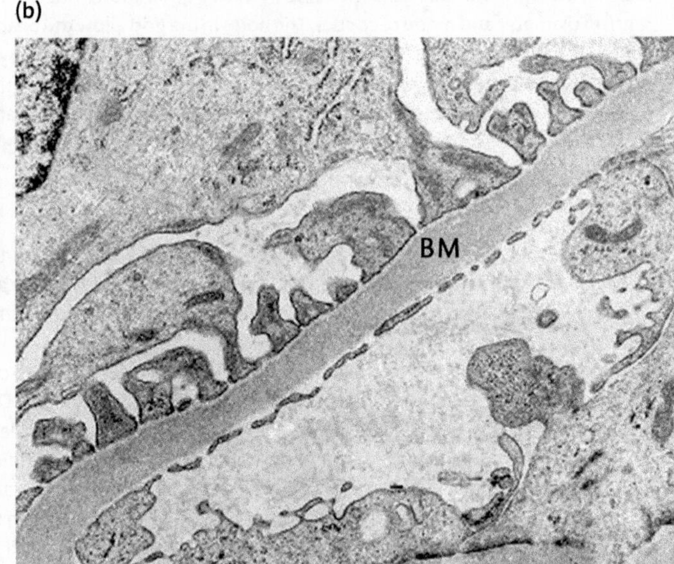

FIGURE 7.20

(a) Electron micrograph of renal glomerular basement membrane (BM) from a patient suffering from membranous glomerular nephritis. (b) Note how its appearance differs in structure from that seen in a healthy person. Courtesy of Dr J.C. Jeanette, Department of Pathology and Laboratory Medicine, University of North Carolina, USA.

Cross reference

You can learn about the glomerular basement membrane and epithelial cell foot processes of the kidney in Chapter 3, 'Kidney disease', of the companion textbook, *Clinical Biochemistry* and in *Histopathology*.

The electron microscope is also used extensively in tumour diagnosis, particularly where there is poor cell differentiation, and it is difficult to see specific features and markers at the resolution of light microscopy. An example of this occurs in the differential diagnosis of malignant melanoma, where the quantity of melanin pigment within the tumour cells is so small that conventional staining and immunocytochemical demonstrations appear negative. With the electron microscope it is possible

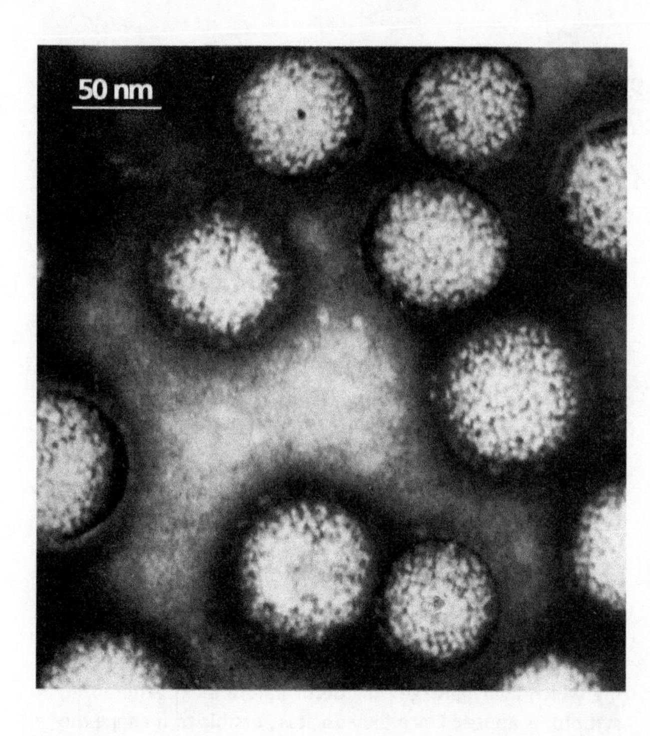

FIGURE 7.21
Electron micrograph of influenza viral particles. Courtesy of H. Cotterill, Manchester Royal Infirmary, UK.

to see very small numbers of melanosomes and even the non-pigmented pre-melanosomes within the cytoplasm of melanoma cells. The development of specific tumour markers and the use of immunocytochemistry have, however, led to less reliance on electron microscopy in tumour diagnosis.

Viral particles are far too small to be seen with a light microscope but are clearly visible using electron microscopy (Figure 7.21). Early identification and diagnosis of viral infections is sometimes essential if the patient is to be diagnosed quickly and receive prompt, effective treatment. This is particularly the case in patients infected with HIV and suffering from AIDS, where their immune response is severely compromised. Electron microscopic examination of suitable samples can be used to identify viral particles by their characteristic morphology and offers a simple and quick method of sample analysis. You can see an electron micrograph of HIV particles escaping from an infected cell in Figure 7.22.

Cross reference
Immunocytochemistry is described in Chapter 13, 'Immunological techniques'.

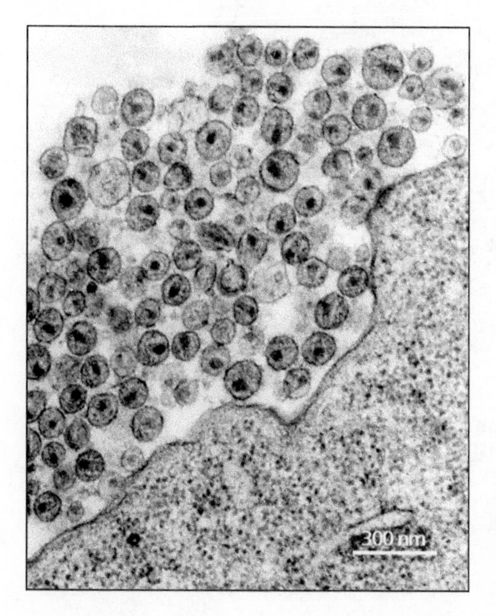

FIGURE 7.22
Electron micrograph of HIV particles being released from the surface of a cell. Courtesy of H. Cotterill, Manchester Royal Infirmary, UK.

CASE STUDY 7.1 Liver biopsy and diagnosing liver cirrhosis

A 44-year old man living rough collapsed and was seen at a local drop-in surgery. The patient was unemployed but had previously owned a business that had gone bankrupt. He admitted to a long history of alcohol abuse, presented overt signs of jaundice, had a temperature of 38.2 °C, exhibited tenderness in the right upper quadrant of his abdomen, and percussion of his flanks indicated abdominal oedema probably due to excessive ascites fluid. He was referred to the local hospital where blood samples were taken for testing. Prothrombin time was prolonged; serum bilirubin concentration was above reference values, and alanine and aspartate transaminase activities greatly increased. Tests for hepatitis B and C virus infections were negative. The patient was provisionally diagnosed with cirrhosis of the liver.

Cirrhosis is a condition in which the liver responds to lesions by producing strands of fibrous tissue, for example collagen, that surround nodules of regenerating cells. The excessive production of scar tissue leads to a loss of cells and tissue architecture, and inflammation, which interferes with liver functions and can lead to liver failure and death. The major causes of liver cirrhosis are prolonged alcohol abuse, infection with the hepatitis C virus and morbid obesity leading to non-alcoholic steatohepatitis, a condition characterised by excessive deposition of fats in the liver.

Symptoms of cirrhosis may initially be generalized, for example nausea, vomiting, and weight loss, making diagnosis difficult, and while serological tests taken together can infer cirrhosis, they are not absolutely accurate. Therefore, a biopsy of the patient's liver was taken and material was prepared for light microscopy. The slides were stained with haematoxylin and counterstained with van Gieson's component (a mixture of picric acid and acid fuschin), which selectively stains connective tissue fibres red. Figure 7.23 shows a section of the liver, clearly showing red-stained collagen fibres enclosing nodules of regenerating cells and surrounded by liver parenchyma cells, thus confirming the diagnosis of liver cirrhosis. This is almost certainly the result of alcohol abuse over a number of years.

It is not possible to reverse cirrhotic damage to the liver. Patients may be asymptomatic for a considerable time during which liver damage may occur before even generalized symptoms appear. Once they do, it is possible to manage the symptoms and complications that arise from the condition. However, if the patient fails to stop the abuse, which is often the case given the addictive nature of alcohol consumption, then liver failure and death result when 80–90% of liver parenchyma cells have died.

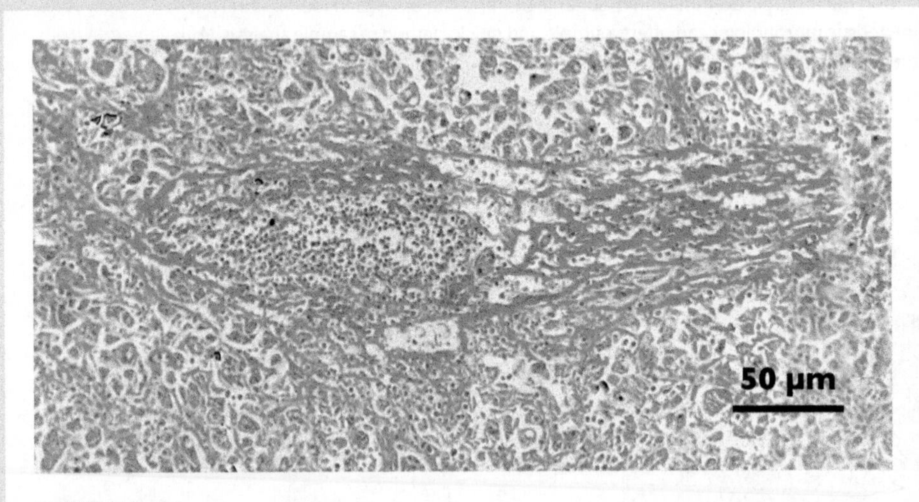

FIGURE 7.23
Liver biopsy material stained with haematoxylin and van Gieson's component. Regions of liver parenchyma cells surround deposits of red-stained collagen fibres that enclose nodules of regenerating cells.

CASE STUDY 7.2 A case of gastroenteritis in infants

Over a roughly 2-week period, a substantial proportion of the 2-3-year-old infants attending a playgroup succumbed to, in most cases, relatively mild gastroenteritis. The major symptoms consisted of watery diarrhoea, but some children presented with mild fever and occasional bouts of vomiting. Treatment consisted largely of isolating children in their respective homes and rehydration therapy using oral rehydration powders. However, the two most severe cases required hospitalization and intravenous hydration. In nearly all cases, the symptoms were resolved within 7 days without the need for additional treatment.

Standard medical microbiological procedures failed to identify any causative bacterial agent. However, transmission electron microscopy identified particles with a surface morphology and size characteristic of adenovirus virions in all faecal samples examined after preparation for electron microscopy and staining with phosphotungstate (Figure 7.24).

Various types of adenoviruses are found in over 50% of the population and often appear to have little clinical effect. However, they are associated notably with eye, respiratory, and gastrointestinal infections. Infants and young children are most susceptible to these infections. Adenoviruses spread easily in crowded play areas and primary schools. Transmission can be by direct contact or touching contaminated toys or surfaces or from adults who have failed to wash their hands correctly after using the lavatory (see Figure 4.12). Whatever the source, infection begins when the virus invades cells of soft tissues, such as the epithelia lining the respiratory or digestive systems. Adenoviruses can cause a number of serious clinical conditions; however, most types of infections are usually mild. For example, the gastroenteritis associated with gut infections is relatively benign and normally resolves within 1-2 weeks.

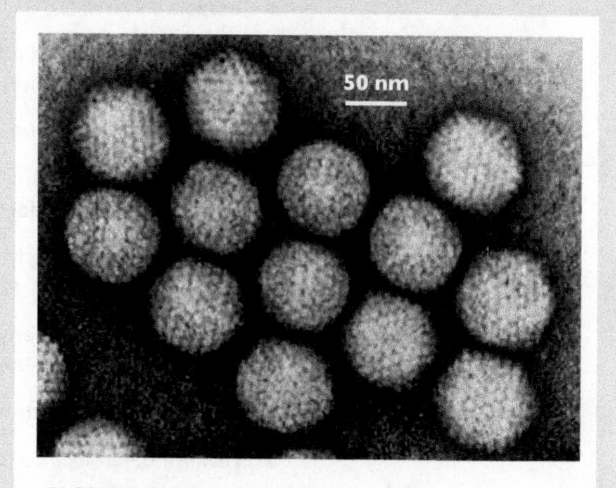

FIGURE 7.24
Transmission electron microscopy of adenovirus particles. These are easily identifiable from their characteristic surface morphology and size. Courtesy Dr G William Gary Jr and the Public Health Image Library, Centers for Disease Control and Prevention, USA.

Some points to consider:

What is the most probable cause of the gastroenteritis? Justify your answer.

Could this outbreak and similar ones be prevented?

Suggest measures to minimize the risk of further outbreaks.

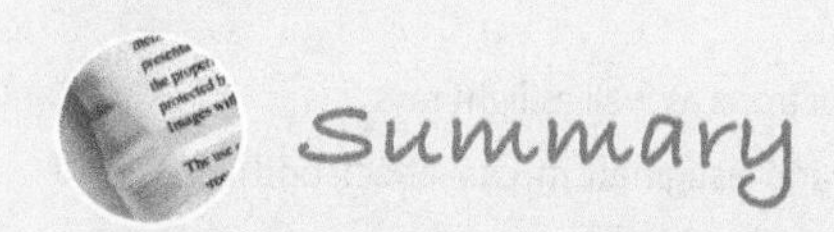

Summary

- Microscopes are an essential part of laboratory testing and diagnosis.
- Microscopes are used in many different disciplines.
- Various types of microscopy exist and are used for different purposes.
- Microscopes must be set up and adjusted correctly to ensure optimum visualization of specimens.
- When significant increases in magnification or resolution are required, electron microscopy must be used instead of light microscopy.

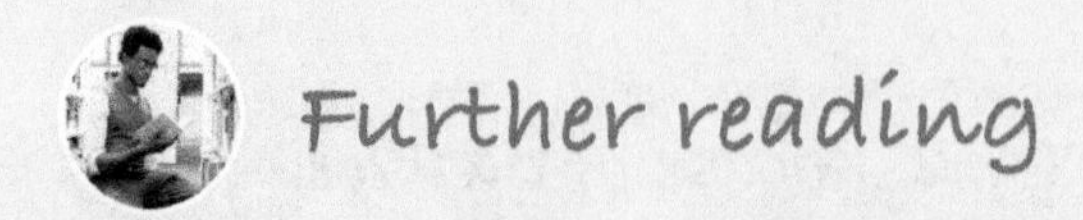

Further reading

- Bozzola J, Russell LD. *Electron Microscopy*. 2nd ed. Boston, MA: Jones and Bartlett, 1998.
- Bradbury HSM. *Introduction to Light Microscopy, Royal Microscopical Handbook*. London: BIOS Scientific Publishers, 1998.
- Cheville N. *Ultrastructural Pathology*. 2nd ed. Oxford: Wiley-Blackwell, 1998.
- Ghadially F. *Diagnostic Ultrastructural Pathology: A Self Evaluation and Self Teaching Manual*. London: Hodder Arnold Publishers, London, 1998.
- Heidelbaugh JJ, Bruderly M. Cirrhosis and chronic liver failure: Part 1. Diagnosis and evaluation. *American Family Physician* 2006; **74**: 756–63.
- Heidelbaugh, JJ, Bruderly M. Cirrhosis and chronic liver failure: Part 1. Complications and treatment. *American Family Physician* 2006; **74**: 767–76.
- Ostroff L, Zeng H. Electron microscopy at scale. *Cell* 2015; **162**: 474–5.
- Pawley JB. *Handbook of Biological Confocal Microscopy*. New York: Springer, 2006.
- Piper T, Piper J. Variable multimodal light microscopy with interference contrast and phase contrast; dark or bright field. *Journal of Microscopy* 2014; **255**: 30–41.
- Rogers K. *Complete Book of the Microscope*. London: Usborne Publishers, 2006.

Questions

1. If the diameter of a virus particle appears to be 5 mm when magnified 100,000×, its 'real' diameter is:
   (a) 1.0 µm
   (b) 0.005 mm
   (c) 50 nm
   (d) 500 µm
   (e) 5 µm.
2. Indicate which of the following statements are **TRUE**:
   (a) Specimens for light microscopes are thicker than those for electron microscopy
   (b) Electron microscopes have a greater resolving power than light microscopes
   (c) Moving electrons have shorter wavelengths and therefore give poorer resolution than light rays
   (d) Electromagnetic lenses can focus beams of electrons as well as light rays
   (e) Chromatic aberration, spherical aberration, and astigmatism can affect both glass and electromagnetic lenses.
3. **(a)** Describe what is meant by chromatic aberration. **(b)** How is this defect in a microscope lens corrected?
4. Describe how you would set up a light microscope illumination system to give Köhler illumination.
5. What two main factors affect the resolution of a microscope?
6. Why is it necessary for the column of an electron microscope to be maintained under very high vacuum?

7. Describe, with the aid of a diagram, how the incident light fluorescence microscope works and can be used in the biomedical science laboratory. You may find it helpful to read relevant parts of Chapter 13, 'Immunological techniques', and Chapter 14, 'Molecular biology techniques'.

8. What is the total magnification of a microscope using a 40× objective lens with a focal length of 2 mm and an eyepiece with magnification of 10×?

Answers to self-check and end-of-chapter questions are available at the end of the book.

8

Electrochemistry

Peter Robinson

Learning objectives

After studying this chapter, you should be able to:

- Define key terms related to electrochemical analyses
- Describe the reactions that occur in a typical electrochemical half cell
- List common substances that may be measured in a clinical laboratory using electrochemical analytical devices
- Describe the Nernst equation and outline its use in potentiometric analysis
- Describe the underlying principles of amperometric electrochemical devices
- Outline the principles and clinical applications of ion selective field effect transistors
- Define and describe the underlying principles of operation of biosensors
- Discuss the development of biosensor technology

Introduction

It is likely that a pH electrode was the first piece of analytical equipment that you used when you first entered a science laboratory. Like so much modern technology, you could use such a device without ever really understanding how it works. However, working within a clinical laboratory you may well have to use several, possibly dozens, of different electrode devices to measure a wide range of **analytes** (the term generally used for the substances being measured) in clinical samples. Ion-selective electrodes, oxygen electrodes, carbon dioxide electrodes, and various biosensors are now found in most clinical settings. It is useful therefore to have sufficient background knowledge to be able to understand the potential and limitations of such devices. The recognition that such electrodes fit into a small number of 'families' should enable you to develop experience of using one electrode and help you understand how other 'close relatives' might also function. Learning a few fundamental principles can therefore save hours of having to 'rote learn' the specifics of each device in isolation. In addition, some knowledge of the principles on which these devices are based is particularly useful when, as can occur, an electrode appears to be malfunctioning.

In many ways electrochemical analysis is ideally suited for use in the clinical laboratory. Indeed, many techniques are also the basis of the point-of-care testing devices described in Chapter 16. Once set up and calibrated, electrochemical devices are simple to use, such that after even the most basic of training you will be able to measure numerous analytes in clinical samples with high accuracy and reliability. Because electrochemical techniques do not rely on any colour formation they are highly suited to measurement of samples with a high background colour, in particular blood. Finally, unlike most other analytical techniques, electrochemical techniques use few reagents. Once an electrode and meter have been purchased the actual running cost per sample is generally low, and this may be of great importance to the financial competitiveness of a clinical laboratory.

Cross reference

You can read about automated clinical laboratory instruments in Chapter 15, 'Laboratory automation'.

The aims of this chapter are to outline the likely uses of these electrode devices within a clinical laboratory and to describe their fundamental principles of operation. General advice on the care and maintenance of electrodes is also included. However, you are always advised to consult the manufacturer for specific information regarding such matters, as electrodes from different manufacturers may have some quite specific requirements.

8.1 Basic concepts and definitions

Electrochemistry is the study of the chemical changes produced by electrical currents and also the production of electrical currents by chemical reactions. For more than 200 years electrochemical techniques have developed through advances in chemistry, physics, electronics and, more recently, biology. The sheer number of electrochemical techniques can be bewildering, and while there may be fundamental differences between some groups of techniques, it is often the case that numerous techniques differ from each other in only minor detail. Sadly, even the naming of electrochemical analytical techniques has been applied with little or no semblance of a systematic approach. As a consequence, the nomenclature surrounding the subject can be quite confusing and, at times, plainly inconsistent!

It is therefore necessary to go back to the basics of electrochemistry before jumping into a detailed description of the devices and techniques themselves.

8.1.1 What is an electrical current?

An electrical current is simply a movement (flow) of electric charge. However, one fundamental inconsistency that might well lead to confusion relates to something as simple as the actual direction of the flow itself. This inconsistency dates back to the mid-eighteenth century, when Benjamin Franklin was studying static electricity. He found that when a wool cloth was rubbed over a paraffin wax block the two materials subsequently attracted each other. Franklin considered that the wool cloth was in some way removing an 'invisible fluid' from the wax block, hence the type of charge that was associated with rubbed wax became known as 'negative' (because it had a deficiency of the 'fluid'), while that associated with the wool became known as 'positive' (because it had an excess of the 'fluid'). This terminology was widely adopted such that when electric batteries were later developed, it was natural to assign the direction of the flow of current to be from the positive to the negative pole. Over a century later, however, J.J. Thomson showed that cathode rays were negatively charged particles, which he called 'corpuscles', and claimed that atoms were built from these corpuscles. The electron had been discovered and it was suddenly realized that it was electrons that carried the current in metal wires. Thus, electrons must be moving in the opposite direction to that of the 'conventional' current. Sadly, it was much too late to change Franklin's naming convention and today it must be accepted that current can be considered to move in either of these two ways. Hence, both of the conventions shown in Figure 8.1 are acceptable.

(a)

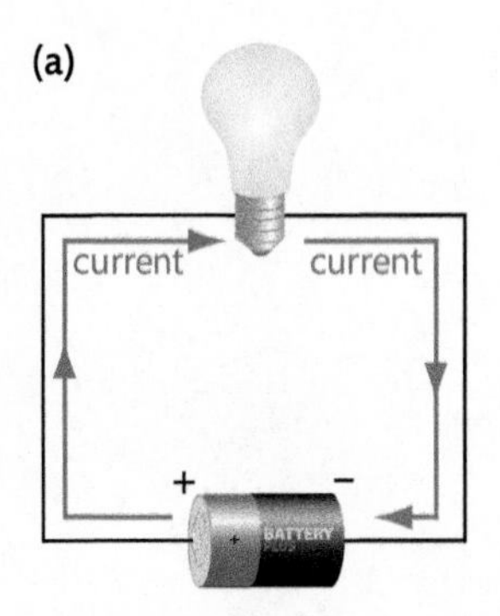

(b)

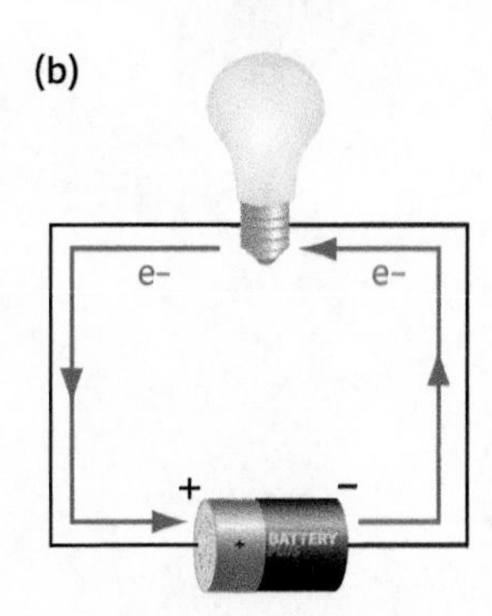

FIGURE 8.1
(a) Conventional and (b) electron flow notation for an electric current showing the apparent inconsistency in direction of current flow.

8.1.2 Ohm's law

A flow of an electrical current only occurs when there is a *potential difference* between two points within a circuit. It is also influenced by *resistive forces* present within the system. These relationships are usually described by **Ohm's law** (Box 8.1), which states that within an electrical circuit the current passing between any two points of a conductor is directly proportional to the potential difference,

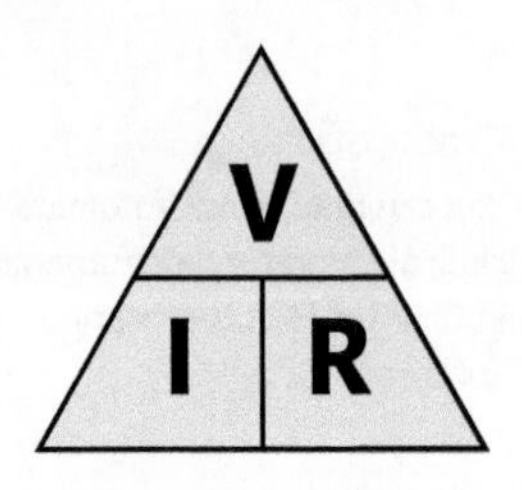

FIGURE 8.2
VIR triangle used to help calculate Ohm's law.

BOX 8.1 Ohm's Law

Ohm's law can be rearranged to describe the current (I), voltage (V), or resistance (R). The easiest way to remember these relationships is to use a VIR triangle, as shown in Figure 8.2. To use this, simply use your finger to hide the value that you want to calculate. The remaining two values then show you how to do the calculation.

that is voltage drop or voltage, between the two points but is inversely proportional to the resistance between them. In mathematical terms, Ohm's law may be written as:

$$I = V/R$$

where I is the current in amperes (often abbreviated to amps, A), V is the potential difference in volts (V), and R is the resistance measured in Ohms (Ω).

SELF-CHECK 8.1

Calculate the voltage required to force a current of 0.2 A through a wire with a resistance of 1200 Ω.

8.2 Principles of electrochemical techniques

Whenever you study electrochemical reactions you will come across the concept of an electrochemical cell that is made up of two electrochemical **half-cells**. Each half-cell contains an electrode, which at its simplest is just a metal wire that is surrounded by an electrolyte. The two half-cells are joined by a shared electrolyte or by a relatively concentrated solution of indifferent ions, often saturated or 4 mol dm^{-3} KCl or KNO_3, called a **salt bridge**.

Consider a relatively simple half-cell consisting of a copper wire dipping into a solution of $CuSO_4$. Within this system, **oxidation** may occur so that the copper electrode may lose copper ions to the solution, leaving behind electrons delocalized on the metal surface:

$$Cu_{(electrode)} \rightarrow Cu^{2+}_{(aqueous)} + 2e^{-}_{(electrode)}$$

This leaves the electrode negatively charged with respect to the surrounding solution. Alternatively, the conditions may favour a **reduction**, where copper ions in solution combine with electrons at the metal surface to deposit copper atoms:

$$Cu^{2+}_{(aqueous)} + 2e^{-}_{(electrode)} \rightarrow Cu_{(electrode)}$$

This produces a deficit of electrons in the copper electrode making it positively charged with respect to the surrounding solution. However, in both cases, a spontaneous potential difference called the **electrode potential** is established across the electrode/electrolyte interface. Hence, when two half-cells are combined, a flow of electrons, that is an electrical current, may occur. Consider, for example, the Daniell cell shown in Figure 8.3, which consists of zinc and copper wires immersed in solutions of $ZnSO_4$ and $CuSO_4$, respectively. If the Zn^{2+} and Cu^{2+} concentrations are approximately equal, then reactions will result in the oxidation of the zinc metal in one half-cell to leave the zinc electrode negatively charged, and the reduction of the copper ions in the other half-cell, which results in the copper electrode becoming positively charged. If the two electrodes are then connected by a wire, electrons will flow from the zinc anode to the copper cathode. The Daniell cell can therefore produce an electrical current and is thus a simple battery.

The salt bridge connecting the two solutions of the half-cells in Figure 8.3 completes the circuit and allows migration of ions between the two compartments but prevents gross mixing of the two

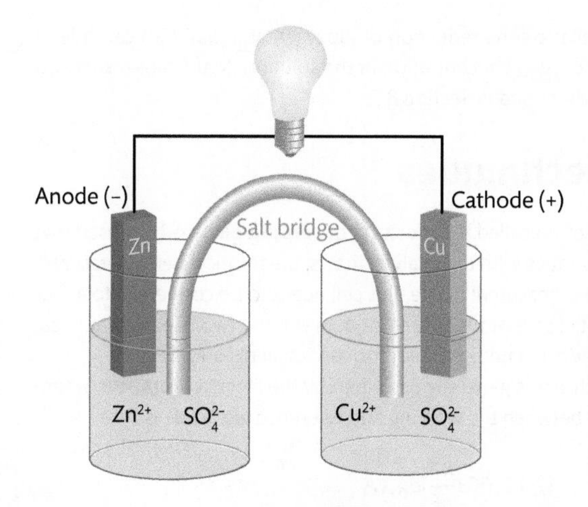

FIGURE 8.3

Daniell cell. See text for details. Note, this is a Galvanic cell and therefore the anode and cathode are negatively and positively charged, respectively, which is the opposite of the electrodes in simple electrophoresis (see Chapter 12).

A cation (pronounced cat-ion) is an ion that has fewer electrons in its electron shells than protons in its nucleus. It thus carries a net positive charge and will move towards the negative electrode during electrolysis.

An anion (pronounced an-ion) is an ion that has more electrons than protons. Thus, it has a net negative charge and will move towards the positive electrode during electrolysis. You might be able to remember that it is the anion that is negatively charged by the mnemonic 'a-n(egative)-ion'.

solutions. The salt bridge might be as simple as a strip of filter paper soaked with a relatively inert electrolyte such as KCl, or a U-shaped glass tube filled with the same electrolyte held within an agar gel.

8.2.1 Schematic representation of electrochemical cells

In 1953 the International Union of Pure and Applied Chemistry (IUPAC) developed a standard notation to represent electrochemical cells that is still accepted today. The IUPAC notation for the Daniell cell shown in Figure 8.3 may be written as follows:

$$-Zn\,|\,Zn^{2+}\,\|\,Cu^{2+}\,|\,Cu^{+}$$

where the single vertical lines represent a phase boundary (electrode/electrolyte interface) and the double vertical lines the salt bridge. Note that according to this notation the negative electrode (**anode**) is placed to the left, the positive electrode (**cathode**) is placed to the right, and electrons would therefore flow from left to right in the connecting wire.

8.2.2 Classification of electrochemical techniques

Electrochemical cells can be categorized as **galvanic** or **electrolytic**. A galvanic cell is one where the reactions occur spontaneously at the electrodes when they are connected externally by a conductor. In an electrolytic cell, however, the electrochemical reactions are not spontaneous and only occur if an external voltage is applied across the two electrodes. A modern rechargeable battery is, in essence, an electrolytic cell when it is being recharged.

SELF-CHECK 8.2

State whether the Daniell cell shown in Figure 8.3 is a galvanic or electrolytic cell.

Electrochemical analytical techniques may be conveniently grouped according to whether their electrochemistry is galvanic (spontaneous) or electrolytic (non-spontaneous). **Potentiometry** involves the use of galvanic cells, and the measurement of the spontaneous potential of an electrode with minimal current flow. The most common applications of potentiometry are the pH electrode and ion-selective electrodes described in Section 8.3. In contrast, **voltammetry** (including amperometry) uses electrolytic cells and the measurement of the current passing through an electrode when an external potential is applied. By far their most common application is the Clark oxygen electrode. This device is described in Section 8.4 and relies on the reduction of dioxygen at a platinum cathode to give a current that is proportional to the oxygen concentration in the solution. Many biosensors also incorporate amperometric detectors, as described in Section 8.5.

8.3 Potentiometric techniques

In potentiometric analysis, an electrode, often called the **working** or **sensing electrode**, is used that generates a spontaneous potential at its surface with the analyte, that is, the solution being measured. However, only potential *differences* can be measured; hence, this cell needs to be connected to a second cell that produces a standard potential so that the *difference* between the two can be evaluated. This second, so-called **reference electrode**, is a necessary component of all pH and ISEs.

The behaviour of potentiometric electrodes is generally described by the **Nernst equation**, where the total potential, E (in volts) developed between the working and reference electrode is:

$$E = E_{constant} + 2.\ 303\frac{RT}{nF}\log_{10} A$$

or simplified as:

$$E = E_{constant} + S/n \log_{10} C$$

where $E_{constant}$ is a constant potential that depends mainly on the reference electrode; 2.303 RT/F (= S) is the Nernst factor or slope (where R is the universal gas constant, T is the temperature in Kelvin, and F is the Faraday constant); n is the number of charges on the ion; A is the activity of the ion; and C is the concentration of the ion.

Potentiometric electrodes actually measure the **activity** of an ion in solution (Box 8.2). In most biomedical situations the concentrations of ions are low and activity is synonymous with concentration.

The Nernst equation shows that the responses of potentiometric electrodes will depend both on the temperature and the number of charges carried by the ion. At 25°C the Nernst factor (2.303 RT/F) becomes 0.059 and thus the equation becomes:

$$E = E_{constant} + 0.059 / n\log_{10} C$$

This means that there will be a 59 mV change in potential for a tenfold change in the concentration of monovalent ion, such as H^+, Na^+, K^+, and Cl^-.

A potentiometric electrode is said to have Nernstian characteristics if it obeys Nernst's law. If the changes in potential relative to concentration are less than the theoretical values, it may indicate an electrode malfunction or interference from other ions.

SELF-CHECK 8.3

Given that the universal gas constant is 8.314 J mol^{-1} per K, T is the temperature in Kelvin (= 273 + T°C), and that the Faraday constant is 96,485 coulomb mol^{-1}, calculate the change in potential in mV for a ten-fold change in the concentration of H^+ when the temperature of the system is 35°C.

BOX 8.2 Activity

Activity is a useful chemical concept. Consider an individual ion in solution. The ability of this ion to participate in any chemical reaction will be reduced if it is surrounded by other ions, which 'shield' it. This is more likely to occur when there are more such ions in solution. Activity, therefore, is a true measure of the ion's ability to affect chemical equilibria and reaction rates; it is often said to be the *effective* concentration of the ion in solution. In most biological situations, where the concentrations of ions are rather low, the values of the concentration and activity are equal. At higher concentrations, however, the activity may become significantly less than the concentration.

8.3.1 pH electrode

Virtually all textbooks, including this one, describe the pH electrode separately from ion-selective electrodes (ISEs), even though it is simply one specific type of ISE that responds to H^+. However, it is much more commonly found in laboratories than other ISEs and more likely to be used by students and thus tends to get a more detailed coverage by authors.

In 1906, Max Cremer found that a thin glass membrane separating the two electrodes of a galvanic cell could produce a potential that was responsive to changes in the concentration of H^+. Three years later, what we would recognize as a glass electrode was constructed, and studied by Fritz Haber and Zygmunt Klemensiewicz. However, this device did not become popular until the 1930s, when reliable amplifiers became available to measure the output from the electrodes. Glass electrodes are still the most convenient and accurate way of determining pH, and the pH electrode/pH meter is a basic item in any science laboratory.

What most people refer to as a glass electrode, as you can see in Figure 8.4 (a), is technically a device, not an electrode. It consists of a thin soft glass membrane, which gives the electrode its name, and an electrically insulating tubular body made of hard glass or epoxy. This structure encloses an internal electrolyte solution (usually 0.1 mol dm^{-3} HCl) and a silver/silver chloride **internal reference electrode**. The silver/silver chloride internal reference electrode is connected to the pH meter. A potential difference must, however, be measured between two points. Hence, a second **external reference electrode** is required, which also needs to be immersed in the same test solution. Frequently, this second electrode is made up of a second silver/silver chloride wire surrounded by a solution of 4 mol dm^{-3} KCl saturated with AgCl. The external reference electrode can be a separate probe, as shown in Figure 8.4 (b), although it is more common to find this external reference electrode built into the glass electrode in a concentric double barrel arrangement, commonly referred to as a **combination electrode,** as you can see in Figure 8.4 (c).

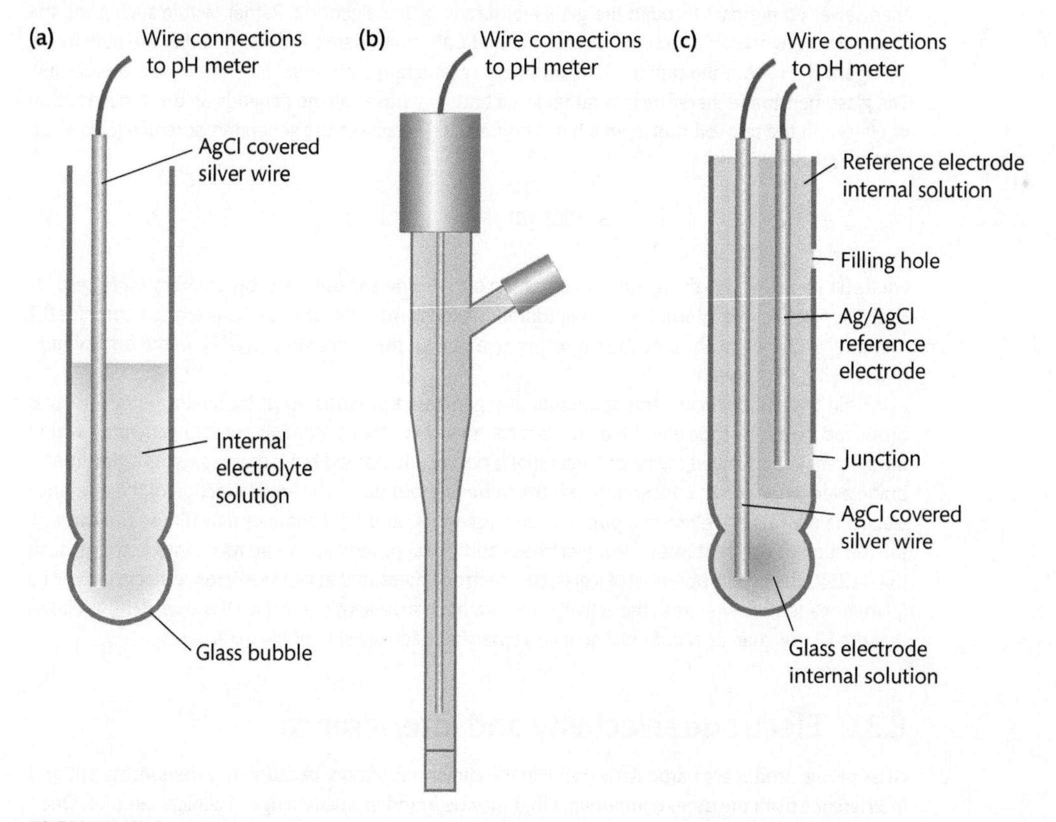

FIGURE 8.4

(a) A pH 'electrode', (b) an Ag/AgCl external reference electrode, and (c) a combination pH electrode. See text for explanations.

Whichever reference electrode is being used, it must also form electrical continuity with the test solution. In many electrodes, the KCl of the reference electrode is an aqueous solution that flows out through a small hole in the electrode body to form a liquid junction. To minimize the loss of KCl, the junction is packed with ceramic, fritted, or fibrous material, although *some* flow of KCl, however small, always occurs and therefore the pH electrode will slightly contaminate the test solution. For most analytical applications this contamination is not significant. The KCl solution lost from the electrode must be replaced. This can be done by refilling through a small hole, high up on the electrode body. Refilling electrodes adds to their maintenance costs and makes them less suited to situations where portability is a factor. To overcome such problems, the internal KCl solution may be gelled, which helps slow the loss. However, as the composition of the internal solution cannot then be restored by refilling, such **gel electrodes** generally have a shorter operational lifetime, although they are easier to use and maintain.

In situations where contamination of the test solution by the KCl of the reference electrode *is* an issue, or where the test solution contains materials that might diffuse into the external reference electrode and affect the electrochemistry, a specially designed reference electrode called the **double junction reference electrode** may be used. This device may also be selected simply to prolong the operational lifetime of the electrode. In a double junction electrode, an additional chamber between the reference electrode and the external solution is introduced. This slows the loss of KCl to the test solution and presents a further diffusional barrier to prevent the inward diffusion of any contaminating materials. Double junction electrodes are, however, more difficult to construct and hence are more expensive.

Modern pH electrodes are robust and reliable devices. However, you might be interested to know why the glass membrane found in these devices generates a potential when separating two solutions. Not all glasses possess this property and it has taken considerable research to optimize the composition of the glass to allow interference-free detection of H^+. Soda-silica glasses, such as Corning 015 glass, consisting of 22% Na_2O, 6% CaO, and 72% SiO_2, are effective for measuring pH. Hydrogen ions themselves *do not* pass through the glass membrane of the electrode. Rather, within such glass, the molecular network of SiO_2 contains both Na^+ and Ca^{2+} coordinated with oxygen that are able to exchange with H^+ from the test solution and thereby generate a potential across the glass membrane. The glass membrane therefore acts rather like a battery whose voltage depends on the concentration of H^+ ions in the test solution in which it is immersed. The size of the generated potential (E) is given by the equation:

$$E = 2.303\,(RT/F)\log_{10}[H^+]_t/[H^+]_0$$

where $[H^+]_i$ and $[H^+]_o$ are the molar concentration of H^+ inside and outside the glass membrane, respectively. In practice, the internal H^+ concentration is fixed at 10^{-1} mol dm^{-3} as the electrode contains 0.1 mol dm^{-3} HCl. Hence, the electrode is responsive only to the concentration of H^+ in the test solution in which it is immersed.

Within the pH electrode, this spontaneously generated potential must be measured against that produced by a reference electrode. In practice, however, other potentials are also generated within the system. The so-called *asymmetry potential* is poorly understood but is present across a glass membrane even when the H^+ concentration is the same on both sides. The liquid junction of the reference electrode may also give rise to a potential because the K^+ and Cl^- it contains may diffuse out through the junction at different rates. However, these additional potentials can be taken into account, such that at 25°C the overall potential of a glass pH electrode measured against a reference electrode will be 59 mV for a tenfold change in the activity of H^+. Rather conveniently, as pH itself is $-\log_{10}[H^+]$, it follows that the 59 mV change in potential actually represents each division of the pH scale.

8.3.2 Electrode selectivity and interference

Glass pH electrodes are particularly useful in the clinical laboratory because they demonstrate little interference from the other components that may be found in a wide range of clinical samples. One exception to this is the interference by monovalent cations, such as Na^+ and K^+. These ions behave in similar ways to H^+ at the glass surface and generate comparable potentials. Under acidic conditions, where the concentration of H^+ in solution is high, the interference is likely to be negligible. At alkaline pH, where there are much lower concentrations of H^+ in solution, interference can be more obvious and the measured pH is lower than the true pH of the test solution. This effect is often called the *alkaline error* or sometimes the *sodium error*, although it is not only Na^+ that interferes in this way.

The careful selection of the glass composition is crucial to minimize such interferences, as it is the composition of the glass that is solely responsible for the **selectivity**. Most commercial pH electrodes have selectivities that are sufficient under all but hyperalkaline conditions, and below pH 12 any interference effects should be negligible even in modestly priced devices. Detailed information about selectivity is available from electrode manufacturers. It is possible to take advantage of this interference: by selecting a glass that has *high* response to Na^+ it is possible to construct an Na^+ selective electrode, as described later.

Cross reference

Measuring Na^+ and K^+ in clinical samples is described in Chapters 1 and 2 of *Clinical Biochemistry*.

8.3.3 Electrode storage and cleaning

The thin glass membrane of a pH electrode is fragile. Care must be taken not to break or scratch it, or to cause a build up of static charge by rubbing it. Many pH electrodes have a plastic casing surrounding the glass electrode to protect them from damage. Even so, the temptation to use the pH electrode as a stirrer must be resisted: this is a common cause of breakage in many laboratories!

It is essential that the outer layer of glass in the pH electrode remains hydrated, so the electrode is normally immersed in a solution at all times. Most manufacturers now supply specific storage solutions for their electrodes. If these are not available, then the electrode should *never* be stored in distilled or deionized water, as this will cause ions to leach out of the glass bulb and render the electrode useless. It is acceptable to immerse combination electrodes in a buffer solution, those pH 4 to 7 are generally recommended, for storage between frequent measurements. For longer-term storage it is generally recommended to keep the electrode immersed in 4 mol dm^{-3} KCl.

The electrode should be washed thoroughly with distilled water after each use. As the electrode is susceptible to contamination or dirt, it is recommended that it is cleaned perhaps every 1-2 months, depending on the extent and condition of use. General cleaning can involve the use of a mild detergent. Any scale deposit on the electrode surface can be removed with a concentrated acid, for example 6 mol dm^{-3} HCl. Biological materials can be removed using a pepsin solution, and oily deposits with acetone or methylated spirits. However, the manufacturer's instructions should always be consulted before using any of these methods.

If an electrode has been allowed to dry out and has lost activity, it is worth contacting the manufacturer first, but a rejuvenating procedure might include soaking it in concentrated hydrochloric or nitric acid, or even boiling the electrode. Such procedures are not guaranteed to be successful and should only be used as a last resort.

8.3.4 pH meter

The glass pH electrodes developed in the early part of the twentieth century were not used on a large scale because of various technical difficulties. The main problem was caused by the large internal resistance of the glass electrodes themselves (typically 50–500 MΩ). This necessitated the use of sensitive and expensive galvanometers, which at the time were only available in a small number of well equipped research laboratories. However, in the mid-1930s Arnold Beckman, an assistant professor of chemistry at the California Institute of Technology, was trying to develop a technique (and technology) to make quick and accurate measurements of the acidity of lemon juice. Beckman's innovative approach was to use a high-gain amplifier made using two vacuum tube valves. The amplified current could then be read with much more reliable and cheaper voltmeters. In 1934, Beckman filed a patent, not actually for the entire pH meter, but rather for the amplifier component alone. Some time later he put the amplifier, a measuring electrode, and a data meter together in a compact walnut case, which he called an *acidimeter* (Figure 8.5(a)). Today we recognize this device as a pH electrode and meter, modern versions of which you can see in Figure 8.5(b).

When using a pH electrode and meter, calibration usually involves the use of two buffered solutions of widely differing pH. It is increasingly common to purchase such buffer solutions ready-made from the electrode suppliers. These are often colour-coded to make the calibration process easier: the most common buffers used are pH 4.0 (red), pH 7.0 (yellow), and pH 10.0 (blue). Calibrating most meters

(a)

(b)

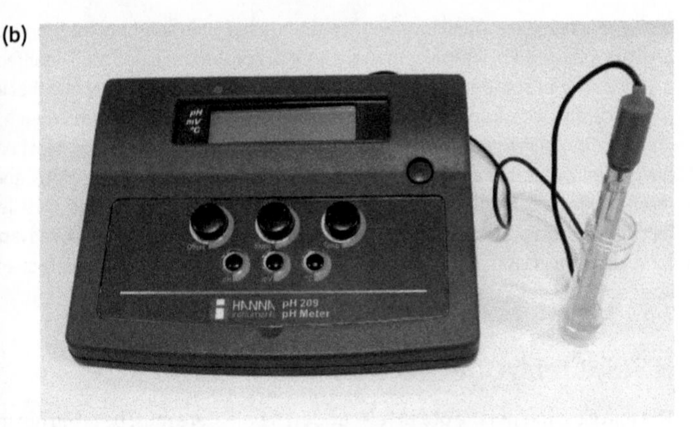

FIGURE 8.5

(a) The first commercial pH meter and probe, called the *acidometer*, developed by Arnold Beckman in the 1930s for the measurement of the acidity of lemon juice. Photograph supplied courtesy of Beckman Coulter. (b) Contemporary modern pH electrode and meter.

initially involves the pH 7 buffer. With the electrode in the buffer solution and after sufficient time to equilibrate, the 'calibrate' dial on the meter should be adjusted to give an output of pH 7.0. The electrode is then removed, rinsed with distilled water, and blotted dry before being placed in a pH 4 buffer (if the test solution is expected to be acidic) or a pH 10 buffer (if the test solution is expected to be basic). This second point is adjusted using the 'slope' control on the pH meter. Once the pH electrode has been calibrated it can simply be rinsed with distilled water and blotted dry before being immersed in the test solution for a rapid and accurate estimation of pH (Box 8.3). Between measurements, the electrode should be rinsed with distilled water and blotted dry of excess liquid.

Should commercial buffer standards not be available you can make standards yourself from recipes available in, for example, the Merck Index or the US Pharmacopeia. Typical standard buffers might include:

- 0.05 mol dm^{-3} potassium hydrogen phthalate (pH 4.01)
- 0.025 mol dm^{-3} disodium hydrogen phosphate + 0.025 mol dm^{-3} potassium dihydrogen phosphate (pH 6.86)
- 0.01 mol dm^{-3} sodium borate (pH 9.18).

BOX 8.3 Expressing mean pH values

Consider the case of measuring the pH of a urine specimen with a pH meter. Measurement of three samples of urine gives results of 5.5, 6.0, and 6.5. You then calculate the mean pH of the urine to be pH 6.0. But is this value appropriate?

This is a remarkably common situation, yet the simple analysis as described above is significantly flawed. pH is a logarithmic scale; hence, in terms of H^+ concentration the three pH values are not equally spaced. Converting the individual pH values into H^+ concentrations gives:

pH 5.5 = 3.16×10^{-6} mol dm^{-3}
pH 6.0 = 1.00×10^{-6} mol dm^{-3}
pH 6.5 = 0.32×10^{-6} mol dm^{-3}

The arithmetic mean of these concentrations yields a H^+ concentration of 1.49×10^{-6} mol dm^{-3}, which equates to a pH of 5.8. This difference (5.8 vs 6.0) might be small enough to ignore, but when pH readings are more widely spread the effects are significant. Try Self-check question 8.4 to see this for yourself.

You should note that many papers are published that report a simple arithmetic mean of pH value. This may have been used either pragmatically (the pH values are very close together and hence errors are small) or through the author's own ignorance. Beware!

Glass pH electrodes are affected by temperature (see the Nernst equation, Section 8.3). Each division of the pH scale will, for example, represent a potential of 54 mV at 0°C, 59 mV at 25°C, and 62 mV at 37°C. Temperature will also affect the thermal characteristics of the electrode and the test solution such that, in practice, the pH electrode will have an **isopotential point** where its potential is 0 and where temperature will not have any effect on the potential. This is often designed to be near pH 7. The further the pH is from this isopotential point, the greater will be the effect of temperature and the more necessary it will be to apply temperature compensation. Most pH meters have a temperature compensation control that needs to be set to the temperature of the calibrating buffers and test solution. More advanced instruments have a temperature probe that is dipped into the calibrating and test solutions alongside the measuring electrode. This ensures automatic correction of any variability resulting from solutions having differing temperatures.

SELF-CHECK 8.4

Given three solutions of pH 5, 7, and 9, respectively, you might expect the mean pH to be pH 7. However, calculate the mean as described in Box 8.3. What is the mean value?

8.3.5 Severinghaus CO_2 electrode

The need to measure the carbon dioxide concentration in clinical samples, especially blood, developed out of necessity during the 1950s' polio epidemics in Europe and the USA. In the USA alone, nearly 60,000 cases of polio were reported in 1952 and over 20,000 children were left with mild to disabling paralysis. Such paralysis often produced respiratory difficulties, which were treated with the use of the negative pressure ventilator, the 'iron lung'. During such treatment it was necessary to measure blood gases accurately and quickly to assess the effectiveness of the therapy.

In 1954, Richard Stow, working at the Ohio State University in Columbus, developed an electrode capable of measuring the partial pressure (concentration) of carbon dioxide or PCO_2 in solution. The device was essentially a modified glass pH electrode. Stow realized that carbon dioxide freely penetrated rubber and that it also acidified water. Hence, he constructed a glass pH electrode and surrounded this with a thin film of distilled water held in place by a thin rubber membrane (actually, in this instance, a section of rubber glove). When this device was dipped into a sample of blood, the carbon dioxide from the blood diffused through the membrane and acidified the water surrounding the pH electrode, resulting in a measurable drop in pH that was proportional to the PCO_2. Stow published and presented his findings, but his device was rather insensitive and unstable, and was never produced commercially. It was, however, soon modified by John Severinghaus who added a solution of $NaHCO_3$ and NaCl between the membrane and the electrode. Severinghaus also preferred a Teflon (polytetrafluoroethylene) membrane, rather than Stow's rubber membrane. In this device, carbon dioxide from the blood sample diffuses across the membrane into the $NaHCO_3$ and NaCl solution, and disturbs the carbonic acid/hydrogen carbonate equilibrium to produce H^+:

$$CO_2 + H_2O \rightleftharpoons H_2CO_3 \rightleftharpoons H^+ + HCO_3$$

The system is allowed to achieve equilibrium, at which time the pH is measured. This is then compared with a calibration curve of pH against known PCO_2 values.

Severinghaus' modifications doubled the sensitivity of the original device, enhanced its stability, and improved the response time. In 1958, Severinghaus described a device incorporating both his PCO_2 electrode and a Clark O_2 electrode, which you can read about in Section 8.4, to form a combined blood gas analyser. Commercial manufacture of the Severinghaus PCO_2 electrode began in 1959 and such electrodes are still found in PCO_2 analysers in clinical laboratories today.

Cross reference

Measuring HCO_3^-, PCO_2, and pH in clinical samples are described in Chapter 6 of *Clinical Biochemistry*.

The Severinghaus electrode is a *gas-sensing electrode* and as it is based on the pH electrode, the electrical output is a *logarithmic function of PCO_2*. At 37°C an output of 62 mV is expected for a tenfold change in PCO_2. The response is also temperature dependent as temperature affects the solubility of dissolved gases, the equilibrium constants within the HCO_3^- layer, and the Nernstian characteristics of the pH electrode itself. It is therefore usual to maintain the temperature of such devices at a constant 37°C. For the most accurate analyses, it may also be necessary to correct for the temperature of the patient who provided the samples, especially if the patient is known to be either febrile or hypothermic.

8.3.6 Ion-selective electrodes

The glass pH electrode is just one of a wide range of ISEs available to the biomedical scientist. Note at the outset that these devices are called *ion-selective* electrodes, not ion-specific electrodes. All of these devices *do* suffer from some interference, that is, responsiveness to other ions, although in many clinical applications this interference may well be minimal. Within the clinical laboratory you may well need to use Na^+, K^+, and Cl^- ISEs to measure the electrolyte balances of samples. Lithium ion ISEs may be used to measure concentrations of Li^+ in samples of serum in cases where this is being used to treat psychiatric illness. In general, ISEs can be divided into three main groups:

- modified glass electrodes
- polymer membrane electrodes
- solid-state electrodes.

Modified glass electrodes incorporate a membrane of specifically modified glass, as their name implies. **Polymer membrane electrodes** have an ion exchange material in an inert matrix, such as polyvinylchloride, polyethylene, or silicone rubber, while **solid-state electrodes** incorporate a membrane consisting of a single crystal of a sparsely soluble salt or a heterogeneous membrane in which the salt is incorporated in an inert binding agent.

The interfering effect of Na^+ on pH measurements was discussed earlier. We noted how the chemical nature of the glass used within a pH electrode is designed to minimize this effect. However, to measure Na^+ in clinical samples, a glass electrode with a *high* Na^+ responsiveness and with an inner reference solution of a fixed Na^+ activity are required. In early Na^+ ISEs, NAS 11-18 glass, comprising 11% Na_2O, 18% Al_2O_3, and 71% SiO_2, was found to be effective, although this has now generally been superseded by a range of lithium-based glasses (Li_2O), sometimes referred to as LAS glasses. Unfortunately, both NAS and LAS glasses are more responsive to Ag^+ and H^+ than they are to Na^+. Fortunately, Ag^+ is rarely encountered in clinical samples and such interference can generally be ignored. In some clinical applications, H^+ interference may also be of little concern. Consider, for example, the analysis of Na^+ in blood where in healthy individuals the plasma Na^+ concentration is maintained within the range 0.135–0.145 mol dm^{-3}. The pH of such samples would be expected to be close to pH 7.4, which equates to a H^+ concentration of 4×10^{-8} mol dm^{-3}, at which concentration little observable interference would be expected. However, in other applications, the interfering effects of H^+ may well be significant. Standardizing the H^+ concentration of all samples and standards to ensure that the electrode is responding to Na^+ rather than to H^+ interference must be considered.

While it is possible to produce a glass electrode that is responsive to K^+ using, for example, NAS 27-4 glass, a more common approach is to use electrodes constructed by incorporating specific ion exchange materials, such as valinomycin, into a polymer membrane. Molecules of the cyclic antibiotic valinomycin have a hexagonal ring structure with an internal space of almost exactly the same diameter as K^+. They can therefore form complexes with the K^+ and preferentially conduct them across a membrane. Similarly, Li^+ ISEs can be constructed from Li^+-selective carriers, such as crown ethers. These are capable of transporting Li^+ across the membrane and releasing the ion, in preference to the many other cations that might be present in a sample.

Various solid-state membranes are also available. Chloride ISEs are constructed using a crystalline membrane that incorporates AgCl in direct contact with the test solution. This membrane slowly dissolves and Cl^- from the electrode diffuses into the test solution to leave behind a surplus of Ag^+. This generates an electrical potential that is proportional to the Cl^- in the test solution. Such electrodes are quite robust and generally have faster response times than glass-based electrodes. However, as such devices rely on the slow dissolution of the membrane itself, they have a finite operational lifetime.

With all ISEs, we have emphasized that the electrode responds to the *activity* of the ion in solution. If the instrument is calibrated with a standard of known concentration then, provided the ionic strengths of the solutions are similar, the concentration of the test solution will be recorded. To ensure that the ionic strengths are similar, an **ionic strength adjustor** can be added to the sample and standards. Ionic strength adjustors contain high concentrations of ions and sometimes pH adjustors, and decomplexing agents or agents to remove chemical species that interfere with the measurement of the ion of interest. If some of the ion is not free in solution, but exists in a complex or as an insoluble precipitate, then ISEs will give a lower value than methods such as atomic absorption spectrophotometry that detect all of the ions present. The ISE results may, however, be more applicable because it is often the *free ions* that are responsible for any clinical or biological effects.

The response of all ISEs is logarithmic and is described by the Nernst equation. Tenfold changes in ion activity give equal increments on the meter scale. ISEs are not intrinsically sensitive and generally show a linear response range of 10^{-1}–10^{-5} mol dm^{-3}. In practice, the minimum detection limit of most commercial devices is 10^{-6} mol dm^{-3} at best. As with pH electrodes, the potential produced from an ISE is temperature dependent, except at the isopotential point, which varies with the type of electrode. It is therefore necessary to have some form of temperature control and/or compensation. All ISEs also require a reference electrode to generate a steady potential against which the varying potential of the working electrode is compared. If either K^+ or Cl^- is measured, then a double junction reference electrode is needed to prevent contamination of the sample by the internal solution of the reference electrode.

Many ions can be measured directly or indirectly by ISEs. One form of indirect measurement is the use of the electrode as the end point indicator of a titration. The electrode can be sensitive to the ionic species being determined or to the titrant ion. Such titrations may be ten times more accurate than direct measurements, as they only require the accurate measurements of a *change* in potential rather than the absolute value of the potential itself.

8.3.7 Ion-selective field effect transistors

A range of ISEs may be constructed by modifying field effect transistors to respond to specific ions. Such **ion-selective field effect transistors** (ISFETs) are available as probe-type electrodes and also as microelectrodes that are suitable for implantation and advanced experimental studies. It is likely that these devices will have an increasing clinical value owing to their robustness and because they can be stored dry, unlike conventional pH electrodes and ISEs.

The electrical transistors found in nearly all electrical devices are **metal oxide semiconductor field effect transistors** (MOSFETs). Within such semiconductor devices, as you can see in Figure 8.6(a), a channel of semiconductor material enables a current to flow from one electrode, called the *source*, to a second electrode, the *drain*. The rate of flow of electrons, that is the *conductivity* of the semiconductor, is changed by altering the electrical diameter of the channel. This is the function of the *gate* electrode, a metallic electrode that is insulated by a layer of silicon dioxide from the channel itself, such that it

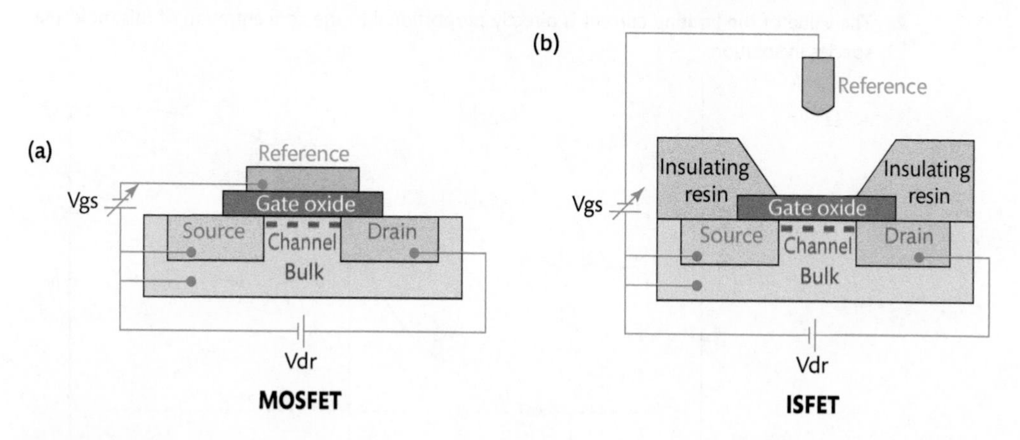

FIGURE 8.6

Schematics showing the structures of (a) metal oxide semiconductor field effect transistor (MOSFET) and (b) an ion-selective field effect transistor (ISFET).

can only influence the source-to-drain current electrostatically. Within such a device, a small change in potential difference between the gate and the body of the semiconductor is able to change the conductivity of the channel and produce a relatively large change in the current from the source to the drain. In this way, the transistor acts as an *amplifier*. This is somewhat like a small tap controlling the flow of water through a pipe: a small turn of the tap can dramatically increase or decrease the flow of water.

As you can see in Figure 8.6(b), the source and drain within an ISFET are constructed in an identical way to the MOSFET. However, the metallic gate of the MOSFET is replaced by an insulating coating capable of interacting with the analyte in the test solution to generate a potential. In a pH-selective ISFET, this may simply be a layer of silicon dioxide. The source and drain leads, as well as the semiconductor itself, are encapsulated so only the gate area is open for contact with the test liquid. The device also incorporates a reference electrode, so that when the device is immersed in a test liquid, an electrical circuit is set up between the reference electrode and gate surface enabling the H^+ concentration in solution to influence the channel and, hence, the current passing from source to drain.

Ion-selective field transistors may be further modified by changing the chemical nature of the gate insulator or by adding an ion sensing membrane layer over the gate insulator. Using such techniques, multifunction ISFETs capable of measuring pH, Na^+, K^+, and Ca^{2+} are formed. It is likely that such devices will become increasingly common for the analysis of clinical samples. ISFET also make eminently suitable **transducers** for incorporation into biosensors, as described in Section 8.5.

8.4 Voltammetric techniques

In all voltammetric techniques, an external voltage greater than that of the spontaneous potential of the electrochemical cell is imposed on the system to force the electrochemical reaction to occur. In simple direct current voltammetry, if an electrode is placed into a solution of an electro-reducible molecule and the potential of the electrode is made progressively more negative a point is soon reached when reduction of the soluble species will begin at the electrode surface and a small current will flow. A graph of current against voltage is called a voltammogram, which you can see in Figure 8.7. The point at which the current begins to flow is E_1. As the electrode potential becomes ever more negative, the current increases dramatically as the molecular species is reduced still further. Eventually, a potential is reached where the molecular species is being reduced at the electrode surface as quickly as it can diffuse towards that surface. At this point, E_2 in Figure 8.7, a *maximum or limiting* current is produced and further increases in potential will not increase the current beyond this value.

The voltammagram shown in Figure 8.7 illustrates two notable properties:

1. The midpoint of the wave, called the half-wave potential ($E_{½}$), is characteristic for each molecular species reduced
2. The value of the limiting current is directly proportional to the concentration of this molecular species in solution.

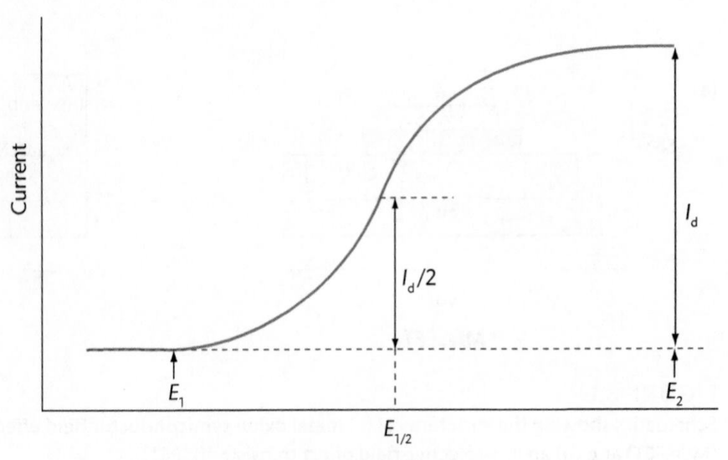

FIGURE 8.7
Typical direct current voltammogram.

These two features enable voltammetric techniques to be used to identify and quantify **electroactive** materials in solution.

More advanced forms of voltammetry, such as pulse voltammetry, differential pulse voltammetry, and alternating current voltammetry all use more complex voltage changes. Stripping voltammetry allows a pre-concentration step to be introduced into the analysis. These advanced techniques are more sensitive than direct current voltammetry, although because of this complexity they rarely find their way into routine laboratory analyses.

Voltammetry is closely related to **polarography** and, indeed, some authors use these terms interchangeably. However, the term polarography should only be used when a dropping mercury electrode is employed as the working electrode. This continuously renews the working electrode surface and ensures that the analysis is not affected by residual material from previous samples contaminating the electrode. **Amperometry** is a much simpler technique. It involves the reduction or oxidation of an electroactive molecular species at a constant applied potential, usually sufficient to generate the limiting current such that the current flowing is proportional to the concentration of the analyte in the test solution.

8.4.1 Clark O_2 electrode

In 1949 Leland Clark, working at the Fels Research Institute, Yellow Springs, Ohio, devised and developed an efficient bubble oxygenator for use in the oxygenation of blood during cardiac surgery. This invention necessitated the development of a device to measure the partial pressure (concentration) of oxygen (PO_2) in the blood of patients undergoing such surgery. Clark duly obliged and in 1956 described a compact oxygen probe that could measure oxygen concentrations in aqueous solutions. The Clark electrode is still widely used today. You can see an example of the device in Figure 8.8. It is an amperometric device that consists of a platinum cathode and a silver anode, both of which are immersed in a solution of saturated KCl. The two electrodes are separated from the test solution by an oxygen permeable Teflon membrane.

When a potential difference of −0.6 V is applied across the electrodes such that the platinum cathode is made negative with respect to the silver anode, electrons are generated at the anode, which reduce oxygen at the cathode.

At the silver anode (positive), silver chloride is formed with the liberation of electrons:

$$4\,Ag + 4Cl^- \rightarrow 4AgCl + 4e^-$$

In contrast, electrons are used at the platinum cathode (negative) to form water, consuming oxygen in the process:

$$O_2 + 4H^+ + 4e^- \rightarrow 2H_2O$$

As oxygen is reduced at the cathode, the concentration of oxygen in the solution of KCl falls, which therefore acts as a sink, so that further oxygen diffuses into the solution from the sample through the oxygen permeable Teflon membrane. As the rate of diffusion of oxygen through the membrane is the limiting step in the reduction process, the current produced by the electrode is directly proportional to the oxygen concentration in the sample. Clark electrodes generally require the sample be stirred or mixed such that gradients of PO_2 are not set up within the sample itself, leading to erroneous (lower) PO_2 values being measured.

Clark electrodes need to be maintained at a constant temperature as the solubility of oxygen in solution is temperature dependent. Similarly, atmospheric pressure also affects solubility. Published solubility tables generally quote dissolved oxygen in pure water, but salting out effects resulting from dissolved solutes may mean that the oxygen saturation values in clinical samples are below those quoted in such tables.

There are many variants of the Clark electrode. It is common to find such electrodes in combined blood gas analysers alongside both Severinghaus and pH electrodes to measure blood PO_2, PCO_2, and pH. Modified oxygen electrodes small enough to be inserted into blood vessels have been developed, but frequently this is avoided because of the dangers of infection or of forming a blood clot. Often it is considered preferable to remove small samples of blood from a warmed earlobe or a finger-tip and measure the oxygen content of the blood using a small probe-type electrode.

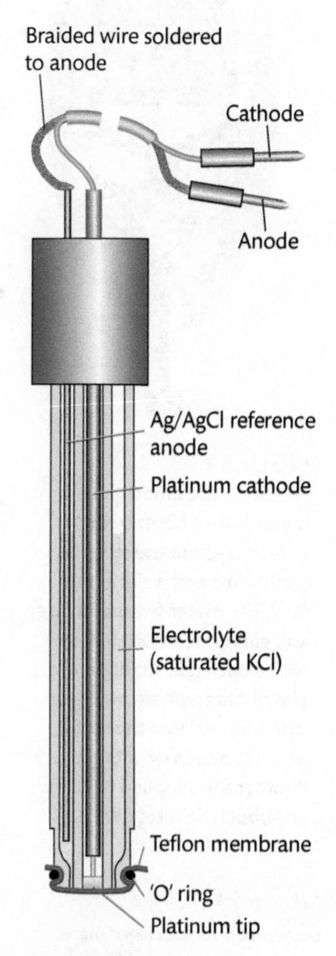

FIGURE 8.8
Clark oxygen electrode.

BOX 8.4 What is the charge of an anode?

You can see that in Section 8.2 the *anode* of a galvanic cell, such as a Daniell cell (Figure 8.3), is negatively charged, whereas in Section 8.4 the anode of a voltammetric cell is positively charged. This is not a mistake, simply the result of rather confusing inconsistency. In both cases, however, the anode is the electrode at which *oxidation* events occur and where electrons therefore become liberated. Similarly, *reduction* events always take place at the *cathode*.

A type of Clark oxygen electrode commonly found in many teaching laboratories is the Rank oxygen electrode (produced by Rank Bros Ltd, UK). This is a versatile instrument and useful for performing enzyme assays in which oxygen depletion or generation can be used as measures of the rate of reaction. Glucose oxidase, amino acid oxidase, and catalase are enzymes whose properties can be studied in this way.

Following its development, it was recognized that the Clark electrode could also be used to measure hydrogen peroxide in solution simply by reversing the potential of the electrodes. If a potential difference of +0.7 V is applied across the electrodes then the platinum anode is made positive with respect to the silver cathode and electrons are generated at the anode and used at the cathode as follows (Box 8.4). Thus, at the platinum anode (now positive), hydrogen peroxide is consumed in an electron releasing reaction:

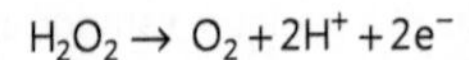

$$H_2O_2 \rightarrow O_2 + 2H^+ + 2e^-$$

and at the now negative silver cathode, electrons are used to reduce silver ions:

$$2AgCl + 2e^- \rightarrow 2Ag + 2Cl^-$$

The flow of current is therefore proportional to the concentration of hydrogen peroxide in solution. Measuring the concentration of hydrogen peroxide is particularly useful in designing many biosensors (Section 8.5). Indeed, approximately 65% of all biosensors incorporate some form of peroxide sensitive transducer.

In this respect, some electrochemical cells differ from the electrophoresis apparatus discussed in Chapter 12, 'Electrophoresis', where the anode is *always* positively charged and the cathode *always* negative.

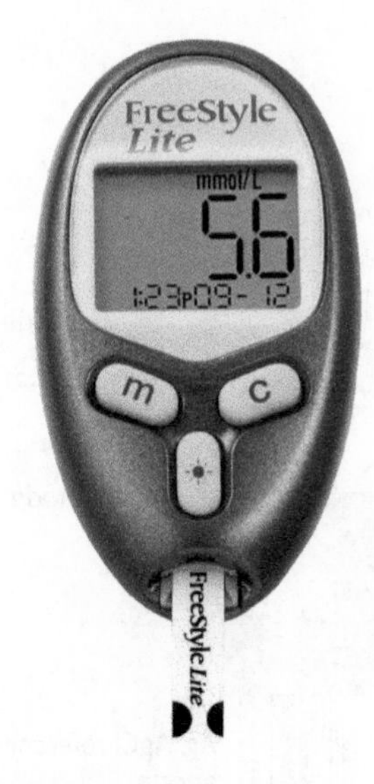

FIGURE 8.9
FreeStyle Lite blood glucose meter from Abbott Diabetes Care. It is about the size of a credit card and weighs only 40 g. The meter features a large and easy to read LED display with a backlight for night use. The photograph shows a test strip inserted into the meter, which is discarded after use. Photograph supplied courtesy of Abbott Diabetes Care.

8.5 Biosensors

Biosensors may be defined as self-contained integrated devices that incorporate a biological recognition element combined with an electrochemical transducer capable of converting its biological or biochemical signal or response into a quantifiable electrical signal. In the majority of biosensors, the biological component is an **immobilized enzyme** that is attached to a membrane or embedded within a structure that holds it in intimate contact with the transducer. This generally allows the enzyme to be re-used and so reduces operating costs.

Much of the technological development of biosensors has been stimulated by the need to determine the concentrations of glucose in clinical samples. In 2000 the World Health Organization estimated the number of people suffering from diabetes worldwide to be 171 million, and predicted this figure would double by 2030. Thus, many companies have made significant investments in research and development programmes that have led to the production of a wide variety of biosensor devices. Within clinical laboratories, a range of high-throughput laboratory analysers are available to assist in the routine analysis of blood glucose and the diagnosis of diabetes. Small hand-held devices are also readily available, which enable diabetic patients to monitor regularly their own blood glucose concentrations (Figure 8.9), and so make any necessary changes in diet or influence their decision to administer insulin as appropriate. Indeed, the *availability* of such portable instruments has been a major influence in the trend towards point-of-care testing within the healthcare sector.

Cross reference

Remember, you can learn more about point-of-care testing in Chapter 16.

8.5.1 Biosensor design and operation

In 1962 Leland Clark (of oxygen electrode fame, as described in Section 8.4) first used the term 'enzyme electrode' to describe a device in which a traditional electrode could be modified to respond to other materials by the inclusion of a nearby layer of enzyme. Clark's ideas became commercial reality in 1975 with the successful launch of the Yellow Springs Instruments model 23A glucose analyser. This device incorporated glucose oxidase, which catalyses the reaction:

$$\text{Glucose} + O_2 \rightarrow \text{gluconic acid} + H_2O_2$$

The device also contained an amperometric peroxide-sensitive electrode (which we described in Section 8.4) to measure the amount of H_2O_2 produced by the reaction. Within the device, the rate of formation of H_2O_2 is a measure of the rate of reaction between glucose and oxygen, which, in turn, depends on the concentration of glucose in solution. Thus, the device can be used to estimate the concentration of glucose in samples.

In enzyme-catalysed reactions, the relationship between substrate concentration, in this case glucose, and reaction rate is not a simple linear one but is described by the **Michaelis–Menten equation**. This is true for the glucose oxidase found within a biosensor. The rate of reaction is only directly proportional to the concentration of glucose when the glucose concentration is low relative to that of the enzyme. However, at higher concentrations of glucose, the concentration of enzyme is the limiting factor, and the change in rate becomes hyperbolic. You can see the Michaelis–Menten curve in Figure 8.10. We can engineer a more linear relationship over a greater range of glucose concentrations by ensuring that the enzyme is enclosed behind or within a membrane through which the glucose must diffuse before it reacts with the enzyme. The effect of the membrane is thus to keep the glucose concentration low compared with that of the enzyme. In this way, the system is diffusionally, rather than kinetically, limited and the response of the biosensor is then more linearly related to the concentration of glucose in solution.

The sensing probe of a device of this type comprises a peroxide-sensitive amperometric electrode covered by a three layered membrane. The first layer is made of porous polycarbonate and limits the diffusion of glucose into the second layer, which contains immobilized glucose oxidase, thereby preventing the reaction from becoming kinetically limited. The innermost, third layer, is of cellulose acetate, which allows only small molecules, such as hydrogen peroxide, to reach the electrode. This

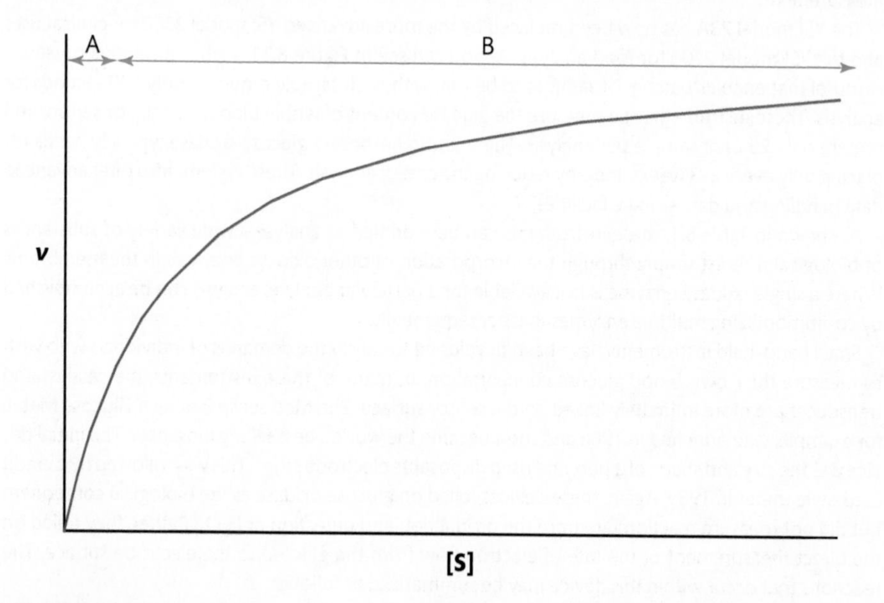

FIGURE 8.10

Michaelis–Menten type response where at low substrate concentrations (Area A) the rate of reaction (*v*) is directly proportional to the substrate concentration [S], whilst at higher substrate concentrations (Area B) the concentration of enzyme becomes the limiting factor such that further increases in [S] have a lesser effect, and eventually the enzyme becomes saturated with substrate.

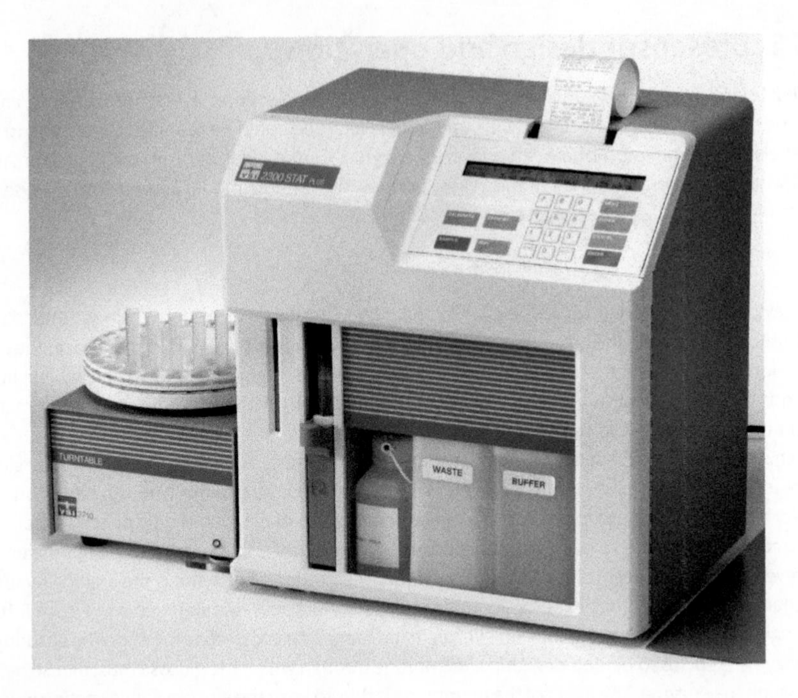

FIGURE 8.11

Model 2300 STAT Plus glucose and lactate analyser. The figure shows the 24-position turntable, which allows batches of samples to be analysed. The buffer reservoir and waste container are both conveniently contained within the analyser. Photograph supplied courtesy of Yellow Springs Instruments Life Sciences.

layer eliminates many larger electroactive compounds that could interfere with the amperometric measurement.

The YSI model 23A has now been replaced by the more advanced YSI model 2300 for clinical uses and the YSI model 2700 for food analysis. As you can see in Figure 8.11, such instruments possess a carousel that enables batches of samples to be run, with each sample requiring only 100 seconds for analysis. These instruments can measure the glucose content of whole blood, plasma, or serum, and require only 25 μl of sample per analysis. The membrane-bound glucose oxidase typically needs replacing only every 2–3 weeks, thereby reducing the cost of analysis. These systems also offer advanced data handling and data storage facilities.

As shown in Table 8.1, these instruments can be modified to analyse a wide variety of substances of biological interest simply through the incorporation of other oxidase enzymes in the membrane. Where a single oxidase enzyme is not available for a particular analyte, analysis can be accomplished by co-immobilizing multiple enzymes that act sequentially.

Small hand-held instruments have been developed to satisfy the demands of individuals who wish to measure their own blood glucose concentration. In many of these instruments, the catalyst and transducer are more intimately linked on the sensor surface. The Medisense Exactech Glucose Meter, for example, was launched in 1986 and soon became the world's best selling biosensor. The initial device was the size and shape of a pen, and used disposable electrode strips. This was followed by a credit card-style meter in 1989. Again, these devices relied on glucose oxidase as the biological component but did not measure reaction rate from the production and detection of H_2O_2. Rather, they relied on the direct measurement of the rate of electron flow from the glucose to the electrode surface. The reactions that occur within this device may be summarized as follows:

$$\text{Glucose} + \text{GOx-FAD} \rightarrow \text{gluconic acid} + \text{GOx-FADH}_2$$

$$\text{GOx-FADH}_2 + \text{Mediator}_{(\text{oxidized})} \rightarrow \text{GOx-FAD} + \text{Mediator}_{(\text{reduced})}$$

and at the electrode surface:

$$\text{Mediator}_{(\text{reduced})} \rightarrow \text{Mediator}_{(\text{oxidized})} + e^-$$

TABLE 8.1 Compositions of enzyme membranes available for analysers with H_2O_2 measuring probes as transducers

Analyte	Enzyme	Reaction catalysed
Alcohol	Alcohol oxidase	Ethanol + O_2 → acetaldehyde + H_2O_2
Glucose	Glucose oxidase	β-D-glucose + O_2 → gluconic acid + H_2O_2
Lactic acid	Lactate oxidase	L-lactate + O_2 → pyruvate + H_2O_2
Lactose	Galactose oxidase	Lactose + O_2 → galactose dialdehyde derivative + H_2O_2
Starch	Amyloglucosidase	Starch + H_2O → β-D-glucose
	Glucose oxidase	β-D-glucose + O_2 → gluconic acid + H_2O_2
Sucrose	Invertase	Sucrose + H_2O → α-D-glucose + fructose
	Mutarotase	α-D-glucose → β-D-glucose
	Glucose oxidase	β-D-glucose + O_2 → H_2O_2 + gluconic acid + H_2O_2

where GOx-FAD represents the flavin adenine dinucleotide (FAD) redox centre of glucose oxidase in its oxidized form, and GOx-$FADH_2$ the reduced form. Within the Medisense device, the **mediator** is ferrocene, although other manufacturers have produced similar devices using alternative mediator technologies. In essence, electrons are stripped from the glucose and then pass from the enzyme to the mediator, which then donates them to the working electrode surface resulting in the generation of an electrical current directly proportional to the rate of oxidation of glucose and hence proportional to the glucose concentration in the sample. As well as a working and reference electrode, later test strips also incorporated a third electrode, known as a background compensation electrode. This is not coated with glucose oxidase and thus measured the non-specific current generated by interfering materials present in the sample. The current generated by this electrode is subtracted from that of the working electrode to produce a response due to glucose alone. Unlike instruments of the Yellow Springs type discussed earlier, these devices do not have any requirement for oxygen. This is advantageous, as the concentration of oxygen in blood can vary, and in some cases may become limiting, such that the glucose measurement is affected.

Medisense, whose only product was their blood glucose meter, was bought by Abbott Diagnostics in 1996, and Abbott-branded devices continued to use and develop this technology for some time. In 1999, however, Therasense marketed a glucose meter that represented the next generation of sensing technology, and integrated the enzyme even more closely with the electrode. Originally developed by Adam Heller at the University of Texas in the 1990s, **wired enzyme electrodes** did not rely on a *soluble* mediator, such as the ferrocene used in the Medisense meter. Rather, the enzyme is immobilized in an osmium-based polyvinyl imidazole hydrogel in which the electroactive osmium groups are firmly held in place within the polymer, and cannot participate in molecular diffusion. Thus, the electrons are not passed from enzyme to electrode through a soluble mediator that moves by diffusion, but by a series of fixed electroactive osmium centres that shuttle the electrons onwards, often called *electron hopping*.

The FreeStyle Lite meter produced by Abbott Diabetes Care and shown in Figure 8.9 incorporates wired enzyme technology. Devices of this type are highly amenable to miniaturization: the Abbott device shown in Figure 8.9 is only the size of a credit card. Its size is largely limited by the dimensions required for a display that can comfortably be read by a diabetic patient, who might well have some visual impairment due to their condition. The electrochemistry is incorporated into a test strip only a few centimetres in length that is inserted into the meter as you can see in Figure 8.9. A sample of blood is obtained using a sterile lancet or similar pricking device and placed on the test strip. Only 0.3 μL blood is needed for analysis, equating to a drop the size of a pin head. Once loaded with sample, the device is able to measure the glucose concentration in approximately 5 seconds and can store the results of up to 400 analyses for comparative purposes. After a single use the test strip is discarded.

You might wonder why such glucose test strips are used once only when the enzyme is immobilized and could possibly be re-used. Re-using the enzyme would require that the user or the device itself be effectively washed between samples to ensure that results are not affected by materials remaining from previous samples. Given that these devices are developed to be used with little training by the general public and in a discrete fashion, it is clear that avoiding such washing procedures by incorporating a single use test strip has great appeal. Further, the test strip is the only part of the device that

Cross reference
Health and safety are discussed in Chapter 4.

is exposed to the blood sample and this can be discarded immediately after use. Thus, it is easier to maintain hygiene and prevent possible contamination from blood-borne diseases such as AIDS and hepatitis. Even so, when using these meters, the test strips together with the lancets used to obtain the sample should be disposed of in accordance with relevant infection control and health and safety guidelines. These may vary slightly from region to region, but, in general, lancets and similar pricking devices need to be disposed of in a properly designed sharps container (available from hospitals and pharmacists), while the test strips should be disposed of as clinical waste.

Cross reference
Measuring renal function is described in Chapter 3 of *Clinical Biochemistry*.

Single measurement devices, as described above, therefore play a key role in both the diagnosis and subsequent management of diabetes. Similar devices are available for the analysis of a number of other clinically important analytes, for example glucose and lactate. Concentrations of lactate in blood may be measured using biosensors to detect the high levels of lactate that typically accompany severe sepsis or septic shock, while creatinine and urea biosensors are used to monitor renal function.

Continuous measuring devices are also becoming available and may well revolutionize the control of certain disease conditions. With respect to diabetes, devices such as the FreeStyle Navigator system from Abbott use the same wired enzyme technology as described earlier but now incorporate this technology into a tiny filament about the diameter of a thin hypodermic needle. This is inserted approximately 5 mm under the skin to measure the glucose level in the interstitial fluid that flows between the cells. The device is designed to remain in place for up to 5 days, during which time it can measure glucose concentration every minute. A wireless transmitter sends the glucose readings to a separate receiver anywhere within a 10-foot range, and this can then issue an early-warning alarm to alert the user to a falling or rising glucose level in time for appropriate action to be taken to avoid a hypo- or hyperglycaemic episode.

All electrochemical techniques, ranging from the earliest pH electrodes to the most advanced biosensors, share common appeal to the biomedical scientist. Electrochemical devices are often small and their portability is an advantage to those who wish to make analyses at the point of care. Such devices are also quite robust and cheap, relative to other analytical technologies, and require limited use of 'wet reagents' in their operation. The simplicity of the devices often means that they can be used with little training, and their selectivity and sensitivity are sufficient for many, although certainly not all, clinical purposes. For these reasons, electrochemical devices are frequently considered to be one of the real workhorses of the clinical laboratory.

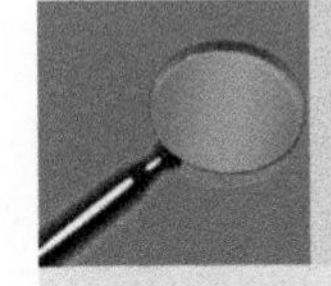

CASE STUDY 8.1 Measuring and interpreting blood lactate level

Lactic acid is the end point of the anaerobic breakdown of glucose and is released into the blood when cells are oxidizing glucose under hypoxic (low oxygen) conditions. As such, blood lactate level is a useful indicator of the oxygen status of tissues. Lactic acid exists as two distinct optical isomers, L-lactate and D-lactate. L-lactate is the most commonly measured level, as it is the form produced in human metabolism; D-lactate is a by-product of bacterial metabolism in the intestinal tract and may accumulate in patients after a gastric bypass or small-bowel resection (short-bowel syndrome).

(a) Philippa, a reasonably fit and healthy 29-year-old, is undergoing training for the London marathon. At the beginning of the training regime she has blood samples taken while exercising on a treadmill; these were analysed immediately with a portable lactate biosensor. After the training regime Philippa repeats the procedure. Results are presented in Figure 8.12. What conclusions can be drawn as to the effectiveness of the training regime?

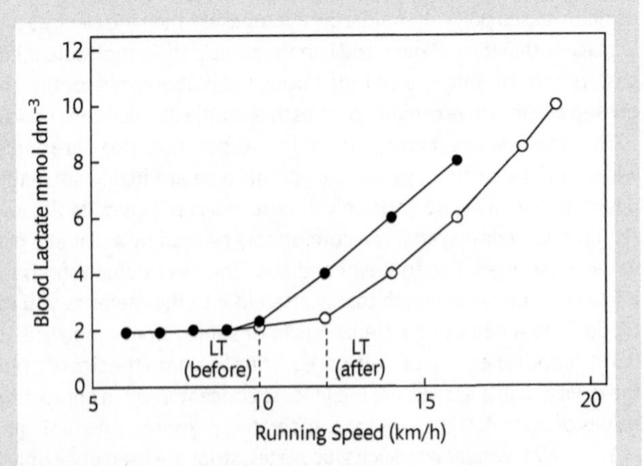

FIGURE 8.12
Philippa's blood lactate concentration when running at various speeds. Closed symbols prior to training regime; open symbols after training regime. LT is the lactate threshold.

(b) Matt has felt unwell all day and goes to hospital reporting a high temperature (40°C) and tachycardia (heart rate 100 beats per minute). A blood sample is taken and sent to the laboratories for analysis using a clinical lactate analyser (see Figure 8.11). Results reveal a blood lactate level of 4.0 mmol dm^{-3}. Is this a cause for concern and what is the likely cause of the lactate elevation?

(c) Why is Matt's blood lactate level of 4.0 mmol dm^{-3} of more concern than Philippa's level of nearly 10 mmol dm^{-3} while training?

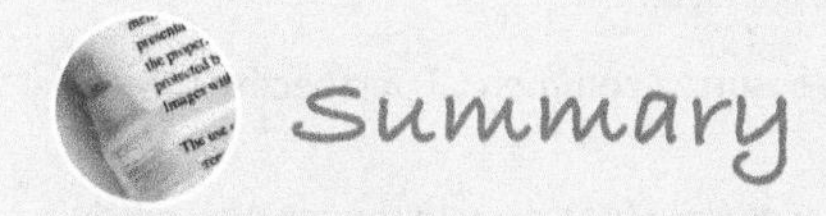

Summary

- All electrochemical techniques involve the use of two electrodes and the measurement of the potential difference between them (potentiometry) or the current that flows between them (voltammetry).
- pH, pCO_2, and ISEs are all potentiometric devices.
- Potentiometry involves the generation of a spontaneous potential that is measured against that produced by a reference electrode as a potential difference.
- The behaviour of these devices is described by the Nernst equation.
- The minimum detection limit of most ISEs is 10^{-6} mol dm^{-3} at best.
- ISFETs can be used to construct miniaturized solid-state potentiometric devices.
- In voltammetric techniques, electrochemical reactions are not spontaneous, but are forced to occur by an imposed external voltage.
- These advanced techniques can be used to both identify and quantify analytes.
- The oxygen electrode has widespread applications in medicine and clinical analysis.
- The hydrogen peroxide electrode is used in about 65% of biosensors.
- Biosensors incorporate a biological component for specific recognition, together with a transducer capable of generating an electrical output.
- Biosensors are widely used in the analysis of glucose concentrations for the diagnosis and management of diabetes, and also have widespread application in the clinical laboratory and in point-of-care testing.

Further reading

- **Brett CMA, Brett AMO. *Electroanalysis. Oxford Chemistry Primers Series, Number 64.* Oxford: OUP, 1998.**

  A concise and well-written introductory text providing a good 'next step' in your reading around the subject. Particularly useful in its coverage of more advanced voltammetric techniques.

- **International Union of Pure and Applied Chemistry. Classification and nomenclature of electroanalytical techniques (rules approved 1975). *Pure and Applied Chemistry*, 1976; 45: 81–97.**

  A well-accepted scheme of classifying electrochemical techniques. Still awaiting an update by IUPAC, hence continues to be current.

- **Luong JHT, Male KB, Glennon JD. Biosensor technology: technology push versus market pull. *Biotechnology Advance*, 2008; 26: 492–500.**

  A review that describes the commercial aspects relevant to the production of and demand for biosensors. The review also describes the current commercial activities of most of the major companies producing such devices.

- **McGrath MJ, Scanaill CN. *Sensor Technologies: Healthcare, Wellness and Environmental Applications.* New York: Apress Media, 2014.**

  Covers sensor technologies and their clinical applications, together with broader applications relevant to wellness, fitness and lifestyle, and the environment. Deals with social and ethical aspects of data security and sharing, and considers possibilities that will occur as sensor networks, cloud-based data storage, and 'big data' analytics merge together.

- **Newman JD, Turner APF. Home blood glucose biosensors: a commercial perspective. *Biosensors and Bioelectronics*, 2005; 20: 2435–53.**

  A thorough review of the commercial development of glucose biosensors, although now missing some of the most recent developments in continuous monitoring.

- **Pretsch E. The new wave of ion-selective electrodes. *Trends in Analytical Chemistry*, 2007; 26: 46–61.**

  An interesting review revealing the significantly enhanced sensitivity and selectivity of modern ISEs.

- **Riddle P. pH meters and their electrodes: calibration, maintenance and use. *The Biomedical Scientist*, 2015; 59: 246–8**

  A concise review of pH meters and their use in a biomedical laboratory.

- **Vashist SK, Zheng D, Al-Rubeaan K, Luong JHT, Fwu-Shan Sheu F-S. Technology behind commercial devices for blood glucose monitoring in diabetes management: a review. *Analytica Chimica Acta* 2011; 703: 124–36.**

  A useful article describing the technology behind a wide range of commercially available glucose-sensing devices.

- **Yoon J-Y. *Introduction to Biosensors: From Electric Circuits to Immunosensors.* New York: Springer, 2013.**

  Covers the underlying circuitry of sensors as well as sensing approaches; providing an applications-based approach to the study of biosensors through hands-on labs, towards the end result of building an antibody-based immunosensor from scratch.

Useful Websites

- www.freepatentsonline.com. A free and searchable database of patents. Very useful for investigating recent developments in sensors and similar devices.

Useful Journals

- *Analyst*, Royal Society of Chemistry.

  The home of premier fundamental discoveries, inventions, and applications in the analytical and bioanalytical sciences.

- *Biosensors and Bioelectronics*, Elsevier.

  The principal international journal devoted to research, design development, and application of biosensors and bioelectronics.

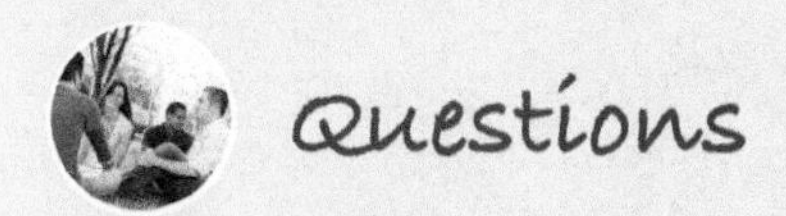

Questions

1. If an electrode is described as an ISE, what does the acronym ISE mean?
   (a) Ion-specific electrode
   (b) Ion-selective electrode
   (c) Ion-sensitive electrode
   (d) Ion-sensing electrode
   (e) Ion standard electrode.
2. Which of the following occur within an electrochemical analytical device?
   (a) The anode always carries a negative charge
   (b) The anode always carries a positive charge
   (c) The anode is the electrode at which the oxidation reactions occur
   (d) The anode is the electrode at which the reduction reactions occur
   (e) None of the above.
3. Which of the following statements are TRUE or FALSE?
   (a) All biosensors contain an enzyme able to recognize the analyte and a transducer that converts the response into an electrical signal
   (b) In simple direct current voltammetry, the limiting current is the highest one that will flow when an analyte is being reduced at the electrode surface.
4. Given that the total potential, *E* (in volts) developed between the working and reference electrodes of an ISE is given by the Nernst equation:

$$E = E_{constant} + 2.303RT / nF \log_{10} C$$

   (a) Calculate the potential developed by a pH electrode and a Na^+ ISE at 37°C.
   (b) Is the accuracy of analysis of these two devices likely to be the same?
5. While your main interest in electrochemistry might be for the analyses of clinical samples, such devices do have many other uses. List some other analytical situations in which these devices might be used, and consider why electrochemical techniques might be favoured in these particular situations.
6. Section 8.3 gives general information on pH electrodes. Having read this section, locate the manufacturers' instructions for the pH electrodes in your laboratory.
   (a) What type of electrodes are you using, for example, combination electrodes, gel electrodes, or double junction reference electrodes?
   (b) Do they all have the same mode of operation, that is, calibration, cleaning, and storage procedures?
7. Section 8.5 describes the Yellow Springs Instruments glucose biosensor in which glucose is measured using the enzyme glucose oxidase and a peroxide-sensitive electrode as the transducer. Suggest two other types of electrode transducers that could, in theory, be used to measure the rate of this reaction and form the basis of a glucose biosensor. Why is a device measuring peroxide preferable to either of these alternative instruments?

Answers to self-check and end-of-chapter questions are available at the end of the book.

9

Spectroscopy

Qiuyu Wang, Helen Montgomery, Nessar Ahmed, and Chris Smith

Learning objectives

After studying this chapter, you should be able to:

- Define the term spectroscopy
- Describe the electromagnetic spectrum
- Discuss the general theory of spectroscopy
- List the major spectroscopic techniques
- Discuss some of the uses of spectroscopy in the biomedical sciences

Introduction

Spectroscopy is the study of the interactions between electromagnetic radiation and matter. Atoms, molecules, and ions can scatter, absorb, or emit electromagnetic radiation in ways that depend upon the chemical composition and structure of the substance. Thus, the pattern of absorption or emission is characteristic for a given substance. This means that spectroscopy can be used as an analytical technique to identify which elements are present in a sample, determine how much of an analyte is present, and even be used to elucidate the structures of molecules. It is a non-invasive technique and samples are not destroyed during spectroscopic analyses. These characteristics give spectroscopic analysis a broad range of applications in the biomedical and biological sciences.

9.1 Radiation and the electromagnetic spectrum

Radiation is the emission or transfer of energy (E) as electromagnetic waves or particles. The particles are often called **photons**. Electromagnetic radiation, such as light, is propagated through space as electric and magnetic fields, which are mutually perpendicular to each other. You can see this expressed diagrammatically in Figure 9.1.

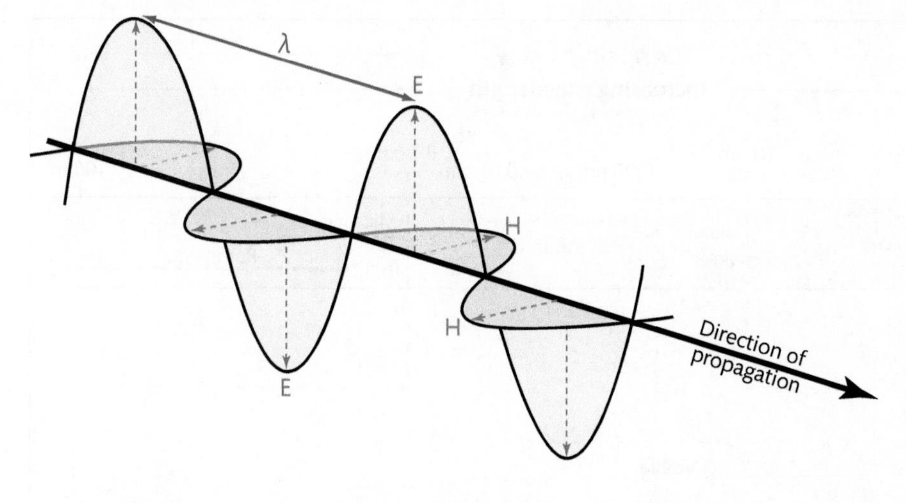

FIGURE 9.1
Propagation of electromagnetic radiation. See main text for explanation.

Two characteristic features of any sort of wave are its wavelength (λ) and frequency (*v*). You can see in Figure 9.1 that the wavelength is the distance between any two adjacent peaks. The frequency is the number of peaks passing a fixed point in a given time, which is normally measured in seconds (s). Wavelength and frequency are interrelated by the velocity of light (*c*) in the following way:

$$\lambda = c/v$$

The energy (*E*) of any type of electromagnetic radiation is described as:

$$E = hc/\lambda = hv$$

where *h* is Planck's constant. The use of these simple equations means that the frequency, wavelength, and energy of electromagnetic waves can be easily interconverted, although care must be taken over units (see Box 9.1).

Thus, the energy of a wave is directly proportional to its frequency: the higher the frequency, the greater the energy. However, the energy is inversely proportional to its wavelength: the shorter the wavelength, the greater the energy. The correlation between wavelength and energy resembles the energy of a spring. Think of squeezing a spring: the more you compress it, which gives it energy, the greater its tendency to jump back to its original, resting length.

The full range of electromagnetic radiation is often called the **electromagnetic radiation spectrum**. It is often subdivided or classified into a number of types of radiation depending upon their wavelengths (or frequencies), as shown in Figure 9.2.

BOX 9.1 Spectroscopy and units

Wavelengths (λ), frequencies (*v*), and the Planck constant (*h*) can all be expressed in one of a number of units. Wavelengths are usually given in nanometres (nm), but some older texts use the angstrom (Å), which is equal to 10^{-10} m or 0.1 nm. Frequencies are normally given in cycles per second or hertz (Hz), which is reciprocal seconds (s^{-1}). However, frequencies may also be described as a wave number (*v'*), which is often expressed as reciprocal centimetres (cm^{-1}), simply because the numbers become easier to manage. Planck's constant has a value of 6.63×10^{-34} J s^{-1}. Again, it can be expressed in alternative units including 6.63×10^{-27} erg s^{-1} and 4.135×10^{-15} eV s^{-1} (eV, an electron volt, is the energy acquired by a single electron accelerating through a potential difference of 1 volt and is equal to about 1.6×10^{-19} J).

The velocity of light (*c*) is 2.998×10^{8} m s^{-1}.

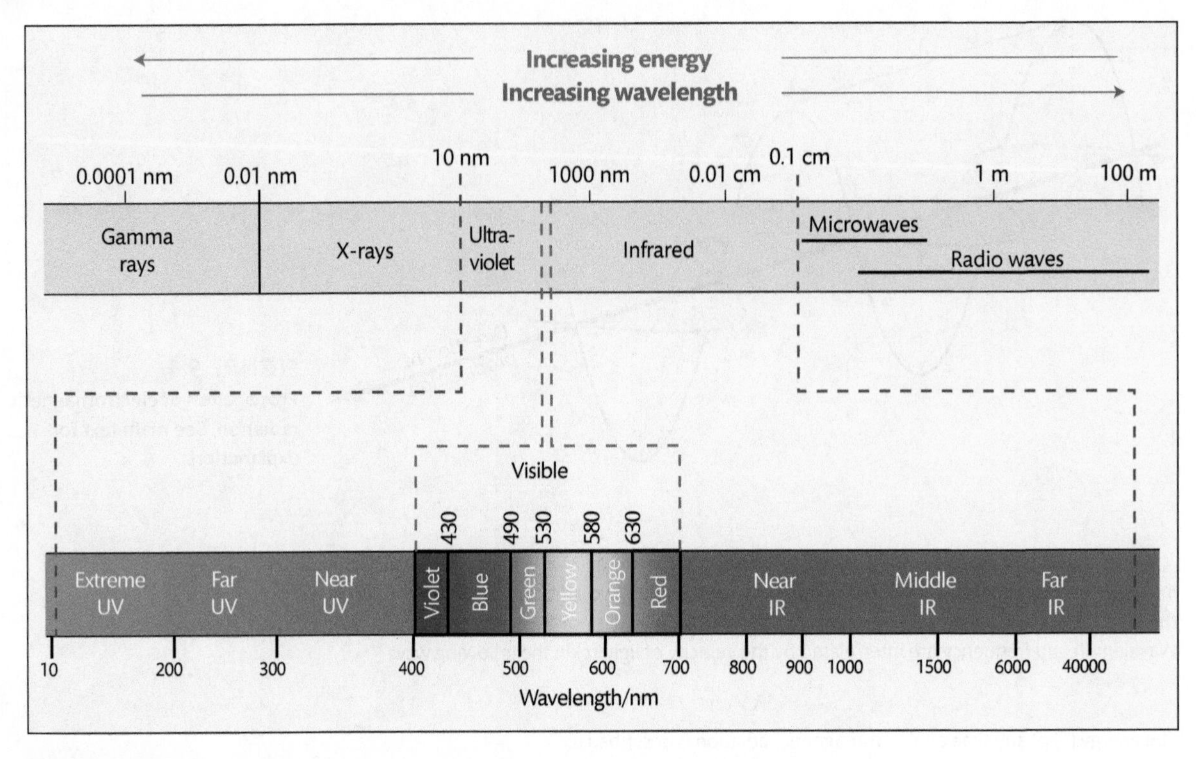

FIGURE 9.2
A portion of the electromagnetic radiation spectrum. UV: ultraviolet; IR: infrared

9.2 Interactions between radiation and matter

Cross reference

You can read about gel electrophoresis, analytical gel filtration, and analytical micro-ultracentrifuges in Chapters 12, 11, and 10 respectively.

When radiation of a particular wavelength interacts with matter, the radiation may be *scattered* or it may be *absorbed*. Scattering, often referred to as Rayleigh scattering, occurs when a photon collides with an atom or molecule, and its direction of propagation is changed. However, its wavelength and frequency, and therefore energy, are not usually altered. Turbidity and nephelometry depend on the scattering of light and are briefly described in Section 9.6. Light scattering can be used to determine the M_r of macromolecules in solution or the sizes of larger suspended particles, such as those of viruses. This technique has been superseded by simpler, more convenient methods, such as gel electrophoresis and analytical gel filtration or by using analytical microultracentrifuges.

The scattering of radiation is also the basis of a number of techniques used to investigate the structures of particles of clinical and biological interest. These techniques include X-ray diffraction, electron microscopy, and neutron scattering. A discussion of these techniques is, however, outside the scope of this chapter.

If a photon of radiation is absorbed by matter, energy is transferred from it to the atom or molecule. This means that the electronic state of the atoms or molecules is affected; their energy is increased from a ground state to an excited one and the atom or molecule is described as being excited. The atom, molecule, or even part of a molecule that becomes excited by absorption is often called a **chromophore**. The excitation energy is usually dissipated as heat when the excited atom or molecule collides with, for example, a solvent molecule. However, some molecules can lose the energy by emitting a photon of lower energy in a process called photoluminescence. If the photon is emitted virtually instantaneously, this is called **fluorescence**. Phosphorescence differs from fluorescence in that the phosphorescent material continues to emit light after removal of the source of excitation, although this slowly diminishes. By contrast, fluorescent materials fluoresce only while excited.

SELF-CHECK 9.1

In fluorescence, which of the photons, absorbed or emitted, has the shorter wavelength?

9.3 Radiation, particles, and quanta

Electromagnetic radiation is often difficult to describe in general everyday terms. This is because it cannot be described solely in terms of the properties of waves, but neither does it possess only the properties of particles. Instead, it has attributes of both. Some of the ways in which radiation interacts with matter, for example the scattering of light, are most easily described by assuming it is a wave. However, other aspects, such as the absorption of light, are more easily envisaged if radiation is regarded as a particle.

The frequency of radiation is related to its energy content by Planck's constant (6.63×10^{-34} J s^{-1}), as described in Section 9.1. The introduction of Planck's constant means that a specific frequency or wavelength has a discrete and fixed amount of associated energy. Equally, when an atom or molecule absorb or emits energy, the amount of energy absorbed or released can only be of discrete or fixed amounts. These 'amounts' are called quanta (singular quantum); a quantum is the smallest individual quantity of energy. Hence, quanta occur in discrete, specific amounts, and not as a continuous range of values. Furthermore, atoms and molecules can possess only specific amounts of energy or be at certain 'energy levels'; so they, too, are quantized.

For a photon to be absorbed, its energy must exactly match the difference in energy between two energy levels in an atom or molecule. The discrete energy levels (quanta) that are associated with atoms and molecules are usually illustrated using an energy-level diagram like the one in Figure 9.3. If you examine this figure you can see that the ground state is indicated by the lower red line (I). However, in an excited state, the energy of an atom or molecule has increased by a fixed amount or quantum. One such excited state is represented by the upper red line (II). You can also see that within each of the electronic states [the ground (I) and an excited (II)], there are different possible vibrational energies, which are represented by the thinner blue-coloured lines. Rotational energies also occur at values between those for vibrations. Our diagram shows only a limited number of these as short green coloured lines; otherwise it would become too cluttered!

A change between any two energy levels is called a **transition**. The vertical arrow in Figure 9.3 shows one of the many possible electronic transitions. You should note that a transition can occur only if the atom or molecule absorbs an amount of energy exactly equal to that between the two relevant states. This is indicated on the energy axis as ΔE. However, transitions can occur not only between electronic states, but also between the various vibrational and rotational states. The sizes of the energy changes decrease from electronic through vibrational to rotational and therefore the wavelengths of the photons required to promote them increase.

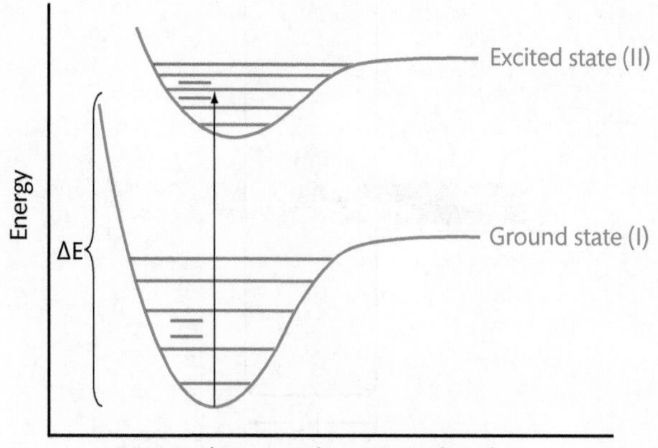

FIGURE 9.3
Schematic energy-level diagram from the ground state (I) to the first excited state (II). The arrow shows a transition from the ground state to the third vibrational level of the first excited state, which requires an input of energy equal to ΔE.

SELF-CHECK 9.2

Do the frequencies of photons (or electromagnetic radiation) increase or decrease with the order of the following transitions: rotational, vibrational, and electronic?

When an excited molecule returns to its ground state, energy is lost and the transition has been from a higher to lower energy level. The energy can be lost as photons, as in fluorescence or phosphorescence (Section 9.7). When radiation is not emitted, however, the transition is called non-radiative decay. Except in the cases of fluorescence or phosphorescence, radiation is only emitted by materials at temperatures that are so high they would normally degrade molecules of clinical or biological interest. Hence, in the biomedical sciences, spectrophotometry is virtually synonymous with measuring the *absorption* of radiation. Figure 9.2 indicates the relative energies associated with different wavelengths. Each of these is used with different types of spectroscopy. All of these different types of spectroscopy have different uses in clinical practice.

These include determining the absorbance, sometimes called optical density, of a substance or its absorption spectra. We will discuss both of these analytical approaches in the following sections.

9.4 Absorbance

Absorbance is a means of quantifying how much light of a particular wavelength is absorbed by a sample. In contrast, absorption spectra show the fraction of electromagnetic radiation absorbed by a material over a *range* of different wavelengths, frequencies, or wave numbers (see Box 9.1).

Spectroscopic techniques can be used to make quantitative measurements. For example, to determine the amount of a substance (the analyte) that is present in a sample. The absorbance of an analyte is proportional to its concentration (c) in a manner described by the Beer–Lambert law. This law describes the absorbance of **monochromatic** light, which is one of a single wavelength, as it passes through a *dilute* solution. The law states that the absorbance is directly proportional to the concentration of the solution and also to the length of the path that the light takes through that solution.

You can see an illustration of the law in Figure 9.4. Light of a specific wavelength and intensity I_0 is entering a sample of concentration c with a path length l or L. Light leaving it has an intensity of I. If the sample does not absorb or scatter any of the light, then I will equal I_0. If, however, some of the light is absorbed by the sample then I will be less intense than I_0, such that the transmitted light (T) is equal to I/I_0. This value is usually multiplied by 100 and expressed as a percentage value. The *logarithm* of 1/T or I_0/I (log (I_0/I)) is the absorbance (A).

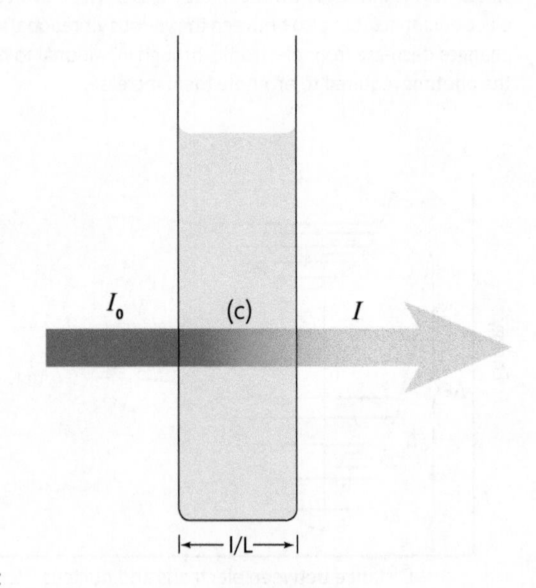

FIGURE 9.4
The absorbance of monochromatic light in accordance with Beer–Lamberts law.

The Beer–Lambert law is summarized by the expression:

$$\log (I_0 / I) = A = \varepsilon cl$$

where ε is often referred to as the molar extinction coefficient.

Absorbances are logarithmic values and so do not have units. In traditional, non-SI units, l is expressed in cm and the concentration in mol L^{-1} or M. This gives ε units of M^{-1} cm^{-1}. However, if SI units are used, then the equivalent Beer–Lambert law expression becomes:

$$\log (I_0 / I) = A = acL$$

where the constant (a) is called the absorptivity, c is in mol m^{-3}, and the path length (L) is in m. This gives absorptivity units of m^2 mol^{-1}. Whatever system of units is used, the generally observed range of transmittance is from 0 to 100%. Absorbance values then vary from 0 to 2.

The absorbance of a given substance will vary according to the wavelength of the light to which it is exposed. When using absorbance for quantitative analysis, however, the absorbance of a solution is normally determined at the wavelength at which the material being investigated absorbs most strongly, that is where the absorbance is at its greatest. This wavelength is often designated as A_λ. At a single wavelength, the absorbances of dilute solutions of a given analyte are directly proportional to the concentration of analyte. This is illustrated in Figure 9.5: note how the graph is only a straight line for lower concentrations, in accordance with the Beer–Lambert law.

Absorbances are usually measured with the solution contained in specialized transparent containers called **cuvettes**, which normally have a standard path length of 1 cm or 0.01 m. This length can be assumed unless you are told otherwise. Disposable cuvettes made of polymethacrylate and polystyrene are commercially available for use at wavelengths of 280–800 and 350–800 nm, respectively. They therefore suffer the disadvantage that they cannot be used to determine the absorbances of solutions of nucleic acids, which have λ_{max} at about 260 nm, and perform poorly with protein solutions that absorb maximally at about 280 nm. Reusable glass, quartz, and fused silica cuvettes are also manufactured. Again, glass cuvettes cannot be used in the ultraviolet (UV) range; their lowest usable wavelength is 340 nm. Cuvettes of quartz and fused silica are the most versatile and can be used throughout the UV and visible ranges, that is, from about 200 to 800 nm.

Given the relationship between absorbance and concentration and because many chemicals absorb light maximally at specific wavelengths, measuring the absorbance of a solution of analyte is extremely useful, as it provides a means of determining its concentrations in patient samples. In theory, if ε (or a) is known for a compound, then its concentration in a sample can be determined simply by measuring its absorbance under the same conditions for which ε (or a) was found. In practice, however, it is more usual to determine the absorbance of that sample and also for a known concentration or a series of concentrations of the analyte. The unknown concentration of an analyte can then be calculated from the ratios of the absorbances given that the concentration of the standard is known.

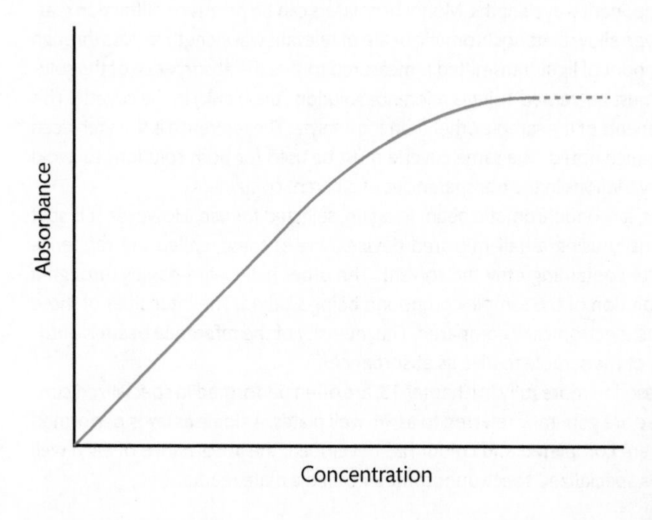

FIGURE 9.5
Graphical illustration of the Beer–Lambert law. See text for details.

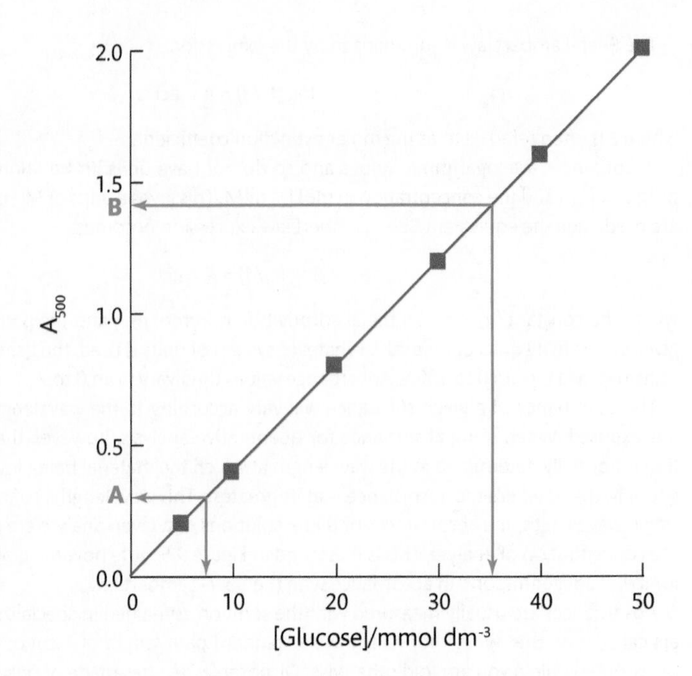

FIGURE 9.6

Standard or calibration curve, relating the concentration of glucose to the absorbance at a wavelength of 500 nm using a glucose oxidase-based test. Two serum samples, A and B, were analysed for glucose using the method at the same time. Sample A was from a healthy individual and B from a poorly controlled diabetic. The absorbances obtained for A and B were 0.30 and 1.40, which correspond to concentrations of 7.0 and 35.0 mmol dm^{-3}, respectively.

Alternatively, a graph or standard curve of absorbance against the known concentration can be plotted from which the unknown concentration can be obtained (Figure 9.6).

SELF-CHECK 9.3

Calculate the concentration of the analyte X, given that a tenfold dilution of the specimen gives an absorbance of 0.458 and a standard solution of the same analyte of concentration 50 μg cm^{-3} had an absorption of 0.625.

9.4.1 Spectrophotometers and colorimeters

Two types of spectrometers are typically encountered. These are single beam and double beam instruments. Both types have a visible light source and either type may have a deuterium lamp that produces UV light so that absorbance at shorter wavelengths can also be measured. You can see outline structures of each type of spectrophotometer in Figure 9.7 (a, b). In single beam instruments, a **monochromator** separates the beam into its component wavelengths. Monochromators can be prisms or diffraction gratings (Figure 9.7 (c)), but either type allow a monochromatic beam of relevant wavelength to pass through the sample in a cuvette. The amount of light transmitted is measured to give the absorbance of the solution. Single beam instruments must be 'zeroed' using a reference solution ('the blank') in the cuvette. This solution contains all the components of the sample other than the analyte. The reference is then replaced with the sample and its absorbance noted. The same cuvette must be used for both solutions to avoid absorbance errors arising from variations in the transparencies of different containers.

Online Resource Centre

To see an online video introducing spectrophotometry, log on to www.oxfordtextbooks.co.uk/orc/fbs

To see an online video demonstrating how the spectrophotometer works, log on to www.oxfordtextbooks.co.uk/orc/fbs

In double beam instruments, a monochromatic beam is, again, selected for use. However, it is split into two beams of equal intensity using a half-mirrored device. One of these, called the reference beam, passes through a cuvette containing only the solvent. The other half-beam passes through a second cuvette containing a solution of the sample compound being studied. The intensities of these light beams are determined and electronically compared. The intensity of the reference beam is automatically subtracted from that of the sample to give its absorbance.

Immunoassays, which we describe more fully in Chapter 13, are often performed in specialized containers that have 96 wells. These are generally referred to as 96-well plates. A single assay is performed in each well. When the assays are completed and colour has developed, the absorbance of each well in the plate is measured using a specialized spectrophotometer called a plate reader.

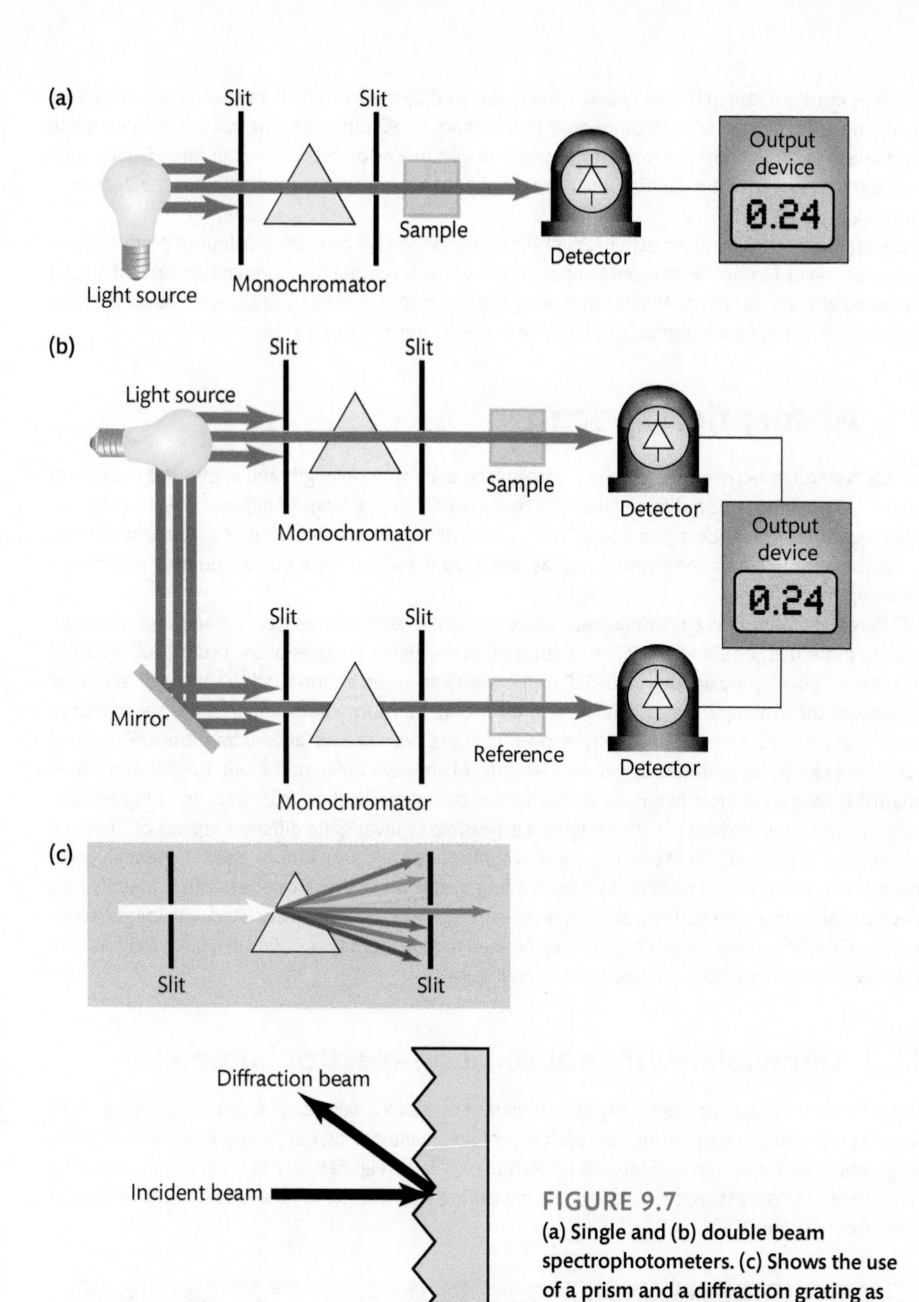

FIGURE 9.7
(a) Single and (b) double beam spectrophotometers. (c) Shows the use of a prism and a diffraction grating as monochromators. See text for details.

The term *colorimetry* is often used when the absorbance of visible, and therefore coloured, light is used. In this case, rather than using a spectrophotometer, which produces light with a limited range of wavelengths, a colorimeter is used. Colorimeters do not produce light of specific wavelengths but use a series of filters that produce narrow bands of light in the visible spectrum.

9.4.2 Clinical applications of measuring absorbances

Spectrophotometers are incorporated in automated analysers in hospital laboratories and used to measure a wide range of analytes colorimetrically. For example, serum creatinine concentrations are determined using the Jaffe's reaction. Creatinine is a breakdown product of creatine phosphate in muscle tissues and its concentration in serum is used to assess kidney function because it is produced

Cross reference

You can read more about automation in Chapter 15, 'Laboratory automation'. Chapters 1-3 of the companion text, *Clinical Biochemistry*, will also provide useful material.

at a fairly constant rate in the body and is freely excreted by the kidneys. Reference ranges for serum creatinine are typically 60–110 μmol dm^{-3} in men and 45-90 μmol dm^{-3} in women. If the filtering mechanism in the kidneys declines, for example owing to loss of nephrons as in renal failure, then concentrations of serum creatinine will increase above these ranges in proportion to the decline in function.

Creatinine reacts with picric acid in an alkaline solution to give an orange-coloured product. After an incubation of 15 minutes at room temperature to allow for colour development, the absorbance is measured at a wavelength of 490–500 nm using the spectrophotometer. The absorbance produced is proportional to the concentration of creatinine in the sample of serum.

9.5 Absorption spectra

An **absorption spectrum** is a graph of absorbance against wavelength and shows the fraction of incident electromagnetic radiation absorbed by a material over a range of *different* wavelengths, frequencies, or wave numbers (see Box 9.1). In a relatively short time, a suitable spectrophotometer can automatically determine absorbances, as described in section 9.4.2, but at a number of different wavelengths.

Different regions of the electromagnetic spectrum can be used to investigate different features of an atom or molecule. Look at Figure 9.2. You can see that the higher energies of UV (usually 200–400 nm) and visible lights (approximately 400–800 nm), collectively often abbreviated to UV-Vis or 'standard' absorption spectroscopy, are associated with electronic transformations. Infrared (IR) spectroscopy uses radiation of 2500–16,000 nm, whose lower energies promote vibrational transitions. Rotational transitions can occur with the longer wavelengths. Microwave radiation (about 10–300 mm wavelengths) is used in nuclear magnetic resonance spectroscopy (Section 9.8). Thus, by using electromagnetic radiation of different wavelengths it is possible to investigate different aspects of the same substance, and gradually build up a picture of its composition and structure. In essence, the absorption spectrum for a compound constitutes a type of 'fingerprint' for it, which can be useful in identifying a specific analyte in a sample. Thus, absorption spectra, particularly those associated with longer wavelengths of the IR and microwave regions, can be used to help identify specific analytes in samples. We will now consider a number of these techniques in turn.

9.5.1 Ultraviolet-visible absorption spectrophotometry

Many compounds absorb electromagnetic radiation of specific wavelengths. Many colourless compounds of clinical or biological interest, such as proteins, nucleic acids, purines, pyrimidines, and some drugs, vitamins, and amino acids absorb non-visible, UV light (Figure 9.2). This is of help in identifying those wavelengths most appropriate to use in measuring absorbances to determine the concentration of an analyte (Figure 9.8).

SELF-CHECK 9.4

A partial absorption spectrum of oxyhaemoglobin is given in Figure 9.8 (a). What would be suitable wavelength(s) to measure its concentration?

9.5.2 Infrared spectroscopy

IR spectroscopy is a useful analytical tool in identifying or partially identifying an analyte. In principle, it is not different from UV-Vis spectroscopy. However, the energies of photons in the IR region of the spectrum vary from approximately 0.25 to about 3.755 J mol^{-1}. Unlike UV and visible light, this is insufficient to produce electronic transitions. Rather, the absorption of lower-energy IR light excites covalently bonded atoms and organic functional groups in molecules, resulting in vibrational and rotational transitions. Such excitation is possible because the covalent bonds in molecules do not behave like rigid, non-elastic structures but more like springs that can be stretched and bent.

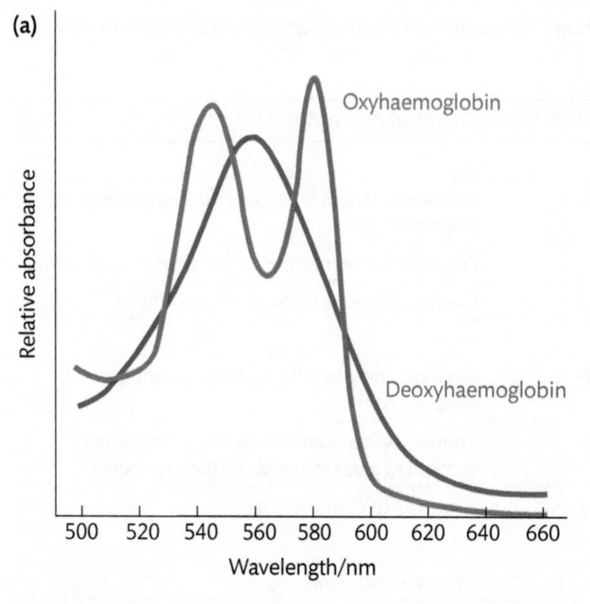

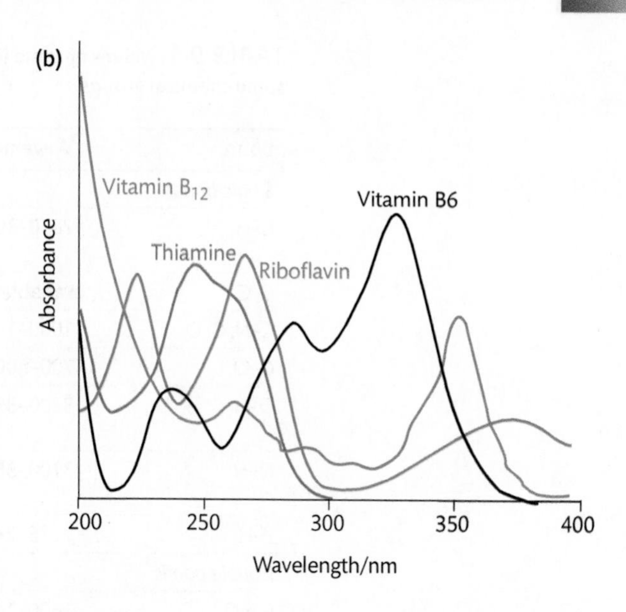

FIGURE 9.8
Ultraviolet-visible absorption spectra of (a) deoxyhaemoglobin and oxyhaemoglobin, and (b) for a number of vitamins.

Thus, molecules of virtually all organic compounds can show a large number of vibrational motions, which are characteristic of the atoms or groups of atoms present. This means IR spectroscopy is especially useful for analysing organic compounds because the different functional chemical groups present absorb IR at characteristic wavelengths. We have listed the characteristic absorption wave number values of some functional groups in Table 9.1. Thus, many of the functional groups in the molecules of an analyte can be relatively easily identified.

Absorption by these groups gives rise to seemingly rather complex spectra when compared with those of the UV–Vis range. You can see what we mean if you examine Figures 9.8 and 9.9. Look at the relative simplicity of the UV–Vis spectrum of the complex protein haemoglobin shown in Figure 9.8 (a) compared with the complex IR spectra of the structurally much simpler compounds shown in Figure 9.9. These spectra are typical of IR absorption spectra in that the absorption is expressed as percent transmittance and wavelengths are given as wave numbers (Box 9.1). However, the very complexity of IR spectra means they provide a 'fingerprint' of an analyte that can be extremely useful in identifying, or at least partially identifying, what it is.

IR spectra can be obtained from gaseous, liquid, or solid samples. Glass cuvettes cannot be used as glass absorbs IR light. Instead, samples are usually examined as thin films sandwiched between two polished salt plates, which are transparent to IR light. Liquids can be applied as films directly between the plates. Solid samples can be examined following their incorporation into thin potassium bromide discs prepared under high pressure. They can be mixed with a solvent and ground to a fine paste, which is spread between the plates. In some cases IR spectra can be obtained directly from finely powdered preparations of samples, although the resolution is sometimes poorer than using salt plates (Figure 9.10 (a) and (b).

SELF-CHECK 9.5

Identify the chemical group that is the one most likely to have produced the trough at 1690 cm^{-1} in the IR absorption spectrum of retinoic acid given in Figure 9.9. Table 9.1 will be of use!

Macromolecules, such as proteins, can dissolve in volatile solvents; the resulting solution can then be spread as a thin film on one of the plates. Evaporation of the solvent then leaves a thin film of protein molecules on the plate for analysis. Unfortunately, organic solvents tend to denature proteins.

TABLE 9.1 Values of some infrared radiation wavelengths characteristically absorbed by some chemical groups

Bond	Wave no.r/cm^{-1}	Chemical group/notes
Single bonds		
C–H	2850–3000	Saturated alkane; presence too numerous to be of diagnostic value
C–C	variable	Presence too numerous to be of diagnostic value
C–N; C–O	1000–1400	Overlap makes assigning signals difficult
C–Cl	700–800	
O–H	3200–3550	Alcohols and phenols; hydrogen bonding will broaden peaks
N–H	3100–3500	Amines; primary amines give several signals, secondary give one peak, tertiary no peaks
P–H	2275–2440	Phosphorus compounds
Double bonds		
C = C	1600–1675	Often a weak signal
C = N	1630–1690	Overlaps with other signals
C = O	1715–1750	Esters
C = O	1680–1740	Aldehydes
C = O	1665–1725	Ketones
C = O	1670–1720	Carboxylic acids
C = O	1630–1695	Amides
N = O	1300–1370 1450–1570	Nitro compounds
P = O	1140–1300	Phosphorus compounds
Triple bonds		
C ≡ C	2120–2260	Alkynes; weak bands
C ≡ N	2210–2260	Nitriles

If a solvent is used to dissolve solids, care must be exercised in its choice; the solvents chosen should not absorb IR light in regions of the spectrum that may be of analytical use. Hence, water cannot be used because it absorbs strongly in the 1600 and 3400 cm^{-1} regions, also it is normally present in high concentrations in samples and would dissolve the salt plates. Commonly used solvents include carbon tetrachloride, chloroform, and tetrachloroethene.

9.5.3 Raman spectroscopy

The difficulties associated with obtaining IR spectra in water can be overcome using Raman spectroscopy, which also probes the vibrational energy levels of atoms and molecules. You can see an outline of a Raman spectrophotometer in Figure 9.11.

The incident, monochromatic, light used in Raman spectroscopy is chosen so that its wavelength is not absorbed by the sample; hence, most of it is simply transmitted. However, a relatively small proportion of the light will be scattered and this can be viewed at an angle of 90° to the sample. It will be noted in Section 9.6 that scattering does not in general alter the frequency of the exciting light. However, if an exceedingly small proportion of the energy of an exciting photon is transferred

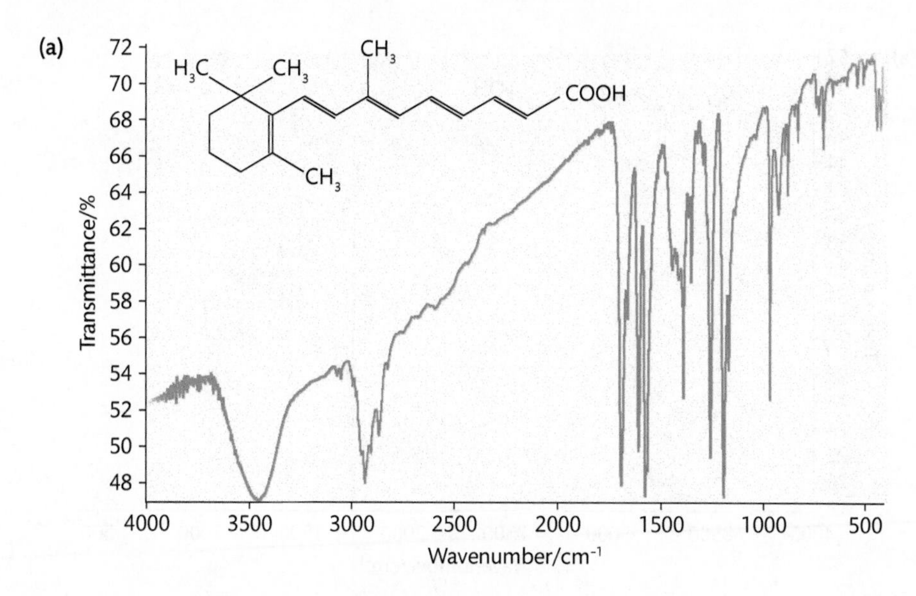

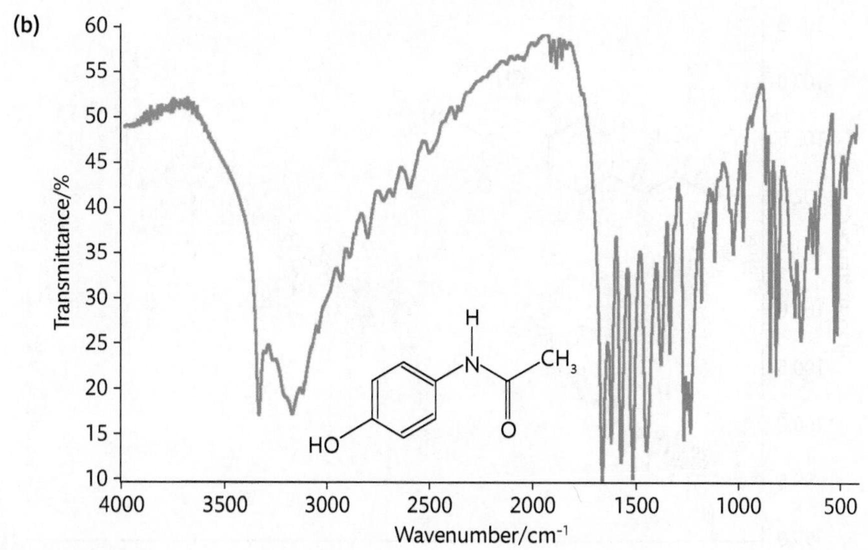

FIGURE 9.9

Infrared spectra of (a) retinoic acid and (b) paracetamol.

to a sample molecule, it can lead to a transition from one vibrational state to another. Hence, a small proportion of the scattered light will have its frequency decreased owing to its involvement in the vibrational transitions. The net result of such changes in frequencies means that a series of vibrational transitions can be observed as a collection of spectral lines in the scattered light collected at 90° to the sample. These spectral lines form the Raman spectrum.

Given that the information gained in Raman spectroscopy is from vibrational transitions, it is similar to that obtained by IR spectroscopy. However, it differs in that the information is independent of the frequency used to excite the atoms/molecules. Thus, any wavelength can be used to produce a Raman spectrum. However, because only a small proportion of the light is absorbed by vibrational transitions, a high-intensity light source is needed; typically a laser is used. Thus, Raman spectroscopy is carried out with a dedicated, specialized spectrophotometer. The sample must also be highly concentrated. In practice, a 2% solution is usually used, which can lead to problems with molecules aggregating or precipitating.

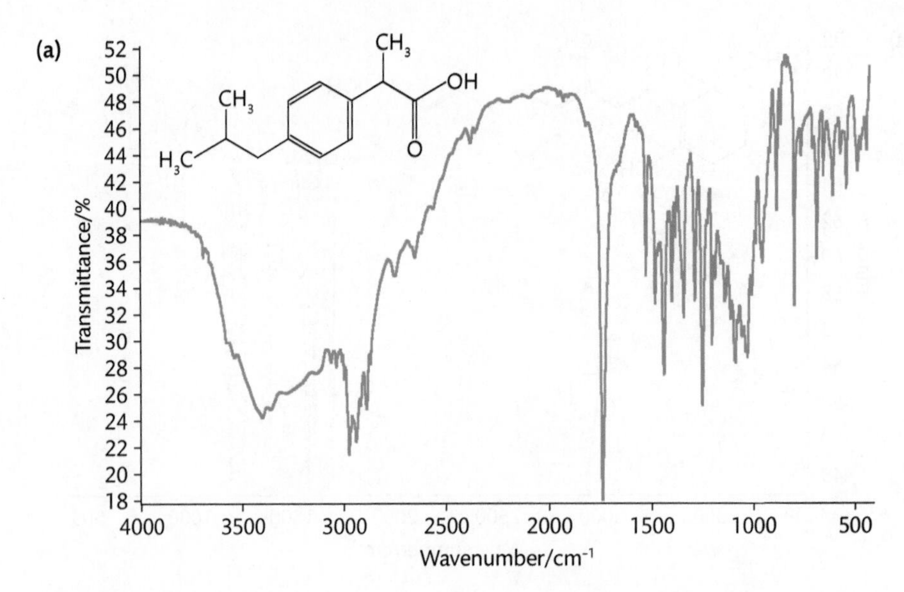

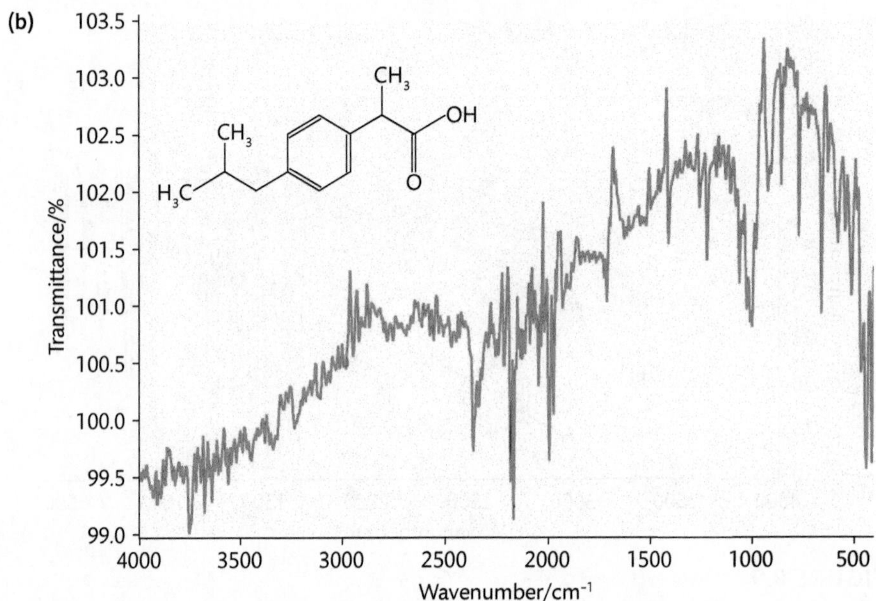

FIGURE 9.10
Infrared spectra of (a) ibuprofen in KBr plates and (b) obtained directly using a powdered tablet.

Clinical applications of infrared and Raman spectroscopy

IR and Raman spectroscopies, often coupled with gas liquid chromatography (GLC; Chapter 11), have been used to study drugs such as the various derivatives of penicillin and amphetamines. Amphetamines are a group of structurally similar molecules that are difficult to differentiate. However, amphetamines in samples can be identified using differences in their IR spectra using GLC–IR. IR spectroscopy is used to monitor pollutants in the environment, particularly harmful chemicals in the air. These include measurement of toxic gases, such as carbon monoxide in garages and formaldehyde produced during manufacture of plastics. There is considerable interest in the use of IR spectroscopy as a non-invasive technique to determine the concentrations of glucose in samples of blood from diabetic patients. The blood glucose is quantified by exposing a region of the skin to IR radiation and analysing the resultant spectrum.

Cross reference

You can read about chromatography in Chapter 11.

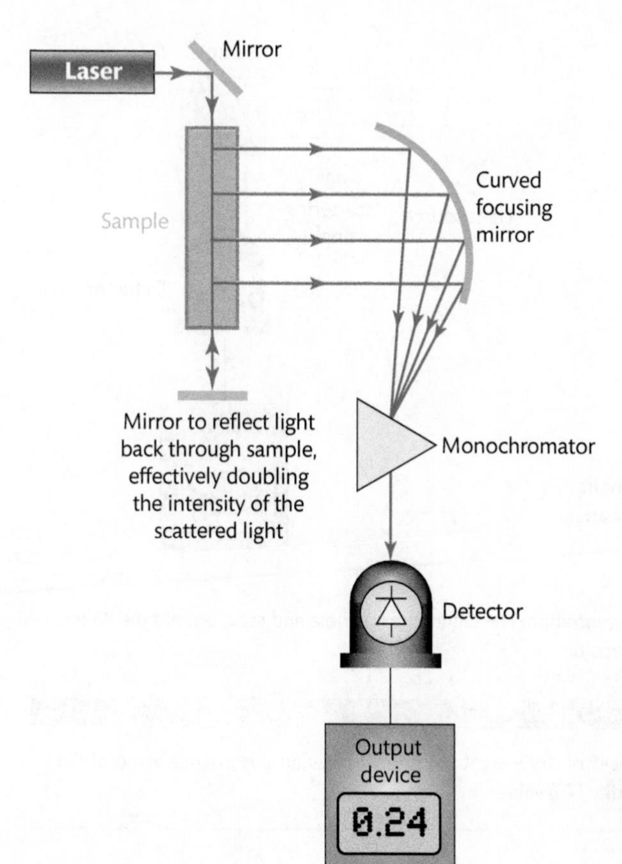

FIGURE 9.11
Outline of a Raman spectrophotometer. See main text for an explanation of its operation.

9.6 Light scattering methods

The two main experimental techniques encountered in the clinical laboratory based on the scattering of light are **turbidity** and **nephelometry**.

Turbidity is the degree of opacity or cloudiness of a fluid caused by particles suspended in it, which causes a beam of light passing through the suspension to scatter. An increase in turbidity reduces the amount of light that can penetrate the suspension, that is, its transmittance is decreased. The amount of light that is scattered depends on a number of factors, including the size and shape of the particles, and their concentration. Turbidimetry is the measurement of turbidity. It relies on the use of a spectrophotometer, which can be used to determine the concentration of the suspension of particles in a turbid sample. The particles are often bacterial cells or precipitated protein. A beam of light is passed through the suspension. Some of the light is absorbed, reflected, and scattered. However, the rest is transmitted through the solution and measured by the spectrophotometer and used to determine the concentration of particles. Turbidity (Tb) can be defined mathematically as:

$$I = I_0 e^{-lT} \text{ or, more simply, } Tb = 2.303 \times 1/l \times \log (I_0/I)$$

where I_0 and I are the intensities of incident light and transmitted beams of light, and 1 is the length of the light path. This relationship only holds for relatively finely dispersed suspensions of absorbances less than 0.3–0.5.

Turbidimetric measurements have some clinical uses. For example, they can be used to determine the number of bacterial cells in microbial suspensions. In some clinical disorders, for example nephrotic syndrome, proteins are lost in the urine. Determining their presence and concentration therefore has medical value. The urine samples are treated with sulphosalicylic acid, which causes any protein present to precipitate as a fine dispersion. The amount of precipitation is measured by its turbidity and

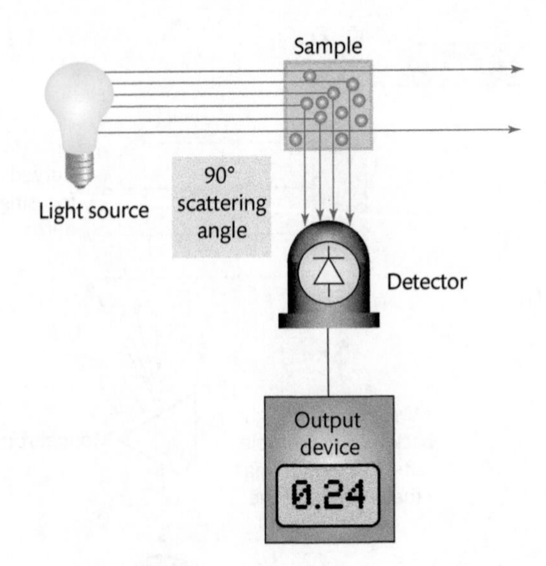

FIGURE 9.12
Schematic illustrating measurement using nephelometry. See text for an outline of the method.

is proportional to the protein concentration. The technique is simple and rapid but has the disadvantages of lacking sensitivity and precision.

SELF-CHECK 9.6

A suspension of *Eschericia coli* cells of dry weight 0.4 g dm^{-3} gives an absorbance value of 0.8. Calculate its corresponding turbidity (Tb) value.

Nephelometry (from the Greek *nephele*, meaning cloud) is a technique in which the intensity of light scattered by a suspension is measured at an angle to the incident beam of light, rather than using the transmitted light to determine the concentration of suspended particles. Thus, unlike turbidity, nephelometry is not measured in a spectrophotometer but uses a nephelometer (Figure 9.12). Nephelometry resembles turbidity but has the advantage of being more sensitive.

The amount of scattered light depends on the size of the particles and their concentration. Thus, for example, the concentration of protein in clinical samples can be determined by comparison with known concentrations. One clinical application of nephelometry is to measure the concentration of rheumatoid factor (RF) in samples of serum. Rheumatoid arthritis is an extremely painful and debilitating chronic condition, which is characterized by symmetrical arthritis, radiological changes to the bone, and the presence of autoantibodies (RF) in the plasma of patients. RF is an immunoglobulin that recognizes and binds to antigenic sites at specific positions of IgG antibodies. The presence of high levels of RF in patients is associated with a poor prognosis.

Cross reference

You can read more about the clinical uses of turbidity and nephelometry in Chapter 13, 'Immunological techniques'.

9.7 Fluorescence and fluorimetry

Fluorescence was introduced in Section 9.2. To recap briefly, when most molecules absorb electromagnetic radiation, the electrons move to a higher energy level before falling back to their ground state, generally releasing energy in the form of heat in doing so. However, in some molecules only part of the energy is lost as heat and the rest is emitted as fluorescent light. The wavelength of this light (λ_{em}) is longer than that used to excite the molecule (λ_{ex}) because, as we noted earlier, the excited molecules lose some of the absorbed energy as heat. The difference between the two wavelengths is known as the Stokes shift.

A fluorimeter or fluorescence spectrophotometer is an instrument used to measure the fluorescence of a given substance. Its components are outlined in Figure 9.13. The source of light produces high-energy radiation of short wavelength. This passes through the first monochromator, which can be used to select an excitation beam of appropriate wavelength that passes through the cuvette.

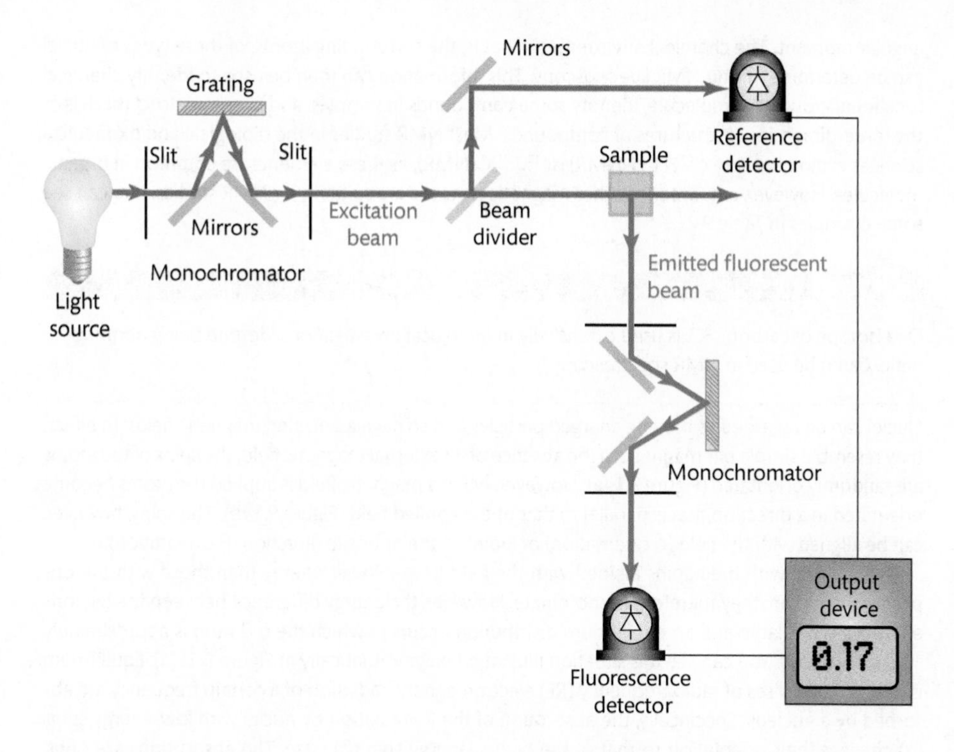

FIGURE 9.13
Outline of a fluorimeter. The main text describes the operational basis of the instrument.

Following its excitation, the sample in the cuvette emits fluorescent radiation in all directions. The second monochromator, which is placed at right angles to the cuvette, only allows emitted light, that is, of the fluorescent wavelength, to fall onto the detector and be measured. The detector produces an electrical signal that is proportional to the intensity of the fluorescent light, which, in turn, is proportional to the concentration of the sample.

Fluorescence has two principal advantages over the use of visible or UV spectrophotometry. First, it is more sensitive, indeed this can be up to a 1000-fold greater. Second, it is also more specific as two wavelengths are used to produce a signal, not one as in spectrophotometry. However, there are also a number of disadvantages to using fluorimetric techniques. They are sensitive to changes in pH and the presence of contaminating molecules in the solution. Furthermore, not all compounds generate fluorescence, although it may be possible to couple non-fluorescent compounds with those that do exhibit fluorescence. For example, proteins can be reacted with fluorescamine to produce a fluorescently labelled complex, which allows the concentrations of minute amounts of them to be quantified. Fluorescent groups can also be coupled to immunoglobulins and this is the basis of immunofluorescence assays, which are described in Chapter 13.

Cross reference
Chapter 13, 'Immunological techniques'.

Sometimes, the emitted fluorescence is less than that expected. This phenomenon is called **quenching** and occurs when substances in the sample absorb the emitted fluorescence or when molecules such as oxygen in the sample interfere with energy transfer.

Fluorimetry can be used to measure the concentration of phenylalanine in samples of serum from newborn infants in screening for phenylketonuria (PKU). This condition is characterized by excessive concentrations of phenylalanine in the plasma and is associated with mental retardation in untreated children. Phenylalanine is reacted with ninhydrin in the presence of the dipeptide leucyl-alanine to give a product whose fluorescence is proportional to the concentration of phenylalanine. The method is calibrated using known standards of phenylalanine.

9.8 Nuclear magnetic resonance spectroscopy

Nuclear magnetic resonance (NMR) spectroscopy provides information on the position of specific atoms within a molecule by exploiting the magnetic properties of certain nuclei. Nuclei with odd masses or odd atomic numbers, for example ^{1}H and ^{13}C, have a magnetic moment, also called spin or angular moment. The chemical environment, that is, the surrounding atoms, of these types of nuclei can be determined using NMR spectroscopy. This information can then be used to identify chemical functional groups in a molecule, identify some compounds in samples, and even construct models of the three-dimensional structures of compounds. Most NMR studies in the biological and biomedical sciences exploit the spin of ^{1}H (^{1}H-NMR) or ^{13}C (^{13}C-NMR), as these elements are common in organic molecules. However, a number of other magnetic isotopes are suitable for NMR studies. You can see some examples in Table 9.2.

TABLE 9.2 Examples of isotopes with odd masses or odd atomic numbers with magnetic properties making them amenable to analysis by nuclear magnetic resonance

Nuclei
^{1}H, ^{2}H
^{13}C
^{14}N, ^{15}N
^{17}O
^{19}F
^{23}Na
^{25}Mg
^{31}P
^{33}S
^{39}K

SELF-CHECK 9.7

One isotope of carbon, ^{12}C, is used extensively in biological investigations, despite being nonmagnetic. Can it be used in NMR spectroscopy?

Nuclei can be regarded as moving charged particles and so have associated magnetic fields; in effect, they resemble simple bar magnets. In the absence of an external magnetic field, the spins of the atoms are randomly orientated (Figure 9.14a). However, when a magnetic field is applied the atoms become orientated in a direction that is parallel to that of the applied field (Figure 9.14b). The spins, however, can be aligned with the field (α orientation) or 'point' in the opposite direction (β orientation).

Those nuclei with their spins aligned with the field have a lower energy than those with the opposite orientation; they therefore predominate. However, the energy difference between the two orientations is not large and an equilibrium distribution occurs in which the α:β ratio is approximately 1.00001:1.00000. You can see this situation illustrated diagrammatically in Figure 9.15(a). Equilibrium is disrupted if pulses of radio frequency (RF) electromagnetic radiation of a certain frequency are absorbed by a nucleus. Specifically, the absorption of the RF radiation by nuclei with low-energy spins (α) changes their orientation to that of the higher-energy spin (β) state. The absorptions are often

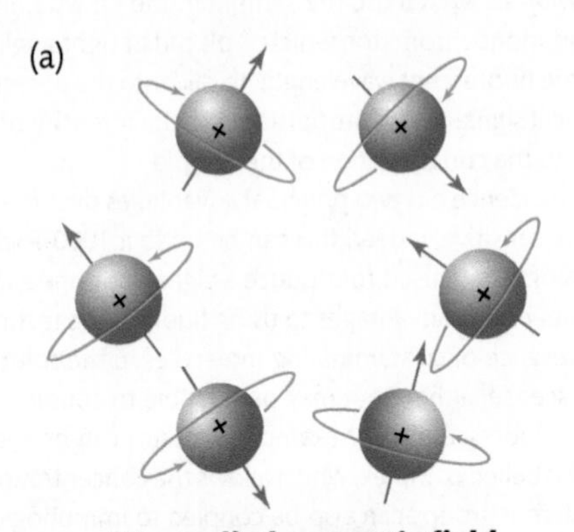

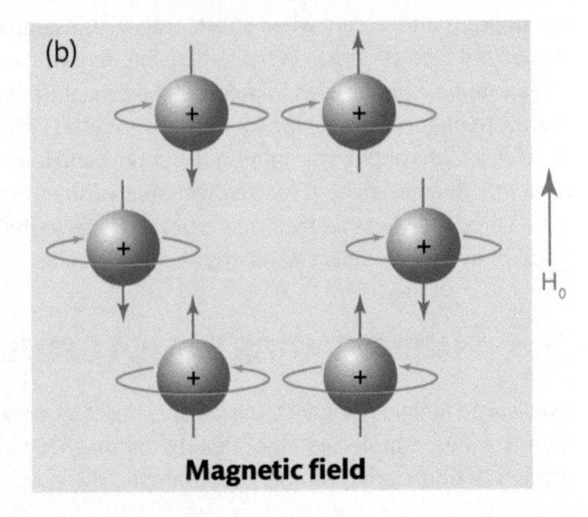

FIGURE 9.14
(a) The spins of atoms are randomly orientated unless (b) they are subjected to an applied magnetic field (H_0).

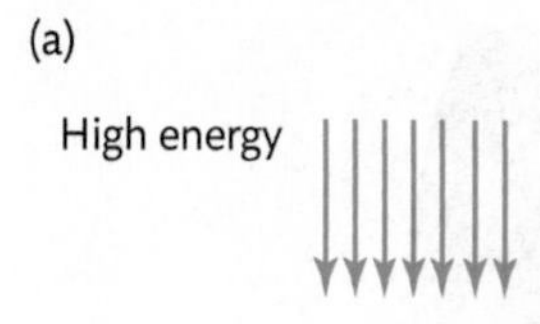

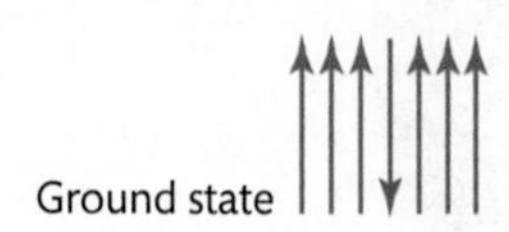

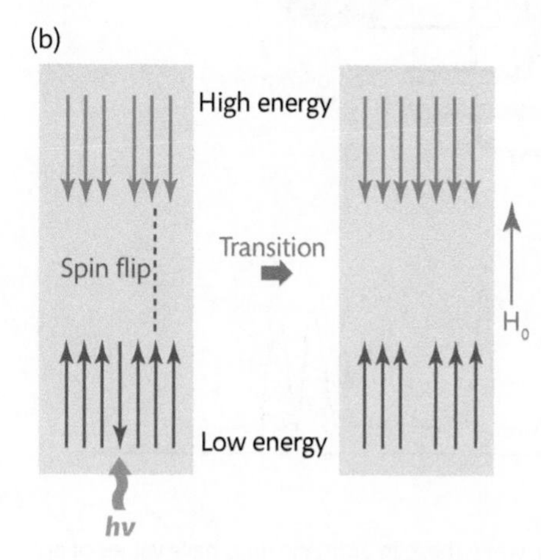

FIGURE 9.15

(a) Shows the non-random alignment of nuclei in the magnetic field H_0, using arrows to represent the direction of spin. (b) Pulses of radiation in the radio frequency region of the electromagnetic spectrum are used to reverse the alignment of nuclear spins from the low-energy spin aligned state (α) to the opposite higher-energy spin state (β). The energy required for this transition depends on the strength of the applied magnetic field.

called **resonances** because the frequency of the RF radiation matches that at which the nuclei spin. The exact resonance or frequency of any given magnetic nucleus depends upon its exact chemical environment. This enables NMR spectroscopy to characterize the structures of chemical compounds. When the nuclei revert back to the equilibrium distribution, that is when those nuclei that have been excited to a high-energy spin state relax to their original low-energy state, RF radiation is emitted, and can be detected and measured. NMR spectra are graphs of intensity against the RF applied.

Figure 9.16 shows the basic components of an NMR spectrophotometer. The sample is placed in the magnetic field and excited by pulses in the RF input circuit. The realignments of the magnetic nuclei in the sample induces a radio signal in the output circuit, which generates a complex output signal. The exciting pulse is repeated as many times as necessary to allow the signal output to be distinguished from any background noise. A mathematical (Fourier) analysis of the output produces the NMR spectrum.

In practice, the resonance frequencies are converted to chemical shifts (δ) to allow results obtained in different experimental conditions to be easily compared. Chemical shifts are defined as the resonance frequency associated with the sample compared with that of a reference compound whose chemical shift is defined as 0, using the expression:

$$\text{Chemical shift } (\delta) = (V_{signal} - V_{reference}) \times 10^6 / V_{reference}$$

You can see from this equation that the scale of the chemical shift is unitless. The multiplication by 10^6 is simply for the convenience of being able to express chemical shifts as parts per million (ppm). Tetramethylsilane [TMS, $Si(CH_3)_4$] is often the reference compound of choice, although other substances, for example 3-trimethylsilyl 1-propane sulfonate (DSS), are used. Whatever the standard, its nature must be specified. You can see examples of NMR spectra in Figure 9.17(a, b).

The frequencies and duration of the applied RF pulses may be varied in an enormous number of combinations allowing different molecular properties of the sample to be investigated. The exact frequency emitted by each nucleus depends upon its molecular environment. For example, protons in methyl

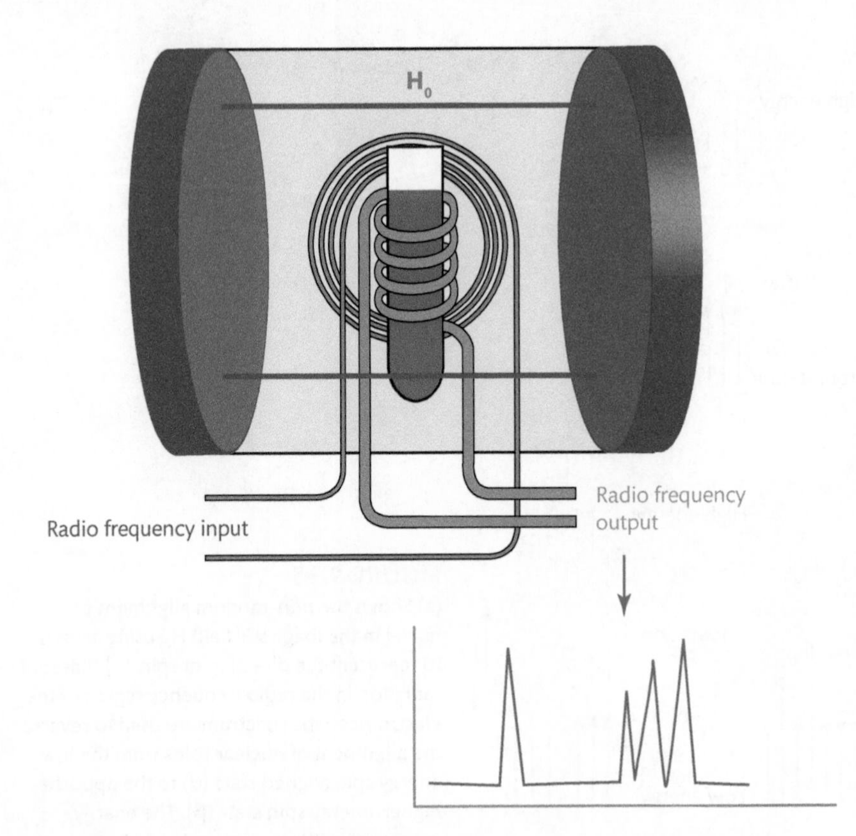

FIGURE 9.16

Basic arrangement of a nuclear magnetic resonance spectrophotometer.

groups ($-CH_3$) have chemical shifts of about 1–2 ppm, while those in aromatic rings have values of approximately 7–8 ppm. Thus, the frequencies differ for each nucleus unless they are chemically equivalent and in identical molecular environments. One-dimensional NMR spectra, like those in Figure 9.17, give valuable information about the structures of the molecules as each of the chemical shifts arises from a particular chemical functional group (Figure 9.18). Also, like the spectra obtained from IR spectroscopy (Section 9.5) the pattern of signals can act like a 'fingerprint' and help identify a particular sample.

SELF-CHECK 9.8

How many peaks would you expect to see in NMR spectra arising from the hydrogen atoms of (a) methyl bromide (CH_3Br) and (b) methyl ethanoate (CH_3COOCH_3)? Explain your answer.

In theory, it should be possible to obtain chemical shifts for all the hydrogen atoms present in a molecule and therefore obtain a one-dimensional NMR spectrum for any molecule. In practice, however, this is not possible for complex molecules such as those of polypeptides and nucleic acids. These molecules contain many hydrogen atoms and therefore the differences in their chemical shifts are smaller than the resolving power of the technique. In other words, the NMR signal is obscured by the presence of so many hydrogen atoms, making it impossible to identify the signal from any one specific atom. Even using extremely strong magnets to increase the frequencies of the radio waves required to change the spin states of the nuclei, which increases the energy of the waves and therefore the resolving power of the technique, does not improve the situation. However, since the 1980s a variety of two-dimensional NMR spectroscopy techniques have been developed, which use a sequence of pulses separated by different time intervals, rather than the single burst associated with one-dimensional NMR. These techniques produce additional peaks that arise from the interactions of protons that are less than 0.5 nm apart in the molecules. These sorts of data, together with other structural information, such as the lengths and angles of covalent bonds, allow the three-dimensional structures of complex molecules, such as those of relatively large proteins, to be predicted.

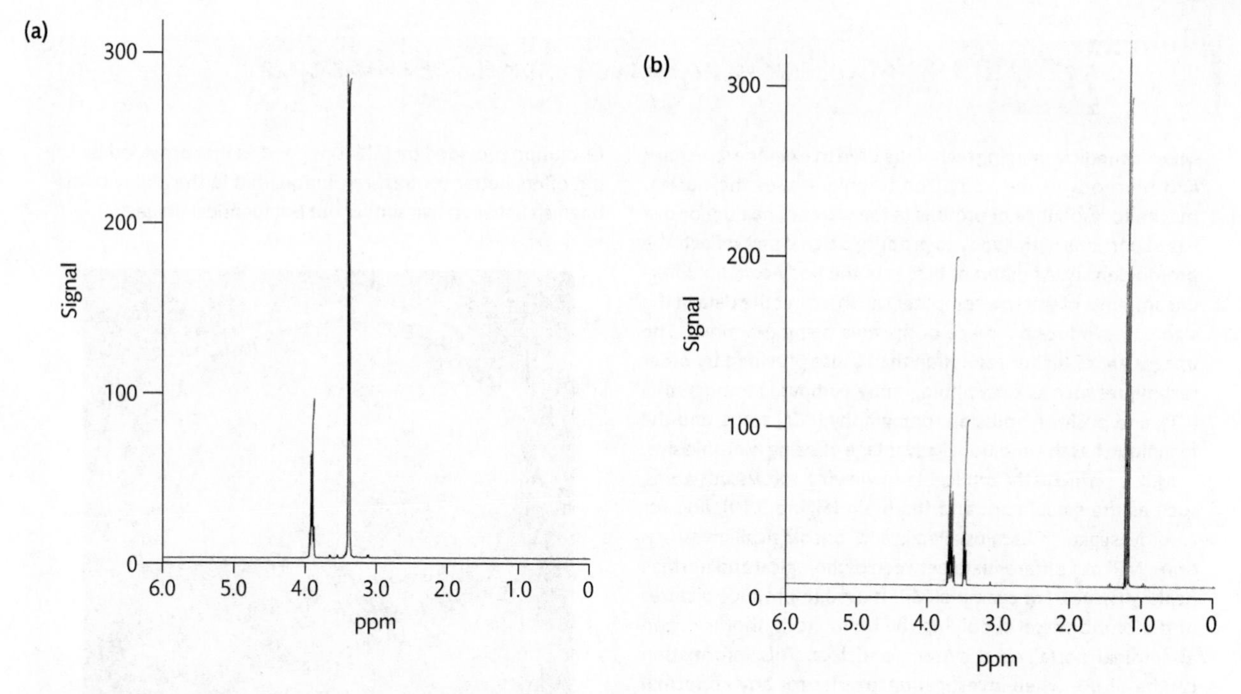

FIGURE 9.17

Nuclear magnetic spectra for (a) methanol and (b) ethanol.

9.8.1 Clinical applications of nuclear magnetic resonance spectroscopy

Magnetic resonance spectroscopy is a non-invasive technique used to investigate neurological disorders such as epilepsy, brain tumours, Alzheimer's disease, and strokes. This technique produces a spectrum or graph that identifies the types of chemicals in the brain (or other organs) and quantifies the amounts of each. The information provided is a useful indicator of the metabolic processes and biochemical changes occurring in diseased tissues. In contrast, nuclear magnetic resonance imaging (MRI; Box 9.2) provides a picture of the anatomical changes that occur in disease conditions.

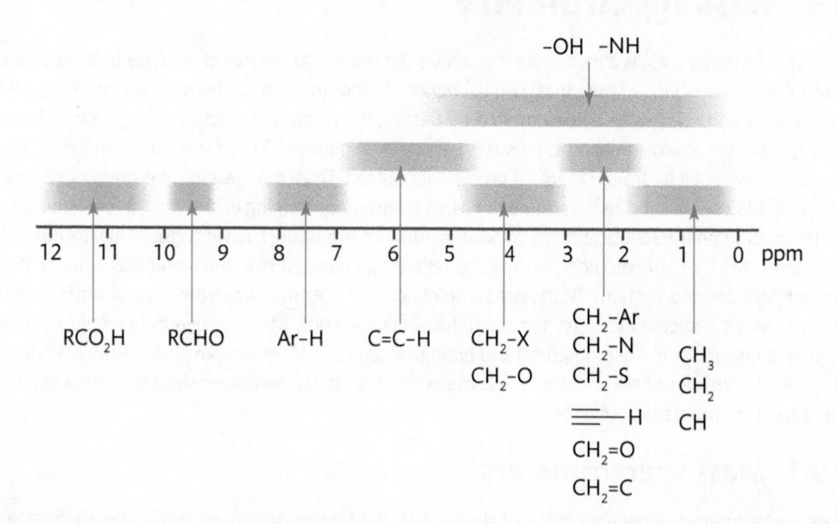

FIGURE 9.18

Typical values of chemical shifts for protons influenced by a single chemical functional group.

BOX 9.2 Nuclear magnetic resonance imaging

MRI is a medical imaging technique used to examine structures and functions of the body. The technique uses the nuclear magnetic resonance of protons in the water of healthy or diseased portions of the body to produce a signal that reflects the proton density. As different tissues of the body contain different amounts of water, a computer can interpret the data in the signal to produce an image of the area being examined. The images are of higher resolution than those provided by other techniques such as x-ray photography, computed tomography (CT), and positron emission tomography (PET) scans, and the technique has the additional advantage of being non-invasive.

MRI is particularly applicable to viewing soft tissues areas, such as the spinal cord and the brain (Figure 9.19), and for cardiovascular, musculoskeletal, and oncological investigations. MRI can differentiate between pathological and normal, healthy tissues. For example, MRI is used to produce pictures of the blood vessel supplying the brain, neck, thoracic, and abdominal aorta, renal system, and legs. This information can be of use when investigating arteries for any abnormal narrowing or stenosis, or dilations of vessel walls or aneurysms. The technique also can be used to identify accurately the location, shape, and orientation of a tumour in a patient prior to radiation therapy.

A major advantage of MRI is that it is harmless to the patient because it uses strong magnetic fields and non-ionizing radiation. In contrast, the ionizing radiation utilized in traditional X-ray techniques may increase the likelihood of developing a malignancy particularly in foetuses. Furthermore, the resolution provided by MRI is as good as that provided by CT, but offers better contrast resolution, that is, the ability to distinguish between two similar but not identical tissues.

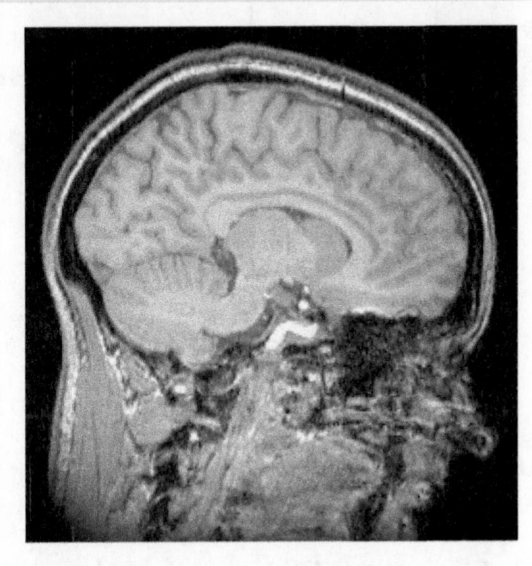

FIGURE 9.19
Nuclear magnetic resonance image (MRI) of a head. The MRI scan shows much more detail than an X-ray investigation. Courtesy of Dr G. Williams, Wolfson Brain Imaging Centre, Cambridge, UK.

9.9 Mass spectrometry

Mass spectrometry (MS) is a technique that allows the molecular masses of analytes to be accurately determined on the basis of their mass (m) to charge (z) ratios (m/z). Analytes must be ionized to allow their masses and charges to be determined by passing them through electric or magnetic fields in a mass spectrometer. Molecular mass is usually expressed in daltons (Da); 1 Da is approximately equivalent to the mass of a hydrogen atom, or proton, or neutron. Thus, the mass in Da is numerically equal to the M_r. MS can be used to determine the M_r of compounds ranging in size from a few hundred, as would occur in pharmaceutical drugs, to several hundred thousand, such as those of macromolecular complexes. MS can also provide information regarding chemical structures and the primary structures of peptides and proteins. Both pure samples and complex mixtures from organic and inorganic sources can be subjected to mass spectrometric analyses. Thus, MS is a powerful technique, with a range of applications in the biomedical and biological sciences. Here, we can give only an overview of the general principles of MS and describe some of the techniques developed since the 1990s and their applications to biomedical science.

9.9.1 Mass spectrometers

Mass spectrometers generally consist of three parts: the ionizer or source region, the analyser, and the detector (Figure 9.20(a) and (b)). Each part is held at high vacuum, of between 10^{-2} and 10^{-10} Pa, to limit collisions between analyte and residual molecules of air, which could reduce the accuracy of the analysis or even cause compounds to fragment. Typically, subfemtomole (<10^{-15}) to low picomole (10^{-12}) amounts of sample are introduced into the mass spectrometer through a staged vacuum lock. Atoms or molecules in the sample need to be ionized by adding a H^+ to them to form a positive ion (M + H^+) or removing one to give a negative ion (M – H^-). Ionization occurs in the source region using one of a number of different techniques. Samples can be liquids, as with electrospray ionization (ESI), where they are dispersed into the gas phase using high-voltage electric currents. Alternatively, solid samples can be mixed with a UV-absorbing matrix compound and irradiated using a UV laser. This forms sample ions in a gas phase, as is the case with matrix-assisted laser desorption ionization (MALDI). Less commonly used ionization techniques are electron ionization (EI), chemical ionization, and fast atom bombardment (FAB). In EI, an electron beam is passed through the vaporized sample causing it to ionize. Samples can be chemically ionized by introducing a gas, such as methane, ammonia, or isobutene, in the source region and allowing it to collide with molecules of the analyte. In FAB, the analyte is mixed with a non-volatile matrix and then ionized by bombarding it with a beam of high-energy atoms, typically of an inert gas, for example argon or xenon. You can see the different uses and some applications of these ionization techniques in Table 9.3.

Whatever the method used, once ionized, the application of a potential difference is used to transfer the sample into the analyser, as you can see in Figure 9.20(b). The analyser uses static or dynamic

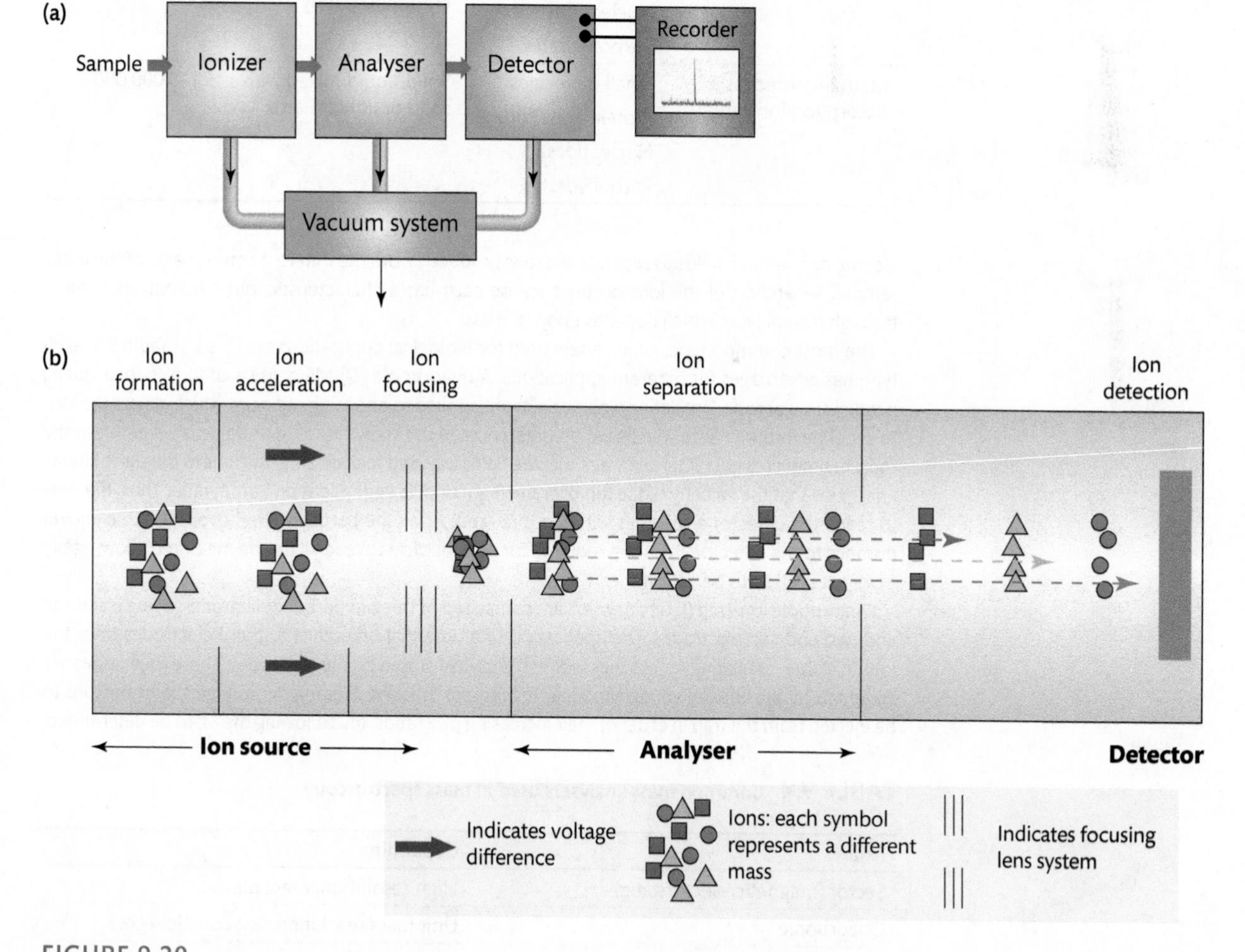

FIGURE 9.20

(a) Schematic showing the general organization of a mass spectrometer. (b) Overview of the operation of a mass spectrometer. An electric potential difference is used to accelerate ions through the system.

TABLE 9.3 Ionization techniques used in mass spectrometry

Ionization method	Analytes	Sample introduction method	Mass range/Da
Electron impact	Small molecules Volatile	Gas chromatography Solid Liquid	1-1000
Chemical ionization	Small molecules Volatile	Gas chromatography Solid Liquid	1-1000
Fast atom bombardment	Carbohydrates Organometallics Peptides Non-volatile	Solid/liquid dissolved in glycerol	1-6000
Electrospray ionization	Small molecules Proteins/peptides Nucleotides Carbohydrates	Liquid High-performance liquid chromatography	1-2,000,000
Matrix-assisted laser desorption ionization	Small molecules Proteins/peptides Nucleotides Carbohydrates	Solid/liquid mixed with a solid or liquid matrix	1-5,000,000

electric or magnetic fields to separate the ions produced from the different components of the initial sample. Separation of the ions occurs because each has a characteristic, but different, movement through the analyser, which depends upon its mass.

The most common types of analysers used for biological applications are listed in Table 9.4; each type has advantages for different applications. A quadrupole (Q) MS consists of four (hence 'quad') parallel circular rods. Oscillating electric fields are applied to opposing rod pairs, which causes the ions to separate in the quadrupole based on differences in the stabilities of their trajectories between the rods. In time-of-flight (TOF) analysers, electric fields are used to accelerate the ions to the same kinetic energy, so that the velocity of the ion over the flight path is dependent on its m/z ratio. Thus, the time to reach the detector also varies with the m/z ratio. As all the particles travel over the same known distance to the detector, the times required for each particle to reach it can be measured allowing the M_r of the particle to be determined.

Quadrupole ion-trap (QIT) analysers are composed of three hyperbolic electrodes: a ring electrode and two end cap electrodes. The application of a potential difference to these electrodes traps the ions in stable, oscillating trajectories within the cavity formed by the electrodes. These trajectories are governed by the applied voltage and the m/z ratio of the ions. Altering the voltage allows the ions to be ejected from the trap in order of their increasing m/z ratios, thus allowing the M_r to be determined.

TABLE 9.4 Common mass analysers used in mass spectroscopy

Analyser	Capability
Sector (magnetic/electrostatic)	High resolution, exact mass
Quadrupole	Unit mass resolution, fast scan, low cost
Time-of-flight	Wide mass range, high throughput
Quadupole ion-trap	MS/MS and MS^n capability
Ion cyclotron resonance	Very high resolution, exact mass

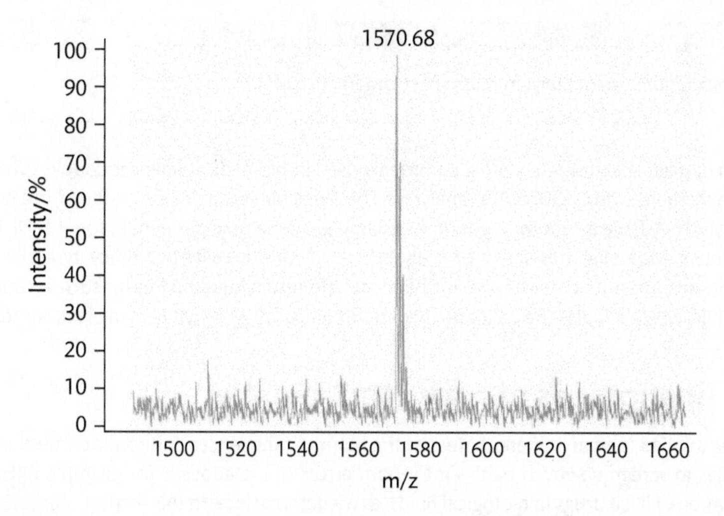

FIGURE 9.21

Mass spectrum of the peptide with the sequence EGVNDNEEGFFSAR, where each letter stands for its usual amino acid residue, and of M_r 1569.65. The value for the ionized peptide $(M + H)^+$ was found to be 1670.68.

Once the ions are separated by the analyser, they are directed towards the detector again because of the applied electric field, which you can see in Figure 9.20(b). The current produced by the ion impacting on, or passing through, the detector surface is recorded. This value is converted into an m/z value and a mass spectrum, typically a graph of m/z against ion signal intensity, produced as you can see in Figure 9.21.

Mass spectrometers that have a single mass analyser can only perform simple M_r determinations, as described earlier. However, mass spectrometers can be designed with two mass analysers arranged in sequence. These instruments are referred to as tandem mass spectrometers. If you look at Figure 9.22, you can see the layout of a tandem MS, which produces spectra called MSMS.

The sample is introduced into the ionizer and the individual components of the mixture separated from one other in the first analyser and a selected ion is transmitted into the collisional-induced dissociation (CID) cell. Here, the ion is fragmented when it collides with molecules of an inert gas, typically helium or argon, present in the CID cell. Fragments of ions from an individual component are analysed in the second analyser to give an MSMS spectrum that is representative of that compound. The fragmentation of the ions from any one individual component of the sample usually occurs in a predictable manner. Thus, the sizes of the fragments produced can be pieced together like a jigsaw to give information about the structure of the component.

In some specific cases, as with QIT instruments, several rounds of mass analysis can be performed in sequence within the same analyser. This analysis is performed by sequentially trapping and fragmenting the ions generated in the source. Specific fragments are trapped and fragmented in turn. This process can be repeated, giving up to ten consecutive rounds of mass analysis. The resulting spectrum is commonly referred to as an MS^n spectrum, where n is equal to the number of consecutive mass analyses.

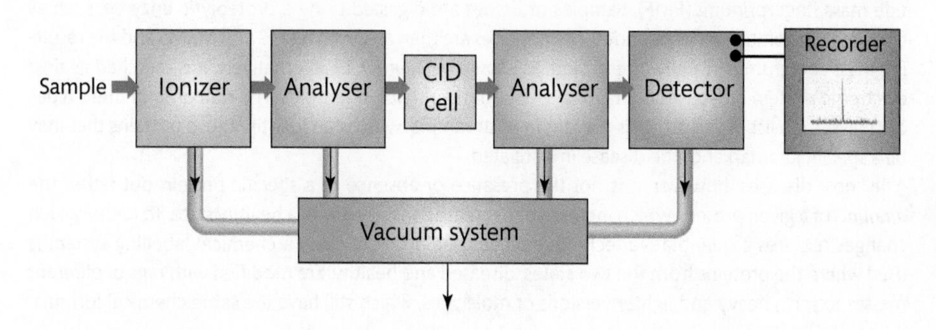

FIGURE 9.22

Schematic showing the general organization of a tandem mass spectrometer. CID: collision-induced dissociation.

SELF-CHECK 9.9

Briefly describe the use of tandem mass spectrometry (MSMS).

The types of mass spectrometers most commonly used in biomedical sciences couple ESI or MALDI ionization methods with Q, QIT, or TOF analysers. These machines can determine the M_r of a wide range of materials. They have a high throughput: for example, one sample per minute for MALDI-TOF mass spectrometers. Moreover, this speed of analysis is coupled to excellent accuracy, typically to 5–100 ppm of the real mass, and sensitivity. Furthermore, the results obtained using such techniques are reproducible, which enables the data obtained to be compared with that held in electronic databases.

9.9.2 Applications of mass spectrometry

MS can be applied to many different areas of the biomedical sciences. It is routinely used in hospital laboratories to screen newborn babies for inborn errors of metabolism, for example PKU, and the determination of illicit drugs in biological fluids, as we describe later in this section.

PKU is a genetic disorder caused by a deficiency of hepatic phenylalanine hydroxylase. This enzyme is necessary for catalysing the conversion of phenylalanine to tyrosine. Thus, in PKU phenylalanine accumulates in the blood and if the condition is left untreated causes mental retardation. PKU cannot be cured but can be treated by restricting the amount of dietary phenylalanine while ensuring an adequate intake of tyrosine. However, the mental retardation caused by PKU is irreversible; hence, its early detection in newborns is essential. MSMS is routinely used to measure the molar ratio of phenylalanine to tyrosine in bloods spots collected from newborn babies. This method of investigating PKU has been shown to be more effective than other types of screening. It is also used to screen for other treatable inborn errors of metabolism.

MS is also routinely used to test for illicit drugs in biological fluids. These test are normally performed on urine samples, as the concentrations of the drugs and their metabolites in urine are relatively high compared with other fluids. Liquid chromatography or gas chromatography coupled to MSMS is a powerful technique in this area, providing selectivity, sensitivity, and reliability. It is also used for the study of drug metabolism, and for the investigation of the dynamics of endogenous biologically active substances, as well as contributing to the development of new therapeutics.

The research applications of MS include the analysis of peptide/protein (proteomics), metabolite studies (metabolomics), and carbohydrate analysis (glycomics). Emerging, although lesser used areas, include mass spectrometric imaging and microorganism identification.

The proteome is all the proteins present in a given cell, tissue, or organism. Proteomics is the study of proteomes. However, the proteome changes in disease cells or tissues compared with healthy ones, both in composition and in the amounts of specific proteins present. MS techniques involving ESI–MS and MALDI–MS can identify and quantify the proteins present in a sample, allowing tissues from normal (reference) and diseased states (patients) to be directly compared. For example, increases in the concentrations of proteins such as amyloid-A protein are associated with the inflammatory response and can be detected in samples of plasma from patients with rheumatoid arthritis. Thus, proteins in a given tissue that change during the disease can be identified as a potential marker for that disease and/or as targets for drug therapy. This is often referred to as biomarker discovery.

Comparison of the proteins in clinical samples is generally done in an indirect manner called peptide mass fingerprinting (PMF). Samples of tissues are digested using a proteolytic enzyme, such as trypsin, to generate a set of peptides. The peptides are then analysed by MS and MSMS and the resulting mass spectrum is visually interpreted, as shown in Figure 9.23; alternatively, it is matched against electronic protein databases containing similar data for known samples. The outcome of these types of analyses is a list of the proteins present in each sample, which can identify those proteins that may be a specific biomarker of the disease investigated.

In most diseases, however, it is not the presence or absence of a specific protein but rather the *amounts* of a given protein, which increase or decrease compared with a healthy state. To identify such changes requires a quantitative technique to be used. In most cases a chemical labelling system is used where the proteins from the two states, diseased and healthy, are modified with tags of different masses to form heavy and lighter versions of molecules, which still have the same chemical formula.

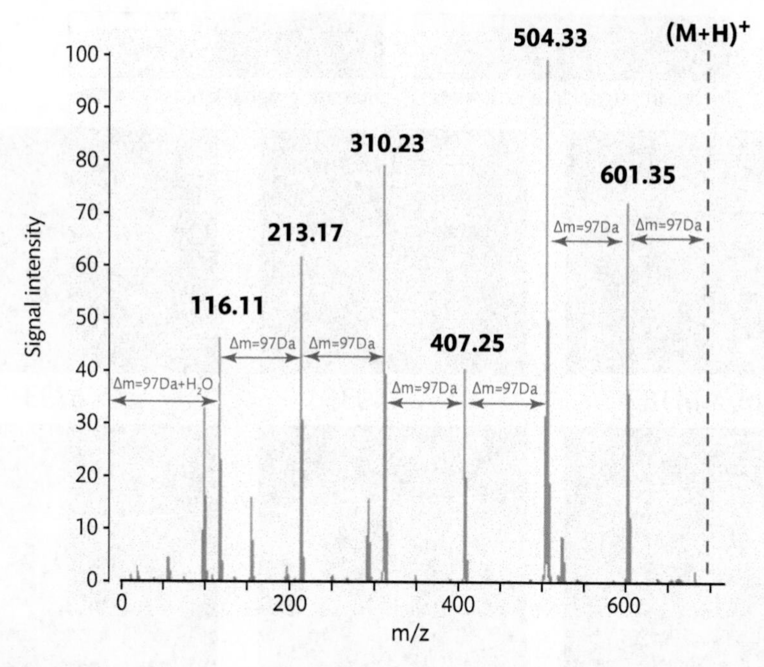

FIGURE 9.23

Tandem mass spectrometry spectrum of a peptide. Note the difference in mass (Δm = 97 Da) between the consecutive peaks in the series is equivalent to the residue mass of the amino acid proline (M_r = 97). Thus, a valid interpretation of the data gives the sequence, PPPPPPP.

The heavy form is produced by incorporating a number of stable isotopes such as ^{2}H, ^{13}C, ^{15}N, or ^{18}O into the protein molecules thus forming a small, but easily measurable, mass difference. The diseased and healthy states are labelled with the light and heavy versions, respectively. The two types are mixed, enzymatically digested, and the resulting peptides analysed by MS.

Analysis of the resulting spectrum allows the quantitative ratio of a particular protein present in samples from patients and healthy individuals to be calculated. Analysis requires examination of the relative signal intensities of the ions from the heavy and light versions of both sets of peptides. A PMF analysis of the peptides allows the proteins in question to be identified. This, combined with the ion ratios, allows the amounts or concentrations of any identified protein in the diseased and healthy samples to be calculated. Those proteins that change significantly may be potential biomarkers or indicators of disease.

In addition to the discovery of potential biomarkers, proteomics also includes the study of protein modifications, many of which have been implicated in disease pathways. These modifications occur post-translationally and may be catalysed by enzymes, for example phosphorylation and glycosylation, or occur non-enzymatically as is the case with glycation. Often proteins are not active until they are enzymatically modified in the cell. Analysis of protein modifications tends to exploit MSMS-based techniques, which are used to identify signature ions or differences in masses characteristic of a particular modification. The fragmentation pattern obtained by MSMS analysis can also identify the specific sequence of the modified peptide, thus identifying the site where the modification has occurred.

The *metabolome* is all the metabolites, which are molecules of small M_r, present in a biological system. These include, for example, many hormones, signalling molecules, and the chemical intermediates of metabolic pathways. It is believed that about 3000 metabolites are essential for normal human growth and development. Studying variations of these metabolites could be helpful in determining if a patient is fighting a disease, reacting badly to a drug therapy, or responding to another form of stress. This type of knowledge could have essential roles in optimizing therapy and management of a disease.

> **Cross reference**
> **You can read about the various forms of chromatography in Chapter 11.**

MS, along with NMR, is a major technique for analysing metabolomes. However, in this role MS is usually combined with a chromatographic separation method such as GLC or high performance liquid chromatography (HPLC), which are described in Chapter 11. GLC and HPLC greatly simplify the analyses by separating the components of complex clinical samples (e.g. urine, serum, cerebrospinal

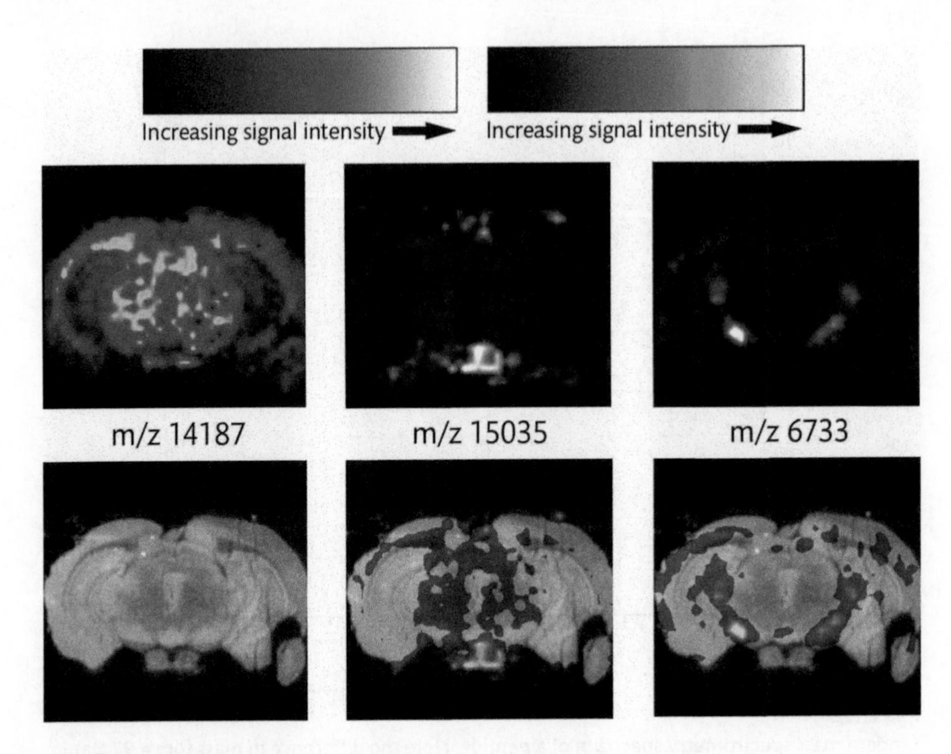

FIGURE 9.24
Mass spectrometric imaging of a section of tissue from a mouse brain. The upper panels show the relative distributions of compounds with m/z ratios of 14,187, 15,035, and 6733, respectively. The corresponding lower panels are tissue sections to allow the sites of the compounds identified by the imaging to be mapped to specific areas of the brain.

fluid) that are typically studied in metabolomics. This simplification means each component can be subjected to MS analysis in turn. This separation, coupled with the speed, high-throughput, accuracy, and sensitivity of MS provides a powerful tool for the validation, evaluation, and monitoring of drug therapies, biomarker discovery, and clinical diagnosis.

Imaging by MS was first used in 1999 and specifically uses MALDI-MS. A section of tissue is placed directly in the source region of the MALDI mass spectrometer. The sample is moved in two dimensions as a laser is fired at it. Specific ions are formed at each point on the section by the action of the laser light. This procedure allows an image of the tissue section surface to be generated using a computer with appropriate software. This image can be manipulated to show the location of any given mass in the tissue section and therefore track, for example, where a drug and its metabolites are accumulating. This is illustrated in Figure 9.24, which shows mass spectrometric images of a mouse brain section highlighted for three different masses.

MS can be used to identify species of bacteria and fungi. Traditional methods for identifying these groups are time consuming and often relatively complicated. Using MS, identification can be carried out using standard proteomics-type techniques, as described earlier or, more elegantly, by an approach called intact-cell MALDI. The organisms are grown on agar plates until mature colonies are visible, then a single colony is transferred into the MALDI source region and irradiated with a laser. As with proteomics, the analysis of these samples provides a pattern or fingerprint of the microorganism being investigated. This fingerprint is then compared with a database of known organisms and so identified provided it has previously been subjected to analysis. This technique has obvious uses in identifying pathogenic organisms in clinical samples.

SELF-CHECK 9.10

Which techniques are typically combined with mass spectrometry to simplify the analysis of complex mixtures?

CASE STUDY 9.1 Fluoroquinolone and kidney failure

A 70-year-old man without any recorded history of kidney disease was given the drug fluoroquinolone to treat a lower urinary tract infection. An automated analyser programmed to perform assays based on the modified colorimetric methods of Jaffe and Berthelot, respectively, determined his serum concentrations of creatinine and urea nitrogen. Prior to treatment, they were measured as 92.5 μmol dm^{-3} (male reference range for creatinine is 60-110 μmol dm^{-3}) and 6.8 mmol dm^{-3} (male reference range for urea is 2.5-7.1 mmol dm^{-3}).

Three days after treatment commenced he presented with a greatly reduced urinary output, and purpura and erythematous skin lesions, which are generally indicative of a hypersensitive response, over his legs and lower abdomen. Following admission to hospital, his serum creatinine and urea nitrogen concentrations were found to be 571.3 μmol dm^{-3} and 56.9 mmol dm^{-3}, respectively. However, his serum electrolyte and bilirubin concentrations were within reference range values; as were aminotransferase, alkaline phosphatase, and lactate dehydrogenase activities. The values are indicative of a 'normal' liver metabolism.

Treatment with fluoroquinolone was stopped immediately. Intravenous fluids and the diuretic furosemide were administered to reduce creatinine and urea concentrations, and prednisone was given to decrease his inflammatory response. During the next month, the patient's serum creatinine and urea nitrogen concentrations returned to reference range values and he fully recovered.

The two most widespread screening tests for assessing renal function are measuring the concentrations of creatinine and urea nitrogen in serum. A failure of renal function leads to increases in both serum creatinine and urea nitrogen concentrations. Acute renal failure can occur rapidly but is potentially reversible depending upon the initial causes or insult if the problem is resolved.

Points to consider:

What is the most likely interpretation of the serum creatinine and urea test results?

What do you consider the primary cause of these results? Explain your answer.

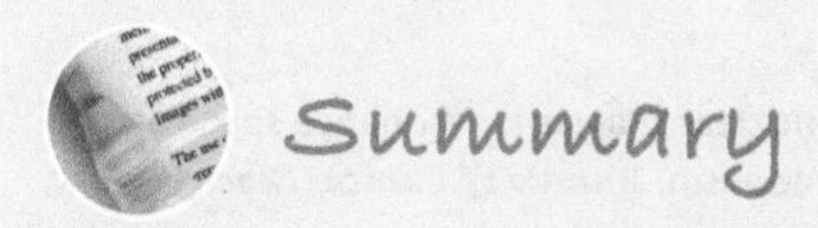

Summary

- Spectroscopy is concerned with the interactions of electromagnetic radiation and matter, particularly the absorption of radiation by matter and, in some cases, its emission, and can be used to quantify molecules of interest in biomedical laboratories.
- Many molecules absorb light of characteristic wavelengths in the visible, UV, or IR regions of the electromagnetic spectrum, and this is used as the basis for determining their concentrations using spectrophotometers and colorimeters.
- The absorbance of light of a specific wavelength by a solution of biological molecules depends on their concentration and the distance travelled by the light through the solution, as described by the Beer–Lambert law.
- IR and Raman spectroscopy are specialized techniques in spectroscopy with applications in identifying biological molecules.
- Light-scattering methods include turbidimetry and nephelometry. Turbidimetry can be used to estimate concentration by measuring the amount of light scattered by a suspension, whereas the more sensitive technique of nephelometry measures the amount of light scattered at an angle to the incident beam of light.
- Some molecules are fluorescent and are able to absorb light of a characteristic wavelength and then emit radiation of a longer wavelength. Fluorescence-based methods are highly sensitive, but are limited to molecules that fluoresce or which bind to one that is.

- NMR utilizes magnetic properties of nuclei within molecules and provides information on positions of atoms within their structures. Its value is in identifying molecules of biological interest and in determining their structures.
- NMR imaging can distinguish between healthy and diseased tissues and is of value, for example, in locating the positions of tumours in patients.
- MS allows us to determine the masses of molecules based on their charge to mass ratios. The technique has wide applications in biomedical science, ranging from detecting phenylalanine when screening newborn babies for PKU and drugs of abuse in urine of suspected individuals, to identifying and sequencing proteins. This sophisticated technique is also the basis of locating the positions of molecules of interest in patients.

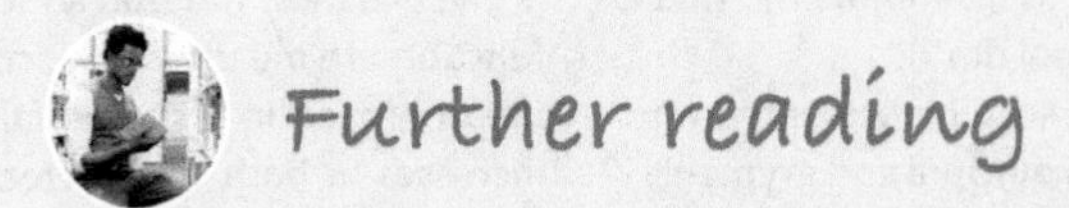

Further reading

- Chary KVR, Govil G. *NMR in Biological Systems: From Molecules to Human*. Berlin: Springer, 2008.
- Gore MG. *Spectrophotometry and Spectrofluorimetry: A Practical Approach*. Oxford: OUP, 2000.
- Hoffman E, Stroobant V. *Mass Spectrometry: Principles and Applications*. Chichester: John Wiley, 2007.
- Honour JW. Benchtop mass spectroscopy in clinical biochemistry. *Annals of Clinical Biochemistry*, 2003; **40**: 628–38.
- Huettel SA, Song AW, McCarthy G. *Functional Magnetic Resonance Imaging*. 2nd ed. Boston, MA: Sinauer Associates, 2009.
- Meng QH. Mass spectrometry applications in clinical diagnostics. *Journal of Clinical and Experimental Pathology*, 2013; **S6**: e001.
- Moolenaar SH, Engelke UFH, Wevers RA. Proton nuclear magnetic resonance spectroscopy of body fluids in the field of inborn errors of metabolism. *Annals of Clinical Biochemistry*, 2003; **40**: 16–24.
- Riddle P. Principles of spectrophotometry. *The Biomedical Scientist*, 2005; **49**: 135–7.
- Smith E, Dent G. *Modern Raman Spectroscopy: A Practical Approach*. Chichester: John Wiley, 2004.
- Stuart BH. *Infrared Spectroscopy: Fundamentals and Applications.* Chichester: John Wiley 2004.
- Stuart B, George B, McIntyre P. *Modern Infrared Spectroscopy*. Chichester: John Wiley, 1998.
- Thomas MJK. *Ultraviolet and Visible Spectroscopy: Analytical Chemistry by Open Learning.* Chichester: John Wiley, 1996.
- Watson JT, Sparkman OD. *Introduction to Mass Spectrometry: Instrumentation, Applications, and Strategies for Data Interpretation*. 4th ed. Chichester: John Wiley, 2007.

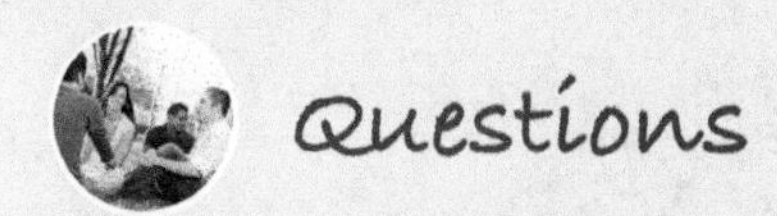

Questions

1. Which of the following requirements must be satisfied to establish a colorimetric method to measure the concentrations of a specific analyte (X)?
   - (a) Pure analyte (X) is needed for the production of a standard or calibration graph
   - (b) Analyte X should absorb light of a particular wavelength or it should be possible to convert it to one that can
   - (c) The wavelength of light at which maximum absorbance by X occurs needs to be determined
   - (d) The relationship between concentration and absorbance of light for X must obey the Beer–Lambert law
   - (e) All of the above criteria must be satisfied.

2. State which of the following are TRUE or FALSE?
   - (a) Spectrophotometers differ from colorimeters in only being able to analyse coloured compounds
   - (b) Tungsten lamps in spectrophotometers are used to measure light in the UV region of the spectrum
   - (c) Light of the desired wavelength can be selected using a monochromator
   - (d) Plastic cuvettes are used when measuring light in the UV region of the spectrum
   - (e) The relationship between concentration and absorbance is not linear when an analyte obeys the Beer–Lambert law.

3. A 1.0 mmol dm^{-3} solution of a compound gives an A_{340} of 0.6. Calculate
   - (a) the molar extinction coefficient (ε) and
   - (b) the absorptivity (*a*) of the compound.

4. What would be the advantage of using a standard curve formed from a range of concentrations of an analyte compared with estimating its value from the absorption of a single solution of known concentration?

5. An aqueous solution of yeast DNA has an absorptivity of 25 m^2 mol^{-1} at 260 nm. Calculate the percentage transmission of a solution of the DNA of concentration 2.0 mol m^{-3} if the absorbance is measured in a standard-sized cuvette.

6. What happens to light as it passes through a turbid suspension contained in a cuvette in a colorimeter?

7. In ^{1}H-NMR spectroscopy, the amino acid alanine [$^+NH_3CH(CH_3)COO^-$] produces two major chemical shifts at 1.5 and 3.8 ppm, respectively, using TMS as standard. Which hydrogens (protons) are responsible for each of these signals?

8. Describe the major components of a mass spectrometer and give an example of each.

9. Give three examples of how mass spectrometry is used in the biomedical sciences.

Answers to self-check and end-of-chapter questions are available at the end of the book.

10

Centrifugation

Qiuyu Wang, Nessar Ahmed, and Chris Smith

Learning objectives

After studying this chapter, you should be able to:

- Describe the basic features of a centrifuge
- Explain the general principles of centrifugation
- Calculate the speeds and forces involved when separating mixtures by centrifugation
- Describe the major types of centrifuges, tubes, and rotors encountered in clinical laboratories
- Discuss the principal techniques used in centrifugal separations
- Outline the precautions to be taken in the safe use of a centrifuge
- List some of the uses of centrifuges in a clinical laboratory

Introduction

Given sufficient time, gravity will cause many particles, such as cells, suspended in a liquid to settle eventually to the bottom of their container. However, this is not usually a practical method of separating or isolating the particles as the time required for them to sediment is normally too long. Indeed, particles that are extremely small in size will not separate at all unless they are rotated at a sufficiently fast speed that the rotational or centrifugal force generated causes the particles to move radially away from the centre of rotation. **Centrifugation** is the term applied to the mechanical process of separating mixtures by applying centrifugal forces. It is one of the most widely used techniques in biochemistry, molecular and cellular biology, and in the biomedical sciences. Applications of centrifugation are many and may include harvesting viruses and cells from growth media, the separation of subcellular organelles, the collection of lipids, and the isolation of macromolecules, such as DNA, RNA, and proteins. For example, centrifugation can be used to separate blood into plasma and blood cells, or obtain cells from body liquids, such as the ascites fluid that accumulates in the abdomen in some pathological conditions. Centrifugation can be applied to a great many clinical and biological materials, so for the sake of convenience in this chapter we will generally refer to them all as *particles*.

Centrifugation is performed in a **centrifuge**. At its simplest, a centrifuge is a mechanical device that produces a centrifugal force by spinning a sample around a central axis. They consist essentially of

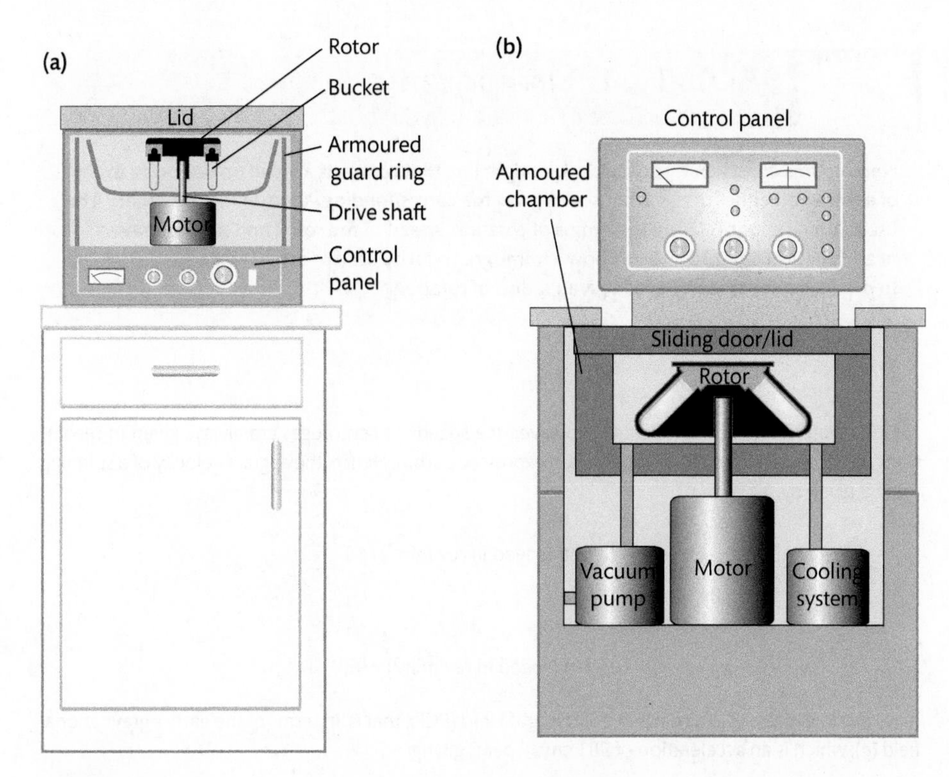

FIGURE 10.1
Outline structures of (a) bench and (b) floor standing centrifuges.

two compartments. One contains a motor whose drive shaft enters the second compartment, called the rotor chamber, where it can be attached to a specialized container called the **rotor**. You can read about rotors in Section 10.3. The rotor contains liquid samples in suitable **centrifugation tubes** or bottles, although in some applications the samples are centrifuged in plastic bags. The centrifugal force generated by the motor-driven rotation of the rotor separates components of the mixture that differ in size and/or density. In general, the mixture is separated into two fractions. One fraction is the **pellet** at the bottom of the tube that contains the sedimented material; the other is the liquid **supernatant** that contains unsedimented material. Whatever the type, all centrifuges operate in essentially similar ways. However, different types are designed for a variety of uses, which you can read about in Section 10.4, and so differ in design details. Centrifuges vary from relatively simple bench-top models to sophisticated floor standing instruments that must be sited with care (Figure 10.1). All types of centrifuges have a brake to slow rapidly and stop the spinning rotor following centrifugation and incorporate a number of safety features, such as locking lids that can be opened only when the rotor has stopped spinning.

10.1 Basics of centrifugation theory

The rate of sedimentation of a particle when spun in a centrifuge depends upon the **centrifugal field** or **acceleration** (G), which is directed outwards from the centre of rotation. The magnitude of this field is described by the equation:

$$G = \omega^2 r$$

where ω is the angular velocity of the rotor in radians per second and the radius of the particle from the centre of rotation (**radius of rotation**) is r in cm (Figure 10.2).

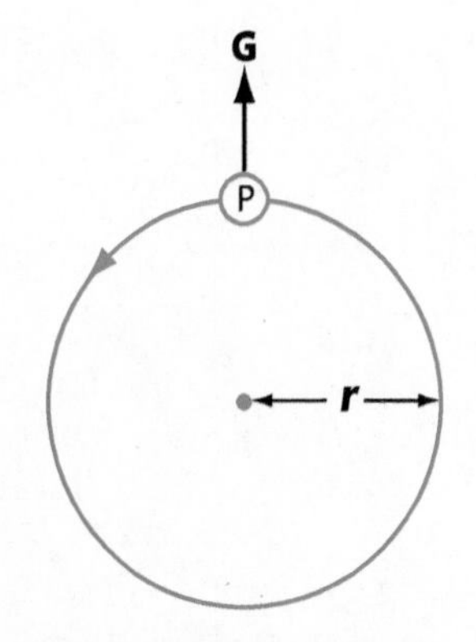

FIGURE 10.2
Diagrammatic representation of the centrifugal field (G) operating on a spinning particle (P), which is a distance of *r* cm from the centre of rotation.

BOX 10.1 Nomograms

Nomograms consist of three scales. A straight line that connects a value on one scale to that of another on either of the other scales gives the corresponding value on the third scale. The use of nomograms relating the radius of rotation, speed of the rotor, and RCF can save a lot of arithmetic! Figure 10.3 shows how a nomogram can be used to estimate the speed needed to produce a specified RCF for a given radius of rotation.

One revolution equals 2π radians. However, the speeds of centrifuges are always given in revolutions per minute (rev min^{-1}), although often expressed as rpm. Hence, the angular velocity of a spinning rotor is given by:

$$\omega = 2\pi \text{ (speed in rev min}^{-1}) / 60$$

So, if $G = \omega^2 r$

$$G = 4\pi^2 \text{ (speed in rev min}^{-1})^2 r / 60^2$$

It is usual to express G as a **relative centrifugal field** (RCF), that is, in terms of the earth's gravitational field (*g*), which is an acceleration of 981 cm s^{-1} per s, giving:

$$\text{RCF (in } g) = 4\pi^2 \text{(speed in rev min}^{-1})^2 r / 60^2 \times 981,$$

which simplifies to:

$$\text{RCF (in } g) = 1.12 \times 10^{-5} \text{ (speed in rev min}^{-1})^2 r$$

Alternatively, the RCF can be determined using a **nomogram** (Box 10.1).

If it is to be repeated successfully, an initial report of an experiment or procedure that involved centrifugation must state the rotor dimensions, so that *r* can be determined, and the speed of the centrifugation, or the RCF and rotor radius plus, in each case, the duration of the centrifuge run.

10.1.1 Sedimentation velocity and sedimentation coefficient

The value of G (that is the RCF) influences the rate of sedimentation or sedimentation velocity (*v*) of a particle, that is, how fast it is moving through the liquid when it is centrifuged. However, *v* also depends upon a number of other factors, including the size of particle, its density and **frictional coefficient**, and the density of the solvent, which are interrelated in the following manner:

$$v = M_r (1 - \bar{v} \rho) \omega^2 r / Nf$$

The mass of the particle measured in grams is m_0, but this is converted into its relative molecular mass (M_r) by the introduction of Avogadro's number (*N*, 6.02×10^{23}). The densities are accounted for by $\bar{v}$ and ρ, which are the **partial specific volume** of the particle (i.e. the change in volume that occurs when the particle is added to a large excess of solvent) and the density of the solvent, respectively. The sedimentation of a particle through the solvent is opposed by a frictional resistance that is equal to the product of its velocity and frictional coefficient (*f*), that is, the *vf* of the particle. Thus, as *v* and *f* become larger, the resistance also increases. The frictional coefficient depends upon the shape and size of the particle. It is larger for asymmetric molecules, for example nucleic acids and fibrous proteins, than for spherical ones, such as globular proteins, of similar M_r.

SELF-CHECK 10.1

Use Figure 10.3 to estimate the speed required to produce an RCF of 50,000 *g* for a rotor of radius 7.67 cm.

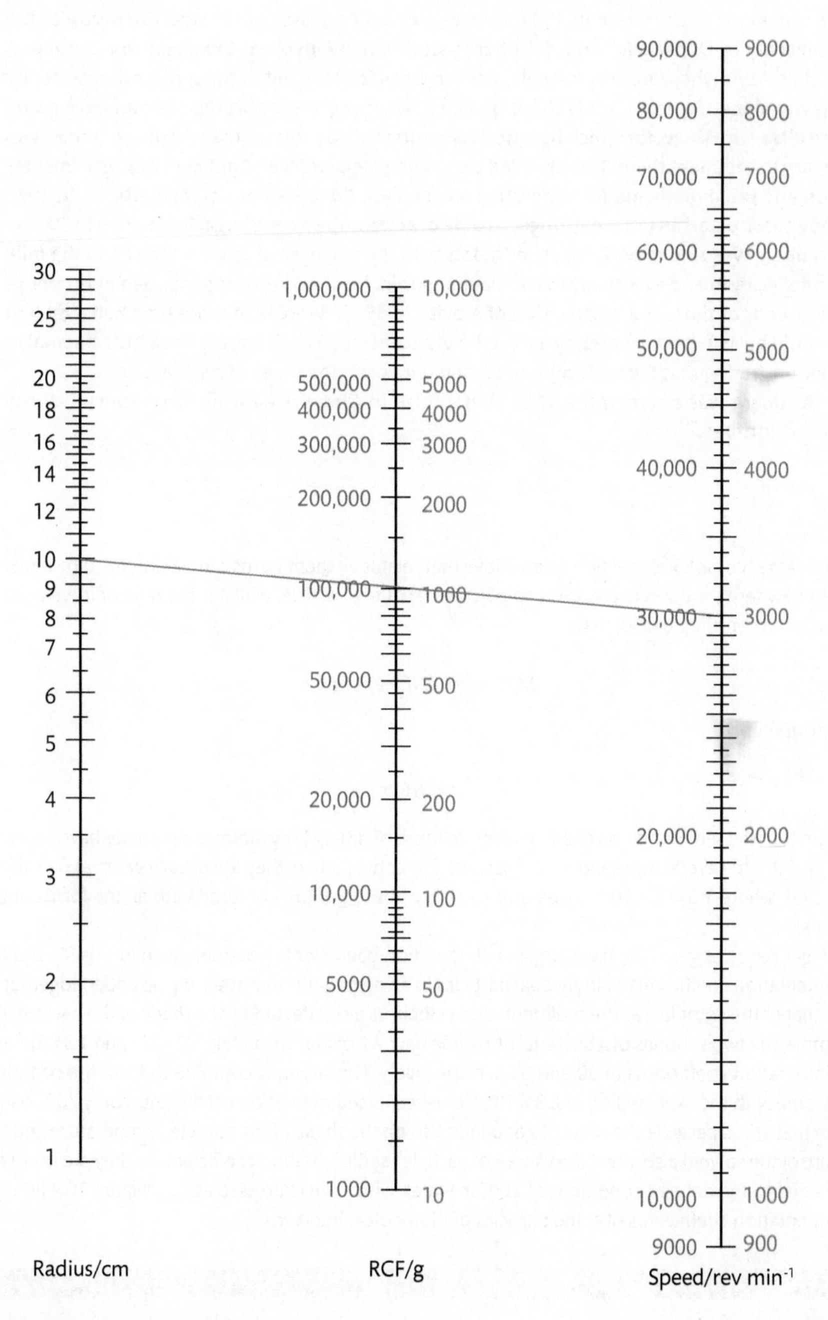

FIGURE 10.3

Nomogram to estimate the speed in rev min^{-1} needed to produce a specified relative centrifugal field (RCF) for a given radius of rotation. Note the line giving the RCF of 100,000 or 1000 *g* for a radius of 10 cm and specified speeds of 30,000 and 3000 rev min^{-1}, respectively.

BOX 10.2 Theodor Svedberg

The Svedberg (S) is a non-SI unit of time, equal to 1×10^{-13} s, used in characterizing the behaviour of particles during centrifugation. It was named after Theodor Svedberg (1884–1971), a Swedish physicist and chemist. Svedberg studied at the University of Uppsala, where he obtained his doctorate in 1907. He was elected Professor of Physical Chemistry at the University of Uppsala in 1912. Svedberg's work largely involved the chemistry of colloids and proteins whose macromolecules are too small to settle out of the surrounding water by gravity alone. To better study these particles, Svedberg and co-workers began developing the ultracentrifuge, for which he is best remembered, to produce the centrifugal forces necessary to sediment them. This involved combining mechanics and optics in the instrumentation and using mathematics to describe and analyse the behaviour of the particles. In 1924 they constructed an ultracentrifuge capable of generating a centrifugal force of up to 5000 *g*. Svedberg was able to perform experiments with the machine to calculate the M_r of the milk protein casein and eventually that of haemoglobin. Later centrifuges generated hundreds of thousands of times the acceleration of gravity. In 1933–34 RCFs of more than half a million *g* could be achieved. Studies based on the ultracentrifuge by Svedberg were instrumental in demonstrating that the molecules of certain pure proteins are all of a single size.

Svedberg was awarded the 1926 Nobel Prize in Chemistry for his development of the ultracentrifuge.

The rather complex expression given above that relates all these terms can be simplified by introducing a new term, the **sedimentation coefficient** (*s*) of the particle, which is the ratio of its velocity to the applied centrifugal field, that is

$$M_r(1 - \bar{v}\rho)/Nf = v/\omega^2 r = s$$

or, more simply:

$$s = v/\omega^2 r$$

The units of the sedimentation coefficient (*s*) are those of time. Many biological particles have values of *s* of 1×10^{-13} to several thousand $\times 10^{-13}$ second. For convenience, they are usually given as **Svedberg units** (S), where 1 S = 1×10^{-13} s (see Box 10.2). Try not to get too confused with all the forms of the letter s!

Bigger particles generally have larger sedimentation coefficients (whether given in s or S!), but the sedimentation coefficients of individual particles in an aggregate cannot simply be added together to give that of the complex, as the sedimentation velocity is also affected by the *shape* of the particle. For example, the two subunits of a bacterial ribosome have M_r of approximately 1.0×10^6 and 1.8×10^6 and sedimentation coefficients of 30 and 50 S, respectively. The subunits combine to form the complete ribosome with the expected M_r of 2.8×10^6. However, its sedimentation coefficient is only 70 S, a situation that arises because the value of *s* depends also on the *shape* of the particle. Temperature and the nature of the solvent also affect the value a of particle's sedimentation coefficient, so they are normally corrected to the defined conditions of 20°C in water, which are expressed as $s_{20,w}$. Figure 10.4 lists the sedimentation coefficients of some particles of biomedical interest.

SELF-CHECK 10.2

An enzyme has M_r 14,400 and a sedimentation coefficient (*s*) of 1.9 S. However, when bound to an inhibitor of only M_r 220, the sedimentation coefficient of the complex was only 1.6 S. How can you account for this change?

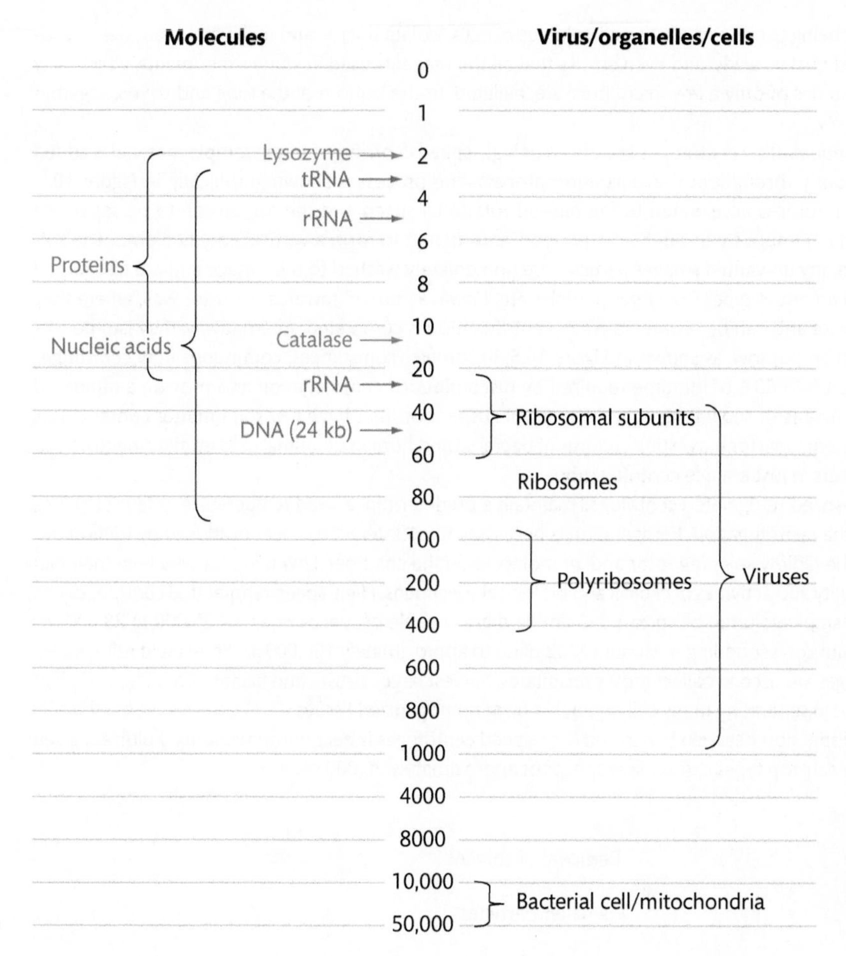

FIGURE 10.4
The sedimentation coefficients (*s*) of some structures (particles) of biological and biomedical interest.

10.2 Types of centrifuges

The main types of centrifuge most likely to be encountered in biomedical or biological sciences laboratories are:

- bench centrifuges
- continuous-flow centrifuges
- high-speed refrigerated centrifuges
- ultracentrifuges.

Bench centrifuges (Figure 10.1a) are the most common type and so the most likely type to be encountered. Small bench centrifuges generally have maximum speeds in the range of 4000–10,000 rev min^{-1} and generate RCFs of 3000–7000 *g*. They are mainly used to collect material that sediments rapidly, such as erythrocytes, yeasts, and others cells, and relatively granular precipitates of, for example, proteins. Larger bench centrifuges develop similar RCFs but can accommodate a greater variety of tubes with sample volumes of 5–250 cm^3. They can also accommodate a number of adapters that allow, for example, the 96-well plates used in enzyme-linked immunosorbent assays (ELISAs) you can read about in Chapter 13 to be centrifuged. Many high-throughput biochemical-based assays that require the rapid yet efficient separation of rather coarse precipitates or cells depend upon bench centrifuges.

Microfuges are small bench-top centrifuges capable of reaching speeds of 12,000–15,000 rev min^{-1} producing RCFs of about 12,000 *g* in only a few seconds. They are used to process samples of less than 1.5 cm^3 contained in Eppendorf tubes. Microfuges are indispensable tools in molecular and cell biology, being used to harvest small volumes of cells, isolate micro- and milligram quantities of proteins and nucleic acids, and even ensure that all the reagents used in some experiments, which may have volumes of only a few microlitres, are 'pelleted' to the bottom of the tube and so mix together thoroughly.

Continuous-flow centrifuges are a form of high-speed centrifuge but the sample material is added continuously throughout the separation process. The process is shown graphically in Figure 10.5. A vertical rotor is accelerated to the desired rotational speed and the suspension to be separated is added continuously through an input port and rotated to form a vertical liquid column. The solvent and any unwanted smaller particles are immediately washed to the surface and are discharged through an outlet pipe. The larger particles are, however, forced towards the rotor wall, where they accumulate and can be recovered when centrifugation is completed. Alternatively, they can be collected in an outflow, as shown in Figure 10.5. In some circumstances, continuous flow centrifuges can save up to 80% of the time required by discontinuous centrifugation and provide a rapid and reliable means of separating large volumes of suspension that have low solid matter content. This makes them ideal for harvesting cultures of bacterial and human or animal cells from fermentors and bioreactors in just a single centrifugation.

High-speed refrigerated centrifuges maintain a predetermined fixed temperature, often set at 4°C, during the centrifugation. Refrigeration is necessary to counteract the heat produced by friction between the rapidly spinning rotor and air molecules in the chamber. Low temperatures help maintain the stability and activities of clinical and biological specimens. High-speed refrigerated centrifuges can accept sample volumes of up to 1.5–2 dm^3 and are capable of speeds of about 25,000 to 28,000 rev min^{-1}, with corresponding maximum RCFs of up to approximately 100,000 g. High-speed refrigerated centrifuges are used to collect most precipitates, harvest larger viruses and bacterial, animal, and plant cells, and in separating many cell organelles from homogenized tissues.

The distinction between bench and high-speed centrifuges is becoming increasingly blurred, given some bench-top types can achieve speeds of approximately 20,000 rev min^{-1}.

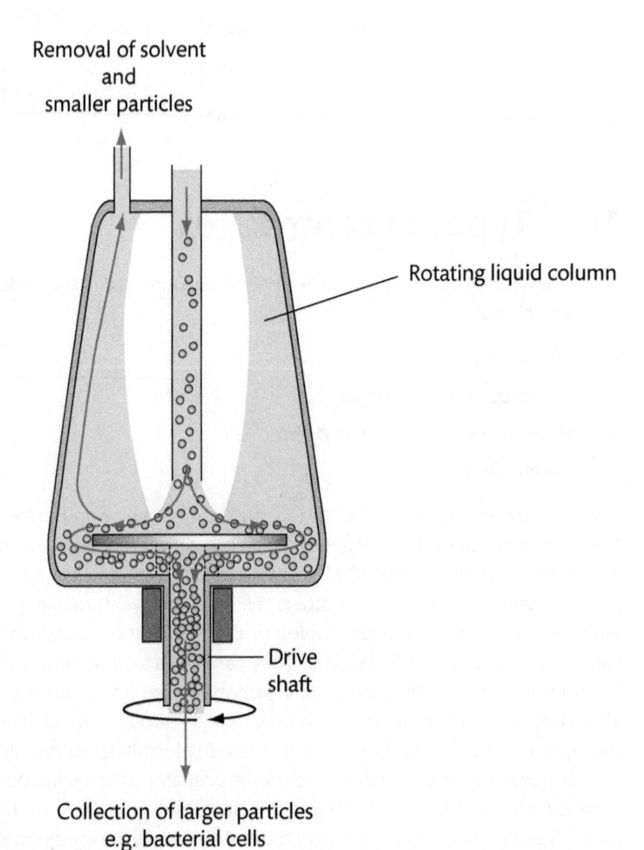

FIGURE 10.5
Outline of a continuous flow centrifuge. See text for details.

BOX 10.3 Analytical ultracentrifugation

Estimating the purity of samples and determining the M_r of molecules is usually performed using chromatographic and electrophoretic methods, which are quick and convenient. You can read about these procedures in Chapters 11 and 12, respectively. However, the M_r obtained by these methods is estimated by comparison with standards of known size. An advantage of analytical ultracentrifugation is that the M_r value and other biophysical properties of macromolecules are given from first principles. Such analytical methods and the analyses of data were relatively time consuming. Also, while high-speed refrigerated centrifuges and preparative ultracentrifuges are bulky floor-standing instruments, large by any standards, analytical ultracentrifuges were even bigger! However, analytical ultracentrifugation can now be performed with microprocessor-controlled table-top micro-ultracentrifuges. These instruments accept small volume samples, up to about 4 cm^3, but can reach speeds of 100,000–150,000 rev min^{-1} and generate RCFs of over 800,000 *g*. This allows analytical experiments to be performed in a fraction of the time previously associated with analytical centrifugation. This, together with contemporary methods of data capture and analysis, means that analytical ultracentrifugation is often the method of choice for determining a range of hydrodynamic properties of biological molecules. These properties may include, for example, determining the numbers of subunits that bind together in complexes, the value of the associated equilibrium constant, and characterizing the conformational changes undergone by a macromolecule on binding with a ligand.

Ultracentrifuges (Figure 10.1b) are not only refrigerated, but the rotor chamber is also evacuated to prevent friction between air molecules and the rotor increasing the temperature of the rotor and sample during centrifugation. They are capable of speeds of about 100,000–150,000 rev min^{-1}, with corresponding maximum RCFs of up to approximately 800,000 *g*. Two types of ultracentrifuge are used: preparative and analytical types. Preparative ultracentrifuges are used, for example, to collect membrane fractions and ribosomes, or to prepare samples of viruses and isolate macromolecules. As with high-speed refrigerated centrifuges, preparative ultracentrifuges can use a variety of types of rotors and so a range of sample volumes can be processed depending upon the size and type of rotor accepted by the centrifuge.

Analytical ultracentrifuges incorporate some form of optical system to observe the sedimentation of the particles, which are often macromolecules or molecular aggregates. This allows biophysical analyses to be performed on the particles during centrifugation. The use of small, table-top micro-ultracentrifuges has made ultracentrifugation particularly popular for investigating the hydrodynamic properties of biological particles, as outlined in Box 10.3.

10.3 Tubes and rotors

During centrifugation the samples are generally contained in specialized tubes. The tubes are made of a variety of materials, including glass, polyallomer, polycarbonate, and polypropylene, and range in volume from less than 1 cm^3 to over a litre. Safe working practice means choosing the most suitable type of tube, rotor, and centrifuge for the separation or experiment to be performed. Most centrifuges can be used with a variety of rotors that accept tubes which differ in volume, or by using a single rotor that has adapters for different tubes. Rotors can be broadly classified into three common types:

- swing out bucket rotors
- fixed angle rotors
- vertical tube rotors.

You can see how the rotors differ from one another in Figure 10.6. In swing out bucket rotors, the tubes are placed in pivoted buckets that move from the vertical to horizontal as the rotor gathers speed. As their name implies, fixed-angle rotors hold the tubes at a constant angle, which generally varies from 14° to 40°.

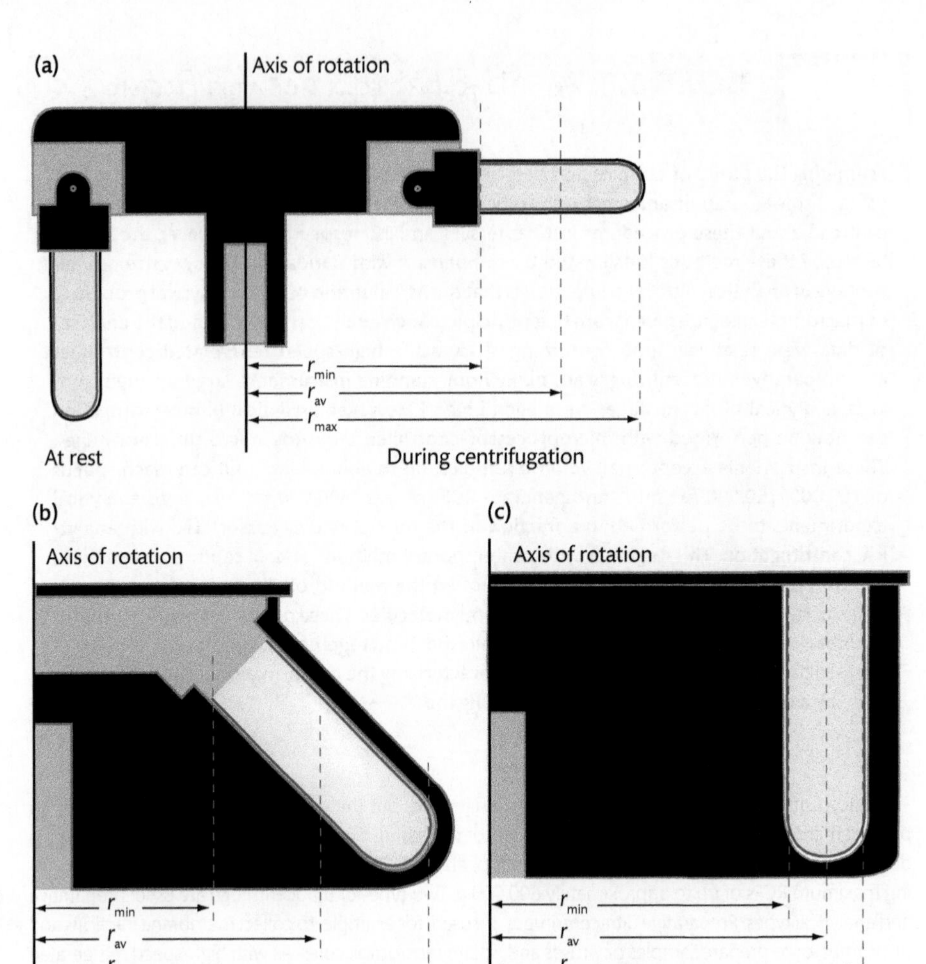

FIGURE 10.6

Schematics showing the basic structures of (a) swing out bucket, (b) fixed-angle, and (c) vertical tube rotors. Note that in each case, the effective radius of rotation varies along the length of the tube from a minimum (r_{min}) to the maximum (r_{max}) value. Relative centrifugal fields are often quoted for the average radius (r_{av}).

Vertical tube rotors are a variant of the fixed angle type, where the tubes are at right angles to the horizontal. During centrifugation, rotors are subjected to considerable stress (see Section 10.5) and must be made of appropriately strong material. Those used in bench centrifuges are generally composed of steel. However, high-speed and ultracentrifuges traditionally used rotors made of aluminium alloy or titanium.

Aluminium is easily corroded especially by alkalis. However, aluminium rotors are given an anodized (aluminium oxide) surface that resists corrosion. Unfortunately, it is thin and easily scratched by sharp metal objects. Titanium is resistant to corrosion and can withstand higher centrifugal forces than aluminium but has the disadvantage that it is much more expensive. Metal rotors have the further drawback of being heavy, not only to handle, but also to accelerate in the centrifuge. The weight of a metal rotor also contributes to the potential hazards involved in using centrifugation (see Section 10.5), as the forces involved are the *product* of the weight of the fully loaded rotor and the RCF applied. If a rotor fails during a centrifugation, then the large forces involved mean nearby persons are in potential danger. Also, considerable damage can be done to the centrifuge and its surroundings.

Rotors made of carbon fibre composites are available, which are lighter than metal ones in weight. These can be accelerated and decelerated more rapidly than metal types, resulting in quicker centrifugations and reduced maintenance costs.

SELF-CHECK 10.3

What is the force (apparent weight) of a rotor of weight 13.6 kg in (a) tonnes and (b) tons when centrifuged at 150,000 *g*? Assume 1 kg is 2.2 lb.

10.4 Separation methods using centrifuges

If you examine Table 10.1 you should notice that many materials of biomedical interest differ in size and/or density. They can therefore be separated or studied by centrifugation. The two most common methods used to separate particles in centrifugation are differential centrifugation and density gradients.

10.4.1 Differential centrifugation

Differential centrifugation, as the name implies, separates particles mainly on *differences* in the sizes of the particles. It is the most common centrifugation technique and is illustrated diagrammatically in Figure 10.7. However, to achieve an effective separation of two particles they must differ in size by an order of magnitude, that is, one must have a diameter ten times larger than the other.

The centrifuge tube is filled with the sample solution and then centrifuged. Depending upon its size and the conditions applied during the centrifugation, any one component of the initial mixture may end up in the supernatant, the pellet, or even be distributed between the two fractions. However, as all the components were initially distributed evenly throughout the sample, a pellet of larger particles will always be contaminated with smaller ones. It is, of course, possible to decant off the supernatant carefully. The pellet can then be resuspended and recentrifuged, which will improve its purity. The supernatant can also be recentrifuged but at a higher speed and possibly longer time to allow smaller particles to be obtained as a pellet.

Separation by differential centrifugation is commonly used in simple pelleting to isolate cells from blood and in obtaining partially pure preparations of subcellular organelles and macromolecules, for example.

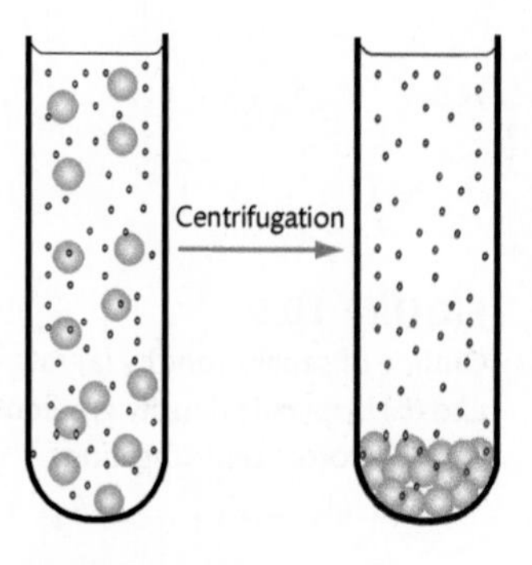

FIGURE 10.7
Outline of separation by differential centrifugation.

10.4.2 Density gradient centrifugations

Density gradient centrifugation is performed in a centrifuge tube that contains a column of liquid whose density increases towards the bottom of the tube. This separates particles in a mixture with greater resolution than discontinuous centrifugation because larger particles are not contaminated with smaller ones. The increase in density can be in discrete steps forming a discontinuous gradient or it can be continuous, as shown in Figure 10.8. Discontinuous or step gradients are suitable for separating whole cells or subcellular organelles and for isolating some types of viruses. Continuous gradients are the more commonly used. Like step gradients they are used in isolating cell organelles and some viruses, and also hormones, enzymes, and other proteins, as well as nucleic acids.

Whether continuous or discontinuous, the gradient is formed by dissolving a suitable solute in an appropriate solvent. Sucrose dissolved in water or buffer is common but, depending on the sample, other materials can be used. The sample to be separated is generally layered on top of the gradient and then centrifuged. Two methods of density gradient centrifugation are routinely used: rate-zonal and isopycnic, as you can see in Figure 10.9.

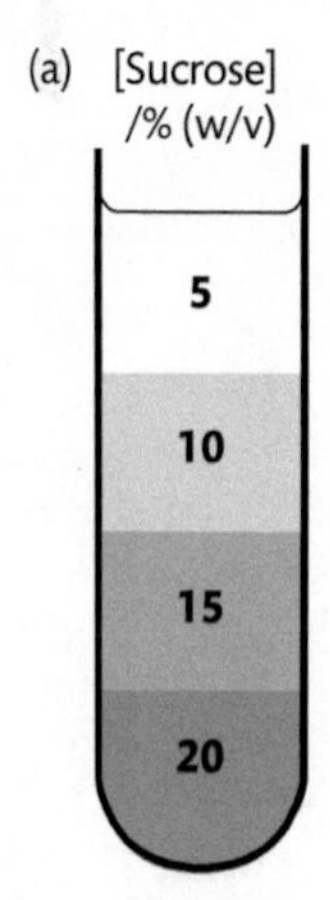

FIGURE 10.8
(a) Continuous and (b) discontinuous density gradients.

TABLE 10.1 Sizes and densities of some materials of biological and biomedical interest

Particle/structure	Size	Density/g cm^{-3}
Haemoglobin	~8 nm	~1.35
Nucleic acids	Variable	~1.6
Viruses	20–50 nm	1.3–1.45
Ribosomes	~25 nm	~1.6
Mitochondria	0.5–2 µm	1.19
Lysosomes	0.05–1.0 µm	1.21
Peroxisomes	0.1–1.5 µm	1.23

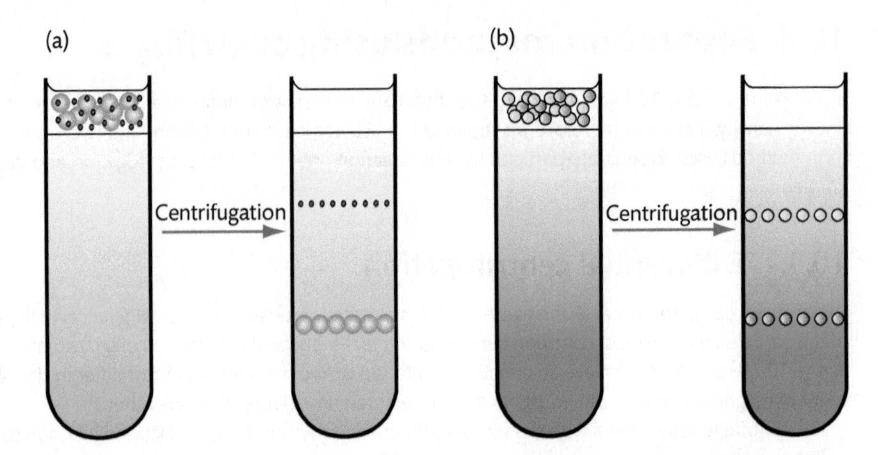

FIGURE 10.9
Outline of separations by (a) rate-zonal and (b) isopycnic density gradient (vertical rotor) centrifugations.

Rate-zonal density gradient centrifugation (Figure 10.9a) separates particles because they sediment through the gradient at different velocities that depend upon their sedimentation coefficients. They thus separate into discrete zones, with each consisting of one type of particle. The zones of the separated particles are stabilized because the gradient slows diffusion and convection currents. The length of the gradient must be sufficient to allow the zones to separate. However, given that the densities of the particles in the sample must exceed that of the densest part of the gradient, the centrifuge run must be terminated before any or all of the zones reach the bottom of the tube. In **isopycnic density gradient centrifugation** (Figure 10.9b), the density gradient is such that it exceeds the densities of all the particles in the sample. Thus, particles will sediment in the gradient to a position that equals their own density. Once all the particles have reached these positions, they will remain there and further centrifugation will not increase the migration of the particles.

A variety of materials are used to form gradients suitable for density gradient separations in biomedical and biological applications. For example, gradients of sucrose and CsCl are commonly used to separate subcellular fractions and nucleic acids, respectively. In many density gradient centrifugations, separation is really achieved by a combination of rate-zonal and isopycnic principles. Rate-zonal centrifugation is often used to purify cellular organelles (Box 10.4), and to isolate proteins, such as immunoglobulins, as all classes of immunoglobulins have similar densities but differ in their masses. Rotors must be selected with care, as the choice affects the efficiency of the separation.

BOX 10.4 Subcellular fractionation

If a cell homogenate is subjected to differential centrifugation, then generally larger particles, such as organelles in this case, will sediment faster than the smaller ones. This is the basis of obtaining fractions enriched in specific subcellular organelles. Thus, centrifugation of the homogenate at 1000 *g* for 10–20 minutes will result in a pellet consisting largely of unbroken cells and larger nuclei. The supernatant can be recentrifuged at 10,000 *g* for 20 minutes, which will mainly pellet mitochondria and the larger types of lysosomes and microbodies. Centrifuging the supernatant at 105,000 *g* for 60–90 minutes will sediment smaller subcellular structures, such as ribosomes and the artificial vesicles formed from fragments of the endomembrane system and the plasma membrane that are often called microsomes. The particles remaining in solution are largely macromolecules and are generally regarded as the soluble fraction.

Obtaining partially purified organelles by differential centrifugation is usually only a preliminary purification step; the fraction is normally purified further using some form of density gradient separation.

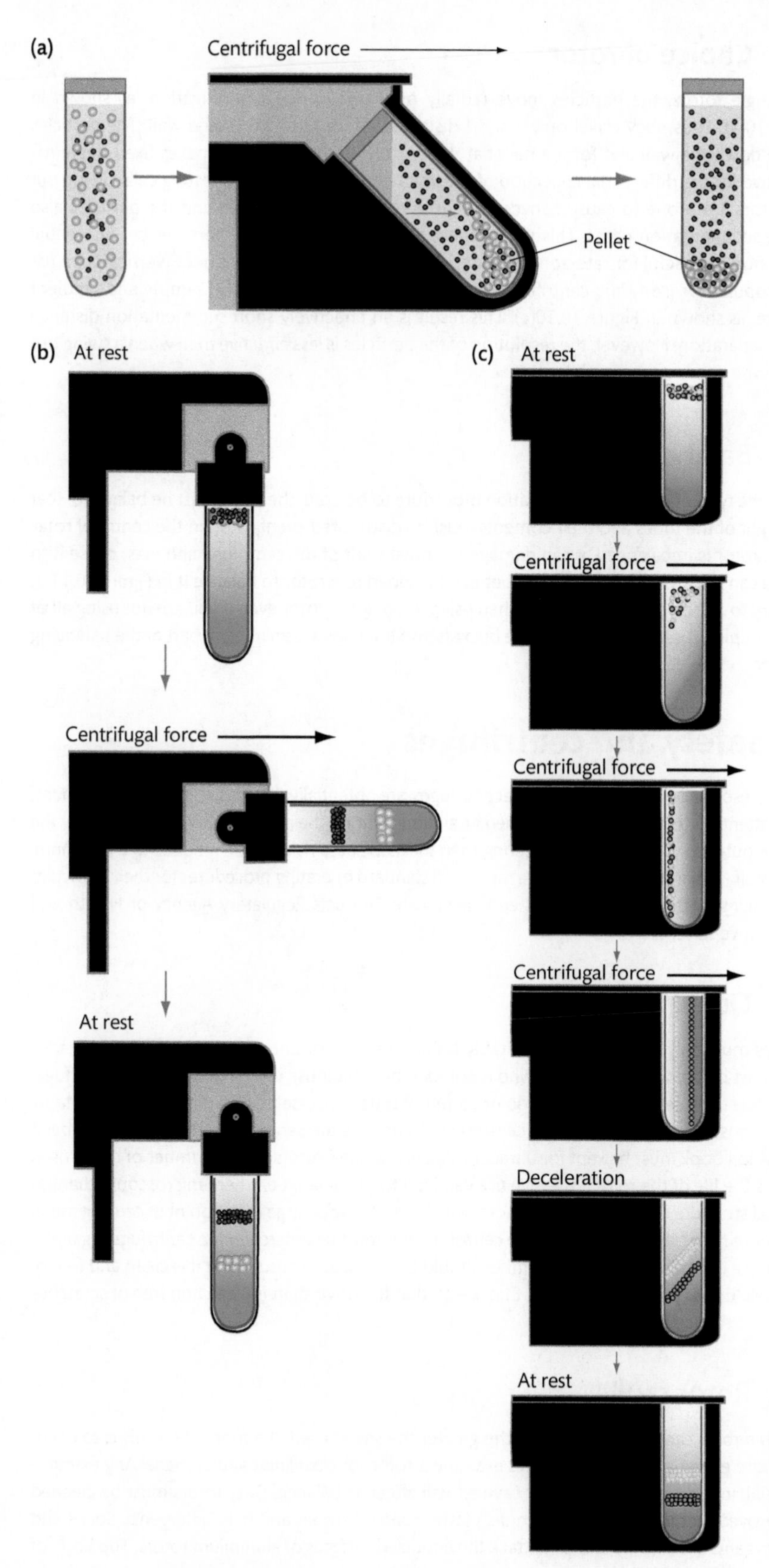

FIGURE 10.10
The behaviour of particles during (a) differential in a fixed-angle rotor, (b) rate-zonal (swing out rotor), and (c) isopycnic centrifugations.

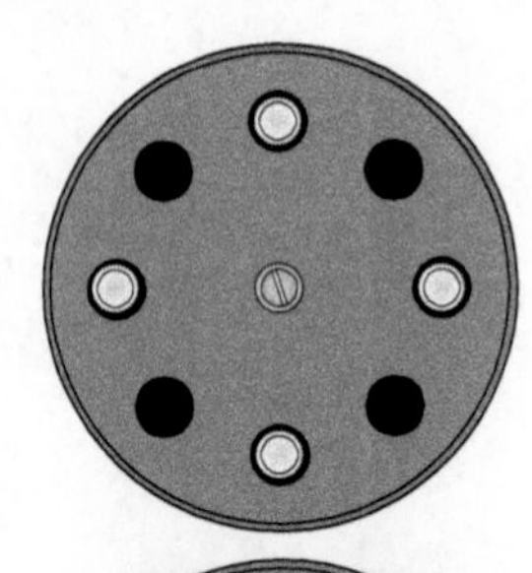

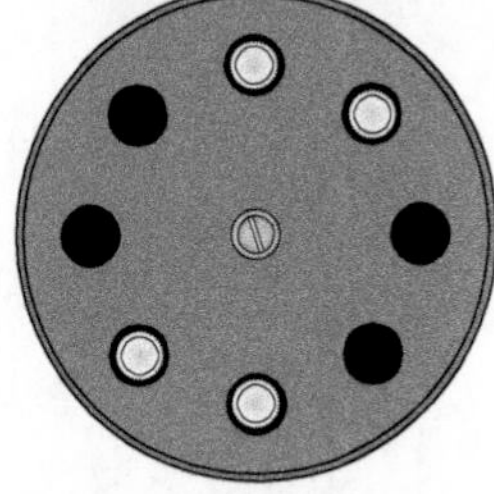

FIGURE 10.11
Surface view showing alternative arrangements for balancing four tubes in a rotor capable of accommodating eight. Filled tubes are indicated by the solid surfaces. Notice how the tubes are distributed symmetrically, with filled tubes always opposite one other.

Cross reference

General aspects of health and safety are described in Chapter 4.

10.4.3 Choice of rotor

In fixed-angle rotors, the particles move radially outwards during centrifugation, as shown in Figure 10.10(a). Thus, they travel only a short distance before hitting the tube wall. The particles then slide down the wall and form a pellet at the bottom of the tube. This makes fixed-angle rotors effective for the differential collection of precipitated materials and harvesting cells. Swing out bucket rotors are prone to cause convection currents during deceleration and the particles also have long sedimentation paths. This makes them inefficient for pelleting particles by differential centrifugation but useful for rate-zonal (Figure 10.10b) and isopycnic techniques. Vertical tube rotors are popular for isopycnic centrifugation. During the centrifugation the sample and gradient reorientate, as shown in Figure 10.10(c). This result is an effectively short sedimentation distance and rapid separation. However, the resolution of the particles is less effective than when a swing out bucket rotor is used.

10.4.4 Balancing rotors

Whatever the type of rotor or centrifugation procedure to be used, the rotor must be balanced; that is, the weight of the tubes and their contents must be distributed evenly around the centre of rotation. If the rotor is imbalanced, it can damage the drive shaft of the centrifuge and even cause it to move. You can see an example of how tubes can be added to a rotor to balance it in Figure 10.11. It is necessary to attach *all* of the buckets when using a swing out rotor even if you are not using all of them for sample tubes. This is because the buckets and their lids are an integral part of the balancing of the rotor.

10.5 Safety and centrifuges

Centrifuges, especially high-speed and ultracentrifuges, are potentially dangerous pieces of equipment. Care and attention to detail must be applied at all times during their use. However, if you follow the procedures outlined in the following sections, then the associated risks will be reduced to a minimum. Note that staff in charge of any centrifuge must read standard operating procedures for their particular machine, taking note of any Medicines and Healthcare Products Regulatory Agency or Health and Safety Executive safety alerts.

10.5.1 Qualified users

Centrifuges must only be operated by suitably trained persons or under their supervision. All staff/students must be trained by someone who is competent in their use before they use any centrifuge. If you notice a fault, report it promptly, do not attempt repairs yourself: only authorized and suitably trained persons may service or repair a centrifuge. Do not use the centrifuge until the fault has been repaired. A log book must be kept for ultracentrifuges and their rotors, as the number of hours used determines the life of the rotor. Excessive use can lead to metal fatigue, where microscopic changes in the metal structure can lead to small cracks, which eventually enlarge to the point of causing metal failure. Where necessary, the details of the centrifugation must be entered in the centrifuge log book. Before using a centrifuge, the rotor chamber should be checked to ensure that it is clean and free of dust or loose debris and condensation. Check also that the drive shaft is clean and free of scratches or burrs.

10.5.2 Rotor faults

Faults with a rotor can lead to its failure: the greater the speed used, the more likely this is to occur and the more extreme the results. Always examine a rotor for *cleanliness* and *damage*. Any extraneous material, including dirt in the rotor cavities, will affect its balance. Dirty rotors must be cleaned by an approved method using, for example, plastic-coated brushes and mild detergents. Soaps and many detergents are alkaline and can attack the anodized surfaces of aluminium rotors. The body of a swing out bucket rotor should never be immersed as hanging mechanisms are difficult to dry and can corrode. Only the buckets should be washed. After cleaning, the rotor must be rinsed with deionized water and air dried with the buckets or cavities pointing downwards. Corrosion, cracks, or dents on any part of the rotor are potentially hazardous and even the most careful design will not protect against its misuse and abuse. When centrifuged, rotors are subjected to stress causing them to be stretched. You can see the effects of stress diagrammatically in Figure 10.12. The first stage of stretching constitutes the elastic limit and the rotor is able to return to its original size when the stress, that is the centrifugation, stops. However, above a certain level of stress, plastic strain occurs and the rotor will not regain its former dimensions. Plastic damage can lead to cracks in the fabric of the rotor and cause its eventual failure. Damaged rotors must *not* be used and must be reported to the appropriate supervisor.

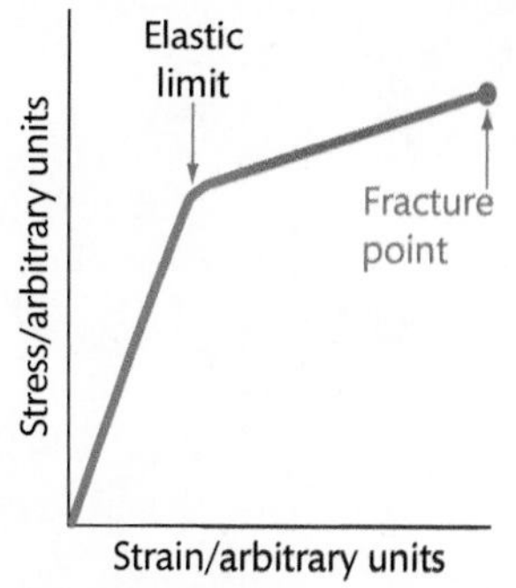

Check that all **O rings**, which are gaskets consisting of a flat rubber or plastic ring used to seal a joint against high pressure, are in place and in good condition. Rotors must never be used if their O rings are missing or perished. Chill the rotor in a refrigerator to the desired temperature before placing it in the rotor chamber as this minimizes the chance of it seizing to the drive shaft. Close the rotor lid when pre-cooling it to prevent condensation in the chambers or buckets.

FIGURE 10.12
Schematic showing the effects of stress on a typical rotor alloy.

10.5.3 Tube checks

Use only correctly fitting tubes and check that they are made of a material that is compatible with the sample solvent. Balance the rotor to within its specified limits as described in Section 10.4. Do not operate the centrifuge without the appropriate rotor lid securely fitted and its seals in place. Never exceed the maximum stated speed for any rotor. Following centrifugation, do not remove the centrifuge tubes or bottles from the rotor with a metal object. Use only the tools provided by the centrifuge manufacturer for that purpose.

10.5.4 After checks

Check the rotor chamber and the rotor for any leaks and clean if necessary. Return the rotor to its normal storage place. Rotors must be stored inverted in a dry environment.

SELF-CHECK 10.4

You need to isolate viral particles from a cell culture lysate when you notice that the aluminium rotor you were going to use has some fine hairline cracks in its surface. You do, however, have a pressing need to prepare the sample. What should you do?

10.6 Examples of clinical centrifugation

In addition to freestanding centrifuges, a number of automated instruments, for example DNA extractors (see Chapter 14), encountered in clinical laboratories include 'built-in' centrifuges to help process specimens. Specialist centrifuges for use in routine clinical laboratories are uncommon. One specific type, the cytocentrifuge or slide centrifuge is described more fully in the *Cytopathology* text of this series. A cytocentrifuge is essentially a low-speed centrifuge with a modified rotor that accommodates microscopic slides, as well as liquid samples. The centrifugal force separates the cells and sediments them as a monolayer onto the slides, while preserving their integrity. This simple technique can be used, for example, to isolate cells from samples of cerebrospinal fluid or fine-needle aspirates directly for cytological examination. Haematocrits are a type of centrifuge that were used to determine the ratio of the relative volumes of erythrocytes and plasma in blood samples but they have been superseded largely by automated blood analysis systems (see Chapter 15), which can determine all haematological indices. The multibucket centrifuges used for serum separations are probably the commonest type found in biochemistry, immunology, and haematology pathology laboratories, although microfuges are used in a range of applications as discussed in Chapter 14, and refrigerated centrifuges find use in specialist complement work (Chapter 13) and some biochemistry assays. For

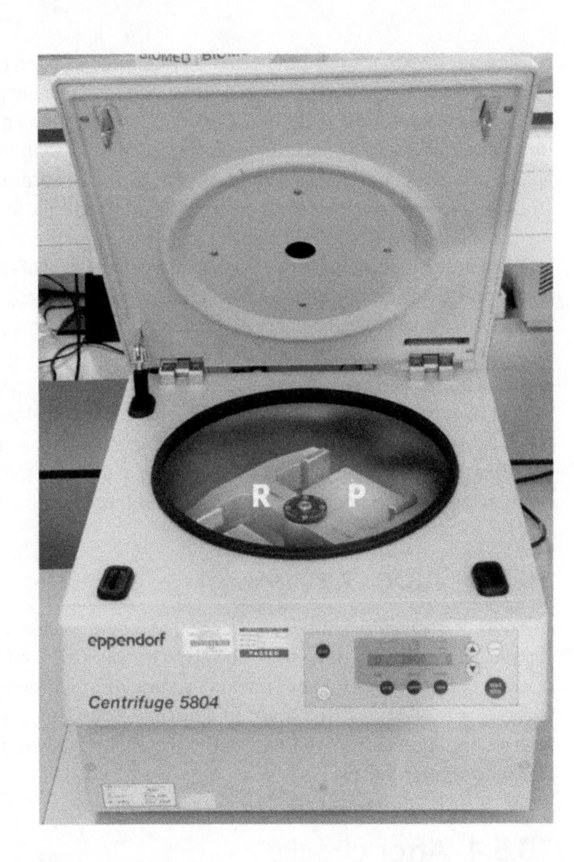

FIGURE 10.13
Centrifuge with a specialized rotor (R) for centrifuging enzyme-linked immunosorbent assay (ELISA) plates. An ELISA plate is indicated (P). Courtesy of Department of Clinical Biochemistry, Manchester Royal Infirmary, UK.

Cross reference

You can read about immunosorbent assays in Chapter 13, 'Immunological techniques'.

Cross reference

You can read more about obtaining blood samples in Chapter 6, and more about blood tubes in Chapter 15.

example, adaptors can allow the 96-well plates used by ELISAs to be directly centrifuged allowing a high throughput of samples (Figure 10.13).

Blood is the most frequently requested of clinical samples that require centrifugation. Once collected, it is added to a blood tube. Most blood tubes are made from a strong plastic like polyethylene terephthalate, and so the blood sample can be directly centrifuged. Many blood tubes also have an inert gel barrier added to them during their manufacture, which aids in the centrifugal separation. The specific gravity of the gel is between those of the blood clot and the serum. During centrifugation the barrier gel moves upward and forms a distinct dividing barrier between the plasma and cells or the serum and clot, respectively (Figure 10.14). Thus, aliquots of the sample can be taken directly from the tube for analysis. Most of the analytes routinely investigated in clinical laboratories remain stable for several days if the tubes are used according to manufacturer's instructions and if stored at 4°C. The introduction of separator gels has enhanced the use of *primary* samples for clinical analyses. A primary sample is one presented to the analyser in the same container in which it was originally collected from the patient. Aliquots may be taken directly from the serum/plasma and directly sampled by the analyser.

Manufacturers generally colour-code blood tubes depending on whether they have a gel barrier, and whether they include an anticoagulant or not, and, if so, what type of anticoagulant is present. For example, tubes with grey-coloured caps contain oxalate and fluoride, and are used for samples to determine blood glucose measurements. The oxalate binds Ca^{2+} and prevents coagulation, while fluoride is an inhibitor of hexokinase, and so prevents glycolysis. The effect of the two additives is to maintain the concentration of blood for several days. Purple-capped tubes also contain an anticoagulant, in this case the metal chelator, ethylenediaminetetraacetic acid. A green cap indicates the tube contains the anticoagulant lithium heparin. These tubes are available with and without gel barriers. The heparin allows the sample to be centrifuged immediately, so whole blood or plasma can be tested. Tubes with red and yellow caps both contain clot activators and are used in most routine chemistry tests on serum. Centrifugation of the former prevents analytes moving between the serum and the clot. The yellow-capped tubes have a gel barrier, which allows for **primary sampling**.

(a)

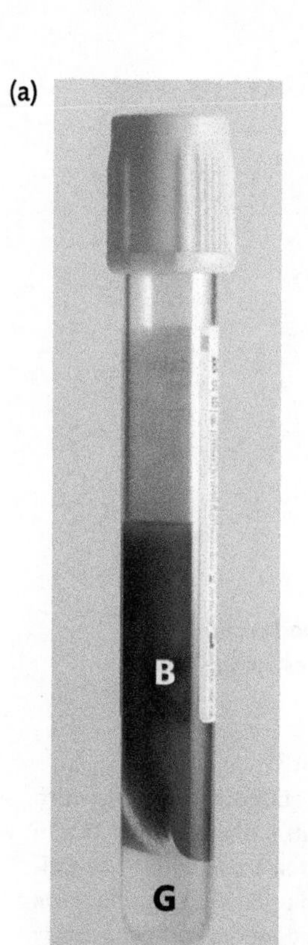

(b)

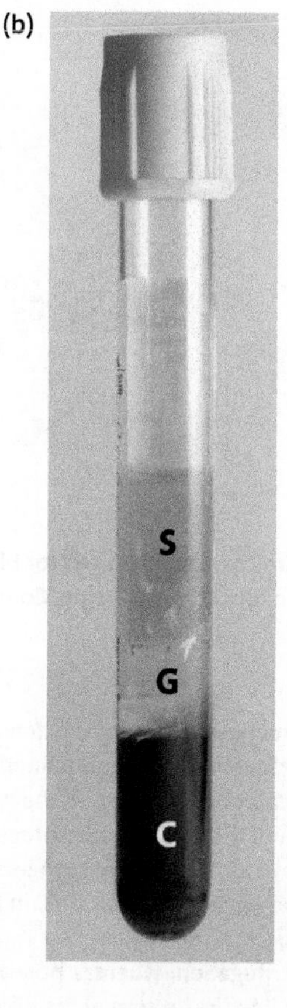

FIGURE 10.14
(a) Blood tube containing a blood sample (B). Note the position of the separator gel (G). (b) Blood tube with a separator gel following centrifugation. Note how the gel (G) has moved up the tube to form a layer between the cells and clot (C) and serum (S). Courtesy of Department of Clinical Biochemistry, Manchester Royal Infirmary, UK.

The automated chemistry analysers used to perform many of the standard clinical tests on patient samples contain centrifuges, which may be standalone devices or occur as discrete units on a tracked system. The tubes are centrifuged in specialized racks. Robots can perform all the operations associated with centrifugation of the samples. Thus, they transfer the sample tubes onto a balance to weigh them and then allocate each tube to an appropriate position on a centrifuge rack. This ensures the final load in each centrifuge rack is balanced before they are transferred to the centrifuge. Blood specimens are centrifuged at 1000 g or higher for 8–12 minutes. Following centrifugation, the robot removes the racks from the centrifuge and may de-cap the tubes before transferring the racks onto a track to transfer to other units for clinical testing. Centrifugation is used extensively in blood banks for a number of procedures (Box 10.5).

Cross reference

The sampling of blood, the centrifugation of clinical specimens, and the use of primary samples are described more fully in Chapters 1 and 2 of *Clinical Biochemistry*.

BOX 10.5 Blood banks

For a number of clinical reasons, blood that contains leucocytes cannot be transfused legally in the UK. Hence, usually within a few hours of collection, leucocytes are removed from blood by filtering it in a process called leucodepletion. The filters used are leucocyte-specific and trap them but not the smaller erythrocytes or platelets. Leucodepletion does not involve centrifugation, but centrifugation is used extensively in blood banks for a number of procedures.

To ensure safety, the blood is tested to check that it is not contaminated with harmful microorganisms and to determine its blood group. Blood must be typed for the ABO and Rhesus (Rh) blood group systems. The ABO system depends on the presence of antibodies in the serum of individuals, which are complementary to the antigens on the erythrocytes. It is these antibodies that determine the identity of the blood group. Thus, individuals of group A have antibodies in their serum to blood group B antigens and those of blood group B possess antibodies to group A. Group AB individuals do not have either type of antibody; however, group O individuals possess both. ABO blood groups are identified for all donors and recipients

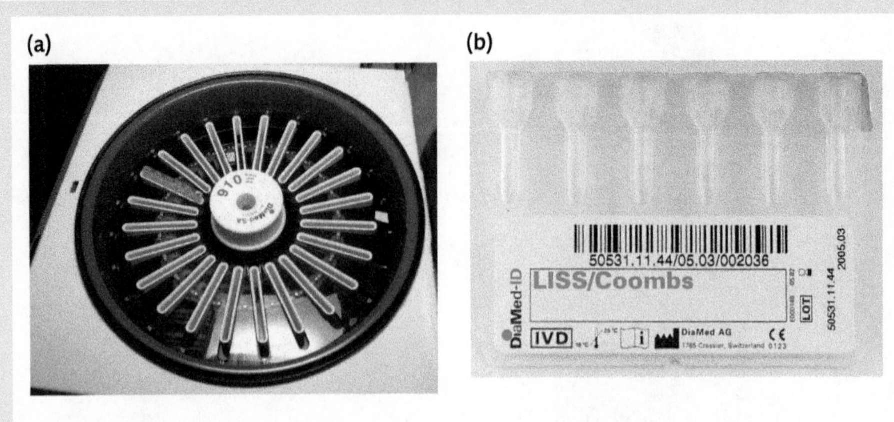

FIGURE 10.15

(a) A centrifuge with a specialized rotor that allows gel cards to be centrifuged for blood group typing as outlined in the text. (b) A typical gel card used in blood group typing. Courtesy of the Manchester Blood Transfusion Service, UK.

to ensure that blood used for transfusion is of the same or a compatible type. If ABO incompatible blood is administered to a patient, the antibodies may mediate an immediate and fatal transfusion reaction.

The terms Rh positive and Rh negative refer to the presence or absence of the Rh(D) antigen. An Rh incompatibility between mother and foetus is a common cause of haemolytic disease of the newborn in white people.

The majority of transfusion laboratories now use gel technology for blood grouping and compatibility testing, which involves some centrifugation. Recent years have seen increasing use of the Diamed typing system to detect haemagglutination. This system uses typing monoclonal antibodies (mABs), which are distributed in a gel contained in individual tubes set in plastic 'cards'. Erythrocytes to be typed are added to the antibodies and the cards are centrifuged using a specialized rotor. Specific blood groups are recognized by the complementary mAB and agglutination occurs. In those tubes where agglutination occurs, the agglutinates remain on top of the gel during centrifugation, whereas non-agglutinated cells sediment through it to the bottom of the tube (Figure 10.15).

In addition, the *production* of blood products, such as erythrocytes, platelets, and plasma, all involve the large-scale centrifugation of blood using dedicated centrifuges.

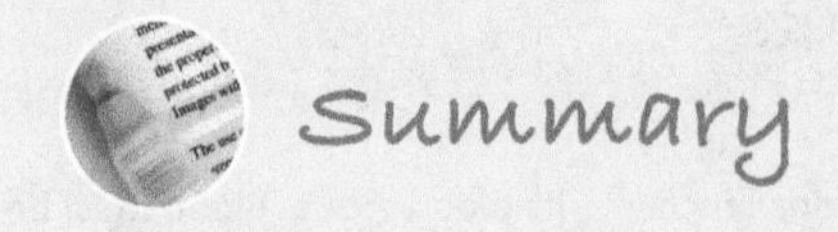

Summary

- Centrifugation is a separation technique conducted in an instrument called a centrifuge. A centrifuge holds a spinning rotor containing samples in tubes that are rotated around a central axis by a motor. This spinning of the rotor generates a centrifugal force that separates the components in the mixture on the bases of differences in size and/or density.
- Particles sediment when subjected to a radial acceleration or centrifugal field (G). This is normally expressed as a relative centrifugal force (RCF) as multiples of the earth's gravitational field (g).
- A variety of centrifugation tubes, rotors, and centrifuges are available, which allows a range of different centrifugation methods to be used as appropriate.

- The major centrifugation techniques used are differential and density gradient centrifugation. The former separates particles on differences in size. Density gradient separation can use rate-zonal and isopycnic centrifugations. Particles that differ in their sedimentation coefficients can be separated by rate-zonal centrifugation, while isopycnic methods rely on differences in the densities of the particles.
- All appropriate safety considerations must be observed when using centrifuges. For example, only clean and undamaged rotors must be used.
- Centrifugation has applications in biomedical science where it is commonly used, for example in separating blood materials or isolating DNA. Clinical laboratories use specialist cytocentrifuges to isolate components onto slides prior to cytological analysis. In research laboratories, centrifugation is used in preparing fractions of cellular constituents such as macromolecules and organelles.

Further reading

- **Goodman, T. Centrifuge safety and security. *American Laboratory* 2007; 39: 20–1.**
  An interesting article detailing design features relating to safety and security that are available on a number of centrifuges.
- **Goodman T. Centrifuge rotor selection and maintenance. *American Laboratory* 2007; 39: 12–14.**
  An interesting, short read that describes the different types of centrifuge rotors, their applications, and also outlines procedures for their safe handling and maintaining their life.
- **Goodman T. Choosing the right centrifuge for your application. *American Laboratory* 2008; 40: 28–29.**
  A short, useful article, whose title says it all.
- **Graham JM. *Biological Centrifugation*. Oxford: Bios Scientific Publishers, 2007.**
  Excellent coverage of the basics of centrifugation. Also includes a number of useful centrifugation protocols.
- **Laue T. Biophysical studies by ultracentrifugation. *Current Opinion in Structural Biology* 2001; 115: 579–83.**
- **Schuck P. Analytical ultracentrifugation as a tool for studying protein interactions. *Biophysical Reviews* 2013; 5: 159–71.**

Questions

1. Calculate the RCF to the nearest 100 *g*, experienced by a particle 8.0 cm from the centre of rotation when it is centrifuged at 3500 rev min^{-1}.
2. Which of the following centrifuges:
   - microfuge
   - ultracentrifuge
   - bench centrifuge
   - high-speed refrigerated centrifuge

   would be your first choice to:

   (a) Isolate bacterial cells from liquid medium

   (b) Collect a fine precipitate of DNA molecules

(c) Isolate mitochondria from a tissue homogenate

(d) Harvest the relatively small poliomyelitis viral particles (diameter 28 nm).

3. Which of the following would not affect the sedimentation velocity of a particle during centrifugation?

(a) Solvent density

(b) Atmospheric pressure

(c) Particle density

(d) Distance of the particle from the centre of rotation

(e) Angular velocity

(f) *Mr* of the particle.

4. A viral particle takes 10 minutes to move 1 cm when centrifuged at 20,000 rev min^{-1}. How long would it take to move the same distance if the speed was doubled?

5. Which would have the larger frictional ratios (*f*) and sedimentation coefficients (*s*): enzyme molecules or DNA fragments of comparable M_r?

6. Non-filamentous influenza viral particles have a sedimentation coefficient (*s*) of 750 S. How far would a virus particle travel in 10 minutes, when centrifuged at 30,000 rev min^{-1} at the r_{av} position of 6.12 cm?

7. Serum albumin has a sedimentation coefficient (*s*) of 4.6 S. Determine its M_r, given its partial specific volume ($\bar{v}$) and frictional coefficient (*f*) are 0.74 $cm^3\ g^{-1}$ and $6.63 \times 10^{-8}\ s^{-1}$, respectively. The density of water at 20°C is 0.998 g cm^{-3}.

8. You are going to perform a centrifugation experiment using CsCl (a corrosive salt) when you spill a *little* of the solution into a rotor bucket. You do, however, have to complete the experiment quickly, as the results are needed in a hurry. Should you ignore this small spillage and proceed with the experiment?

Answers to self-check and end-of-chapter questions are available at the end of the book.

11

Chromatography

Qiuyu Wang, Nessar Ahmed, and Chris Smith

Learning objectives

After studying this chapter, you should be able to:

- Define the term chromatography
- Discuss the general basis of chromatographic separations
- List some of the major chromatographic techniques
- Describe the theoretical bases of specific chromatographic techniques
- Discuss the relative efficiencies of some chromatographic methods
- List some of the uses of chromatography in the clinical and biological sciences

Introduction

Chromatography was first used by the Russian botanist Mikhial Tsvet, who used it to separate plant chlorophyll and carotenoid pigments. Tsvet first gave a written description of the method in 1903, although he did not coin the term **chromatography**, which he adopted from the Greek words chroma, colour in reference to the plant pigments, and graphein, meaning to write, until 1906. Chromatography is now the collective term for a family of analytical techniques used to separate the components of mixtures of molecules for their identification and possible estimation of their concentrations in the original mixture. A major difficulty that is often encountered when investigating clinical and biological samples is the presence of substances that can interfere or be confused with the substance or analyte under investigation. This means it is often necessary to separate and isolate a particular analyte from other contaminants in the mixture to allow its concentration to be determined. The principal uses of chromatographic methods are to remove contaminating materials from the analyte of interest, to aid in the identification of unknown analytes, and to estimate the concentration of analyte in the sample.

All chromatographic techniques are based on differences in the relative affinities of the molecules in a mixture of two different and **immiscible phases**, one of which is **mobile** and the other **stationary**. Mobile phases may be liquids or gases that flow over or through the stationary phase, which can be a solid or an immobilized liquid. Different combinations of phases can be used. For example, in column chromatography, particles of a hydrated gel are used as the stationary phase and a liquid moving through the gel is the mobile phase. In many cases these have been developed into extremely

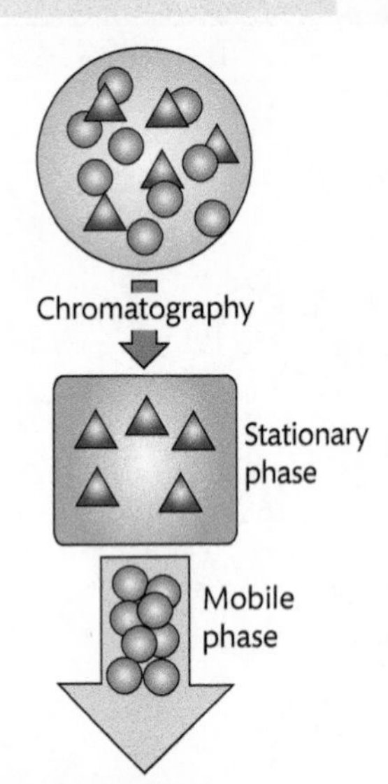

FIGURE 11.1
Schematic illustrating the separation of two substance using differences in their relative affinities for two phases in chromatography.

high-resolution techniques by forcing the liquid through the column under high pressure. These latter methods are usually referred to as high-performance liquid chromatography or HPLC. Whatever the specific type of chromatography, different substances in a mixture can partition differentially between the two phases and so are separated from one another. You can see this schematically in Figure 11.1. Even complex mixtures of closely related molecules can be separated if an appropriate combination of phases is selected.

Most types of chromatography are used in biomedical science laboratories to assist in analysing and purifying analytes from a variety of clinical samples. You will be introduced to a number of these applications in the chapter.

11.1 Partition or distribution coefficient

The basis of all forms of chromatography is the **partition** or **distribution coefficient** (K_d), which describes the way in which a substance distributes at equilibrium between two immiscible phases. For example, the K_d of a substance X, which distributes between the two immiscible phases A and B, is given by:

$$K_d = \text{The concentration of X in A/The concentration of X in B}$$

given that the distribution of X between A and B is at equilibrium. You can see an illustration of this in Figure 11.2.

Partition coefficients are normally determined at 20 or 25°C. Often, one of the two phases is water and the other a hydrophobic solvent such as octanol. This gives an octanol–water partition coefficient, K_{OW}. Note that the K_{OW} is not the same as the ratio of the solubility of the substance in octanol to that in water because the two phases are not fully immiscible! At equilibrium, octanol contains 2.3 mol dm^{-3} of water and the water phase 4.5×10^{-8} mol dm^{-3} of octanol. The measured values of K_{OW} for organic compounds vary from 10^{-3} to 10^{7}. This is an enormous range. Hence, they are often quoted as their logarithmic value (log K_{OW}), which condenses the range from −3 to 7. The K_{OW} value describes the relative hydrophobicity/hydrophilicity of the compound in question: the lower the value, the more hydrophilic the compound; the larger it is, the more hydrophobic.

SELF-CHECK 11.1

What are the units of K_d, K_{OW}, and log K_{OW}?

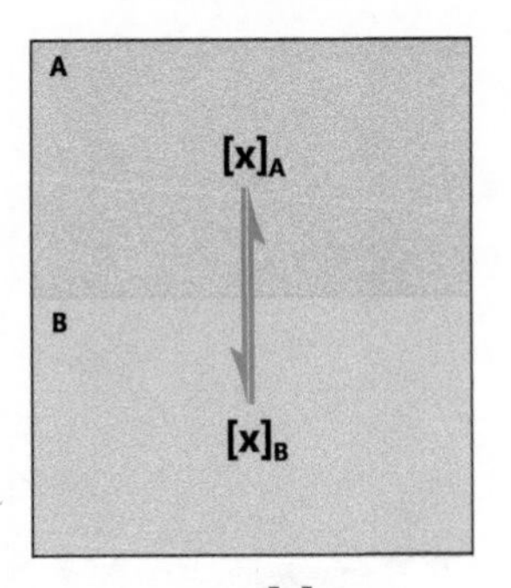

$$k_d = \frac{[x]_A}{[x]_B}$$

FIGURE 11.2
The definition of the partition coefficient for a solute (X) following its equilibrium partitioning between the two immiscible phases, A and B.

In chromatography, the mobile and stationary phases must be chosen such that the molecules to be separated differ in their partition coefficients. This, in turn, depends upon them differing in some particular physical property. If you examine Table 11.1 you will see some of the properties upon which different types of chromatography rely.

TABLE 11.1 **A summary of the differences in physical properties of molecules in the mixture that can be used as a basis for their separation by various forms of chromatography**

Property	Chromatographic techniques
Solubility and polarity	Paper chromatography Thin-layer chromatography
Size (mass)	Gel-filtration chromatography
Charge	Ion-exchange chromatography
Hydrophobicity	Hydrophobic chromatography Gas–liquid chromatography
Ligand binding	Affinity chromatography Metal chelation chromatography

Polar molecules have an unequal distribution of electrons owing to differences in the electronegativity of the atoms from which the molecule is composed. For example, water is a polar molecule, as oxygen has a greater electronegativity than hydrogen. This means the electrons are more attracted to the oxygen leaving less of an electron density around the hydrogens. As a result, the oxygen acquires a slightly negative charge and the hydrogens have an equal but positive one (Figure 11.3).

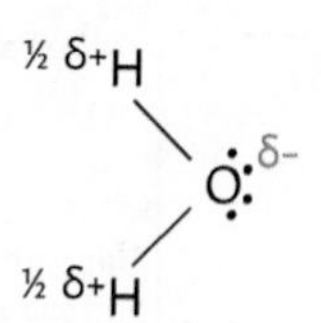

FIGURE 11.3
Structure of a water molecule.

The polarity of molecules influences their solubility and **adsorption** on to surfaces. Thus, in chromatographic methods based on differences in solubility and polarities, the particles partition to different extents between two immiscible liquids, and so separate. This is often described as 'differential partitioning'. We will describe in Section 11.2 how differences between the solubilities and adsorption properties of molecules are exploited in paper and thin-layer chromatographies. Differences in the *sizes* of molecules can also be exploited in gel-filtration chromatography. Molecules may become charged under certain conditions. These charges can be used as a basis for their separation, as in ion-exchange chromatography. Hydrophobic and gas–liquid chromatographies exploit differences in the partitioning of hydrophobic molecules between two phases. Finally, differences in how molecules bind to ligands are utilized as the basis for their separation in affinity and **metal chelation chromatography** (Sections 11.2–11.5).

The major types of chromatography likely to be encountered in biomedical science laboratories are those grouped together as:

- planar techniques
- column chromatographies
- gas–liquid chromatography.

Each of these and some of their uses will be discussed in turn.

11.2 Planar chromatography

Planar or open-bed chromatography uses a stationary phase consisting of a thin, essentially two-dimensional, sheet. The two major types you are likely to encounter are paper and thin-layer chromatography (TLC). In paper chromatography, a sheet or a narrow strip of paper binds the stationary phase. In TLC, a film of solid particles, bound together for mechanical strength with a binder such as calcium sulfate, is coated on a supporting plate and also binds a stationary liquid phase. Both are examples of partition chromatography because molecules distribute themselves between bound and mobile liquid phases, although adsorptive effects to the particles in TLC also contribute to the separation.

11.2.1 Paper chromatography

In paper chromatography, the stationary phase consists of a layer of water or other polar solvent molecules that are bound to the cellulose fibres of the paper, even although it may appear dry. Materials to be separated are applied to the paper, generally as a series of discrete spots. One end of the paper is then dipped into the mobile liquid phase, which moves through the paper by capillary action. The molecules to be separated are differentially partitioned between the stationary and mobile phases and so move along the paper at different rates and separate. Paper chromatography is a useful method for assessing the purity of substances but is now seldom used in clinical laboratories, having been superseded by TLC.

11.2.2 Thin layer chromatography

TLC relies on the differential partitioning of substances between stationary and moving liquid phases. In this respect, it resembles paper chromatography. A solvent is used to suspend an adsorbent material, which is then bound to an inert glass, plastic, or aluminium plate as a thin layer and allowed to dry. The thin layer of solvent remaining in the bound adsorbent forms the stationary phase (hence the name). A range of materials are suitable for forming the adsorbent layer and, indeed, mixtures of materials can be used. Some of them are listed in Table 11.2; the choice depends on the material to be analysed.

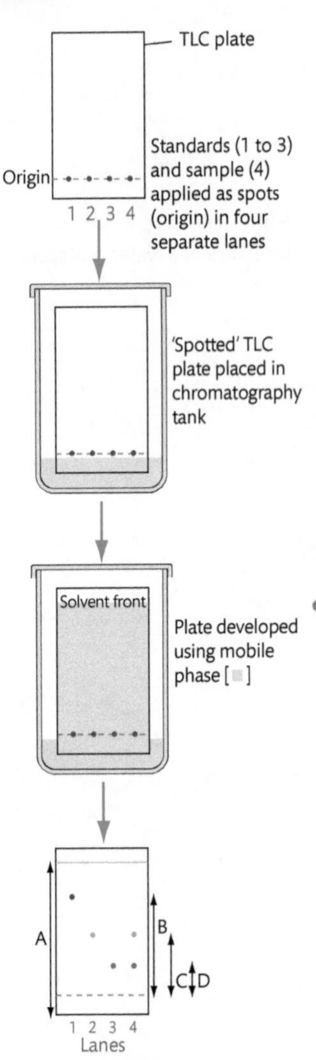

FIGURE 11.4
General procedure for thin layer chromatography (TLC). See text and Method box 'Generalized procedure for TLC'. In this case, lanes 1, 2, and 3 have been used to 'run' substances of known identity ('standards'). Lane 4 was used for an unknown sample. On the basis of its migration, the unknown would appear to contain standards 2 and 3.

TABLE 11.2 **Suitable adsorbents for thin-layer chromatography and their applications**

Adsorbent	Used in separating
Aluminium oxide	Steroids
Calcium phosphate	Proteins
Cellulose	Proteins
Magnesium carbonate	Porphyrins
Silica gel	Steroids, amino acids, lipids

Again, a mobile liquid phase migrates through the thin layer by capillary action. A wide range of solvents may be used as mobile phase, including water, ethanoic acid, ether, hexane, methanol, and benzene, which can be mixed together in different combinations and proportions as appropriate. A general procedure for TLC is outlined in the Method Box 'Generalized procedure for TLC' and illustrated in Figure 11.4.

METHOD Generalized procedure for TLC

- **The adsorbent material is prepared by mixing it with a binding agent such as calcium sulfate or plaster of Paris (10%) and water to form a slurry. The binding agent helps hold the adsorbent to the plate.**
- **The slurry is applied to the plates as a thin uniform layer using a plate former and allowed to set.**
- **The plates are activated by heating in an oven at 100–120°C to remove free liquid, with the remainder forming a thin layer bound to the adsorbent particles. The final thickness of the adsorbent layer is between 0.25 and 2 mm.**
- **A 1–20 μdm^3 (μL) sample is applied to the plate as a concentrated spot or band using a micropipette or microsyringe and allowed to dry.**
- **The TLC plate is placed in a tank that contains the mobile phase and developed for 30–90 minutes, as shown in Figure 11.4. The atmosphere within the tank must be saturated with solvent vapour in equilibrium with the mobile phase to minimize evaporation from the plate surface and prevent the composition of mobile phase changing because of losses of its volatile components. The mobile phase travels up the plate by capillary action. Development is stopped when the moving edge of the mobile phase (usually referred to as the *solvent front*) approaches the top of the plate (Figure 11.4).**
- **Following their separation on the TLC plate, the positions of the separated components of the sample can be determined and the presence of the analyte of interest confirmed and its concentration in the sample possibly estimated.**

As with all chromatographic techniques, the substances to be separated partition differentially between two phases. Initially, the stationary phase consists of the liquid retained by the adsorbent layer during its drying. Separation is achieved because as the mobile phase carries sample and standards up the thin layer, the more polar molecules in them will associate with the thin layer. In contrast, the relatively hydrophobic substances will be more soluble in the mobile phase and be carried further along the plate. However, the mobile phase may be a complex mixture and contain polar liquid molecules. As these move through the thin layer by capillary action, they may bind to the adsorbent material and will form what is essentially a new stationary phase, which may assist in the separation. TLC plates have a large surface area. Hence, significant adsorption of sample molecules to the particles can also take place. This ability is often due to the presence of hydroxyl groups (–OH) on their surface, which can form hydrogen bonds with molecules in the sample. The more polar the molecules in the sample, the greater the strength of the bonds and the more adsorption occurs. Adsorption to the particles can increase the effectiveness of separation. It is apparent that the factors that affect separation in TLC are complex!

Following its development, it is necessary to detect the separated components on the plate. A number of methods are available to achieve this, including the use of:

- ultraviolet (UV) light
- specific colouring reagents
- autoradiography
- sulfuric acid

Examining the TLC plate under UV light will show the positions of compounds that absorb or fluoresce in the presence of UV light. For example, nucleotides and nucleic acids absorb UV light, and so appear as dark spots. Spraying of plates with specific colouring reagents can be used to stain certain compounds. For example, ninhydrin will react with amino acids to form purple-coloured complexes, which will indicate their location on the plate. Alternatively, the TLC plates can be sprayed with sulfuric acid and then heated to 110°C. Most organic compounds are charred by this procedure and show up as brown-black spots or bands. Radioactively labelled compounds will present as dark spots or bands on X-ray film after the TLC plates have been subjected to autoradiography (Box 11.1)

It is usual to calculate the **Rf** (relative to front) value, which is the distance travelled by any component relative to that moved by the solvent using the expression:

$$R_f = \text{migration distance of analyte/migration distance of solvent front}$$

In theory, R_f values are identical for any given substance analysed under the same chromatographic conditions. However, given that experimental conditions vary, it is more usual to also 'spot' known reference compounds, in addition to the samples to be analysed, and subject them all to TLC. Component(s) in a mixture can then be tentatively identified by comparing the R_f value(s) to those of the standards. You can see this diagrammatically in Figure 11.4, where the unknown compounds in the sample (lane 4) have migrated the same distances (C, D) relative to the front (A) as the standards in lanes 2 and 3.

Cross reference

See the *Histopathology* book in the same series, and Chapter 12, 'Electrophoresis', in this volume.

SELF-CHECK 11.2

Is it possible to have an R_f value of 1.2 for a particular analyte? What effect would changes in temperature have on the TLC? What difference would an increase in room temperature make to R_f values?

BOX 11.1 Autoradiography

Autoradiography is a simple and sensitive method of producing a photographic image of a specimen containing radioactively labelled material(s). The image is formed by placing an X-ray film directly in contact with a chromatogram, histological preparation, or electrophoresis gel. These are then wrapped in a material which prevents external light fogging the film. However, radiation from the labelled materials impinges on the film causing the deposition of silver grains in its emulsion. The length of exposure varies depending upon the intensity of the radiation source: it can be as little as a few minutes or up to several days. The package is often stored in a freezer at −70°C during longer exposures to reduce the diffusion of the molecules in the sample. Once the film is developed, it forms a radiograph, whose patterns of deposited silver record the location(s) of radioactive substance(s) in the specimen. The intensity of the developed spots or bands on the film is proportional to the dose of radiation and thus the concentration of the labelled material.

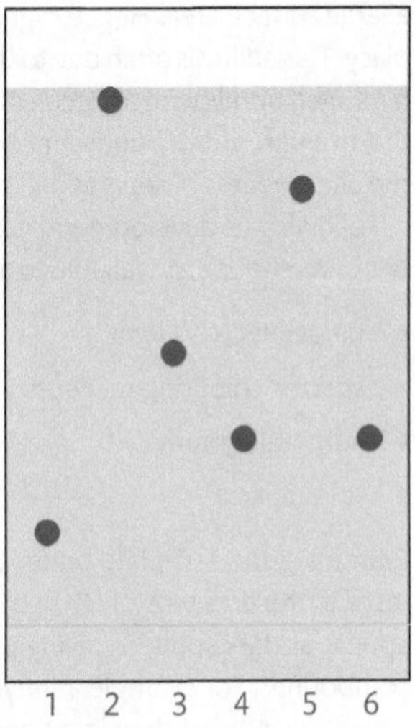

1 2 3 4 5 6

FIGURE 11.5

Thin layer chromatographic separation of morphine (lane 1), cocaine (2), nicotine (3), methadone (4), heroin (5), and a clinical sample (6) using silica gel as the adsorbent material and a mobile phase consisting of ethanol:dioxin:benzene:ammonium hydroxide (4:40:50:5).

TLC has a number of advantages over some other chromatographic techniques, particularly over paper chromatography. One particular advantage is its versatility. If you examine Table 11.2 you will see it can be applied to the separation of a number of differing types of substances. Moreover, separation is quick, often requiring as little as 30–90 minutes, it is simple to perform and is inexpensive. Hence, it is often the first method of choice, particularly when the components of the mixture are known to be structurally similar. Its major disadvantage is that it is not especially suitable for quantitative measurements, for example for determining the amounts of each substance in a mixture. However, TLC is used successfully to identify a number of substances of clinical interest, including drugs, lipids, amino acids, and vitamins. Figure 11.5 illustrates its use in separating a number of drugs of abuse.

Once separated using TLC, analytes can be located on the plate using UV light, and any of potential interest can be extracted from the plate and identified using gas chromatography–mass spectrometry.

SELF-CHECK 11.3

Calculate the R_f value of the drugs and the unknown analyte used in Figure 11.5. What is the most likely identity of the unknown analyte?

11.3 Column chromatography

In column chromatography, the stationary phase is organized into a column by packing it into a glass or steel cylinder (Figure 11.6), which is in equilibrium with a liquid mobile phase. Equilibrium is achieved by using a gravity-based system or motorized pump to move the mobile phase through the column. The sample is added to the top of the column as a band, which is then flushed through the column by the mobile phase. Individual components in the sample are separated within the column and eluted from it at different times as the mobile phase leaves the column. The liquid leaving the column and carrying the (hopefully) separated molecules is often called the eluant. It is normally collected in discrete fractions in a series of tubes, either manually or automatically using a fraction collector. The different fractions can be analysed and an elution profile or chromatogram constructed by plotting a graph of the amounts of substances eluted against time, eluant volume, or fraction number, as shown in Figure 11.6. Hopefully, different fractions will contain different components of the original mixture and so it will at least have been partially separated.

Whatever the form of column chromatography, it is necessary to be able to detect the analytes of interest as they leave the column in the elution buffer. If the analytes are coloured then this is not a problem, but many materials of biomedical interest are colourless. However, most proteins and nucleic acids are colourless but absorb light at about 280 and 260 nm, respectively. This is also the

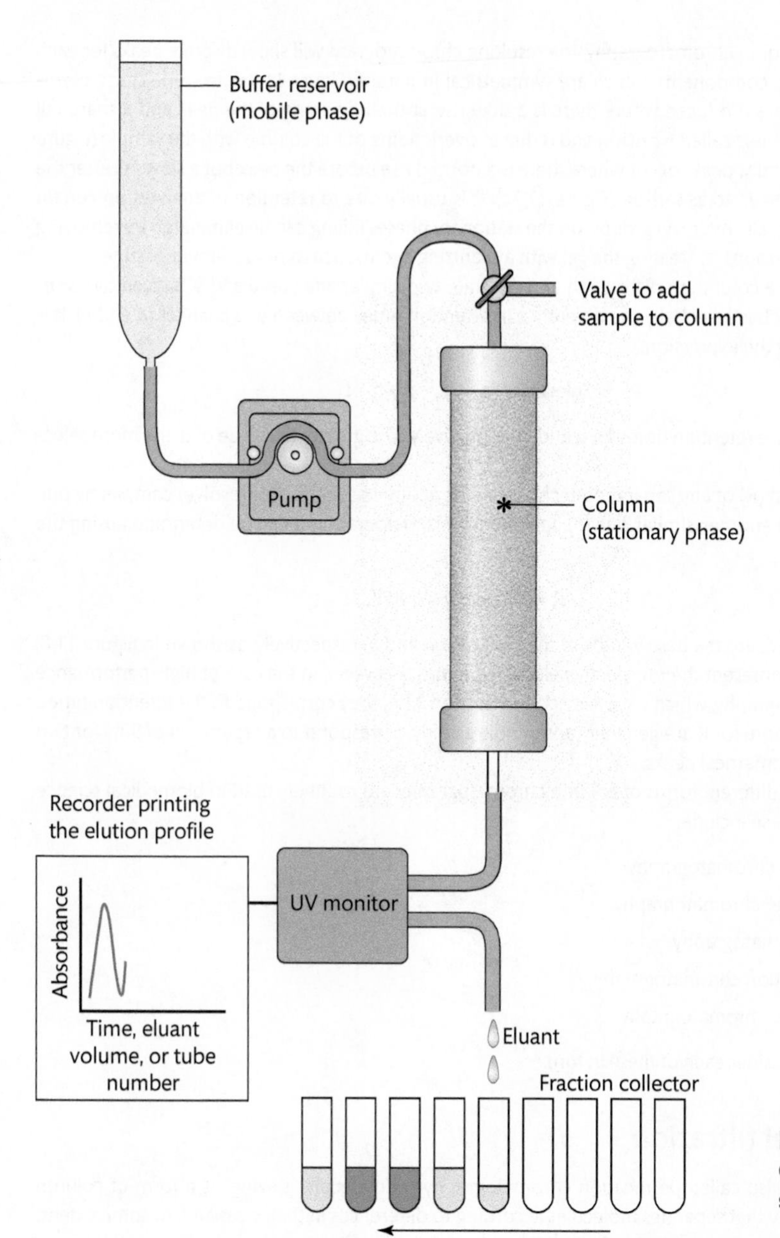

FIGURE 11.6
Outline of a general column chromatography system (see text for details). UV: ultraviolet.

case for many other compounds of clinical interest. Thus, they can all be detected by monitoring the column using UV light, which will indicate when they are being eluted.

It is also possible to exploit the biological activities of some substances to monitor their separation. For example, in the cases of enzymes the fractions of eluant can be analysed for the reaction they catalyse. Carbohydrates are often detected by monitoring the refractive index of the eluant, which changes when a sugar is present.

Cross reference

See Chapter 9, 'Spectroscopy', and Chapter 13, 'Immunological techniques'.

The movement of any one analyte through a column is described by its individual retention time (t) or elution volume (V_e). The retention time for an analyte is the period between addition of this substance to the column and its maximum concentration in the eluant. The term V_e is the volume of mobile phase required to elute a particular analyte from the column. It is used particularly when referring to separations by gel filtration (see Section 11.3.1). The time taken for an analyte to pass through the column without any interaction with the stationary phase is the dead time (t_o) and the volume of mobile phase required for its elution is the void volume (V_0).

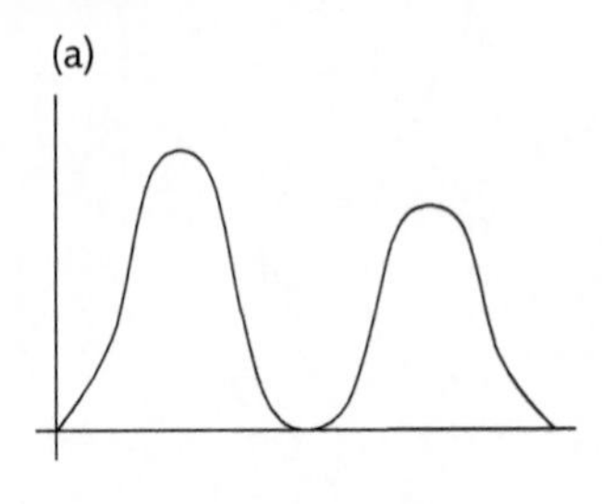

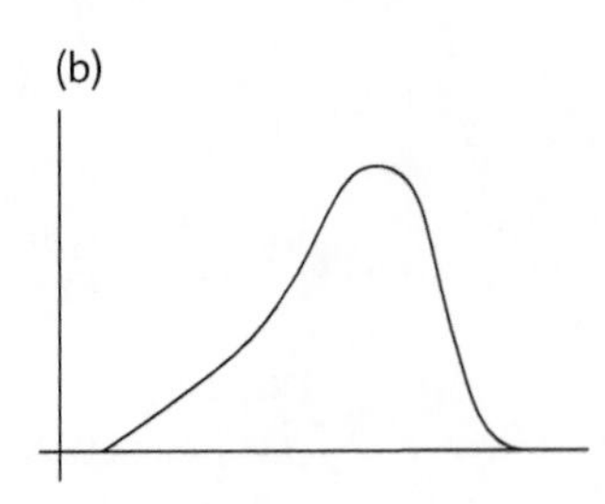

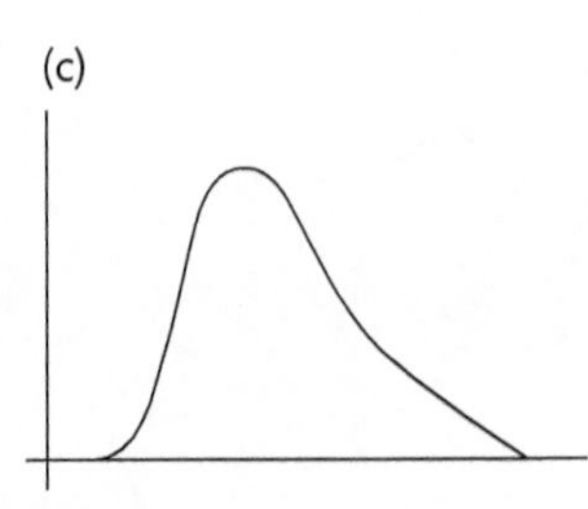

FIGURE 11.7
(a) Shows the symmetrical peaks following an ideal separation of two components by column chromatography. Components isolated by column chromatography showing (b) fronting and (c) tailing.

Following column chromatography, the resulting chromatogram will show discrete peaks for each of the separated components, which are symmetrical in nature (Figure 11.7a). In some cases, asymmetrical peaks are produced where there is a slow rise at the beginning of the peak and a sharp fall after the peak. This is called **fronting** and is due to overloading of the column with the sample (Figure 11.7b). Occasionally, peaks occur where there is a normal rise before the peak but a slow fall after the peak; this is referred to as **tailing** (Figure 11.7c). It is usually due to retention of analytes on certain binding sites, usually hydroxyl groups, on the stationary phase. Tailing can be eliminated by removing these hydroxyl groups by treating the gel with a silanizing agent, such as hexamethyldisilazine.

The ability of a column to distinguish and separate two similar analytes *a* and *b* is given by its selectivity (α), which is essentially the difference in retention times between two analytes (*a* and *b*). It is calculated using the expression:

$$\text{Selectivity } (\alpha) = (t_b - t_0)/\ t_a - t_0$$

where t_a and t_b are retention times for *a* and *b*, respectively. The greater the value of α, the more selective the column.

The resolution (R) of any separation technique is its ability to separate (or resolve) completely one analyte (*a*) from another, similar one (*b*). In column chromatography, it can be determined using the expression:

$$R = 2(t_{Rb} - t_{Ra})/\ W_a + W_b$$

where W_a and W_b are the base widths of the peaks for *a* and *b*, respectively, as shown in Figure 11.8, and t_{Ra} and t_{Rb} represent their respective elution volumes. However, in the case of high-performance liquid chromatography, which we will discuss in Section 11.4, they correspond to the retention times. Values of 1 or more for R are generally acceptable as they correspond to a separation of 98% for two consecutive symmetrical peaks.

A number of different forms of column chromatography are routinely used in biomedical science laboratories. These include:

- gel filtration chromatography
- ion-exchange chromatography
- affinity chromatography
- metal-chelation chromatography
- hydrophobic chromatography

We will now consider each of these in turn.

11.3.1 Gel filtration

Gel filtration, also called permeation chromatography or 'molecular sieving', is a form of column chromatography that separates molecules according to differences in their sizes and, to some extent, shapes. The column is composed of polymeric materials hydrated in the form of microscopic porous

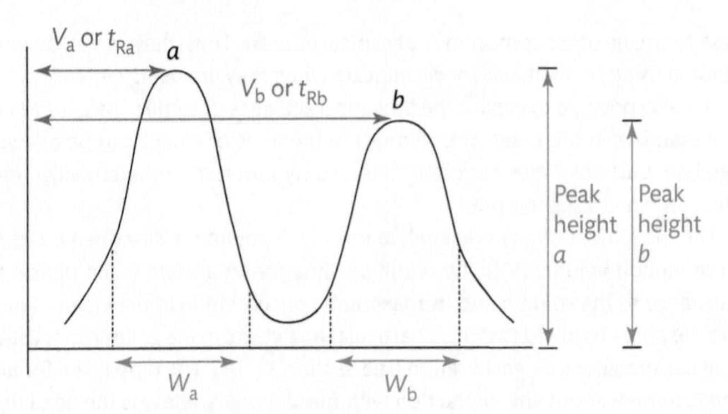

FIGURE 11.8
Definitions of elution volumes or times, peak heights, and peak widths.

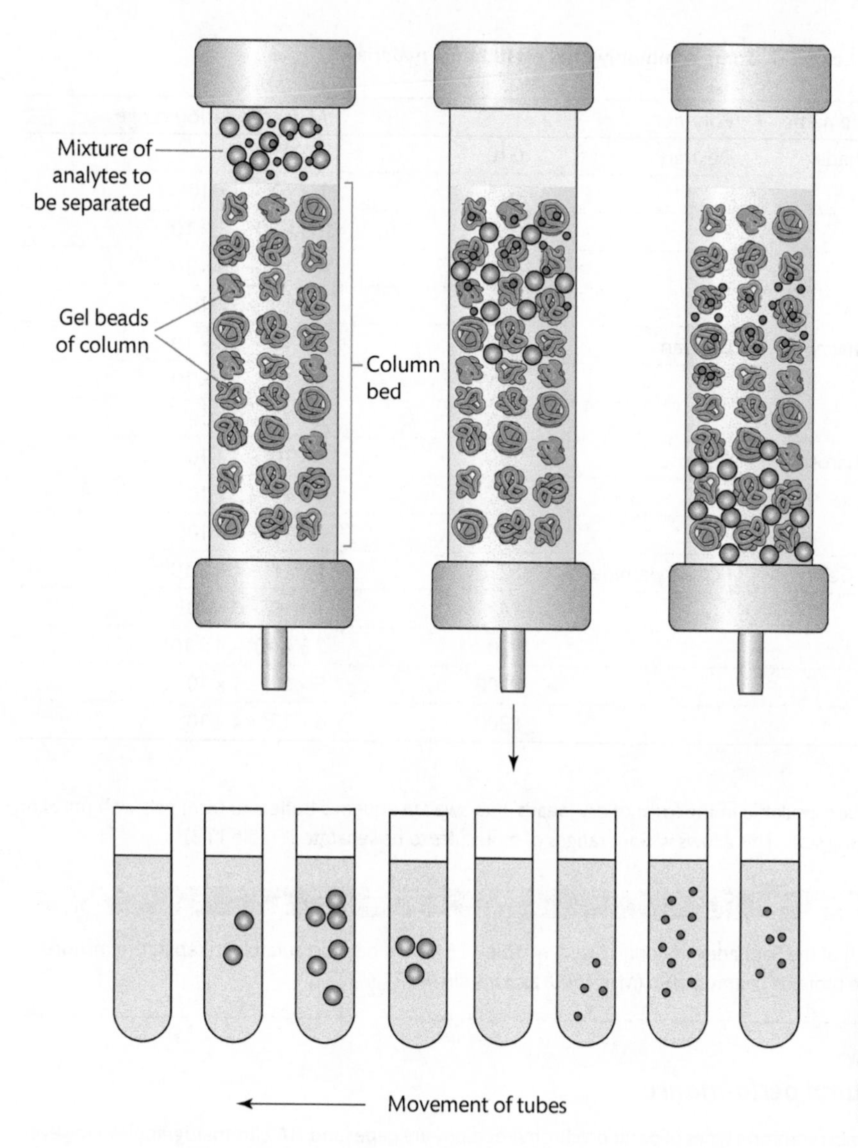

FIGURE 11.9
Schematic outline of separation by gel filtration. The separation of two differently sized particles in a mixture is illustrated. See text for details.

beads. These are in equilibrium with a buffered solution suitable for use with the molecules to be separated, which usually has a constant composition and pH. This is an **isocratic elution**.

Molecules are washed out of the column in order of *decreasing* size. You can see this diagrammatically in Figure 11.9. Each bead is permeated by a three-dimensional network of pores that can only be penetrated by molecules of a specific size range. Molecules bigger than the pore dimensions are completely excluded from the gel particles and will pass through the spaces between the beads as they are washed through the column. The volume of liquid required to elute such molecules from the column is its void volume (V_0). Smaller molecules that can enter the pores of the beads completely, however, will be distributed between the solvent that is both inside and outside the beads and will therefore be eluted by a volume equal to the total volume of the column (V_t). Molecules between these two extreme sizes will be eluted by intermediate volumes. Hence, if a mixture of differently sized molecules is applied to the column the largest ones will pass through the column faster than the smaller ones. Thus, the larger molecule will require a lesser volume of eluant than the smaller ones to move it through the column.

A variety of commercial materials are available for gel filtration. Sephadex is a commonly used material and consists of particles of dextran that have been modified to give varying degrees of cross-linking and therefore different pore sizes. Other substances used include agarose and polyacrylamide.

TABLE 11.3 Some commonly used gel filtration materials

Trade name	Polymer		M_r fractionation range
Sephadex	Dextran	G10	<700
		G25	$1 \times 10^3 - 5 \times 10^3$
		G50	$1.5 \times 10^3 - 3 \times 10^4$
		G100	$4 \times 10^3 - 1.5 \times 10^5$
		G200	$5 \times 10^3 - 6 \times 10^5$
Sephacryl	Dextran	S200	$5 \times 10^3 - 2.5 \times 10^5$
		S300	$1 \times 10^4 - 1.5 \times 10^6$
		S400	$2 \times 10^4 - 8 \times 10^8$
Sepharose	Agarose	2B	$1 \times 10^4 - 4 \times 10^6$
		4B	$6 \times 10^4 - 2 \times 10^7$
		6B	$7 \times 10^4 - 4 \times 10^7$
Bio-Gel	Polyacrylamide	P2	$1 \times 10^2 - 1.8 \times 10^3$
		P6	$1 \times 10^3 - 6 \times 10^3$
		P30	$2.5 \times 10^3 - 4 \times 10^4$
		P100	$5 \times 10^3 - 1 \times 10^5$
		P300	$6 \times 10^4 - 4 \times 10^5$

They are available in the form of dry 'beads' that swell in aqueous buffers to form gels with pores of differing sizes. This allows various ranges of molecules to be separated (Table 11.3).

SELF-CHECK 11.4

Which of the Sephadex materials listed in Table 11.3 would be most suitable to separate a mixture of the proteins haemoglobin (M_r 64,500) and insulin (M_r 11,470)?

Column performance

The most common types of partition chromatography are paper and TLC chromatographies. However, gel filtration is also a partition method, in which separations are based on differences in size. Thus, molecules to be separated by gel filtration must have different distribution coefficient (K_d) values for their successful separation. In gel filtration, the K_d calculated for a given type of molecule represents that fraction of the stationary phase available to it. In practice, K_d is difficult to determine in relation to gel filtration, and it is usually replaced by a K_{av} value as there is a constant relationship between K_{av} and K_d. A value of K_{av} for a solute can be determined using the equation:

$$K_{av} = (V_e - V_o)/(V_t - V_o)$$

where V_e is the elution volume of the solute. Molecules must have values of K_{av} that differ significantly if separation is to be effective.

While components of a mixture are moving through a gel filtration column, their separation is hindered by the diffusion of their molecules, which tends to mix them together. This can restrict the resolution and lead to an incomplete separation, with the individual components overlapping with one another as they leave the column. The net result is a less efficient separation, which you can see diagrammatically in Figure 11.10.

The efficiency of a column is a measure of the diffusion of the sample component during separation. A column can be thought of as consisting of a number of adjacent zones called theoretical plates. The length in the column (L) is the height of a theoretical plate (H) multiplied by their number (N). The

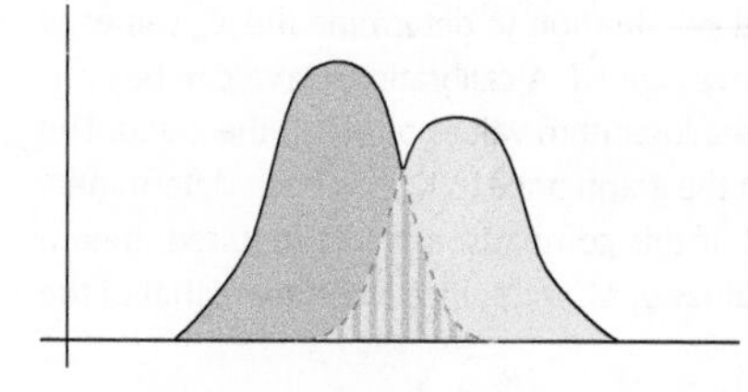

FIGURE 11.10
Overlapping peaks resulting from the incomplete separation of two components in a mixture during column chromatography.

solute reaches equilibrium between the mobile and stationary phases within each plate; thus, N is a measure of the efficiency of the column. It can be calculated using the formulae:

$$N = 16\,(t/W)^2 \text{ or } 5.54\,(t/W_h)^2$$

where W is the peak base width and W_h the peak width at half the peak height as shown in Figure 11.8, and t is the retention time. If W and W_h are expressed in units of time, then N is unitless.

The ability of a column to separate (resolve) different substances increases with the length of the column, L, as this increases N. Calculating the values of N for different columns can be used to compare their relative efficiencies. Indeed, the simplest way to increase the efficiency of a gel filtration step in purification would be to use a column of increased length. There are obvious practical limitations to the maximum size of L that can be used!

The number of theoretical plates is related to the surface area of the particles forming the stationary phase. Smaller particles have a relatively greater surface area than larger ones, thus the smaller the size of the particles of the stationary phase, the more effective the resolution of the column. If the size of N is calculated for a column of known length, the size or height equivalent to a theoretical plate (HETP) can be calculated using the formula:

$$\text{HETP} = L/N$$

and is normally expressed in μm. The smaller the value of HETP, the better the chromatographic performance. Thus, the HETP value can be used to assess changes in the efficiency of the same column under different chromatographic conditions.

SELF-CHECK 11.5

Calculate the HETP of the column of length 64 cm that eluted a protein with a V_e of 90 cm^3 using a suitable buffer at a flow rate of 45 $cm^3\ h^{-1}$. The base width of the eluted protein peak was equivalent to 30 cm^3.

Applications of gel filtration

Gel filtration chromatography has a number of applications including:

- the purification of analytes
- desalting samples
- determining the M_r of biological molecules (analytical gel filtration).

Gel filtration has been used to purify molecules of biomedical interest by separating them from larger or smaller contaminating molecules. For example, it has been applied to the purification of a large range of proteins, including enzymes, hormones, and antibodies, as well as nucleic acids, polysaccharides, and viruses.

Desalting is the process of removing unwanted small molecules, particularly ions, from preparations of macromolecules. An example is the removal of phenol and ammonium sulfate from partially purified nucleic acids and proteins, respectively. A column of Sephadex G-25 is often used because the macromolecules elute in the void volume, while the smaller contaminant molecules are retarded on the column. This process is faster and more effective than the alternative desalting process of dialysis.

The M_r of molecules can be estimated using analytical gel filtration to determine the K_{av} values of molecules of known M_r and similar type to that being investigated. A calibration curve can be constructed by plotting the K_{av} values on the *y*-axis against the logarithm values of M_r on the *x*-axis. The M_r of the unknown molecule can then be estimated from the graph once its K_{av} has been determined. This method allows different columns to be compared. If this comparison is not required, then a simple graph of elution volumes against the logarithm values of M_r will suffice to estimate that of the unknown analyte.

11.3.2 Ion-exchange chromatography

Ion-exchange chromatography uses a stationary resin in the form of a charged gel that is able to bind ions of the opposite charge, and a mobile liquid phase that is able to release selectively the bound ions, as we explain in this section. Ion-exchange resins are insoluble porous cross-linked polymers that contain chemical groups of a particular charge. The extent of cross-linking of the resin influences the size of its pores. Polymers commonly used include those of cellulose, polystyrene, and agaroses modified by the attachment of charged functional groups, such as sulfonate ($-SO_3^-$) and ammonium derivatives ($^+NR_3-$). These are regarded as strong exchangers because they are ionized over a wide range of pH values. In contrast, carboxylate ($-COO^-$) and diethylammonium [$-H^+ N(CH_2CH_3)_2$] are weak exchangers because they are ionized only within a narrow range of pH.

Positively charged resins are *anion* exchange resins, while those with a negative charge are *cation* exchangers. In both cases, ions of the opposite charge can bind to the column. Ion exchange occurs rapidly and is an equilibrium process, which you can see in Figure 11.11 (a). The strength of binding depends on the extent of ionization of the analytes in the mobile phase. The more highly charged an ion, the tighter it will bind to oppositely charged groups on the column. The total capacity of a column, which is the maximum number of exchangeable ions it can bind, must not be exceeded. If it is, the 'excess' charged analyte molecules will simply pass through the column without binding. Manufacturers normally give total capacities of ion-exchange resins as milliequivalents of exchangeable groups per mg of dry resin.

The ion-exchange column must initially be equilibrated in a buffer whose composition and pH will (a) stabilize the analyte of interest and (b) allow it to bind to the column. Once bound, the components of the mixture need to be released (eluted) sequentially (Figure 11.11b). This may use an isocratic buffer system, which differs in composition and pH from the one used to equilibrate the column. More often, a buffer of gradually changing pH or one whose concentration of salt progressively increases is used (Figure 11.12), and this is referred to as **gradient elution**. The former changes the charge of the bound molecules so they are released from the column in to the mobile phase. The effect of increasing the concentration of salt is that its ions displace the bound components from the column.

SELF-CHECK 11.6

Give the most probable order of elution of the following amino acids alanine ($^+NH-CH(CH_3)\ COO^-$), glutamate ($^+NH_3CH(CH_2CH_2COO^-)COO^-$), and lysine ($^+NH_3CH(CH_2CH_2CH_2CH_2{}^+NH_3)COO^-$) from a cation exchanger gel column at pH 7.

11.3.3 Affinity chromatography

Affinity chromatography relies on a specific binding between the analyte of interest and some type of ligand, which you can see diagrammatically in Figure 11.13. The ligand is immobilized by covalently attaching it to the column matrix (Box 11.2). Dextrans, agarose, and polyacrylamide have all been used as matrices in affinity chromatography. The ligand is often attached to the matrix using a chemical with an extended structure. This forms a spacer arm that separates the ligand from the bulk of the column preventing the matrix interfering with the specific binding of ligand and analyte.

When the sample is added to the column, only those molecules that can bind to the ligand, that is, those of the analyte, will be attached to it. The column is then washed with buffer to remove the non-binding impurities. It is then possible to release the analyte of interest and elute it from the column in a pure form. The release of the bound molecule requires that the affinity between them and the

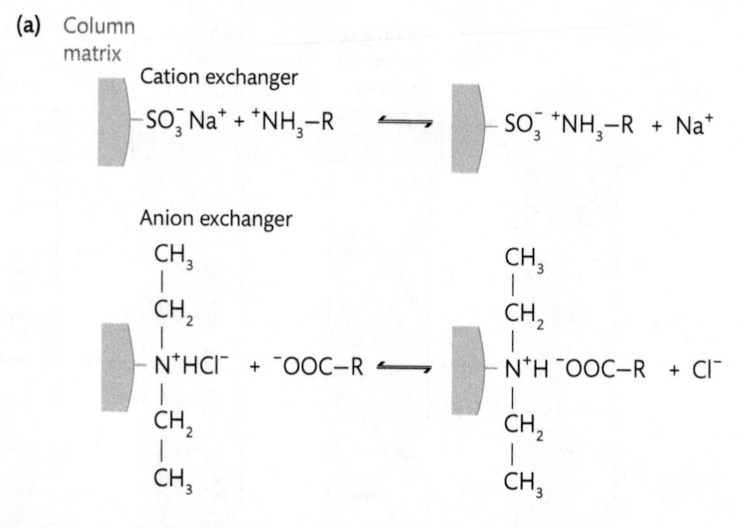

FIGURE 11.11

(a) Schematic illustrating the binding and release of ions from gels used in ion-exchange chromatography. (b) Binding of an anionic analyte from a mixture of materials and its release from an anion exchange resin.

ligands be reduced or that molecules be added to the eluting buffer that have a greater affinity than the bound ligand and so can displace the analyte from the ligand. For example, bound enzymes can be eluted from the column with solutions of substrate, coenzyme, or an inhibitor that competes for the enzyme's binding sites. If binding requires the presence of a metal ion, ethylenediaminetetraacetic acid (EDTA), which chelates and removes free metal ions, can be added. Often, changing the pH of the mobile phase can alter the conformation of bound proteins, allowing their release.

Key points

Chelation **is the binding and therefore removal from solution of metal ions by substances such as some amino acids and EDTA.**

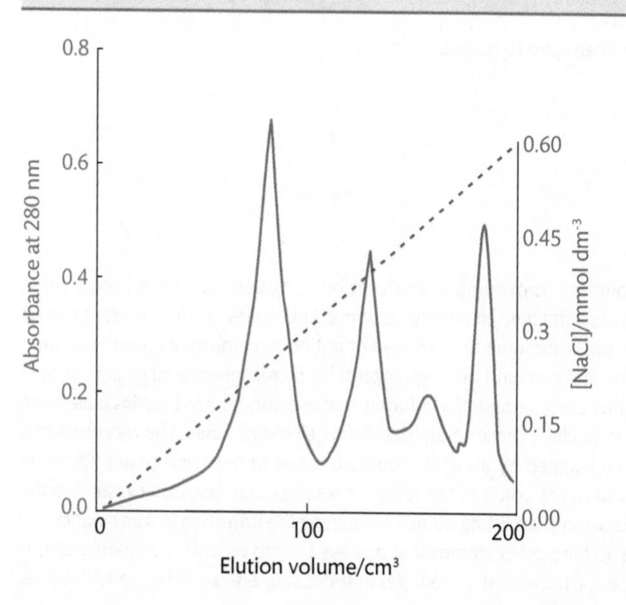

FIGURE 11.12

Sequential release of four proteins from an ion-exchange column using a gradient of increasing salt concentration.

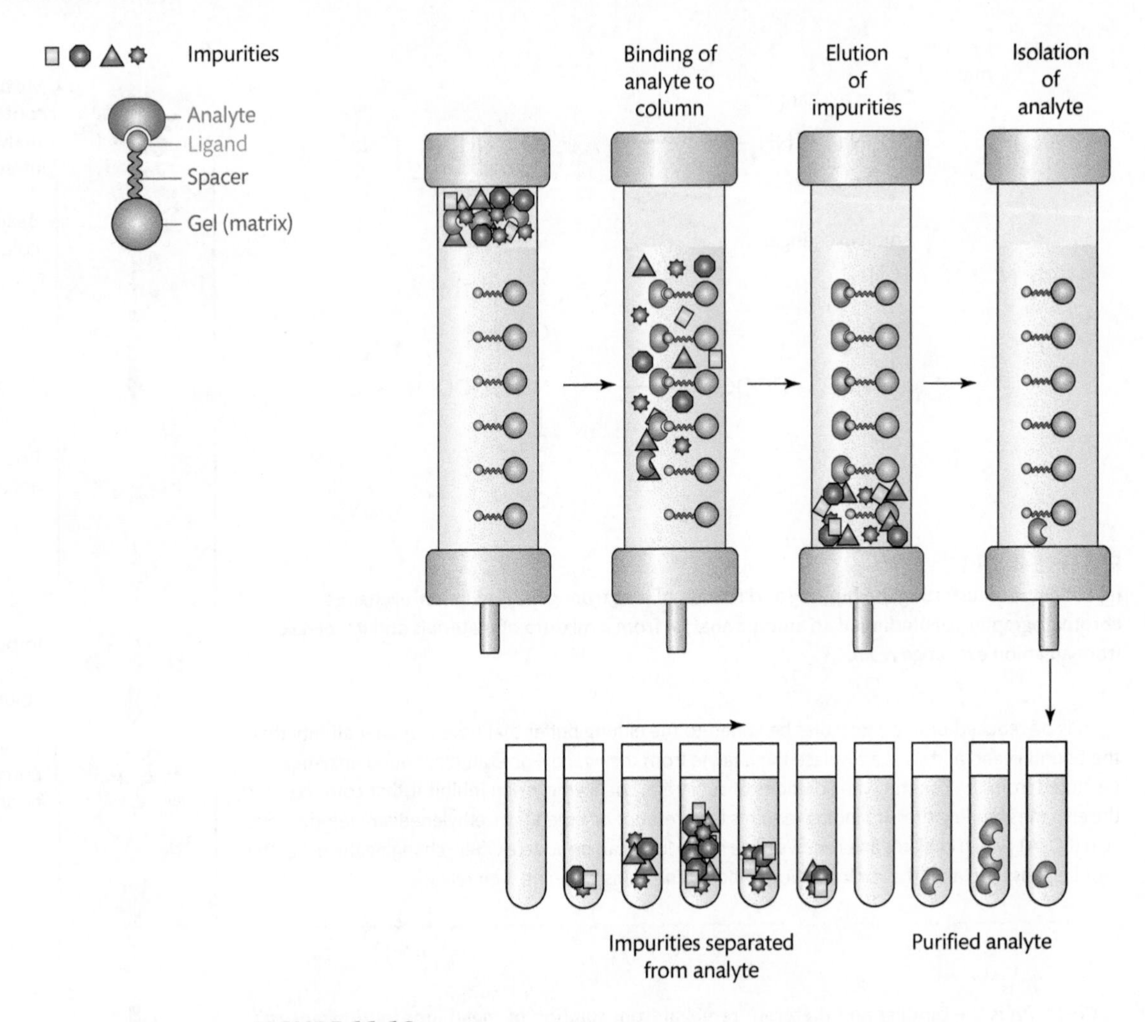

FIGURE 11.13

Outline of the use of affinity chromatography in purifying an analyte. See text for details.

Affinity chromatography has been used to isolate:

- enzymes
- antibodies/antigen
- lectins
- receptor proteins
- nucleic acids.

Enzymes can be isolated and purified from complex mixtures by using their substrates, coenzymes, or competitive inhibitors as ligands attached to affinity columns. Antibodies and/or antigens can be purified by immobilizing either the complementary antigen or antibody, respectively, on the column matrix. Lectins are glycoproteins that can bind other glycoproteins by recognizing all or part of their carbohydrate component. Affinity chromatography columns that use immobilized lectins have been used to isolate glycoproteins, particularly those solubilized from cell membranes. The receptors of a number of hormones have been isolated by using immobilized forms of the appropriate hormone as an affinity ligand. Once bound to the column, the receptor is released in ways that resemble the purification of enzymes with, for example, eluting buffers containing the hormone in free solution.

Nucleic acid strands are able to bind other strands that possess a complementary sequence of nucleotides, as explained in Chapter 14. Individual nucleotides or short, single-stranded oligonucleotides

BOX 11.2 Selecting a matrix for affinity chromatography

An ideal matrix for affinity chromatography will have at least some of the following characteristics. It must:

- **form a stable link between the ligand to the matrix**
- **be stable during binding of the macromolecule and its subsequent elution**
- **not interact with the macromolecules generally to avoid non-specific adsorption**
- **allow the mobile phases to flow freely through the matrix readily**
- **be a hydrophilic polymer with pore sizes large enough to allow free passage of molecules.**

can be immobilized for use as affinity ligands. These have been used to extract nucleic acids with the complementary nucleotide sequence or to isolate nucleic acid binding proteins that are capable of recognizing and binding the nucleotide or oligonucleotide.

11.3.4 Metal-chelation chromatography

Metal-chelation chromatography can be regarded as a form of affinity chromatography that exploits differences in the strengths of binding of different amino acids to the immobilized ions of some heavy metals. Metal ions, such as Cu^{2+}, Zn^{2+}, Hg^{2+}, and Ni^{2+}, are immobilized by chelation with aminodiacetic acid, which, in turn, is covalently attached to suspended agarose gel. The suspension of agarose is then poured into a tube to form a column. The side chains of a number of amino acid residues found in proteins, including imidazole groups of histidine, thiol groups of cysteine, and indole groups of tryptophan, can bind to the metal ions. When mixtures of proteins are washed through the column, some will bind to the metal ions and be retained. The bound proteins are eluted sequentially by, for example, adding free amino acids to the mobile phase, which can displace the bound proteins. Thus, mixtures of proteins are at least partially separated.

11.3.5 Hydrophobic interaction chromatography

Hydrophobic interaction chromatography can be regarded as a form of affinity chromatography that is based on differences in the binding of hydrophobic sites on the surfaces of protein molecules to immobilized hydrophobic ligands attached to a solid support matrix in the form of a column. Hydrophobic interaction chromatography is generally used to separate mixtures of proteins. Proteins often have hydrophobic sites on the surfaces of their molecules due to the presence of certain amino acid residues, such as those of valine, leucine, isoleucine, and phenylalanine. However, the hydrophobic sites of different proteins vary in their nature and area, so the strengths of binding of different proteins to the column vary.

The bound proteins are eluted from the column by reducing the hydrophobic interactions between sample molecules and the stationary phase by, for example, eluting the column with a solution of a non-ionic detergent, such as Triton X-100, or by changing the pH or temperature of the eluting mobile phase.

11.4 High-performance liquid chromatography

Increasing the length of a column and its number of theoretical plates improves its ability to resolve mixtures into their components (Section 11.3). However, long and densely packed columns of small particles have a greater resistance to the flow of the mobile phase. To overcome this resistance, and gain the advantages of faster and increased resolution, columns have been

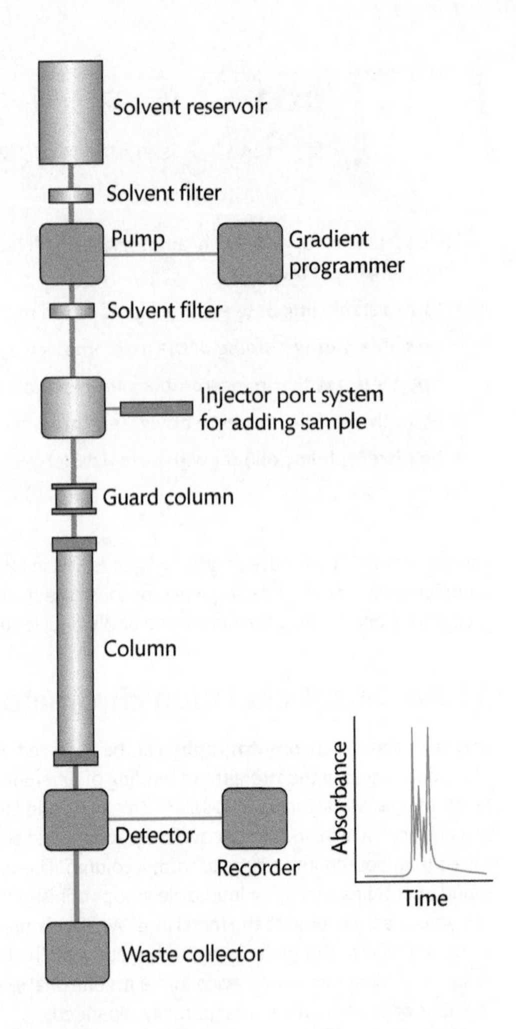

FIGURE 11.14
General components of a high-performance liquid chromatography system.

developed that can withstand the relatively high pressures produced when high-pressure pumps are used to force the mobile phase through the column. The pumps used should give continuous flows of at least 10 cm^3 per minute. This group of techniques is called **high-performance liquid chromatography** (HPLC). A typical HPLC system is shown in Figure 11.14. The columns are contained in tubes made of stainless steel that can withstand pressures up to 5.5×10^7 Pa. Columns typically have lengths of 15–50 cm and are normally 1–4 mm in diameter. The stationary phase consists of tightly packed small particles of material giving a correspondingly large surface area. Three types of material are used in forming these particles:

- microporous
- pellicular
- bonded.

Microporous particles have diameters of 5–10 μm and contain pores with diameters of less than 2 nm. Pellicular particles are porous and coated onto an inert solid core, such as a glass bead of about 40 μm diameter. Bonded particles have a stationary phase that is chemically joined to an inert support, such as silica. Columns packed with pellicular material offer efficiencies but only accept samples of small volumes compared with those with microporous material. Hence, the latter are generally preferred. Bonded phases give excellent chromatographic separations, and are chemically and mechanically stable, giving long-lasting columns.

Partition chromatography usually uses a liquid stationary phase coated onto an inert microporous or pellicular support. A problem with such supports is the tendency of the stationary liquid phase to be

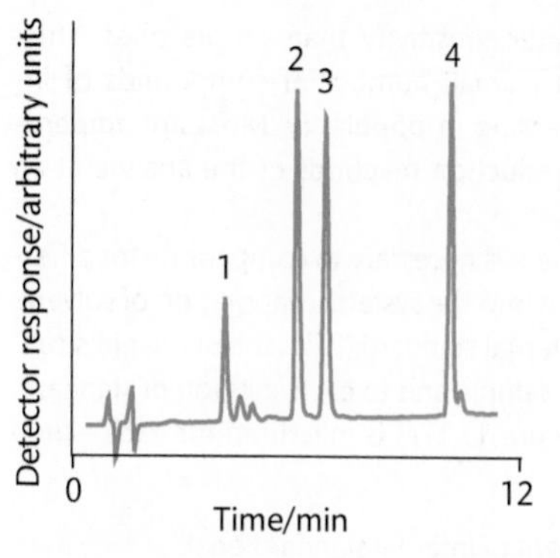

FIGURE 11.15

Use of reverse-phase high performance liquid chromatography to separate a mixture of proteins and protein hormones. Chromatogram courtesy of Varian, Inc.

washed away by the mobile phase. However, bonded phases have been developed where the liquid stationary phase is chemically joined to the silica to overcome this problem.

The choice of mobile phase depends on the nature of the analytes to be separated. The polarity of the solvent chosen should be such that effective partitioning of components in the sample occurs between the two phases. All the solvents used must be of high purity because contaminants can affect the column and also interfere with the detection system. Solvents also require degassing, not only because gas in the solvent can adversely affect the separation of the sample components, but also because it can interfere with the continuous monitoring of the eluant by, for example, absorbing UV light. Unfortunately, some solvents, for example, acetone, also absorb at these wavelengths. The mobile phase should not react with the stationary phase nor break the chemical bonds linking it to its supporting material, as this will, of course, decrease the effectiveness of purification and/or interfere with the detection of the components leaving the column.

The correct application of samples onto HPLC columns is necessary to obtain the best resolutions. The sample is injected onto the column using a microsyringe. There are two ways of applying the sample: stop flow and loop injection systems. In the former, the pump is switched off so that the pressure inside the column falls. The sample is then applied and the pump then switched back on. Loop injection systems allow the sample to be added while there is still a high pressure in the column. Repeated additions of impure samples can cause the column to lose its resolving power, which can be prevented by using a guard column. This is a short column of material similar to that of the main one, sited between the sample injector system and the analytical column. It is able to retain contaminating material, so protecting the main column and extending its lifespan.

HPLC methods generally involve adsorption and partition effects. Partition chromatography exploits differences in the solubilities and partition coefficients of polar substances as the bases for separating polar substances between stationary and mobile phases. The two usual techniques of partition chromatography are normal phase and reverse-phase liquid chromatography. In normal phase, the stationary phase is a polar substance whereas the mobile phase is a non-polar solvent. These are reversed in reverse-phase liquid chromatography (Figure 11.15), hence the name! Here, the stationary phase is a non-polar substance and the mobile phase is a polar solvent. This is the commonest type of HPLC. In reverse-phase liquid chromatography, non-polar molecules dissolve in the non-polar stationary phase to varying extents, whereas polar molecules pass rapidly through column within the mobile phase.

However, HPLC can also be used with ion-exchange (Figure 11.16), exclusion, and affinity chromatographies.

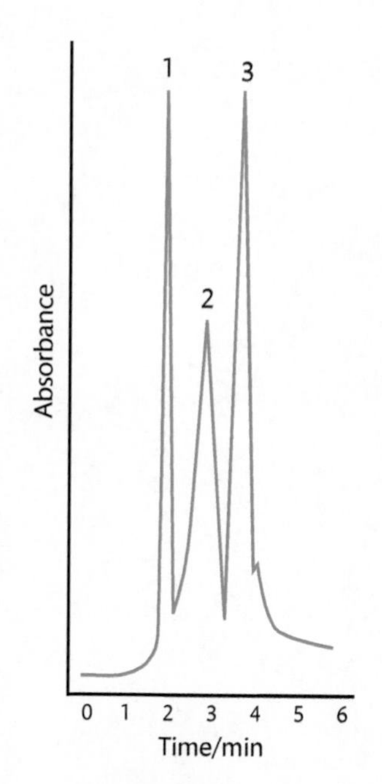

FIGURE 11.16

Separation of a mixture of proteins using a high-performance liquid chromatography system with an ion-exchange column.

11.4.1 Detector systems used in HPLC

Whatever the form of HPLC, a sensitive method is required to detect the separated components, including the analytes of interest, as they are eluted from the column. Often the eluant passes through a spectrophotometric cuvette or flow cell, which are described in Chapter 10, and its absorbance in the UV region of the spectrum is continuously monitored to detect proteins and nucleic acids, which absorb maximally at about 280 and 260 nm, respectively. However, other detection systems are deployed. These include fluorescence and electrochemical detectors.

Based on absorbance, fluorescence detectors have greater sensitivity than simple ones. Their applications are, however, limited owing to the relatively small number of compounds of interest that fluoresce. Electrochemical detectors are increasing in popularity. Most are amperometric: they measure a current produced by oxidation-reduction reactions of the analyte at an electrode.

Cross reference

See Chapter 8, 'Electrochemistry'.

Whatever the method of detection, calibration of the system is necessary to compensate for differences in the volumes of samples, variations in their injection into the system, evaporation of solvent, and other irregularities. This is usually achieved using an internal standard (IS) that is similar in structure to the analyte. The same quantity of IS is added to the sample and to each solution of standard. You can see this in Figure 11.17. The peak height ratio (Figure 11.17a) is measured for each eluted analyte or standard concentration using the expression:

$$\text{Peak height ratio} = \text{Height of analyte peak}/\text{Height of internal standard peak}$$

A graph of peak height ratio on the *y*-axis against the corresponding concentrations of standards on the *x*-axis will give a straight line graph as you can see in Figure 11.17(b).

SELF-CHECK 11.7

A method to determine the concentration of drug X in clinical samples was established using HPLC. This method was calibrated using a range of standard concentrations of X and an IS. A serum specimen from a patient was analysed and the chromatogram produced is shown in Figure 11.17(a). Use this chromatogram and the accompanying calibration curve (Figure 11.17b) to determine the concentration of X in the serum of the patient.

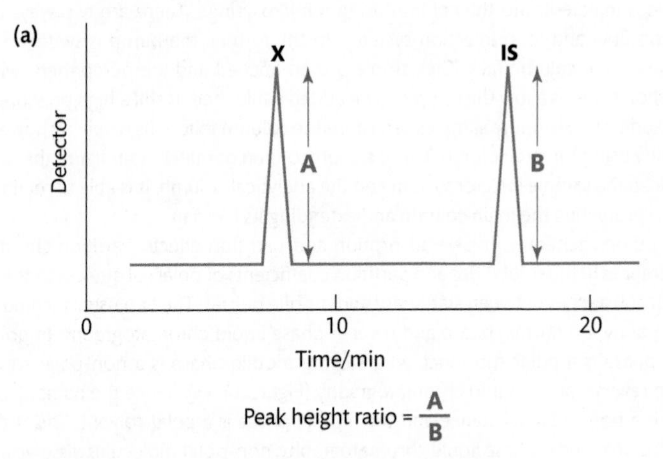

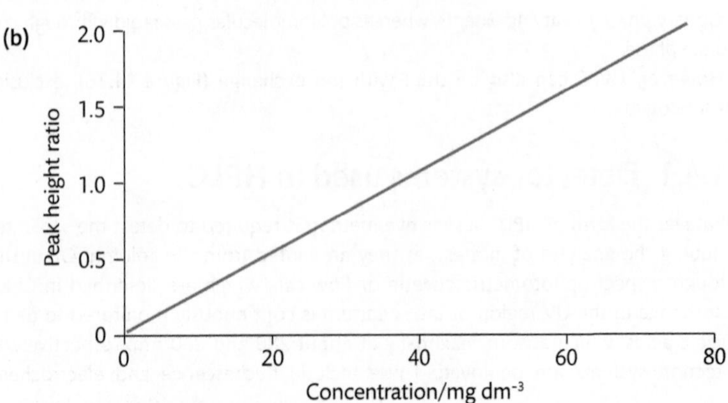

FIGURE 11.17

(a) Schematic showing the separation of an analyte X (or a standard of fixed concentration) and an internal standard (IS). The ratio of the heights of their peaks give a peak height ratio for that concentration. Calculating peak height ratios for a number of concentrations of the standard gives a calibration graph as shown in (b), which can be used to determine the concentration of an analyte in a sample.

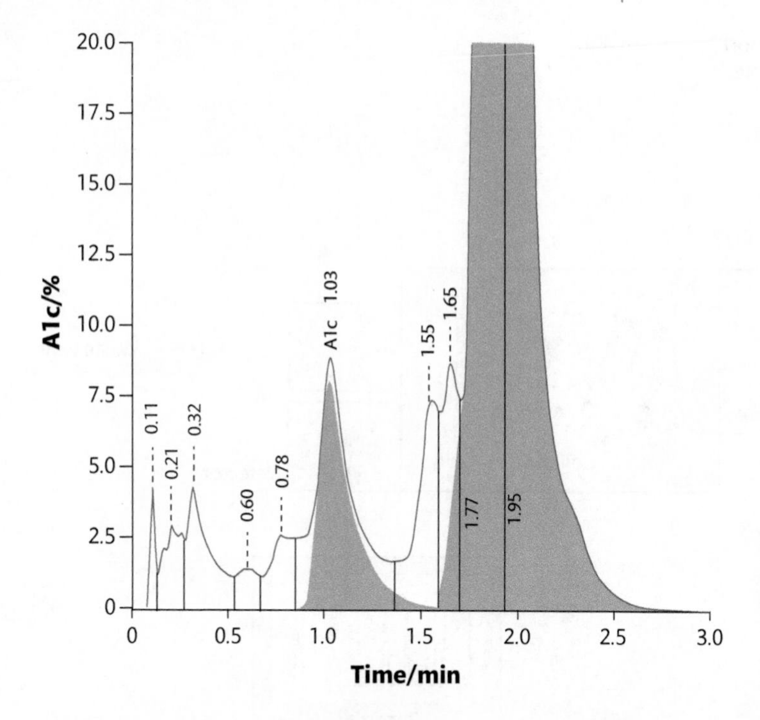

FIGURE 11.18
Use of high-performance liquid chromatography to separate and estimate the proportion of glycated haemoglobin (HbA_{1c}) to total haemoglobin. Courtesy of Department of Clinical Biochemistry, Manchester Royal Infirmary, UK.

11.4.2 Applications of high-performance liquid chromatography

The major advantages of HPLC are its speed, wide applicability, and sensitivity. The versatility of HPLC ensures it has many applications in biomedical science. It has been used to separate and analyse a wide range of materials of interest, including amino acids, proteins, nucleotides, nucleic acids, drugs, and vitamins. For example, ion-exchange resins are used in HPLC to determine the proportion of glycated haemoglobin (HbA_{1c}) to the total haemoglobin in the blood of diabetic patients (Figure 11.18). In erythrocytes, glucose binds to haemoglobin non-enzymatically and the extent of this glycation depends on the degree and duration of hyperglycaemia. Measurements of HbA_{1c} give an estimate of average glycaemia over the preceding 2 months. In non-diabetics, the proportion of HbA_{1c} is less than 6%, whereas that in diabetic patients is often in excess of 9%. A high relative proportion of HbA_{1c} reflects a poor control of blood glucose concentration. It is well established that such patients are more susceptible to long-term micro- and macrovascular complications.

Cross reference

You can read more about diabetes in Chapter 11, 'Diabetes mellitus and hypoglycaemia', in the companion text *Clinical Biochemistry*.

11.5 Gas-liquid chromatography

Gas-liquid chromatography (GLC) partitions molecules between a stationary phase consisting of a microscopic layer of liquid or polymeric material contained within a glass or metal tubing, called a column, and a mobile gas phase. The mobile phase is forced continuously through the column and carries the components of the mixture being analysed. Samples for analysis are injected into the system where they are vaporized and introduced into the gas phase. The gas is heated to comparatively high temperatures to keep the components of the sample in the gas phase, allowing them to be carried along in the mobile phase. Those components that have a greater affinity for the stationary liquid phase have their movement through the column retarded. Those more soluble in the gas are carried through the column faster. The partitioning of the components of the mixture (sample) between the stationary and mobile phases causes different compounds to elute from the column at different retention times.

A typical GLC system is shown in Figure 11.19. The columns have capillary diameters between 0.03 and 1 mm and may be as long as 100 m. Two types of capillary columns are used: wall-coated open tubular and support-coated open tubular. In the former, a liquid stationary phase is coated directly

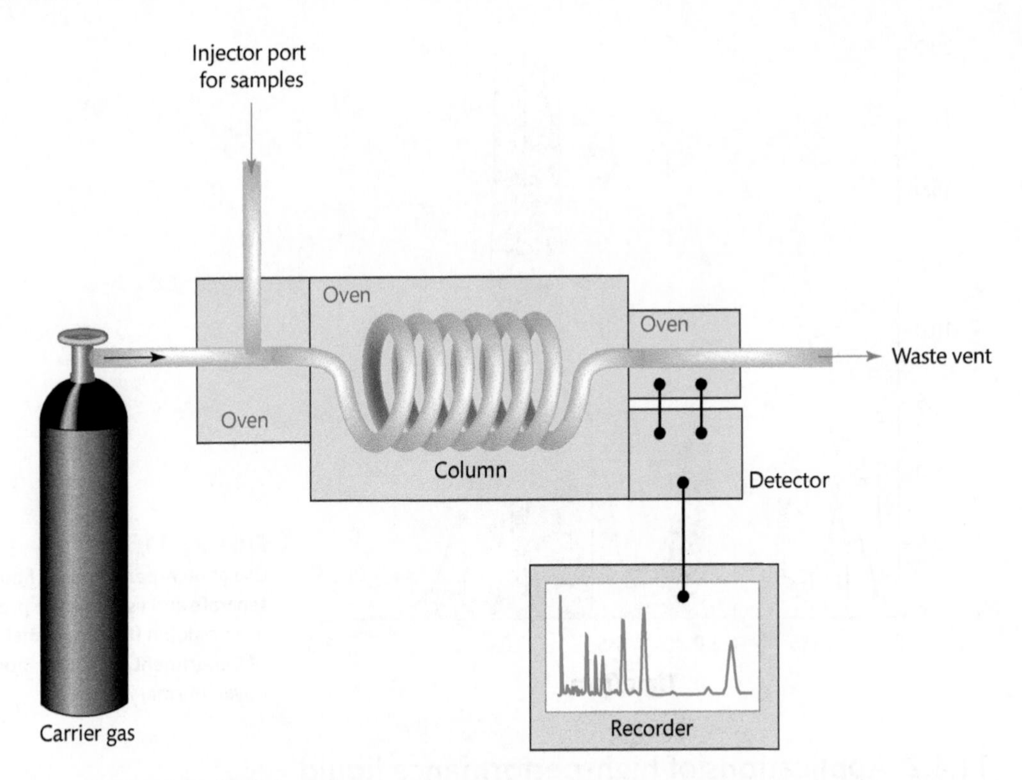

FIGURE 11.19
Typical outline of a gas-liquid chromatography (GLC) system.

onto the wall of the capillary column, whereas the latter uses a liquid stationary phase coated onto a support material which is then bonded to the walls of the column. Stationary phases must be non-volatile and not enter the gaseous phase at the temperatures used. They are usually high boiling point organic compounds such as silicone greases.

The mobile phase is usually an inert gas such as helium or argon, or a nonreactive one, such as nitrogen. The gas is forced through the column at a rate of 40–80 cm^3 per minute. The column is contained within an oven that maintains the column at higher than ambient temperature during separation. This is necessary because, in general, the higher the temperature, the greater the volatility of molecules. Temperature therefore influences the separation because higher temperatures speed up the movement of more volatile molecules through the column. Temperatures may be set at constant values between 50 to 250°C to give isothermal conditions. However, for some applications it may be necessary for it to be pre-programmed to increase by, for example, 10°C every minute.

The sample for analysis is dissolved in a solvent, such as acetone or methanol, and injected into the GLC system using a microsyringe. The injection region of the column is usually at a higher temperature than the column itself. This ensures the rapid and complete vaporization of the sample. Wall-coated open tubular columns have a relatively small amount of stationary phase and so only small volumes of samples can be analysed. The sample injection port has a splitter system, which allows a small fraction of injected sample to reach the column and the remainder is vented as waste. The greater capacity of support-coated open tubular columns means there is no need for a splitter system and the sample is injected directly into the column.

Cross reference
You can read about mass spectrometry in Chapter 9, 'Spectroscopy'.

In some ways, GLC resembles column chromatography. However, it differs from other forms of chromatography in that the molecules must be *volatile*, that is, have low boiling points, if they are to move through the column and be separated. GLC is therefore particularly suitable for separating hydrophobic molecules or ones of low polarity, which therefore have low boiling points.

Molecules that are non-volatile can, however, often be converted to volatile derivatives prior to their separation. For example, long-chain fatty acids can be converted to their methylesters and monosaccharides to trimethylsilyl compounds. Both of these types of derivatives have much lower boiling points than their parent compounds and can be analysed by GLC (Figure 11.20).

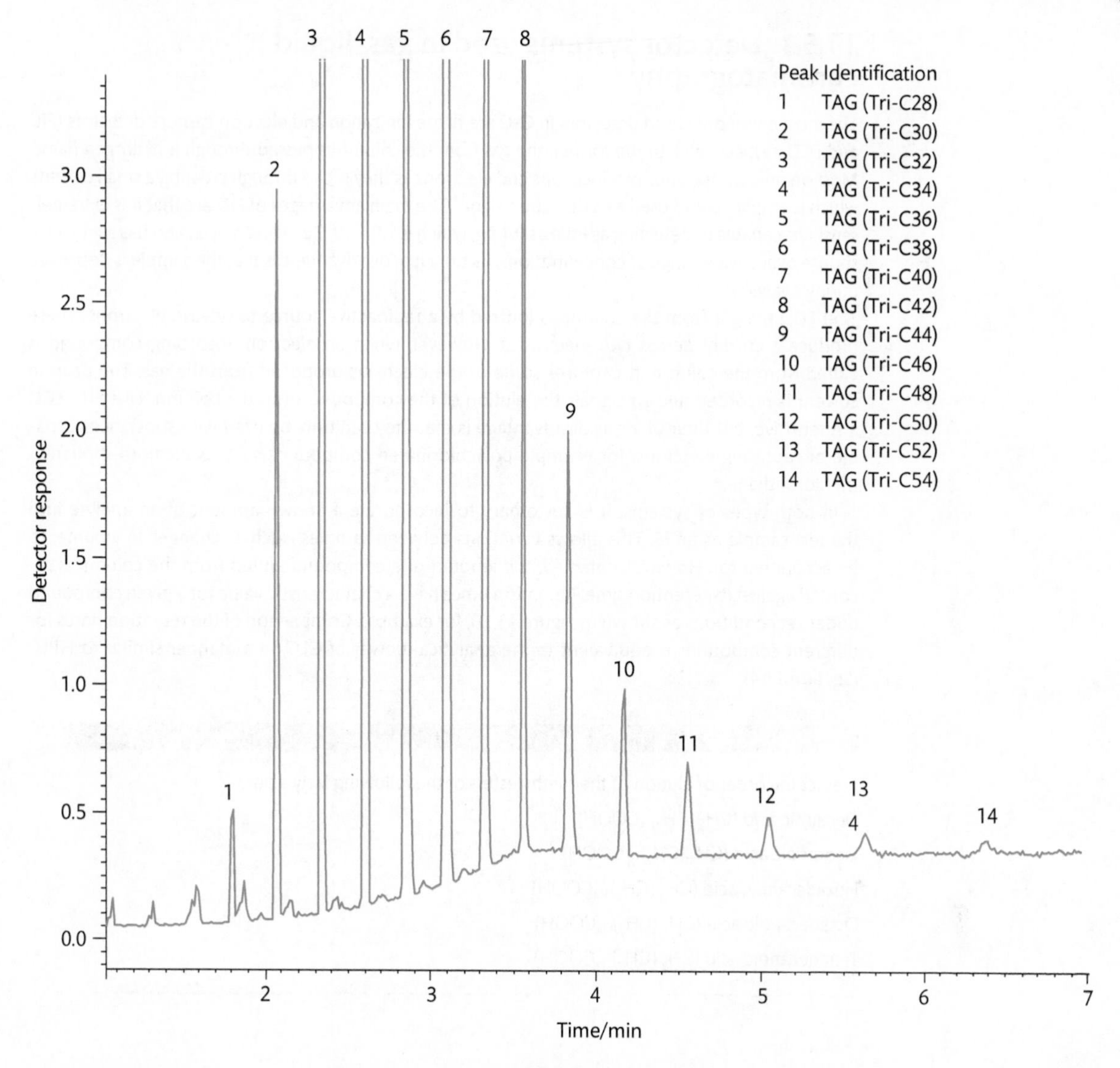

FIGURE 11.20

Separation of a number of triacylglycerols (TAGs) by gas-liquid chromatography. The lengths of the fatty acids in each TAG are given as the number of carbon atoms present. Courtesy of Varian, Inc.

Like TLC (Section 11.2), GLC can be combined with other analytical techniques. For example, clinical samples can be subjected to GLC and the separated components identified by tandem mass spectrometry. The peroxisomal diseases adrenoleukodystrophy, Zellweger syndrome, and Refsum's disease, although rare, are extremely debilitating and are associated with high concentrations of very long-chain fatty acids (VLCFAs) in the blood of patients. Thus, GLC-tandem mass spectrometry is used, for example, to investigate the VLCFA composition of samples of plasma from patients suspected of having these diseases.

GLC is thus a versatile technique and can be used to analyse a wide range of substances including lipids, alcohols, phenols, and a number of drugs. The major advantages offered by GLC in separating such materials are:

- excellent resolution of a wide range of materials
- reproducibility of separations
- excellent sensitivity of its analyte detection systems.

11.5.1 Detector systems used in gas-liquid chromatography

The most commonly used detectors in GLC are flame ionization and electron capture detectors (FID and ECD, respectively). In the former, the gas from the column is passed through a hydrogen flame. Most organic compounds produce ions and electrons as they move through creating a small current, which is amplified and used as a detection signal. The main advantages of FID are that it is extremely sensitive, capable of detecting as little as 0.1 pg, which is 1.0×10^{-13} g of an analyte, and has a linear response over a wide range of concentrations. Its principal disadvantage is that the sample is destroyed during analysis.

In ECD the gas from the column is ionized by a radioactive source to release electrons. These produce a current across two electrodes. However, when an electron-absorbing compound is eluted from the column, it captures some of the electrons produced from the gas. The drop in current is recorded and so signals the elution of the compound from the column. Like FID, ECD are sensitive, but their obvious disadvantage is that they can only be used with substances capable of capturing electrons, for example polychlorinated compounds such as dichloro-diphenyl-trichloroethane.

In both types of systems, it is customary to incorporate a known amount of an analyte into the test sample as an IS. This allows variations between samples, such as changes in volume, to be accounted for. However detected, the amount of a compound eluted from the column is recorded against its retention time. Retention times have a characteristic value for a given compound under set conditions as shown in Figure 11.20, for example. Comparison of the retention times for different compounds is equivalent to the analytical power of GLC, in a manner similar to HPLC (Section 11.4).

SELF-CHECK 11.8

Predict the order of elution of the methylesters of the following fatty acids:

Decanoic acid ($CH_3(CH_2)_8COOH$)

Eicosanoic acid ($CH_3(CH_2)_{18}COOH$)

Hexadecanoic acid ($CH_3(CH_2)_{14}COOH$)

Octadecanoic acid ($CH_3(CH_2)_{16}COOH$)

Tetradecanoic acid ($CH_3(CH_2)_{12}COOH$)

CASE STUDY 11.1 Investigating galactosaemia

Joe, a child of 7 months of age, was suspected of having galactosaemia. His urine sample and three known sugars, glucose, fructose, and galactose, were subjected to TLC. The following R_f values were obtained:

Glucose	0.42
Galactose	0.35
Fructose	0.39
Urine sample	0.35

(a) Which sugar is most likely to be present in Joe's urine sample?

(b) Is the diagnosis supported?

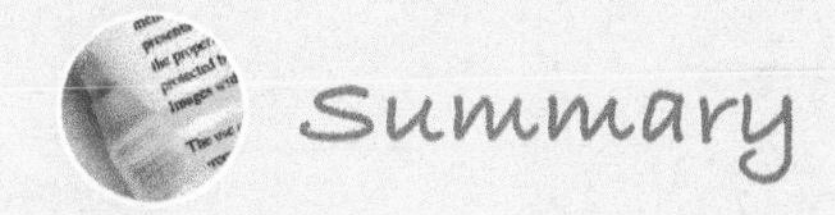

Summary

- Chromatography is a set of techniques that separate mixtures whose components differ in their distribution or partition coefficients.
- Chromatographic techniques can also be useful in identifying individual components of the mixture and, in some instances, in providing quantitative data.
- Planar chromatographies include paper and thin layer chromatographies. Both separate mixtures using a mobile liquid phase moving through a bound stationary liquid. TLC is a simple, quick, and inexpensive technique, suitable for qualitative but not quantitative analyses. It is used to separate a wide range of components, which can often be identified by their R_f values using those of known standards.
- Column chromatographies comprise a range of techniques in which an immobilized solid forms a cylinder of material, through which a mobile liquid phase is allowed to percolate. Molecules in a mixture are applied to the column and separate owing to differences in their sizes as in gel filtration chromatography, their charges in ion-exchange chromatography, hydrophobicities in hydrophobic chromatography, and ligand binding in affinity and metal chelation chromatographies. Whatever the type of column chromatography, components of the mixture differentially distribute between the stationary and mobile phases. Thus, individual components leave the column and can be identified by their elution volumes or retention times. An effective separation of the components of a mixture is indicated if they appear as discrete peaks on the chromatogram that do not overlap or show tailing or fronting.
- Specialized types of column chromatography materials are used to increase the surface area of the stationary phase, allowing the mobile liquid phase to be forced through the column at high pressure, as is the case in HPLC. This gives an improvement in resolution over low-pressure systems.
- GLC separates mixtures of volatile compounds using a stationary liquid phase and a mobile gaseous phase. Partitioning of the individual components of the mixture between the two phases effects their separation. Individual components leaving the column are detected by their absorbance, fluorescence, or electrochemical properties, and identified by their retention times.

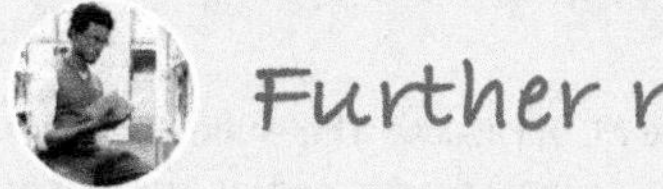

Further reading

- **Blum F. High performance liquid chromatography. *British Journal of Hospital Medicine* 2014; 75: C18–21.**
- **Dong M. *Modern HPLC for Practising Scientists.* Harlow: John Wiley, 2006.**
- **Hagel L. Gel filtration: size-exclusion chromatography. *Methods of Biochemical Analysis* 2011; 54: 51–91.**
- **Lundanes E, Reubsaet L, Greibrokk T. *Chromatography: Basic Principles, Sample Preparations and Related Methods.* Weinheim: Wiley-Blackwell, 2013.**
- **Jungbauer A, Hahn R. Ion-exchange chromatography. *Methods in Enzymology* 2009; 463: 349–71.**
- **Stuart B. *Gas Chromatography.* (Coordinating author E Prichard). Cambridge: Royal Society of Chemistry, 2003.**
- **Wall PE. *Thin Layer Chromatography: A Modern Practical Approach.* Cambridge: Royal Society of Chemistry, 2005.**
- **Urh M, Simpson D, Zhao K. Affinity chromatography: general methods. *Methods in Enzymology* 2009; 463: 417–38.**

Three Useful Journals

- *Biomedical Chromatography*
- *Chromatography Today*
- *Journal of Chromatography B: Biomedical Sciences and Applications*

Questions

1. Chloramphenicol, fluorouracil, and paracetamol are three therapeutic drugs. Their log K_{OW} values are 1.14, −0.85, and 0.48, respectively. List them in order of their *increasing* water solubility.

2. In chromatography, the separation of a mixture of molecules is determined by their differential distributions between two different phases. Which **ONE** of the following terms best describes this relationship?
   - **(a)** Retardation factor
   - **(b)** Partition coefficient
   - **(c)** Elution volume
   - **(d)** Void volume
   - **(e)** Electrophoretic front.

3. Which of the following statements concerning TLC is **FALSE**?
   - **(a)** The adsorbent material often used is silica gel
   - **(b)** The mobile phase moves up the thin layer by capillary action
   - **(c)** The R_f value is always less than 1
   - **(d)** TLC is a simple, relatively inexpensive technique
   - **(e)** The separation of molecules is based on differences in their sizes.

4. Predict the probable order of elution from a column of Sephadex G-100 of the following proteins: carbonic anhydrase (M_r 30,000), catalase (M_r 250,000), chymotrypsinogen (M_r 23,250), fibrinogen (M_r 330,000), haemoglobin (M_r 64,500), and insulin (M_r 11,470). You may find it helpful to consult Table 11.3 when attempting this question.

5. Gel exclusion chromatography was used to determine the M_r of catalase. The elution volumes (V_e) of five proteins of known M_r from a column of Sephacryl were determined, as was that of catalase. Chromatography was carried out at 4°C using a dilute buffer of pH 7.4. Under these conditions, the structures of all the molecules concerned are preserved. Use the data in the following table to determine M_r of catalase.

Protein	M_r	V_e/cm^3
Hexokinase	102, 000	24.7
Aldolase	158,000	22.7
α-Amylase	199,000	21.6
Fibrinogen	330,000	19.4
RNA polymerase	450,000	18.0
Catalase	?	20.7

6. Amino acids are polar molecules and mixtures of them cannot be resolved by GLC. Explain a general strategy for overcoming this limitation.

7. Which **ONE** of the following properties of molecules generally forms the basis for their separation by affinity chromatography?

(a) Size

(b) Volatility

(c) Charge

(d) Shape

(e) Polarity.

8. Galactosyl transferase catalyses the reaction:

$$\text{UDP-galactose} + \text{acceptor} \rightarrow \text{UDP} + \text{galactose-acceptor}$$

The enzyme binds to a UDP-agarose affinity column in the presence of Mn^{2+}. Suggest a plausible method of eluting it from the column.

9. Two molecules, X and Y, were separated using HPLC. They had retention times of 3 minutes 30 seconds with a base width of 38 seconds, and 5 minutes and 30 seconds and base width 42 seconds, respectively. A molecule Z, which was completely excluded from the column, had a retention time of 1 minute and 30 seconds. Determine (a) the selectivity of the column and (b) the resolution of X and Y. Comment on each value.

Answers to self-check and end-of-chapter questions are available at the end of the book.

12

Electrophoresis

Qiuyu Wang, Nessar Ahmed, and Chris Smith

Learning objectives

After studying this chapter, you should be able to:

- Explain the underlying principles of electrophoresis
- Describe the uses of the major types of electrophoretic separation techniques
- Describe how the positions of separated components may be determined following electrophoresis
- Explain the underlying principles of isoelectric focusing
- Discuss the uses of pulsed field and two-dimensional gel electrophoresis

Introduction

Electrophoresis is the movement of charged molecules or **ions** in an **electric field** (E). Ions that differ in their charge and/or mass will move at different rates such that if they have a common starting point they will separate over time as some will move faster than others. As a result, the components of a mixture will be separated. Thus, electrophoresis is a separation technique. However, it can also be adapted to provide analytical data, such as the size (M_r) of molecules. Many biological molecules, such as amino acids, peptides, proteins, nucleotides (nt), and nucleic acids, possess groups that can ionize, and give the molecules an overall negative or positive charge. Hence, electrophoresis has many applications in the biomedical sciences: in separating proteins and nucleic acids, in assessing purity of biological samples, and in estimating the sizes of molecules.

Online Resource Centre

To see an online video demonstrating electrophoresis, log on to www.oxfordtextbooks.co.uk/orc/fbs

12.1 Principles of electrophoresis

If a potential difference or voltage (V) is applied between two electrodes, it produces an electric field or **potential gradient** (E). The value of E can be calculated by dividing the applied voltage by the distance (d) between the electrodes:

$$E = V/d$$

Values of E are normally quoted in units of V cm^{-1}. When an E is applied to molecules carrying a charge of q, the force produced is equal to Eq; this force drives the molecules towards the electrode of opposite charge. For example, the positive electrode or anode will attract negatively charged molecules or anions; whereas the negative electrode or cathode attracts positively charged ones or cations (Figure 12.1).

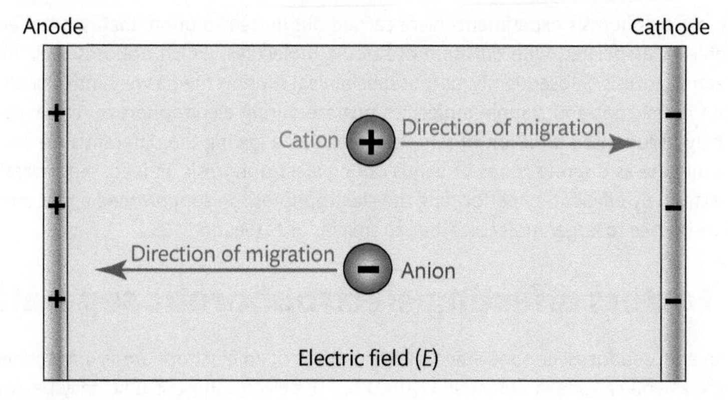

FIGURE 12.1
Schematic showing the movements of anions and cations in an electric field.

SELF-CHECK 12.1

Calculate the value of the electric field (E) when a potential difference of 200 mV is applied between two electrodes 20 cm apart.

During electrophoresis, the applied potential difference, usually expressed in volts (V), is normally kept at a constant value. However, in some cases, electrophoresis is performed using a constant current, and the voltage will vary with the electrical resistance of the system. Occasionally, electrophoresis is carried out under conditions of constant power, which is the *product* of the current and voltage.

The strength of E can be increased by simply increasing the voltage applied between the electrodes. An increase in voltage increases the force that the molecules experience, making them move more quickly. As a result, an increase in voltage shortens the time needed to complete electrophoresis experiments. However, it is not always practical to use high voltages because of their heating effects. Heating can increase the diffusion of molecules in the sample and may also cause convection currents, both of which decrease the efficiency of separation. Higher temperatures can also result in the thermal denaturation of some types of biological molecules, such as proteins and nucleic acids.

12.1.1 Electrophoretic mobility

Cross reference

You met frictional coefficients in Chapter 10, 'Centrifugation'.

The movement of ions subjected to an E is opposed by a frictional resistance, which depends, in turn, upon the frictional coefficient (f) of the ions.

The velocity (v) of the ion is given by:

$$v = Eq/f$$

However, the movement of an ion is often described by its **electrophoretic mobility** (μ), that is, its velocity per unit of electric field. Electrophoretic mobility can be obtained by dividing v by E:

$$\mu = v/E$$

Combining these equations, gives the expression:

$$\mu = q/f$$

Thus, in theory, it should be possible to obtain the value for f of a molecule from its mobility in an electric field if its net charge is known, which should make it possible to obtain information about its shape and size. In practice, this has not proved possible because the simple equations given above do not adequately describe electrophoretic processes. For example, interactions between migrating molecules and the medium in which electrophoresis is performed all complicate the system. However, what is clear is that during electrophoresis, molecules that differ in their values of μ will separate from one another.

Initially, electrophoresis experiments were carried out in free solution, that is, a buffer, but this poses problems associated with diffusion of sample molecules, which opposes their movement during electrophoresis. Subsequently, porous mechanical support media were introduced, through which both buffer ions and sample molecules migrate during electrophoresis. This type of technique greatly reduces the diffusion of sample molecules, allowing the different substances in the sample to migrate as discrete zones or bands during electrophoresis. Indeed, the separations can be made extremely efficient by performing the electrophoresis in support media that offer greater frictional resistance to larger molecules than to smaller ones (Section 12.2).

12.2 Factors affecting electrophoretic separations

The equipment used for electrophoresis typically consists of an electrophoresis unit connected to a power pack, as you can see in outline in Figure 12.2. The electrophoresis units may be arranged to hold the supporting material in either a horizontal or a vertical arrangement. The power pack provides a flow of direct current to the electrodes in each compartment of the unit. Buffer solution in the compartments and in the supporting medium that links them then carries electricity between the two electrodes to generate a flow of electricity through the system.

The electric field generated during electrophoresis is associated with the current that is carried by the buffer connecting the electrodes, although a relatively small proportion of it is also conducted by the ions of the sample. If the ionic strength of the buffer is increased, the proportion of current carried by it will also increase, whereas that carried by the sample ions will decrease. Consequently, their velocities will slow. If the ionic strength of the buffer is decreased, the converse happens; the proportion of the current carried by the ions of the sample will increase and their rate of migration will speed up. However, although a low ionic buffer decreases the overall current and causes less heat production, there is greater diffusion of the sample due to an increase in separation time and therefore some loss of resolution or separation may occur. Thus, the E chosen is always a compromise between the time required to separate the molecules and the resolution obtained.

During electrophoresis, all ions are subjected to the same E and separation occurs because different substances have different electrophoretic mobilities. Individually, the mobility of charged ions is influenced by a number of factors including:

- their sizes and shapes
- the buffer in which electrophoresis is performed
- the supporting medium.

The sign of the charge (+ or −) of an ion will determine whether it moves towards the anode or cathode, as described in Section 12.1. However, the mobility of an ion during electrophoresis at constant voltage depends solely on the ratio q/f, irrespective of the direction in which it is moving. For mixtures of molecules with *similar conformations*, for example globular proteins or linear DNA molecules (Chapter 14), then the variables affecting their mobility are restricted to the individual charges and f values. The value of f depends to a great extent upon the mass of the molecule. Thus, each type of molecule in the mixture will migrate in a given electric field at a velocity proportional to its mass-to-charge ratio. The shape of molecules also influences their mobility. For example, globular proteins migrate differently to fibrous proteins of the same mass-to-charge ratio.

During electrophoresis, a buffer is used to saturate the supporting medium. The choice of buffer can affect the migration of particles in a number of ways. Its pH influences the ionization of some chemical groups within the sample molecules, in particular that of amino and carboxyl groups of polypeptides, and so determines the overall charge of the ions. pH is temperature dependent, which is a major reason for performing electrophoresis under conditions in which the temperature remains constant.

The ionic composition of the buffer is also a factor to be considered. Ions in the buffer, such as Na^+, K^+, Mg^{2+}, and Cl^-, and those of phosphate, can all interact with charged groups on molecules of the sample. This results in them becoming surrounded with ions of the opposite charge, and decreases both their overall mass-to-charge ratio and mobilities.

In general, electrophoresis is carried out in buffers of low ionic strength that are compatible with the solubilities of biological macromolecules. Such compatibility reduces the types of interaction described to minimal levels. In some cases, however, the binding of buffer constituents to sample

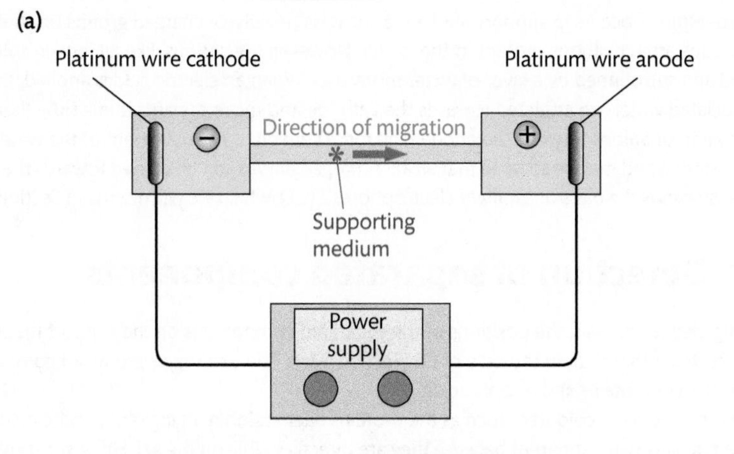

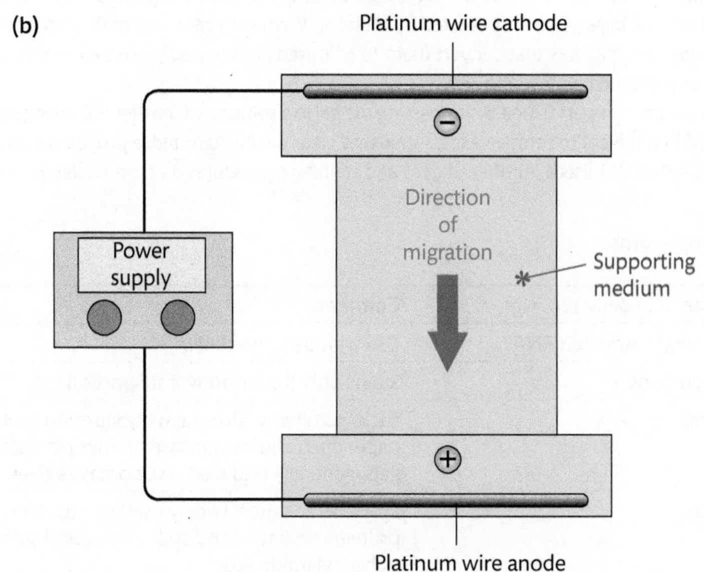

FIGURE 12.2

Schematics showing (a) horizontal and (b) vertical electrophoresis systems. Note how in each system, the electrodes are positioned in their own buffer compartments and the buffer completes the electric circuit between them. The electrodes in both systems are made from a platinum wire.

molecules can be exploited. For example, borate buffers are used when separating carbohydrates as their components bind to carbohydrates. This converts neutral carbohydrates to charged complexes, which can then be subjected to electrophoresis.

The support medium used in electrophoresis is usually inert, although it can lead to **molecular sieving**, adsorption, and **electro-osmosis**, each of which can influence the rate of migration of sample ions. Some support media consist of gels that contain pores (Sections 12.6 to 12.8) that restrict the flow of ions, particularly those of large M_r, during electrophoresis. Smaller ions experience a lower frictional resistance as they migrate through the gel and so travel faster and therefore further than larger ones for a given time. This effect is referred to as molecular sieving and, as we shall see in Sections 12.6 to 12.8, allows separation of molecules based on their size alone.

Adsorption is the retention of sample molecules onto the *surface* of the support medium. Adsorption slows the migration of sample molecules and causes tailing of the sample so that it moves in the shape of a comet, that is, in an extended area, rather than as a distinct zone. Thus, the individual components of a mixture do not separate into distinct bands and the resolution of separation is impaired.

Cross reference

We discuss adsorption in the context of chromatography in Chapter 11, 'Chromatography'.

Electro-osmosis occurs in support media that possess negatively charged groups on their surface. These groups attract cations present in the buffer. However, the cations, like all ions in solution, are hydrated and surrounded by a layer of water molecules. When an electric field is applied, the cations and associated water are attracted towards the cathode and move towards it in a flow that opposes the migration of anions towards the anode. The force generated by movement of the water is called electro-osmosis and may be so large that weakly charged anions can be carried towards the cathode! Electro-osmosis is the basis of capillary electrophoresis (CE), which we will discuss in Section 12.9.

12.3 Detection of separated components

Following electrophoresis, the positions of the separated components on the support medium need to be detected, other than in the case of CE (Section 12.9). The two major groups of compounds we will consider are proteins and nucleic acids.

Online Resource Centre
To see an online video demonstrating haemoglobin electrophoresis, log on to www.oxfordtextbooks.co.uk/orc/fbs

If the substances are coloured, such as the proteins haemoglobin, myoglobin, or the cytochromes, then their positions are apparent because they are directly visible on the gel. However, most proteins are colourless and, in general, are detected by **staining** them with coloured or fluorescent dyes or by reacting them with reagents that convert them to coloured complexes. These treatments allow them to be seen as coloured bands in the gel.

Normally, staining is performed by immersing the gel in a solution of the dye. Following staining, the gel is soaked in a solvent to remove excess stain and clear the background: a procedure usually called **destaining**. Table 12.1 lists a number of dyes and staining procedures that are available.

TABLE 12.1 **Common stains used in electrophoresis**

Stain	Substance(s) detected	Comment
Acridine orange (Section 12.8)	DNA, single-stranded RNA	Use with denatured RNA
Alcian blue (Section 12.7)	Glycoproteins	Stains only the carbohydrate portion
Amido black	Proteins	Lacks sensitivity, gives high background stain with paper and cellulose acetate, shrinks polyacrylamide gels, generally replaced by Coomassie dyes
Coomassie blue dyes (Section 12.7)	Proteins	Probably the most widely used dye to detect proteins on native and sodium dodecyl sulfate polyacrylamide gels
Ethidium bromide (Section 12.8)	DNA RNA	Sensitive for double-stranded nucleic acids, used with agarose and polyacrylamide gel electrophoresis
Enzyme activities (Section 12.7)	Enzymes	Sensitive, specific for enzyme activity, can be time consuming
Ninhydrin (Section 12.4)	Amino acids, peptides, proteins	Sensitive reagent, mainly used with paper electrophoresis
Ponceau S (Section 12.5)	Proteins	Used with cellulose acetate and starch gels, stains rapidly
p-Anisidine/ethanol and sodium periodate/acetone (Section 12.4)	Sugars	Two-reagent technique for detecting sugars following paper electrophoresis
Periodic acid–Schiff reaction (Section 12.7)	Glycoproteins	Stains only the carbohydrate portion, lacks sensitivity
Pro-Q emerald 300 (Section 12.7)	Glycoproteins	Sensitive fluorescence stain for detecting sugar portions of glycoproteins
Silver staining (Section 12.7)	Proteins	More sensitive than Coomassie dyes but more expensive and staining can be variable
Sypro orange and red (Section 12.7)	Proteins	Sensitive fluorescence dyes for detecting proteins on polyacrylamide gels

Following staining, scanning densitometry may be used to estimate the amounts of the separated proteins or nucleic acids present. The densitometer passes a narrow beam of light from a laser through the gel and measures the proportion of light transmitted by each band of protein. The central processor of the densitometer or a computer to which it is connected can then calculate the amount of each protein present from the amount of light it absorbs: the smaller the amount of light transmitted by the gel, the greater the amount of sample present. These data are often presented as graphs of absorbance against the distance each protein migrated during electrophoresis, as you can see in Figure 12.3. However, the estimated amounts of each protein must be treated with caution. Absorbance is only linear up to a particular concentration of protein, in accordance with the Beer-Lambert law described in Chapter 9. Furthermore, equal amounts of different proteins do not stain to the same extent. Thus, estimates of the amounts of proteins present are, at best, semiquantitative.

Cross reference

You can read about the Beer-Lambert law in Chapter 9, 'Spectroscopy'.

Gel documentation systems are now replacing densitometers. These consist of a computer linked to a video imaging system enclosed in a small dark compartment containing a white or ultraviolet (UV) light transilluminator. Transilluminators are instruments that project a narrow beam of visible or UV light through a translucent sample, such as an electrophoresis gel, to allow the stained bands of the separated components to be observed and photographed. Images of the gel can be stored on the computer and the intensities of the protein bands analysed.

SELF-CHECK 12.2

Examine Figure 12.3 and deduce from the densitometric scan which protein is present in serum in the greatest concentration.

How the separated components are detected and analysed depends upon their nature and the type of electrophoresis performed. Although the principles and points outlined in Sections 12.1 and 12.2 underpin all types of electrophoresis, there are major differences between the various techniques. These include the types of support media used, which, in turn, depends upon the sizes and nature of the molecules to be separated, and whether the equipment has a vertical or horizontal geometry.

The support media used in electrophoresis include:

- paper (cellulose)
- cellulose acetate
- starch
- polyacrylamide
- agarose.

Horizontal electrophoresis systems (Figure 12.2a) are often used in paper, cellulose acetate, or agarose gel electrophoresis (AGE). In contrast, polyacrylamide electrophoresis generally employs a vertical system (Figure 12.2b).

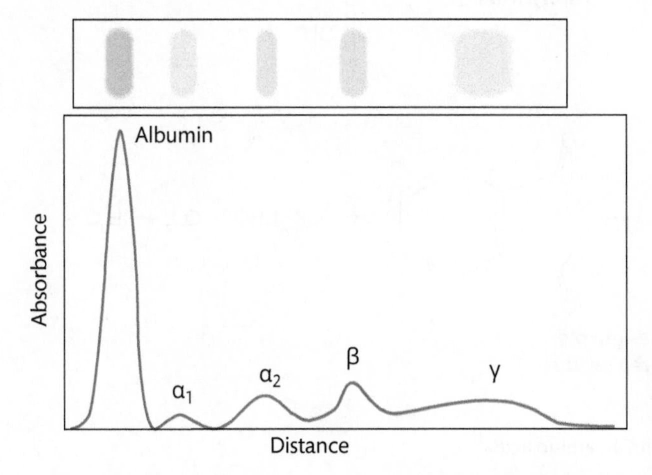

FIGURE 12.3
Densitometric scan of proteins, in this case serum proteins, separated by electrophoresis.

We will now review the different types of support media used in electrophoresis in more detail, starting with paper electrophoresis.

12.4 Paper or cellulose electrophoresis

Paper is composed of the polymer of β-glucose called cellulose (Figure 12.4). It was first used as a support medium in the 1950s in both vertical and horizontal geometries. Paper electrophoresis is largely used to separate small molecules, many of which, for example, amino acids and sugars, are of biomedical importance. However, the use of paper in electrophoresis has a number of flaws. These include its fragility, particularly when soaked in buffer, making it difficult to handle. Furthermore, a number of biological molecules, for example those of proteins, interact with the hydroxyl groups of the cellulose retarding their mobility and resulting in tailing (Section 12.2).

Following electrophoresis, the dye ninhydrin can be used to detect amino acids because it reacts with their amine groups to give a blue-purple-coloured complex (Table 12.1; Figure 12.5). Imino acids, proline, and hydroxyproline are secondary amino acids and react to yield yellow-orange-coloured products.

A number of reagents are available to detect sugars following their electrophoresis. For example, the electrophoresis paper can be immersed in a solution of *p*-anisidine/ethanol followed by one of sodium periodate/acetone, which converts the sugars to coloured derivatives. This method can detect as little as 20 μg of a variety of sugar types, including polyols, pentoses, hexoses, sugar acids, amino sugars, and their acetyl derivatives.

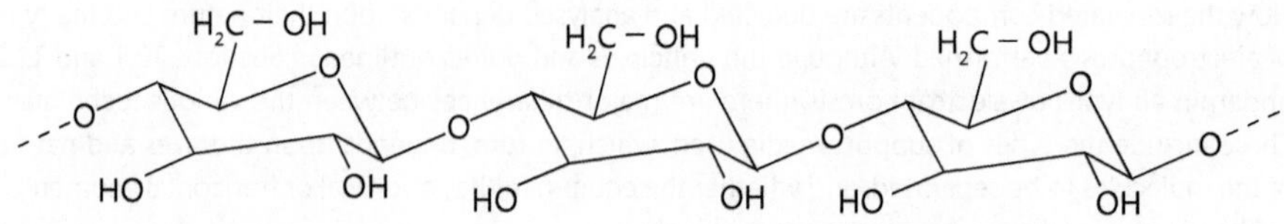

FIGURE 12.4
Structure of cellulose.

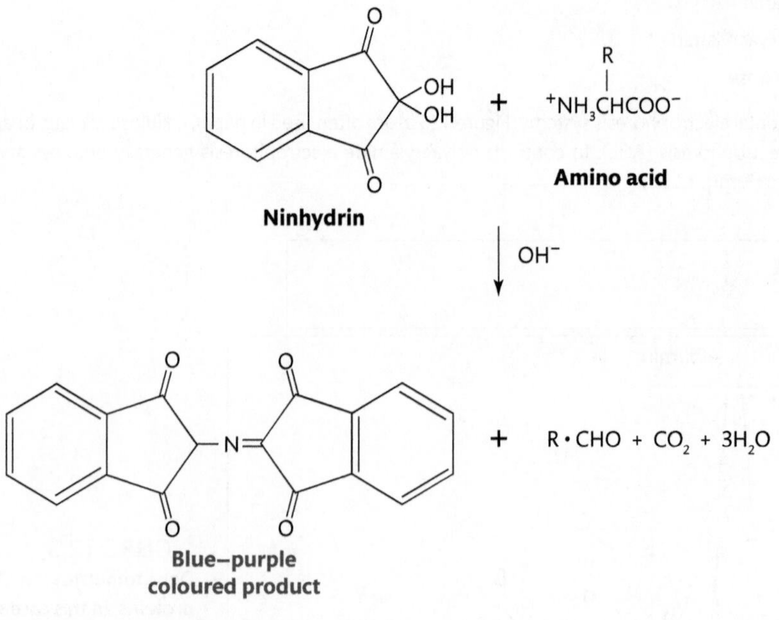

FIGURE 12.5
The reaction of ninhydrin with an amino acid.

Cellulose contains a relatively small number of carboxyl groups, but these can ionize in many buffers and result in electro-osmosis. Both tailing and electro-osmosis hinder resolution and the separation of mixtures. For these reasons, other support media have largely replaced paper.

12.5 Cellulose acetate electrophoresis

The use of cellulose acetate in the late 1950s overcame some of the disadvantages of paper described in Section 12.4. Cellulose acetate is made by acetylating the hydroxyl groups in cellulose (Figure 12.6). The absence of hydroxyl groups reduces interactions with sample ions and produces a structure that is less hydrophilic and therefore holds less water, both of which increase resolution. The pores inside cellulose acetate are uniform and of sufficient size to allow passage of all except the largest molecules.

Single strips of cellulose acetate, typically 2.5 × 12 cm, are used for single samples, although wider ones are available to allow the analysis of multiple samples simultaneously. The strips are first fully wetted in electrophoresis buffer, and placed on a glass or plastic sheet in a horizontal geometry. Buffer-soaked filter papers are used to connect the strip to the electrode compartments (Figure 12.7). Care must be exercised to ensure the strips never dry out. The pH of the buffer used (8.6) ensures proteins have a negative charge and move towards the anode. A small, 1 or 2 µL volume of sample is applied as a 1 cm-wide band to the cathode end of the strip.

Electrophoresis is carried out for about 45 minutes and as different proteins have different mass-to-charge ratios, they migrate at different rates and separate. Following electrophoresis, proteins are fixed in their positions in the cellulose acetate by immersing the strip in a solution of trichloroacetic acid (TCA), which denatures proteins and reduces their diffusion. The strips are stained to detect the proteins, often with the dye Ponceau S (Table 12.1 and Figure 12.8). The negatively charged dye not only binds to positively charged amino groups of the protein, but it also binds through hydrophobic interactions with non-polar parts of the protein.

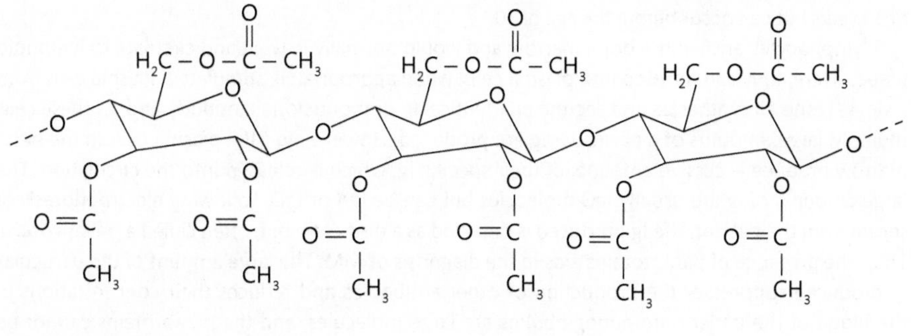

FIGURE 12.6
Structure of cellulose acetate.

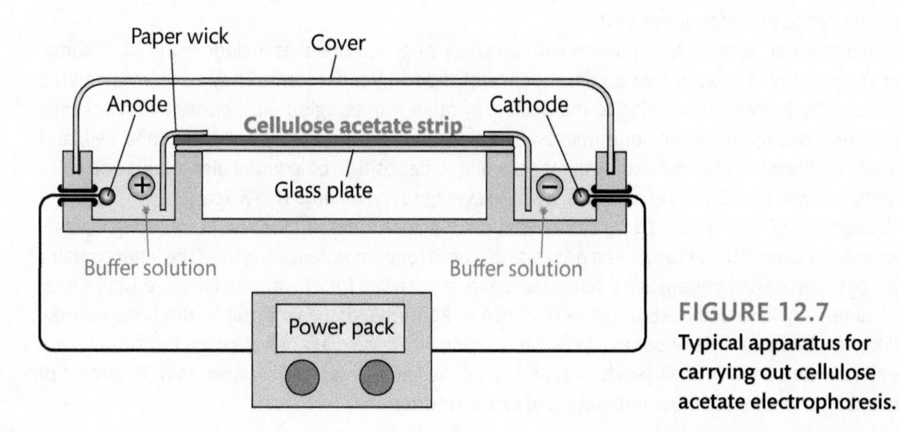

FIGURE 12.7
Typical apparatus for carrying out cellulose acetate electrophoresis.

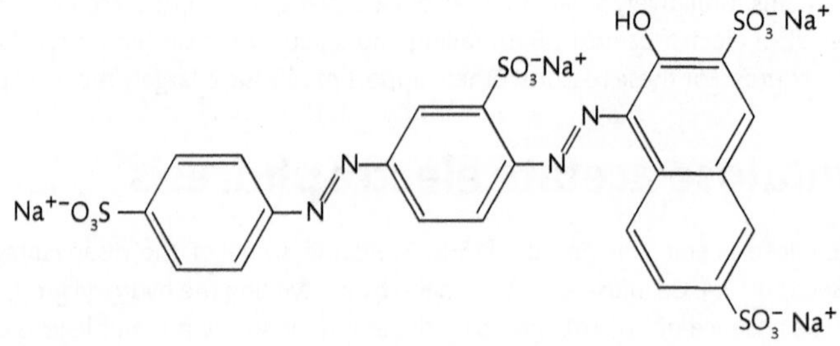

FIGURE 12.8
Structure of Ponceau S.

> **Cross reference**
> You can read about spectrophotometry in Chapter 9, 'Spectroscopy'.

Alternatively, the strips can be rendered transparent by immersing them in a mineral oil whose refractive index equals that of cellulose acetate. The positions of the proteins can then be determined spectrophotometrically.

Cellulose acetate electrophoresis lacks the resolution associated with gel-like materials, such as starch, polyacrylamide, and agar. However, it has the advantages that it is rapid, simple, and relatively inexpensive, and is used in pathology laboratories to analyse serum proteins, for example in the investigation of multiple myeloma (MM).

> **Cross reference**
> You can read more about Bence Jones protein in Chapter 18, 'Specific protein markers', in the book, *Clinical Biochemistry.*

12.5.1 Multiple myeloma and paraproteins

MM is the cancerous proliferation of immunoglobulin (Ig)-producing plasma cells located in the bone marrow. It has an incidence of four in 100,000 worldwide. The condition is more frequent after the age of 50 years but can occur before the age of 30.

B lymphocytes arise in the bone marrow and would normally leave it and circulate to lymphoid tissues. Here, they can develop into plasma cells, when appropriately stimulated. Plasma cells from a single clone all synthesize and secrete an identical Ig or monoclonal antibody and, as these cells multiply, large amounts of a particular Ig are produced. However, in MM, plasma cells in the bone marrow produce excessive amounts of their specific Ig, which is released into the circulation. The large amounts of Ig are usually IgG molecules but can be IgA or IgD. Following electrophoresis of serum from the patient, the Ig produced is detected as a discrete band, often called a **paraprotein**. Thus, the presence of paraproteins aids in the diagnosis of MM. The large amount of the particular Ig produced suppresses the production of other antibodies and reduces their concentrations in the blood of the patient. Immunoglobulins are large molecules, and the paraproteins cannot be removed from the blood by the kidneys and so increase the viscosity of the blood. However, in some patients, a defect in the plasma cells results in the production of only light chains of the Ig being produced, which can be filtered by the kidneys and lost in the urine. These can be detected and referred to as Bence-Jones protein.

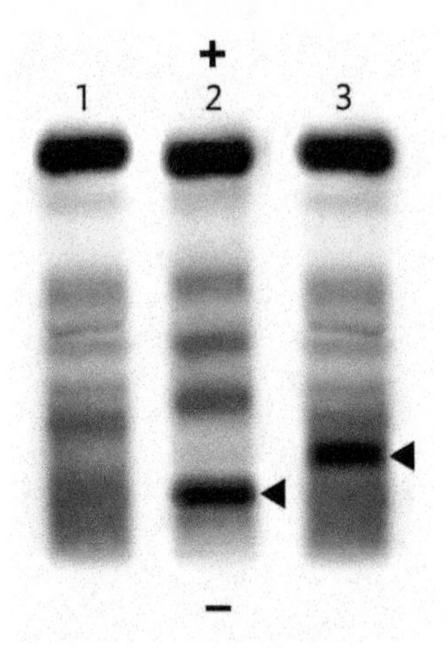

FIGURE 12.9
Separation of serum and plasma proteins by cellulose acetate electrophoresis. Lane 1 shows the separation of a sample of serum from a healthy individual. Lane 2 is serum from a patient with multiple myeloma. The position of the separated paraprotein is indicated. Lane 3 shows the separation of plasma from a healthy individual. The position of fibrinogen is indicated.

Patients suffering from MM present with a variety of symptoms that include bone pain, coma, and retinopathy, the latter two a consequence of blood hyperviscosity. They have an increased susceptibility to infections owing to the decline of other immunoglobulins; present with anaemia owing to a decline in other bone marrow functions, such as erythrocyte, white blood cell, and platelet syntheses; and suffer kidney damage owing to deposition of Ig molecules. About one-third of patients with MM die of renal failure, and the average survival time is 2–3 years.

A diagnosis of MM is assisted by demonstration of a paraprotein in the serum following electrophoresis. It is essential that serum and not plasma is used for paraprotein analysis as the latter contains fibrinogen, which can present as a discrete band and be mistaken for a paraprotein (Figure 12.9). Other clinical tests include demonstrating the presence of abnormal malignant cells in the bone marrow following a biopsy. X-rays of bones can also be used to aid in the diagnosis as characteristic holes are seen in affected areas of skull, pelvis, ribs, and vertebrae in patients suffering from MM. Treatment of patients involves the use of radiotherapy and cytotoxic drugs.

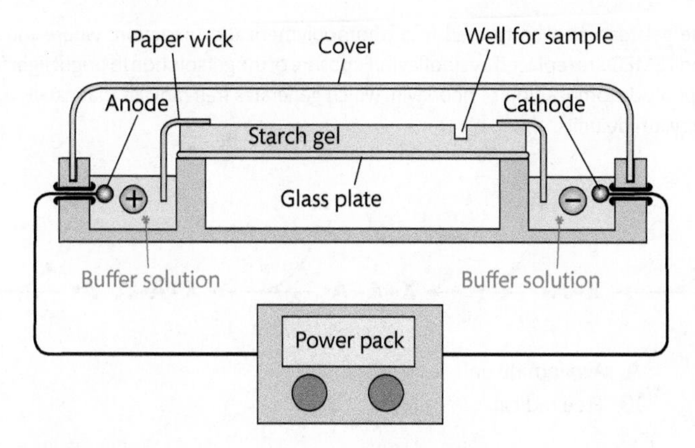

FIGURE 12.10
Schematic showing typical apparatus for starch gel electrophoresis.

12.6 Starch electrophoresis

Starch was first used in the late 1950s as a molecular sieve in gel electrophoresis to separate proteins. The usual type of apparatus is shown in Figure 12.10. As you can see, it has a horizontal geometry, which holds a slab of starch connected to the buffer compartments with wicks, rather like the system encountered in cellulose acetate electrophoresis (Section 12.5).

Starch gels are normally prepared using hydrolysed potato starch. The starch is suspended in the electrophoresis buffer and heated until a transparent solution forms. The solution is then poured into a suitable mould and solidifies to form a gel as it cools. Slots are cut into the gel into which the samples for electrophoresis are added. Following electrophoresis, the gel can be sliced into a number of layers using a fine wire, which provides several replicate gels, each of which can be analysed using different detection methods. For example, different slices of gels can be examined for different enzyme activities (see Section 12.7), particularly to demonstrate the presence of isoenzymes, such as those of serum lactate dehydrogenase.

Cross reference

You can read about isoenzymes in Chapter 7, 'Clinical enzymology and biomarkers', in the accompanying textbook, *Clinical Biochemistry*.

The major drawbacks of using starch as a support medium in electrophoresis are that, like cellulose, starch contains some negatively charged groups that can interact with sample cations, and also lead to electro-osmosis. Also, the pore sizes of starch gels vary with the concentration used. This is a major disadvantage to its use as a sieving medium. If the concentrations of starch are too extreme, the gel is too soft to handle or excessively stiff to use, which restricts the range of sizes of molecules that can be analysed.

12.7 Polyacrylamide gel electrophoresis

Polyacrylamide gel electrophoresis (PAGE) separates mixtures of molecules of proteins or nucleic acids through a gel composed of polyacrylamide. The gel is formed *in situ* between two glass plates that are clamped to form a sandwich, with the plates being held apart by plastic spacers that also retain the reaction mixture used to form the gel.

Polyacrylamide gels are formed by polymerizing acrylamide units into long chains. The polymerization is initiated by free radicals generated in the mixture by a reaction between ammonium persulfate and *N,N,N',N'*-tetramethylethylenediamine (TEMED) as shown in Figure 12.11 (a, b). In Figure 12.11(a) the free radical is represented as X·, and the acrylamide unit as A. The free radical X· reacts with A to form X–A·, which is still reactive as it has an unpaired electron. Thus, it will react with another molecule of A to form XAA· and so polymers of acrylamide units are built up in a stepwise fashion. As the chains grow, they are cross-linked by bisacrylamide (*N,N'*-methylene bisacrylamide), which consists of two acrylamide units linked by a methylene group. The cross-linking converts the solution to a solid gel, the structure of which is shown in Figure 12.11(b), and consists of a three-dimensional network of pores.

Acrylamide gels can also be formed in a photopolymerization reaction, where the ammonium persulfate and TEMED are replaced by riboflavin. Exposure of the gel solution to bright light for 2–3 hours leads to the photodecomposition of riboflavin, which generates free radicals that catalyse polymerization of the acrylamide units.

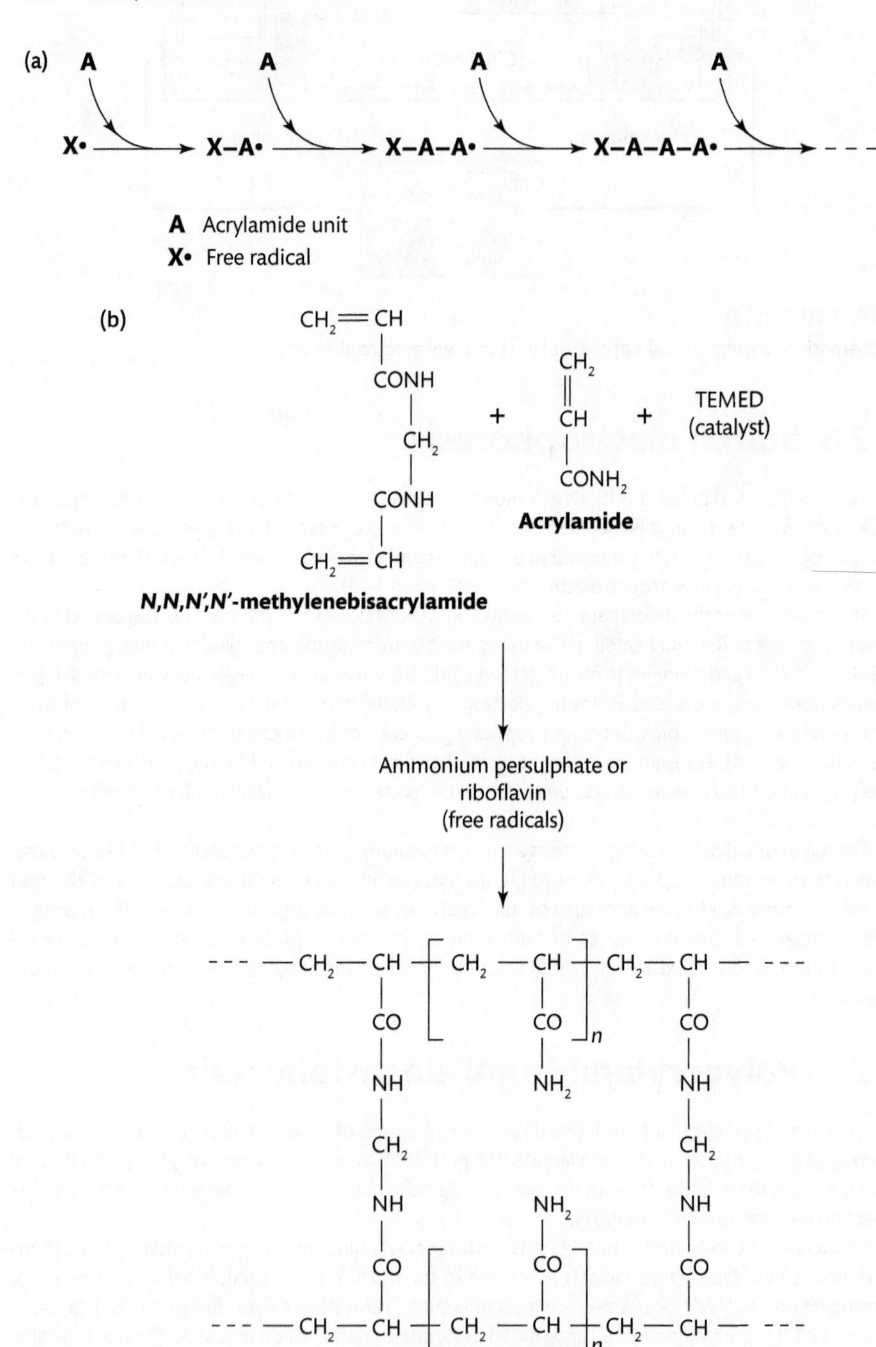

FIGURE 12.11
(a) Schematic illustrating the polymerization of acrylamide. See text for details. (b) Structure of polyacrylamide.
TEMED: *N,N,N',N'*-tetramethylethylenediamine.

Separation during PAGE is based not only on differences in the electrophoretic mobilities of the different biological molecules, but also on the molecular sieving effects of the pore size. The sizes of the pores in the gel can be varied by altering the concentrations of acrylamide and bisacrylamide. Typically, gels consist of 3–30% concentrations of acrylamide. Those with a low percentage of acrylamide have larger pores and are more suitable for separating larger proteins or nucleic acids, as described in Chapter 14. Smaller molecules move through the gel easier than larger ones and so the relative rates of migration are the smallest ones fastest and the largest slowest.

There are a number of elementary safety precautions that should be observed for electrophoresis (see Health & Safety Box, 'Some elementary precautions for electrophoresis').

HEALTH & SAFETY

Some elementary precautions for electrophoresis

- **Mouth pipetting must not be used in preparing any of the solutions used for gel formation, staining, blotting, or detection.**
- **All acrylamide solutions should be prepared in a fume cupboard. Unpolymerized acrylamide and bisacrylamide are a skin irritant and a potential neurotoxin, respectively. Thus, safety gloves must be worn throughout all handling procedures and a face mask when weighing the dry powder. Mercaptoethanol has a powerful and unpleasant odour, and should be handled only in a fume cupboard.**
- **Acrylamide begins to polymerize when TEMED and ammonium persulfate are added to it. The ammonium persulfate solution should be freshly prepared and it and the TEMED added to the acrylamide solution immediately before it is poured into the gel apparatus.**
- **Do not touch the electrophoresis chamber or wires while electrophoresis is in progress. Voltages may be routinely as high as 200–300 V and shocks may be fatal.**
- **Ethidium bromide is a mutagen. Always wear gloves when handling solutions or agarose gels containing the compound. Dispose of the final electrophoresis buffer containing ethidium bromide according to directions from your instructor and local health and safety procedures.**
- **Avoid looking into UV light during the detection of DNA fragments on the agarose gel. Wear UV-protective goggles.**

SELF-CHECK 12.3

What basic safety precautions must you take when you prepare and handle a polyacrylamide gel?

Polyacrylamide gel electrophoresis occurs in a vertical gel (Figure 12.2b). A photograph of a typical commercial vertical apparatus for carrying out vertical gel electrophoresis is shown in Figure 12.12(a), which is shown schematically in Figure 12.12(b). The typical dimensions of a polyacrylamide gel are 12 × 14 cm with a thickness of 1–2 mm, although gel dimensions can vary enormously.

Whichever method is used to polymerize the acrylamide, a plastic comb or well former is inserted between the plates into the polymerization mixture before it sets. The spacers and comb are removed following polymerization. The indents in the gel formed by the comb form wells to which samples are subsequently added.

The buffer provides the electrical contact between the electrodes and the gel. After assembly, the lower electrophoresis tank buffer surrounds the gel plates helping to keep them cool. Polyacrylamide gel electrophoresis can use continuous or discontinuous buffer systems.

12.7.1 Continuous buffer systems

In a **continuous polyacrylamide gel electrophoresis** system, the same type of buffer is used to dissolve the sample as that present in the gel and in the compartments of the electrophoresis equipment. Sucrose or glycerol are added to the samples, which are loaded directly into the wells. This increases their densities, ensuring they settle quickly to the bottom of the wells. The samples also contain a

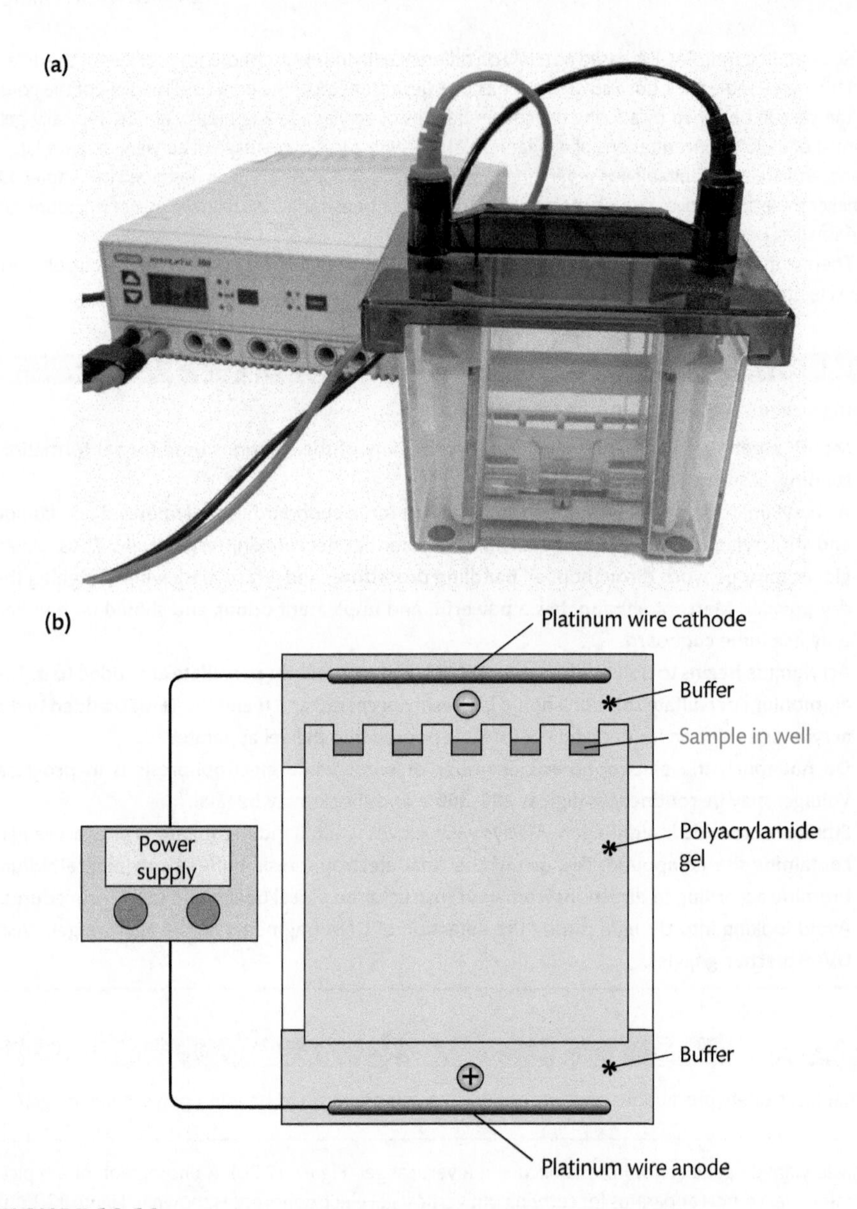

FIGURE 12.12
(a) Photograph of a typical commercial gel electrophoresis system for performing polyacrylamide gel electrophoresis. (b) Schematic showing typical apparatus for performing polyacrylamide gel electrophoresis.

so-called **tracking dye**, usually bromophenol blue, which is a small molecule and passes through the pores in the gel to act as a marker for the electrophoretic front.

Electrophoresis is performed for a defined period ranging from an hour to overnight, usually until the tracking dye has neared the bottom of the gel. The current is then switched off.

SELF-CHECK 12.4

If 6 cm^3 of a 30% combined acrylamide and bisacrylamide solution was made up to 18 cm^3 with the other gel-forming reagents, what is the concentration of acrylamide in the final gel?

Following electrophoresis, the gel is removed from the glass plates. The separated proteins are fixed in position usually using a mixture of TCA and methanol; the gel is then stained to reveal the positions of the separated proteins.

12.7.2 Discontinuous buffer systems

In a **discontinuous polyacrylamide gel electrophoresis**, the buffers used to dissolve the samples and those present in the gel and reservoir compartments of the equipment differ in terms of their composition and pH. In addition, the gel is formed in two parts. The main separating or resolving gel is made as described above. However, it occupies only part of the space between the glass plates and does not contain wells. Once this has polymerized, a much smaller stacking gel is formed above it, which generally has a buffer of lower pH to the resolving gel. The wells for samples are formed in the stacking gel, which typically consists of only 4% acrylamide or even less. The large size of its pores allows the sample molecules to move through it freely during electrophoresis.

Samples are added to the wells of the stacking gel generally in the same type of buffer used to form the resolving gel. When an electric field is applied to the system, the molecules of proteins or nucleic acids become sandwiched between various ionic species of the buffer, which concentrates them as a thin band within the stacking gel. Although the pH is higher in the resolving gel, which would tend to increase the mobilities of the ions, its pore size is smaller and separation of the various molecular species occurs as described above for continuous systems. However, because the proteins or nucleic acids have entered the resolving gel as narrow concentrated bands, discontinuous systems have *enhanced resolution* over continuous systems.

PAGE can be performed under non-denaturing conditions as described earlier, in which case the technique is referred to as native PAGE. Native PAGE, although useful for many applications, is not generally applicable for determining the M_r of molecules or of resolving molecules that have similar sizes, because mobility under these conditions is influenced by both charge and size. However, when it is unnecessary to preserve the structures and functions of biological molecules, PAGE can also occur in denaturing conditions, such as in the presence of the detergent sodium dodecyl sulfate (SDS).

12.7.3 Sodium dodecyl sulfate electrophoresis

The most widely used PAGE technique is known as **sodium dodecyl sulfate polyacrylamide gel electrophoresis** (SDS-PAGE). SDS-PAGE is particularly useful in analysing proteins to assess their purity or to determine their Mr and was extensively used in sequencing DNA molecules, as described in Chapter 14, before being replaced by CE (Section 12.9).

Cross reference

The sequencing of nucleic acids is outlined in Chapter 14, 'Molecular biology techniques'.

SDS-PAGE, as the name implies, uses the anionic detergent SDS (Figure 12.13), which is added to samples and also incorporated in the acrylamide gel and added to the electrophoresis buffers. Samples are boiled for several minutes or kept at 37°C overnight in a buffer containing SDS. The SDS strongly binds to molecules and disrupts the hydrophobic interactions that are largely responsible for maintaining their structures. In the case of protein samples, β-mercaptoethanol is often added to reduce the disulfide bonds that help stabilize the protein structure and which hold the monomers of some quaternary proteins together. Urea is added to nucleic acid samples to denature them to single-stranded forms, suitable for analysis by SDS-PAGE as, for example, in the diagnosis of some genetic diseases such as cystic fibrosis, which is discussed in Box 12.1.

In general, proteins bind SDS in quantitative amounts with one SDS molecule binding to every two amino acid residues; furthermore, the amount of SDS bound overwhelms the native charges carried by the proteins. As the negative charges of the sulfate groups of SDS repel each other, the protein molecules adopt rod-like structures (Figure 12.15). This means that all proteins in the presence of SDS and β-mercaptoethanol have a similar rod-shaped conformation and possess similar mass-to-charge ratios. Thus, the negatively charged proteins will move towards the positive electrode. Furthermore, all the protein molecules will now have the same charge per unit length and therefore travel with the

$$CH_3-CH_2-CH_2-CH_2-CH_2-CH_2-CH_2-CH_2-CH_2-CH_2-CH_2-CH_2-SO_4^-Na^+$$

FIGURE 12.13
Structure of sodium dodecyl sulfate (SDS).

BOX 12.1 Investigating cystic fibrosis

Cystic fibrosis (CF) is the most common fatal, inherited homozygous recessive disorder of white people. It affects approximately one in 2000 people in the UK, and about one in 20 carry a defective gene. The condition results from mutations in the gene *CFTR* that encodes the CF transmembrane conductance regulator protein. Thus, there is reduced permeability to Cl^- in the apical membranes of epithelial tissues that line the lungs and other organs, and viscous mucus accumulates in the lungs, digestive tract, and associated organs and epididymis. Patients suffer chronic respiratory disease, malabsorption, cirrhosis, and electrolyte disturbances.

Possible cases of CF are investigated by applying pilocarpine to the skin of patients to stimulate a flow of sweat, which is then sampled. Concentrations of Cl^- in the sweat greater than 70 mmol dm^{-3} are indicative of CF. Unfortunately, the sweat test is not always reliable in the first 6 weeks of life or in adulthood, and some patients with CF test normally. Several screening tests for CF based on its effects on pancreatic function are available. One measures immunoreactive trypsin (IRT) in dried spots of blood. The amount of IRT increases significantly in infants with CF compared with healthy ones. However, after the first few weeks of life pancreatic insufficiency develops and IRT decreases, and the test cannot be used.

Identifying mutations in *CFTR* of patients with CF can not only confirm diagnosis, but can also assist with assessing the carrier status of their relatives. It also provides an accurate and rapid prenatal diagnosis of CF once the parental genotypes are confirmed. Unfortunately, over 1300 mutated forms of *CFTR* have been reported, which makes their identification in individual cases difficult. The most common CF mutation is called ΔF508 and is the deletion of the codon in *CFTR* that specifies a phenylalanine residue at position 508 in the amino acid sequence of the encoded protein. Molecular techniques offer relatively simple ways to detect this mutation. The gene is hydrolysed and the relevant portion amplified to increase its concentration using specific enzymes. The most common technique produces a DNA fragment from healthy individuals that is 98 base pairs (bp) in size. However, if the ΔF508 mutation is present, the product is only 95-bp long. SDS-PAGE can separate the appropriate gene fragments, as shown in Figure 12.14, and detect the deletion. The test is not totally specific for ΔF508 given that another 3-bp deletions in this region of the gene would also result in the same difference in size. Mutations other than ΔF508 can also be identified because they produce differing banding patterns following SDS-PAGE.

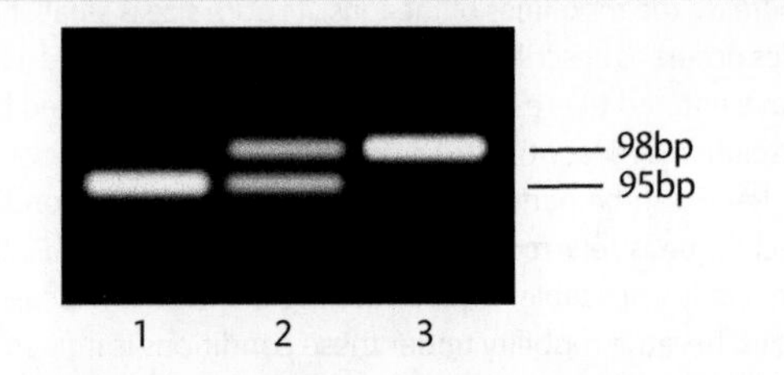

FIGURE 12.14
Use of electrophoresis in detecting the ΔF508 mutation in cystic fibrosis. Lane 1 shows the position of the 95-base pair (bp) fragment associated with the mutation. Lane 2 has the two bands that occur in the heterozygote condition, the normal 98-bp fragment and the 95-bp one. Lane 3 is the normal condition where only a single 98-bp fragment is observable.

Cross reference
See Chapter 14, 'Molecular biology techniques'.

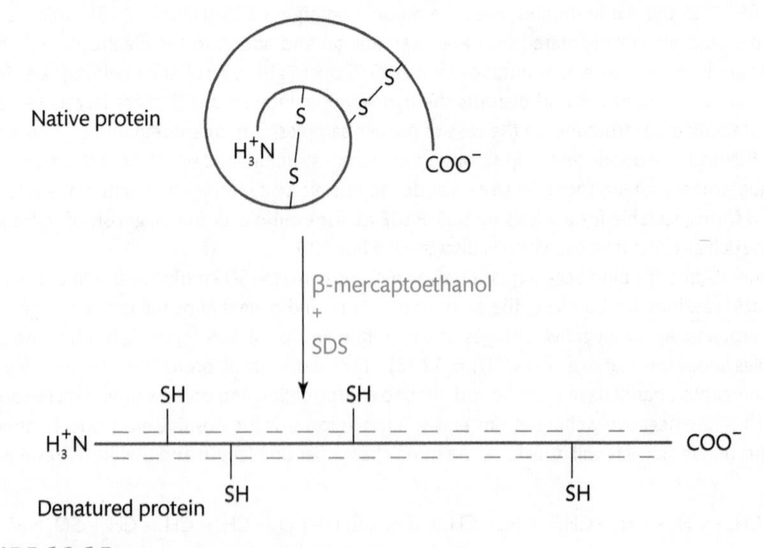

FIGURE 12.15
Schematic showing the denaturation of proteins by β-mercaptoethanol and sodium dodecyl sulfate (SDS).

BOX 12.2 Coomassie dyes

Coomassie dyes (Table 12.1), which are sometimes called Coomassie brilliant dyes, were introduced in the 1960s, following electrophoresis, as stains to detect proteins. They are particularly used following native PAGE and SDS-PAGE. The Coomassie dyes were originally developed as acid woollen dyes (they stain wool in acid conditions). They were named to commemorate the 1896 British occupation of Kumasi, which was formerly called Coomassie, in Ghana, West Africa.

Coomassie dyes constitute a group of chemically related compounds. The first to be developed was Coomassie Blue R-250; the R stands for reddish tint of the dye, while the 250 is an indicator of the strength of the dye on a scale used by the original manufacturer, Imperial Chemical Industries. Coomassie Blue G-250, G for greenish, and Coomassie Violet R-150 followed later. The structures of R-250 and G-250, which are the most commonly used for staining proteins in electrophoretic gels, are shown in Figure 12.16(a, b). Coommassie R-250 generally gives better resolution and is more often used than G-250, even though the latter has greater sensitivity.

Coomassie dyes can be added to native (non-denatured) proteins, rather than SDS, prior to electrophoresis. The name of this technique is Blue Native or BN-PAGE. Like SDS, Coomassie dyes impart a net negative charge to the proteins, allowing them to separate during PAGE. Only negatively charged proteins would migrate without the presence of the dye (or SDS) during native PAGE.

The Bradford method is one of the most common used to determine the concentrations of proteins in solution. Coomassie dyes are one of the components of the Bradford reagent. Coomassie Brilliant Blue G-250 can bind to proteins in acid solution and, when it does so, its wavelength of maximum light absorbance shifts from 465 to 595 nm. Equal amounts of Bradford reagent are added to solutions containing known concentrations of proteins, as well as to the samples whose concentrations are unknown. The absorbances of the resulting coloured solutions are then measured in a spectrophotometer. The data from the known solutions are then used to construct a calibration graph from which the concentration of protein in the samples can be determined.

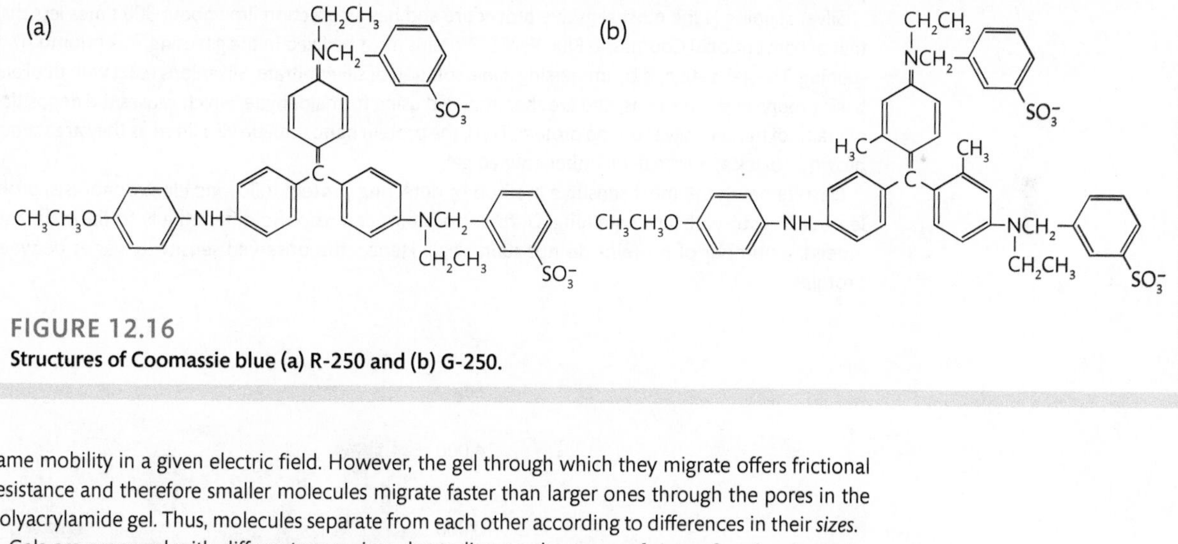

FIGURE 12.16
Structures of Coomassie blue (a) R-250 and (b) G-250.

same mobility in a given electric field. However, the gel through which they migrate offers frictional resistance and therefore smaller molecules migrate faster than larger ones through the pores in the polyacrylamide gel. Thus, molecules separate from each other according to differences in their *sizes*.

Gels are prepared with different pore sizes depending on the range of sizes of molecules to be separated. For example, 15% polyacrylamide gels are typically used to separate molecules with M_r ranging from 10,000 to 100,000. Gels with a lower concentration of acrylamide and therefore larger pore sizes are used for larger molecules. For example, 5% acrylamide gels separate molecules with *Mr* in the range of 60,000–350,000.

12.7.4 Visualizing proteins following separation

Perhaps the most commonly used dyes to stain proteins in polyacrylamide gels are the Coomassie types (Table 12.1 and Box 12.2). Coomassie dyes bind non-specifically to virtually all proteins through hydrophobic and van der Waals interactions, and because the negative charges of their sulfonate groups make ionic interactions with arginine, lysine, and histidine residues of the

Cross reference

You can read more about spectrophotometry and standard graphs in Chapter 9, 'Spectroscopy'.

proteins. Gels are stained by immersing them for 2–4 hours in a solution of the Coomassie brilliant blue dye in a mixture of methanol and TCA, until the gel is a uniformly blue colour. The solvents used precipitate the proteins and so fix their positions in the gel. Following this immersion, the gels require destaining by gently shaking them in a mixture of methanol and acetic acid until the background is clear, which usually takes 4–24 hours. The proteins will then appear as darkish blue bands against a light background. The inclusion of positively charged resins speeds up the destaining time considerably.

SELF-CHECK 12.5

What is the structural difference between Coommassie R-250 and G-250?

Coomassie Blue R-250 has a detection limit of about 0.3–1.0 μg of protein per band on the gel (Figure 12.17). If greater sensitivity is required to detect even smaller amounts of proteins, then Coomassie blue in a *colloidal* form may be used. A solution of colloidal and molecularly dispersed Coomassie blue in equilibrium is prepared by adding methanol and ammonium sulfate to a solution of the dye in dilute phosphoric acid. During staining, only the molecular form enters the gel and stains the proteins preferentially, while the colloidal form is unable to enter the gel and so background staining does not occur.

The fluorescent Sypro orange and red dyes have comparable sensitivities to colloidal Coomassie dye and can detect about 1–2 ng of protein (see Table 12.1). Stained proteins can be seen on the gel by illuminating them with light of approximately 300 nm. Although both fluorescent dyes have similar staining properties, Sypro orange is the more sensitive, but the red form gives less background interference.

Silver staining is the most sensitive procedure and has a detection limit about 300 times less than that of conventional Coomassie Blue R-250. Proteins must be fixed in the gel using TCA prior to silver staining. The gel is stained by immersing it in a solution of silver nitrate. Silver ions react with thiol and basic groups in the proteins, and are then reduced using formaldehyde, which causes the deposition of grains of metallic silver on the protein. Thus, the protein bands can be visualized as they are stained brown or black against a light amber-stained gel.

Despite being the most sensitive method of detecting protein following electrophoresis, problems can occur with silver staining. Different proteins can respond differentially to the stain and, indeed, a number of proteins do not stain at all. Hence, the observed sensitivity varies between proteins.

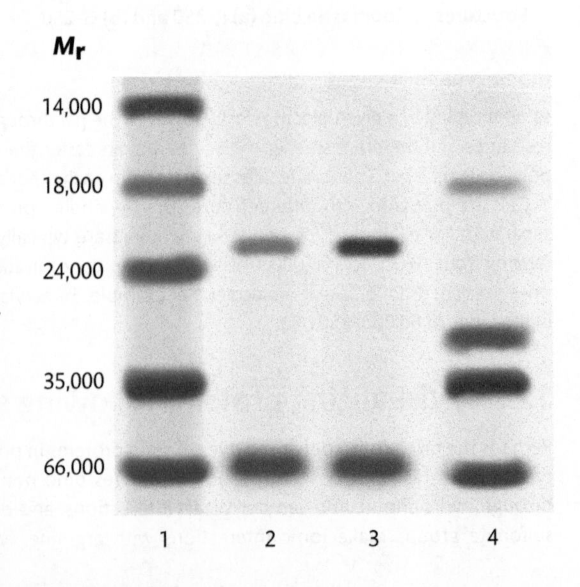

FIGURE 12.17

Photograph of a polyacrylamide gel showing the separated proteins stained with Coomassie blue R-250. Lane 1 shows the positions of proteins of known M_r. Lanes 2–4 show the positions of a number of proteins of differing M_r from three separate samples. Courtesy of Dr M. S. Ahmad, School of Healthcare Science, Manchester Metropolitan University, UK.

SELF-CHECK 12.6

Calculate the approximate lowest sensitivity in ng of silver staining to detect the presence of proteins following PAGE and SDS-PAGE.

Proteins can also be detected by labelling them with radioisotopes such as ^{125}I and their positions on the gel are detected by autoradiography (see Box 11.1, 'Autoradiography').

12.7.5 Detection of glycoproteins

Glycoproteins can be detected on the gels using Alcian blue stain or the periodic acid–Schiff (PAS) reaction (Table 12.1). The former is now little used. You can see an outline of the PAS procedure in Figure 12.18. The periodic acid oxidizes sugars that contain hydroxyl groups on adjacent carbon atoms in the carbohydrate part of the glycoprotein to a dialdehyde product. The aldehydes subsequently react with Schiff reagent to form a coloured product. However, the method lacks sensitivity, being only able to detect bands with at least 2.5 ng glycoprotein. Thus, the weakly coloured pink or light-red bands are often difficult to see on the gel.

The Pro-Q emerald 300 dye also reacts with periodate-oxidized carbohydrates but gives a bright green fluorescent signal when excited, with light of wavelength 300 nm (see Table 12.1). This sensitive fluorescence stain can detect as little as 0.5 ng of glycoprotein. However, rather than using these methods to detect glycoproteins, blotting techniques using lectins as sensitive and specific probes of the carbohydrate portion of the molecule are often employed.

If the proteins are enzymes, then their position in the gels can be detected by exploiting their catalytic activities. This method of detection is particularly applicable to native PAGE and starch gel electrophoresis (Section 12.6) because these support media are permeable to the substrates, cofactors, and dyes required to detect enzyme activities. For example, hydrolases can be detected using p-nitrophenylphosphate. This artificial substrate is hydrolysed by the enzyme to phosphate and p-nitrophenol:

$$\text{p-nitrophenylphosphate} + H_2O \xrightarrow{\text{Hydrolase activity}} \text{p-nitrophenol} + \text{phosphate}$$

The substrate is colourless, but p-nitrophenol is yellow coloured. Hence, the positions of hydrolases on the gel show as bright-yellow bands.

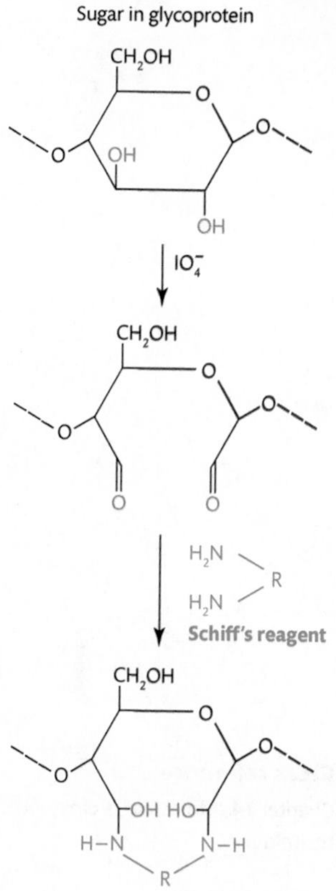

Sugar in glycoprotein

CH_2OH, OH, OH, O

IO_4^-

H_2N, R, H_2N

Schiff's reagent

CH_2OH, OH HO, H–N, N–H, R

Pink-to-red-coloured product

FIGURE 12.18
Schematic showing the periodic acid–Schiff reaction of sugars.

Cross reference
You can read about the blotting of gels and the use of probes in Chapter 14, 'Molecular biology techniques'.

SELF-CHECK 12.7

Why is it not possible to detect the positions of enzymes following SDS-PAGE by monitoring their catalytic activities?

Whatever staining procedure is used, it may cause hydration or dehydration of the gel, and so alter its size. Such changes must be taken into account when measuring the mobilities of the separated components. Thus, the relative mobility of, for example, a protein can be determined using the expression:

Relative mobility = (Migration distance of protein/migration distance of tracking dye) × (Length of gel *before* staining/length of gel *after* staining)

If the M_r of an unknown protein was being estimated by SDS-PAGE, then proteins of known size would be included in the electrophoresis as shown in Figure 12.17. A calibration graph would then be drawn of the logarithm of size (M_r) on the *y*-axis against relative mobility on the *x*-axis. The M_r of the unknown protein may be estimated by determining its relative mobility using the calibration graph.

12.8 Agarose gel electrophoresis

Agarose is one of the components of agar, which, in turn, is a mixture of polysaccharides isolated from some seaweeds. It is a linear polysaccharide consisting of units of agarobiose, that, in turn, consist of alternating galactose and 3,6-anhydrogalactose residues linked together by alternating α1-4 and β1-3 glycosidic bonds (Figure 12.19).

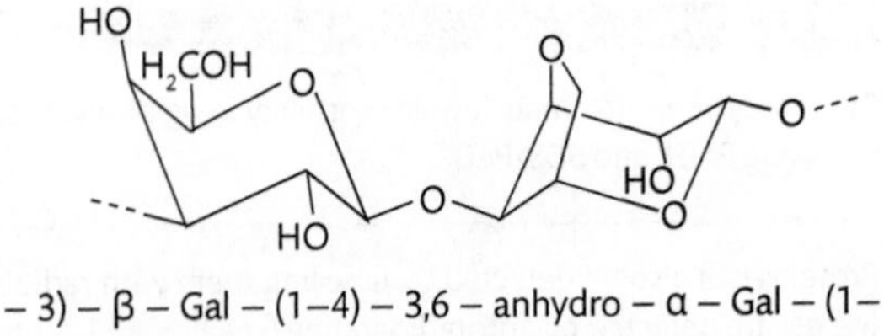

HO
H_2COH
O
O
O
HO
O
HO
HO
O

- 3) – β – Gal – (1–4) – 3,6 – anhydro – α – Gal – (1–

FIGURE 12.19
Structure of agarose. Gal: galactose; anhydro: anhydrogalactose.

Agarose is able to gel because hydrogen bonding within and between the agarose polymers forms a three-dimensional lattice containing a continuous system of pores. The size of the gel pores is determined by the concentration of agarose. A low concentration of agarose gives a gel with larger pores and vice versa. The use of an appropriate concentration allows mixtures of differently sized molecules to be resolved during electrophoresis because the smaller molecules are able to move faster through the gel than the larger ones. Like all supporting media, the gel also limits the diffusion of the molecules, which improves separation.

Molecules that differ in their M_r by as little as 1% can be resolved on agarose gels. In addition to its ability to separate molecules with a wide range of M_r with high resolution, other advantages of using agarose include its lack of toxicity and relative ease of use. Agarose is generally used in electrophoresis to separate nucleic acids as we discuss below, and is also used in isoelectric focusing (IEF) to separate proteins (Section 12.10).

AGE is often used to separate and analyse mixtures of nucleic acid molecules, particularly fragments of DNA. These are usually of large size and their separation requires gels with larger pores. The size of double-stranded DNA is often expressed as the number of bp or kilobase pairs (Kbp). However, even a fragment of DNA as small as 1 Kbp has a M_r of 620,000. In the case of RNA and single-stranded DNA, size is usually given as the number of nucleotides. You can read more about the structures of nucleic acids in Chapter 14.

Cross reference

Chapter 14, 'Molecular biology techniques'.

Every nucleotide in DNA and RNA includes a negatively charged phosphate group. Thus, the mass-to-charge ratio for all nucleic acid molecules is identical. In this respect they resemble polypeptides in the presence of SDS (Section 12.7; Figure 12.14). This means that DNA and RNA molecules will move towards the anode with identical mobilities when subjected to an electric field. Hence, in theory, they should not separate during electrophoresis. However, in agarose electrophoresis, the gel matrix resists the movement of DNA and RNA molecules: larger fragments have the most difficulty in getting through the pores, while the smaller fragments move faster because they experience the least resistance. Indeed, the mobility of very large nucleic acids may be blocked completely. Therefore, separation of DNA and RNA molecules by AGE depends solely on differences in their sizes, like proteins in SDS-PAGE.

The use of appropriate concentrations of agarose with different pore sizes allows nucleic acids of varying ranges of size to be resolved. Those most frequently used include 0.3% agarose gels, which are used to separate fragments of DNA 5–60 Kbp in size, 0.8% for 0.5–10.0-Kbp fragments, and 1–2% agarose gels for 0.1–3.0-Kbp fragments. Using an appropriate concentration of agarose means the distance moved by a limited range of nucleic acid fragments is inversely proportional to the logarithm of their sizes, as was the case for SDS-PAGE described in Section 12.7. Whatever the concentration of agarose to be used, the gel is formed by dissolving dry agarose in buffer and boiling the mixture until a clear solution forms. The solution is poured into the gel former and a plastic comb inserted in it before it sets. Once the solution cools, it sets to form a rigid gel. Removal of the comb leaves wells into which samples can be added. The gel is placed in the electrophoresis tank and covered by the buffer (Figure 12.20a, b).

Samples for electrophoresis are, as in PAGE (Section 12.7), prepared by dissolving them in a buffer that contains sucrose or glycerol to increase their densities above that of the electrophoresis buffer. The samples are loaded directly into the wells using a micropipette and settle quickly to the bottom of each one. Also, as in PAGE, a dye, most commonly the small molecule bromophenol blue, is normally included in the sample buffer. The dye moves rapidly through the gel in advance of all but the smallest DNA fragments. It therefore acts as a marker for the electrophoretic front allowing electrophoresis to be stopped before the samples migrate so far that they run out of the end of the gel!

(a)

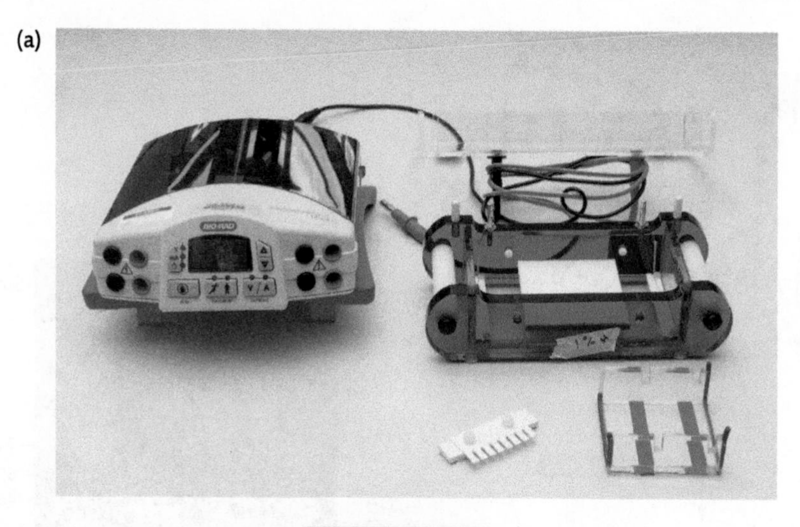

(b)

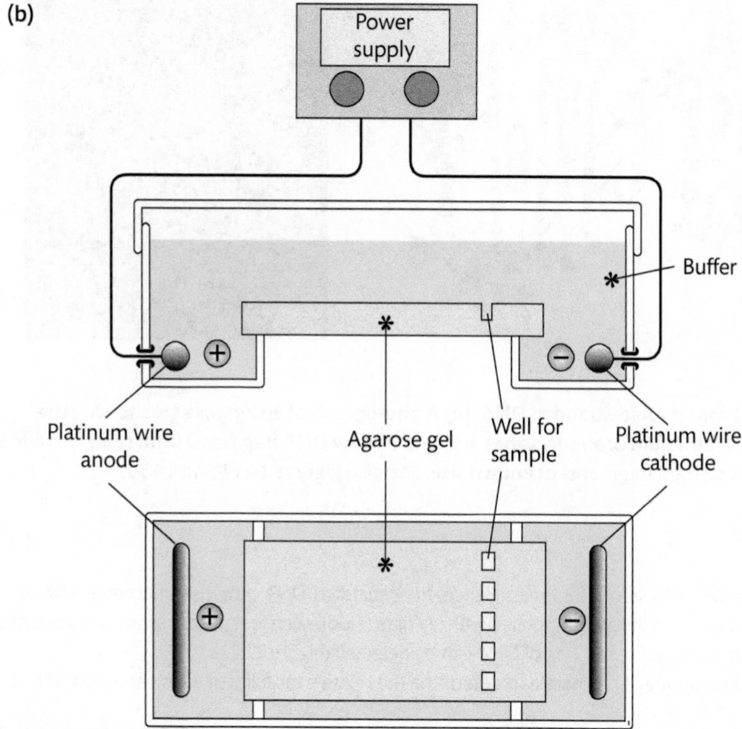

FIGURE 12.20

(a) Photograph of a typical commercial gel electrophoresis system for performing agarose gel electrophoresis. (b) Schematic showing side and top views of typical apparatus for performing agarose gel electrophoresis.

12.8.1 Staining nucleic acids in agarose gels

Ethidium bromide (Table 12.1; Figure 12.21) is the stain of choice for double-stranded DNA molecules that have been separated by AGE. It is normally added to the electrophoresis buffer and so is present throughout the electrophoresis run. As ethidium bromide consists of relatively small, planar molecules, it binds to all the DNA molecules present by intercalating between their stacked base pairs (Figure 12.22a). In this position, the ethidium bromide enhances the normally weak fluorescence of the purine and

FIGURE 12.21

Structure of ethidium bromide.

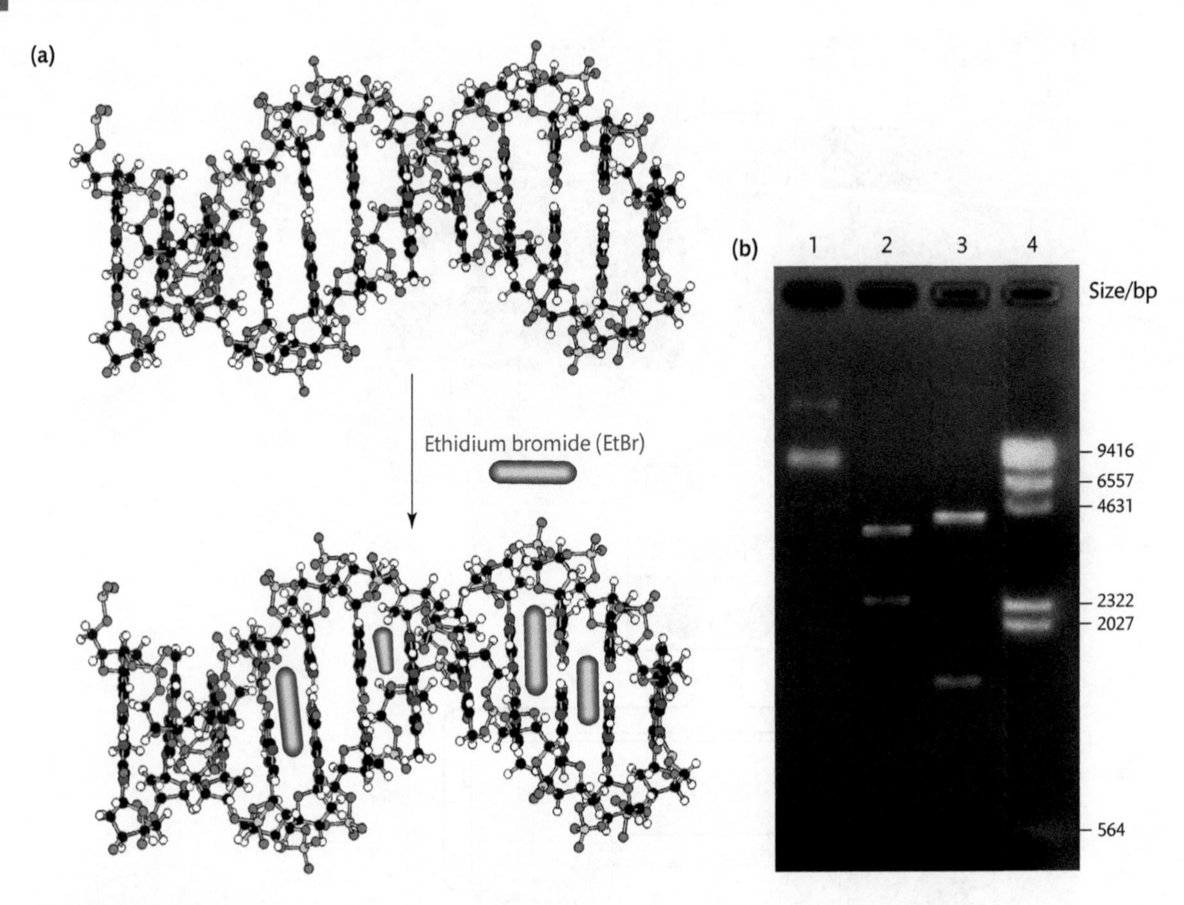

FIGURE 12.22

(a) Intercalation of ethidium bromide into double-stranded DNA. (b) A photograph of an agarose gel showing the separated DNA molecules stained with ethidium bromide. Lanes 1, 2, and 3 show DNA fragments from samples stained with ethidium bromide. Lane 4 shows stained fragments of known size. See also Figures 14.17 and 14.30. bp: base pairs.

pyrimidine bases by about 25 times. Hence, the separated DNA molecules fluoresce with an intense orange colour when the gel is irradiated with UV light. Fluorescence stains are sensitive means of detecting the fragments: as little as 1 ng of DNA can be detected (Figure 12.22b).

Following staining, it is possible to determine the relative mobility of each band from the formula:

Relative mobility = Distance moved by band from origin/distance moved by electrophoretic front from origin

As the ethidium bromide is usually added to the electrophoresis buffer prior to electrophoresis, it is not necessary to correct for changes in the size of the gel during staining. It is possible to determine the sizes or M_r of unknown DNA fragments by including in the electrophoresis fragments of DNA of known size (Figure 12.22b). A calibration graph of the logarithm of size (in Kbp or M_r) on the *y*-axis against relative mobility on the *x*-axis can then be drawn. The values of the unknown nucleic acids can then be determined by using their relative mobilities and the calibration graph. This is essentially the same method as used to estimate the M_r of proteins using SDS-PAGE (Figure 12.17 and its associated text).

AGE can be used to separate molecules of RNA, as well as DNA. Single-stranded molecules of RNA are able to double back on themselves. Consequently, *intrachain* double-stranded regions can form that will bind ethidium bromide, allowing their detection. However, determining the sizes of RNA molecules requires the electrophoresis be performed in the presence of denaturants, such as

FIGURE 12.23
Structure of acridine orange.

formaldehyde, glyoxal, or methylmercuric hydroxide, which are compatible with agarose. These denaturants bind to bases in the RNA and so prevent double-stranded regions forming.

The dye, acridine orange, can detect the single-stranded RNA molecules following electrophoresis (Table 12.1; Figure 12.23). Its binding involves a complex mixture of electrostatic interactions and the intercalation of dye molecules between bases of the single-stranded polymer. When excited by blue light, RNA-acridine orange complexes fluoresce orange but DNA-acridine orange emits yellow-green light. However, in many experiments, RNA is radio-labelled and so can be detected by autoradiography.

SELF-CHECK 12.8

Explain why ethidium bromide is not a suitable stain for single-stranded RNA molecules.

Determining the order of bases of a DNA molecule (sequencing) also involves separating single-stranded nucleic acids, in the case of DNA. Chapter 14 gives a description of sequencing, which involves separating fragments of *single*-stranded DNA by electrophoresis. The methods initially developed used SDS-PAGE gels containing urea, but contemporary techniques use CE (Section 12.10). Urea, like SDS, is a denaturant and together they ensure the DNA chains remain single-stranded during electrophoresis. Following their separation, the different lengths of DNA can be interpreted to give the equivalent sequences of bases in the DNA sample. You can see examples of sequencing DNA using gel electrophoresis and CE in Figures 14.19 and 14.20, respectively.

12.8.2 Pulsed field gel electrophoresis

AGE is unable to resolve molecules of DNA larger than 15 Kbp as molecules of this size and above tend to migrate as a group during electrophoresis. Pulsed field gel electrophoresis (PFGE) is an electrophoretic technique able to separate such large DNA molecules. It does this by periodically altering the magnitude of the applied electric field across the agarose gel. Thus, during PFGE, two different electric fields are applied alternately at different angles to the gel for a defined period of time, for example, 60 seconds (Figure 12.24).

Each field causes the DNA molecules to become stretched and to move in one direction through the gel. The application of the alternate *field* then causes the DNA molecules to change direction. Even though all the DNA molecules are relatively large, the smaller ones are able to realign faster when the direction of *field* changes. This enables the smaller molecules to move faster through the gel than larger ones, leading to separation of the mixture.

PFGE is used in genotyping or genetic fingerprinting. It is also used in epidemiological studies of pathogenic organisms, for example in identifying subtype strains of *Listeria monocytogenes* in clinical samples.

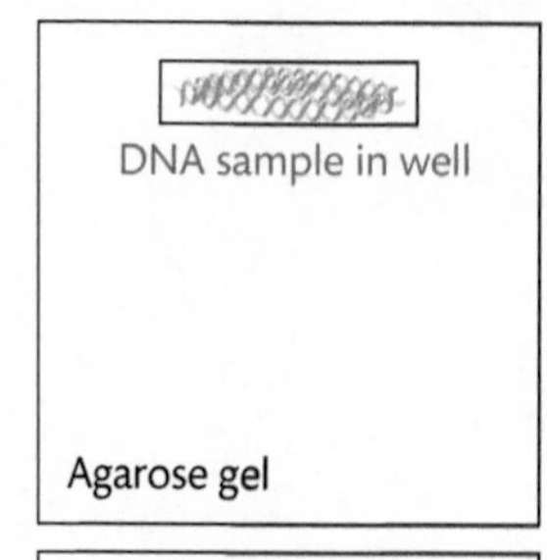

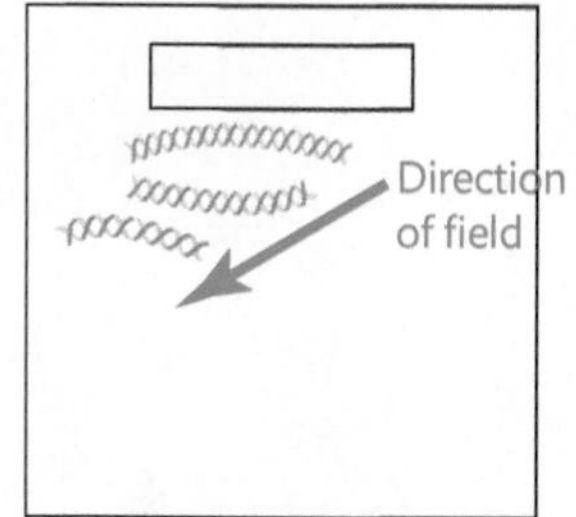

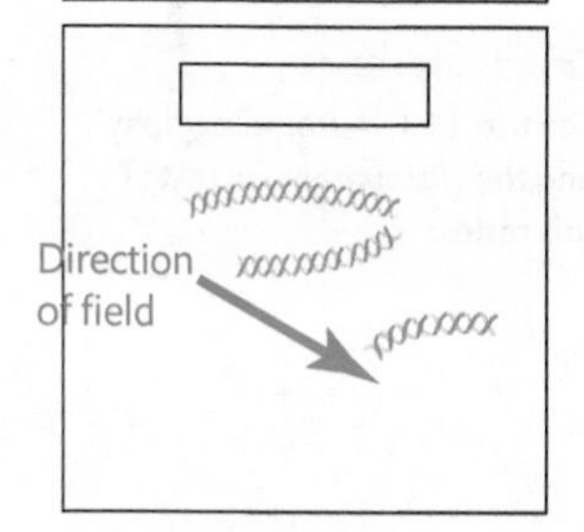

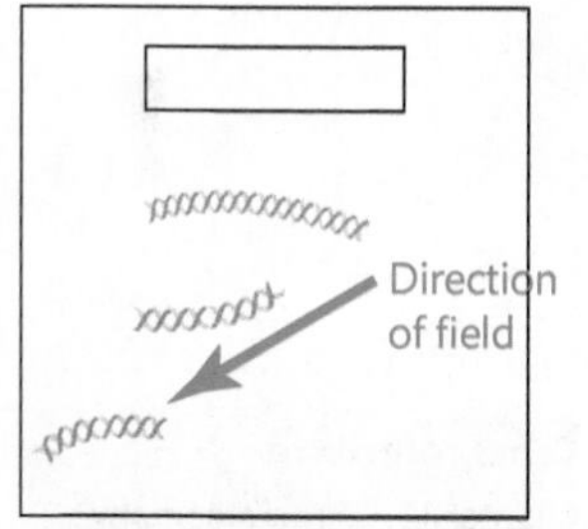

FIGURE 12.24
Schematic showing the operation of pulsed field electrophoresis. See text for details.

12.9 Capillary electrophoresis

CE, as the name implies, separates the components of samples in narrow bore tubes, whose internal diameters are only 25–100 µm, but they may be 40–100 cm in length. Like all electrophoresis, charged molecules move in an electric field. However, because capillary tubes have a high surface area to volume ratio, the high voltages used and resistances encountered allow the considerable heat generated to dissipate rapidly. Hence, convection currents and diffusion of sample

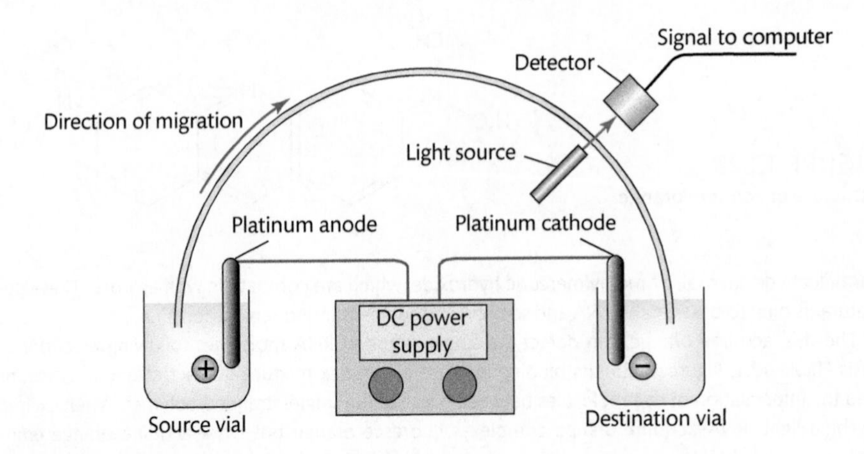

FIGURE 12.25

Schematic showing typical apparatus for performing capillary electrophoresis. DC, direct current.

Cross reference

See Box 11.1, 'Autoradiography', and the *Histopathology* text of this series.

components are minimized, and separation can be carried out in solution without the need for the support media necessary in other forms of electrophoresis.

CE combines the high resolution of electrophoresis with the speed of high-performance chromatography, which you can read about in Chapter 11. For this reason, CE is also called high-performance CE.

Figure 12.25 shows a typical arrangement for CE. The capillary tubes are made of fused silica and coated externally with a polymer to provide mechanical strength. Samples are injected into the capillary by one of two methods:

- the inlet of the capillary tube is removed from the anode reservoir and immersed in the solution of sample; a high voltage is then applied across the tube, which attracts sample ions into the tube
- the sample can be forced into the capillary tube by pressurizing the sample container with an inert gas.

Whatever the method, a major advantage of the technique is that sample volumes as small as 5–10 nL can be analysed.

Cross reference

Chapter 11, 'Chromatography'.

Once the sample has entered the tube, a high voltage is applied across the capillary tube and the constituent molecules in the sample migrate along its length at different rates and so separate. Typical voltages used range from 7 to 10 kV, but they can be as high as 50 kV, depending upon the samples to be separated.

The basis of separation in CE is rather different from that encountered in other types of electrophoresis. Normally, charged particles would move towards the electrode of the opposite charge. However, in CE, electro-osmosis causes all the charged molecules to be carried towards the cathode because the electro-osmotic flow is greater than their electrophoretic mobility. Electro-osmosis occurs in the capillary tubes owing to a net negative charge on the fused silica surface. This attracts cations, which migrate to the inner wall forming an electrical double layer. The cations, in turn, are surrounded by water molecules. When a potential difference is applied to the column, the cations migrate towards the cathode. As these cations are solvated and clustered at the walls of the capillary tube, they drag the rest of the solution with them including the anions (Figure 12.26).

Cross reference

Chapter 9, 'Spectroscopy', and Chapter 8, 'Electrochemistry'.

Although all components of the sample, both charged and neutral, migrate towards the cathode, cations reach it first because they move by a combination of electro-osmotic flow and electrophoretic migration. Cations are followed by neutral molecules and then anions. Thus, the components of the sample are all separated in typically 10–20 minutes. As they flow towards the cathode, the individual components are detected by UV or visible spectrophotometry, fluorescence, or using electrochemical detection methods.

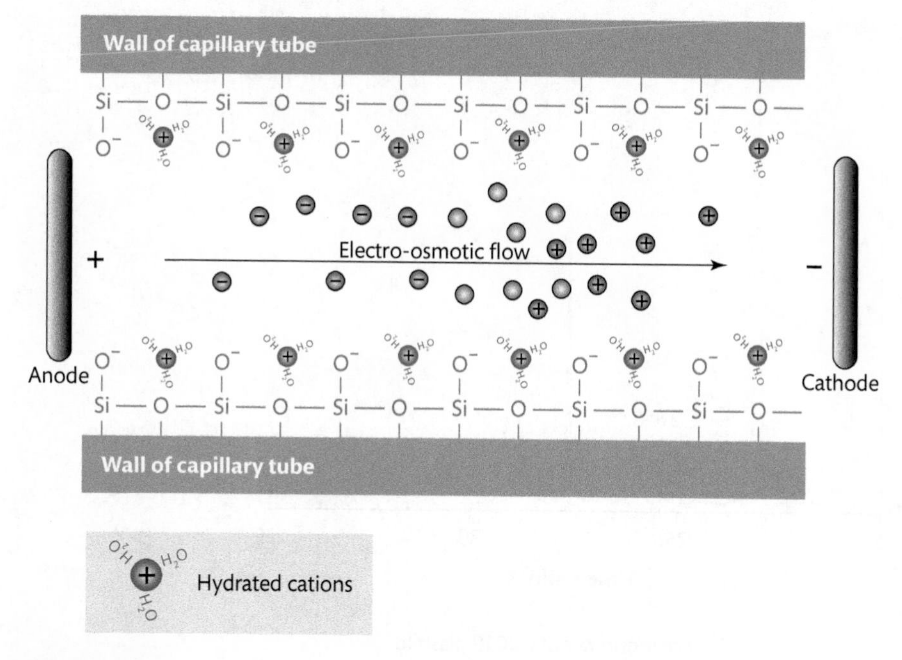

FIGURE 12.26
The role of electro-osmosis in capillary electrophoresis. See text for details.

Amino acids, peptides, proteins, DNA fragments, drugs, and even metal ions can all be separated by CE. A major use is in sequencing DNA molecules and assessing plasmid homogeneity (Box 12.3). The main application of CE in biomedical laboratories is to analyse serum proteins and to monitor disease. For example, CE is used in the detection of paraproteins in the sera of patients as their presence indicates MM, a malignant proliferation of B cells in the bone marrow (Section 12.5). Haematology laboratories use CE to separate haemoglobins for the qualitative analysis of haemoglobin S in diagnosing sickle cell anaemia.

BOX 12.3 Separating plasmids using electrophoresis

Plasmids are extrachromosomal, double-stranded DNA molecules found in bacteria, yeasts, and some other eukaryotic cells. A specific type of plasmid can occur in several forms that differ from one another in conformation and size. One conformation is the supercoiled covalently closed circular (ccc) structure. In this most compact form, the circular DNA double helix is interwoven, rather like a twisted rubber band. Loss of this compact structure happens if one of the DNA strands breaks, which allows the circular molecule to relax with a loss of coiling producing the open circular (oc) or nicked conformation. If both strands of the DNA are hydrolysed, then the closed circular structure is lost and a linear plasmid formed. Further structural heterogeneity results from the ability of plasmids to dimerize.

Plasmids are used as vectors in the production of recombinant DNA molecules (Chapter 14, 'Molecular biology techniques'). Thus, both purity and heterogeneity of preparations of plasmids must be assessed before their use as a vector. Pharmaceutical-grade plasmid DNA is of adequate homogeneity if more than 90% of the preparation is in the form of ccc molecules.

AGE is the traditional method for assessing the homogeneity of plasmid DNA. However, AGE suffers some drawbacks: it cannot be automated and is only semi-quantitative. Furthermore, the assignment of bands to the different plasmid conformations is difficult, as their mobility changes with electrophoretic conditions. However, CE is a more useful technique for the routine analysis of plasmid preparations (Figure 12.27). It has high resolution, sensitivity, and reproducibility, in addition to being readily automated.

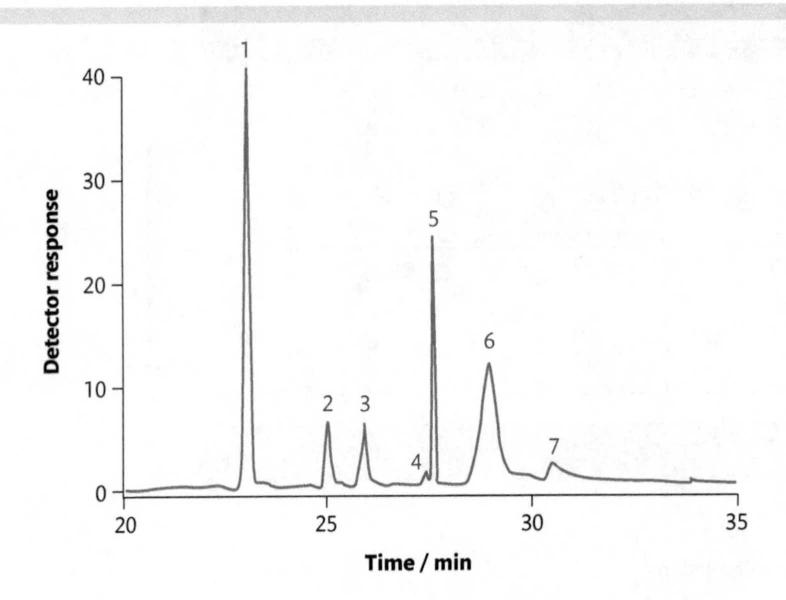

1. ccc monomer of pUC19 plasmid
2. ccc dimer of pUC19 plasmid
3. Linear monomer of pUC19 plasmid
4. Linear dimer of pUC19 plasmid
5. Marker of 5.7 Kbp size
6. oc monomer of pUC19 plasmid
7. oc dimer of pUC19 plasmid

FIGURE 12.27

Separation of forms of the plasmid pUC19 by capillary electrophoresis. Courtesy of Dr Martin Schleef, PlasmidFactory, Bielefeld, Germany. See also Figure 14.28.
ccc: covalently closed circular; oc: open circular.

12.10 Isoelectric focusing

IEF is used to separate molecules, usually peptides and proteins, according to differences in their isoelectric points (*pI*). The *pI* of a molecule is the pH at which it lacks a net charge. The charge on a protein (or amino acid or peptide) is pH dependent. If the pH of the solvent is below the *pI* of the protein, then the protein will be positively charged and will move towards the cathode. Conversely, if the pH of the buffer is above the *pI*, then the protein will be negatively charged and will move towards the anode. If the pH, however, is equal to that of the *pI* of the protein it will not migrate because it carries no charge. Thus, separation occurs in a gradient of pH values formed using a mixture of **ampholytes**. Ampholytes are synthetic low *Mr* polyamino-polycarboxylic or polysulfonic acids that differ from one another in the value of their *pI* (Figure 12.28).

In most IEF systems, focusing is carried out in horizontal, large-pore gels (e.g. 4% acrylamide, or agarose gels of low concentration) on glass or plastic plates or in horizontal glass tubes. The gel stabilizes the pH gradient and the large pores reduce any molecular sieving. Thus, for example, gels can be prepared as thin layers between two glass plates, using a solution containing riboflavin and acrylamide, as described in Section 12.7. The solution also contains a range of ampholytes. Different ranges of pH, for example 3–10, 7–8, or 2–12, are available. Naturally, a range is chosen that encompasses the *pI* values of the molecules to be separated. Once polymerization of the gel has occurred, one of the glass plates is removed and thick strips of wetted filter paper are used as wicks to connect the gel to the electrode

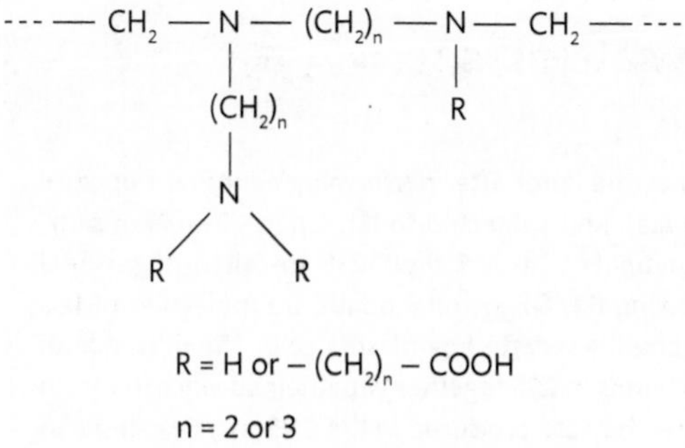

$$\text{---} - CH_2 - N - (CH_2)_n - N - CH_2 - \text{---}$$

$$N(-R)_2 \text{ attached via } (CH_2)_n \text{ to the first N; R on the second N}$$

$R = H$ or $-(CH_2)_n - COOH$

$n = 2$ or 3

FIGURE 12.28
General structure of an ampholyte.

compartments. The anode compartment contains phosphoric acid and the cathode chamber sodium hydroxide. At this stage, the pH is constant throughout the gel. However, when a potential difference is applied, the ampholytes migrate and form an appropriate pH gradient within the gel. The current is then switched off and the samples applied to the gel as a concentrated, salt-free layer on its surface. Alternatively, the sample can be added directly to the gel preparation solution. When the current is switched back on, the proteins are subjected to an electric field and migrate either towards the anode or the cathode depending upon their charge and position in the pH gradient (Figure 12.29). Once they reach a position where the pH of the gradient is equal to their *pI*, they will no longer have a net charge and will cease moving (Figure 12.29). Given that different proteins differ in the value of their *pI*, they will move to different parts of the pH gradient and so separate.

Following focusing, the gels are treated with TCA, which removes the ampholytes and fixes the proteins in the gel. The positions of the proteins on the gel can then be found by staining them, as described in Sections 12.3 and 12.7.

IEF can resolve proteins that differ in their *pI* by a pH value of as little as 0.01: it is, therefore, a high-resolution technique. It can also be used to determine the *pI* of a protein if a mixture of

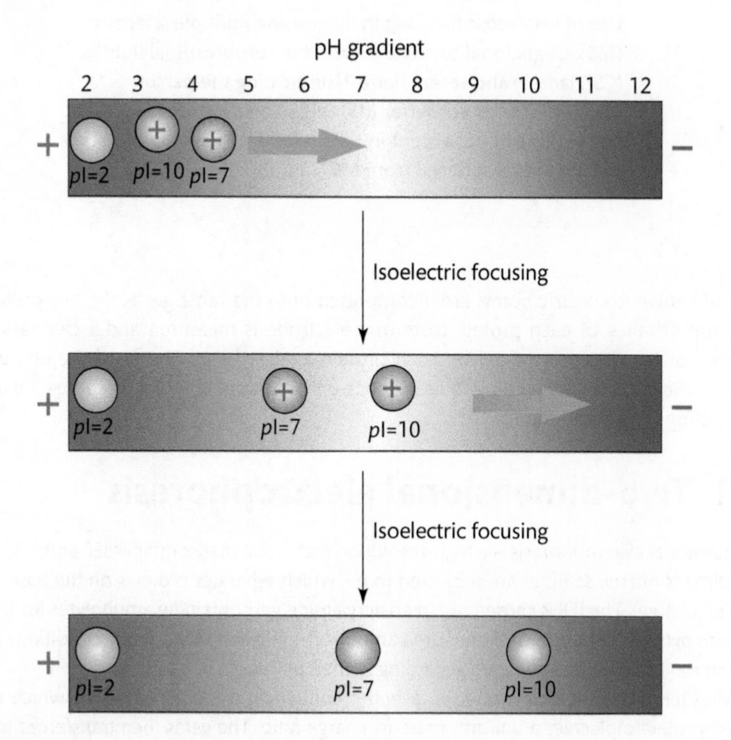

FIGURE 12.29
Outline of the operation of isoelectric focusing. See text for details.

BOX 12.4 Diagnosis of multiple sclerosis using IEF

Multiple sclerosis (MS) is an autoimmune condition where antibodies or immunoglobulins (Ig) attack nerves within the central nervous system (CNS) causing their demyelination. These antibodies are produced within the CNS and are not found in blood. The disease is commoner in females than males and its onset is typically during young adulthood. MS affects between two and 150 per 100,000 people.

In the majority of patients, MS can be diagnosed using IEF. Samples of serum and cerebrospinal fluid (CSF) are obtained from patients, the latter after performing a lumbar puncture (see Figure 6.8), and subjected to IEF. Up to 79–90% of samples from patients with MS show 2–5 so-called oligoclonal bands following IEF. Oligoclonal bands are molecules of IgG or their fragments secreted by plasma cells. The presence of oligoclonal bands in CSF together with their absence in serum indicates that they are produced in the CNS, as you can see in Figure 12.30. It is therefore normal to subtract bands in serum from bands in CSF when investigating CNS diseases.

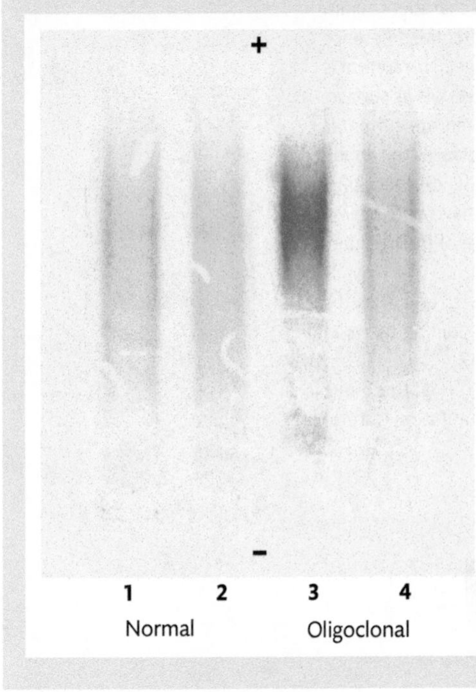

FIGURE 12.30

Use of isoelectric focusing in diagnosing multiple sclerosis (MS). Oligoclonal bands are absent in cerobrospinal fluid (CSF, lane 1) and serum (lane 2), indicating the person concerned does not suffer MS. Oligoclonal bands are present in CSF (lane 3) but absent in serum (lane 4), which indicates the individual suffered from MS.

Cross reference

You can learn more about immunoglobulins in Chapter 13, 'Immunological techniques'.

proteins of known isoelectric points are incorporated onto the same gel as the test protein. After staining, the distance of each protein from one electrode is measured and a calibration graph constructed of the distance moved by each protein against its *pI*. The *pI* of the test protein is obtained as the value corresponding to its distance on the graph. Box 12.4 illustrates the use of IEF in investigating clinical conditions.

12.11 Two-dimensional electrophoresis

Two-dimensional electrophoresis is a high-resolution technique that combines IEF and SDS-PAGE. In the first dimension, the samples are subjected to IEF, which separates proteins on the basis of differences in *pI* (charge). The IEF is carried out in an acrylamide gel containing ampholytes for making an appropriate pH gradient and 8 mol dm^{-3} urea and a non-ionic detergent. These conditions denature the protein molecules, which separate according to their *pI* values.

Following IEF, the gel is equilibrated in a buffer containing the detergent SDS, which gives the denatured protein molecules a uniform mass-to-charge ratio. The gel is then transferred to the top of an SDS polyacrylamide gel. The proteins are now subjected to SDS-PAGE, which forms the second dimension by separating the proteins according to differences in their sizes. The gels are usually silver stained to show the separated components as described in Section 12.7. Large (20 × 20 cm) 10% acrylamide two-dimensional gels are often used in this technique and have been reported to resolve up to 10,000 proteins in samples isolated from cells or in tissue extracts (Figure 12.31).

Two-dimensional electrophoresis can be used to detect and compare changes in gene expression. For example, in two different tissues where one is healthy and the other is diseased. An extract from each of these tissues is subjected to two-dimensional electrophoresis and the gel patterns of the two can be compared. Such a comparison may reveal the presence or absence of a particular protein in the diseased state, which may ultimately be identified and its gene expression investigated. Thus, this technique can increase understanding of the disease.

Proteomics is that branch of science that studies the complete complement of proteins encoded by a genome. It includes identifying and quantifying all the proteins produced (expressed) by a cell, tissue, or organism during the development of a disease. Two-dimensional electrophoresis aids proteomic studies in a number of ways, particularly when combined with other analytical techniques. For example, separate two-dimensional 'maps' can be produced using extracts from a cell or tissue from a healthy individual and from a patient, respectively. The gels are difficult to compare because of the large amounts of data (the number of spots). However, there is commercially available software that can compare the two different two-dimensional maps, and analyse these data both quantitatively and qualitatively. Any proteins that differ between the two maps are of clinical interest and require further analysis. Thus, they may be subjected to peptide mass-fingerprinting. The spots of interest are cut out of the gel and digested with trypsin, which hydrolyses the polypeptide at the carboxyl terminal of

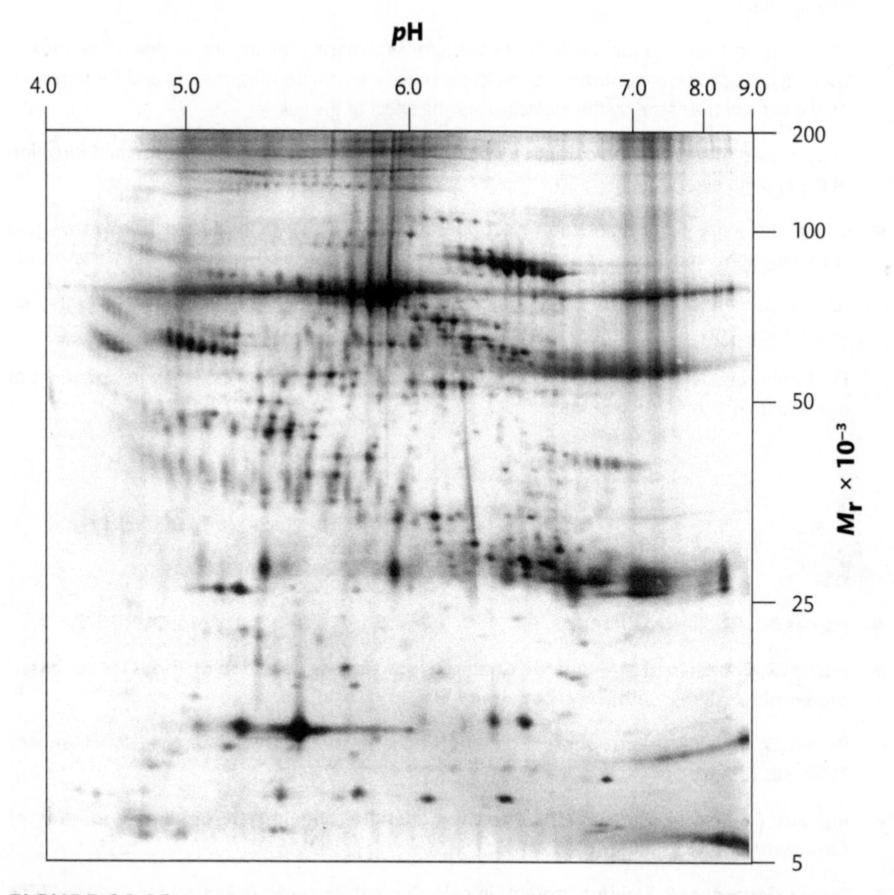

FIGURE 12.31

Photograph of a two-dimensional electrophoretic separation of proteins in a sample of cerebrospinal fluid. Courtesy of P.R. Burkhard, Faculty of Medicine, University of Geneva, Switzerland.

Cross reference

You can read more about mass spectrometry in Chapter 9, 'Spectroscopy'.

each arginine and lysine residue to produce a number of peptides. Matrix-assisted laser desorption ionization-mass spectrometry is then used to give quantitative measurements of the peptides. The set of peptides produced is highly specific for any particular protein; hence, the pattern of peptides produced from the protein of interest can be compared with those in an electronic database and so identified. In a similar way, it is possible to explore the effects of exposure to a particular drug or treatment regime, or toxin.

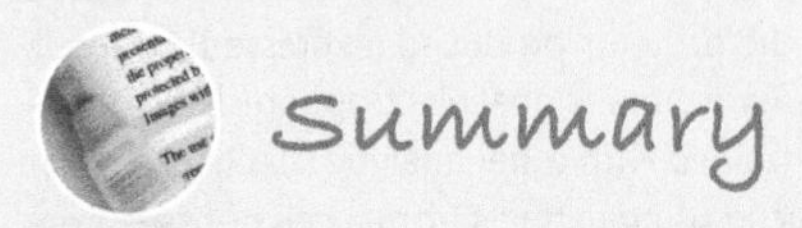

Summary

- Electrophoresis is the movement of charged molecules or ions in an electric field. Separation of the ions is influenced by their nature, the support media used, and the potential difference used.
- Electrophoresis is performed using different types of support media, some that provide only physical support such as paper and cellulose acetate, while others are gels, for example starch, polyacrylamide, and agarose. The latter offer improved resolution because of their molecular sieving effect.
- SDS-PAGE and AGE are the methods of choice for separating proteins and nucleic acids, respectively. In both procedures, mixtures of molecules have uniform negative charges and are separated on the basis of their sizes by the molecular sieving effect of the gel.
- PFGE is used to separate larger molecules of DNA by periodically altering the size and direction of the applied electric field.
- CE combines the high resolution of electrophoresis with the speed of high-performance liquid chromatgraphy, and is used to separate analytes such as amino acids, proteins, DNA, and drugs.
- IEF is a specialized technique where molecules are separated on the basis of differences in their isoelectric points.
- Electrophoretic techniques have applications in biomedical science, including investigation of diseases such as MM, CF, and MS.

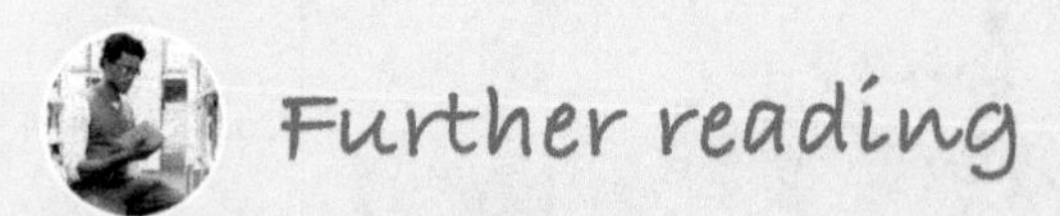

Further reading

- **Hames BD. *Gel Electrophoresis of Proteins: A Practical Approach*. Oxford: OUP, 1998.**
- **Harris LR, Churchward MA, Butt RH, Coorssen JR. Assessing detection methods for gel-based proteomic analyses. *Journal of Proteome* 2007; 6: 1418–25.**
- **Perrett D. Capillary electrophoresis in clinical chemistry. *Annals of Clinical Biochemistry* 1999; 36: 133–50.**
- **Righetti PG. Electrophoresis: the march of pennies, the march of dimes. *Journal of Chromatography A* 2005; 1079: 24–40.**
- **Sasse J, Gallagher SR. Staining proteins in gels. *Current Protocols in Molecular Biology* 2009; 10: 10.6.1–10.6.27.**
- **Westermeier R. *Electrophoresis in Practice*. Weinheim, Wiley-VCH, 2006.**

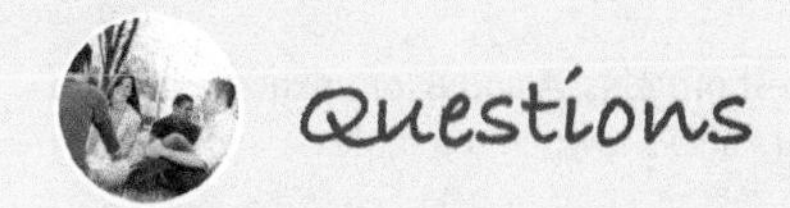

Questions

1. You normally separate fragments of DNA by electrophoresis using a potential difference of 100 mV. The distance between the electrodes in your apparatus is 150 mm. However, this apparatus is unavailable and the larger system available does not have a protocol or manufacturer's instructions. The distance between its electrodes is 200 mm. What potential difference is necessary to produce the electric field you would normally use?

2. Why are glycerol and the dye bromophenol blue added to gel loading buffers?

3. A protein was subjected to native PAGE and centrifugation to determine its size. Native PAGE indicated that the molecules had a M_r of 64,000, which was confirmed by the centrifugation studies (Chapter 10). However, its M_r was found to be only 16,000 when subjected to SDS-PAGE. Are these data compatible? If so, explain your answer.

4. When DNA molecules are separated by AGE, which of the following statements best describes their migration?

   (a) Shorter molecules move faster than longer ones because they have a greater density

   (b) Shorter molecules move faster than longer ones because they can move through the pores in the gel more easily

   (c) Longer molecules move faster than short ones because they are subjected to more frictional resistance

   (d) Longer molecules move faster than shorter ones because of their larger mass

   (e) Longer molecules move faster than short ones because they have a greater mass-to-charge ratio.

5. Which of the following are used to stain proteins and which to stain nucleic acids?

   (a) Acridine orange

   (b) Amido black

   (c) Coomassie blue R-250

   (d) Ethidium bromide

   (e) Silver stain.

6. A number of standard proteins of known M_r were subjected to SDS-PAGE. The bromophenol blue tracking dye moved a distance of 57.0 mm. The distances moved by the proteins were measured. The distances and their M_r are given in the accompanying table. A protein of unknown M_r was also included in the experiment and the distance it migrated is also shown in the table. Staining and destaining did not affect the dimensions of the gel.

$M_r \times 10^{-3}$	Migration distance/mm
14.3	50.0
18.4	43.0
24.0	30.0
34.7	20.0
66.0	10.0
?	38.0

   Calculate the relative mobilities of each of the proteins and plot a suitable graph of these data. Use this graph to determine the M_r of the unknown protein.

7. A number of standard proteins of known *pI* were subjected to IEF. Following staining, the distances of the proteins from the anode edge of the gel were measured. The distances and their

*p*I values are given in the accompanying table. Two test proteins, A and B, of unknown *p*I were also included in the experiment and their distances are also shown in the table.

Standard proteins	*p*I	Distance/mm
	3.50	4.5
	4.55	17.0
	5.20	33.5
	5.85	53.0
	6.55	69.0
	6.85	79.5
	7.35	90.0
	8.15	103.0
	8.45	116.5
	8.65	125.0
	9.30	140.0
Test proteins		
A	?	85.5
B	?	31.5

Plot a suitable graph of these data and determine the *p*I values of A and B.

Answers to self-check and end-of-chapter questions are available at the end of the book.

13

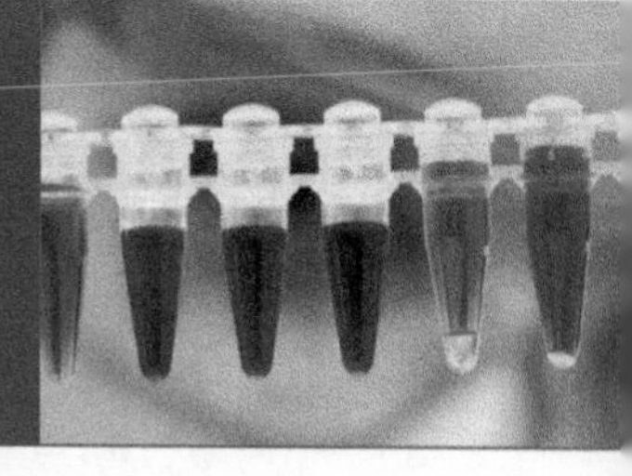

Immunological techniques

Christine Yates and Qiuyu Wang

Learning objectives

After studying this chapter, you should be able to:

- Outline the components and functions of the immune system
- Describe the *in vivo* production of antibodies
- Explain the principles of antigen–antibody reactions
- Discuss the uses of immunological techniques in the clinical laboratory
- Discuss the uses of monoclonal antibodies in the clinical laboratory

Introduction

A grasp of the underlying principles of the immune system and the fundamentals of immunoassay are essential to the understanding of routine immunological diagnostic assays. Immunological assays are techniques that depend upon the visualization of interactions between antibodies and antigens to detect and measure the concentrations of a wide range of substances in health and disease. They are relative newcomers to the clinical laboratory having developed from basic agglutination tests (Box 13.1). The applications were revolutionized in the mid-1970s with the development of monoclonal antibodies. The rapid growth in our understanding of the immune system and the utility of antibodies as diagnostic markers and reagents has led to the development of a wide range of applications that assist in diagnosis and treatment.

Immunological techniques largely depend upon the visualization of antibody–antigen reactions, detecting the formation of **immune complexes** or the attachment of labelled antibodies to cells or an artificial surface by a variety of methods, which we describe in this chapter. A range of immunological techniques have been developed to detect and measure the concentrations of cellular and humoral elements of the immune system in health and disease and, in some cases, assess their biological activities. They have also been used to raise antibodies to other analytes, such as hormones and tumour markers, increasing the range of immunoassays available to biomedical scientists.

BOX 13.1 Beginnings of research into the immune system

The discovery of antigen-antibody reactions took place at the beginning of the twentieth century. Michael Heidelberger measured antibodies using a precipitation reaction in the 1930s producing a globulin fraction of 70% purity and established the existence of complement (Box 13.2). In the decades that followed, electrophoretic and ultracentrifugation studies increased our knowledge of immunoglobulins and antibody activities in the immune response. In 1972, Rodney Porter and Gerald Edelman received the Nobel Prize in Medicine for their work on the structure of immunoglobulin molecules. These studies were the beginning of our understanding of the immune response and led the way for the techniques that we take for granted in clinical laboratories of the present day.

13.1 Outline of the immune system

The immune system comprises a number of physical barriers, lymphoid organs, cells, and humoral components that work in harmony to protect the body from infection. Its primary function is to destroy or disable microorganisms, for example viruses, bacteria, fungi, and protozoa, and parasites, such as helminth worms, which are present in the environment.

Lymphoid organs are divided into two types: primary and secondary organs. Primary organs consist of the bone marrow and thymus, where the cells of the immune system are generated and undergo maturation. Secondary lymphoid organs consist of the spleen, lymph nodes, mucosa-associated and gut-associated lymphoid tissues. These are essential for the development of the immune response and cells circulate between the tissues and lymphoid organs in the blood and lymph and maintain the process known as **immune surveillance**. The cells and humoral elements of the immune system are listed in Table 13.1.

TABLE 13.1 Cells and humoral elements of the immune system

Cells	Myeloid
	• Neutrophils
	• Eosinophils
	• Macrophages
	• Mast cells
	• Dendritic cells
	Lymphoid cells:
	• Lymphocytes
	• Plasma cells
	• Natural killer cells
Humoral elements	**Antibodies**
	Complement (Box 13.2)
	Mannan-binding lectin
	Proteolytic enzymes, cytokines
	Adhesion molecules

BOX 13.2 Complement

The **complement system** is part of the innate immune system and, as such, it is not adaptable and does not change over the course of an individual's lifetime. **Plasma** contains over 40 complement proteins. They are often abbreviated as C and given a number, for example, C1q, C2, C3; although others, such as MBL, factors B, D, and so on, obviously do not follow this rule. Complement proteins work in a sequential cascade fashion; individual proteins in the sequence activating several molecules of the next component. The effect of such a cascade is to magnify greatly the final effect. The cascade may initially follow one of three pathways:

- classical pathway, which is activated by an **antibody**
- the pathway activated by MBL
- the alternative pathway, which is activated by direct contact with bacteria.

However, all three pathways converge in a common final pathway that destroys the cells of the pathogen. For example, complement kills bacteria by piercing their cell surface membrane. A further function of the complement system is the disposal of immune complexes that are produced naturally as the immune system combats infection.

An inappropriate activation of such a powerful process could lead to many potential problems; hence, control of the complement pathway is strict, and includes a number of inhibitory proteins, for example, C1-inhibitor, factor I, and factor H.

13.1.1 Innate immunity

Innate immunity is that present at birth. It is non-specific, does not develop an immunological memory, and is most effective against **pyogenic bacteria**, fungi, and multicellular parasites. Physical barriers, such as the skin, that present an impermeable surface and secretions that wash mucosal tissues, are the first line of defence of the innate systems. Humoral components such as complement (Box 13.2), mannan-binding lectin (MBL), and proteolytic enzymes, which are present in blood, mucous, saliva, and other body fluids, are also part of the innate response. The cells of the innate system are **neutrophils**, **eosinophils**, **mast cells**, and **natural killer cells**. All of these factors combine to disable and dispose of potentially pathogenic organisms.

13.1.2 Acquired immunity

Acquired immunity is also known as the adaptive response. Unlike innate immunity, acquired immunity is not present at birth but develops throughout life. The response to microorganisms is specific. Acquired immunity has the ability to develop an immunological memory so the responses increase in intensity with repeated exposure to a microorganism.

The active components of the adaptive immunity are **lymphocytes** derived from the bone marrow. These produce antibodies in response to protein **antigens** on the surfaces of viruses and bacteria. The antibodies bind specifically to the organisms carrying their complementary antigen and interact with complement and receptors on **granulocytes** and mononuclear **phagocytic cells**, leading to the death of the target organism.

13.1.3 Cytokines

In both innate and acquired immunities, chemical messengers called **cytokines** are produced by cells. These allow the cells to communicate with each other, and coordinate and control their growth, differentiation, and functions. **Chemokines** and **adhesion molecules** are also instrumental in the movement of cells between the different lymphoid organs.

13.2 Synthesis of antibodies *in vivo*

Antibodies are glycoproteins that are able to bind to antigens, usually proteins that are normally bound to cell surfaces. They belong to a group of proteins called immunoglobulins, which are so called because they have a role in acquired immunity and appear in the globulin fraction when **serum** proteins are separated by electrophoresis (see, e.g., Figures 12.3 and 12.9). Immunoglobulin molecules are formed of two heavy and two light chain polypeptide chains joined together by disulfide bonds. The structure of their molecules can be thought of as Y-shaped. It consists mainly of a relatively conserved constant region. However, there is a region at the end of each of the two arms of the Y called the variable region, which forms the antigen **binding site** and which varies in different types of antibodies (Figure 13.1).

Cross reference

You can read about electrophoresis in Chapter 12, 'Electrophoresis'.

Immunoglobulins may be divided into one of five classes depending largely upon their structures and which of the five main types of heavy chain are present. There are two types of immunoglobulin light chain, named lambda and kappa. The different light chains have no particular function except to contribute to the diversity of the variable region. Immunoglobulins that exist as multiples of immunoglobulin G molecules are joined together by a protein called the J chain.

The five classes of immunoglobulins are:

- immunoglobulin G (IgG)
- immunoglobulin A (IgA)
- immunoglobulin M (IgM)
- immunoglobulin E (IgE)
- immunoglobulin D (IgD).

IgG is the most abundant immunoglobulin in samples of serum. It may be subdivided into four subclasses. IgG is most effective when it associates with its antigen. It can then bind to complement proteins, which results in the recruitment and activation of phagocytes. An interesting aspect of IgG activity is its ability to cross the placenta and give the newborn baby some protection against infection during the first few months of life.

IgA is found in relatively low concentrations in serum in a form that resembles two IgG molecules joined together. It is secreted generously in saliva, tears, and other fluids that bathe the mucosal surfaces limiting microbial entry to the body through these vulnerable routes.

IgM is the first antibody produced in the **primary immune response** to infection. It is formed of the familiar four polypeptide chains but in a structure where they are repeated five times. This gives it multiple antigen-binding sites. IgM is a potent activator of complement and phagocyte recruitment.

IgE occurs at low concentrations in serum samples from healthy individuals but this concentration increases in patients with parasitic infections or allergies. IgE principally interacts with mast cells that produce substances capable of deactivating parasites and which contribute to the development of allergic reactions.

IgD is found on the surface of immature **B lymphocytes** (Section 13.2). It is normally present in very low levels in serum and its function is not fully understood.

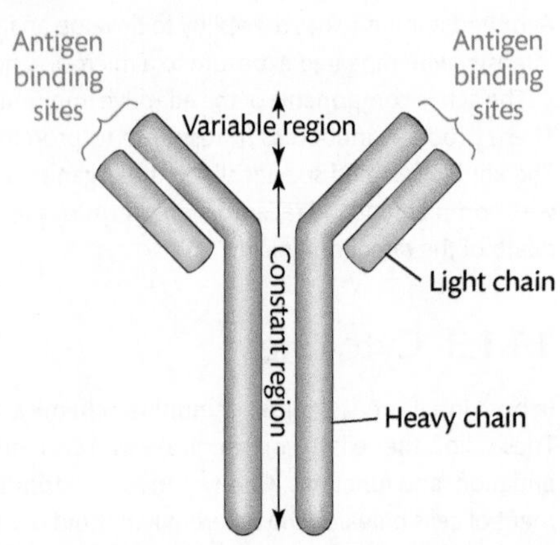

FIGURE 13.1
Structure of an immunoglobulin G (IgG) molecule.

SELF-CHECK 13.1

(a) List the primary lymphoid organs. (b) Define *innate immunity*. (c) State the major activity of an antibody.

13.2.1 Polyclonal and monoclonal responses

Antibodies are produced by **plasma cells** that are activated through receptors on the surface of T and B lymphocytes. These cells recognize specific parts of antigens called **antigenic determinants** or **epitopes**. The recognition sites of antibodies interact with the complementary epitope surface rather like pieces of a jigsaw puzzle. Plasma cells that produce and secrete antibodies develop from B lymphocytes. Each type of plasma cell is a **clone** and uniquely produces a single, specific type or **monoclonal antibody**. However, antigens are complex molecules that potentially have multiple epitopes on their surfaces. They therefore have the potential to stimulate the production of many different types of antibodies. Most infections would elicit the production of multiple antibodies, which is known as a polyclonal response. Measurable monoclonal responses are usually confined to clinical cases where a plasma cell tumour leads to the overproduction of a single clone of cells, or when plasma cells are manipulated under laboratory conditions.

When an antigen is first encountered, the immune system mounts a primary response within 5–10 days, usually producing an IgM antibody, whose concentration increases then falls to a low level. However, if the same antigen (epitope) is encountered again, *memory* plasma cells respond by producing a secondary response. This is usually in the form of an IgG antibody that is produced more rapidly than the initial IgM and is of a higher intensity and specificity.

Cross reference

You can read more about immunoglobulins in the relevant chapter of the *Clinical Immunology* book of this series.

13.2.2 Associated disease states

Diseases of the immune system are varied but are generally categorized by an inability to mount an effective immune response or by an inappropriate activation of the immune system. Individuals with **immunodeficiency** disorders are unable to raise an appropriate immune response. This may be due to an abnormality in the development of the cells of their immune system, a defect in communication between the immune cells, HIV infection, or a fault in another essential process, such as the complement pathway.

Autoimmune disorders occur when immune responses are directed against an individual's own tissues or specific organs. This is normally avoided by the development of **immune tolerance** mechanisms during the maturation of T and B cells. Individuals may be genetically predisposed to developing **autoimmune disease**, but it usually takes the added trigger of an environmental stimulus, such as a viral infection, to start the chain of events that leads to the development of disease. In some cases, an autoantibody can be identified as a diagnostic marker, which may or may not have direct pathological activity. The autoimmune disorders are divided into organ-specific, where a particular organ is targeted, for example the pancreas in type 1 diabetes, or non-organ-specific, where antibodies are directed at cell components such as nuclear proteins as in systemic lupus erythematosus (SLE).

In multiple myeloma (see Figure 12.9 and associated text), a clone of malignant plasma cells produces excessive amounts of a single immunoglobulin, which may or may not behave as an antibody. This may inhibit the production of other immunoglobulins, effectively creating an immunodeficient state and making the individual particularly vulnerable to infections.

Allergic symptoms develop in susceptible individuals when an antibody reaction is elicited by a normally harmless environmental substance or food. The presence of the allergen, which may be inhaled pollen or an ingested food substance, stimulates the production of IgE. The molecules of IgE cross-link receptors on mast cells, leading to the release of histamine, and other factors that produce the allergic reaction.

Cross reference

You can read about diseases of the immune system in the *Clinical Immunology* volume of this series.

SELF-CHECK 13.2

(a) Describe the structure of an immunoglobulin. (b) What is the main function of a plasma cell? (c) Name the two factors that may combine to predispose an individual to develop an autoimmune disease.

13.3 Production of polyclonal and monoclonal antibodies *in vitro*

Polyclonal antibodies are heterogeneous pools of antibodies that arise from a **secondary immune response** to an antigen; that is, they are directed against multiple epitopes. They are produced by injecting patients with antigenic material that may be from the same species or a different species: a procedure known as immunization. A variety of animals are used to raise antibodies such as the **antihuman globulins** used in immunological assay systems (Section 13.5). Antigens are injected into the patient accompanied by a carrier called an **adjuvant**. It is unclear how adjuvants work, but it is thought they slow down the release of the antigen in the patient, resulting in a series of small 'hits' to the immune system. Examples of adjuvants include Freund's adjuvant, Syntex adjuvant, Gerbu adjuvant, saponin, and aluminium hydroxide, which is one of the few licensed for use in humans. Protocols vary depending on the response that is required and the routes of administration, but the procedures take into account the size of the patient, the form of the antigen, whether it is particulate, cellular, or soluble material, and the amount of antigen available.

Polyclonal antibodies may be raised in human volunteers but the range is limited by the choice of acceptable antigens and other ethical considerations. An example of human antibody production is given in Box 13.3. Animal sources are also controlled by stringent legal requirements but offer a wider range of possible antigens and antihuman options in antibody production (Figure 13.2). Polyclonal antibodies are versatile and easy to produce; they may recognize a range of epitopes on a single antigen, which is often an advantage in immunoassay. However, drawbacks to their use include limited reproducibility, cross-reactivity, low quantity, and the high cost of maintenance of laboratory animals.

Monoclonal antibodies are homogenous pools of antibodies that are raised from one clone of B cells, which produce antibodies directed against one specific epitope. The technology was developed in the 1970s by Georges Köhler and César Milstein, and is considered one of the most significant contributions to the biological sciences. B lymphocytes from the spleen of mice immunized with antigen are fused with myeloma cells. The resulting **hybridoma** forms a single clone of cells that produces a monoclonal (pure) antibody. You can see this procedure in outline in Figure 13.3. The hybridoma cells may be grown in culture for many years raising large quantities of antibody to almost any antigen.

Monoclonal antibodies are specific and may be produced in unlimited supply maintaining high quality. They are invaluable reagents for research and diagnostic purposes, and have been developed to treat some immune disorders and cancers. However, high levels of investment are required for their

BOX 13.3 The production of polyclonal anti-D antibodies in human volunteers

Approximately 15% of the UK population have the rhesus-negative blood group, which is determined by the absence of the rhesus D antigen on their erythrocytes. When a rhesus-negative woman conceives a rhesus-positive baby there is a risk that the mother may produce anti-D antibodies to her unborn child. This may have implications when she has a second child, as the secondary immune response will produce IgG antibodies that can cross the placenta and enter her baby's bloodstream. The antibodies will react with the fetal blood cells, damaging them and causing anaemia, jaundice, and an enlarged liver and spleen. This is known as haemolytic disease of the newborn, and potentially could affect 70,000 babies per year in the UK. However, this life-threatening condition can be averted by immunizing mothers with antibodies to the D antigen during their pregnancy. Thus, the anti-D antibodies can attach to D-positive cells, destroying them before they can illicit a reaction from the mother's immune system.

The anti-D antibodies are obtained by raising a polyclonal response in male human volunteers who donate plasma regularly. The volunteers are injected with rhesus-positive blood cells, which stimulate their immune systems to produce anti-D antibodies. Research into producing a monoclonal source of the antibody is continuing.

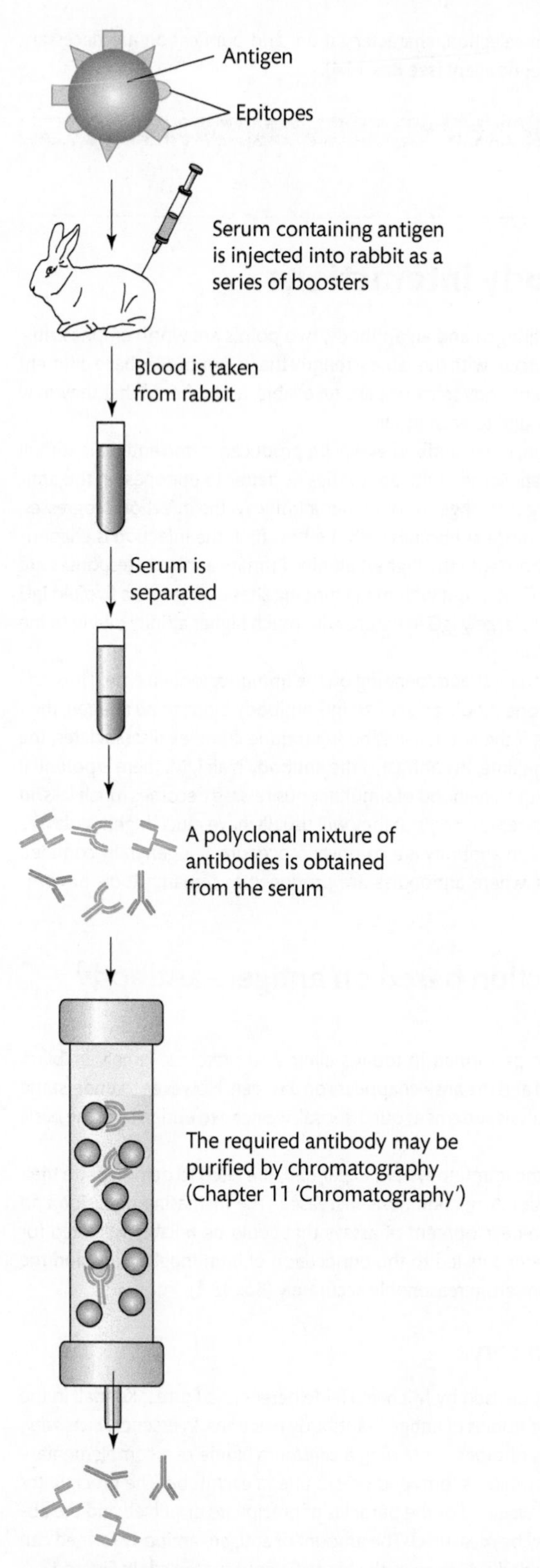
Antigen
Epitopes
Serum containing antigen is injected into rabbit as a series of boosters
Blood is taken from rabbit
Serum is separated
A polyclonal mixture of antibodies is obtained from the serum
The required antibody may be purified by chromatography (Chapter 11 'Chromatography')

FIGURE 13.2

Schematic showing the production of polyclonal antibodies in a rabbit. See main text for details.

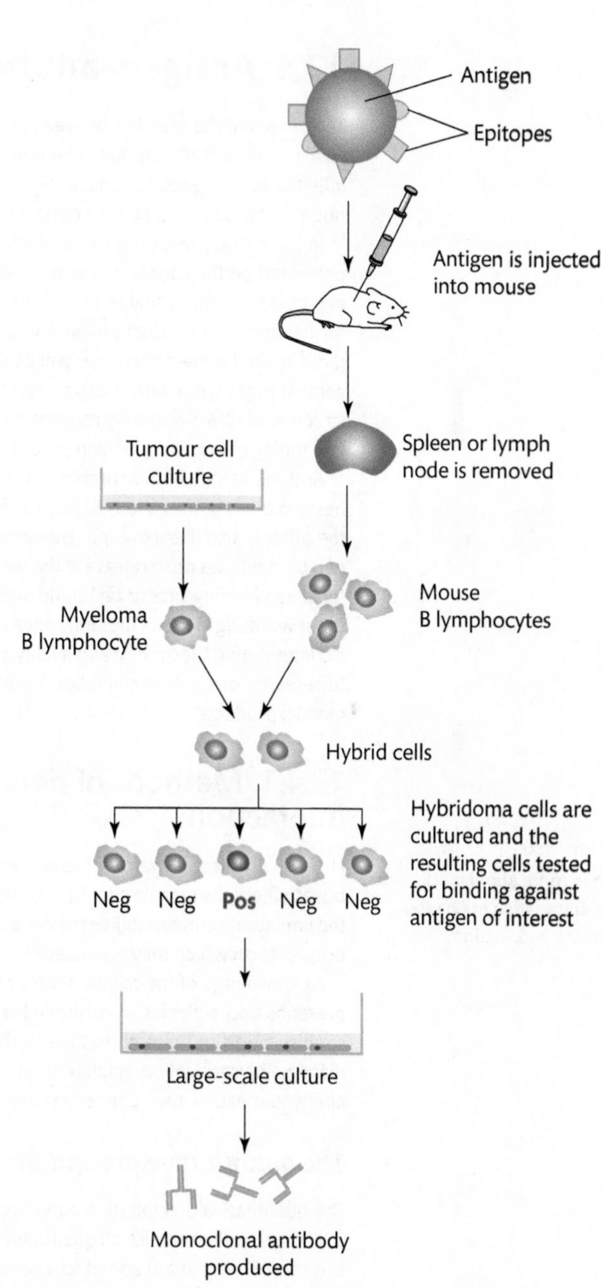
Antigen
Epitopes
Antigen is injected into mouse
Tumour cell culture
Spleen or lymph node is removed
Myeloma B lymphocyte
Mouse B lymphocytes
Hybrid cells
Hybridoma cells are cultured and the resulting cells tested for binding against antigen of interest
Neg
Neg
Pos
Neg
Neg
Large-scale culture
Monoclonal antibody produced

FIGURE 13.3

Schematic showing the production of monoclonal antibodies in a mouse. See main text for details.
Neg: negative; Pos: positive.

production as sophisticated methods of selection, characterization, and purification are necessary, especially if the final product is a therapeutic agent (see Box 14.4).

SELF-CHECK 13.3

How are adjuvants thought to work?

13.4 Antigen-antibody interactions

To appreciate the reaction between an antigen and an antibody, two points are worth emphasizing. First, not all antigen-antibody reactions occur with the same strength; that is, antibodies have different affinities for antigens. Second, antigen-antibody reactions are reversible, and differ in that they may not have the same capacity to remain bound to each other.

In the primary response to an infection, many antibodies will be produced to the antigenic stimuli presented on the surface of the pathogen. Some of the antibodies fit better to epitopes on the antigen than will others, that is, their binding is stronger or of higher affinity. As the infection progresses, further antibody production will favour those antibodies with the best fit. If the infection is encountered again the memory B cells will be those with the highest affinity. Primary antibody responses are generally IgM types with relatively low affinities but with many binding sites owing to the fivefold IgG structure of IgM. Secondary responses are largely IgG in nature with much higher affinity owing to the fine tuning of immunoglobulin production.

Antibodies vary in their number of binding sites depending on the immunoglobulin type. Thus, IgG has two binding sites and IgM has 10. If one binding site of an IgG antibody binds to an antigen, then the other is also likely to bind. However, if the antigen-antibody immune complex disassociates, the two binding sites must release at the same time. In contrast, if the antibody is an IgM, there is potential for all ten binding sites to be bound and the likelihood of simultaneous release becomes much less. In other words, IgM molecules may have lower **affinity** but they will usually have much higher **avidity**.

Determining the affinity and avidity of an antibody is a complex procedure. It is generally confined to research or commercial laboratories, where antibodies are produced as research tools or commercial products.

13.4.1 Methods of detection based on antigen-antibody interactions

Cross reference

You can read more about black boxes and automation in Chapter 15, 'Laboratory automation'.

High-quality, fast turnaround assays are performed in today's clinical laboratories, largely in 'black boxes' where the sample goes in one end and the answer appears on a screen. However, to understand the limitations and capabilities of the analysis systems at our disposal, we need to understand the basic principles on which they are based.

As knowledge of the components of the immune system expanded, the need to demonstrate their presence and activities in controlled laboratory conditions increased. The increasing precision and sophistication of research has led to the development of assays that could be reliably validated for diagnostic uses. Early precipitation experiments led to the purification of immunoglobulins and the ability to measure their concentrations in serum reasonably accurately (Box 13.1).

The quantitative precipitation curve

The quantitative precipitation curve was devised by Michael Heidelberger and Forrest Kendall in the 1950s, and is the basis for all quantitative studies of antigen-antibody reactions. In essence, *increasing* aliquots of antigen are added to a series of tubes containing a *constant* volume of a complementary antibody. Antigen-antibody immune complexes form and precipitate in each tube. The precipitates may be separated by centrifugation and weighed or the particles of precipitate dispersed and the absorbance of the resulting suspension may be measured. The amount of antigen-antibody formed can then be plotted against that of antigen added, which gives the type of graph you can see in Figure 13.4. The Heidelberger curve varies according to the species from which the antibody or antigen is derived. In the initial tubes, the amount or concentration of antibody is in excess of that of antigen. However,

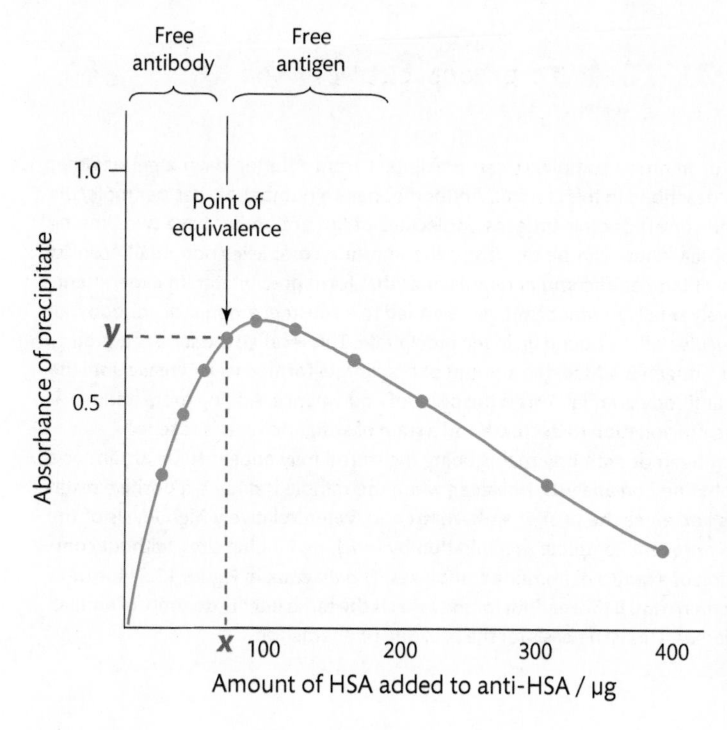

FIGURE 13.4
Precipitation curve for human serum albumin (HSA). An increasing amount of HSA is added to a fixed concentration of anti-HSA. Free antibody is present in solution prior to the point of equivalence. However, after the point of equivalence, antigen will be in excess. See text and Box 13.4 for details.

in the later tubes, the antigen is in excess and so the amount of precipitation will decrease because of the formation of soluble complexes. These arise because, although the binding sites of the antibody molecule are blocked with antigen, they are unable to form cross-links, as explained in Box 13.4.

This type of experiment illustrates the basic principles of all immunological test systems that depend upon the formation of immune complexes. An equivalence of antigen and antibody is reached just before maximum precipitation occurs (Box 13.4). There are points of *equal* precipitation on both sides of the curve shown in Figures 13.4 and 13.5: calculations based on readouts from this type of curve must take this into account. The method can be adapted to quantify the amounts of antibody in samples of serum, to determine antigenic valency, and is, indeed, the basis of all other precipitation reactions. Light-scattering methods, such as nephelometry and turbidimetry (Section 13.5) have, however, largely replaced precipitation methods, making the quantitation of immune complexes more accurate and more sensitive.

Other immunological techniques visualize immune complexes in media such as gels or observe the **agglutination** of cells or particles with antigens attached to plastic surfaces, or are based on the attachment of antibodies to thin sections of solid organs or sheets of tissue-cultured cells.

SELF-CHECK 13.4

(a) Define the meanings of avidity and affinity in relation to antigen–antibody binding. (b) At what point is equivalence reached on the quantitative precipitation curve?

13.5 Immunological techniques used in clinical laboratories

Immunological techniques depend in some way upon visualizing and quantifying interactions between antigens and antibodies to detect and measure the concentrations of analytes in clinical samples. We will discuss them under the following headings:

- immunodiffusion techniques
- agglutination techniques

BOX 13.4 To precipitate or not

Antigen–antibody or immune complexes can precipitate from solution or in a gel or other formats, which are described in this chapter. Antibodies have a number of sites per molecule that can bind to the corresponding antigens. Molecules of IgG antibodies have two binding sites; in the case of IgM there can be ten. Thus, the immune complexes potentially consist of large aggregates or lattices of combined molecules that form precipitates. In experiments when a comparatively small amount of antigen is added to a constant amount of antibody, all of the antigen molecules will be bound up in the precipitate. This is called the state of antibody excess. When more antigen is added, the amount of precipitate formed will increase until the ratio of antigen to antibody is equal. This is the point of equivalence. Adding more antigen at this point will cause precipitation to decrease and a state of antigen excess is reached.

Sometimes the antigen or antibody that is being measured may appear to be absent or a negative result is obtained on analysis. However, when the sample is diluted a positive result is obtained. This is known as the **prozone phenomenon**, when relatively high levels of the antigen or antibody prevent complete precipitation by blocking binding sites, without completing the formation of a lattice of immune complexes. The diagram in Figure 13.5 illustrates this situation. At points A and B the readout for the assay is the same but the quantity of antigen present is very different. This is the basis for the prozone phenemenon.

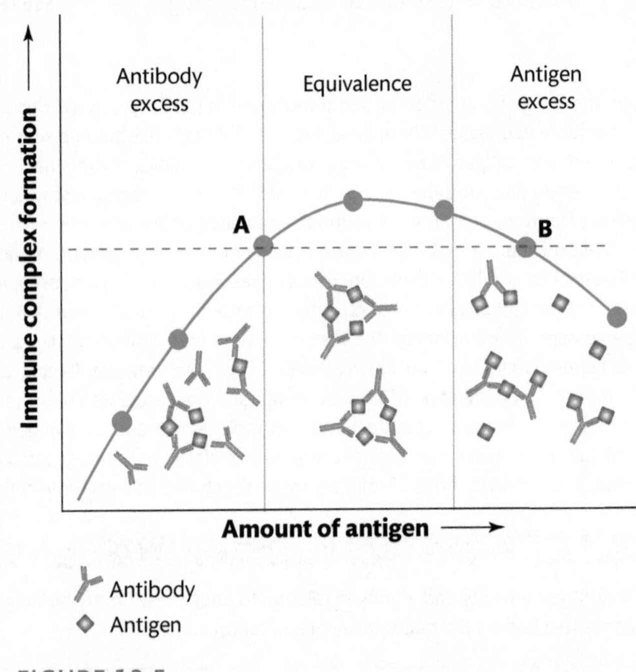

FIGURE 13.5
Prozone phenomenon. The basis of the prozone phenomenon is that at points A and B the results for the assay are the same, even though the quantities of antigen present are different. See text for details.

- immunoelectrophoresis techniques
- immunoassays
- nephelometry and turbidimetry techniques
- enzyme-linked immunosorbent and enzyme-linked immunospot assays
- flow cytometry and fluorescence-activated cell sorting
- multiplex techniques
- microarrays
- complement assays
- indirect and direct immunofluorescence techniques
- immunocytochemistry.

13.5.1 Immunodiffusion techniques

Immunodiffusion techniques depend upon the diffusion of antigens and antibodies from holes or wells cut into an agarose gel. The antibodies employed may be monoclonal or polyclonal, depending on the design of the testing protocol. Where antibody meets antigen in the gel, recognition, and binding occur, forming a white line of precipitated immune complexes.

The gel may be inert or may contain defined quantities of known antibody or antigen. Polyethylene glycol may also be added to enhance precipitation or Coomassie blue may be used to stain and clarify the lines of precipitation. The incubation time and temperature employed depends on the type of the immunoglobulin employed in the method. Incubation times range from 24–72 hours, although with some IgG antibodies, reactions can be observed after only 4 hours if incubated at 37°C.

The use of gel diffusion was a significant development in the transfer of immunology methods from the research area to the clinical laboratory. The techniques require smaller volumes of reagent and have diverse applications in diagnosis.

Single radial immunodiffusion

In single radial immunodiffusion (SRID), antigen of unknown concentration is added to the wells in a gel that contains uniformly distributed known antibody or **antiserum**, which is serum that contains known antibodies and so may be used as an analytical reagent. The gel is usually incubated at room temperature for 24–48 hours. The antigen diffuses through the gel and on binding with its complementary antibody forms a precipitate in the shape of a ring that surrounds the well and moves outwards from it. Movement of the ring stops at a diameter where the point of equivalence is reached, that is the antigen and antibody are present in equal quantities. The final diameter of the ring is directly proportional to the concentration of the antigen present (Figure 13.6). This is measured using a calibrated eyepiece that magnifies the ring of precipitation or an electronic reader. A **standard curve** can be constructed using known concentrations of antigen in a number of wells to allow the semi-quantitation of unknown antigen samples. The method can be reversed by incorporating antigen in the gel to quantify the amount of antibody in a sample.

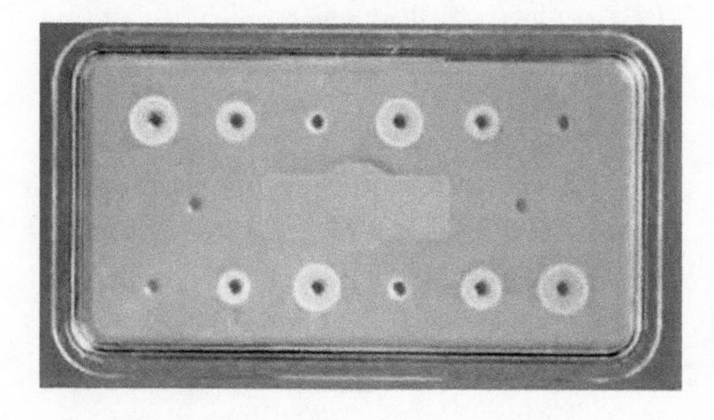

FIGURE 13.6
Photograph of a single radial immunodiffusion experiment, showing precipitation rings of varying diameter relative to the quantity of antibody in the sample. Courtesy of D. Raine, Immunology Department, Hull Royal Infirmary, UK.

Cross reference

The *Clinical Immunology* volume of this series has further details about complement assays and the practical applications of gel diffusion methods.

SRID has limited applications in present-day clinical laboratories as the determinations are not sufficiently accurate and are operator dependent, being easily affected by the delivery of sample into the wells and inattention to the test protocol. The method is also time consuming and inappropriate for the analyses of large numbers of samples. This method is, however, routinely used to measure the components and inhibitors of the complement system in rare immunodeficiency disorders. In these cases, the number of samples is likely to be low and few other analytical options are available.

Online Resource Centre

To see an online video demonstrating single radial immunodiffusion, log on to www.oxfordtextbooks.co.uk/orc/fbs

Double diffusion and Ouchterlony double diffusion techniques

Double diffusion techniques involve placing antigen and antibody in opposing wells of gel and observing the precipitation of bands where immune complexes form. The most common application is in the Ouchterlony format, where a number of antigens or antibodies can be compared around a single well of known antigen or antiserum. In the Ouchterlony double diffusion method, a number of wells are cut in the gel such that one is placed centrally and the rest are equidistant from it. Figure 13.7 outlines the Ouchterlony double diffusion technique. If a known antibody is placed in the central well and a variety of unknown antigens placed in the surrounding ones, then precipitation lines will appear between them that may be one of three characteristic patterns:

1. Line of full identity
2. Line of non-identity
3. Line of partial identity.

A line of full identity is a continuous one and indicates that the antigens reacting with the antibody are all identical. In contrast, a line of non-identity is cross-shaped; thus, the antigens reacting with the antibody are different. Lastly, a line of partial identity appears as a single spur on a line indicating that the antigens reacting to the antibody have some activity in common.

The Ouchterlony double diffusion technique may be used to differentiate the different antinuclear antibodies in autoimmune disease. However, it is not a quantitative method and its use in clinical laboratories is becoming less frequent as automated methods have replaced such time-consuming manual techniques.

SELF-CHECK 13.5

(a) What happens when an antigen meets its corresponding antibody in an agarose gel? (b) What term is used to describe the situation when high concentrations of analyte prevent the formation of an antigen-antibody precipitate?

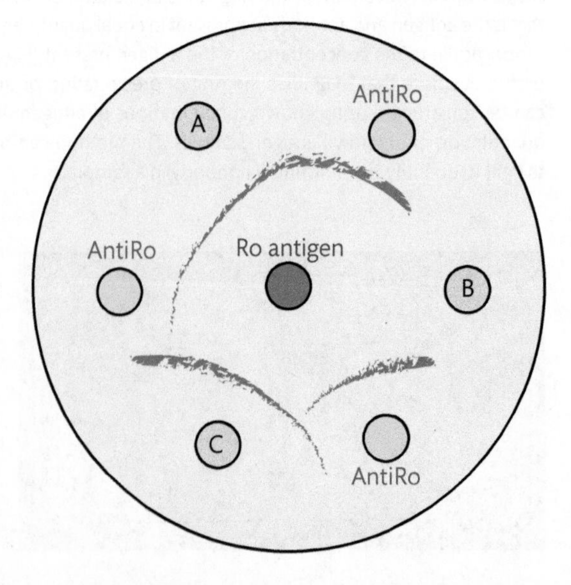

FIGURE 13.7

Schematic showing the results of an Ouchterlony double diffusion assay. Wells in the gel plate were loaded with three test sera, A, B, and C. The central well contained a purified nuclear antigen (Ro), and alternate surrounding wells contained known anti-Ro. Following incubation, the line of precipitate formed to give the results shown. Thus, sample A is positive for anti-Ro as it shows a line of full identity with the Ro antiserum. Sample B is negative as there is no line of precipitation and sample C shows some reactivity with the Ro antigen but cannot be fully identified as anti-Ro. This latter type of reaction can be caused by impurities in the antigen preparation.

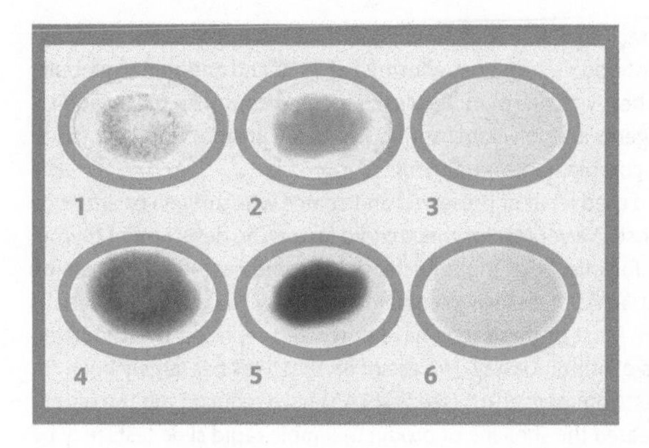

FIGURE 13.8

Schematic showing the agglutination of erythrocytes and latex particles on a slide. On the left erythrocytes (pink) and latex particles (blue) are clumped or agglutinated, giving a grainy appearance. On the right the suspension is smooth in appearance because the cells and particles are not agglutinated.

13.5.2 Agglutination techniques

Visible agglutination, that is, the binding together of particles, may be observed in a liquid matrix on mixing antibodies complementary to an antigen carried on the surfaces of cells or particles, as you can see in Figure 13.8. Erythrocytes and latex particles can be used to give a semi-quantitative assay for estimating the concentrations of antibodies. This is achieved by preparing a range of serum dilutions from high to low concentration. Equal quantities of each dilution and appropriate cells or particles are then mixed on slides or in microwell plates. The strength of the antibody is expressed as a **titre**, which is the dilution at which agglutination is still, but only just, visible, as you can see in Figure 13.9.

In blood group serology, the agglutination of erythrocytes with known antisera is used to identify ABO and **Rhesus blood groups**. When donor blood is selected for transfusion into an individual, a cross-match procedure is performed to determine the compatibility of the donated blood with that of the recipient. Incompatibility is indicated if agglutination occurs when the donor's blood is mixed with the recipient's plasma. The ABO blood group are antibodies, which are usually of the IgM type. They will agglutinate erythrocytes of an opposing ABO group within a few seconds at room temperature. However, IgG antibodies have fewer binding sites and their agglutination reactions are more

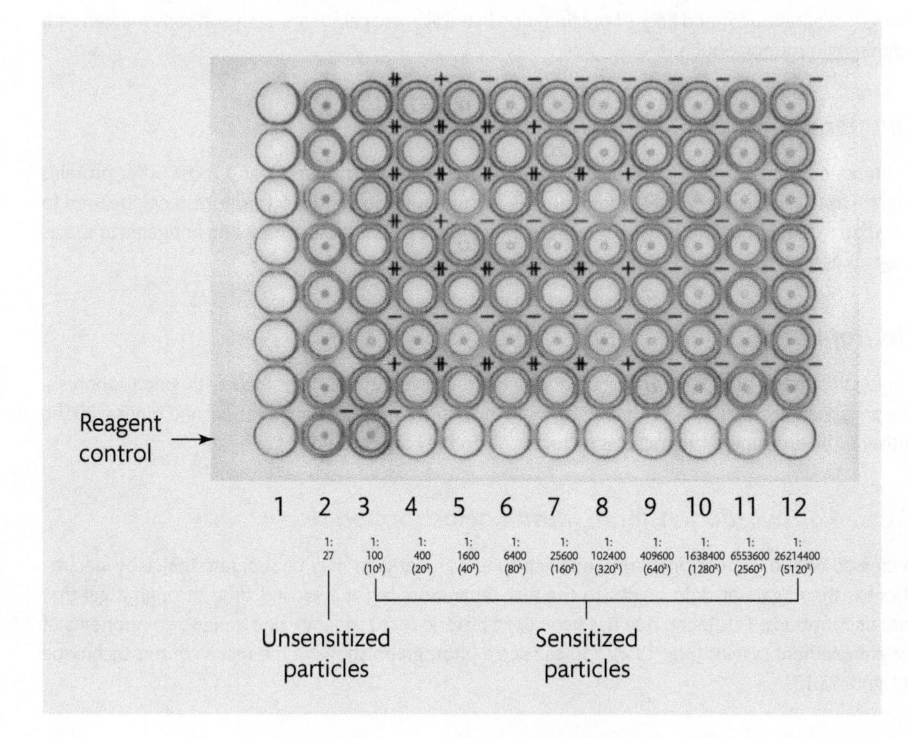

FIGURE 13.9

Agglutination of particles in a multiwell plate. Particles coated with thyroid antigen have been mixed with varying dilutions of test sera. Non-agglutinated cells settle in a small pellet at the bottom of the well. Agglutinated particles form a smooth layer on the well and partially agglutinated or weak reactions settle as a ring in the well.

Cross reference

You can read more about blood transfusion in the accompanying text of the series, *Transfusion and Transplantation Science*.

temperature dependent. They may attach to cells and bind complement but not achieve visual agglutination. In the case of IgG Rhesus antibodies, tests are performed at 37 °C and antihuman globulin antibodies are used that attach to the heavy chains of molecules of IgG and complete the agglutination reaction. Antihuman globulin reagents are polyclonal antibodies raised in laboratory animals to specific immunoglobulin isotypes and complement components.

In the past, animal erythrocytes were used fresh or preserved and coated with antigen or antibody in a range of assays, for example the Rose Waaler test for rheumatoid factor; the detection of thyroid antibodies; and the Paul Bunnell test for infectious mononucleosis. In some cases sera containing antibody required heat inactivation at 56 °C to destroy complement activity that could potentially haemolyse the cells during incubation. Most of these tests have now been replaced by automated analysis, nephelometry, or other forms of immunoassay. The use of erythrocytes has largely been replaced by polystyrene or latex particles. Commercial kits (see Box 14.1) have reduced the use of reagents obtained from animals and increased the shelf life of products. Simple rapid slide tests may be used to detect rheumatoid factor, C reactive protein, and a wide range of other immunological assays that are varied in quality and utility. However, these kits are not suitable for high-volume work in contemporary clinical laboratories.

13.5.3 Immunoelectrophoretic techniques

Diffusion methods are confined to systems that use few reactants and where the precipitation lines can be easily observed. However, the resolution of *multiple* antigens and antibodies can be achieved by combining immunodiffusion with electrophoresis.

Cross reference

Electrophoresis is described in Chapter 12.

Proteins in serum and urine, including immunoglobulins, can be separated by electrophoresis and visualized by staining them with one of a number of different dyes (Table 12.1). The quantities of the separated proteins can be determined by densitometry. This is the method of choice when detecting and monitoring monoclonal paraprotein, which is a measure of tumour load and a feature of myeloma and other B-cell malignancies.

The simplest application of antibodies and electrophoresis is immunofixation. Antisera to the range of immunoglobulin isotypes present in the sample is added to proteins separated by electrophoresis, which identifies the light chain type of the immunoglobulin paraprotein. However, the early methods were unsatisfactory. Adding a range of antisera to immunoglobulin isotypes following their electrophoretic separation formed lines of precipitation in the agarose gel, where each antibody reacted with a different immunoglobulin.

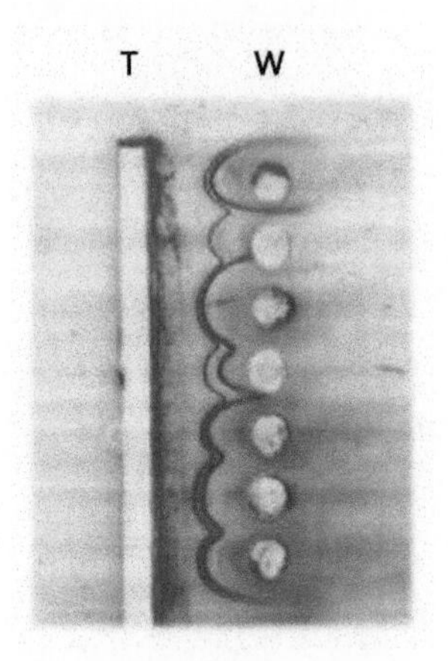

FIGURE 13.10
Photograph showing the results of immunoelectrophoresis. Samples have been placed in wells (W) and antibody in a trough (T). Following electrophoresis, arcs of precipitation have formed and been stained with Coomassie blue dye. See text for details. Courtesy of D. Raine, Immunology Department, Hull Royal Infirmary, UK.

Counter immunoelectrophoresis

As immunoglobulins migrate towards the cathode during electrophoresis and most other proteins migrate to the anode, a combination of double immunodiffusion and electrophoresis can be used to speed up the reaction. This is particularly useful in the preparation of antisera and antigens for use as reagents as their specificities can be assessed quickly.

Electrophoresis in gels containing antibodies

Single radial immunodiffusion, which we described above, can be combined with electrophoresis. The precipitate forms in rocket-shaped zones as the antigen moves through the gel and binds with the antisera. The area under the rocket is proportional to the concentration of antigen.

Crossed or two-dimensional immunoelectrophoresis

In crossed or two-dimensional immunoelectrophoresis, antigens may be separated once by electrophoresis then again at right angles to the first separation, but the second time through a gel that contains antibody. This technique has been used historically to separate and analyse components of the complement system (Box 13.2). You can see a photograph showing the results of this technique in Figure 13.10.

13.5.4 Immunoassays

The term *immunoassay* is broadly applied to methods of analysis that use antibodies to detect and determine the concentration of any substance present in serum and urine in health and disease. This is made possible by attaching a detectable label, which is usually an enzyme, radioisotope, or a fluorescent tag to an antibody or antigen. Immunoassays are flexible and this has led to the development of a number of manual and automated techniques, which are used in clinical biochemistry, virology, histology, cytology, and immunology laboratories. Many immunoassays are now available for diagnostic purposes using a wide range of tests on large high-throughput analytical platforms. Clinical tests are available to measure the concentrations of, for example, hormones, tumour markers, autoantibodies, allergens, and for helping to diagnose many infectious diseases. Biomedical scientists are often challenged in validating selected assays for clinical use, bearing in mind the wide variation in analytical and clinical sensitivity and specificity, and the quality of technical performance.

As it makes no difference in principle whether the concentrations of an antibody or antigen are being measured by the assay, the unknown substance is usually referred to as the *analyte* and the labelled substance *reagent.*

Early immunoassays were based on the principle of saturation analysis where excess labelled antigen (the analyte in this case) is reacted with a limiting amount of antibody (reagent). When known aliquots of unlabelled antibody are added they will inhibit the binding of labelled antigen; determining the amount of bound labelled antigen indicates the amount of unlabelled antigen that has been added. In this way an inhibition standard curve can be drawn to estimate the amount of analyte by the level of inhibition it induces. Ideally, the antibody concentration should be relatively low; however, this can intensify the effects of other factors, including the accuracy of the method for detecting the labelled reagent, the speed of reaction, and the concentrations, avidities, and affinities of the reactants. The development of immunoassays was initially limited by the cross-reactivity of polyclonal antibodies raised in animals, and by the difficulty in producing large quantities of reliable antisera. However, with the advent of commercially available monoclonal antibodies (see Box 14.4) such limitations have been largely overcome, producing high-affinity, specific reagents of good quality that give accurate and reproducible results. A schematic that outlines the use of radioactive labels is given in Figure 13.11.

Immunoassays may be classified using a number of factors:

- choice of label, whether radioisotope, enzyme, or **fluorochrome**
- method of signal detection
- competitive or non-competitive assay
- whether the assay requires the separation of bound and unbound reactants.

Radioisotope labels

Radiolabelled immunoassays with radioisotopes, such as ^{125}I, ^{3}H, and ^{32}P, are rarely performed in clinical laboratories because of the stringent health and safety restrictions that apply to their use.

Such safety regulations define the location where an experiment may be performed, the monitoring of staff, the storing of reagents, and disposal of waste, as well as requiring a trained radioisotope safety officer to oversee all such activities. These limit the practicality of the technology in routine clinical laboratory tests.

Cross reference

You can read more about health and safety in Chapter 4.

A number of technical aspects to the use of radioisotopes also pose problems, especially when measuring the relatively low concentrations of analytes in serum and urine. Measuring the activities of radioisotopes requires low background levels and isotopes with a radioactive half-life that is short enough for the assay protocol but long enough to give a reasonable shelf life. The radioactive signal can be counted using a γ-scintillation counter or liquid scintillation counter, which is sensitive and accurate. However, the signal of an isotope is determined by its rate of radioactive decay; to obtain useful results the isotope must bind to millions of reagent molecules to give a count that is statistically significantly above the background.

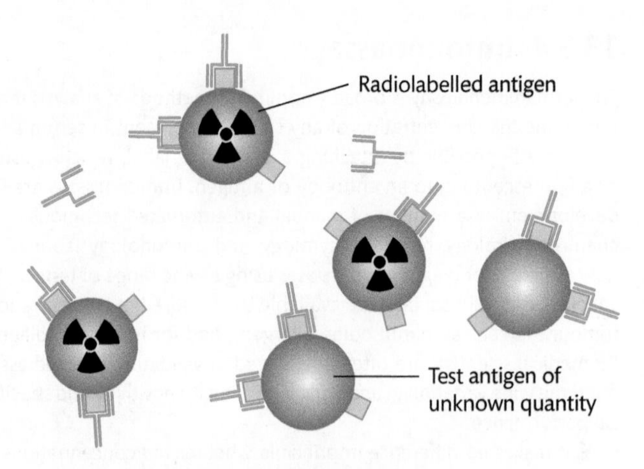

Radiolabelled antigen
Test antigen of unknown quantity
Antigens compete to bind antibody
Separation stage

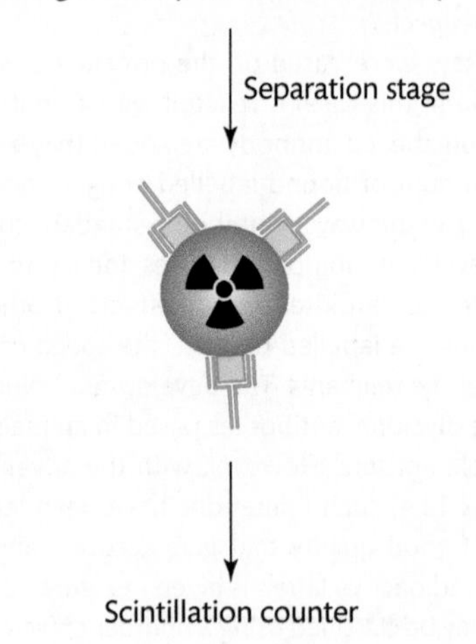

Scintillation counter
The number of radiolabelled antigens bound to antibody can be used to calculate the concentration of test antigen

FIGURE 13.11
Schematic showing a competitive radioisotope labelled immunoassay. See text for details.

Key points

Theoretically, a minimum of 600 molecules of labelled reactant (equivalent to 10^{-21} mole) must be detected in an analysis to obtain a statistically relevant result.

Enzyme labels

Attaching an enzyme to a reactant provides a convenient way of generating a sensitive signal. In many cases, a chromogenic substrate is used, which the enzyme can convert into a coloured product. As a single enzyme molecule is capable of converting many molecules of substrate to product, a detectable colour signal is readily available. Unfortunately, the photometers or spectrophotometers used to measure the intensity of the colour are limited by the optical quality of the cuvettes and microtitre trays used in analyses. The most common enzymes used in immunoassays are alkaline phosphatase and horseradish peroxidase, together with chromogenic substrates such as tetramethylbenzidine and 2,2′-azino-*bis*-ethylbenzthiazoline sulfonic acid.

BOX 13.5 Biotin and avidin

Biotin is a coenzyme that is synthesized by many plants and microorganisms, particularly those present in the flora of the gastrointestinal tract. Avidin is found in egg white and can also be isolated as streptavidin from *Streptomyces avidinii*.

Biotin can bind to antibodies, enzymes, and many other proteins without significantly affecting their biological activities. Avidin can bind to antibodies and be linked to a wide variety of enzymes, fluorochromes, and radioisotopes. Avidin also has an extremely high affinity for biotin; indeed, when combined, they may be considered for practical purposes to be linked by the equivalent of a covalent bond. Avidin has four binding sites for biotin. If a range of proteins are biotinylated, and all four sites on the avidin molecule are used to recognize and bind to them, this gives many possible combinations of complexes. Many variations of biotin-avidin systems are available commercially for use in immunoassays.

Some immunoassays combine the advantages of both radioactive and enzyme labels by using an enzyme-labelled antibody or antigen with a radioisotope-labelled substrate. Thus, the action of the enzyme releases the radioactive label, which can then be isolated and accurately measured.

Fluorescent labels

Fluorescent molecules absorb radiant energy but then emit energy as light of a longer wavelength because energy is lost in the process. Using fluorescent labels in immunoassays presents a number of problems, such as high background fluorescence and quenching of the emitted light by other components in the assay system. The advantage of using fluorescent labels, for example fluorescein and rhodamine β-isothiocyanate, in immunoassays is the sensitivity they give to the system.

Whatever type of label is used, radiolabelled, enzyme, fluorescent, all can be combined with the **avidin-biotin** system outlined in Box 13.5.

Cross reference

Fluorescence is described in Chapter 9, 'Spectroscopy'.

Competitive and non-competitive assays

Immunoassays using labelled reagents may be subclassified into homogenous and heterogeneous types. Homogenous assays are quicker and easier to perform. Heterogeneous assays require an extra step in the procedure to remove unbound antigen or antibody from the test system. They may be competitive or non-competitive.

Competitive immunoassay is when the unknown analyte competes with a similar labelled antigen to bind with an antibody. The quantity of detectable labelled reagent is therefore inversely proportional to the concentration of analyte because the greater the amount of labelled reagent that is able to bind means there are fewer binding sites available for the analyte. In non-competitive or sandwich immunoassays, the unknown analyte is bound to an antibody that is usually attached to a solid phase. Thus, when a labelled antibody is attached to the *bound* analyte it forms a type of molecular 'sandwich'. In this case, measurements of the signal from the labelled antibody are directly proportional to the quantity of the analyte.

Separation of bound and unbound reactants

Incomplete separation of the labelled component from the other components of the immunoassay mixture is a potential error in the test result. Separation may be achieved in a number of ways:

- adsorption onto activated charcoal
- precipitating the antibody with 50% saturated solution of ammonium sulfate
- precipitation with polyethylene glycol
- using a secondary binding antibody such as antihuman globulin (AHG)

- attaching a component to a solid phase, such as plastic tubing or a microtitre plate, so that unbound components can be washed away
- attaching a component to a solid phase in the form of beads or particles so that the bound reactant can be separated by simple centrifugation.

Cross reference

Automation is described in Chapter 15, 'Laboratory automation'.

Adsorption and precipitation methods have largely been replaced in clinical laboratories by the effective use of antibodies in solid phase assays that can be adapted for use on automated analysers with the capacity for high volumes of tests.

13.5.5 Nephelometry and turbidimetry

Cross reference

You can read some more about nephelometry and turbidimetry in Chapter 9, 'Spectroscopy'.

Nephelometry and turbidimetry techniques are based on the absorbance or deflection of monochromatic light by immune complexes in suspension. They differ in that turbidimetry measures the amount of light *absorbed* by the complex, while nephelometry measures the amount of light scattered at an angle to the original light path.

The most advanced instruments are fully automated, incorporate laser light sources, and can detect quantities of analyte. Rate nephelometry is a frequent method of choice in clinical immunology laboratories; this measures the amount of immune complex production against time. As explained earlier (Section 13.4), the principles of the quantitative precipitation curve must be taken into account. This is especially true regarding the phenomenon of antigen excess, where the antibody reagent may be completely consumed by relatively high amounts of antigen in the sample, leading to an inaccurate result.

The concentrations of immunoglobulins in samples of serum are normally measured by nephelometry or turbidimetry in the clinical laboratory. An antihuman globulin reagent is employed and the heavy chain of the immunoglobulin that is being measured becomes the antigenic target. A wide range of analytes, such as complement, rheumatoid factor, β,2-microglobulin, and α,1-antitrypsin, may be measured by these methods when investigating B-cell malignancies, autoimmune and liver diseases, and immunodeficiency.

SELF-CHECK 13.6

(a) List three ways in which immunoassays may be classified. (b) What is the main practical disadvantage of using radioisotope-based immunoassays? (c) Name three separation techniques used in immunoassays.

13.5.6 Enzyme-linked immunosorbent assays

Enzyme-linked immunosorbent assays (ELISAs), also known as enzyme immunoassays, are one of the most common immunoassay techniques used in clinical laboratories. The general technique has many applications in immunology, virology, and other analytical areas. It is easy to perform manually, but there are many automated ELISA platforms. This is a highly commercially driven area that has expanded rapidly. Clinical laboratories have a wide choice of readily available commercial kits for a multitude of different analyses. ELISAs are commonly used in the diagnosis of autoimmune and infectious diseases, such as when investigating possible HIV and hepatitis B viral infections, and a range of autoantibodies, complement components, and immunoglobulin subtypes.

The basis of ELISA technology is that unknown antigens are attached to a solid phase in the form of a plastic multiwell plate. Areas of the plate that might non-specifically bind antibody are prevented from doing so by blocking them with an inert protein. Enzyme-labelled antisera that can detect the antigen are added to the wells. After a period of incubation the unbound antibody is washed away. Removing the excess antibody or reagent by repeated washings at each stage of the procedure is essential to the technique as residual reagent or analyte in the well can potentially interfere with the next step of the process. Washings are usually performed using an automated plate washer or as part of a fully automated procedure. Following washing, a chromogenic substrate is then added. The enzyme catalyses the release of a coloured product. Colour development is stopped after a defined period of incubation using substances that prevent enzyme activity, usually by changing the pH of the solution. Colour

Cross reference

More on ELISAs is given in Chapter 14, 'Molecular biology techniques'.

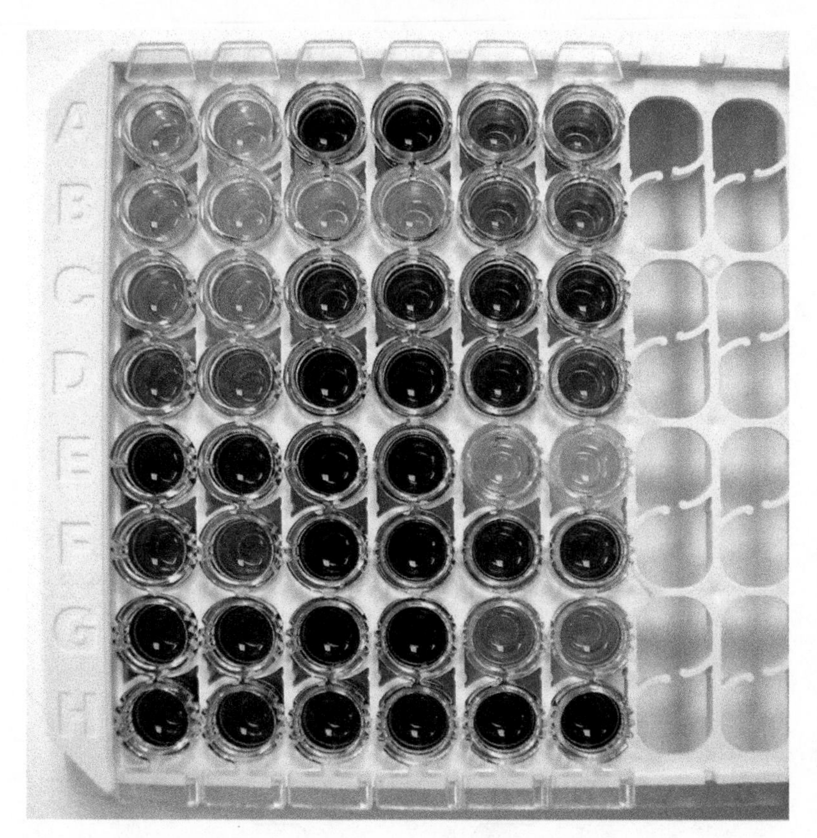

FIGURE 13.12
Photograph showing the final stage of an enzyme-linked immunosorbent assay with different intensities of colour development in the wells. The colour developed in an assay using a biotinylated antigen with an avidin-linked peroxidase and 2,2′-azino-*bis*-ethylbenzthiazoline sulfonic acid substrate. Courtesy of D. Raine, Immunology Department, Hull Royal Infirmary, UK.

intensity is measured as the absorbance or optical density at an appropriate wavelength on a specially adapted spectrophotometer designed to monitor ELISA plates. The intensity of colour development is directly proportional to the quantity of antigen bound to the plate (Figure 13.12).

Online Resource Centre
To see an online video demonstrating the ELISA technique, log on to www.oxfordtextbooks.co.uk/orc/fbs

Some ELISAs use a standard curve to quantify the level of antibody in patient sera. This employs a range of solutions containing known concentrations of antibody, which are measured at the same time as the unknown serum. The absorbances obtained on measuring these solutions are plotted against their known antibody concentrations producing a standard curve. In many cases, there is no recognized international standard for the antibodies measured by ELISA, so often results are expressed in arbitrary ELISA units. If a recognized standard does exist, the reagents in commercial kits should be calibrated to it and the results may be expressed as international units.

Fluorescent labels and other detectable probes that are used in similar ways with multiwell plates are usually referred to as ELISAs because they use similar protocols, even although enzymes are not involved. Modifications to the basic ELISA protocol include:

- sandwich ELISA
- competitive binding ELISA
- automated procedures.

Sandwich ELISA, outlined in Figure 13.13, is a sensitive and robust variation of the basic ELISA protocol. A known antibody is bound to the wells of the microtitre plate at a predetermined concentration. Diluted serum containing unknown concentrations of the analyte is then added to the wells. The bound reagent will capture the analyte within the wells. Following washings, a second antibody is added. The second antibody, which has an enzyme label attached to it is the detection antibody, and binds to the immobilized analyte. The rest of the procedure is as described above for the basic technique. Competitive binding ELISA is commonly used for detecting small molecules with limited numbers of epitopes. The sera containing unknown quantities of analyte are added to wells that

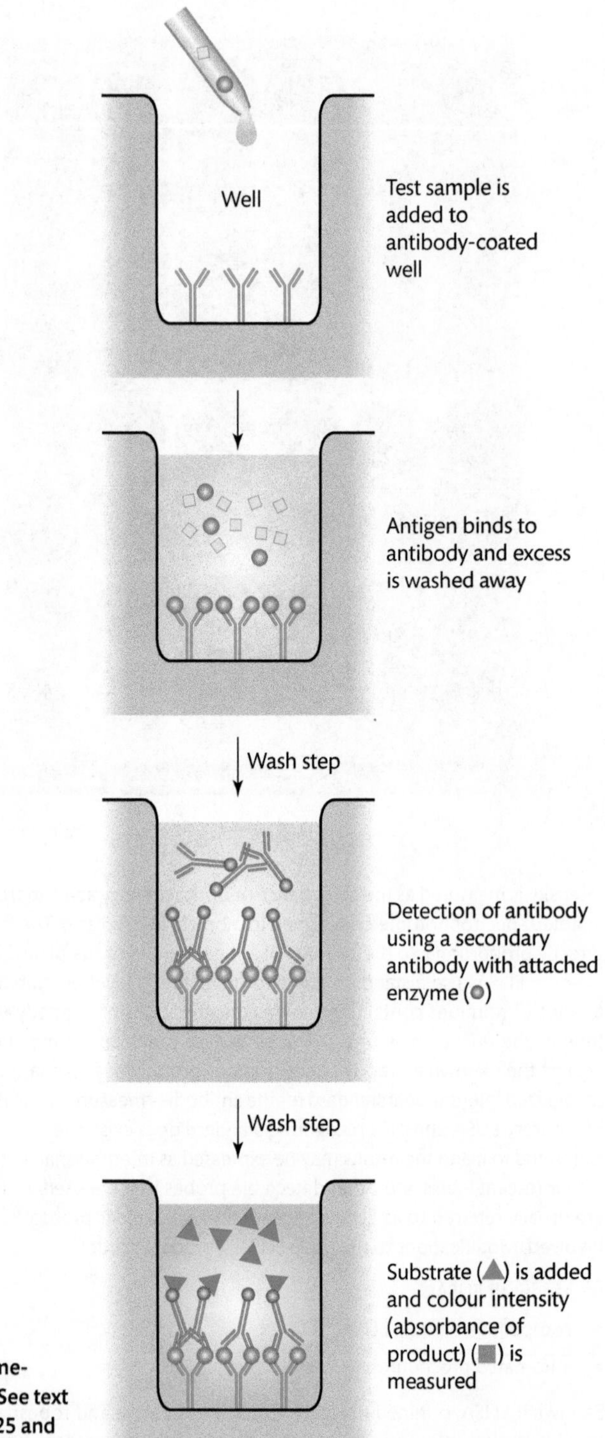
Well
Test sample is added to antibody-coated well
Antigen binds to antibody and excess is washed away
Wash step
Detection of antibody using a secondary antibody with attached enzyme (●)
Wash step
Substrate (▲) is added and colour intensity (absorbance of product) (■) is measured

FIGURE 13.13
Outline of the sandwich enzyme-linked immunosorbent assay. See text for details. See also Figure 14.25 and its associated text.

have been coated with a capture antibody, along with an enzyme-labelled reagent of similar antigenic material. If the colour change of the chromogenic substrate is less than that obtained from a well containing only labelled reagent, then the difference is proportional to the quantity of the analyte.

Commercially developed automated techniques based on the principle of the sandwich ELISA use individual plastic wells or nitrocellulose sponge discs as the solid phase. However, when the enzyme comes into contact with its substrate, a light-emitting fluorescent product is formed. The fluorescence is measured using an ultraviolet light source and its value can be compared with a standard curve to determine the amount of analyte in the sample of serum. This type of analysis is the basis for determining IgE values of serum and a wide range of allergens and autoantibodies. Other adaptations, using a range of absorbent materials, attach antibodies to the solid phase such that the colour change produced forms a line or dot if the result is positive. This technique underlies home pregnancy kits and a range of rapid techniques that are used in laboratories, including autoantibody typing and the diagnosis of malaria.

Online Resource Centre
To see an online video demonstrating the ImmunoCAP procedure, log on to www.oxfordtextbooks.co.uk/orc/fbs

Key points

ELISA techniques depend upon the purity and quality of the immunological reagents and robust binding to the solid phase.

Enzyme-linked immunospot assays

Enzyme-linked immuno*spot* (ELISPOT) assays are an adaptation of the ELISA technique, which were developed to measure antibodies produced by plasma cells. The cells are isolated from immunized mice and added to antigen-coated microtitre tray wells where they bind to antigen and are captured on the solid phase. The cells are then removed by washing and enzyme-labelled AHG is added to the wells, followed by a substrate suspended in agarose gel. The appearance of coloured spots in the gel shows where the plasma cells are located in the wells.

Modifications to the original ELISPOT method have been made to investigate the production of cytokines by **T lymphocytes**. Stimulated T cells are isolated and added to wells containing a monoclonal antibody to the cytokine of interest. The procedure is the same as the original ELISPOT but uses a labelled antibody directed against the cytokine. The sites of colour development indicate the locations of cells secreting the cytokine, as you can see in Figure 13.14. If the spots are counted, it is possible to estimate the proportion of T cells secreting the cytokine present, based on the number of cells originally added to the plate.

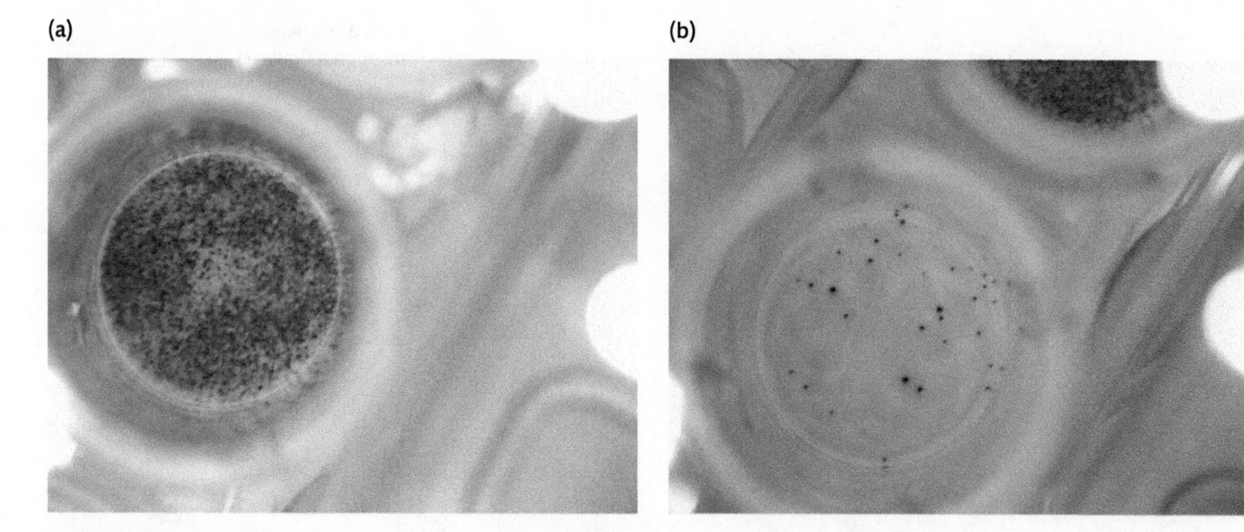

FIGURE 13.14

(a) Positive control for an enzyme-linked immunspot (ELISPOT) assay. (b) Sample of a patient showing a positive ELISPOT result. Courtesy of D. Spradbery, Immunology Department, Hull Royal Infirmary, UK.

Commercial variations on the ELISPOT have been developed that have increased the reproducibility of the assays and made them more suitable for use in clinical laboratories. They are used for the detection of interferon secretion by T lymphocytes that have been sensitized to *Mycobacterium tuberculosis.* This is of particular importance in the diagnosis of immunodeficient individuals who may have latent or subclinical tuberculosis.

SELF-CHECK 13.7

(a) Why are washing steps essential in ELISAs? (b) Which components of the immune system are detected by ELISPOTs?

Online Resource Centre

To see an online video demonstrating flow cytometry, log on to www.oxfordtextbooks.co.uk/orc/fbs

13.5.7 Flow cytometry and fluorescence-activated cell sorters

Flow cytometry is a method of counting and evaluating cells or particles by passing them in a thin stream through a laser beam. A fluorescence-activated cell sorter (FACS) is able to separate the cells as they are identified. Figure 13.15 outlines their operation.

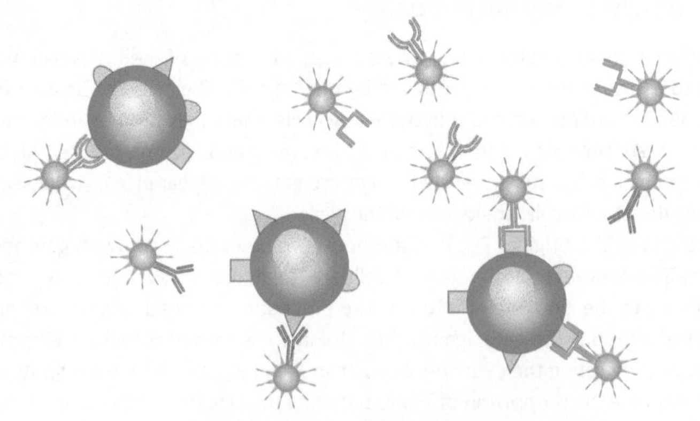
Lymphocytes are mixed with a range of fluorochrome-labelled antibodies

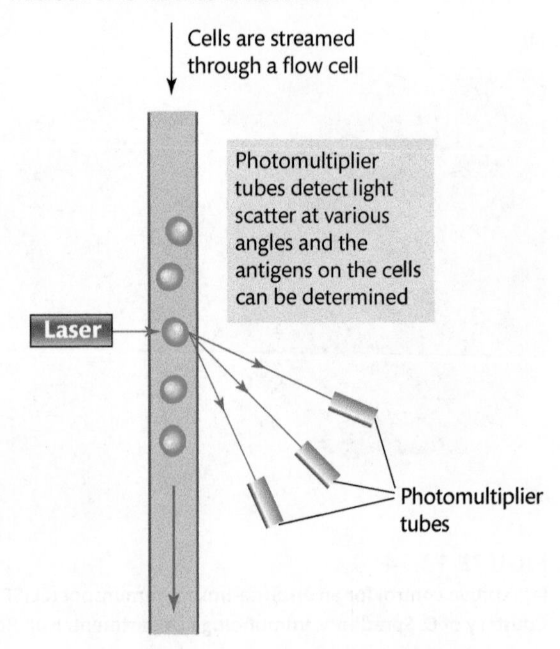
Cells are streamed through a flow cell
Photomultiplier tubes detect light scatter at various angles and the antigens on the cells can be determined
Laser
Photomultiplier tubes

FIGURE 13.15
Basic principles of flow cytometry. See text for details.

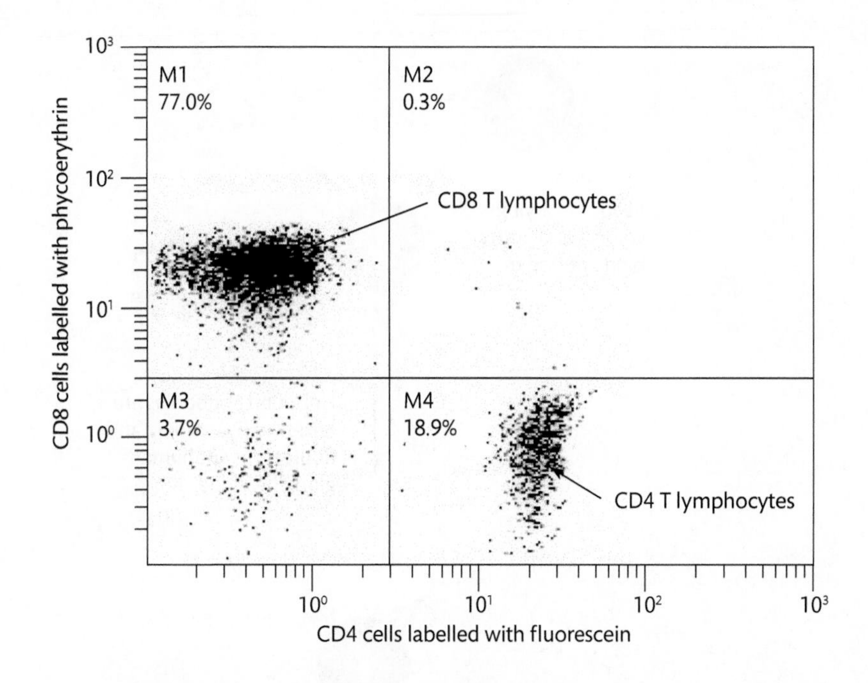

FIGURE 13.16
A dot plot showing the separation of CD4 and CD8 T lymphocytes in a sample from a patient with HIV. Courtesy of D. Spradbery, Immunology Department, Hull Royal Infirmary, UK.

Flow cytometers are incorporated into analysers that count and identify leucocytes in routine haematology tests. More complex flow cytometers with multiple lasers are used to differentiate subpopulations of a range of cells. This is an important tool in the **immunophenotyping** of cells when classifying leukaemias and in the classification of immunodeficiency diseases.

Different types of leucocytes, which may superficially appear similar when stained and examined by light microscopy (see Figure 7.1a), have, however, many differences in their cytoplasm and on their surfaces allowing them to be distinguished and identified. Some surface markers are termed *clusters of differentiation* (CD). Many CD markers have been identified and numbered. For example, we refer to CD4 and CD8 lymphocytes when monitoring cell counts in patients with AIDS.

Cells are incubated with monoclonal antibodies directed against a predetermined panel of CD markers or intracellular components. The antibodies are labelled with fluorescent dyes called fluorochromes that will vary in their response to lasers of differing wavelengths. As the dilute suspension of cells is forced through a fine tube they pass through the laser beam. Sensitive photomultiplier tubes detect the extent of light *scattering*, which will indicate the size and granularity of the cells. They also measure the fluorescence emission: this indicates which of the labelled antibodies is attached to the cells. The data are electronically stored for analysis and can be displayed in the form of a graph or histogram if only one antibody is used and as a two-dimensional dot plot if two or more antibodies have been used, as you can see in Figure 13.16.

The use of complex analysers with multiple lasers, and the wide variety of fluorochromes and monoclonal antibodies, have increased the use of this versatile technology. Flow cytometers are able to process large numbers of cells and calculate the number of cells carrying a particular marker. This is necessary in the routine testing of patients who are being treated for HIV infection. Along with FACS analysis, flow cytometry has added enormously to our knowledge of the cells of the immune system.

Cross reference

***Clinical Immunology* in this series gives more information on flow cytometry.**

13.5.8 Multiplex technology

Multiplex technology is an advanced development in the clinical laboratory, which makes it possible to measure multiple analytes in one assay. There are a number of variations on the basic principle but all allow complex panels of antibodies and antigens to be evaluated simultaneously.

Microscopic polystyrene beads or microspheres of approximately 5 µm diameter with differing ratios of fluorescent dyes incorporated into them form the solid phase of the assay. The combination of the dyes can be varied to produce up to a hundred different coloured beads; each bead can then have a different antigen or antibody attached to it. The beads are added to a sample of serum

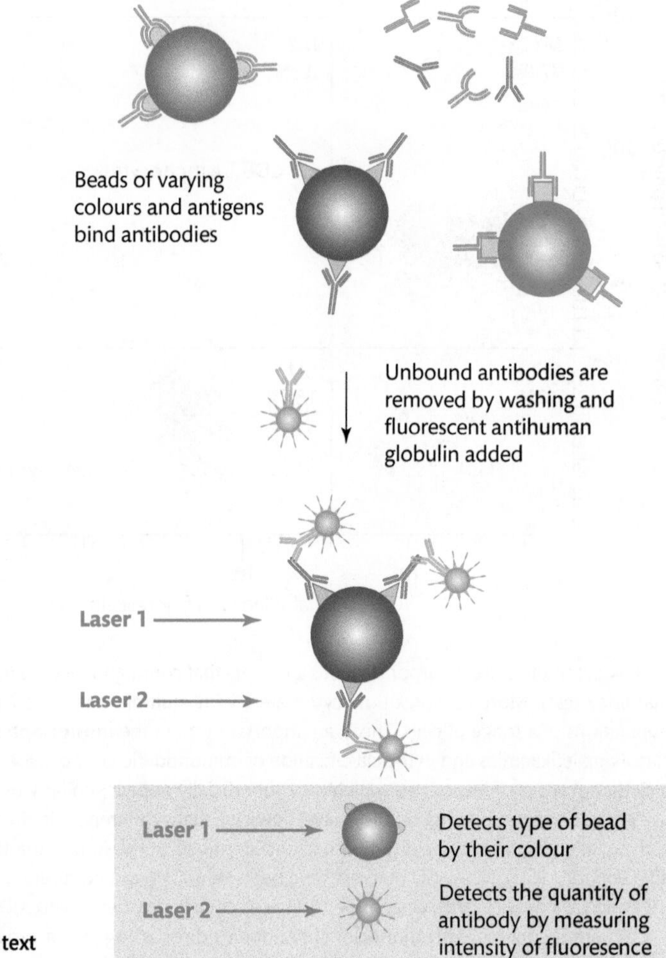
Beads of varying colours and antigens bind antibodies
Unbound antibodies are removed by washing and fluorescent antihuman globulin added
Laser 1
Laser 2
Laser 1
Detects type of bead by their colour
Laser 2
Detects the quantity of antibody by measuring intensity of fluoresence

FIGURE 13.17
Basic principles of multiplex analysis. See text for details.

that potentially contains antibodies or antigens that will attach to an appropriate bead. A secondary fluorochrome-labelled antibody, which recognizes and binds to the primary one, is then added. The beads are then analysed using flow cytometry, which we described earlier. As the beads pass through a flow cell, they pass through two laser beams: one detects the bead and the other excites the fluorochrome. The data obtained will identify the bead and determine the type and quantity of antibody or antigen bound to the surface of the bead (Figure 13.17).

The potential of this technology to assist clinical laboratories and research facilities is impressive, providing extensive data on a wide range of analytes.

13.5.9 Microarrays

Typically, microarrays are used for DNA analysis where they are known as 'DNA chips'. However, their use has widened with the introduction of microarrays to measure cytokines. Glass slides with minute wells etched into their surface are loaded with a range of monoclonal antibodies that are used to bind ('capture') the cytokine of interest. When the serum to be analysed is added to the microarray, the capture antibodies will bind to some of the proteins in the serum. The analysis is performed very much like an ELISA, a further antibody that has an attached fluorescent label is added. The slides are read using an array reader with compatible image analysis software. The presence of a fluorescence signal indicates the positions of the bound cytokines.

Cross reference

For more information on microarrays see Chapter 14, 'Molecular biology techniques'.

SELF-CHECK 13.8

(a) What name is given to the cell surface markers that are detected by flow cytometry? (b) What is the main advantage of multiplex technology over other antibody-based methods of analysis?

13.5.10 Complement assays

Severe or chronic infections will cause a continuous activation of complement (Box 13.2) and increased concentrations of circulating immune complexes, which are a feature of some autoimmune diseases. Both types of condition increase consumption of complement proteins and therefore may lower their concentrations in the blood. Complement proteins are also labile and care must be taken to prevent their activation in the sample tube prior to testing. This means samples should be tested promptly or frozen immediately until required for testing. Some procedures may require the use of the anticoagulant ethylenediamintetraacetic acid in samples or the addition of a protease inhibitor.

Complement components C3 and C4 are present in low, but relatively stable and measurable, quantities in normal sera. Both are routinely analysed by nephelometry and other automated immunoassay systems as a first line complement screen. If a complement immunodeficiency is suspected, many of the other proteins can be determined. A number of complement proteins can be measured by SRID.

Haemolytic assays are sometimes used when investigating complement deficiencies to determine which activation pathway is affected. Erythrocytes are used to demonstrate the presence of complement by creating an environment in which the complement system pierces the erythrocyte membrane and releases haemoglobin into solution. The procedure was referred to as the *total haemolytic complement test* or CH100. However, it is difficult to determine the final end point of **haemolysis**, and therefore the working end point is usually accepted as occurring when 50% of the erythrocytes are lysed (CH50).

The classical pathway is investigated using sheep cells that have IgG antisheep antibodies attached to their surfaces. They are incubated at 37°C and observed for haemolysis at regular intervals until 50% haemolysis is reached. You can see an example of this in Figure 13.18. This test may be performed in test tubes, multiwell plates or on SRID gels containing erythrocytes. The time and detection systems vary with method of choice. A number of in-house protocols and commercial kits are available.

If 50% lysis is achieved under comparable conditions to a 'normal' serum control, the complement activity of that pathway is normal.

The alternative pathway is tested in the same way as the classical, but antibody is not required to activate complement. Rabbit, guinea pig, or chicken cells that are susceptible to direct haemolysis by human complement may all be used.

In virology laboratories, complement fixation tests are used to detect the presence of microorganisms such as mycoplasma, those causing Q fever and psittacosis, and a range of influenza viruses. The antibody to be detected is incubated with an appropriate antigen in the presence of complement. Sheep erythrocytes that have been coated with anti-sheep IgG are then added and after a further

FIGURE 13.18
The CH50 test (to determine where 50% of the erythrocytes are lysed) for haemolytic complement activity. The photograph shows erythrocytes in the gel. Circles of haemolysis can be observed where complement activity is present. Courtesy of D. Raine, Immunology Department, Hull Royal Infirmary, UK.

incubation, the suspension is centrifuged. If there is still free complement in the mixture the erythrocytes will be haemolysed, that is, the complement will have interacted with the IgG and pierced the cell surface membrane. This means that the test is negative as the complement has not been bound by an antigen–antibody reaction during the first incubation.

13.5.11 Indirect immunofluorescence and immunocytochemistry

Indirect immunofluorescence (IIF) and immunocytochemistry techniques differ from those already described in that the solid phases necessary to detect the analytes are sections of solid organs, or cells grown in tissue culture that have been fixed to a glass or plastic slide.

Indirect immunofluorescence

IIF is a commonly used technique in clinical immunology laboratories to detect and identify antibodies in the serum of individuals with autoimmune diseases. Many autoantibodies may be detected using this technique by employing a wide range of substrates that may be sections of animal tissues or cultured cell lines. Table 13.2 shows examples of antibodies and tissues used in investigating autoimmune diseases. The AHG used is normally directed at IgG; however, anti-IgA is used for some cases.

The principle of the technique is outlined in Figure 13.19. A range of animal specimens are prepared by cutting sections of frozen tissue and fixing them in an alcohol–acetone mixture. Cultured cell lines used include **Hep2**, which are human epithelial-type cells derived from a carcinoma, that have large nuclei and a rapid rate of division. This allows observation of many nuclear antibody patterns. The tissue or cultured cell substrate is fixed to a glass slide, which is then flooded with dilute serum, which may contain antibody. Following a period of incubation, the serum is washed away leaving antibodies attached to their target antigens. Fluorescently labelled polyclonal AHG is then added to the slide and will bind to the antibodies attached to the antigens. After further incubation, the excess labelled antibody is removed by washing. As with ELISAs (see Section 13.5.6) the washing steps are essential to ensure excess unbound components are removed (Figure 13.20). The sections or cells are then immersed in mounting medium and a cover slip added.

Online Resource Centre

To see an online video demonstrating indirect immunofluorescence, log on to www.oxfordtextbooks.co.uk/orc/fbs

The slides are then examined using fluorescent microscopy and the patterns formed by the location of antibody in the tissues or cells recorded. Figure 13.21 shows examples of the types of staining patterns. Depending on the results obtained in the IIF panel of tests, immunology laboratories will follow up positive results with more specific assays such as ELISA for subtypes of **antinuclear antibodies**, liver antibodies, and vasculitis-related antibodies.

Cross reference

Chapter 7, 'Microscopy', has further information on fluorescence microscopy.

The amount of antibody present can be estimated by testing a range of dilutions of serum on the slide and expressing the result as a titre, that is, the dilution at which fluorescence can still be observed in the cells.

The most common fluorochromes employed in IIF are fluorescein, rhodamine, and phycoerythrin.

Many antibodies can be identified by IIF. Antinuclear antibodies include a number of antigenic targets and create a variety of patterns on the nuclei of Hep2 cells, for example smooth homogenous staining of

TABLE 13.2 Tissue substrates used in the detection of autoantibodies by indirect immunofluorescence

Cell/tissue	Source	Target antigen	Disease state
Hep2 cells	Human epithelial cells	Nuclear components, e.g. nucleolus, centromere	SLE
Liver	Rat, mouse	Mitochondria, nuclear components	PBC
Oesophagus	Primate	Tissue transglutaminase	Coeliac disease
Neutrophils	Human	ANCA	Vasculitis

ANCA: antineutrophil cytoplasmic antibody; SLE: systemic lupus erythematosus; PBC: primary biliary cirrhosis.

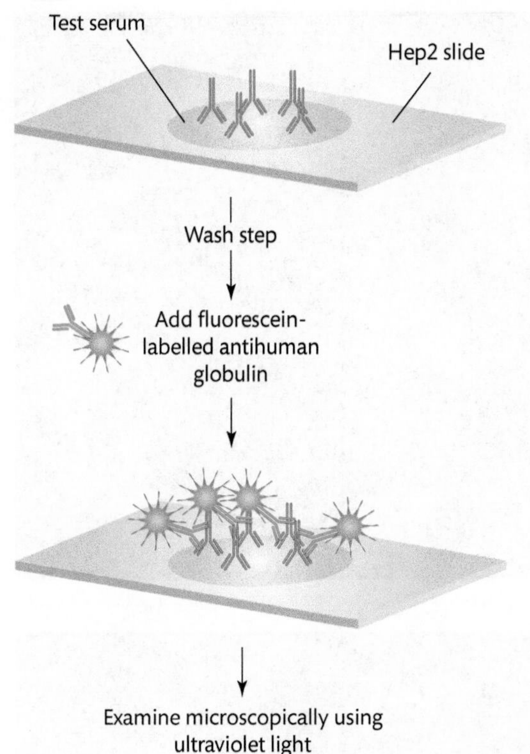

FIGURE 13.19
Outline of indirect immunofluorescence. See text for details.

the nucleus is seen in SLE. Rat or mouse kidney and liver sections will reveal the presence of mitochondrial antibodies, as seen in primary biliary cirrhosis. Pancreatic islet cells and sections of adrenal gland, ovary, testes, and oesophagus are commonly used when investigating for autoantibodies such as those for antitissue transglutaminase, which occur in coeliac disease. Human neutrophils are used to demonstrate the antineutrophil cytoplasmic antibodies that are a feature of some forms of vasculitis.

The drawback of IIF in routine clinical laboratory testing is that it is extremely operator dependent and time consuming. The method relies on the expertise of individuals in slide preparation and microscopy. Higher than normally expected numbers of staff are also required to achieve an appropriate

Cross reference

The *Clinical Immunology* volume of this series covers an extensive range of autoimmune disorders and their associated antibodies.

(a)

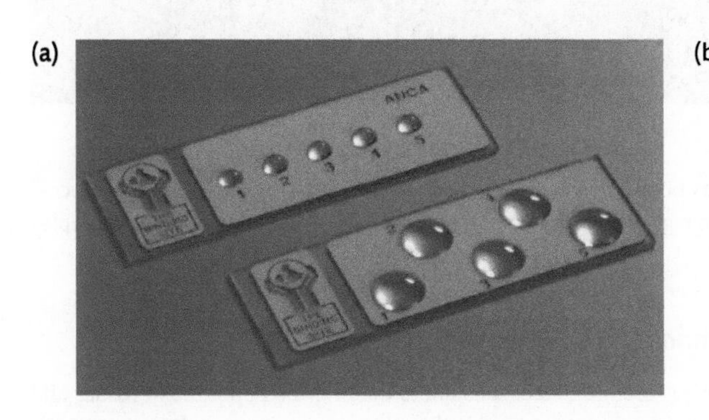

(b)

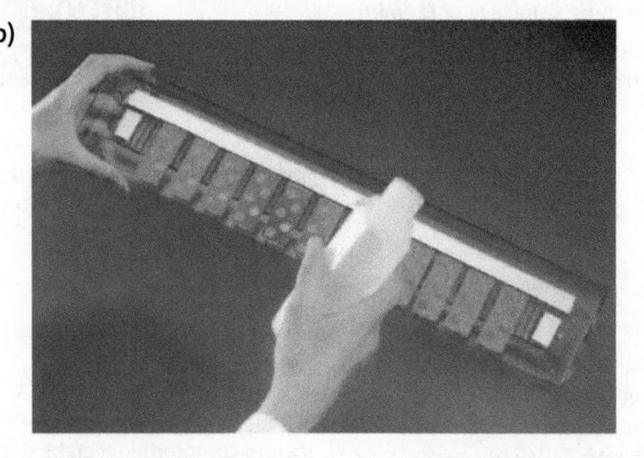

FIGURE 13.20
(a) Photograph showing commercial indirect immunofluorescence slides that have tissue sections attached to them and diluted serum placed in the wells prior to incubation. (b) Photograph showing the careful washing of the indirect immunofluorescence slides. Courtesy of D. Raine, Immunology Department, Hull Royal Infirmary, UK.

(a)

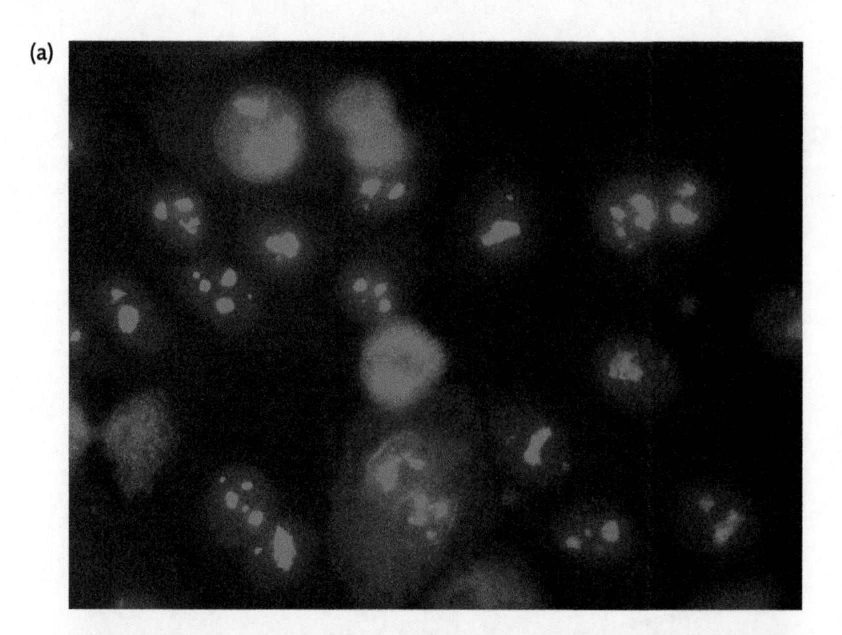

(b)

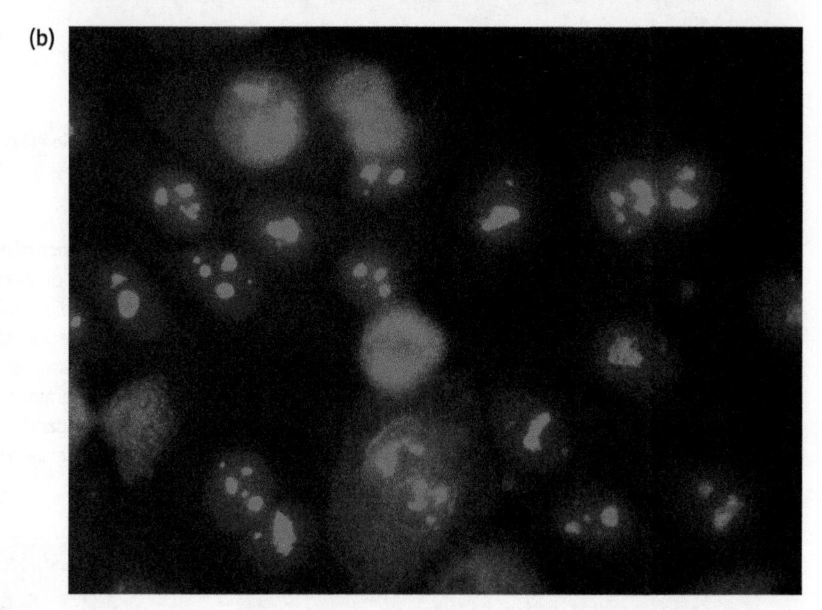

FIGURE 13.21
Photographs of indirect immunofluorescence patterns. (a) Nucleoli of Hep2 cells are indicated by the presence of an antinucleolar antibody. (b) Diffuse, granular staining of kidney tubules indicates the presence of mitochondrial antibodies. Courtesy of D. Raine, Immunology Department, Hull Royal Infirmary, UK.

throughput of samples. This has been alleviated to some extent by using automated slide processors, but eventually many of the high-volume antibody tests are likely to be performed by multiplex technology-based assays.

Cross reference

You can read more about cytological and histological investigations in Chapter 6, 'Samples and sample collection', and Chapter 7, 'Microscopy'.

You can learn more about DIF and immunocytochemistry in the *Histopathology* book of this series.

Direct immunofluorescence

Direct immunofluorescence (DIF) is used to detect substances that may be deposited in tissues and organs during some disease states, for example some skin disorders can be diagnosed when antibodies and complement can be demonstrated in all layers of the epidermis. A biopsy of the patient's tissue is removed from a site adjacent to a skin lesion and fixed on a glass slide. Antihuman globulin with a fluorescent marker is added to the slide, which will bind to antibodies present in the tissue sample. The procedure is the same as IIF in that excess antihuman globulin is removed by washing, and then the tissue is examined for the fluorescence that will demonstrate the presence of autoantibodies.

Immunocytochemistry

Immunocytochemistry is similar to the fluorescence methods described above, in that direct and indirect methods can be used. The techniques were developed to identify tumours by accurately targeting tumour cell surface markers. This technology has been automated and is a growing area in laboratories that specialize in histological and cytological investigations. Although fluorescent labels can be used, the methods are often based on colour development using enzyme-labelled antibodies and substrates similar to those employed in ELISAs.

SELF-CHECK 13.9

(a) What is the difference between direct and indirect immunofluorescence? (b) Name two examples of fluorochromes that may be used when detecting antibodies by IIF.

CASE STUDY 13.1 Rubella and pregnancy

A 32-year-old carer in a playgroup became pregnant. Late in her first trimester, several infants in the group developed German measles, a disease caused by the rubella virus. This virus can cross the placenta and harm unprotected fetuses by causing congenital rubella syndrome (CRS), which is associated with defective hearing and vision, and mental retardation, or may even lead to the death of the fetus. The woman consulted her general practitioner, who sent samples of her blood for antibody testing. The woman was found to possess rubella-specific IgM but not IgG antibodies.

Some points to consider:

What does the presence of IgM and absence of IgG antibodies indicate?

Is the fetus at risk of CRS?

What advice should the woman's counsellor offer her?

CASE STUDY 13.2 Delayed-type (type IV) hypersensitivity to cheap jewellery

A 14-year-old girl visited a local car boot sale where she bought an old but fashionable retro matching paste ring and bracelet. Initially, she wore both with no ill effects. However, she developed a severe dermatitis where both her finger and wrist had been in contact with the jewellery. Her general practitioner advised that she had had an 'allergic' reaction to the jewellery and had developed an allergic contact dermatitis. The girl was told not to wear the ring and bracelet again or, indeed, similar types of old or cheap jewellery. Topical treatment with 1% hydrocortisone cream resolved her dermatitis.

In the past, cheap jewellery often contained the metal nickel. Transition metals, like nickel, can react with skin proteins to form hapten-protein like complexes, which are antigenic. Cells of Langerhans in the skin can present these complexes to helper T lymphocytes, which become sensitized. Further contact with the jewellery leads to a delayed-type hypersensitive allergic (that is type IV) reaction.

The nickel content of jewellery sold within the European Union is now regulated strictly and the incidence of these types of cases has rapidly declined over recent years.

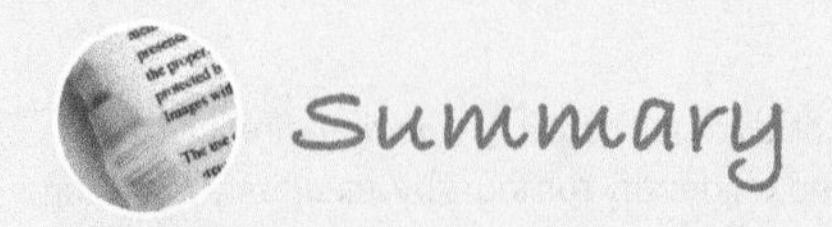

Summary

- The immune system consists of the interaction of specialized organs, cells, and humoral components that protect the body from infection.
- Innate immunity is present at birth and is immediate acting but non-specific. Acquired immunity is slower to develop, more specific, and has immunological memory.
- Immunoglobulins, in the form of antibodies, are essential components of the immune system, which have been adapted as diagnostic tools.
- The formation of antibody–antigen immune complexes and their behaviour form the bases of immunoassays.
- The use of radioisotopes, enzymes, and fluorochromes to label antibodies and antigens was an important advance in the development of immunoassay.
- The use of monoclonal antibodies in immunoassays has revolutionized laboratory medicine.
- Advances in areas such as flow cytometry and multiplex technology will have a lasting impact on clinical laboratories that were previously dominated by manual methodology; small-scale manual immunoassays are being overtaken by automated analytical methods in all specialities in clinical laboratories.

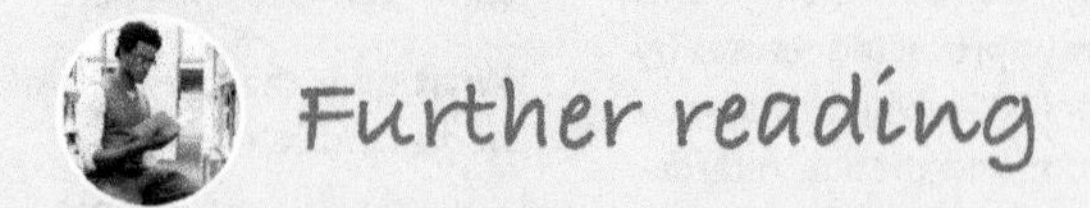

Further reading

- **Cruse JM, Lewis RE. *Historical Atlas of Immunology*. Abingdon: Informa Healthcare, 2005.**

  An interesting and well illustrated account of the development of our knowledge of immunology.

- **DePriest AZ, Black DL, Robert TA. Immunoassay in healthcare testing applications. *Journal of Opioid Management* 2015; 11: 13–25.**

  This review highlights the performance characteristics of immunoassay, with special emphasis on prescription drug classes and testing at the point of care.

- **Hay FC, Westwood OMR. *Practical Immunology*. 4th ed. Oxford: Blackwell Science, 2002.**

  Practical Immunology is a useful bench textbook for research scientists and gives insight into the origins of immunological techniques and increases the understanding of biomedical scientists.

- **Murphy KM. *Janeway's Immunobiology*. 8th ed. London and New York: Garland Science, 2011.**

  A comprehensive text that gives a detailed, well-written account of the immune system, beginning with innate immunity, then moving to adaptive immunity, and ending with applied clinical immunology.

- **Peakman M, Vergani D. *Basic and Clinical Immunology*. 2nd ed. London: Churchill Livingstone, 2009.**

  This excellent book describes the immune system and associated disease states. It is well laid out and relevant to clinical laboratory practice.

- **Price CP, Newman DJ. *Principles and Practice of Immunoassay*. 2nd ed. Oxford: Macmillan, 1997.**

  Comprehensive account of the development and practice of immunoassay in the clinical laboratory. Covers all aspects of methodology, standardization, and quality.

- **Sompayrac LM. *How the Immune System Works*. 5th ed. Harlow: Wiley-Blackwell, 2015.**

  Brilliant, readable account of how our immune system works; recommended to anyone who wants to learn about immunology. Well written, humorous, and explained in simple terms; difficult concepts are presented in a highly imaginative way making them easier to remember.

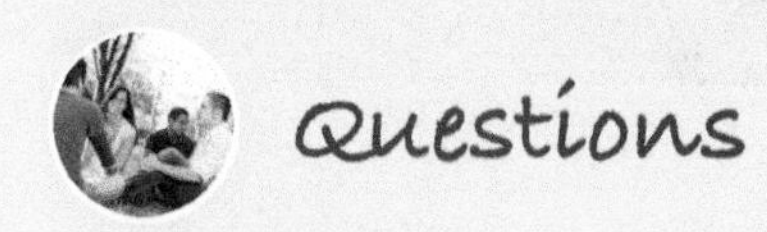

Questions

1. Select three characteristics of a high-affinity antibody from:
   (a) An antibody with more than two binding sites
   (b) An antibody produced in the secondary immune response
   (c) Is most likely to be IgG
   (d) Forms immune complexes that do not easily dissociate
   (e) Produced by memory B cells that have refined 'best fit' antibodies.

2. Which of the following statement(s) is/are **TRUE**.
   (a) Acquired immunity is present at birth
   (b) Complement can kill pathogens without the involvement of antibodies
   (c) Antibodies are produced by T lymphocytes
   (d) Genetic predisposition alone commonly causes autoimmune diseases
   (e) The precipitation curve is not affected by the source of the antibody or antigen.

3. Discuss the difference between a polyclonal antibody response and a monoclonal response.

4. What are the drawbacks of using polyclonal antibodies?

5. Give a short description of the three types of antibody labels that may be used in immunoassays with examples of assays that use them.

Answers to self-check and end-of-chapter questions are available at the end of the book.

14

Molecular biology techniques

Qiuyu Wang, Nessar Ahmed, and Chris Smith

Learning objectives

After studying this chapter, you should be able to:

- Outline the structures of nucleic acids and proteins
- Describe the general functions of biological macromolecules
- Explain the basic molecular biology encountered in clinical laboratories
- Discuss the principal techniques of molecular biology used in clinical tests
- Outline the uses of molecular biology techniques in clinical laboratories

Introduction

The term **molecular biology** is thought to have been first used by Warren Weaver in a 1938 report to the Rockefeller Foundation. Molecular biology is the study of biological phenomena in terms of the physical and chemical properties of the molecules of organisms. It can be described generally as the study of the structures, functions, and interactions of biological macromolecules. Two major types of macromolecules common to all cells are the nucleic acids, deoxyribonucleic acid (DNA) and ribonucleic acid (RNA), and proteins. DNA and RNA carry biological information. DNA contains the genetic information of all cells and some viruses, that is to say, it is the material of genes. RNA only holds the genetic information of some viruses. However, it is present in all cells in a number of forms, including messenger RNA (mRNA), transfer RNA (tRNA), and ribosomal RNA (rRNA) molecules that are necessary to allow the genes to be expressed and form proteins. Proteins perform most of the activities in cells. They catalyse metabolic reactions, perform mechanical work in muscles, cilia, and flagella, act as sensors, perform regulatory and control functions, and also constitute some of the major structural features of the cell.

The definition of molecular biology given above is rather limited in that it fails to describe how the subject overlaps with other areas of biology, chemistry, microbiology, and computational biology, in particular biochemistry, genetics, immunology, genomics, proteomics, and bioinformatics. The techniques of molecular biology have reached sophisticated levels, and are used extensively in the biological and biomedical sciences. Many of these techniques have been automated and are performed by robots, which makes them particularly suitable in the diagnosis of clinical conditions and monitoring treatments. This chapter will provide an overview of the scope of molecular biology and introduce you to the structures, properties, and functions of some of the molecules involved. It will describe some of the experimental techniques of molecular biology and outline a number of their applications in the biomedical sciences.

14.1 Structure and properties of macromolecules

Macromolecules are ones that are large, that is, have M_r in excess of approximately 5000. Biological macromolecules are polymers, formed by covalently linking much smaller units together in condensation reactions (Figure 14.1). Within the polymer, such units are often referred to as *residues*. Linking two units gives a dimer, three a trimer, and so on, to eventually give a polymer, which may contain hundreds or thousands or millions of residues.

Cross reference

You can read about automation in Chapter 15, 'Laboratory automation'.

14.1.1 Nucleic acids

The complete hydrolysis (see Section 14.4) of any nucleic acid gives:

- pentose (five-carbon) sugars
- pyrimidine and purine bases
- phosphate groups.

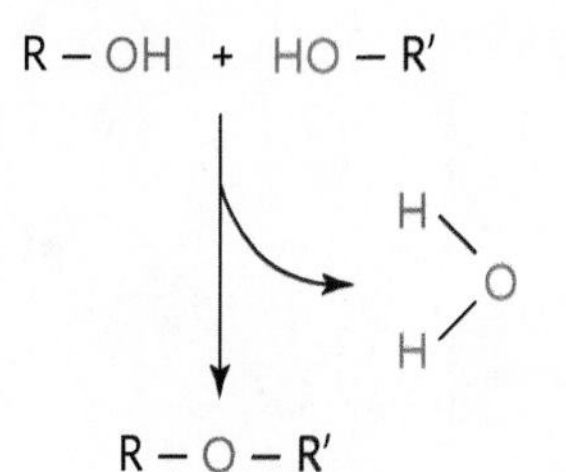

FIGURE 14.1
A general condensation reaction.

The sugars are ribose (Rib) in the case of RNA and deoxyribose (dRib) in DNA. DNA contains the purine bases adenine (A) and guanine (G), and the pyrimidines, cytosine (C) and thymine (T). Ribonucleic acid also contains A, G, and C but the pyrimidine uracil (U) rather than T (Figure 14.2). Minor amounts of other, chemically related, bases are also released. The phosphate groups are negatively charged at physiological values of pH.

The combination of a base and sugar forms a **nucleoside** (Figure 14.3a). Note that the atomic positions in sugars and bases need to be distinguished, and this is achieved by adding a prime (') to the sugar positions as you can see in Figure 14.3(a). Bases are attached to the 1' carbon atoms of the sugars. The addition of a phosphate(s) to a nucleoside gives a **nucleotide** (Figure 14.3b). Phosphates may be bonded to the 5' or 3' positions giving nucleoside 5' phosphate or nucleoside 3' phosphate, respectively. Table 14.1 summarizes the nomenclature of the major bases, nucleosides, and nucleotides.

Both DNA and RNA are polymers of nucleotides hence their alternative name of **polynucleotides**, and both are similar in structure. Phosphodiester bonds link the nucleotides together. If you examine Figure 14.4, you will see these bonds link the 3' and 5' carbons of adjacent sugar residues giving the link directionality: the convention when writing or drawing them is to place the 5' to the left and the 3' to the right (as in Figure 14.5).

This means the *sequence* of bases also runs in a 5' to 3' direction forming the primary structure of the nucleic acid, which differs in nucleic acids from different sources. Primary structures of nucleic acids can be described by one of several conventions, which you can see in Figure 14.5(a, b).

Native DNA exists as two complementary strands held together by hydrogen bonds. The strands form the famous double helical or duplex structure first described by Crick and Watson, and constitute the secondary structure of DNA. You can see this in Figure 14.6. The strands run in opposite directions, an arrangement described as antiparallel. The bases occur as complementary pairs, with A hydrogen bonded to T and G with C. In each case, a purine base is complementary to a pyrimidine meaning the lengths of each base pair (bp) are approximately the same and the strands can line up as a double helix of regular diameter. Two hydrogen bonds hold each A–T pair together unlike the G–C combination that has three, making it the more stable of the two pairs. The sequence of bases of one strand automatically gives that of the other. RNA is single stranded. However, it can form double helical regions, when the strand double backs on itself. Its base pairing rules are similar to those of DNA, but A pairs with U and G with C. It is also possible to form heteroduplexes that consist of complementary strands of DNA and RNA.

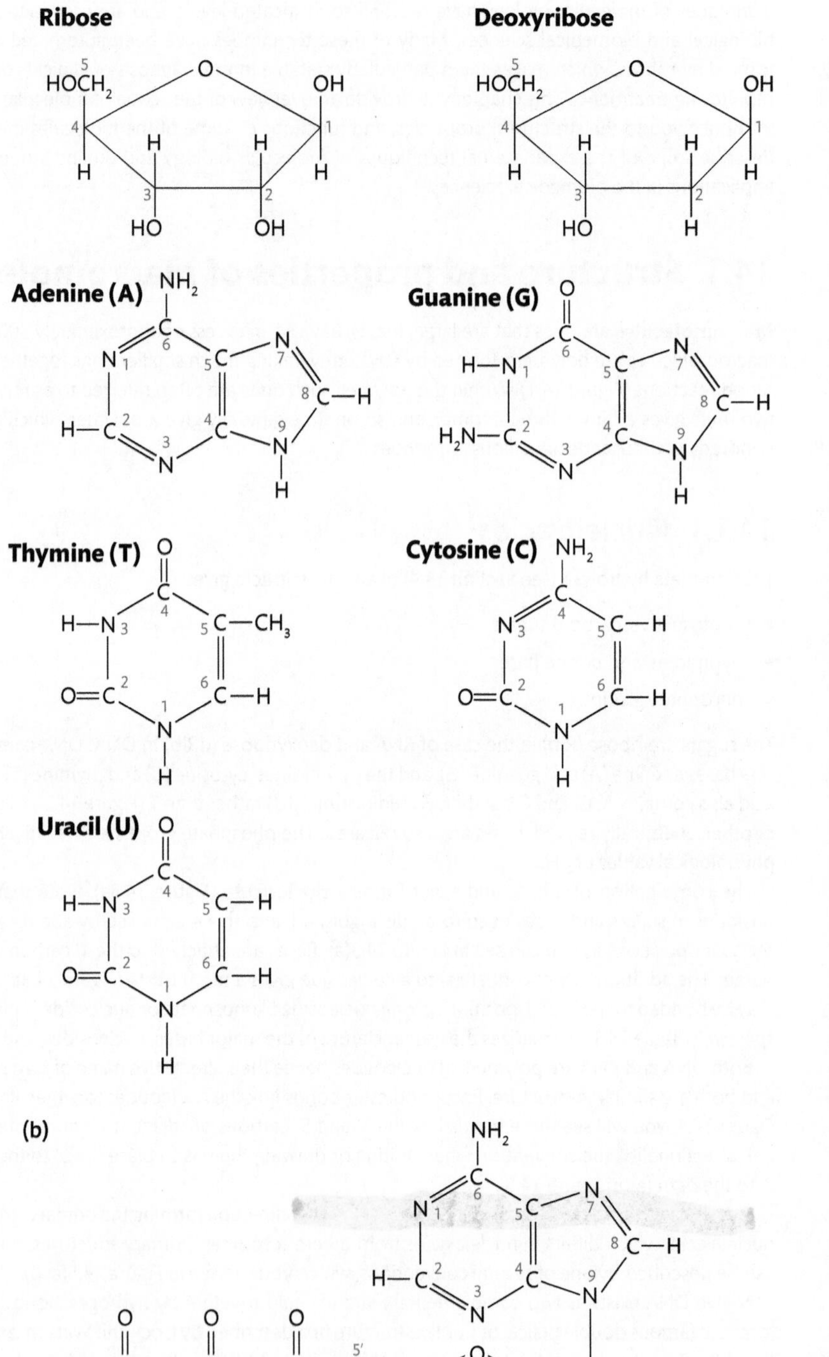

Ribose

5 HOCH2 O OH 4 H H 1 H 3 2 H HO OH

Deoxyribose

5 HOCH2 O OH 4 H H 1 H 3 2 H HO H

Adenine (A)

NH2 C 6 N 1 5 C N 7 8 C—H H—C 2 4 C 9 N 3 N H

Guanine (G)

O C 6 H—N 1 5 C N 7 8 C—H H2N—C 2 4 C 9 N 3 N H

Thymine (T)

O C 4 H—N 3 5 C—CH3 O=C 2 6 C—H 1 N H

Cytosine (C)

NH2 C 4 N 3 5 C—H O=C 2 6 C—H 1 N H

Uracil (U)

O C 4 H—N 3 5 C—H O=C 2 6 C—H 1 N H

FIGURE 14.2
Structures of the sugars and bases of nucleic acids.

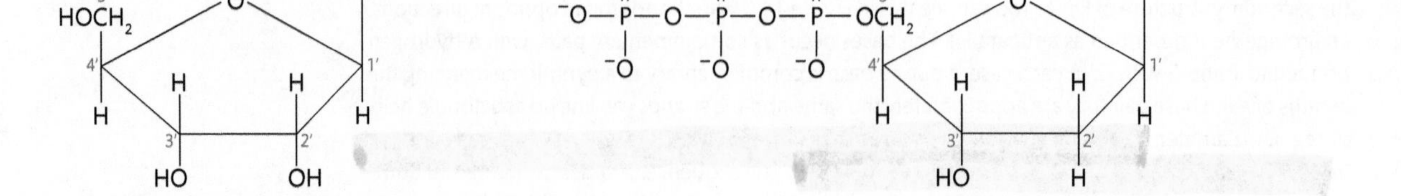

(a)

Uracil

O C 4 H—N 3 5 C—H O=C 2 6 C—H 1 N

Ribose

5' HOCH2 O 4' H H 1' H H 3' 2' HO OH

(b)

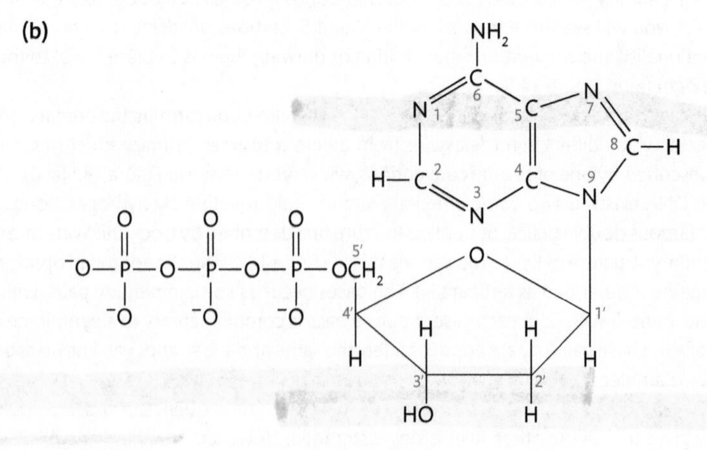

FIGURE 14.3
Structures of (a) the ribonucleoside uridine and (b) the nucleotide deoxyadenosine 5'-triphosphate. See also Table 14.1.

TABLE 14.1 Standard nomenclature of the major nucleic acid bases, nucleosides, and nucleotides

Bases	Ribonucleoside	Ribonucleotide*,'
Adenine (A)	Adenosine	Adenosine 5'-monophosphate (AMP)
Guanine (G)	Guanosine	Guanosine 5'-monophosphate (GMP)
Uracil (U)	Uridine	Uridine 5'-monophosphate (UMP)
Cytosine (C)	Cytidine	Cytidine 5'-monophosphate (CTP)
Bases	Deoxyribonucleoside	Deoxyribonucleotide (as 5'-monophosphate)
Adenine (A)	Deoxyadenosine	Deoxyadenosine 5'-monophosphate (dAMP)
Guanine (G)	Deoxyguanosine	Deoxyguanosine 5'-monophosphate (dGMP)
Thymine (T)	Deoxythymidine	Deoxythymidine 5'-monophosphate (dTMP)
Cytosine (C)	Deoxycytidine	Deoxycytidine 5'-monophosphate (dCTP)

*Where the attachment for the phosphate is not specified, the 5' can be assumed by default. Other sites, for example C-3', must be denoted.

'Other nucleotides with two (di), three (tri), and, rarely, four (tetra) phosphates are known. For example, adenosine 5'-triphosphate (ATP), deoxythymidine 5'-diphosphate (dTDP).

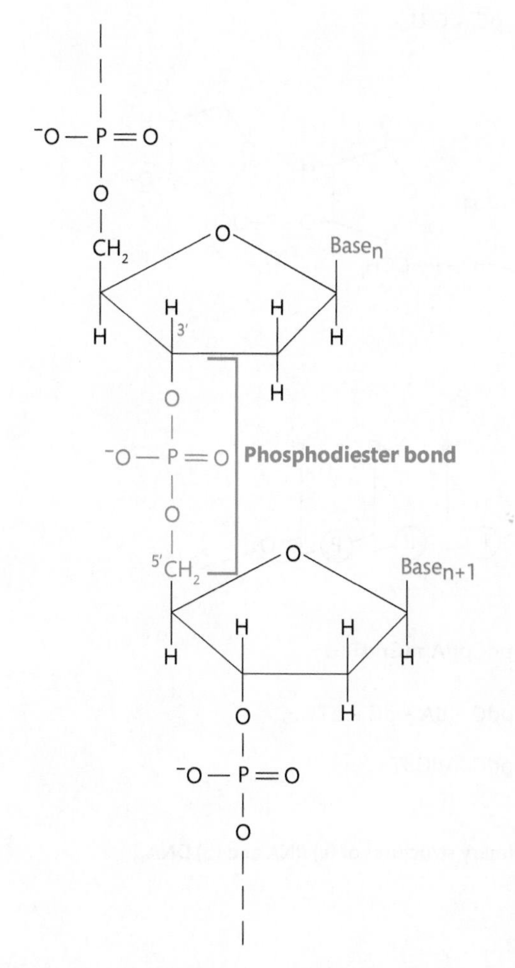

FIGURE 14.4
Structure of a phosphodiester link.

(a)

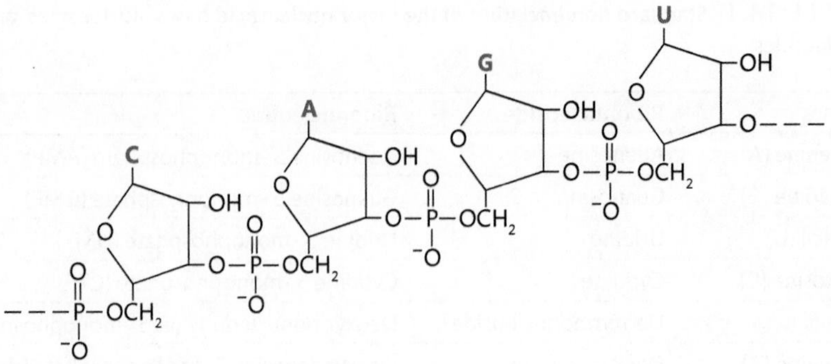

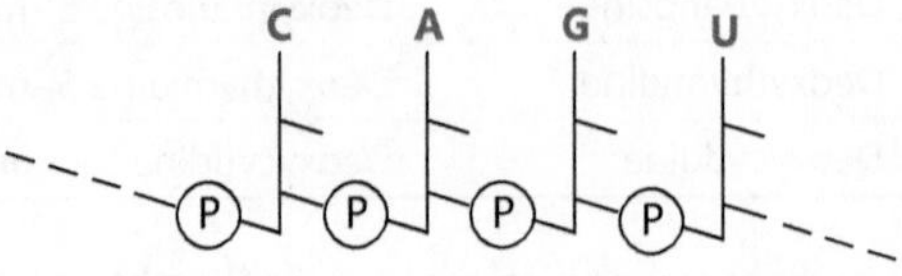

. . . pCpApGpU . . .

. . . pC – A – G – U . . .

. . . pC A G U . . .

(b)

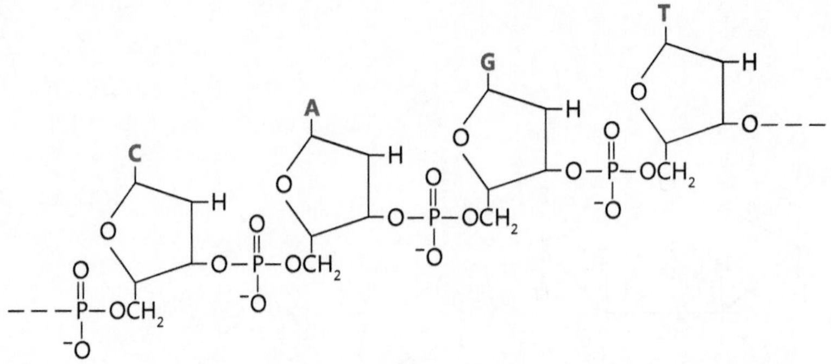

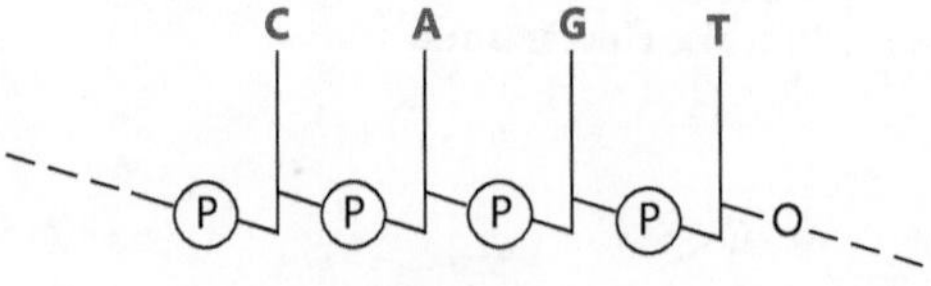

. . . pdCpdApdGpdT . . .

. . . pdC – dA – dG – dT . . .

. . . pdCdAdGdT . . .

FIGURE 14.5

Various conventions used in showing the primary structures of (a) RNA and (b) DNA.

Introduction info

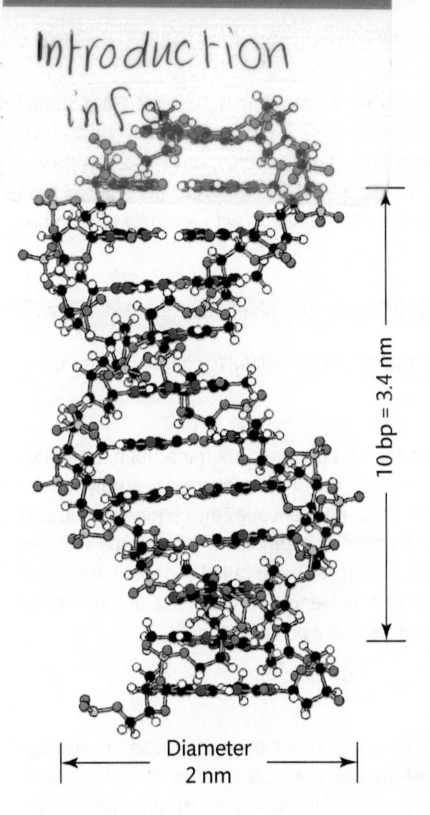

FIGURE 14.6
Computer-generated model of double-stranded DNA. bp: base pairs.

Complementary base pairing is the basis for replicating genes and for allowing their expression. During replication, each strand of DNA acts as a separate template for the synthesis of a new complementary strand. **DNA-dependent DNA polymerases** (usually abbreviated to DNA polymerase or DNA pol), assisted by a variety of other proteins, copy each template to give a new complementary strand. Thus, the two new DNA molecules each consist of one old (parental) and one newly synthesized strand; hence, the process is called semi-conservative replication. Molecules of RNA are formed when DNA-dependent RNA polymerases use a single strand of DNA as a template to catalyse the formation of a complementary RNA copy. This process is called **transcription**.

Cross reference

You can learn more about replication and transcription of DNA in Papachristodoulou *et al.* (2014) and Craig *et al.* (2014), which are listed in the Further Reading section.

SELF-CHECK 14.1

A strand of DNA has the sequence ATGCCGTTAGGAGTA. Give the complementary sequence to it, written in the 5' to 3' direction.

All chromosomes contain a single molecule of DNA. In prokaryotic cells, the genetic material usually occurs as a single chromosome containing a single circular DNA molecule. However, mutual repulsion between the negative charges of the phosphate groups opposes the bending of the molecule. In addition, bacterial chromosomes have M_r of at least 2×10^9, and are over 1 mm in circumference and need to be folded to fit into bacterial cells that are only approximately 1 µm diameter. The charges on the phosphate groups are neutralized by binding of basic, that is positively charged, proteins to the DNA. This makes the bending and elaborate folding of DNA possible. Bacteria, and some yeasts, which are eukaryotes, also contain extrachromosomal DNA structures called **plasmids**. Like the bacterial chromosome, plasmid DNA is circular and double stranded. However, it is much smaller with M_r of 2×10^6–20×10^6, which corresponds to 3000–30,000 bp. Bacterial plasmids normally contain genetic information for proteins that confer a specialized and sometimes protective phenotype on the organism, often, for example, imparting antibiotic resistance. Plasmids are self-replicating and their DNA is duplicated before every cell division and copies of the plasmid DNA segregate to each daughter cell, assuring continued propagation of the plasmid through successive generations of the host cell.

The paired chromosomes of eukaryotes are much larger than bacterial ones. For example, a human haploid cell contains DNA molecules of total length approximately 3×10^9 nucleotides divided between 23 chromosomes (22 autosomes, and a sex chromosome, X or Y). Each eukaryotic chromosome contains a linear DNA molecule many millimetres long. These can only fit into a nucleus by being even more elaborately folded than bacterial types, following their combination with a variety of chromosomal proteins.

SELF-CHECK 14.2

Calculate the total length of DNA in a diploid human cell. You may find it helpful to consult Figure 14.6.

The sequences of bases in the nucleotides of a gene that encodes a protein are transcribed to eventually form a mRNA molecule. The mRNA is then used in the synthesis of the corresponding polypeptide in a process called translation; the sequence of bases in the mRNA encodes the sequence of amino acid residues in the polypeptide. Translation of mRNA occurs on ribosomes and requires the activities of many other nucleic acids and proteins. Each of the groups of three bases called codons in the mRNA specify each of the 20 types of amino acids used in translation. Hence, the sequence of bases in the gene (DNA) determines the sequence of amino acid residues in the polypeptide.

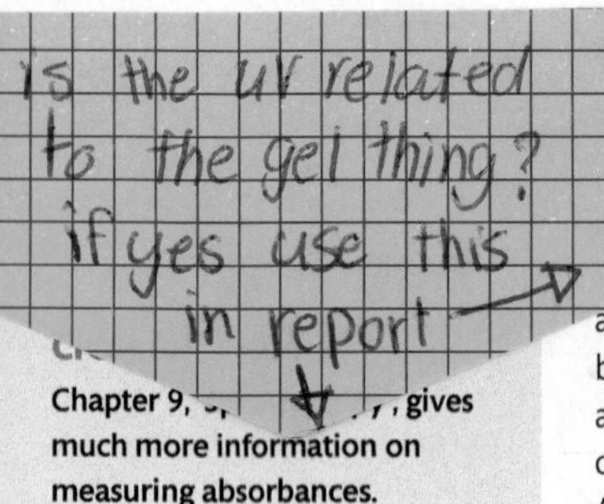

Is the UV related
to the gel thing?
if yes use this
in report

traviolet absorption by nucleic acids

e presence of the purine and pyrimidine bases (A, T, G, C, and U) means nucleic acids absorb sig- icant amounts of light in the ultraviolet (UV) region of the spectrum (see Chapter 9), with maximal sorbance occurring at about 260 nm (Figure 14.7). Hence, solutions of pure double- and single- anded DNA and RNA of concentrations 50 and 40 μg cm^{-3}, respectively, have absorbances of 1.0 at 260 nm. Proteins, the major contaminants in isolated DNA samples, also absorb in the UV region, but maximally at 280 nm (Figure 14.7). Thus, the ratio of the absorbances, A_{260}/A_{280}, is often used to assess the purity of isolated samples of nucleic acids. The typical A_{260}/A_{280} for isolated DNA with little contaminating protein is about 1.8. Lower ratios indicate increased contamination with proteins. The A_{260}/A_{280} ratio of extracted RNA is generally 1.8–2.0.

Chapter 9, ... gives much more information on measuring absorbances.

SELF-CHECK 14.3

A 10 μl sample of a purified DNA solution was diluted to a final volume of 1 cm^3. The A_{260} of the diluted solution was 0.55. What is the concentration of DNA in mg cm^{-3}?

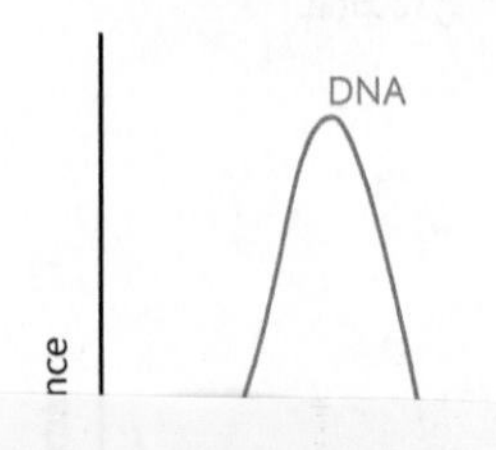

REMEMBER

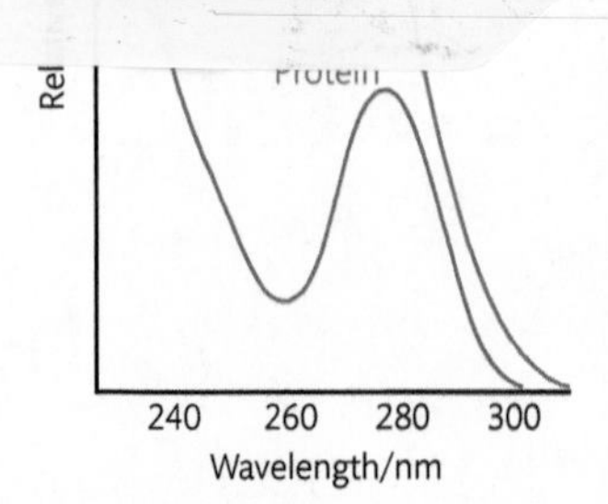

FIGURE 14.7
Ultraviolet absorption maxima of DNA and proteins.

Separating the double helical structure of purified DNA with denaturing agents, such as heat, alkali, and organic solvents, to form two single strands causes its absorption at 260 nm to increase markedly. This hyperchromic effect arises because the transition from an ordered double-helical structure to unpaired single stranded DNA means base–base interactions are now at a minimum. The resulting change in the properties of the rings of the bases means they can absorb more UV light. If the strands are allowed to reassociate, or **reanneal**, to give the original double helical structure, the UV absorption returns to its initial value. Thus, the processes of DNA disassociation and reannealing under various experimental conditions can be followed by monitoring the absorbance at 260 nm.

SELF-CHECK 14.4

Would you expect RNA molecules to show a hyperchromic effect?

14.1.2 Proteins

Proteins are polymers of amino acid residues, which all have a common structure but differ from one another in the nature of their side chains, which you can see in Figure 14.8.

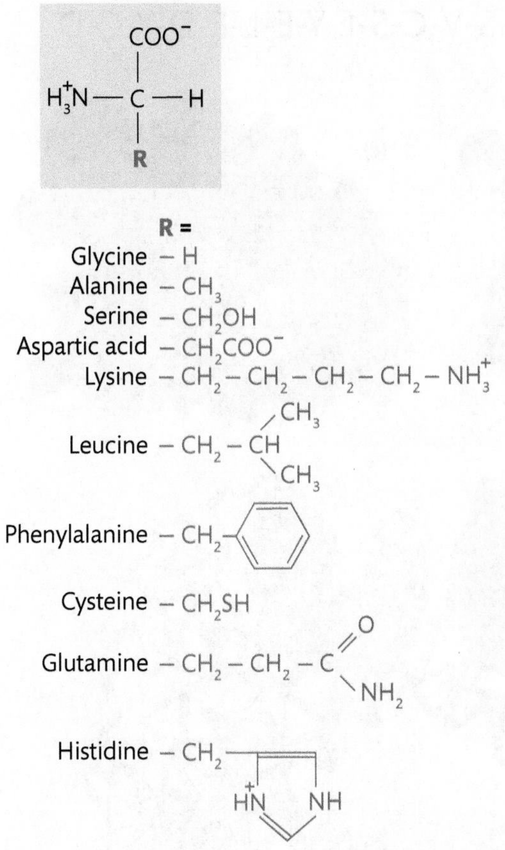

COO^-
H_3^+N-C-H
R

R =
Glycine $-H$
Alanine $-CH_3$
Serine $-CH_2OH$
Aspartic acid $-CH_2COO^-$
Lysine $-CH_2-CH_2-CH_2-CH_2-NH_3^+$
Leucine $-CH_2-CH(CH_3)_2$
Phenylalanine $-CH_2-$(benzene ring)
Cysteine $-CH_2SH$
Glutamine $-CH_2-CH_2-C(=O)NH_2$
Histidine $-CH_2-$(imidazole ring, $H\overset{+}{N}$, NH)

FIGURE 14.8
Structures of some α amino acids.

Individually, proteins have more complex structures than nucleic acids because they are formed from 20 different types of amino acids, which are linked together by peptide bonds (Figure 14.9).

The structures of most proteins are subdivided arbitrarily into hierarchical levels. The four main levels are the primary, secondary, tertiary, and quaternary structures, which you can see illustrated in Figure 14.10, although other structural levels have also been described. Note, like any classification, this is an artificial division. The primary structure of a protein is the sequence of its amino acid residues or side chains. The secondary structure is the regions of local folding of the polypeptide chain.

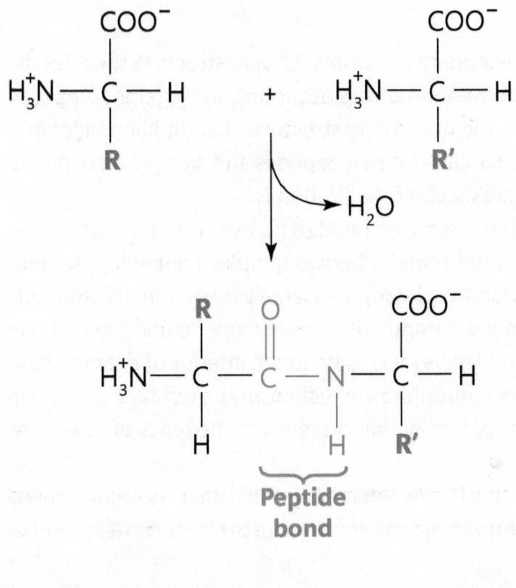

$H_3^+N-CH(R)-COO^- + H_3^+N-CH(R')-COO^- \rightarrow H_3^+N-CH(R)-C(=O)-NH-CH(R')-COO^- + H_2O$

Peptide bond

FIGURE 14.9
Formation of a peptide bond by condensing two amino acids.

(a) G-I-V-E-E-C-C-A-S-V-C-S-L-Y-E-L-E-D-Y-C-D

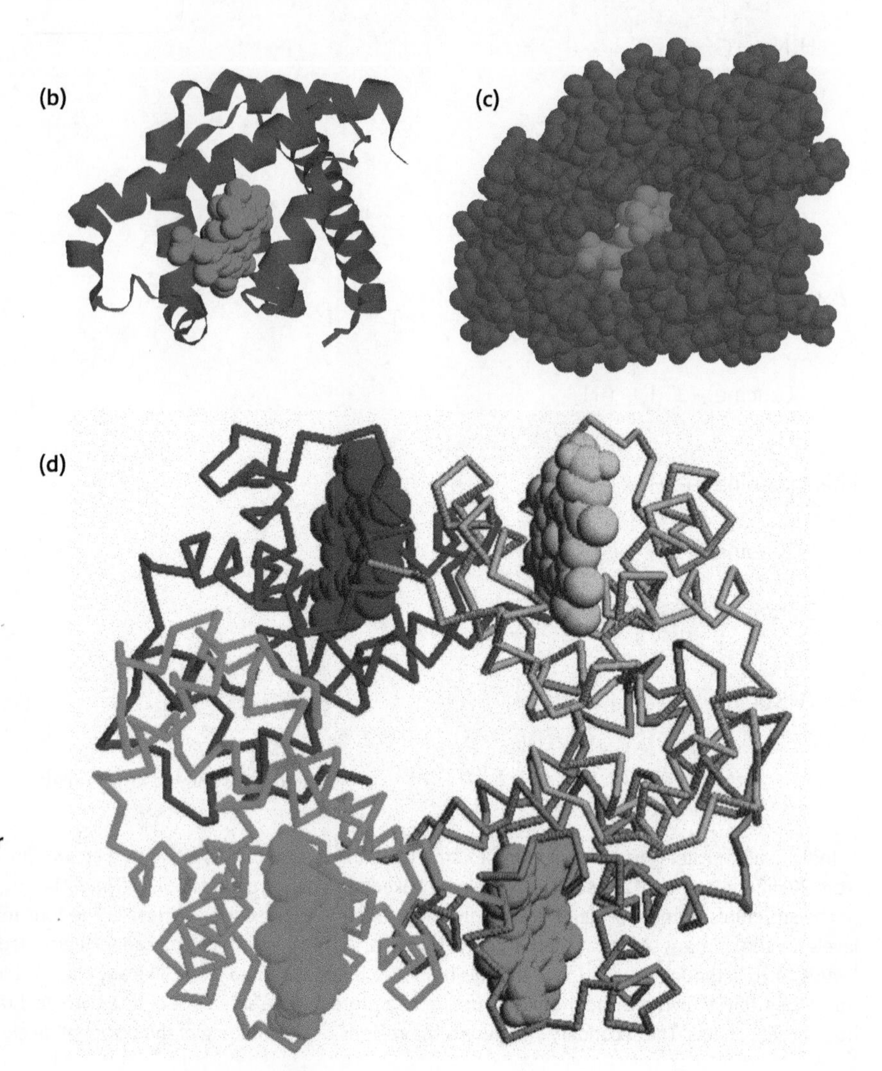

FIGURE 14.10

Examples of the four main structural levels of proteins. (a) Primary structure of the A chain of insulin denoted using the standard one-letter abbreviations for each of the amino acid residues. (b) Secondary structure (largely α-helical) of a myoglobin molecule. The red portion is a haem group. (c) Tertiary structure of myoglobin. (d) Quaternary structure ($\alpha_2\beta_2$) of haemoglobin.

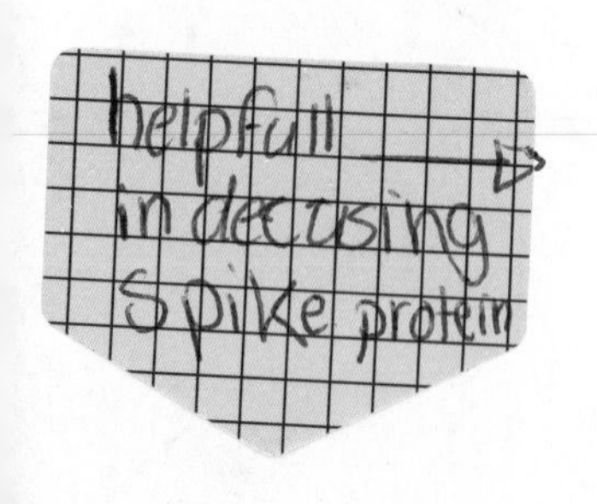

The famous α helix and β sheet are examples of secondary structures. Tertiary structure specifies the positions of all the atoms in the polypeptide. Quaternary structures occur only in proteins composed of more than one polypeptide chain. For example, the quaternary structure of adult haemoglobin is $\alpha_2\beta_2$, meaning its molecules are composed of two so-called α polypeptides and two β polypeptides. Individual polypeptides of a quaternary structure are usually called subunits.

Protein molecules are only marginally stable. The energy of a folded (native) protein is only relatively slightly below that of an unfolded or denatured protein; hence, samples containing proteins have to be treated with care. The structure of any protein is determined largely by its primary structure, that is, the sequence of its side chains. In an aqueous environment, the polypeptide folds so that side chains that cannot interact favourably with water molecules segregate to the interior of the molecule. This is referred to as the **hydrophobic effect**. The hydrophobic effect ensures each type of protein has a unique conformation that determines the properties of that protein and the types of molecules with which it can interact.

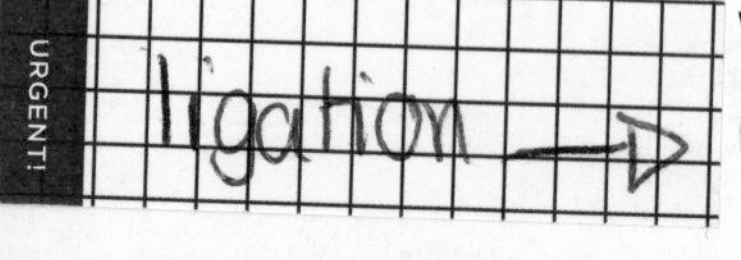

Proteins generally function by recognizing and specifically interacting with other molecules called ligands. Ligands are normally small, that is, they are not macromolecules, but proteins can also bind to relatively small areas of other proteins and nucleic acids. Protein molecules have discrete sites on their surface that are complementary to the ligand, which allows the protein to recognize and specifically bind to it in a reversible manner. The protein–ligand complex can dissociate back to the free molecules or it can induce a change in the conformation of the protein, which induces some biological event to occur. For example, in the case of enzymes, the ligand would be a substrate and the event would be the catalysis of a reaction that converts it to product. If the protein were a receptor and the ligand a hormone, then changes would be stimulated in target tissues.

14.2 Molecular biology-based techniques

The complementary binding of pairs of bases in nucleic acids offers a convenient way of recognizing and isolating specific base sequences within a fragment of a DNA or RNA molecule or of isolated nucleic acid molecules. Similarly, proteins specifically recognize and bind to a variety of different types of molecules. The abilities of macromolecules to bind complementary substances is the basis of many routine clinical techniques, for example fluorescence *in situ* hybridization (FISH) and enzyme-linked immunosorbent assays (ELISAs), as we will describe later (Section 14.8).

Throughout this chapter you will encounter many molecular biology techniques where the specific binding properties of macromolecules has been exploited in recognizing, isolating, and analysing molecules in biological and clinical samples. In all cases, the specificity of recognition is such that the techniques are extremely sensitive. They therefore require relatively small amounts of sample compared with more traditional chemical techniques and can be performed on volumes as small as a few microlitres; hence, minimal amounts of reagents are required, saving the health services' money. Experimental molecular biology kits are also available (Box 14.1).

14.3 Isolation of nucleic acids

The study of nucleic acids and their analysis or use in clinically related procedures, for example preparing the **probes** (Box 14.2) used in FISH described in Section 14.8, often involves their isolation from cells and tissue samples. The nucleic acid purified for clinical purposes is usually DNA rather than RNA. Chromosomal DNA can be purified from bacterial or eukaryotic cells and this is generally described as *total* DNA. However, plasmids (Section 14.1) can also be isolated from bacterial and some types of eukaryotic cells, such as yeasts. In general, the separation of plasmids follows the same basic protocols

BOX 14.1 Experimental molecular biology kits

Many of the techniques of molecular biology are well established and can be performed with experimental kits that can be purchased from commercial suppliers. Indeed, experimental kits are central to research in molecular biology and a number of the protocols described in this chapter make use of them. For example, you can buy kits to extract DNA or RNA from tissues (Section 14.3) and synthesize complementary DNA (cDNA) from the isolated mRNA (Section 14.9). Such kits provide all the reagents and other materials needed for the particular experiment, descriptions of the underlying method, and easy to read and follow protocols. Reasonably 'good' results can usually be obtained using a kit. Indeed, in many cases, experimental results are easier to obtain than with the traditional method of purchasing the materials separately and preparing individual solutions of the reagents. This saves time, but kits are generally the more expensive option. They are also inflexible in that their protocols are difficult to modify for different circumstances. Furthermore, their extensive use carries the danger of lowering one's general laboratory skills. You will never fully understand an experimental method if your depth of knowledge is that 3 μL of the reagent in the red tube must be added to 5 μL of sample and then heated to 30°C for 5 minutes! A thorough understanding of any experimental procedure will enable you to use it to its full effect, know what to do when it goes wrong, as it will on some occasion, and appreciate when a result is aberrant or outside the experimental limits, artefactual, or simply incorrect. The best advice is to learn and understand the technique before using the corresponding kit.

BOX 14.2 Probes

Probes are labelled molecules that are complementary to the molecules one is studying. For example, if a solution of DNA is heated at 100°C or exposed to pH values above 13 the complementary bp are disrupted and the double helix dissociates into its two single strands. Complementary single strands of DNA, however, readily reform double helices by a process called annealing, hybridization, or DNA renaturation, if they are allowed sufficient time at temperature at or below 65°C. Hybridization reactions can also occur between any two single-stranded complementary polynucleotides forming DNA-DNA, RNA-RNA homoduplexes, or RNA-DNA heteroduplexes. These specific hybridization reactions are the basis of sensitive methods widely used to detect and characterize specific nucleotide sequences in both RNA and DNA in samples. In such cases, the probe could be a single-stranded DNA molecule that has been labelled in some way to identify and quantify the resulting hybrid (sample nucleic acid–DNA probe).

as those used to obtain chromosomal DNA, although at some stage, procedures must be used to separate the plasmid from the bulk of the cellular DNA. This section will discuss general methods for isolation and purification of chromosomal DNA from bacteria and eukaryotic cells; the isolation of plasmids is described in Section 14.9.

Chromosomal DNA is difficult to isolate in an intact and undamaged form because of the large size and fragile nature. However, several isolation procedures have been developed to purify it from cells. Indeed, a number of them are commercially available in kit form. Used correctly, these procedures produce preparations of DNA that are stable and of large M_r, and relatively free of contaminating RNA and proteins. Whatever its source, a number of points must be considered in the isolation of DNA. These include:

- pH
- temperature
- ionic strength
- nuclease activities
- mechanical stress.

The isolation of DNA requires it be released into a medium of appropriate pH. This is necessary to preserve the interactions and chemical bonds that stabilize the molecules. For example, hydrogen bonding between the complementary strands is stable at pH 4–10, while the phosphodiester linkages become unstable below pH 3 and above 12. Furthermore, the *N*-glycosidic bonds joining purine bases (adenine and guanine) to the deoxyribose sugars are hydrolysed at pH values of 3 or less. Phosphodiester linkages and *N*-glycosidic bonds are stable at temperatures up to 100°C. However, temperatures of 80–90°C destabilize the hydrogen bonds holding the double helix intact causing it to unwind. Care must be taken in choosing an appropriate ionic strength for solutions used in DNA extractions. DNA is most stable and soluble in salt solutions. However, concentrations less than 0.05 mol dm^{-3} weakens the hydrogen bonding between complementary strands. Enzymes, such as deoxyribonucleases or DNases (see Section 14.4) are present in all cells and can degrade DNA during its purification if their action is not inhibited.

Native DNA molecules are highly asymmetric, being relatively long (mm to cm), but only 2 nm in diameter. The extraction of DNA involves removing the stabilizing basic proteins, which leaves the DNA prone to breakage by shearing forces generated by routine laboratory procedures, such as grinding, shaking, stirring, and pipetting. Hence, all extraction procedures must be performed with care. It is not, however, possible to isolate chromosomal DNA without fragmenting the molecules: isolation procedures always result in preparations containing DNA molecules whose lengths are much shorter than those found in chromosomes. However, damage to the secondary (double helical) structure does not usually occur.

The isolation of DNA and purification from cells and tissues are essential for many other molecular biological techniques. A number of methods have been developed for extracting and purifying DNA from different cells, tissues, and organisms, which are often modified by different laboratories to accommodate different types of samples. Here, we will give simplified protocols used in isolating DNA from bacteria and animal samples.

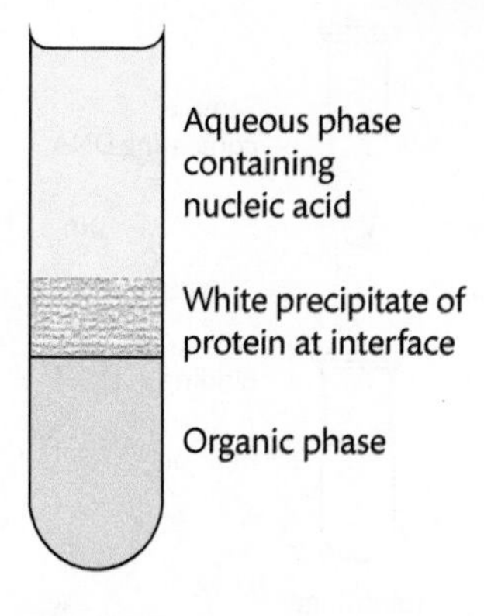

FIGURE 14.11
Final centrifugation stage in the extraction of DNA.

14.3.1 DNA extraction from bacteria

DNA can be extracted from bacteria following the generalized procedure outlined below. Its isolation from specific species of microorganisms may require modifications to the procedures.

The cell wall and cell surface membrane of the bacterial cells must be disrupted, often using lysozyme and sodium dodecyl sulfate (SDS), to release the DNA. Lysozyme catalyses the hydrolysis of glycosidic bonds of cell wall peptidoglycans and lyses the cell wall, while the SDS disrupts the hydrophobic interactions that stabilize the membrane. Hence, DNA, together with other cellular components, is released into a medium in which it is soluble and protected from degradation. The medium of choice is usually a saline solution, buffered to pH 8, that contains ethylenediaminetetraacetic acid (EDTA). The DNA is soluble in the salt solution. The EDTA chelates divalent metal ions, such as Ca^{2+}, Mg^{2+}, and Mn^{2+}, removing them from free solution so they cannot form salts with the phosphate groups of DNA and so inhibits the DNases that require their presence for activity. The relatively high pH also reduces nuclease activities. A secondary action of SDS is to act as a denaturant of DNases and other proteins.

Basic proteins bound to native DNA must be dissociated from it during the extraction. The presence of detergent, the high concentration of salt, or the addition of sodium perchlorate to the medium, and its mildly alkaline (pH 8) nature all reduce electrostatic interaction between the two. The DNA can now be isolated from other soluble cellular components by removing the proteins and then precipitating the DNA in a relatively pure form. The solution is deproteinized by adding a mixture of chloroform–isoamyl alcohol followed by centrifugation. Following centrifugation, three layers are produced, as shown in Figure 14.11: an upper aqueous phase that contains the DNA, a lower organic layer, and a compact band of denatured protein at the interface between the aqueous and organic phases. Chloroform causes surface denaturation of proteins. Isoamyl alcohol is used because it reduces foaming and stabilizes the interface between the aqueous phase and the organic phase where the protein collects. The upper aqueous phase containing nucleic acids is removed from the centrifugation tube. The ionic nature of DNA is exploited in its purification because the addition of an organic solvent to the aqueous medium reduces its polarity so it is no longer a solvent for nucleic acids. Thus, the addition of ethanol precipitates the DNA as threadlike material that can easily be collected from the medium. Any remaining proteins contaminating the isolated DNA are removed by dissolving it in a saline medium and repeating the chloroform–isoamyl alcohol treatment until denatured protein no longer collects at the interface. While RNA does not normally co-precipitate with DNA during these procedures, it could still be present as a minor contaminant. Ribonucleases (RNases) may be added to the preparations during the procedure after the first or second deproteinization step to digest any RNA present. Its removal sometimes makes it possible to remove additional proteins using chloroform–isoamyl alcohol.

If highly purified DNA is required, several deproteinization and alcohol precipitation steps may be necessary. In general, up to 50% of cellular DNA can be recovered by this procedure, with an average yield of 1–2 mg DNA per gram of wet packed cells.

Phenol and chloroform are potentially dangerous materials and their use in extracting DNA from samples is often time consuming. Extraction procedures based on the reversible binding of DNA to columns of silica overcome these disadvantages because phenol and chloroform are not used. Thus, these methods are safer than the organic solvent extraction methods described previously. They are also easier to perform because there is less handling and manipulation of the sample. Such methods rely on the property of DNA to bind to silica in the presence of high concentrations of chaotropic salts, such as sodium iodide, guanidine hydrochloride, or sodium perchlorate, but to be released when their concentrations are reduced. The procedure is outlined in Figure 14.12. The column is contained within a small tube that can be centrifuged in a microfuge to drive the various solutions through it, speeding up the purification considerably. High concentrations of salts dehydrate the DNA and favour the formation of hydrogen bonds between it and the silica. Thus, if the DNA is extracted in a medium of high salt concentration, the sample can be added directly to the column. The column is then extensively washed with alcohol-based solutions to remove the impurities. The DNA can then be recovered in a pure form by washing the column with a buffer containing a low concentration of salt, for example a 2-amino-2-hydroxymethylpropane-1,3-diol-(Tris)–EDTA buffer or even nuclease-free water.

Cross reference

Centrifuges, including microfuges, are described in Chapter 10, 'Centrifugation'.

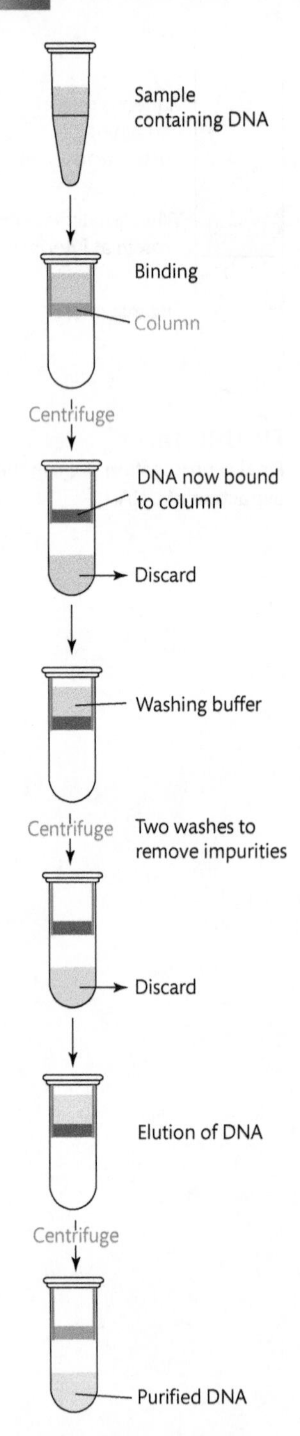

FIGURE 14.12
Column-based purification of DNA. See text for details.

14.3.2 DNA extraction from animal cells

The major steps for isolating DNA from animal cells, such as clinical samples, resemble those used for bacteria described above. The tissue must be disrupted, the cellular components solubilized, and the DNA separated from the other released cellular components. Care in the initial collection of animal tissue samples greatly reduces the problems encountered during DNA isolation. Ideally, samples should be immediately preserved in a freezer at –80°C or in liquid nitrogen. Excellent samples of DNA can be prepared from blood, heart, liver, kidney, stomach, intestine, cerebrospinal fluid, and skeletal muscle. Clinical samples are often blood and its cells should be separated from the plasma prior to freezing.

The simplest means of disrupting cells to isolate DNA is to boil them in water or to use an alkaline extraction medium. High temperatures (100°C) or pH values above 9 disrupt cell membranes, releasing the DNA into solution. However, the DNA obtained by these methods is denatured, but active degradation in low ionic strength solutions is stimulated by metal ions also released from the cells. Chelating resins have been used to remove metal ions from solutions, thus making feasible the rapid isolation of DNA. It should also be noted that because of the high alkalinity (pH 10–11) of the chelating resin, samples prepared with it may degrade much more rapidly than those prepared in other ways such as isolation using proteinase K with detergent SDS or with formamide. However, denatured DNA is amenable to some molecular biology techniques, for example polymerase chain reaction, but not others such as **recombinant DNA** cloning (Section 14.9), Southern **blotting**, or restriction fragment analysis (Section 14.7).

Extraction buffers used to isolate animal DNA are usually based on Tris and contain EDTA to chelate divalent cations such as Ca^{2+} and Mg^{2+} and thus inhibit nuclease activity, and a Na^{+} or K^{+} salt to stabilize the nucleic acids in an isotonic medium. A protease, usually proteinase K, is often added to digest cellular proteins. An anionic detergent, usually SDS, is normally included to solubilize cellular membranes and denature proteins, as is formamide, an ionizing solvent that dissociates DNA–protein complexes and denatures the released proteins. However, its presence significantly reduces the activity of proteinase K. Both the pH of the isolation buffer and the specific protective agents or detergents included may need to be modified to give optimal conditions for specific samples. Once released into solution, the DNA can then be purified by the methods outlined earlier for bacterial samples.

Robotic workstations are available that automatically purify genomic DNA from tissue samples. Figure 14.13 shows a typical instrument that can, for example, simultaneously process eight or 16 samples of 5–10 cm^3 of whole blood with yields of up to 350–500 μg of DNA per sample. The DNA obtained is sufficiently pure for its immediate use in many of the techniques described in the remainder of this chapter.

FIGURE 14.13
Robotic workstation capable of extracting DNA from numerous samples simultaneously.
Courtesy of S. Smith, Regional Genetics Services, St Mary's Hospital, Manchester, UK.

14.3.3 Isolation and purification of RNA

Typical mammalian cells contain approximately 10^{-5} µg of RNA, of which 80–85% is rRNA, the remaining 15–20% is largely low M_r tRNA and small nuclear RNA molecules, which can be isolated by density gradient centrifugation, anion-exchange or high-performance liquid chromatography, or gel electrophoresis. mRNA constitutes 1–5% of the total cellular RNA and is heterogeneous in both size, from several hundred bases to many kilobases in length, and sequence. However, most eukaryotic mRNAs have a tract of several 100 adenylate residues at their 3' termini, the so-called poly(A) tail, that is exploited to allow their purification by affinity chromatography with a column of oligo(dT) cellulose. The yield of total RNA depends on the tissue or cell source.

RNAs are chemically more reactive than DNA because they have hydroxyl groups at both the 2' and 3' positions. They are therefore more susceptible to hydrolysis by contaminating nucleases. Ribonucleases are released from cells during purification procedures and are also present on the skin. Unlike many DNases, RNases do not require divalent cations for activity and are not inactivated by EDTA or other chelators in buffer solutions, and are often resistant to prolonged boiling and mild denaturants. Thus, glassware and material are often autoclaved to denature any RNases present. Disposable gloves must always be worn. To avoid contamination with RNases during work with RNA, instruments, such as automatic pipettes should be reserved for that specific use and all procedures carried out in restricted areas or fume cupboards.

Cross reference

You can read about centrifugation, chromatography, and gel electrophoresis in Chapters 10, 11, and 12, respectively.

When isolating RNA from tissues such as pancreas or the gastrointestinal tract, that are rich in digestive enzymes, it is best to cut the dissected tissue into small pieces and add them quickly to liquid nitrogen. These fragments are stored in a freezer at about −70°C or used immediately for RNA extraction. Extraction requires the cells or tissues to be disrupted in denaturing solutions that contain phenol, and guanidinium thiocyanate or guanidinium isothiocyanate, that help disrupt the cells while solubilizing their components, and denature endogenous RNases simultaneously. The addition of chloroform followed by centrifugation separates the sample into a colourless upper aqueous phase, an interphase, and a lower organic phase. RNA remains exclusively in the aqueous phase and can be recovered by precipitating it with alcohol and then washing the precipitate with 75% ethanol before redissolving it in RNase-free water or 0.5% SDS solution.

The Health & Safety Box 'Precautions when isolating DNA' highlights some of the safety procedures necessary when using the harsh chemicals used in extracting DNA. Naturally, similar precautions are essential when isolating RNA from samples because, for example, guanidinium thiocyanate and isopropanol are harmful if ingested or absorbed through the skin or if the alcoholic fumes are inhaled.

HEALTH & SAFETY

Precautions when isolating DNA

- **Experimental work with bacterial cells presents a potential biohazard. Some bacteria recommended for use, for example strains of *Salmonella* and *Haemophilus*, are pathogenic and can cause disease. Eukaryotic tissues carry the danger of harbouring pathogens.**
- **Benches and other work surfaces must be washed with a 10% solution of bleach before and after DNA extractions. Do not pipette any solutions by mouth. Always wear a suitable laboratory coat, protective goggles, and disposable gloves, and wash hands well with hot water and soap when leaving the laboratory, and before eating or drinking.**
- **Chloroform is a volatile liquid that irritates the skin, eyes, mucous membranes, and respiratory tract. It is also carcinogenic and can cause liver and kidney damage. Work involving chloroform must always be performed in a fume cupboard.**
- **Ethanol is volatile and flammable. There must be no naked flames in the laboratory during its use.**
- **Isoamyl alcohol can cause tissue damage following inhalation, ingestion, or topical absorption.**
- **Lysozyme is an hydrolytic enzyme that can damage mucous membranes.**
- **SDS is an irritating toxic detergent that can severely damage eyes. It can also cause tissue damage following inhalation of its dry powder form or ingestion of the powder or solution.**

14.4 Hydrolysis of nucleic acids and restriction endonucleases

If you re-examine the structures of Rib and dRib in Figure 14.2 you will see that the only difference in their structures is the presence of the 2'-OH on the former but only a 2'-H on dRib. The name, in fact, says this: *deoxy* means *without oxygen*. The presence of this additional hydroxyl group means RNA is much more susceptible to alkaline hydrolysis than DNA. Mild acid hydrolysis releases purine bases from DNA.

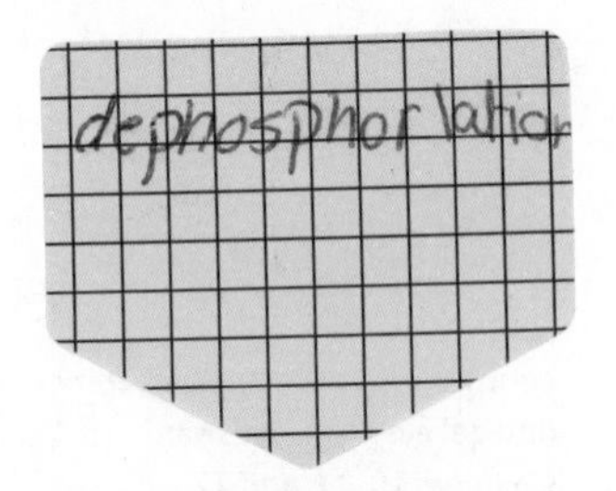

The alkaline or acid hydrolyses of nucleic acids degrades susceptible bonds in a random fashion. However, enzymes are available that can catalyse the hydrolysis of nucleic acids in a more specific manner. These enzymes are called nucleases or phosphodiesterases, and they catalyse the hydrolysis of phosphodiester bonds, although they differ considerably in their specificities. Exonucleases catalyse the sequential removal of terminal nucleotides; endonucleases hydrolyse internal phosphodiester bonds. Some nucleases can only use DNA or RNA, respectively, as substrates, while others can degrade both. Some nucleases are only active against single-stranded nucleic acids, but others can catalyse the hydrolysis of double-stranded structures. Look again at the structure of the phosphodiester link shown in Figure 14.4. Note that it can be cut on one of two sides. If the hydrolysis occurs on the 3' or *a* side then the product is a 5'-phosphate derivative; hydrolysis on the 5' or *b* side gives a 3' prime product. Figure 14.14 summarizes both hydrolyses. Again, different nucleases differ in *a* or *b* specificity. Some nucleases can only catalyse the hydrolysis of phosphodiester bonds associated with nucleotides that contain only purine or only pyrimidine bases. **Restriction endonucleases** (REs) are even more specific and only catalyse the hydrolysis of specific phosphodiester bonds within or near to short specific base sequences in double-stranded DNA. REs are classified into three different groups: I, II, and III. Type II REs are the most widely used in molecular biology.

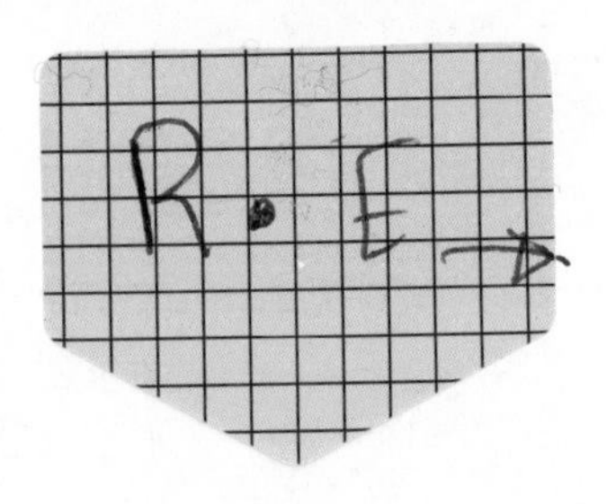

FIGURE 14.14

Definition of *a* and *b* hydrolyses of a nucleic acid. Note how each type of hydrolysis form a different product.

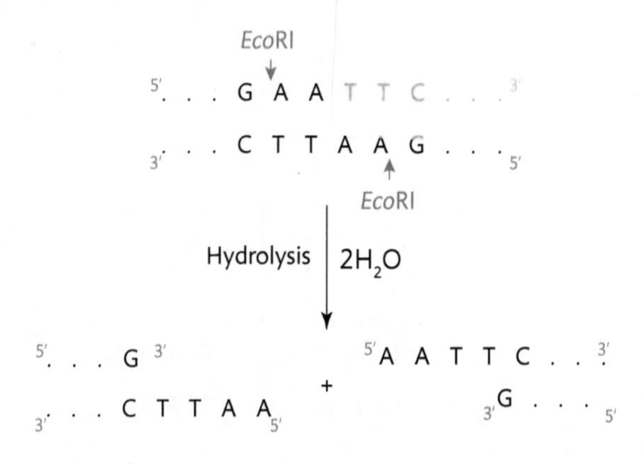

FIGURE 14.15
Palindrome recognized and hydrolysed by the restriction enzyme, *Eco*RI.

14.4.1 Restriction enzymes

Type II REs are endonucleases produced by bacteria that typically recognize specific 4–8 bp sequences, called restriction sites, and then hydrolyse both strands of the DNA at this site at the same position. For example, the RE *Eco*RI recognizes the sequence GAATTC and hydrolyses it at the positions indicated in Figure 14.15. If you examine Table 14.2, which lists a number of REs, notice how the sites of hydrolysis occur at corresponding bonds within identical sequences of bases in both of the strands when they are read in the 5' → 3' direction. Restriction sites are therefore inverted repeats and are called **palindromes** because the sequence of the site is the same on each DNA strand. Different REs recognize and hydrolyse different palindromes. Different species of bacteria synthesize different REs, which protect them from viruses by degrading viral DNA. The bacterial DNA is not degraded because each RE has a corresponding enzyme that recognizes the same palindrome and modifies it, so it is no longer recognized by the RE (Box 14.3).

Many restriction enzymes make staggered cuts in the two DNA strands at their recognition site and produce products that have a single-stranded portion at each end, as is the case with *Eco*RI. Look carefully at the products of the reaction. Note how they have overlapping strands that are often called 'sticky ends'. Such REs are used extensively in recombinant DNA technology, which we will describe in Section 14.9. However, some REs have actions that form products with blunt cut ends, as you can see is the case with *Eco*RV (Figure 14.16).

SELF-CHECK 14.5

Which REs in Table 14.2, besides *Eco*RI, produce products with sticky ends?

BOX 14.3 Restriction endonuclease: sources and nomenclature

REs are produced by bacterial cells. They function in protecting the bacterium from attack by viruses by degrading the viral DNA. The host bacterial DNA is protected from hydrolysis because each RE has a corresponding methyl transferase with an identical palindrome specificity. Thus, the bacterial palindromes, but not the viral, have methyl ($-CH_3$) groups attached to bases near to the palindrome that prevents their recognition by the REs. Several hundred REs have been isolated from different bacterial species and characterized. The nomenclature adopted for them consists of a three-letter abbreviation that identifies the bacterial source, followed by a letter representing the strain of the species (if necessary), and, lastly, a Roman numeral designating the order of discovery of that enzyme when more than one RE has been isolated from that strain. Thus, *Eco*RI was the first RE to be isolated and characterized from the R strain of *Escherichia coli*.

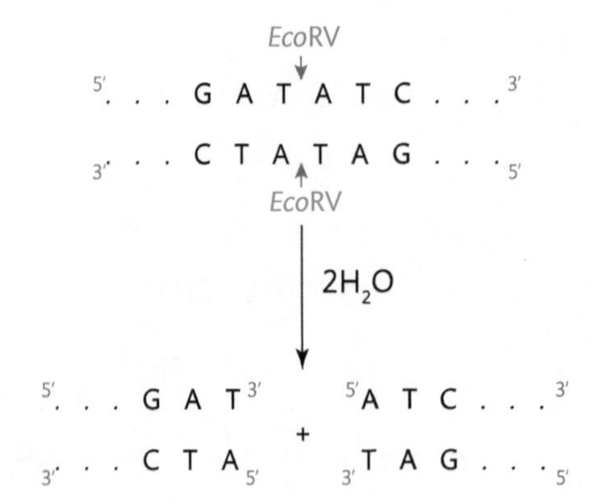

FIGURE 14.16
Reaction catalysed by *EcoRV*.

TABLE 14.2 Examples of restriction endonucleases (REs) and their palindrome specificities (the red arrows indicate the sites of hydrolysis)

Restriction endonuclease (bacterial species)	Palindrome recognized and sites of hydrolysis
AluI (*Arthrobacter luteus*)	↓ . . . AGCT . . . . . . TCGA . . . ↑
BamHI (*Bacillus amyloliquefaciens* H)	↓ . . . GGATCC . . . . . . CCTAGG . . . ↑
EcoRI (*Escherichia coli* RY 13)	↓ . . . GAATTC . . . . . . CTTAAG . . . ↑
EcoRV (*Escherichia coli* J62 pLG74)	↓ . . . GATATC . . . . . . CTATAG . . . ↑
HaeIII (*Haemophilus aegptius*)	↓ . . . GGCC . . . . . . CCGG . . . ↑
HindIII (*Haemophilus influenzae* Rd)	↓ . . . AAGCTT . . . . . . TTCGAA . . . ↑
KpnI (*Klebsiella pneumonia*)	↓ . . . GGTACC . . . . . . CCATGG . . . ↑
MspI (*Moraxella* species)	↓ . . . CCGG . . . . . . GGCC . . . ↑
PstI (*Providencia stuartii* 164)	↓ . . . CTGCAG . . . . . . GACGTC . . . ↑
PvuII (*Proteus vulgaris*)	↓ . . . CAGCTG . . . . . . GTCGAC . . . ↑
TaqI (*Thermus aquaticus*)	↓ . . . TCGA . . . . . . AGCT . . . ↑

Uses of restriction endonucleases

The abilities of REs to recognize palindromes in DNA and restrict hydrolytic events to those sites means they can be used to degrade molecules of DNA into a specific number of fragments in a reproducible manner. Thus, REs are invaluable tools in molecular biology and used extensively to:

- hydrolyse specifically large DNA molecules
- produce physical maps
- convert circular plasmids in to linear DNA molecules
- produce recombinant DNA molecules
- investigate restriction fragment length polymorphisms (RFLPs).

REs can be used to hydrolyse large DNA molecules into smaller fragments that are more amenable to analysis. The fragments are readily separated and their size estimated by agarose gel electrophoresis, as you can see in Figure 14.17 (Section 14.5 and Chapter 12; you may also wish to look at Figure 12.22b). It is unlikely that the sets of fragments produced from any two different DNA molecules using the same RE will be the same. Hence, the pattern of DNA products is likely to be unique and so can be considered a 'fingerprint' of the DNA substrate.

REs can be used to produce physical maps of DNA molecules. An understanding of the genetics, metabolism, and regulation of an organism requires knowledge of the precise arrangement of its genetic material. The construction of a RE map for a DNA molecule provides some of this information as it will show the sites that can be cut by different REs and the number of fragments obtained after each digestion. Two characteristics for every DNA fragment produced by an RE digest are known: the

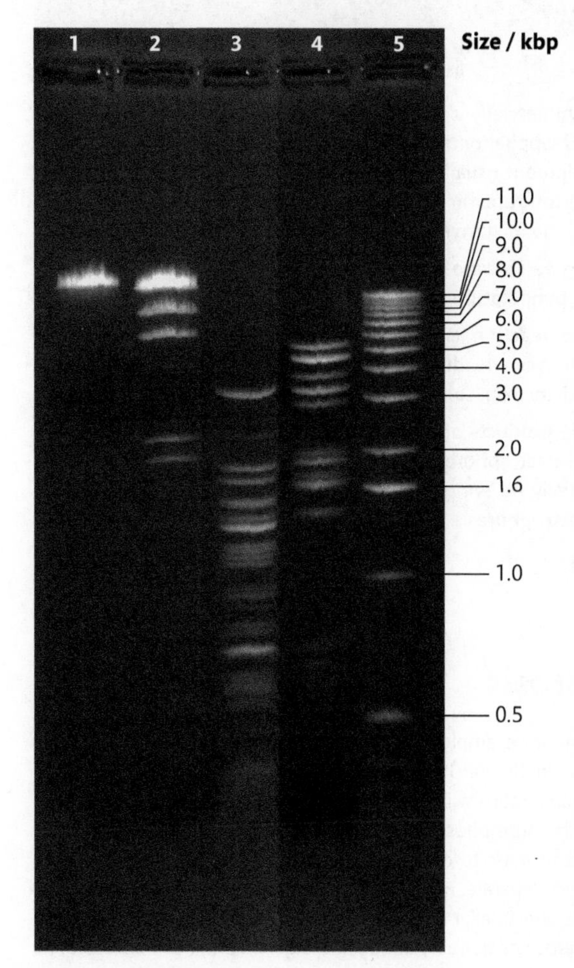

FIGURE 14.17
Separation of DNA fragments by agarose gel electrophoresis. Lane 1 shows a stained sample of DNA isolated from λ bacteriophage. Lanes 2, 3, and 4 show the fragments obtained by digesting the bacteriophage DNA with the restriction endonucleases *Hind*III, *Taq*I, and *Eco*RV, respectively. Lane 5 shows the positions of fragments of DNA of known sizes ('ladder'). Note how each of the enzymes has produced a different set of products by hydrolysing the DNA at different palindrome sites. Courtesy of N. Shaheen, School of Healthcare Science, Manchester Metropolitan University, UK. See also Figure 12.22 (b).

kbp: kilobase pair.

nature of the fragment ends (from the known specificity of the individual RE) and its approximate M_r (from electrophoresis separation). Other information about the fragments, for example the sequence of their bases (Section 14.6), can be obtained using other experimental methods. (See the Method Box 'Experimental procedure for performing a restriction endonuclease digest'.)

The circular DNA molecules present in many plasmids can be converted into linear molecules by the action of some REs. Note that it is necessary that the plasmid have only a single recognition site for the RE being used; otherwise, digestion will fragment it! REs are valuable tools in the construction of recombinant DNA molecules. Thus, once a circular plasmid has been 'linearized', it is capable of binding to other fragments of DNA produced by the actions of the same RE on different samples irrespective of the source of the DNA. This gives molecular biologists a way to cut and splice different pieces of DNA together to produce recombinant DNA molecules, which can be transferred into bacterial cells and cloned as described in Section 14.9. The use of REs not only opens plasmids for insertion of a DNA fragment, but can also destroy a phenotype. For example, if they cleave at a site that specifies resistance to a particular antibiotic, then when the recombinant plasmid is added back to a suspension of bacteria, those that are transformed by taking it up can be identified by their susceptibility to the antibiotic.

The digestion of DNA using REs and the separation of the digest fragments by electrophoresis can be used for clinical purposes, such as the diagnosis of inherited disorders, for example sickle cell anaemia, using RFLPs, which we describe in Section 14.7.

METHOD Experimental procedure for performing a restriction endonuclease digest

- **Commercially available REs are usually dissolved in an appropriate buffer and posted in vials surrounded by dry ice. Disposable gloves must be worn when removing the enzyme container. All REs are heat labile and should quickly be stored at −20°C in a buffer containing 50% glycerol. The solution of enzyme should be removed from the freezer just before use and stored on ice.**
- **A typical RE digest is performed in a microcentrifuge tube (Chapter 10). The mixture normally contains approximately 1 µg or less of DNA and about 1 unit of RE activity in the appropriate incubation buffer.**
- **REs have optimal reaction conditions for activity, especially regarding the composition and pH of the reaction buffer, and incubation temperature. The information for commercially available enzymes is readily available from the supplier or from published literature. The total reaction volume is usually between 20 and 50 µL. The contents of the tube are mixed by gently tapping it or briefly centrifuging the mixture in a microcentrifuge.**
- **Incubations are most often carried out at the recommended temperature, usually 37°C for about 1 hour.**
- **The reaction is then stopped by adding EDTA, which removes the divalent metal ions essential for nuclease activity from solution.**
- **The products of digestion are often analysed by agarose gel electrophoresis (Section 14.5 and Chapter 12). They can usually be temporarily stored at 0–4°C until subjected to electrophoresis.**

14.5 Gel electrophoresis

Electrophoresis is the movement of ions, for example nucleic acids and proteins, in an electric field. You can read more about electrophoresis in Chapter 12. Here, we will briefly highlight the main points. Particles that differ in their mass-to-charge ratios will move at different speeds in the field and thus separate from one another. However, because phosphate groups are regularly spaced along DNA and RNA molecules, they all have the same mass-to-charge ratio and will all move at the same speed for a given electric field and so would not separate. However, performing electrophoresis in agarose or polyacrylamide gels means plasmids and smaller DNA and RNA molecules, and proteins can all be separated. The concentration of agarose or polyacrylamide in the gel determines the sizes of its pores. Using appropriate concentrations therefore allows mixtures of different-sized molecules to be resolved during electrophoresis because the smaller molecules are able to move faster through the gel than larger ones.

In many types of gels, the distance moved by a limited range of nucleic acid fragments is inversely proportional to the logarithm of their sizes. The size of DNA fragments is usually expressed in bp or kilobase pairs (kbp) but can be given in M_r. In general, the lower the concentration of agarose in the gel, the larger the DNA molecules that can be separated. However, gels with concentrations of agarose below 0.3% are too fragile for ordinary use and this therefore forms a practical lower limit. Gels of this concentration allow the analysis of linear double-stranded DNA fragments of 5–60 kbp (M_r up to 150 $\times 10^6$). Gels of 0.8% agarose can separate DNA in the range of 0.5–10 kbp; and 2% agarose is used to separate smaller DNA fragments of 0.1–3.0 kbp. In addition to the DNA molecules of analytical interest, fragments of DNA of known size are also subjected to electrophoresis, forming a so-called 'DNA ladder'. The sizes of unknown fragments can then be estimated by comparing their mobilities with those of known size in the ladder.

Cross reference
Chapter 12, 'Electrophoresis'.

Online Resource Centre
To see an online video demonstrating electrophoresis, log on to www.oxfordtextbooks.co.uk/orc/fbs

Once stained, separated fragments of DNA in the gel are usually detected using ethidium bromide as a stain (Figure 12.21). Ethidium bromide consists of relatively small, planar molecules, which can bind to DNA by intercalating between the stacked bp, which enhances the normally weak fluorescence of the purine and pyrimidine bases by about 25 times (Figure 12.22a). Thus, when irradiated with UV light the separated fragments fluoresce with an intense orange colour giving a sensitive means of detecting the fragments: as little as 1 ng of DNA can be seen (Figure 14.17).

DNA fragments containing 1000 or fewer bp can be separated by SDS polyacrylamide gel electrophoresis (SDS-PAGE). Such gels are also widely used to separate mixtures of proteins, as you can see in Figure 14.17. Normally, proteins are separated following their denaturation using β-mercaptoethanol and SDS. β-Mercaptoethanol reduces the disulfide bonds that help stabilize protein structure and which hold the subunits of some quaternary proteins together, while the negative charges of the SDS bound to the proteins repel each other. In these conditions, protein molecules adopt rod-like structures. Proteins bind SDS in a size-dependent manner and as SDS overwhelms the native charges of the proteins, all proteins in the presence of SDS have similar mass-to-charge ratios. Polyacrylamide, like agarose, forms gels whose pore size is determined by the concentration used. Hence, if the electrophoresis is carried out in a gel also saturated with SDS (hence SDS-PAGE), smaller proteins or very small DNA fragments will migrate faster through it than larger ones and separate from one another. The higher the concentration of acrylamide, the greater the resolution of proteins of lower M_r; the lower the concentration the better the resolution of larger ones. Following separation, proteins can be detected using one of a variety of different staining procedures (Table 12.1). As with agarose electrophoresis of nucleic acids, it is normal to treat and separate a mixture of proteins of known M_r to allow the sizes of unknown proteins to be estimated.

14.6 DNA sequencing

Sequencing of DNA, that is, determining the order of its bases, is mainly performed using the chain termination or 'dideoxy' method first devised by Sanger in the 1970s. Sequencing methods are largely automated and performed by robotic machines whose results are analysed by computers. However, the basic procedure is relatively straightforward. The DNA must be pure and hydrolysed to give a series of suitably sized overlapping pieces. Each fragment is amplified (Section 14.9) before being independently sequenced. As the fragments have overlapping regions of bases, it is possible to assemble them in the appropriate order to give the complete sequence.

The dideoxy method has two key features. One is that it exploits the specificity of DNA polymerase, the enzyme that copies DNA strands to produce a new complementary strand. This enzyme catalyses the extension of a polynucleotide chain by forming new 3'-5' phosphodiester bonds. Bonds are formed when the enzyme adds a new nucleotide to the terminal 3'-OH. The base sequence of the new strand is determined by that of the template. The second feature exploited in the method is the use of dideoxynucleoside triphosphates (ddNTPs). These lack a 3'-OH and, once a ddNTP is added to the end of a newly forming chain, its growth ceases because the absence of a 3'-OH prevents the addition of a further nucleotide. You can see this diagrammatically in Figure 14.18.

Dideoxy sequencing occurs in a reaction mixture that contains a sample of the single stranded DNA to be sequenced (the template), an oligonucleotide primer that is complementary to part of the

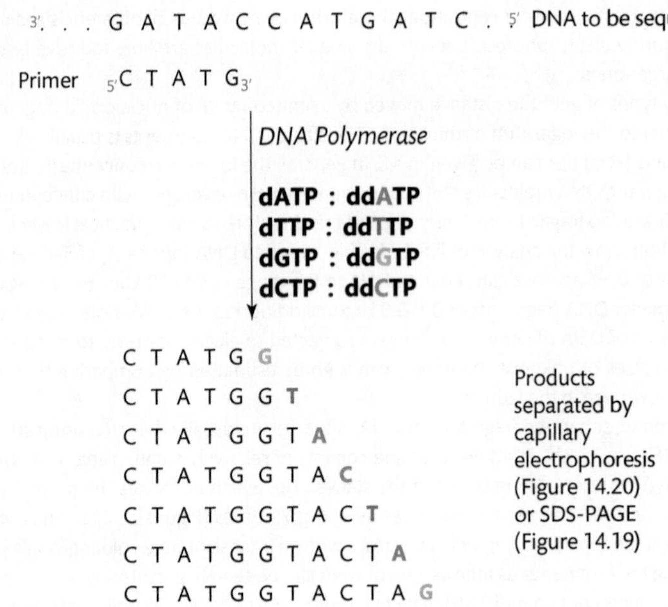

$3'$. . . G A T A C C A T G A T C . $5'$ DNA to be sequenced

Primer $_{5'}$C T A T G$_{3'}$

DNA Polymerase

dATP : ddATP
dTTP : ddTTP
dGTP : ddGTP
dCTP : ddCTP

C T A T G G
C T A T G G T
C T A T G G T A
C T A T G G T A C
C T A T G G T A C T
C T A T G G T A C T A
C T A T G G T A C T A G

Products separated by capillary electrophoresis (Figure 14.20) or SDS-PAGE (Figure 14.19)

FIGURE 14.18
Outline of the dideoxy method for sequencing DNA. See text for details. SDS: sodium dodecyl sulfate.

template, DNA polymerase, the four deoxynucleotides, dATP, dTTP, dGTP, and dCTP, and relatively small amounts of each of the four ddNTPs. The DNA polymerase then catalyses the addition of nucleotides and extends the length of the DNA primer. The newly formed extension is complementary to the template. However, during extension of the primer the enzyme will sometimes use a ddNTP rather than the normal dNTP. When this happens, extension of the growing chain will cease to occur. If the mixture contains an appropriate ratio of the respective ddNTPs to dNTPs, then when the reaction is completed the mixture will contain a set of new but incomplete DNA chains complementary to the template. Extension of each will have started at the same point, but because the ddNTP was selected in a random fashion by the enzyme, they will differ in length by a single base.

The newly formed fragments in the reaction mixture can then be separated on a SDS-PAGE gel containing urea (Section 14.5) or by capillary electrophoresis. Each technique can separate DNA molecules that differ by only a single base in length. The denaturants SDS and urea ensure the DNA chains remain separate during electrophoresis. Following separation, the different lengths of DNA form a densely packed 'ladder' of bands on the gel, as you can see in Figure 14.19. These ladders can be interpreted to give the equivalent sequences of bases in the template. Each of the four ddNTPs is tagged with a different dye. A laser light passes along the gel and each dye fluoresces with a different coloured light that identifies the particular base used in terminating that length, for example yellow for G(uanine), red for T(hymine). The length of the DNA fragments and their colours, which identify each terminal base, are analysed by a computer to give the base sequence of the template.

Cross reference
Chapter 12, 'Electrophoresis'.

SDS-PAGE gels cannot resolve more than 500–800 bases in a single electrophoresis run, which limits sequencing experiments. However, modern DNA sequencers separate the fragments using capillary electrophoresis (Chapter 12) and automatically detect dye fluorescence and record the data output as a fluorescent peak trace (Figure 14.20). The use of such automation means many thousands to millions of bases can be sequenced in an hour.

SELF-CHECK 14.6

Complete the sequence of bases shown in Figure 14.20.

The application of automated methods means that DNA sequencing is largely a routine and rapid procedure. Indeed, the genomes of many of the model organisms studied routinely in the laboratory have been sequenced. These include a number of different yeasts, the nematode *Caenorhabditis elegans*,

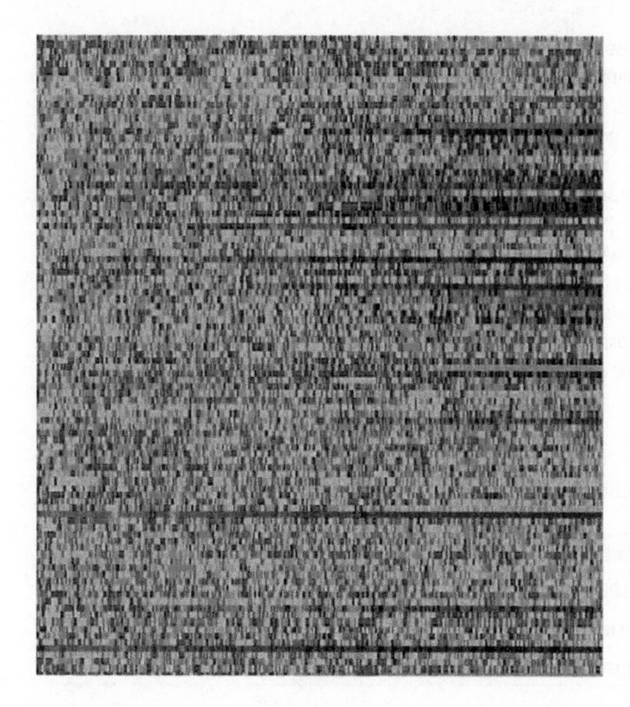

FIGURE 14.19
Portion of a sequencing gel. The colours identifying each base are green for adenine (A), blue, cytosine (C), yellow, guanine (G), and red, thymine (T). See text for details. Courtesy of Wellcome Trust Sanger Institute, Cambridge, UK.

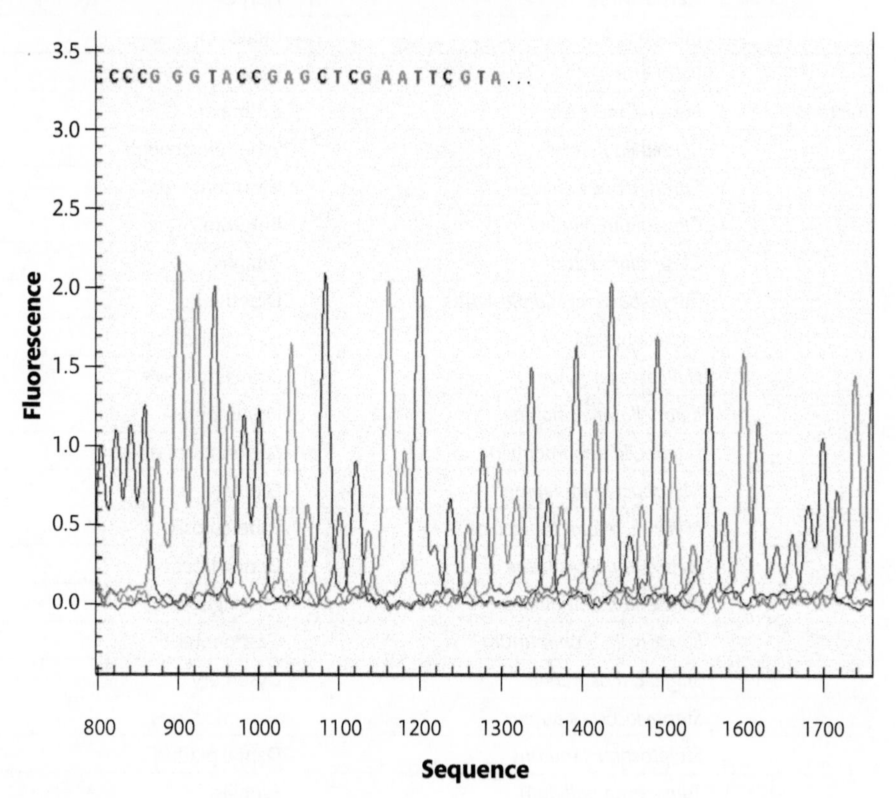

FIGURE 14.20
Trace showing the sequencing output (capillary electrophoresis) from a Beckman Coulter CEQ 8000 automated sequencer. Part of the corresponding base sequence is shown. Courtesy of Dr Patricia Linton, School of Healthcare Science, Manchester Metropolitan University, UK.

the Drosophila fruit fly, the mouse, dog, and chimpanzee, and, finally, yet of great clinical importance, that of humans. The complete sequences of many hundreds of parasitic viruses, for example the smallpox and Epstein–Barr viruses, large numbers of pathogenic bacteria, such as those responsible for cholera, tuberculosis, syphilis, gonorrhoea, Lyme disease, and stomach ulcers, as well as parasitic protozoa like *Plasmodium falciparum*, have also been fully determined (Table 14.3). Knowing these sequences should help produce more rapid and effective diagnostic tools, such as in designing specific probes (Box 14.2) for use in FISH (Section 14.8) and polymerase chain reaction (PCR; Section 14.9) methods. It is anticipated that analysis of their genomes will also provide useful information about their virulence and indicate more effective treatments for the clinical conditions they cause.

Sequencing specific genes can also be used to determine if a person is a carrier of a genetic disease or to confirm a provisional diagnosis, particularly in those conditions where the inheritance does not follow simple Mendelian rules, for example, Fragile X syndrome. This is the most common cause of mental retardation in the UK, affecting one in about 4100 males, but is less common in females, at

TABLE 14.3 Examples of pathogens whose genomes have been completely sequenced

Group	Virus or organism	Disease
Viruses	Enterovirus 100,	Meningitis
	Immunodeficiency viruses 1 and 2	AIDS
	Papillomavirus 16	Cervical cancers
	Herpes virus	Herpes
	Influenza A	Influenza
	Poliovirus	Polio
Bacteria	*Bacillus anthracis*	Anthrax
	Bordetella pertussis	Whooping cough
	Campylobacter jejuni	Gastroenteritis
	Clostridium botulinum	Botulism
	Clostridium tetani	Tetanus
	Corynebacterium diphtheriae	Diphtheria
	Escherichia coli	Food poisoning
	Helicobacter pylori	Stomach ulcers
	Klebsiella pneumoniae	Pneumonia
	Legionella pneumophila	Legionnaires disease
	Mycobacterium leprae	Leprosy
	Mycobacterium tuberculosis	Tuberculosis
	Neisseria gonorrhoeae	Gonorrhoea
	Neisseria meningitidis	Meningitis
	Salmonella typhimurium	Gastroenteritis
	Shigella dysenteriae	Dysentery
	Staphylococcus aureus	Skin infections
	Streptococcus mutans	Dental plaque
	Treponema pallidum	Syphilis
	Vibrio cholerae	Cholera
	Yersinia pestis	Bubonic plague
Protozoa	*Plasmodium falciparum*	Malaria
	Trypanosoma brucei	Sleeping sickness

approximately one in 8000, and who are generally less severely affected. The condition is associated with repeated copies of a trinucleotide, CGG, which occurs in the *FMR1* gene. Most people have a stable copy of *FMR1* containing about 30 repeats. Those with 45–55 copies have an increased chance of passing on an even larger number to their children. Individuals with 55–200 are said to possess a *premutation* because their children may have more than 200 CGG repeats, which is the full mutation associated with Fragile X syndrome. Sequencing *FMR1* or relevant portions of it will directly determine the number of repeats present and confirm the status of the patient.

Next-generation sequencing (NGS) refers to non-Sanger-based high-throughput sequencing technologies. Millions or billions of DNA strands can be sequenced in parallel, yielding substantially more throughput and minimizing the need for the fragment-cloning methods that are often used in Sanger sequencing of genomes. Commonly used NGS technologies include Ion TorrentTM, Roche 454, and Illumina (Solexa) sequencing. The diversity of NGS methods, ranging from DNA sequencing to RNA sequencing has facilitated the investigation of the genetic material of patients. These technologies have accelerated the identification of new disease-causing genes and are now entering molecular diagnostic laboratories.

14.7 Blotting techniques

Electrophoresis is a convenient way of separating nucleic acid and protein molecules. Unfortunately, it is difficult to identify specific molecules in the gel because probes (Box 14.2) do not readily diffuse into gels. Also, gel matrix materials, like agarose or polymerized polyacrylamide, are fragile and can be difficult to handle. These difficulties can be overcome by the common laboratory procedure of blotting, in which nucleic acids or proteins in the gel matrix are transferred from the gel to, for example, a piece of nitrocellulose paper or a nylon membrane. The positions of the molecules of interest on the paper/membrane can be identified and their analysis facilitated because they are bound to the surface of the paper or membrane. Three major types of blotting techniques are used, namely:

- Southern blotting
- Northern blotting
- Western blotting.

The first blotting technique devised was Southern blotting by E. M. Southern, hence its name. The nomenclature of Northern and Western blotting were given with reference to it. These techniques allow the detection of specific DNA fragments, RNA molecules, and individual proteins, respectively. Thus, while genes or fragments of DNA can be detected by Southern blotting, the expression of specific genes is usually monitored by Northern or Western blotting, which detects the specific mRNA or protein molecules produced in the cell. In general, Southern and Northern blotting methods are insufficiently sensitive to detect the amount of nucleic acids obtainable from a single cell. However, PCR, which we describe in Section 14.9, can be used to amplify the nucleic acid in the sample to amounts that are detectable.

14.7.1 Southern blotting

Southern blotting follows gel electrophoresis of double-stranded fragments of DNA, which are separated on the basis of differences in their sizes. The DNA fragments present in the gel are then denatured with alkali and the resulting single-stranded molecules transferred using a buffer onto a nitrocellulose filter or nylon membrane by blotting, as shown in Figure 14.21.

This procedure preserves the pattern of distribution of the DNA fragments in the gel by forming a replica of the gel on the filter paper or membrane. The filter is then incubated under hybridization conditions with a specific radiolabelled DNA probe. The DNA restriction fragment that is complementary to the probe hybridizes with it, and its location on the filter can then be revealed, often by autoradiography (Box 11.1, 'Autoradiography'). However, autoradiography suffers from a number of disadvantages, including possible health hazards, the inconvenience of handling a radioisotope and the relatively short half-life of ^{32}P. The digoxygenin (DIG) system provides a sensitive alternative to autoradiography for analysing blots and is simple to use. Analysis involves adding a probe labelled

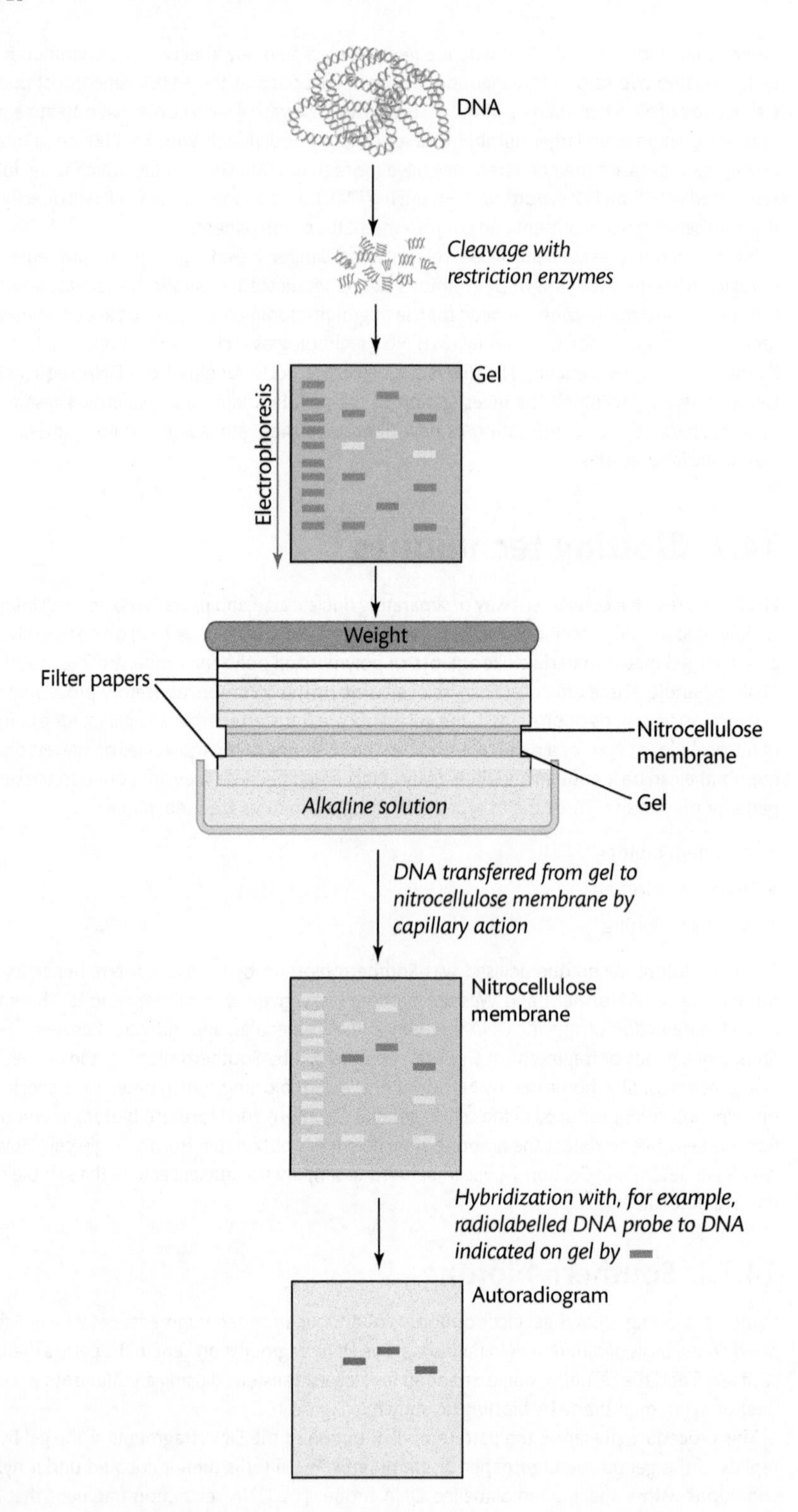

FIGURE 14.21
Outline of Southern blotting, which is explained in the general text.

with DIG to the nitrocellulose filter or membrane. An anti DIG-antibody tagged with an enzyme, such as horseradish peroxidase (HRP) or alkaline phosphatase (ALP) is added to the reaction mixture. The addition of substrates that are broken down by the enzyme to give a chemiluminescence product means that even the binding of a single enzyme molecule is amplified because catalysis produces many molecules of product. The chemiluminescence can be detected using a high-sensitivity detection device, such as a cooled charge-coupled device (CCD) camera system.

DIG-labelled DNA or RNA probes can be used to analyse the products of all types of hybridization reactions, including *in situ* hybridization and ELISAs (Section 14.8), microarrays (Section 14.10), and Northern blots (Section 14.7.2).

Restriction fragment length polymorphisms

Southern blotting is sufficiently sensitive to detect a single specific product in a complex mixture of DNA fragments as is formed when the entire human genome is hydrolysed with an RE. Southern blotting can detect and identify the genomes of specific viruses and microorganisms and the presence of mutations, or **polymorphisms**, in samples of DNA from patients, produced, for example, by the deletion or the insertion of short DNA sequences; most single-base changes, however, cannot be detected in this way unless they form or break a palindrome that can be recognized by an RE. Sickle cell anaemia is a common disorder among some populations inhabiting or originating from malarial-infested regions. Patients typically present with severe joint pain and a microscopic examination of their blood will show the presence of sickle-shaped erythrocytes. The disorder arises from a single point mutation in the gene for β globin, one of the two types of polypeptide that make up haemoglobin. A thymine substitutes for an adenine base resulting in a valine residue being incorporated at position 6 in the globin molecule rather than a glutamate, forming haemoglobin S. β Globin genes are inherited from each parent in a Mendelian fashion. Hence, individuals can be homozygous for adult haemoglobin (Hb^A/Hb^A) or for haemoglobin S (Hb^S/Hb^S), or heterozygous (Hb^S/Hb^A) and form a mixture of adult and S haemoglobins. The condition can be diagnosed using fetal material obtained by amniocentesis using the RE (Section 14.4), *Mst*II. The normal globin gene has three restriction sites that can be hydrolysed by *Mst*II. However, the sickle mutation destroys one of these sites, resulting in a product that is larger in size. Thus, electrophoresis of the globin gene fragments followed by blotting produces different patterns for the hetero- and homozygotes, as shown in Figure 14.22.

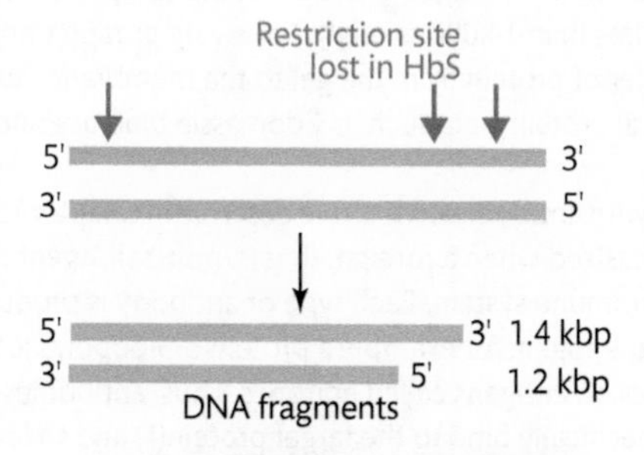

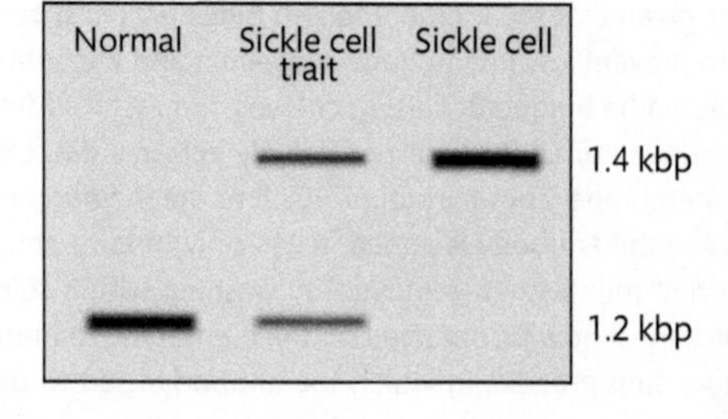

FIGURE 14.22
Application of restriction fragment length polymorphism analysis using the restriction endonuclease *Mst*II and Southern blotting to diagnose sickle cell anaemia. See text for explanation. kbp: kilobase pair; HbS: sickle haemoglobin.

14.7.2 Northern blotting

The expression of a particular gene can be followed by detecting its corresponding mRNA by Northern blotting. This procedure is similar to Southern blotting, but electrophoresis is performed on a sample of mRNA. The RNA sample is often the total RNA purified from cells or tissues. It is denatured by treatment with agents, for example formaldehyde, that disrupt the hydrogen bonds between any intrastrand bp, which ensures that all the RNA molecules have an unfolded, linear conformation. The sample is then subjected to gel electrophoresis and RNA molecules separated on the basis of differences in their sizes. As in Southern blotting, the denatured RNA molecules are transferred to a sheet of nitrocellulose or nylon membrane, which is then probed with a labelled DNA that is complementary to the RNA of interest. Molecules of RNA that hybridize to the probe on the paper/membrane are located by detecting the bound probe by autoradiography or by the DIG system described earlier. The sizes of the hybridized RNA molecules can be estimated by reference to RNA standards of known size that are subjected to electrophoresis side by side with the experimental sample.

Northern blotting is widely used to compare the amounts of a particular mRNA molecule in different cells grown in varying conditions. For example, it can be used to show if a mutation in a gene increases or reduces the amount of normal-sized mRNA produced. It can also be adapted to demonstrate if the mutation has resulted in the transcription of abnormally short mRNA molecules by testing the blotted paper/membrane with a series of shorter DNA probes, each complementary to short portions of the mRNA; this would also demonstrate which part of the normal RNA molecule is missing.

14.7.3 Western blotting

Western blotting (immunoblotting) combines several techniques to detect specific proteins in a sample. Like other blotting techniques, Western blotting gives information about the size of the protein or the relative amounts of protein produced in different cells, or the same cell in different conditions.

You can see an outline of this procedure in Figure 14.23. As with Southern and Northern blots, gel electrophoresis is used to separate native or denatured proteins, which are then transferred to a polyvinylidene difluoride (PVDF) or nitrocellulose membrane. Nylon membranes are not used in Western blotting because they are cationic and strongly bind acidic proteins. Thus, under many conditions they bind so much protein that detection procedures colour the membrane strongly and it is difficult to identify the proteins of interest. PVDF also binds proteins strongly but gives light background staining after analysis. However, nitrocellulose membranes are the most common choice for general use. They are the cheapest alternative and bind adequate amounts of protein, largely owing to hydrophobic interactions. However, they are relatively fragile and do not stand up well to repeated probings and do not bind proteins of M_r less than 14,000 strongly. Following protein transfer, the uniformity and overall effectiveness of transfer of protein from the gel to the membrane can be checked by staining the membrane with a general protein stain such as Coomassie blue or Ponceau S dyes (Table 12.1).

In Western blotting, the membrane is probed with antibodies, which we describe in Chapter 13. To briefly recap, antibodies are glycoproteins synthesized when a foreign, that is, non-self, agent such as an infectious microorganism, stimulates the immune system. Each type of antibody is produced in response to the presence of a macromolecular antigen, for example a protein or lipopolysaccharide. Individual antibodies recognize specific sites on antigens called epitopes. Thus, antibodies can be used in Western blotting because they can specifically bind to the target protein(s) and so locate their positions on the membrane. However, given that the membrane also binds proteins, but in a non-specific fashion, steps must be taken to prevent any interactions between it and the antibody, otherwise detection of the target protein would be hindered. This is achieved simply by adding an inexpensive protein, for example bovine serum albumin (BSA) or non-fat dry milk in a dilute solution of detergent such as Tween 20, to the membrane. These proteins attach to the membrane and block non-specific binding sites. Hence, when the antibody is added, it can only bind to epitopes of the target protein. The excess BSA or dried milk is then removed by washing with a suitable buffer. The bound protein–antibody complex must now be detected on the membrane to pinpoint its position. Traditional detection uses a two-step process, in which the antibody used to detect the protein is called the *primary* antibody. The membrane is rinsed to remove unbound primary antibody, and a *secondary* antibody, which is species-specific to a portion of the primary antibody

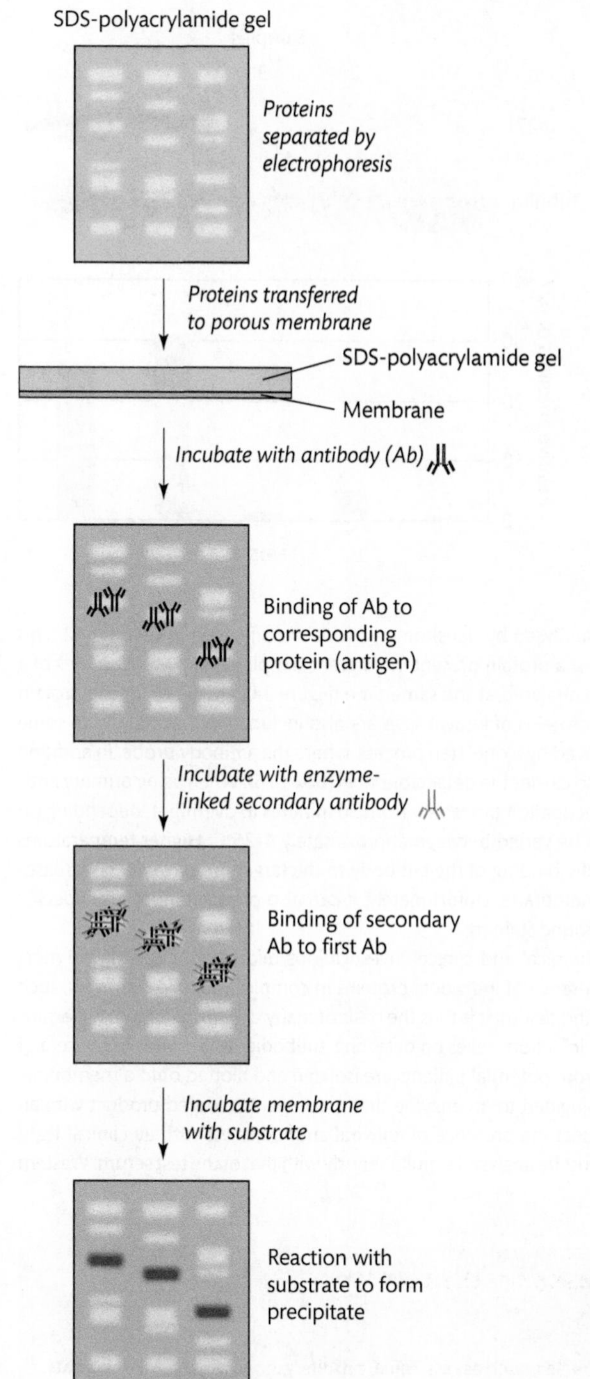

FIGURE 14.23
Outline of Western blotting. See text for details. SDS: sodium dodecyl sulfate.

is added. This secondary antibody is usually linked to biotin or to a reporter enzyme such as ALP or HRP whose activities produce a coloured signal allowing the position of the bound protein to be identified (Figure 14.24). Commonly, a HRP-linked secondary antibody is used in conjunction with a chemiluminescent agent. The reaction product produces a luminescent signal, whose intensity is proportional to the amount of protein. The light may be detected on photographic film or by using a CCD camera that captures a digital image of the Western blot.

Cross reference
Chapter 13, 'Immunological techniques'.

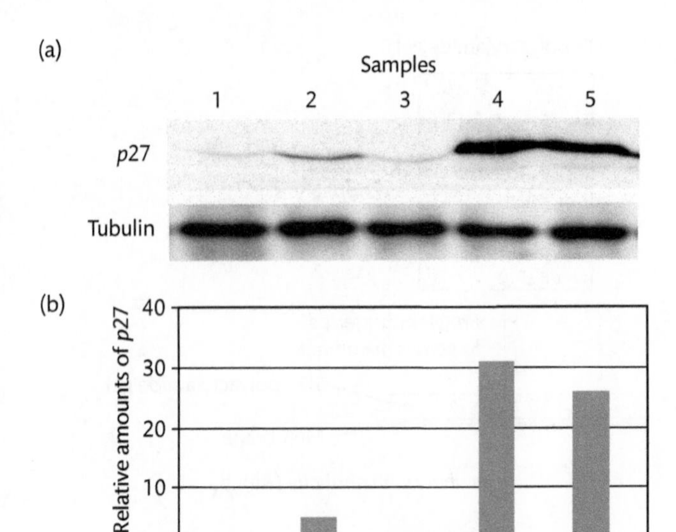

FIGURE 14.24
(a) Picture of a Western blot showing the presence of the protein *p27* in five different samples. In each case, the product of the housekeeping gene, tubulin, has also been analysed as a control. As each sample contains the same quantity of tubulin, any experimental variation in the treatments of the samples can be eliminated. (b) Densitometry to show the relative expression of *p27* in each of the samples.

Images of Western blots can be analysed by densitometry, which is explained in Chapter 12. This allows us to compare the amounts of a protein present in different samples, if known quantities of a so-called housekeeping protein are analysed at the same time (Figure 14.24). The M_r of the protein of interest can also be estimated if proteins of known sizes are also included in the analysis. In some applications, detection can be achieved by a one-step process, where the antibody probe, in addition to recognizing the target protein, also carries the detectable label (Box 14.4). When using primary antibodies or the one-step approach, incubation times vary from 30 minutes to overnight, depending on the temperature. Temperatures can be varied between approximately 4–25°C. Higher temperatures are associated with enhanced specific binding of the antibody to the target protein, which increases the signal from the target on the membrane. Unfortunately, it can also produce more non-specific binding, which increases the background staining.

Cross reference
Chapter 12, 'Electrophoresis'.

Western blotting is used in biochemical and clinical investigations because it is one of the most specific means of identifying the presence of individual proteins in complex biological samples, such as blood, serum, saliva, urine, cells, and tissues; it is thus the basis of many diagnostic assays. For example, the serodiagnosis of HIV type 1 infections relies on detecting antibodies to the virus in the sera of potential patients. Serum proteins from potential patients are isolated and blotted onto a membrane. Antihuman immunoglobulin G conjugated to an enzyme that produces a coloured product with an appropriate substrate is used to detect the presence of antiviral antibodies. As with all clinical tests, positive and negative control sera must be analysed simultaneously with that of the test serum. Western

BOX 14.4 Antibodies for Western blotting

The results of Western blotting depend on the quality of antibody probe used, in particular how specific it is for the target protein.

Both monoclonal and polyclonal antibodies (see Box 13.4) against a large number of target proteins are readily obtainable from commercial sources. If, however, the protein is a novel one, the antibody must be produced in house, although commercial companies are available that will also do this for a fee! In either case, small amounts of the target protein must be purified to stimulate antibody production. A large number of species, such as rats, mice, rabbits, guinea pigs, chickens, goats, sheep, and donkeys, are available in which to raise antibodies. Generally, monoclonal antibodies take longer to develop than polyclonal ones and are therefore more expensive (Figures 13.2 and 13.3). Monoclonal antibodies are, however, usually more specific to their targets and bind more strongly to their epitope than polyclonal types. This means that fewer non-specific bands are detected on the membrane and the background staining is of a lower intensity.

blotting also provides improved serological diagnoses for conditions such as Lyme disease, some forms of spongiform encephalopathy, congenital toxoplasmosis, syphilis, and in patients with rheumatic disease, and is the basis for many home diagnostic tests, for example for pregnancy testing kits.

SELF-CHECK 14.7

List the principal uses of Southern, Northern, and Western blottings. State the major use of each technique.

14.8 Enzyme-linked immunosorbent assays and fluorescence *in situ* hybridization

ELISAs and FISH are two sensitive groups of techniques that rely upon the respective abilities of proteins (antibodies) and nucleic acids to recognize specifically and bind to complementary molecules, as we described in Section 14.1. Each of the techniques forms the basis of many clinical assays.

14.8.1 Enzyme-linked immunosorbent assays

Like Western blotting, ELISAs rely upon the ability of antibodies to recognize and bind to their complementary antigens. This makes them powerful reagents for detecting and quantifying analytes of clinical interest. The use of antibodies in this respect is called **immunoassay**. Immunoassays are used in all branches of the biomedical sciences. Thus, ELISAs use an antibody to which an enzyme has been attached to measure the amount of a specific antigen in a sample. The enzyme used is one that will convert a colourless substrate into, for example, a coloured soluble product that can be measured spectrophotometrically. HRP and ALP are the most frequently used enzymes. The simplest ELISA format (Figure 14.25) is to allow a protein antigen to absorb onto the wells of a plastic microtitre plate. Aliquots of known concentrations of the antigen are also applied to other wells of the plate. An enzyme-labelled antibody that binds to the antigen is then added to the wells. Wells containing a large amount of antigen will bind more antibody and therefore more enzyme. The substrate for the enzyme is then added and, after a limited period, the reaction is stopped and the amount of product formed determined. The amount of product formed with the known concentrations of antigen can be used to form a standard curve from which the concentration of the unknown sample can be determined. Given each enzyme molecule forms many thousands of molecules of product, even relatively tiny amounts of antigen can be detected. Other labels that can be used in immunoassays include fluorescent labels and bio- and chemiluminescent labels.

Cross reference

See Chapter 13, 'Immunological techniques'.

Online Resource Centre

To see an online video demonstrating the ELISA technique, log on to www.oxfordtextbooks.co.uk/orc/fbs

The development of ELISAs in 1971 by Engvall and Perlmann revolutionized medicine by replacing many awkward to perform and time-consuming diagnostic tests. Since then, the basic protocol has been subjected to many different adaptions that allow non-protein antigens to be measured or which have increased sensitivity. The specificity of antibodies allows biomedical scientists to measure the concentration of an analyte, such as a steroid hormone or autoantigens in patients with autoimmune diseases, in biological fluids such as plasma, which contain hundreds of other biomolecules, some of which are similar in structure to the analyte being measured. ELISAs are widely used in diagnosing cases of malaria in Africa, and home pregnancy testing kits in the developed world.

Cross reference

You can read about mutations in Chapter 18, 'Inherited metabolic disorders and newborn screening', in the companion volume, *Clinical Biochemistry*.

Fluorescence in situ hybridization

In FISH, single-stranded DNA molecules chemically linked to a fluorochrome are used as probes to detect DNA samples that have sequences complementary to them; you can see this diagrammatically in Figure 14.26(a). Fluorochromes are molecules or parts of molecules that absorb light of one wavelength, often in the UV range, and re-emit most of the energy but at a longer wavelength. Thus, the fluorochrome appears bright against a dark background when viewed with a fluorescence microscope, giving a technique that can be sufficiently sensitive to detect single DNA molecules in microscopic sections of karyotypes, cells, or tissues.

Cross reference

Fluorescence is described more fully in Chapter 9, 'Spectroscopy'.

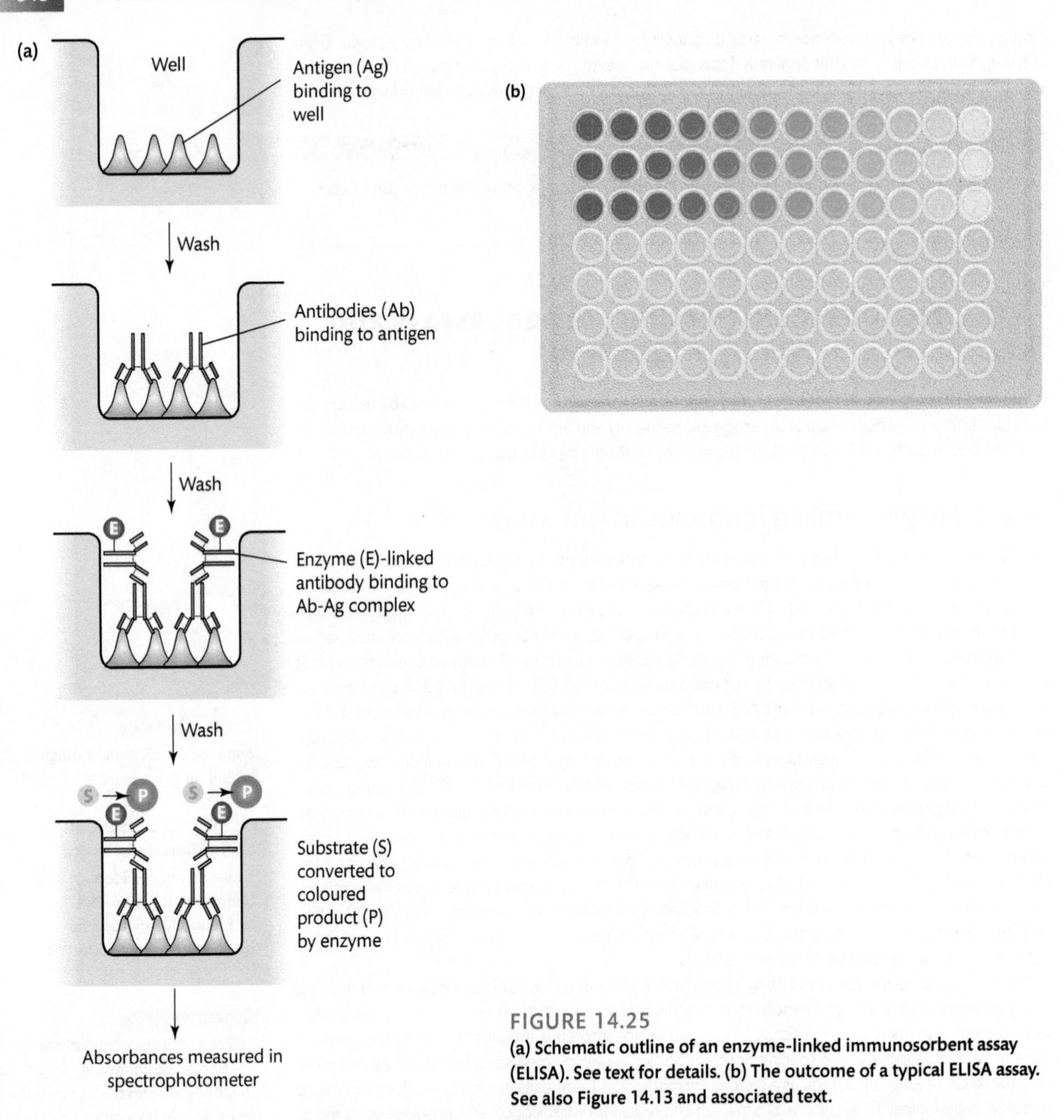

FIGURE 14.25

(a) Schematic outline of an enzyme-linked immunosorbent assay (ELISA). See text for details. (b) The outcome of a typical ELISA assay. See also Figure 14.13 and associated text.

FISH techniques are used for diagnosis. For example, the analysis of metaphase chromosomes is a major tool in cancer cytogenetics. Thin sections of the tumour obtained from biopsies are treated to separate the DNA strands, which are then hybridized *in situ* with fluorochrome-labelled probes (Box 14.2) for known mutated forms of cancer-associated genes. When the slides are viewed using a fluorescence microscope, any fluorescent areas in chromosomes from cells in the metaphase indicate the presence of a mutation. For example, patients with DiGeorge anomaly exhibit a range of signs, including cardiac malformations, hypoparathyroidism, and hypocalcaemia. The majority of patients have a deletion or partial monosomy of chromosome 22. These genetic lesions are detectable *in utero* if cells obtained by amniocentesis are analysed using FISH with an appropriate probe (Figure 14.26b).

The Quadruple Test for Down's syndrome calculates the potential risk of a pregnant woman carrying a fetus with Down's syndrome. When the test indicates a higher risk of Down's syndrome, FISH or PCR (Section 14.9) tests are available from many laboratories in the UK to help clarify the situation.

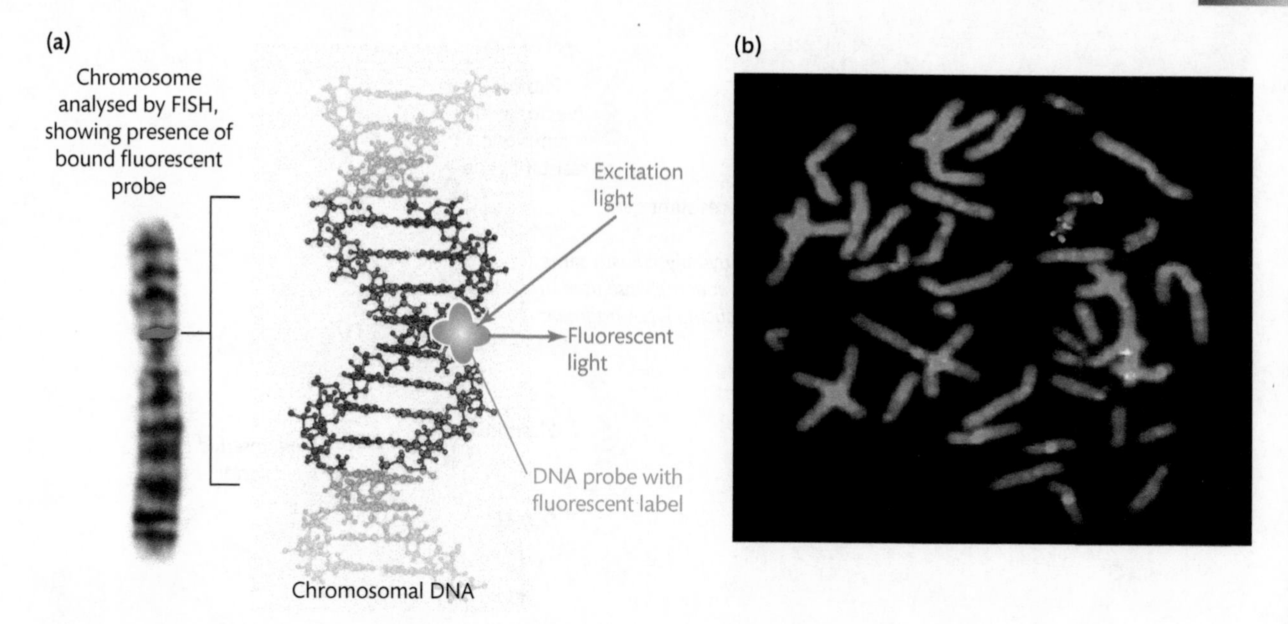

FIGURE 14.26
(a) Schematic to outline the binding of a fluorescently labelled DNA probe to target specific chromosomal DNA sequences. (b) Picture of a fluorescence *in situ* hybridization (FISH) analysis confirming DiGeorge anomaly (DGA). The TUPLE probe fluorescing red shows the 22q11 region that is deleted in DGA. The ARSA green signal is diagnostic for a region at the end of the q arms of chromosome 22, confirming the deletion to this chromosome. Courtesy of Paul Virgo, Department of Immunology, Southmead Hospital, Bristol, UK.

14.9 DNA cloning

Gene or **DNA cloning** is the preparation of many identical copies of a DNA molecule, that is, the amplification of a particular piece of DNA. Cloning can be accomplished by preparing recombinant DNA in a cloning vector or by using PCR.

14.9.1 Recombinant DNA

DNA recombination is the combining of a piece of the DNA one is interested in into another type of DNA to form a recombinant DNA molecule. A recipient cell is then induced to assimilate and replicate the recombinant DNA. Many copies of the recombinant DNA may be produced by the progeny of the recipient cell; thus, the DNA of interest is cloned. If the inserted fragment is a functional gene that encodes a specific protein then the protein could be produced in the host cell when the gene is transcribed and translated. This process is used for the large-scale production of proteins valuable in the clinical sciences, for example insulin, somatostatin, and other hormones, but which are difficult or expensive to prepare by other methods. You can see an overview of the steps involved in cloning DNA by recombination in Figure 14.27.

The carriers of the DNA to be cloned, often called vehicles or **vectors**, are usually plasmids (Section 14.1) or bacteriophage DNA (Boxes 14.5 and 14.6). The recombinant DNA molecule is formed by hydrolysing the vector with an RE to break both its strands at a specific point (as you can see in Figure 14.27). The DNA of interest is prepared as a fragment using the same RE as used to cleave the vector. The fragment and the vector are allowed to anneal at their complementary overlapping ends. DNA ligase is then used to catalyse the covalent joining of the ends of the restriction fragment and vector DNA together to form the recombinant DNA. Ligation is most efficient when the ends overlap, but the

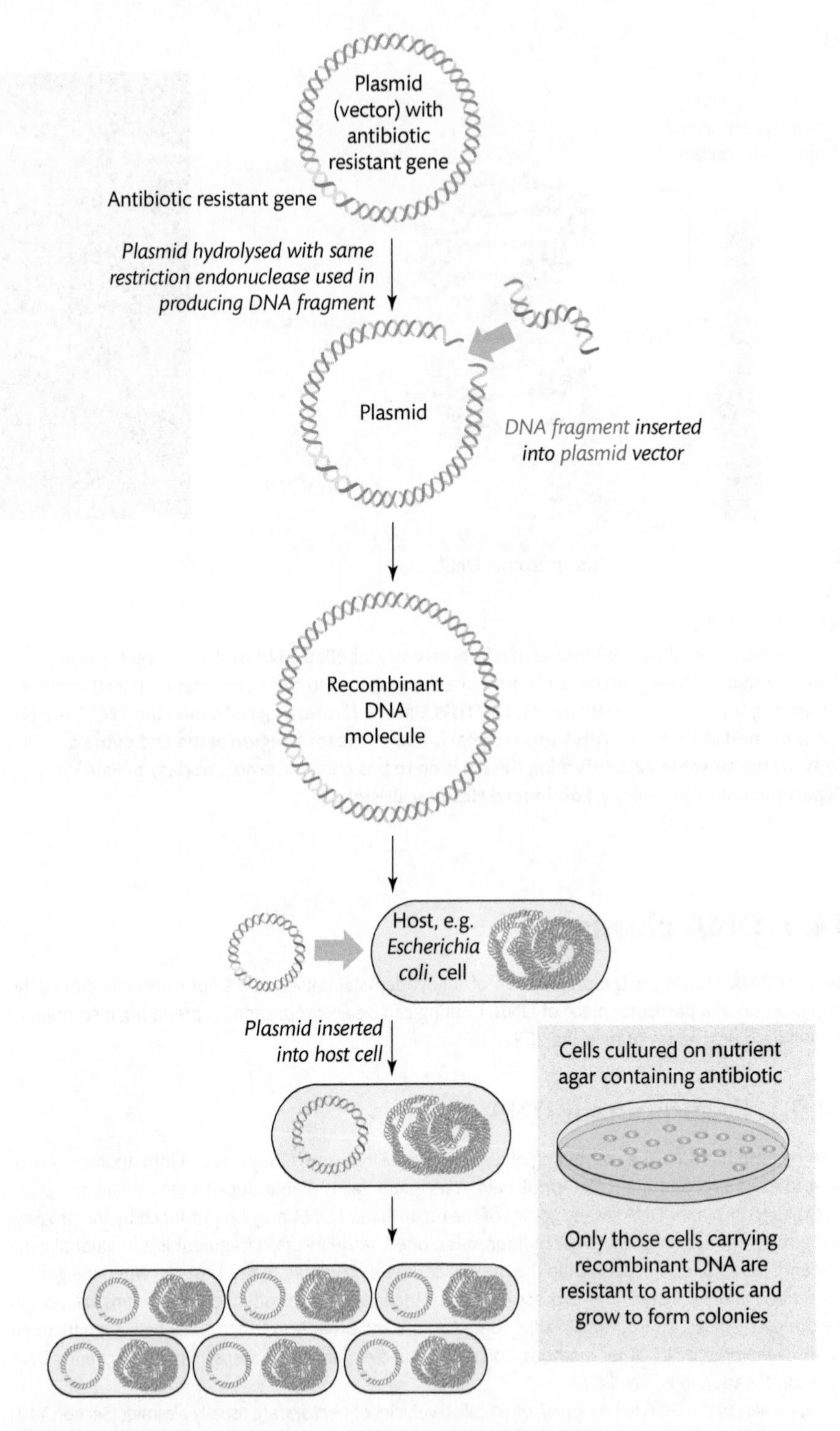

FIGURE 14.27

Overview of the steps involved in a typical recombinant DNA cloning experiment. A plasmid is used in this case as a vector. The plasmid and the piece of DNA of interest are spliced to form a recombinant DNA molecule. Bacterial cells are then transformed when they take up the recombinant DNA. Only transformed cells can grow in an antibiotic-rich medium because of the antibiotic resistance gene on the plasmid. Growth of the bacteria amplifies the DNA of interest.

BOX 14.5 Ideal characteristics of a cloning vector

The ideal cloning vector should have a number of properties, including the following:

- **the vector should possess only a single site for a specific RE**
- **the vector should contain identifiable markers so that it is possible to screen transformed cells for uptake; at least two selective markers are desirable, one to confirm the insertion of foreign DNA into the plasmid, and the second to allow the presence of the plasmid in the cell to be confirmed**
- **the vector should generally be small relative to the fragments of chromosomal DNA, making its isolation and purification easier and reducing the number of potential sites for attack by REs**
- **the vector should be capable of being replicated rapidly by transformed cells to ensure an efficient cloning of the recombinant DNA.**

DNA ligase from bacteriophage T4 can join fragments and vectors that have been produced using REs that produce blunt-ended DNA, although this is a less efficient means of ligating the DNA together.

A major breakthrough in recombinant DNA technology was the isolation of mutant strains of *Escherichia coli* that are not able to digest foreign DNA. These strains are suitable as host organisms for replicating recombinant DNA. The recombinant DNA must now be taken up by such a host cell in a process called **transformation**. For example, *E. coli* cells can be made transiently permeable to DNA and then mixed with recombinant vector DNA. Under certain conditions, a small fraction, about one in 10,000, of the cells will take up the recombinant plasmid. The transformed cells can be distinguished from the others by using, for instance, a plasmid containing a gene that confers resistance to the antibiotic ampicillin. Thus, if the cells are grown in an ampicillin-containing medium, then only the transformed cells will be viable and can be selected. Replication of the plasmid during the growth of the transformed cells ensures that the inserted DNA is cloned.

It is sometimes necessary to isolate plasmid DNA from cells to allow it to be characterized and manipulated. We introduced general methods for purifying DNA earlier (Section 14.3). However, several methods for isolating plasmid DNA have been developed to separate plasmid from chromosomal DNA once the cells have been disrupted. Plasmid DNA differs from chromosomal DNA in having a tight circular conformation, whereas the chromosomal DNA usually consists of linear fragments formed when the chromosome is sheared. The plasmid DNA is still much smaller than these fragments. These structural differences can be exploited to separate the two types of DNA. For example, plasmid DNA can be specifically adsorbed on to nitrocellulose microfilters and hydroxyapatite columns or the chromosomal DNA can be selectively precipitated by using buffers of extreme pH, high temperature, or other denaturing conditions. Finally, centrifugation can be applied to separate them because their size difference means they differentially sediment.

Plasmid vectors based on the naturally occurring F plasmid of *E. coli* are used to clone DNA fragments of 300,000 to 1 million nucleotide pairs. Unlike smaller bacterial plasmids, the F plasmid and its derivative, the bacterial artificial chromosome (BAC; Box 14.6) is present in only one or two copies per *E. coli* cell. The fact that BACs are kept in such low numbers in bacterial cells may contribute to their ability to maintain large cloned DNA sequences stably. Because of their stability, ability to accept large DNA inserts, and ease of handling, BACs are now the preferred vector for building DNA libraries of complex organisms, including those representing the human and mouse genomes.

cDNA cloning

An alternative cloning strategy is to begin the process by selecting only those DNA sequences that are transcribed to give an mRNA molecule. These sequences are presumed to correspond to protein-encoding genes. This is done by extracting the mRNA from cells as outlined earlier (Section 14.3). A complementary DNA, or cDNA, is then made of each mRNA using reverse transcriptase, an enzyme prepared from retroviruses, which synthesizes a DNA molecule complementary to an RNA template. The single-stranded cDNA is converted into a double-stranded molecule using DNA polymerase. Each double-stranded DNA molecule is then inserted into a vector and cloned in the normal manner. Each

BOX 14.6 Vectors

Plasmids are small, only 3–20% the size of the bacterial chromosome, meaning they can carry inserts only 1–20 kbp in size. Their relatively small size also makes them comparatively easy to isolate from the bulk of the DNA present in the cell (Sections 14.3 and 14.9). Also, they have high multiple copy numbers, up to 500 plasmids per cell in some cases, and so the inserted DNA of the recombinant plasmid may be greatly amplified. The presence of selectable markers, such as the genes for antibiotic resistance or the *lac Z* gene, means that bacterial cells transformed by the recombinant DNA can be easily detected because those cells that do not take up the vector will be unable to grow in a medium containing the antibiotic. Thus, a number of cloning vectors have been derived from plasmids.

The pUC vectors were introduced in the early 1980s and are based on fragments of DNA derived from naturally occurring *Escherichia coli* plasmids. Figure 14.28 shows the structure of the pUC18 plasmid. The pUC19 vector is identical except its genes are organized in the opposite direction. The pUC vectors have an origin of replication allowing them to be replicated in high copy numbers. They contain the gene for β-lactamase, which confers resistance to the antibiotic, ampicillin (*Amp*^{R}) and allows transformed cells, which have acquired the plasmid, to be identified. Additionally, pUC18/19 also contains a portion of the *E. coli lac* operon, which allows cells containing the plasmid to synthesize the amino terminal portion of β-galactosidase. pUC vectors may be used with appropriate host cells that have been modified to express the gene for only the carboxyl terminal portion of the enzyme. In the presence of a suitable inducer, for example isopropylthio-β-galactoside (IPTG), those cells containing a pUC plasmid will produce the two fragments of enzyme and possess β-galactosidase activity. Such cells are able to hydrolyse the artificial substrate 5-bromo-4-chloro-3-indolyl-β-galactoside (X-gal) to form a blue-coloured product. Thus, colonies of the cells will appear blue when grown on nutrient agar containing IPTG and X-gal. However, pUC18 also contains a multiple cloning site or polylinker region near the *lac* gene, which contains the recognition sites for a number of different REs (Figure 14.28). The formation of a recombinant plasmid means the inserted fragment of DNA in this region prevents the formation of an active β-galactosidase. Hence, transformed cells possessing the recombinant DNA will be unable to hydrolyse X-gal and their colonies will appear white and so are easily identified.

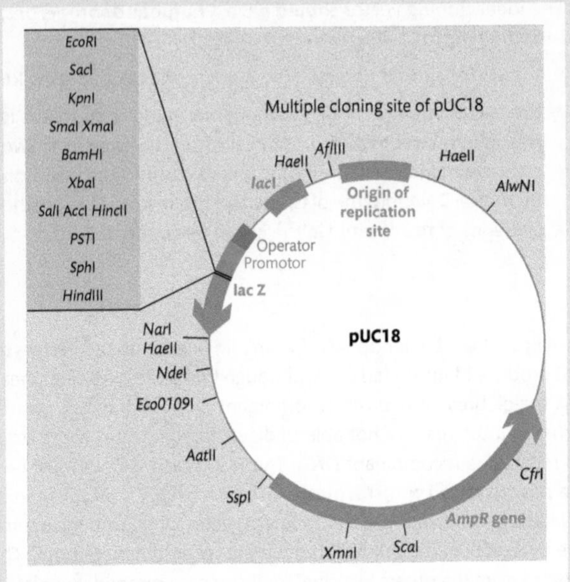

FIGURE 14.28
Outline of the structure of the pUC plasmid.

The F plasmid or factor contains 'fertility genes', which enables a bacterium possessing it (F$^+$ cells) to produce sex pili during bacterial conjugation and act as genetic donors to those cells lacking the factor (F$^-$ cells). Thus, the plasmid and its genetic information are transferred between bacterial strains. Unlike smaller plasmids, only one or two copies of the F plasmid are present in each bacterial cell. Plasmid vectors based on the F plasmid of *E. coli*, called bacterial artificial chromosomes (BACs), can be used to clone pieces of DNA as large as 300–10^3 kbp. The low numbers of BACs in bacterial cells are thought to contribute to the stability of the recombined BAC, despite the large sizes of the DNA inserts. This stability and their relative ease of handling means BACs are useful vectors for building DNA libraries of, for example, the human genome.

The ability of bacteriophages or phages to replicate inside bacterial cells has allowed a number of them to be developed as cloning vectors. Examples of phages used include the Enterobacteria phage λ, or lambda phage, the filamentous bacteriophage M13, bacteriophage Mu (phage Mu), and phage P1. Phage P1 has been modified to form phage artificial chromosomes that, like BACs, can be used to clone relatively large pieces of DNA. Cosmids are a type of hybrid plasmid that contain part of the DNA from the lambda phage to which a suitable origin of replication from other phages or plasmids and a gene for selection, such as antibiotic resistance, have been added. Thus, they can be used to clone DNA inserts as large as 30–50 kbp of DNA and so can be used to build genomic libraries.

clone prepared in this way is called a cDNA clone and the entire collection of clones derived from one mRNA preparation constitutes a **cDNA library**.

The major advantage of using cDNA molecules is that uninterrupted coding sequences of a gene are cloned. Eukaryotic genes usually consist of short coding sequences of DNA called exons that are separated by normally much longer non-coding sequences called introns. During the production of a specific mRNA, those portions of it that were formed by transcribing the introns are removed and the coding sequences (from the transcribed exons) are spliced together to give a continuous sequence. However, the mRNA transcripts of many genes are subject to alternative splicing, so one gene can encode a number of alternatively spliced and therefore different mRNAs. Thus, a cDNA library often contains many of the alternatively spliced mRNAs produced from a given cell line or tissue.

14.9.2 Polymerase chain reaction

Mullis devised the PCR in 1983 and was awarded the Nobel Prize in Chemistry in 1993 for this work. PCR is an elegantly simple **in vitro** method to increase or amplify the number of relatively short strands of specific DNA fragments in a sample. The fragments are normally 0.5–5.0 kbp in length, although longer ones up to 40 kbp can also be amplified. The DNA fragments may be complete genes, although they more usually comprise only a portion of one. Thus, the technique complements the recombinant DNA cloning strategy.

PCR is used to replicate a DNA sample (template DNA) using a DNA polymerase. This enzyme copies the template to produce new complementary DNA strands. DNA polymerases cannot begin to synthesize a strand *de novo*; they can only extend an existing piece of DNA. Thus, two primer DNA strands are needed to initiate the copying process. Primers are artificial oligodeoxynucleotides less than 50 nucleotides long that are complementary to sequences that flank the region of the template DNA of interest. Hence, the primer determines the beginning of the region to be amplified. Primers are usually made to order by commercial suppliers who must be supplied with the required sequence. Some DNA polymerases can proofread, that is, correct any mistakes in the newly formed strand, which ensures that the fidelity of the sequence is preserved. Other polymerases do not, however, have this property.

The PCR consists of a series of cycles, each of which doubles the number of molecules of the DNA template. Prior to the first cycle, the DNA is often heated to 90–96 °C for 5–10 minutes in a 'hot start' designed to break the hydrogen bonds connecting the two strands and ensure the template and primer DNA molecules are fully separated. This separation is called **melting**. Each of the subsequent cycles consists of three identical steps, as shown in Figure 14.29. In the first step, double-stranded DNA is heated for 0.5–3.0 minutes at 94–96°C, which is normally sufficient for melting. In the second step, the temperature is reduced to 50–65°C for 0.5–6.0 minutes during which the primers bind or anneal to their complementary sequence of the templates. The primer must be present in amounts that are in excess of the target DNA, otherwise its strands will simply rejoin. The design of the length of the primer requires careful consideration. Primer melting temperature increases with the length of the primer. The optimum length for a primer is generally 20–40 nucleotides, which will have melting temperatures of 56–75 °C. If primers are too short they will anneal at random positions on the relatively long template and result in a non-specific amplification. However, if the primer is excessively long, the corresponding melting temperature would be above 80°C and this could reduce the activity of the polymerase. The third step is the extension of the primer by the polymerase. This usually requires 0.75–2.0 minutes at 72°C. The extended portions of DNA are complementary to the template strand.

Online Resource Centre

To see an online video demonstrating how to perform a polymerase chain reaction, log on to www.oxfordtextbooks.co.uk/orc/fbs

The high temperatures used in PCR mean that DNA polymerases from thermophilic organisms are preferred. The Taq polymerase from *Thermus aquaticus*, often abbreviated to Taq pol, or simply Taq, is widely used, although it has the disadvantage of lacking proofreading capabilities and therefore can introduce errors (mutations) of 1 in 400–500 nucleotides in the newly formed DNA. Polymerases such as *Pwo* or *Pfu*, obtained from Archaea, have proofreading mechanisms that significantly reduce mutations and are used in 'long-range' PCR of up to about 30 kbp.

The first cycle results in two double-stranded DNA molecules, which are usually overextensions of the target sequences. Each is composed of one of the original strands plus the newly formed complementary strand and associated primer. Thus, the amount of DNA present has been doubled. Cycles of PCR are usually repeated 20–30 times, with each cycle doubling the amount of DNA present (Figure 14.30). Thus, the yield of DNA increases exponentially with each cycle: after 30 cycles the original amount is

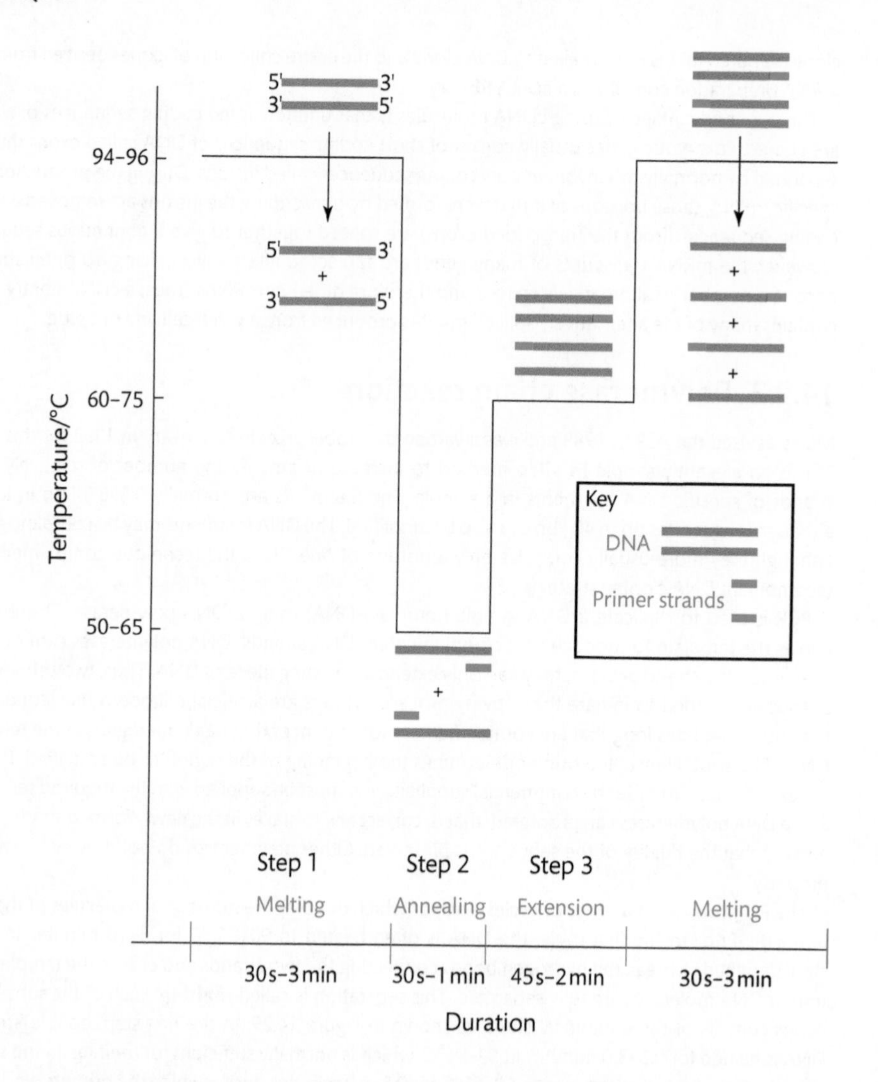

FIGURE 14.29

Three steps of the polymerase chain reaction. See text for details.

amplified over 2^{30} or 10^9-fold. A PCR experiment normally terminates with a 10-minute incubation at 72°C to ensure that all of the new DNA molecules are fully extended by the polymerase.

Cloning by PCR is extremely sensitive to contamination and extremely stringent conditions are necessary to prevent unwanted template DNA being co-amplified with the sample. The products of PCR are identified by determining their base sequences (Section 14.6) and/or their size using agarose or PAGE against standards of known size (Section 14.5).

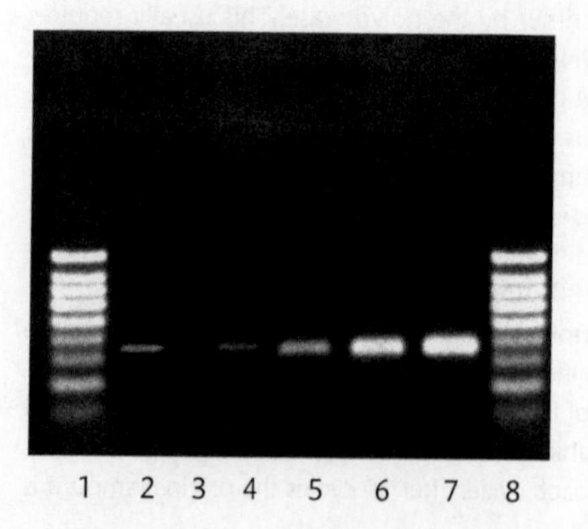

FIGURE 14.30

Agarose gel electrophoresis separation of a polymerase chain reaction (PCR) amplification. Lane I shows the positions of fragments of DNA of known sizes. Lane 2 shows a 390-base pair fragment of DNA, which forms the positive control. Lanes 3–7 show the results of amplifying this DNA by 10, 20, 25, 30, and 35 cycles of PCR. Lane 8 shows the results of the negative control, where the DNA was replaced by distilled water in the amplification.

The great advantages of PCR are the increase in sensitivity it provides by amplifying the amount of DNA present in a sample and the ease with which its step can be automated in a **thermocycler**. These are instruments that can be programmed to heat and cool the reaction tubes to the appropriate temperatures, for the desired times and the required numbers of cycles. Samples of DNA isolated from even a single cell or that present in samples many years old can be amplified and then analysed.

In addition to its application in many areas of molecular biological research, PCR has four major clinical uses:

1. The identification of infectious disease organisms for diagnostic purposes
2. The detection of variations and mutations in hereditary diseases
3. Detecting acquired mutations that lead to cancers
4. Tissue typing.

PCR is especially useful in diagnosing diseases caused by organisms that are difficult or impossible to culture. Its ability to amplify small amounts of DNA means that PCR-based tests can identify sources of infection more accurately, reliably, rapidly, and cheaply than previous methods. For example, PCR is the basis of sensitive and specific tests for *Helicobacter pylori*, the main causative agent of stomach ulcers. Three different sexually transmitted disease agents, Herpes, papilloma viruses, and Chlamydia, can be detected on a single swab using PCR tests. Indeed, even specific strains of papilloma viruses that predispose individuals to cervical cancer can be identified. Diagnostic tests are also available for the viruses involved in AIDS, viral hepatitis, and viral meningitis. In 2002, 47% of meningococcal infections in England and Wales were diagnosed using PCR tests for meningococcal DNA in clinical samples. Bacterial infections in middle ear fluid from children suffering otitis media are detectable by PCR, indicating an active infection, even when standard culture methods for the bacterium fail. The Lyme disease bacterium, *Borrelia burgdorferi*, is often difficult to diagnose accurately using its general symptoms, but PCR can amplify, and therefore identify, its DNA in samples of body fluids. Cloning by PCR has largely replaced Southern blotting for diagnosing genetic diseases.

Reverse transcriptase PCR

Reverse transcriptase PCR (RT-PCR) is a technique that allows an *RNA* molecule to be amplified but as complementary *DNA* copies, whose preparation was described above. The first step is to hybridize an oligodeoxynucleotide primer to the target RNA molecule. This primer may be specifically designed to hybridize to the mRNA from a specific target gene. However, oligo(dT) can be used as a universal primer for RT-PCR of mRNA molecules because it binds to the endogenous poly(A) tail they all possess. An RNA-dependent DNA polymerase or reverse transcriptase, obtainable from retroviruses, forms a cDNA to the target RNA molecule by extending the primer. The cDNA can be amplified by PCR, generally using gene-specific primers to generate copies of it. The amplification of contaminating genomic DNA can be minimized by treating the RNA preparation with RNase-free DNase, which will degrade it. RT-PCR is a sensitive and versatile procedure. It is used to:

- clone the 5' and 3' termini of mRNA molecules
- generate large cDNA libraries, that is, copies of the mRNA molecules present in cell or tissue extracts
- identify mutations (polymorphisms) by analysing mRNA molecules transcribed from the defective gene.

Real-time polymerase chain reaction

Real-time PCR is also called quantitative real-time PCR (qPCR), and kinetic PCR. Real-time PCR is a technique based on PCR that not only amplifies the target DNA, but also simultaneously quantifies the amount present. The amount can be expressed as the absolute number of copies or as the relative amount of target DNA compared with a so-called normalizing gene. Real-time PCR follows the general principles of PCR outlined earlier. However, the amount of amplified DNA is determined as it accumulates in the reaction in 'real time' following each cycle by one of two common methods. One

Cross reference

The automated use of PCR is discussed in Chapter 15, 'Laboratory automation'.

method uses a fluorescent dye that intercalates into double-stranded DNA. The other method uses modified deoxyoligonucleotide probes that fluoresce when hybridized with their cDNA. In both cases, the amount of fluorescence is proportional to the amount of DNA formed.

Real-time PCR is often used in combination with RT-PCR to determine how many molecules of a specific mRNA are present in a sample. This gives the relative expression of its gene at a particular time, or in a particular cell or tissue type.

14.10 DNA microarrays

DNA microarrays are glass microscope slides, nylon filters, or silicon chips to which large numbers of DNA fragments have been attached (Box 14.7). Each fragment is an oligodeoxynucleotide whose sequence is a probe for a specific gene. The densest DNA microarrays contain probes for all of the approximately 21,000 human genes packaged into an area smaller than a postage stamp! Thus, despite their small size, microarrays allow thousands of hybridization reactions to be performed simultaneously. Hybridization reactions use mRNA molecules isolated from the cell and used directly or more stable cDNA copies (Section 14.9) of them. In either case, their sequences are complementary to the probes of the microarray. Each sample mRNA or cDNA is radiolabelled with ^{32}P (or ^{33}P) or tagged with a fluorescent dye. The labelled samples are allowed to hybridize to the probes on the microarray. Note how the probes differ from those we described for other techniques, such as FISH and blotting, in not carrying the label, which in this case is attached to the sample molecules. The microarray is thoroughly washed to remove any unbound (i.e. non-hybridized) mRNA or cDNA molecules. The targets can be identified by autoradiography or by its fluorescent label. Usually, the fluorescent cDNA from the experimental samples is labelled, for example with a dye that fluoresces red when excited, and is mixed with a reference sample of cDNA fragments labelled with a dye that shows green fluorescence. Thus, if the amount of a specific mRNA expressed from a particular gene is increased relative to that of the reference sample, the resulting spot on the microarray is red. Conversely, if the expression of the gene is decreased in comparison, the spot is green. However, If there is no change compared with the reference sample, the spot is yellow (Figure 14.31). These initial hybridization intensities require sophisticated computerized analyses to normalize, summarize, and analyse the differential gene expression patterns. However, using such an internal reference allows the expression of many genes to be accurately assessed.

Since their introduction in 1995, DNA microarrays have proved useful in analysing the expression of numerous genes by allowing the concentrations of *all* the mRNA molecules produced by a cell to be simultaneously monitored. Hence, we can now identify and study the patterns of gene expression that underlie both normal and pathological cell physiology and investigate the differential expression of genes as cells grow, divide, and differentiate. For example, microarrays are used to identify and classify tumours based on the patterns of gene expression they exhibit. This should help identify new targets in tumours for chemotherapy and the development of novel anticancer drugs. Microarrays can also be used to identify quickly disease-causing microorganisms by hybridizing the DNA isolated from infected tissues to a microarray of genomic DNA sequences from potential pathogens. At the time

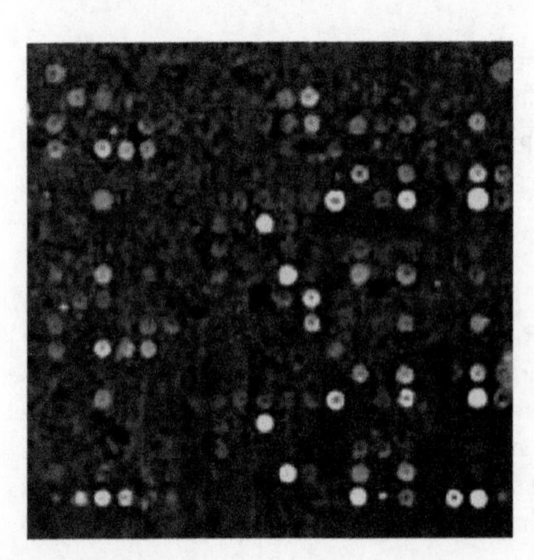

FIGURE 14.31

Photograph showing a portion of a microarray following an experiment on gene expression. Genes whose expressions have been increased during the experiment (upregulated) compared with the control are indicated by red spots. In contrast, those genes whose expression decreased (downregulated) appear as green spots. Yellow spots show genes whose expression was unchanged during the experiment.

BOX 14.7 Preparation of DNA microarrays

DNA microarrays are prepared using longer fragments of DNA produced by PCR (Section 14.9) and then spotted onto the slides by a robot. Shorter artificially synthesized oligodeoxynucleotides are attached to the surface using techniques similar to those that are used to etch circuits onto computer chips. In both cases, the sequence and position of every probe on the chip is known. Thus, by detecting its location on the microarray any nucleotide fragment that hybridizes to a probe can be identified as the product of a specific gene.

of writing, microarrays require rigorous operating procedures, making them difficult to use in everyday clinical applications. Furthermore, they are expensive to use, both in the cost of the microarrays themselves and in their need for highly trained personnel. It is generally expected that as their costs diminishes, microarray analyses will be powerful diagnostic tools in the clinical sciences. However, as the costs of DNA sequencing continues to fall, this more robust technique is likely to be the method of choice given that the two techniques have many similar biomedical applications, for example in **personalized medicine** (see Box 14.8).

Cross reference

You can read about antibody microchips in Chapter 13, 'Immunological techniques'.

SELF-CHECK 14.8

Huntington disease, myotonic dystrophy, and spinobulbar muscular dystrophy are all associated with increased numbers of trinucleotide repeats in their associated genes. Which method(s) could conveniently be used to confirm the status of potential carriers or patients?

BOX 14.8 Personalized medicine

Personalized medicine is the use of molecular biological evidence to ensure the right patient receives the correct therapy in the most appropriate way at the right time. The "one-size-fits-all" traditional medical practice sometimes misses its mark because each patient's genome differs slightly in sequence from that of others in ways that can affect disease development and responses to treatment. Thus, personalized medicine relies on using an individual's genetic and epigenetic information and/or other molecular or cellular analyses to predict that person's risk for a particular disease and to initiate preventative treatment before the disease presents or to target specific medical treatment during all stages of care. The approach depends on multidisciplinary healthcare teams and integrated technologies to optimize healthcare strategies. Advances in personalized medicine will create a more unified approach to treatment specific to an individual and their genome.

Personalized medicine uses recent advances in information and life science technologies. Molecular biological techniques such as gene sequencing can reveal mutations that influence diseases ranging from cystic fibrosis to diabetes to cancer. Two examples can be used to illustrate the usefulness of this approach. First, detecting a mutation that increases the risk of developing type 2 diabetes means the person in question can be advised to adopt a lifestyle that will lessen the chances of the condition developing later in life. Second, if it is known that an advanced non-small cell lung cancer in a patient is due to the c.2573T>G (L858R) mutation in *EGFR*, then targeted therapy using iressa (gefitinib) and/or tarceva (erlotinib) alone or in combination with chemotherapy would be the most beneficial approach for the patient.

Personalized medicine provides better disease prevention, diagnoses that are more accurate, safer drug prescriptions, and more effective treatments for diseases and conditions that diminish health. Its implementation does, however, face a number of challenges such as funding policies, and patient privacy and confidentiality. Its further integration into clinical practice requires overcoming several barriers in education, accessibility, and regulation, and government, medical insurance companies, and individual approaches to funding.

CASE STUDY 14.1 Sequencing as a test for mutations in BRCA1 and BRCA2

A 33-year-old woman of Ashkenazi Jewish ancestry has a serious family history of breast and ovarian cancers. Her maternal grandmother died of ovarian cancer at the age of 56 years and her mother developed breast cancer aged 52 but has survived for 10 years after diagnosis. Genetic testing can identify mutations that make humans more likely to develop certain diseases such as cancers. For example, *BRCA1* and *BRCA2* are human tumour suppressor genes: mutations in these genes are associated with a significantly increased risk of breast and ovarian cancers, and are responsible for 90% of hereditary breast cancers. Gene sequencing or PCR-based methods are currently the gold standard for identifying such mutations. The woman was referred to a hospital high-risk clinic and her blood samples were, given her maternal clinical history, tested for *BRCA1* and *BRCA2* mutations. Sequencing showed her *BRCA1* gene is normal; however, she is positive for the *BRCA2* 6174delT mutation (Figure 14.32).

The woman has not shown any clinical signs or symptoms of developing breast or ovarian cancer when the genetic testing was performed. Some points to consider.

What does the presence of the normal BRCA1 and the BRCA2 6174delT mutation imply about the woman's health? Justify your conclusion.

What advice should the woman's counsellor offer her?

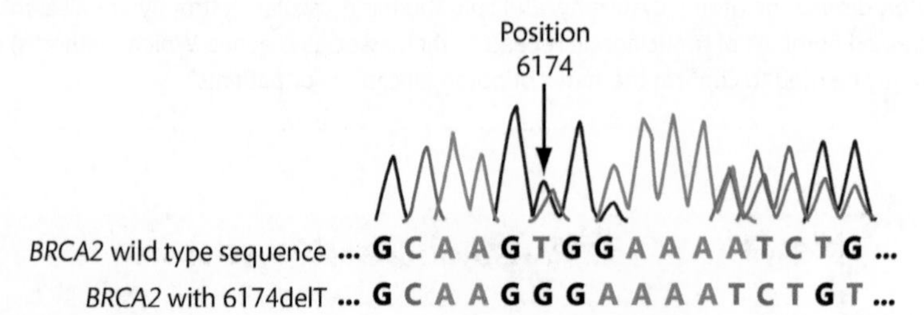

FIGURE 14.32

Nucleotide sequences showing the deletion of a deoxythymidine residue (T) at position 6174 of *BRCA2*, that is a 6174delT mutation.

CASE STUDY 14.2 FISH and HER2 in breast cancer

A 42-year-old woman consulted her general practitioner regarding a lump she had found in her right breast. She was referred to a specialist clinic, where a mammogram indicated the lump was a potential tumour. An aspirate was removed from the lump using a fine needle (**fine-needle aspiration**). Cytological examinations of the fluid confirmed a malignancy.

Fine-needle aspiration biopsy was used to obtain material from the lump. Unfortunately, histological examination proved inconclusive in staging the malignancy and the presence of HER2. However, FISH analysis gave strongly positive results for HER2, as seen in the accompanying photograph (Figure 14.33).

FIGURE 14.33
Section of breast tumour stained using fluorescent *in situ* hybridization indicating the presence of *HER2*, a result diagnostic for treatment with the monoclonal antibody herceptin (trastuzumab).

Treatment with the monoclonal antibody, herceptin (trastuzumab) was initiated as part of an approved chemotherapy regime. The condition of the patient improved, although she felt nauseous and suffered from diarrhoea and headaches, which are common side effects of the therapy.

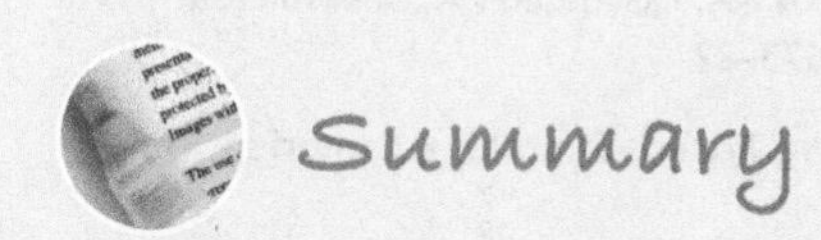

Summary

- Molecular biology is generally concerned with the structures, functions, and interactions of the two major groups of macromolecules, the nucleic acids DNA and RNA, and proteins. DNA and RNA carry biological information. Proteins perform many of the activities in cells and form some of their structural features.
- Nucleic acid molecules or fragments of molecules are able to interact and recognize one another: adenine hydrogen bonds to its complementary bases of thymine in DNA and uracil in RNA, and guanine hydrogen bonds with its complement, cytosine. Protein molecules have discrete sites on their surfaces that are complementary to specific small ligands or to relatively small areas on the surfaces of other proteins and nucleic acid molecules. Thus, proteins also generally function by recognizing and specifically interacting with other biological molecules.
- Nucleic acids can be purified from organisms using a variety of techniques. REs can digest the isolated DNA into smaller-sized fragments suitable for analysis and for use in a number of techniques of clinical interest.

- Molecular biology techniques used in biomedical science that rely on the complementary binding nucleic acids to provide a convenient way of recognizing and isolating specific base sequences within fragments of DNA or RNA molecules include the sequencing of isolated DNA, Southern and Northern blotting, FISH, the cloning of DNA by recombination and PCR technologies, and DNA microarray analysis.
- One group of proteins used extensively in clinical tests are the immunoglobulins, or antibodies. The binding of antibodies to their complementary antigens forms the basis of Western blotting and immunoassay techniques such as ELISAs, which are used in all branches of the biomedical sciences.
- The chapter discusses the underlying basic principles behind these techniques and provides general descriptions of their associated experimental protocols. In addition, we also highlight a number of essential precautions that must be observed if the reagents are to be handled and the procedures carried out safely.

Further reading

- Brown TA. *Gene Cloning and DNA Analysis: An Introduction*. 5th ed. Oxford: Blackwell, 2006.
- Craig N, Cohen-Fix O, Green R, Greider C, Storz G, Wolberger C. *Molecular Biology: Principles of Genome Function*. 2nd ed. Oxford: OUP, 2014.
- Dreier J, Störmer M, Kleesiek K. Real-time polymerase chain reaction in transfusion medicine: applications for detection of bacterial contamination in blood products. *Transfusion Medicine Reviews* 2007; **21**: 237–54.
- Hanna WM, Rüschoff J, Bilous M, Coudry RA, Dowsett M, Osamura RY, et al. HER2 *in situ* hybridization in breast cancer: clinical implications of polysomy 17 and genetic heterogeneity. *Modern Pathology* 2014; **27**: 4–18.
- Kricka LJ. Stains, labels and detection strategies for nucleic acid assays. *Annals of Clinical Biochemistry* 2002; **39**: 114–29.
- Lo YMD (ed.). *Clinical Applications of PCR*. 2nd ed. Clifton, NJ: Humana Press, 2006.
- Maraqa L, Donnellan CF, Peter MB, Speirs V. Clinicians' guide to microarrays. *Surgical Oncology* 2006; **15**: 5–10.
- Mikhailovich V, Gryadunov D, Kolchinsky A, Makarov AA, Zasedatelev A. DNA microarrays in the clinic: infectious diseases. *BioEssays* 2008; **30**: 673–82.
- Nicholl DST. *An Introduction to Genetic Engineering*. 3rd ed. Cambridge: Cambridge University Press, 2008.
- Papachristodoulou D, et al. *Biochemistry & Molecular Biology*. 5th ed. Oxford, OUP, 2014.
- Perez EA, Cortés J, Gonzalez-Angulo AM, Bartlett JMS. HER2 testing: current status and future directions. *Cancer Treatment Reviews* 2014; **40**: 276–84.
- Salto-Tellez M, Gonzalez de Castro D. Next-generation sequencing: a change of paradigm in molecular diagnostic validation. *Journal of Pathology* 2014; **234**: 5–10.
- Sambrook J. *Molecular Cloning: A Laboratory Manual*. Cold Harbor: Cold Spring Harbor Laboratory Press, 2001.
- Sargeant R. In situ hybridization: applications in research and disease. *The Biomedical Scientist* 2015; **59**: 416–18.
- Speers DJ. Clinical applications of molecular biology for infectious diseases. *Clinical Biochemist Reviews* 2006; **27**: 39–51.

- Volpi EV, Bridger JM. FISH glossary: an overview of the fluorescence *in situ* hybridization technique. *Biotechniques* 2008; **45**: 385–6.
- Williams SA, Slatko BE, McCarrey JR. *Laboratory Investigations in Molecular Biology.* Sudbury: Jones and Bartlett, 2007.
- Wiltgen M, Tilz GP. DNA microarray analysis: principles and clinical impact. *Hematology* 2007; **12**: 271–87.
- Yan M, Schwaederle M, Arguello D, Millis SZ, Gatalica Z, Kurzrock R. HER2 expression status in diverse cancers: review of results from 37,992 patients. *Cancer Metastasis Review* 2015; **34**: 157–64.

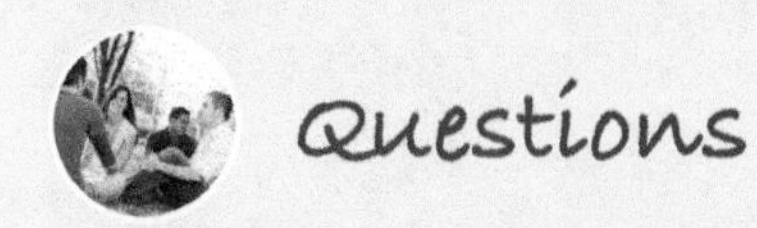

Questions

1. Which of the following statements is/are **FALSE?**
   - **(a)** RNA contains ribose residues
   - **(b)** RNA contains deoxyribose resides
   - **(c)** RNA contains the bases adenine, guanine, uracil, and cytosine
   - **(d)** RNA molecules do not possess any double-stranded structure
   - **(e)** RNA contains the bases adenine, guanine, uracil, and thymine.
2. Which of the following statements is/are **FALSE?**
   - **(a)** Proteins are largely concerned with the storage and transmission of genetic information
   - **(b)** Proteins can form structural elements of cells and tissues
   - **(c)** Proteins form when amino acid residues are joined by phosphodiester bonds
   - **(d)** Proteins can act as catalysts
   - **(e)** A protein composed of 336 amino acid residues would contain 337 peptide bonds.
3. How does SDS help disrupt the plasma membrane?
4. How is contaminating RNA removed from a DNA preparation?
5. What is the major use of ethanol in the preparation of samples of DNA?
6. Pancreatic ribonuclease is an endonuclease that shows *b*-type specificity at nucleotide residues on the 5′ side of pyrimidine nucleotides. List the possible terminal nucleotides of its products.
7. Which of the following base sequences would not be potential recognition sites for REs? Explain your choice.

(a) 5′ . . . GAATTC . . . 3′	**(b)** 5′ . . . CATATG . . . 3′
3′ . . . CTTAAG . . . 5′	3′ . . . GTATAC . . . 5′
(c) 5′ . . . GACGGAT . . . 3′	**(d)** 5′ . . . CAATTG . . . 3′
3′ . . . CTGCCTA . . . 5′	3′ . . . GTTAAC . . . 5′
(e) 5′ . . . CATTAG . . . 3′	
3′ . . . GTAATC . . . 5′	

8. Why is PAGE unsuitable for analysing plasmids?
9. If the proteins you wished to analyse by Western blotting are **(a)** acidic (anionic) and of relatively low M_r, and **(b)** relatively hydrophobic and of M_r 65,000, respectively, what type of membrane would give optimal binding?
10. Outline a test based on Western blotting that would help in the diagnosis of infection with HIV. Assume that a sample of serum is available from the patient.

Answers to self-check and end-of-chapter questions are available at the end of the book.

15

Laboratory automation

Tim James

Learning objectives

After studying this chapter, you should be able to:

- Identify the major benefits of automation
- List the types of specimens that are processed in a high-volume automated laboratory
- Describe the common automation units including those associated with the pre- and post-analytical processes
- Recognize the role of automation in the efficiency and productivity of all biomedical disciplines

Introduction

Laboratory automation has been defined by the Association for Laboratory Automation, a leading organization for laboratory automation, as 'a multi-disciplinary strategy to research, develop, optimize and capitalize on technologies in the laboratory that enable new and improved processes and products'. All clinical laboratories utilize automation albeit to a variable extent, dependent on the discipline and requirements of the service. The application of this automation has produced improvements in efficiency and effectiveness of testing and improved consistency of analysis.

The investigations and analyses undertaken by hospital laboratories can be categorized in a number of ways. Traditionally, this has been based on the association of any given laboratory test with a defined biomedical science discipline, and conventionally these are haematology, clinical biochemistry, immunology, microbiology, histology, and cytology. Automation is evident in each of these disciplines and you can find discipline-specific descriptions of automation throughout the accompanying volumes in this book series. However, the availability of analytical systems with increased sophistication in terms of both robotics and underlying information technology has yielded a model of laboratory operation that crosses the traditional discipline boundaries.

Figure 15.1 outlines the basic pathway that a sample may undergo within most clinical laboratories. The sample will arrive in the laboratory and be registered, usually by booking the specimen into the laboratory computer system. The sample is then prepared for analysis, presented to analysers, and undergoes analysis. Following analysis the sample will be archived, stored, and will eventually, after a defined period of retention, be disposed of in an appropriate and safe manner.

Automation of the central analytical step usually using instrumentation of varying complexity is apparent in all hospital laboratories. However, the preceding steps and those following analysis can also be automated. While the timescale and complexity of steps may vary between each of the individual disciplines, the underlying process is the same. Consequently, there has been a convergence of automation and technology across the conventional disciplines resulting in the development of what is termed the total laboratory automation or the **blood sciences laboratory**. This chapter will review automation in the context of general automation features and their associated benefits.

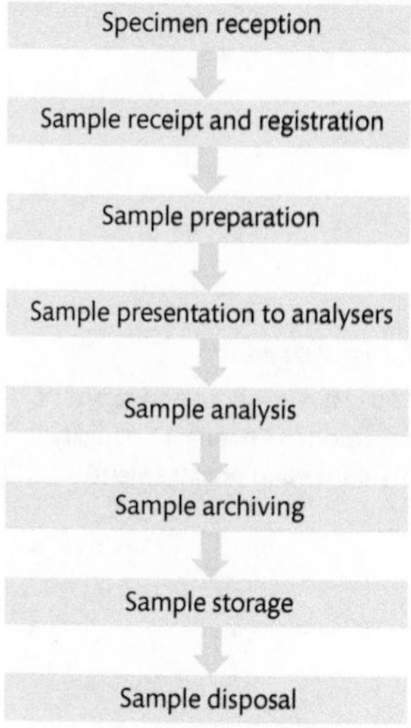

Specimen reception

Sample receipt and registration

Sample preparation

Sample presentation to analysers

Sample analysis

Sample archiving

Sample storage

Sample disposal

FIGURE 15.1

The basic steps undertaken for most samples in clinical laboratories utilize a common process path. Most of these steps can be automated.

15.1 Benefits of automation

The most apparent benefits of full automation compared with manual or semi-automated analysis are the improved speed and efficiency in processing samples to produce test results for clinical management. It has been estimated that growth in workload of clinical laboratories is between 5% and 10% per annum. The management of this expanding workload has only been possible through greater utilization of automation of increasing sophistication and capacity. However, the benefits are much broader than this, as detailed in Table 15.1. The ability of automation to process samples using a common procedure has resulted in simpler, more consistent sample handling and this has been associated with a reduction in processing errors. A consistent finding in laboratory audits is that the majority of errors occur in the pre-analytical phase. This is the step most associated with manual sample handling. Equally, manual laboratory procedures may involve several steps, where the sample may need to be manipulated and analysed. Misidentification of samples and errors can occur at each of these steps even when there are significant quality measures in place. Evidence suggests that automated processes, whether they are pre-analytical, analytical, or post-analytical, are systematic and associated with lower error rates. Automation has produced a number of improvements with respect to health and safety.

Cross reference

Audit is discussed in Chapter 17, 'Quality assurance and management'.

Manual handling of uncapped, open samples can expose staff to a range of hazardous organisms. Automating laboratory processes reduces the number of occasions an individual may need to handle specimens and thereby reduces the risk of exposure to these hazards. Equally, in a laboratory setting where several thousand specimens are manipulated, there is a risk of repetitive strain injury. Replacing the manual manipulation of samples by an equivalent automated process will remove this risk. For example, many automation systems can uncap and recap tubes, steps that if undertaken repeatedly by the same member of staff carry the risk of straining hand, wrist, and shoulders, and temporary or permanent musculoskeletal injury.

Cross reference

You can read more about health and safety in Chapter 4.

TABLE 15.1 **General benefits of automation**

Benefit	Impacts
Efficiency benefits	Improved sample throughput Reduced sample splitting Reduced sample retrieval requirements Reduced staff time required for manual sample handling
Improved health and safety	Reduced exposure of staff to clinical specimens and therefore clinical infection Reduced risk of repetitive strain injury
Error reduction	Reduced risk of samples being mis-aliquoted into the wrong secondary containers Reduced risk of samples being misplaced as the sample location can be monitored at all stages of the laboratory process
Improved patient care	Reduced volume of blood and number of tubes collected Reduced and more consistent turnaround times for test results resulting in improved patient management

A consideration discussed widely in laboratories is the time taken for a test result to be available to the clinician after the specimen is taken from the patient. This is commonly referred to as the turnaround time (TAT). Most laboratories use some measure of TAT as a quality indicator of efficiency. The TAT can be measured in two ways, the *total* TAT and the *within laboratory* TAT. Both can be influenced by laboratory automation. The total TAT is the time taken from the collection of the sample to the time the report is available to the clinician to interpret the result. The within laboratory TAT is the time taken from the receipt of the specimen in the laboratory to the laboratory authorization of the result.

TAT requirements vary with clinical setting and the particular investigation, for example the TAT requirement for glucose analysis in an unconscious patient with diabetes will need to be rapid and is measured in minutes. In comparison, the TAT requirement for cholesterol testing undertaken to assess cardiovascular risk in a healthy, ambulatory patient can be longer and will be measured in hours. Reduced and more consistent TATs are apparent with increased levels of automation and this may improve the timeliness of clinical decisions and patient treatment.

Cross reference

You can read about accuracy and precision in Chapter 17, 'Quality assurance and management'.

SELF-CHECK 15.1

How might a reduced test turnaround time help patient care?

Sequential generations of automated analysers have produced more consistent and reproducible results, as evidenced by reduced imprecision. They are more robust, as measured by the proportion of time the analyser is in operation. The related changes and benefits have included:

- a reduction in the specimen volume required for individual assays
- a reduced usage of reagent
- increased analyser capacity relative to physical size.

The lower specimen volumes required by newer automated analysers compared with manual techniques and older analysers translates into the advantage that a greater number of tests may be undertaken from a single clinical sample. This is particularly important in children and neonates where blood collection may be difficult, and is also beneficial in other patient groups where blood collection is challenging, such as patients with learning difficulties or those with fragile veins (e.g. the very elderly).

The reduced reagent volume has a great number of associated advantages. Total reagent costs will be reduced if less is used and less chemical waste is generated, thereby reducing the environmental impact. Reagent storage requirements, both on the analyser itself and within laboratory storage facilities, are reduced if lower volumes are required. This further reduces the space requirements for both room temperature and refrigerator storage.

Modern automated instrumentation is often considered physically large, but the analytical capacity, as measured either as samples or tests per hour, relative to their size has increased significantly. Therefore, while most laboratories have been modified in design and layout to accommodate automation, the total laboratory space requirement has often remained the same. The components utilized within analysers have become increasingly sophisticated, and particularly with respect to liquid handling, the accuracy and precision of sample and reagent delivery is both highly precise and robust. The consequences of these improvements are that total assay imprecision has improved and methods are highly reproducible. Most instrument manufacturers have a programme of monitoring to identify patterns of analyser component failure. This informs component supplier requirements and subsequent instrument design. Increasingly remote diagnostics, where a manufacturer will monitor critical instrument parameters electronically using the internet, may even predict analyser deterioration before it actually impacts on test results. The best design features utilized within automation systems are maintained and copied by competitors, and therefore it is possible to see common features in a range of different suppliers' instrumentation. The overall analyser component failure rate has reduced and this has produced automated instrumentation that is operational for a greater proportion of time. This reduces demands on both laboratory operators and service engineers.

Automation allows flexible use of staff time as large automated systems require less human intervention for operation and a greater number and proportion of laboratory investigations can be undertaken by automated systems. This could be considered to be an opportunity to reduce overall operating costs, but the majority of laboratories will deploy any staff released by automation and redirect them to analytically more demanding laboratory areas or to improving laboratory quality. However, as automation hardware is expensive, its introduction into clinical laboratories requires careful consideration of the clinical necessity and staff saving against capital expenditure.

Key points

TATs are key indicators of laboratory performance. Turnaround time requirements vary with respect to investigation/test and clinical context. Automation can lead to significantly reduced and more consistent TATs.

15.2 Collection of suitable samples and delivery to the laboratory

The majority of the samples processed in a **core automated laboratory** will be blood specimens taken from a vein by phlebotomists. A smaller proportion, usually less than 10%, of the laboratory investigations are undertaken on other clinical specimens, for example urine. Clinicians often require a number of assays to be undertaken simultaneously and therefore multiple assays may be required on a patient at the same time. In the UK the average test to request ratio of a typical hospital laboratory is between 6 and 8; however, this will vary depending on the local arrangements and requirements as described in Box 15.1.

Cross reference

Blood sampling is described in Chapter 6, 'Samples and sample collection'.

Key points

A laboratory test may be considered an individual investigation or assay, for example measuring blood glucose concentrations. A request is regarded as one or more tests asked for by the clinician at a single point in time; for example, a patient who is tired all the time may require a number of different tests to assess possible causes. A request is not the same as a profile, which is usually a number of clinically or metabolically related tests grouped together, for example a liver function profile, which will include several tests associated with this organ's functions.

It would be convenient if all blood tests could be processed using a single specimen tube. However, many blood constituents require specific preservation conditions to stabilize the test component. Consequently, a number of different tube types are required depending on the combinations of tests required. Commonly available specimen tubes used in routine diagnostic laboratories on automated systems are presented in Table 15.2 together with commonly observed colour coding systems used to identify the different additives.

BOX 15.1 How to calculate the test to request ratio

The average test to request ratio is simply the number of tests divided by the number of requests. A laboratory that undertakes 4,742,200 tests on 750,555 requests per annum therefore has a test to request ratio of 6.31.

TABLE 15.2 Commonly used collection tubes and associated colour coding (see also Table 6.2)

Tube type	Additive	Commonly used tube cap colours
Serum	Clot activator	Red, brown
Heparinized plasma	Lithium heparin	Green, orange
EDTA blood	Potassium EDTA	Lavender, pink
Citrated plasma	Sodium citrate	Blue, green
Fluoride oxalate plasma	Sodium fluoride, potassium oxalate	Grey, yellow

EDTA, ethylenediaminetetraacetic acid.

A serum collection tube contains no anticoagulant and the sample will form a clot: this sample can be centrifuged to yield serum. Serum requires time, usually about 30 minutes at room temperature, for the clot to form. To reduce the clotting time, many serum tubes are manufactured containing an additive to accelerate clotting so that the sample may be centrifuged more rapidly. Serum is suitable for a wide range of tests, including the vast majority of biochemistry, immunology, and virology assays. The tube often contains a gel that has a density that is greater than that of serum but lower than that of the clot. Therefore, when the sample tube is centrifuged, the contents are distributed into three components: a clot at the bottom of the tube; the gel in the middle, forming a barrier; and the serum on the top (see Figure 10.14). The majority of automated instrumentation analyses this serum directly from the original sample tube, a process called 'primary sampling', as distinct from transferring the serum into a secondary container. Many automation systems can produce a secondary aliquot of the specimen, but it is more convenient in terms of efficiency not to do this. Most automation systems benefit significantly from primary sampling.

SELF-CHECK 15.2

Why is it necessary to use a number of different specimen tubes when undertaking laboratory investigations?

Potassium ethylenediaminetetraacetic acid (EDTA) is an anticoagulant by virtue of its ability to chelate Ca^{2+}, which is essential for the clotting process. Blood collected into a tube containing EDTA will not clot and a whole-blood sample is produced that preserves the cellular constituents. This tube type is most frequently used for the analysis of standard haematology investigations such as the full blood count and erythrocyte sedimentation rate. It is also a suitable specimen tube for a range of other investigations including the analysis of glycated haemoglobin, a common test used for the diagnosis and monitoring of diabetes mellitus.

Sodium citrate, like potassium EDTA, binds Ca^{2+} and prevents coagulation. This tube may be centrifuged to yield citrated plasma for coagulation studies. A critical factor for this tube type is the ratio of blood to anticoagulant, which is essential for the accurate analysis of coagulation parameters. To ensure the correct volume of blood is added to each collection tube an indicator line is marked onto the side of the tube as a guide.

Lithium heparin is an effective anticoagulant as it inhibits the prothrombin to thrombin step of clotting. Centrifugation of this sample type produces heparinized plasma which is used for many investigations undertaken in clinical biochemistry. The advantage of this sample type over serum is the ability to centrifuge almost immediately after collection. An additional advantage of lithium heparin blood is that the plasma volume obtained is greater than the comparative yield from an equivalent volume of blood converted into serum. While only a small effect, this can be beneficial for paediatric specimens and their naturally smaller sample sizes. It is common for lithium heparin tubes to contain a gel to enable primary sampling in a similar manner to that described for serum.

The fluoride-oxalate blood collection tube contains two additives: sodium fluoride, which binds Mg^{2+}, a co-factor of several enzymes associated with the glycolytic pathway thereby inhibiting glycolysis, and potassium oxalate, which binds Ca^{2+} and prevents coagulation. This tube is usually associated with the analysis of glucose and lactate, assays that are adversely affected by continued glycolysis, which will occur in the other specimen tubes.

The development of laboratory automation systems has been helped considerably through standardization of specimen container shape and size, and manufacturers produce the range of tubes described above in a standard format with respect to their physical dimensions and capping mechanism. The improved standardization of tube shape and size has improved sample presentation to analysers, the manipulations needed to remove and replace caps, and their transfer by robotic arms. The two most common diameters of specimen tube are 13 and 15 mm, and the commonly encountered tube heights are 75 and 110 mm. Many laboratories attempt to standardize the size of tube used by the automated system so, for example, all the laboratory users are supplied with tubes of one size, for example, 13 × 75 mm.

When the sample tubes have been collected from the patient they may be transported to the laboratory. The process of conveying the specimen from clinical to laboratory area within a hospital has been automated through the use of pneumatic air tubes. These tubes connect the clinical areas to the laboratory, and convey the specimen safely and promptly to enable faster total turnaround times.

Key points

The application of automation to clinical laboratories was greatly enhanced with the introduction of separator gels that are incorporated into the blood collection tubes during manufacture. These gel-containing tubes allow processing of the plasma and serum samples through all steps of the analytical process, from collection through to disposal, in a single vessel, removing the requirement for specimen transfers.

15.3 Specimen reception considerations in the core automated laboratory

Samples arrive in the clinical laboratory with either paperwork in the form of a request card or with an electronically linked request that is conveyed with a label on the specimen tube. A request card is the traditional manner for making a laboratory request and contains information about the patient, the specimen, and the investigations required. This information is conventionally transcribed by manual data entry into the laboratory computer system. Electronic requesting is in widespread use and has improved efficiency and data quality compared with a manual request as the entry process is partly or completely electronic. Whatever method of sample registration is utilized, the interface between the specimen reception area and the automated instrumentation is a critical factor in overall laboratory efficiency. Automation alone does not produce the benefits detailed in Section 18.2; it is its utilization in the overall laboratory process that is critical.

Specimen reception is the usual location for an adhesive sample bar code to be applied to each sample tube thereby allowing its unique identification and routing through the laboratory automation. An alternative process exists in which the bar code is applied at the point of collection in the clinical or phlebotomy area. The accurate reading of the bar code label is required at many points in the automation process from the specimen reception workstations, as well as all of the interconnected devices and analysers. Consequently, the bar code quality in terms of clarity to bar-code readers needs to be high enough to allow consistent reading and identification at each step. The format may be a simple number (numeric) or may allow inclusion of one or more letters into the specimen number (alpha-numeric). The key aspect of the format is to provide unique identification for each specimen.

The interface between specimen reception and automated systems can be described using three basic models presented in Figures 15.2 (a–c):

- a traditional interface model
- a semi-automated interface model
- a totally automated interface model.

Model 1, Figure 15.2(a), represents the traditional laboratory in which the specimen reception process includes the step of sorting specimens into a range of racks appropriate and suitable for presentation

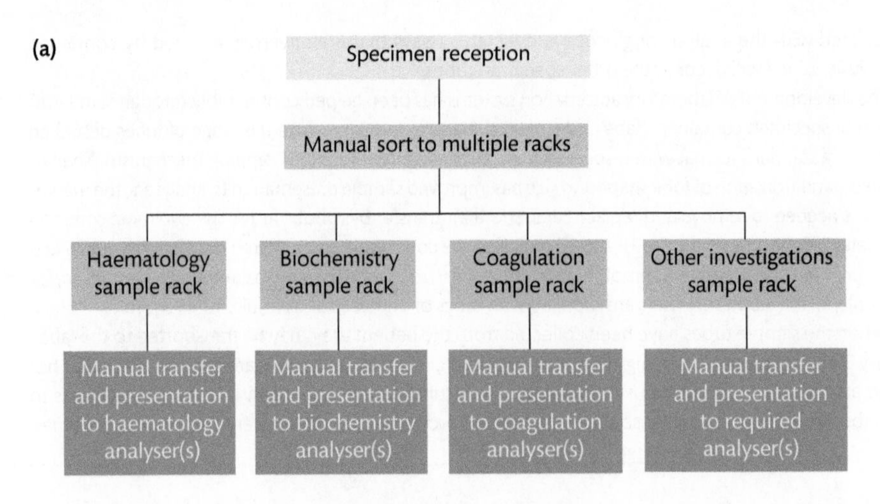

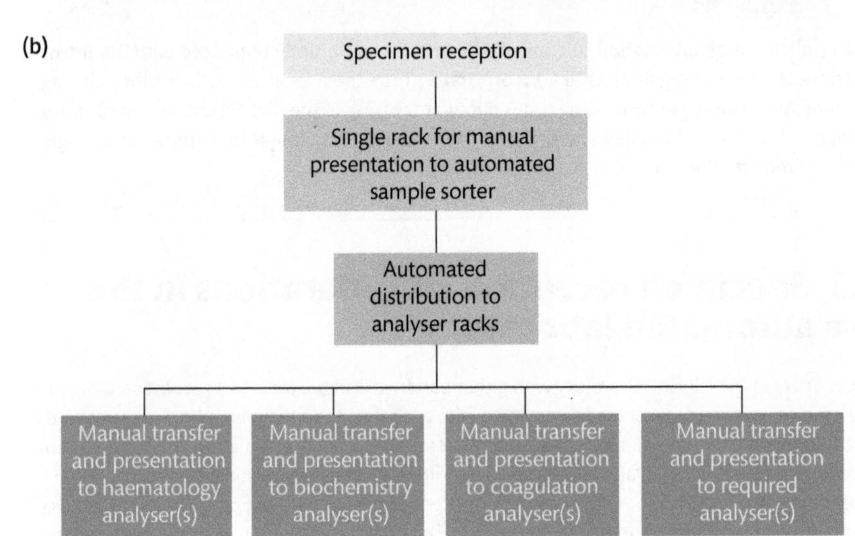

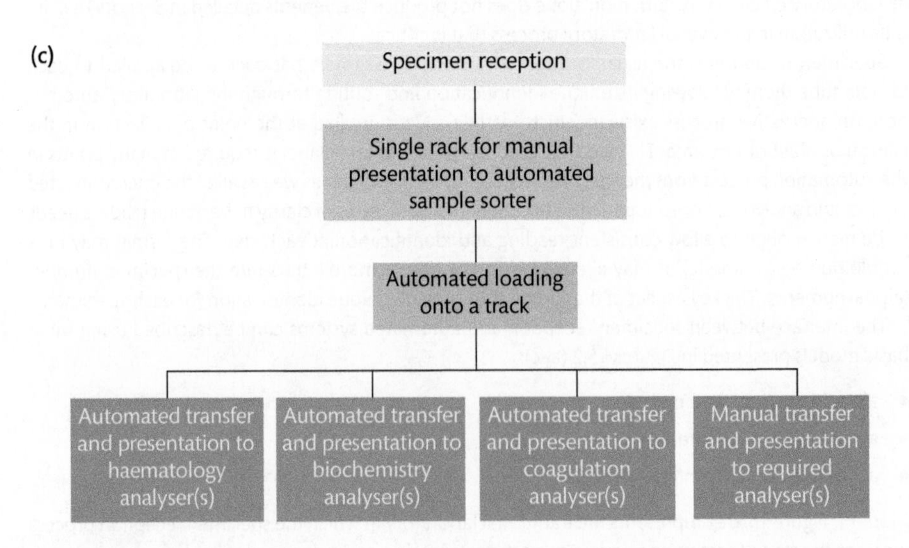

FIGURE 15.2
Specimen reception; automation interface models. (a) Samples are sorted and presented to individual analysers manually. This is complex with respect to the processes on specimen reception. (b) Samples are placed into a single rack that is presented to a standalone automated sample manager that distributes the samples to analyser racks. Racks are removed and presented to the analysers manually. (c) Samples are placed into a single rack that is loaded onto the input rack of an automated tracking system with automated sorting, distribution, and presentation to analysers.

to different analysers or laboratory areas. The disadvantage of this approach is that the specimen reception workstation may be relatively cluttered and that multiple manual steps may be required to present samples to analysers. Model 2, Figure 15.2(b), a model in which the specimen reception workstation is simplified by using a single sample rack into which the bar coded specimens are placed. The single rack is then presented to a standalone automated sample management system that will then distribute the specimens into different analyser racks. The subsequent presentation to the analysers is manual. This model may be particularly beneficial if a laboratory utilizes analysers from more than one manufacturer that cannot be connected through a single tracking system. The system is also beneficial if laboratory space and/or layout is constrained and analysers are in different laboratory rooms and areas. Standalone automation can be utilized to improve the efficiency of multidisciplinary laboratories but is most often found in clinical biochemistry laboratories. Figure 15.2(c), represents a system in which both the sorting and conveyance of samples to analysers is achieved with a track system. Again, the work of the specimen reception is simplified by having only one rack in which to place all the specimens. In reality, specimen reception areas usually operate a mix of these models depending on the automation available (primarily whether the laboratory uses standalone sample management systems as opposed to tracks, as described in Section 15.4), the testing repertoire, and laboratory organization. Most automated laboratories review their specimen reception processes when new or different automation is introduced, with the primary aim of streamlining all the steps, reducing variation, and removing unnecessary steps.

Cross reference

Automation is also explored in Chapter 2, 'Automation' in the companion text, *Clinical Biochemistry*.

SELF-CHECK 15.3

Specimen reception is the only place where bar code labels are applied to specimen tubes. True or false?

15.4 Tracked automation systems and the core automated laboratory

A tracked automation system in a core automated laboratory can be considered as a group of modules or units each with a specific function. The number and type of units/modules within the automation system is dependent on the repertoire of assays and the number of samples/tests requiring analysis. There is also a general requirement for instrumentation back-up, particularly if continuous analysis is required. As no two clinical laboratories are identical, a range of different combinations of instrument can be found connected to tracking systems to meet the requirements of the specific laboratory. You can see an outline design of an automated tracked system in Figure 15.3.

The modules/units of an automated tracking system can be categorized as:

- modules that prepare samples for analysis
- modules that analyse
- modules that undertake post-analytical processes.

Preparation modules include sample input systems, centrifugation units, and tube decappers. The introduction of samples into an automated system requires the specimens to be placed into a rack that can then be loaded onto a sample manager or input module connected to the track. The rack capacity varies depending on the design of the system and may be as low as ten or as high as 100. Most systems also offer the facility to introduce one or a group of samples that can be given priority to ensure they have a faster turnaround time. Increasingly, automation systems employ hoppers into which large numbers of specimens may be placed and fed onto the track, thereby eliminating the need for placing samples into racks. To provide a flexible approach to sample loading some systems have both a bulk specimen loader and a rack-based system for sample presentation, thereby providing greater flexibility. The subsequent movement of the samples around the automation system may use pushing and pulling mechanisms to position small racks of samples or more commonly robotic arms that can lift and convey individual specimens (Figure 15.4). The robotic arm usually moves the sample into an individual specimen holder that is positioned on the track and is referred to

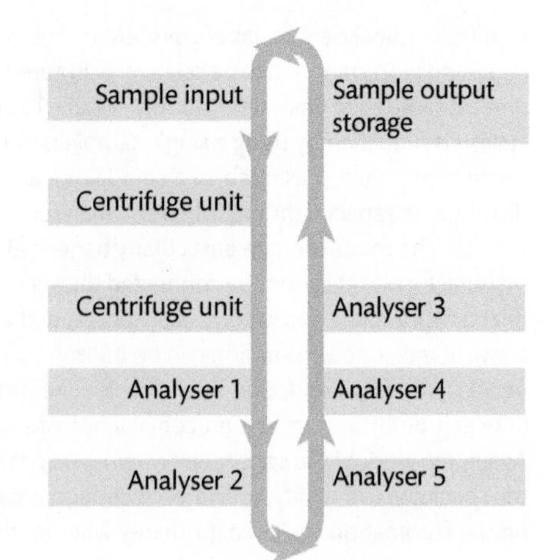

FIGURE 15.3

A schematic of an example tracking system and layout. The modules and units for any given system will vary depending on local laboratory requirements.

as a carrier or a puck. This carrier, or puck, moves around the track and will be directed to the modules and analysers as required for each particular sample. As the specimen progresses around the track the position of the tube may be monitored by re-reading the bar code at bar code-reading stations at set intervals around the track. This enables a constant check to be made on the most appropriate route for the specimen tube to take. Alternatively, each of the pucks/carriers can be identified with a radio frequency tag that is monitored as it progresses around the track. When the sample is placed into each puck an association is established between them and this will be maintained until the sample is removed.

The majority of investigations on blood specimens require the plasma or serum component of blood to be obtained, for which a centrifugation step is required. This can be achieved with a

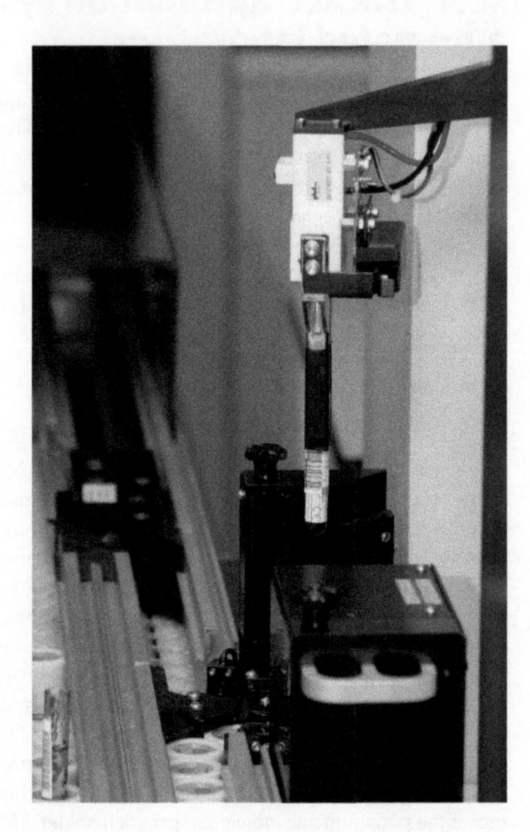

FIGURE 15.4

An example of a robotic arm placing a sample onto a tracking system.

centrifuge unit integrated into the automation system. The centrifuge unit will weigh specimen tubes and load them into centrifuge buckets in an even manner to achieve a balanced load. The buckets are then placed into the centrifuge and spun for a defined period of time, usually set between 8 and 10 minutes, with application of a force of approximately 1000 *g*.

Cross reference

You can read more about centrifugation in Chapter 10.

SELF-CHECK 15.4

How can the progress of a specimen tube on an automation system be followed?

The effect of the centrifugation is to bring the serum or plasma to the upper part of the collection tube where it can be sampled. Following centrifugation, the buckets are lifted from the centrifuge and the spun samples removed. While some analysers have probes that can penetrate the rubber caps of the tubes to remove the plasma/serum/blood samples, most require the caps to be removed. Automated decappers provide this function and these can be found in three general formats:

- as part of a standalone pre-analytical system
- as part of an integrated centrifuge-decapper module
- as a distinct device fitted onto a track system.

The general movement of a decapper is a hold, twist, and lift movement. Once removed, the caps are dropped into a waste receptacle. The waste receptacle may have varying levels of sophistication to alert the operator to the requirement to periodically empty the waste container. Once the cap has been removed the tube proceeds to the next stage of the automation process. In some instances, it may be necessary to produce aliquots of the primary sample as part of the standard automated process on the particular instrument, or because the specimen requires splitting for analysis across several analytical platforms. Sample aliquoting may also be required for referral to another laboratory.

SELF-CHECK 15.5

How do automated decappers reduce the risk of musculoskeletal injury to staff?

Each sample will be routed to one or more of the analysers connected to the track. In many systems the sample remains on the track and the analyser sample probe will extend out and over the track to a pre-set, defined position. The sampling probe will then be lowered until it detects the surface of the liquid sample, it will progress beneath the surface to a pre-set depth and liquid will be sampled and transferred into the analyser. This sample presentation is sometimes called 'point in space' sampling. Alternatively, a robotic device can remove the sample from the track and place it within a distinct and separate sampling area from which the analyser can sample.

The organization of the analytical modules reflects the balance of testing across each of the standard automated chemistry, immunoassay, haematology, coagulation, and urinalysis analysers. Table 15.3 illustrates some common testing areas undertaken on such systems. Each of these individual analysers will generally have a large test capacity ranging from a few hundred tests per hour through to several thousand. In Figure 15.3 the analysers may be the same or any combination of analyser type. Further detailed descriptions of the attributes of the individual analyser types can be found in the discipline-specific books in this series.

Post-analysis samples may be presented to a storage system. This may hold a variable number of specimens, usually in numbered racks within which the specimen is allocated to a specific, *x y* coordinate-based position. The storage capacity may be many thousands of samples with multiple units to provide greater capacity and these are typically refrigerated to reduce specimen evaporation and deterioration. Samples may be re-accessed for further analysis if re-runs or additional tests are required and in these circumstances the samples may be retrieved manually or automatically reloaded onto the track, and re-routed to the designated analyser. Automated disposal of the specimens from

TABLE 15.3 Typical automated analyser units and examples of the associated test repertoire

Traditional discipline	Analyser unit	Assays for:
Clinical biochemistry	General chemistry	Sodium, potassium, chloride, bicarbonate, urea, creatinine, calcium, phosphate, albumin, total bilirubin, alanine transaminase, alkaline phosphatase, C-reactive protein, glucose, total cholesterol, HDL cholesterol, triacylglycerol, uric acid, magnesium, total protein, creatine kinase, aspartate transaminase, lactate dehydrogenase, gamma-glutamyl transferase, amylase, lipase, iron, transferrin, conjugated bilirubin, lactate, paracetamol, salicylate, ethanol, zinc, ammonia, bile acids, angiotensin converting enzyme
	Immunoassay	Thyroid stimulating hormone, total and free thyroxine, total and free tri-iodothyronine, prostate-specific antigen, alpha-feto protein, carcinoembryonic antigen, β-human chorionic gonadotrophin, Ca-125, Ca 19-9, cortisol, testosterone, oestradiol, progesterone, androstendione, follicle-stimulating hormone, adrenocorticotrophin hormone, growth hormone, insulin-like growth factor-1, drugs of abuse assays
Haematology	Haematology analyser	Full blood count, red cell count, white cell count, haematocrit, haemoglobin
	ESR analyser	ESR
	Slide maker	Preparation of blood films
	Immunoassay	Vitamin B_{12}, serum folate, red blood cell folate, erythropoetin
	Coagulation	Pro-thrombin time, activated pro-thrombin time, international normalized ratio, fibrinogen
Immunology	Immunoassay	Autoantibodies, allergens, complements C3 and C4, immunoglobulins
Virology	Immunoassay	Hepatitis A markers, hepatitis B markers, hepatitis C markers, HIV, CMV, rubella, syphillis

HDL, high-density lipoprotein; ESR, erythrocyte sedimentation rate; CMV, cytomegalovirus.

storage systems after a predefined period of time also reduces staff exposure to samples and leads to staff efficiencies. Practical considerations for large automation systems include the requirements for services. These include:

- suitability of space, including the structural integrity of flooring as the systems may be very heavy, walking space around the instrumentation, and access for routine use and maintenance
- water, which may be both a normal tap water quality requirement and/or deionized from a deionization system
- power supply as large tracking systems will require a specific, dedicated power supply; it is standard practice to protect the power supply to each unit with an uninterruptible power supply in case of power failures
- drainage of waste as the waste drainage is best achieved with floor drainage rather than pumps
- information interchange because most automated systems have a complex array of computers controlling different aspects of the system operation.

Key points

The majority of modules connected to a track will be conventional automated analysers that utilize a particular technology suitable for the analyses being undertaken. The other modules undertake sample processing functions, such as centrifugation, archiving, and storage.

15.5 Automation in wider laboratory settings

The biomedical science disciplines most often associated with automation are clinical biochemistry and laboratory haematology. However, all of the biomedical science disciplines, even those that have traditionally been associated with the manual skills and dexterity of biomedical scientists, such as histology, cytology, and microbiology, now have significant automation in use within clinical laboratories. The benefits of automation in any of the laboratory areas mirror those detailed in Section 15.1. Many laboratory processes require a series of independent manual steps and at each one there is potential for sample misidentification. Automation of the more traditionally manual areas of biomedical science is often associated with a reduction in the number of these manual steps at which this misidentification may occur. This can therefore be seen as a significant improvement in patient safety, as it reduces laboratory errors. The introduction of automation in any of the laboratory areas also requires a reassessment of laboratory operation. This often involves moving towards a laboratory work flow that is continuous, rather than as a batch mode, and may also involve laboratory opening for longer periods of the day to maximize the benefits of automation.

In histology, several discrete stages of the preparation of histology slides have been automated. Automated tissue processors for preparation of tissues for sectioning, automated staining instruments, and automated cover slippers are all used routinely. The benefits of these individual automated devices are consistency and efficiency. Linking each of these individual discrete automated elements together into integrated single systems using robotics is developing. This type of automation is also an integral part of automated liquid-based cytology.

Cross reference

You can read more about liquid based cytology in the companion volume, *Cytopathology*, in this book series.

While virology has utilized immunoassay technology and analysers (as described earlier) since the 1980s, the automation of bacteriology processes have only been evident since around 2005. The application of clinical samples to growth media within plates, the allocation of the plates to appropriate incubation conditions, and subsequent retrieval for visual assessment of colony growth has conventionally been undertaken manually. Modular instrumentation can be used to automate these processes (Figure 15.5). Staff work at designated workstations where they are presented with the pre-labelled plate sets for inoculation and sample application. Once prepared the plates can then be conveyed automatically to incubators. After defined incubation periods the plates are automatically re-presented for review. At each stage a digital image may be taken, which enables comparison of

FIGURE 15.5

The Kiestra system: an example of advanced automation in the bacteriology laboratory.

plate growth at different time points and provides greater flexibility to when and where plates are read and interpreted.

Automated systems in blood transfusion laboratories can be used for the majority of routine testing, including blood grouping and subgrouping, antibody screening, and cross-matching. These systems can be highly reliable, sensitive, and specific.

In molecular diagnostics, the polymerase chain reaction (PCR) is a powerful analytical technique with a wide range of clinical applications in infectious diseases, cancer and genetic disorders. The instrumentation used for PCR has a range of designs related to the thermocycler, the assay features, and the detection system, and an increasing sophistication with respect to automation. Real-time PCR, in which the amplified genetic material is quantified after each cycle, is becoming a standard application. Next-generation sequencing offers even greater sophistication and power and the greatest challenge with this latest technology is the amount and complexity of the data generated. The automated instrument platforms have a growing repertoire of applications.

Cross reference

PCR is described in Chapter 14, 'Molecular biology techniques'.

In several clinical laboratory areas there has been a high degree of professional expertise in pattern recognition, for example in reviewing histology and blood slides, or identification of colony formation on a microbiological plate. The use of digital images in such disciplines can be considered an aspect of automation and allows remote review of images. It is also possible to develop pattern recognition software and this is an emerging field.

SELF-CHECK 15.6

Which of the following laboratory areas can benefit from automation?

1. Coagulation
2. Immunology
3. Microbiology
4. Virology
5. Haematology.

Summary

- All laboratory disciplines utilize laboratory automation to improve efficiency of processing.
- The extent of laboratory automation use will depend on the size and complexity of the clinical laboratory service.
- Key aims of automation are process efficiency and consolidation.
- Typical clinical laboratory workload growth of between 5% and 10% per annum has been managed by increasing levels of automation, rather than by increasing staff numbers.
- The design of modern automation systems aims to minimize human intervention in the total analytical process.
- The test turnaround times for large automated track systems are generally considered to be faster and more consistent than separate analytical units.
- Good design features within automation systems are maintained and copied by competitors and, therefore, it is possible to see common features in a range of different suppliers' instrumentation.

- Pre- and post-analytical processes such as centrifugation, decapping, and recapping of specimen tubes, aliquoting, archiving, and storage of samples can be integrated into a single automated process by using tracking systems.
- Automation is prevalent well beyond the highly automated biomedical science areas, and many microbiology, histology, cytology, and blood transfusion laboratories have a range of automated instrumentation.

Further reading

- **Angeletti S, De Cesaris M, Hart JG, Urbano M, Vitali MA, Fragliasso F, Dicuonzo G. Laboratory automation and intra-laboratory turnaround time: experience at the University Hospital Campus Bio-Medico of Rome. *Journal of Laboratory Automation* 2015; 20: 652-8.**
- **Hawker CD. Laboratory automation: total and subtotal. *Clinical and Laboratory Medicine* 2007; 27: 749-70.**
- **Horowitz GL, Zaman Z, Blanckaert NJ, Chan DW, Dubois DW, Dubois JA, et al. Modular analytics: A new approach to automation in the clinical laboratory. *Journal of Automation Methods and Management in Chemistry* 2005; 2005: 8-25.**
- **Melanson SEF, Lindeman NI, Jarolim P. Selecting automation for the clinical chemistry laboratory. *Archives of Pathology and Laboratory Medicine* 2007; 131: 1063-9.**
- **Rhoads DD, Novak SM, Pantanowitz L. A review of the current state of digital plate reading of cultures in clinical microbiology. *Journal of Pathology Informatics* 2015; 6: 23.**
- **Sarkkozi L, Simson E, Ramanathan L. The effects of total laboratory automation. *Clinical Chemistry* 2000; 46: 751-6.**
- **Wheeler M. Overview of robotics in the laboratory. *Annals of Clinical Biochemistry* 2007; 44: 209-18.**
- **Zaninotto M, Plebani M. The "hospital central laboratory": automation, integration and clinical usefulness. *Clinical Chemistry and Laboratory Medicine* 2010; 48: 911-17.**

Useful Website

- www.slas.org/

Questions

1. Why do blood specimens require centrifugation on an automated system?
2. Sodium fluoride is a suitable preservative for glucose because:
   - **(a)** It inhibits coagulation
   - **(b)** It promotes glycolysis
   - **(c)** It inhibits the formation of thrombin from prothrombin
   - **(d)** It inhibits glycolysis
   - **(e)** It promotes coagulation.
3. What is the average test to request ratio for a laboratory that undertakes 606,531 requests and 5,365,852 tests per annum.

4. How do automation units reduce the risk of infection to laboratory staff?

5. True or false? If a re-analysis of a test is required samples may be retrieved from the automation system's sample storage area.

6. True or false? The test turnaround time may be measured in two ways, the total turnaround time and the within laboratory turnaround time.

7. The gels added to blood collection tubes to enable primary sampling have:
   - **(a)** A density higher than serum
   - **(b)** A density higher than plasma
   - **(c)** A density higher than red blood cells
   - **(d)** A density higher than a blood clot
   - **(e)** A density between serum and plasma.

Answers to self-check and end-of-chapter questions are available at the end of the book.

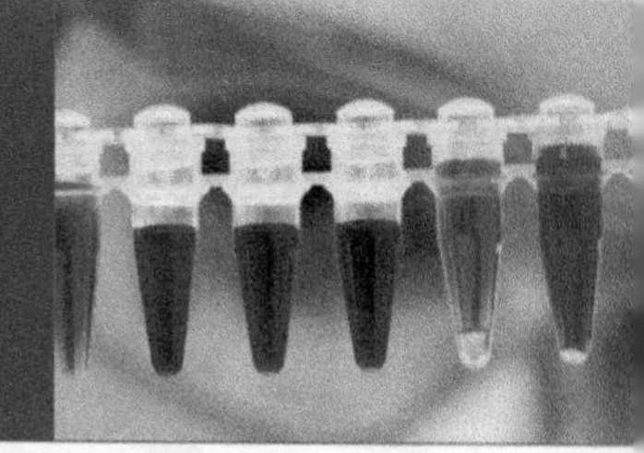

16

Point-of-care testing

Jan Still, Lynda Petley, and Garry McDowell

Learning objectives

After studying this chapter, you should be able to:

- Describe the basis and regulations that underpin point-of-care testing (POCT)
- Discuss the requirements for POCT
- Discuss the differences between internal quality control, external quality assessment, and quality assurance in POCT
- Outline the equipment procurement process
- Describe the audit information required to minimize risk in the provision of a POCT
- Discuss the provision of a patient focused POCT service

Introduction

Point-of-care testing (POCT) is the provision of a diagnostic pathology testing service outside the traditional clinical laboratory setting and physically closer to the patient. This type of testing may be carried out by healthcare professionals in the ward, clinic, or general practitioner (GP) surgery, by high street pharmacists, or by the patients themselves. This chapter describes POCT and its impact on the provision of pathology services, and outlines those aspects that must be considered when providing an effective POCT service.

POCT is not a new concept, but earlier technologies paved the way for new devices in response to the demand for a more patient-centred health service. Thus, urine dipstick testing has been an accepted practice in healthcare for many years and blood glucose meters were first introduced in the 1970s. Technological advances over the years have widened both the test repertoire and software capabilities available to POCT devices, with blood gas analyser manufacturers leading the marketplace in terms of quality standards for POCT. Indeed, POCT blood gas analysis is a well-established practice, having been developed in response to clinical requirements. Modern POCT now consists of a variety of devices from simple hand-held meters and test kits, some available for purchase and use by patients, to more sophisticated portable and bench top analysers, which provide a variety of tests in all disciplines of pathology.

This chapter introduces the basic concept of rapid testing for the benefit of the patient at the point of care. It illustrates how POCT is governed and regulated, and describes the requirements for a well-managed, reliable POCT service, and how such a service can be established and maintained. The training and competence of POCT practitioners is discussed, and the dangers associated with a lack of or poor quality training are highlighted.

16.1 Standards and guidelines

CE

FIGURE 16.1
The CE mark consists simply of the initials in the form shown. The marking on the device must be clearly visible, legible, and indelible throughout the expected life of the device. For complex products, the initials are followed by the identification number of the body involved in the control of its production. All the letters and numbers must have substantially the same vertical dimensions of not less than 5 mm. They may be in any colour provided it is legible and unlikely to be confused with any other markings.

Every area of pathology laboratory testing is subject to regulation, inspection, and accreditation, including the POCT service. The aim of regulation is to ensure a safe and reliable service, offering quality-assured, consistent, and accurate results from *in vitro* diagnostic devices (IVD) that are fit for purpose and used by suitably trained, competent individuals. The organization, management, and documentation of the POCT services are defined to ensure best practice and the competency of users. POCT requires separate or additional requirements to the traditional types of analytical investigations because most or all of the analytical processes take place away from the direct overview of the clinical laboratory. Both standards and guidelines are necessary to ensure that POCT is performed to an appropriate level of quality, thereby ensuring the safety of the patient. Standards define what must be complied with, whereas guidelines are recommendations for *best* practice.

The European-wide directive, 'In vitro diagnostic medical devices' (Directive 98/79/EC; applicable to member states of the European Union (EU) and now also part of UK law) regulates all POCT IVDs, including analysers, meters, single-use kits, associated reagents, and consumables. This directive was amended in February 2006 and deals specifically with the safety, quality, and performance of IVDs on the European market. All manufacturers must meet the essential requirements listed in Annex 1 of the directive to ensure that IVDs do not compromise the health and safety of patients or users, and perform at the levels claimed by their manufacturers. IVDs that meet the requirements of the directive are denoted by the display of the **CE mark** (Figure 16.1). It is a requirement for all POCT IVD devices used in community or clinic settings to have this mark.

Pathology laboratories are subject to inspection and subsequent accreditation to standards produced by the International Organization for Standardization (ISO); its regulatory document for clinical laboratories is ISO 15189:2012 'Medical laboratories—Requirements for quality and competence'. A POCT-specific adjunct to this document is ISO 22870:2006 entitled 'Point-of-care testing (POCT)—Requirements for quality and competence', which is a comprehensive guide to all aspects of POCT and how it should be organized as a quality-assured, well-managed, and accurately documented service.

Accreditation bodies incorporate the ISO requirements into their own standards for medical laboratories. All UK laboratories are inspected to these standards every 4 years by the Clinical Pathology Accreditation (CPA) UK Ltd, now merged with United Kingdom Accreditation Service (UKAS). If its standards are met, the laboratory will be issued with a Certificate of Accreditation. This is a mark of quality, safety, and performance for that laboratory.

Any failure to achieve the necessary standards through deficiencies or non-compliances found during a laboratory inspection will result in the laboratory being given a period of time in which to rectify them. Continuing unresolved failures may result in the loss of or non-award of a Certificate of Accreditation. This, in turn, may have an adverse effect on the laboratory's ability to recruit and retain professional staff or to provide its analytical services to users or other healthcare providers. A number of other professional organizations have produced their own guidelines, including the Institute for Biomedical Science's 'Point of Care Testing Guidance on the Involvement of the Clinical Laboratory (2004)' and the Medicines and Healthcare products Regulatory Agency's (MHRA) 'Management and use of IVD point of care testing devices'.

These guidelines, like the ISO standards, are intended for laboratory use but the MHRA has also produced more informal leaflets to guide POCT users, such as its 'Point of care testing: top 10 tips'. These cover tests such as those to estimate the concentrations of glucose and cholesterol in blood samples and for urinalysis, and are useful for practice nurses and other non-laboratory-based users.

Cross reference

More detail about accreditation and international standards can be found in Chapter 17, 'Quality assurance and management'.

16.2 Advantages and limitations of point-of-care testing

From the advent of medicine, successful clinical outcomes have relied upon accurate diagnosis followed by appropriate intervention. Consequently, in recent years laboratory services have become more centralized to help increase efficiency by economies of scale. This has resulted in clinical laboratory testing often being remote from the clinician and patients, which may lengthen the turnaround times for some important or urgent tests.

16.2.1 Advantages of point-of-care testing

POCT can redress many of the problems associated with traditional testing in the clinical laboratory. For example, the clinician can obtain pathology results during the consultation. Having access to the information required to make a diagnosis, or change the management of the patient during the consultation offers many benefits, which we will discuss in the rest of this section.

Cost benefits

The cost per test in POCT is often higher than when the same analysis is conducted in a laboratory. However, the impact of a rapid result may offset this extra unit cost. The following list describes some of the financial benefits that can be achieved by POCT:

- timely interventions improve patient outcomes, potentially reducing the length of stay as an inpatient
- it can reduce the inappropriate administration of antibiotics and other prescription drugs, and the use of blood products
- POCT can determine a patient's immune state and help decide whether expensive drugs require purchasing; the efficient and timely management of chemotherapy drugs eliminates wastage
- individuals who are less experienced can competently perform POCT, enhancing their role and releasing other professional staff for more appropriate duties
- the enhanced disease management benefits the patient and reduces costly complications
- over-the-counter testing kits, for example fertility monitoring kits and pregnancy testing kits, cut costs associated with their provision by the health service
- diagnostic testing carried out on finger-prick capillary blood samples provides a cost saving as trained phlebotomists, syringes, sample tubes, and pathology request forms are not needed
- the use of POCT devices in, for example, the prison service may reduce the need for costly secure transport of prisoners to the local hospital for treatment, ensuring that only those with a genuine need are referred
- travel and parking costs for patients are reduced; with POCT both the test and the visit to the clinician to review results can be completed in a single visit.

Estate management

The use of POCT has implications for the management of hospital and human resources, and affects the strategic management of many hospital departments. POCT can affect the whole patient experience where pathology results may be a rate-limiting factor. This contributes to the efficient use of bed space and clinician time, and can help shorten waiting lists.

In outpatient departments, the use of a 'one-stop shop' approach reduces repeat visits and minimizes the flow of people visiting the hospital site (see Case study 16.1). Fewer visits may also reduce cross-infections as patient contact is reduced.

Point-of-care devices sited within the hospital reduce the requirements for transportation of samples to the laboratory. They also reduce pressure on the clinical laboratory through workload reductions and can provide some temporary 'back-up' service in the event of laboratory analyser failure. The introduction of local units that use POCT in satellite laboratories can reduce accident and emergency (A&E) department admissions; however, maximum benefit may require redesign of the whole patient pathway as time to clinical decision making can also be a limited factor. This enables patient triage and observation to be performed in a community hospital or 'polyclinic' setting.

Management of acute conditions

The management of rapid-onset or acute conditions often involves the use of 'bench-top' POCT analysers in a 'rapid response laboratory' setting. In these situations, the test results give a baseline from which the clinician can monitor, refer, or discharge the patient. Diagnostic testing is also performed by the healthcare professional on the sample that they obtained, reducing the frequency of samples becoming lost or the result being attributed to the wrong patient. This assumes correct, relevant information on the sample is correctly entered into the analyser.

Smaller sample volumes are required for POCT compared with conventional laboratory samples. This is better for the patient as it minimizes blood loss related to phlebotomy. Furthermore, the analysis of fresh samples reduces inaccuracies caused by sample deterioration.

Management of chronic conditions

Cross reference

The management of the chronic disease diabetes mellitus is described in Chapter 13 of the accompanying volume, *Clinical Biochemistry*.

The management of longer-term or chronic conditions does not require the same level of complexity of devices as required in the acute setting. In chronic conditions, both diagnosis and prognosis are already established, and so only appropriate tests required to manage the patient's condition are necessary. However, the accurate monitoring of chronic diseases enables better treatment and reduces the burden of complications caused by poor patient management.

The patients may be able to monitor themselves at home using a POCT device, increasing their understanding of their own disease and giving them more responsibility in their own disease management.

Chronic conditions can also be monitored in a GP surgery, health centre, or even local pharmacy, releasing hospital resources for the diagnosis and treatment of acute conditions or patients.

Patient satisfaction

Cross reference

Arterial blood sampling is described in Chapter 6, 'Samples and sample collection'.

Patient satisfaction may be often overlooked in the design of health service provision but must always be a major consideration. The provision of an efficient POCT service, delivered at the right time in an appropriate setting improves the patient experience. Test results provided during the consultation reduce worry and stress of patients who are likely to be concerned about their condition by allowing them to ask about the implications of their results and to make better informed lifestyle choices. Patient experiences are illustrated in Case study 16.2.

Meeting government directives

The National Health Service (NHS) throughout the UK is expected to be cost-effective and may be required to achieve government-defined health promotion or prevention targets. For example, in England, patients presenting at A&E departments must be treated within 4 hours of arrival. POCT may help in this situation by offering a rapid pathology diagnostic service within the department, although the introduction of POCT should be part of a whole-systems approach to evaluate the overall patient journey. Other targets include the 18-week pathway, where the patient can expect to wait no longer than 18 weeks from referral to the start of their treatment. While the laboratory may provide an efficient service, any requirement for laboratory testing will result in repeat appointments with the clinician guiding the treatment plan. The use of POCT may eliminate the need for repeat appointments and so speed the patient's progress.

Recent trends require the primary care sector to offer outpatient services to relieve the burden on secondary care (hospital) departments. This has necessitated a different approach in the provision of pathology services and POCT may prove a more effective way of providing test results.

SELF-CHECK 16.1

Indicate which of the following statements are true or are false:

(a) POCT is useful in managing chronic and acute conditions

(b) Patients do not like POCT

(c) Patients must not perform their own self-checks.

16.2.2 Limitations of point-of-care testing

Analysis using POCT is often seen as a quick and easy alternative to standard clinical laboratory testing. However, a report from the British In Vitro Diagnostics Agency in 1997 found that hospital *managers* were twice as confident in the usefulness of POCT as *clinicians*. This may be because managers concentrate on factors such as rapid turnaround times and lower costs, while clinicians have a much better understanding of what can go wrong in practice. A POCT solution may be sought when the problem is really about issues such as transport to the laboratory or speed in obtaining test results. If these difficulties are addressed, then the laboratory could, in fact, be the most cost-efficient, as well as reliable, option. There are many potential and actual limitations in POCT, of which some are readily apparent, but others may only become evident in practice. Where POCT devices are in use, safeguards and protocols must be rigorously adhered to in order to maintain the reliability of results and ensure patient safety. The clinical laboratory must always be used for the confirmation of unexpected or spurious results.

If a particular POCT test becomes commercially available, there must be a clear clinical need that its use, rather than continuing with the laboratory-based analysis, will benefit the patient. There is little advantage in producing a test result in minutes if a doctor will not see it in hours or even days. Easy availability may result in 'over-testing' and duplication of effort by continuing requests to the laboratory for the same test, resulting in overall increased costs. Infrequent use of a POCT device will result in increased user error and wastage.

Before introducing any POCT device, consideration must be given as to who is going to do the testing, where and when. Nursing or clinical staff must have the necessary time to do the analysis and documentation, or patient care may be compromised. Staff will require training, competency checks, and annual update training. This all takes time and resources. Siting of the equipment must be considered from a health and safety perspective and for the convenience of the user. Maintenance and troubleshooting of equipment may divert laboratory staff from other duties, while poor maintenance may result in breakdowns, inaccurate results, and increase the risks of infections. Equipment may be damaged by misuse, especially if unauthorized users can gain access to it.

Unless staff complete written logbooks or data are automatically downloaded from the equipment, information may be lost. Thus, there may be a lack of a permanent record of POCT results. Users may not be correctly identified owing to personal passwords being stolen or simply passed on to others. Errors in **transcription** may occur when test results are written into the patient's notes, and printouts may become damaged, lost, or degrade with time. There is often a lack of traceability: staff in a hurry or pressed for time may omit entering patient identifiers into the analyser so that reports cannot be traced to the correct patient. Reports and printouts that are identifiable may be left in view, leading to issues of patient confidentiality.

When a POCT result is obtained, there may be a lack of 'thinking time' for consideration of that result and for reflection on further action, which may lead to rushed and inappropriate decisions. This can have serious consequences for the patient, such as a misdiagnosis, which may result in **litigation** (see Section 16.11).

The main source of error with any item of POCT equipment is, quite simply, the user. Table 16.1 shows some of the most frequently encountered errors in POCT. Constant observation and the monitoring of all POCT devices is necessary to avoid these errors and ensure that POCT provides a safe and cost-effective addition to the laboratory service. Further information on the use of standard operating procedures (SOPs) for POCT may be found in Section 16.9, but Case study 16.3 describes what may happen when an SOP is not followed and the potential outcomes for a patient.

SELF-CHECK 16.2

State the major advantage(s) of using POCT.

TABLE 16.1 Sources of error in point-of-care testing

Stage	Error
• Pre-analytical	• Wrong patient • Wrong type of sample container • Wrong type of sample (urine or blood?) • Wrong time for sampling (fasting? pre- or post-dose for drugs) • Poor sampling techniques resulting in poor-quality samples affected by factors such as haemolysis or the introduction of air bubbles in blood gas samples • Delays in analysis and failure to mix samples
• Analytical	• Failure to follow instructions or standard operating procedure • Lack of training • Lack of maintenance or servicing • Failure to carry out quality control or appreciate when it is incorrect • Failure of calibration • Failure to identify the patient on the analyser • Failure to understand error codes or warning messages
• Post-analytical	• Failure to check unexpected results with the laboratory • Failure to understand 'out-of-range' flags or implausible results • Failure to consult reference ranges or making inappropriate assumptions about them • Entering results into the wrong patient's notes • Transcription errors when entering results into a patient's notes • Clinical decisions made on poor quality or incorrect results • Delay in acting on results • Loss of results

16.3 Point-of-care committee and point-of-care policy

The POCT committee and POCT policy provide the cornerstones for the POCT service. Establishing a multidisciplinary POCT committee will help ensure that all POCT processes are compliant with the trust's own POCT policy and with national guidelines, and ensures good **clinical governance**. The committee also ensures effective management in controlling POCT staff training and the use of resources, and so contributes to the production of reliable results that will benefit the patient. The committee must be accountable to the trust board, and maintain links with other relevant committees. Indeed, there may already be an established committee, such as a medical devices committee, or medical equipment and procurement committee, whose aims and interests are broadly similar and relevant. It may prove effective to incorporate POCT into this existing structure, cutting down on the number of separate meetings and combining established committee members with new recruits. A typical organizational chart for POCT clinical governance resembles that shown in Figure 16.2.

Clinical laboratories have key functions in the promotion, management, and regulation of a POCT service. The initial impetus for setting up a POCT committee may come with the appointment of a POCT manager or coordinator, or quality manager. This person is usually a senior laboratory scientist, and will be the driving force behind the service. The committee should meet on a regular basis and must be as multidisciplinary as possible. This allows for shared knowledge and experience, and a wide cross-section of opinions. The committee will need a chairperson who is committed to the concept of POCT and willing to be its figurehead. This person must have authority, respect, and drive. A senior clinician or pathologist may be ideally suited for this role. Other members must include the POCT coordinator,

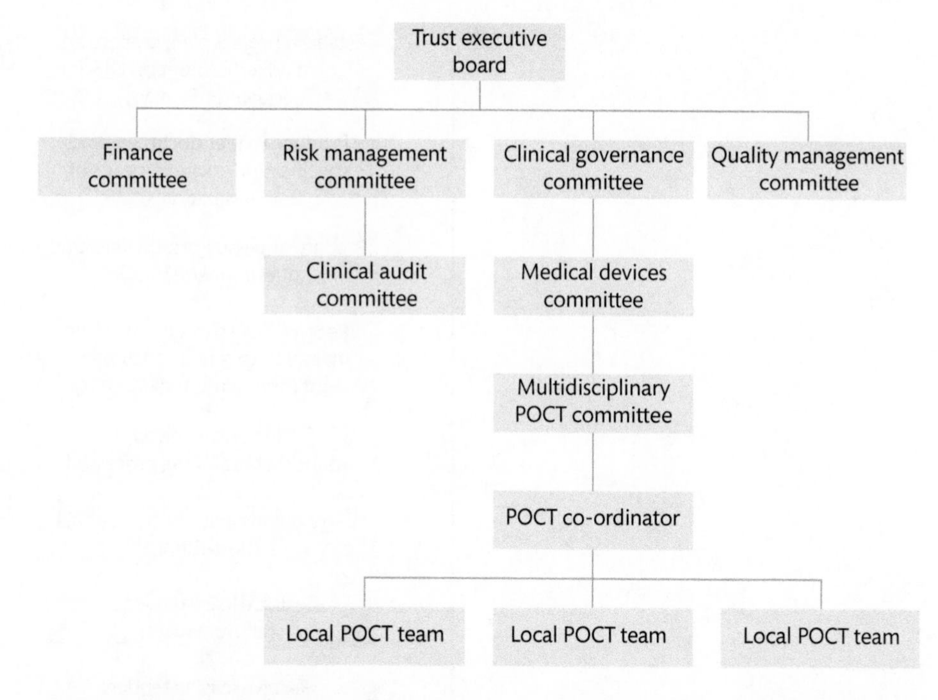

FIGURE 16.2
Clinical governance and organizational chart for point-of-care testing (POCT).

quality manager, and representatives from each of the pathology disciplines. Clinical representatives from surgical and medical directorates, as well as nursing and midwifery staff, will involve those most likely to be active users of POCT devices in wards and departments. Given that the committee may make decisions regarding the purchase or replacement of POCT devices, a representative from finance to give advice on economic considerations and funding of the service can be helpful. A committee member from the NHS Logistics Department to supply information when selecting and acquiring devices is also useful. Pharmacists who are involved in the purchase of relevant items, such as blood glucose strips and pregnancy tests, and also in their storage and issue to wards, should also be invited members.

Clinical governance and **clinical risk** departments should also be represented on the POCT committee and an advisor on training is desirable. Devices may require repair and maintenance, and laboratory staff and biomedical and clinical engineering departments will have relevant input and experience. Valuable insights may come from patient representatives, and, if possible, GPs and community nurses. The committee should aim to reflect the interests of all POCT service users. Secretarial support will be required for taking minutes and preparing agenda items.

Once the committee is appointed, it must agree and develop draft **terms of reference**. These will define the clinical governance, in terms of scope and responsibility of the committee; how it is constituted; accountability and direction; and how the committee relates to the healthcare trust structure.

Once formed, the committee must now develop and publish a trust-wide POCT policy with regard to the elements contained in the standards and guidelines, which we discussed in Section 16.1. The policy must be robust, concise and comprehensive, and provide clear guidance for users. It is recommended that the policy and the minutes be published electronically and be easily accessible. The policy should be subject to an annual review, with an updated version produced at least every 3 years. The policy must be ratified by the trust board and be accessible both in hard copy and electronically by all staff. The policy should define the organization, management, and funding for POCT, and include the extent to which patient self-testing is supported. The document should refer to European and national legislation, and to the accreditation standards of the laboratory and the trust.

The procedures and required documentation for selecting and procuring equipment, including the submission of a detailed business case, must be clear and relevant (Figure 16.3). These documents should show that consideration has been given to the processes involved in the ordering, supply and storage of reagents and other consumables, and the cost of both internal quality control (IQC) and external quality assessment (EQA). The POCT committee should ensure that procedures are in place for the safe location and decontamination of equipment. POCT devices must not be purchased, loaned, or installed without prior approval by the POCT committee.

Week 1
Establish project plan, prepare and post advertisement on OJEU (open for 37 days)
Produce tender document and specification, include means of evaluation
Prepare and issue pre-qualification questionnaire (PQQ)
Receive PQQ shortlist and issue invitation to tender documents (out for minimum of 40 days)
Tenders received, round 1 short-listing processed
Carry out evaluations before final short-listing
Issue lease tender (allow 40 days)
Receive leasing tenders
Issue intent to award (10 day standstill period applies)
Completion of award, including posting of award into OJEU
Week 36

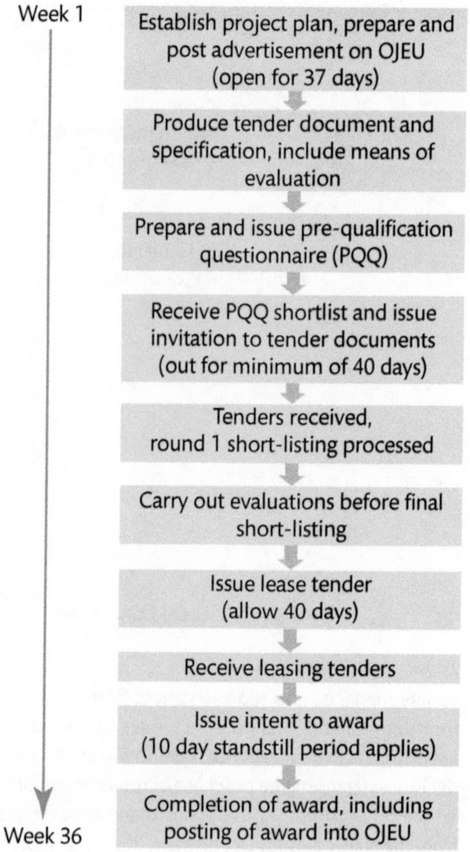

FIGURE 16.3
Flow chart summarizing the steps involved in approving the commissioning of a new point-of-care testing device. OJEU: *Official Journal of the European Union.*

Quality objectives are necessary to ensure correct performance and to maintain safety standards, and these must be regularly evaluated and reviewed. The POCT committee must have the authority to act in the event of poor performance or similar risk issues, including unauthorized use, password theft, or wilful misuse. Action may include withdrawal of the service, or the recommendation of disciplinary action for serious or repeated breaches.

Training objectives and methods of assessing competency must include the provision of SOP for each POCT IVD or test, including risk assessments. You can read more about some aspects of training in Section 16.8. The policy must define the various staff grades authorized to carry out the POCT procedure, and detail their responsibilities. The policy must define how records of all POCT results are kept and stored. This may include electronic transmission from POCT sites direct to an electronic patient record. Such connectivity must be available for all POCT IVDs. Connectivity is discussed in more detail in Section 16.7. Finally, mechanisms for peer review and shared experience must be defined, with an annual review report of the POCT service provided for the trust board.

SELF-CHECK 16.3

Which of the following records would be essential when preparing for a forthcoming accreditation inspection of the POCT service?

(a) Staff sickness records

(b) POCT policy

(c) POCT committee minutes

(d) Laboratory SOP

(e) POCT SOP.

16.4 Procurement

The procurement of POCT devices is best managed by the clinical laboratory in coordination with the supplies department, once the source of funding has been agreed. Funding could be from a specified budget or from specific capital funds, charitable funds, or as part of a defined project. Procurement must be undertaken in accordance with the hospital trust policy and procedures for the procurement of equipment and services, and the process must also comply with the hospital trust standing financial instructions, public procurement law, and EU legislation.

Compliance with all the procedures and legislation ensures that best value for money is achieved and procurement is in line with the trust's strategic and business plans.

The guidelines for purchasing equipment were discussed in Section 16.3. Before the decision to procure any POCT equipment is taken, the POCT manager must undertake the following:

- a cost-benefit analysis
- a patient benefit analysis
- an equipment evaluation
- a connectivity evaluation
- site evaluation.

Cost-benefit analysis is a technique for deciding the economical outcome of a change of practice. It is derived from the sum of the costs of an action subtracted from a sum of the benefits. While the costs are usually simple to identify, the benefits of moving a test from the clinical laboratory to the point of care can be subjective and difficult to forecast, as the impact of such changes are often far reaching. Patient benefit analysis, as the term implies, is an assessment of the impact that a change has in practice on patient outcomes. The analysis must consider the effect on the whole patient experience, as small alterations in the patient pathway may have unforeseen negative effects. Speeding the patient through one area may cause a bottleneck in another, if this has not also been improved by similar efficiencies.

An evaluation of equipment for POCT use is described in Section 16.5. However, in addition to assessing the POCT device itself, the stability, and credibility of the supplier must also be examined to ensure the long-term viability of the support, maintenance, and availability of consumables throughout its life. Connectivity evaluation requires the involvement of the hospital information technology (IT) department from the outset in the purchase of any POCT device. This is necessary to ensure that the hospital network can support any IT solutions offered by the manufacturers. Health and safety aspects must be included in the site evaluation (see Section 16.6); hence, it can be useful to invite the laboratory health and safety officer to visit and offer advice prior to any purchase.

The results gathered from all the evaluations must answer the following questions before proceeding with the purchase of the POCT device:

- How much extra cost will be incurred during the lifetime of the new device?
- What will be the effects on the laboratory, the POCT service provision, and the staff involved?
- What will be its impact on the patient experience and outcome?
- Will the new technology offer a safe and reliable service?
- Is a complete audit trail in place that provides information when required?
- Is there an appropriate infrastructure and back-up from the provider?

The results obtained from these analyses and evaluations can be used to formulate a business case to identify the requirement for the goods. A robust business case must include a service level agreement that outlines the support offered by the clinical laboratory. This will vary according to resources and service needs but should include a series of core requirements so that the evaluation of the whole business case can be considered. The roles of the laboratory will include the maintenance, quality control (QC), and assessment of the device, and include the training and competency assessment of staff expected to use it. The financial information to be considered should include the funding stream for the device(s) including delivery, the costs of consumables over its lifetime, maintenance and service contracts (see Section 16.6), expenses associated with training staff, the disposal of clinical waste, and a depreciation and replacement strategy.

Case study 16.4 shows what may happen if POCT devices are procured without taking into account all of the factors that a POCT committee or coordinator may consider.

All contracts from the public sector, which are valued above a defined threshold for the whole lifetime costs, need to be advertised in the *Official Journal of the European Communities*, to comply with EU legislation.

16.5 Evaluating a point-of-care device

Evaluations are required to ensure that any equipment purchased for POCT is fit for purpose and suitable for safe and effective use at the location for which it is intended. All medical IVD kits and reagents in use within member states of the EU now carry a CE mark (Section 16.1), which indicates their performance has been tested and verified in accordance with the claims of the manufacturer. This does not imply that a particular device will meet the needs and requirements of day-to-day clinical use. It must be borne in mind that what works beautifully in a manufacturer's demonstration in a well-ordered laboratory may well be unsuitable for a busy critical care area. It is essential to carry out a thorough evaluation of the device prior to its purchase, to ensure that the users will have confidence in their ability to use it in a safe and competent manner.

16.5.1 Evaluating the operation of POCT equipment

In a **competitive tender** situation, each proposed device should be placed in the POCT site for a set period. This will enable users to familiarize themselves with the device, comment upon its design, performance, and how well it fits their requirements. Planning is essential to ensure that such trials take place with a minimum of delay, so that if more than one device is under consideration, users can make valid comparisons between them. Staff must be trained by the manufacturer to an agreed set of training objectives and allowed free access to the equipment. Results obtained during trials should not, however, be used for clinical decision-making. If the equipment under trial requires an additional blood sample to be taken from the patient, then approval from the hospital ethics committee is required with informed patient consent. This procedure is required because the additional duplicate sample will not be used for the direct benefit of the patient, and he or she has the right not to have their blood used in such a way.

Users should have access to evaluation forms that they can complete after using each piece of equipment. They should be asked to score each separate device for overall ease of use. This may include what sample types are required and how they are handled, and how passwords and patient identifiers are entered. The speed of use and time from obtaining a sample to acquiring results can be assessed, as can the quality of reports and ease of data recall. Instructions, help menus, and manuals must be clear and easily understandable. QC and maintenance procedures should be evaluated by staff undertaking them, and trouble-shooting interventions should be concise and relevant.

Evaluation forms should be identical for each item and collated at the end of the trial. Laboratory staff should also have access to the equipment, or a separate trial should be arranged within the clinical laboratory. The preferences and comments of all concerned should be carefully considered, to help avoid reaching unsound conclusions. It would be unwise to purchase any item that none of the users liked, on the basis of other criteria, such as economy. Evaluations can avoid expensive mistakes.

16.5.2 Diagnostic kits

Reports indicating the relative performances and costs may also be available for specific diagnostic kits; comparative reports are particularly useful in product selection. Having provisionally chosen a kit, the views, and preferences of their users must be sought with regard to its ease of use and suitability for purpose. Potential sources of user error must be identified. Considerations to take into account include the conditions for kit storage, the sizes of the packs, possible wastage, and, inevitably, costs. It may be financially beneficial to consolidate the purchasing of diagnostic kits to a single supplier if possible. Both IQC and EQA are necessary to ensure batch-to-batch conformity. Precision must be assessed by performing at least 20 replicate tests and calculating the coefficient of variation for the results, which should be less than 5%.

16.5.3 Acceptance testing

Once delivered, the device must be checked by the appropriate department to ensure it is electrically safe. Its acquisition must be entered into the asset register of the trust and it requires a clear label. The supplier should install the device at its working site (Section 16.6) and thoroughly check it to ensure its performance is as specified. Manufacturer-defined QC checks must be carried out and replicate tests performed on laboratory samples as described above for diagnostic kits. An appropriate statistical analysis of the data is essential to ensure that the results concur with those of the clinical laboratory instrument routinely used in testing such samples. The acceptance forms must only be signed when everything is satisfactory.

Evaluation is not a one-off procedure. Continuous evaluation through IQC, EQA, and audit is necessary to ensure that the product performs satisfactorily throughout its life.

SELF-CHECK 16.4

You have just taken delivery of a new POCT analyser. Which of the following points would you check to ensure that it was ready for placement in the ward?

(a) The electrical acceptance testing was acceptable

(b) The ward clerk was aware that it was ready to be installed

(c) All spares, consumables, and peripheral items associated with the analyser have been delivered.

16.6 Use of point-of-care devices

The servicing, acquisition, and storage of spare parts for any POCT device should have been considered during its procurement process in consultation with the POCT committee. Without proper forethought and planning, additional problems may become apparent when the device is implemented. Insufficient space, lack of adequate facilities such as light, ventilation, and power, or an inability to store and provide spares can hamper the POCT service, cause inefficiencies, and lead to waste.

16.6.1 Siting of equipment

POCT equipment should be located in safe, designated sites, close to the clinical area, and where the risks to the health and safety of patients and staff are minimal. The recommendations of the manufacturer should act as a guide to the siting of all POCT appliances.

The site must conform to all health and safety, and fire safety regulations. There must be sufficient space for the equipment and consumables, and room to carry out maintenance procedures. Work surfaces should be at a convenient height and capable of supporting the weight of the instrument. If it produces a significant output of heat, then air conditioning and ventilation may be required. The equipment should be arranged in an effective manner with consumables; documents, including user manuals; spares; centrifuge; refrigerators; or freezers readily to hand. Adequate lighting, power supplies, and network ports must be provided. The mains supply should be on a protected circuit and the equipment provided with an integral uninterruptible power supply. Facilities for the disposal of wastes, including sharps, clinical, and non-clinical waste are necessary. Hand-washing and possibly eye-washing provisions should be readily accessible.

These points must be considered prior to purchase, and included in any plans for new or modified premises to prevent problems only being discovered later, when they may prove costly and difficult to remedy.

16.6.2 Stock control

Appropriate stock control ensures that the POCT service does not fail through a lack of essential supplies, which are used economically preventing wastage. Commercial solutions are available that use barcodes or chip recognition systems to track supplies and to re-order them automatically. Supplies should be delivered to a secure central point for collection.

Immediately on arrival, new stock should be checked to ensure that it is as ordered, has been delivered in the right quantity, and has a long shelf-life. Supplies must be stored in accordance with the manufacturer's instructions to prevent wastage. Thus, for example, there must be adequate freezer or refrigerator space for the volume of stock ordered; regular temperature checks must be carried out and logged to ensure supplies are kept in optimum conditions. Batch and lot numbers of stock should be recorded, together with details of when these are brought into use. Stock rotation is necessary to ensure that reagents and consumables are used before their expiry dates. Out-of-date stock, particularly reagents and diagnostic kits, may deteriorate and produce invalid results. Never write on unopened packs because manufacturers will not accept the return of incorrect stock if the packages or boxes are marked. If they must be marked, it is better to use removable stickers. Regular inventories are necessary to ensure that the recorded levels of stock are accurate. It is all too easy to return an empty box to the store and confuse the system!

16.6.3 Maintenance and servicing

All maintenance and servicing of POCT devices must be carried out in accordance with their manufacturer's instructions to ensure they remain in good working order. Lack of or poor maintenance may result in incorrect results, safety risks, equipment failure, and invalidation of the manufacturer's warranty. All equipment should be entered into an asset register that should detail all maintenance procedures and how frequently they must be carried out. The manufacturer, supplier, laboratory staff, or clinical engineering staff may carry out maintenance, and responsibility for it should be noted in the register. Maintenance logbooks should be provided. These must be updated at each service or this information recorded electronically in the analyser itself. The logbook provides information about what has been done and when, and thus helps pinpoint any recurrent problems. Any POCT device that does not perform to its specification must not be used until the problem has been rectified. Serious faults may require the return of the equipment to the manufacturer for repair or replacement. It may be necessary to withdraw the POCT service while such repairs are in progress, with the clinical laboratory providing the necessary back-up service.

Users should know who to contact and how to report breakdowns. They should be aware of alternative arrangements for dealing with samples whilst analysers are 'down'. Manufacturers should specify response times and frequency of service calls.

All newly purchased equipment has a warranty period, which is usually its first year of use. The manufacturer must carry out servicing during the period of the warranty. Care must therefore be taken not to invalidate the warranty by misuse, adaptation, or unauthorized interference with the equipment. Service agreements should include provision of a back-up instrument if required. If equipment is required urgently, it may be possible to arrange an immediate exchange policy.

For small items, such as blood glucose meters, it is sensible to have a supply of spares in stock for their rotation or for immediate use to replace defective instruments. This arrangement can be stipulated in their procurement specification. However, it is not sensible to have spare parts for large and expensive analysers, so any repairs must be swift and efficient. Returning equipment to the manufacturer for repair may be expensive once the costs of the inspection fee, spare parts, labour, and delivery are included. It is often economical to take out a service contract for a fixed fee, which covers the costs of several repairs.

Cross reference

Aspects of health and safety are covered in Chapter 4.

Items that are returned for repair require a decontamination certificate stating whether the equipment has been exposed to body fluids, infected material, or other biological or radioactive hazards. It must also describe the decontamination process used to render it safe. In-house repairs also require a decontamination certificate.

'Repair and return' requests must state the equipment type, its serial number, the date, the site where it is used, the reason it requires repair, and a name and address to which it must be returned when working. Unless the equipment is under warranty, the manufacturer may require an order number to guarantee payment. Items must be securely packaged and dispatched as soon as possible. Valuable items should be sent via recorded delivery.

SELF-CHECK 16.5

If you find that reagents are not being stored appropriately on a regular basis, what could be done to solve this recurring problem?

16.7 Connectivity

Connectivity facilitates data capture without the requirement for manual transcription. Hence, data from POCT systems may be saved in the laboratory information system (LIS), in the hospital information system (HIS) as part of the patient's electronic record, or to 'standalone' servers with a robust back-up system. However, prior to 2000, POCT connectivity had been unregulated and, consequently, of variable quality. In January of that year, the Connectivity Industry Consortium was formed and devised a document describing the standards required for POCT connectivity. This document formed the basis for the subsequent development of POCT connectivity requirements, including plug-and-play facilities, which allow the bidirectional control of multiple devices using a single data capture platform. Bidirectionality is the transfer of information from the analyser to the laboratory and back again. This allows the POCT coordinator to send the same information to many devices, for example to update the set-up, or to a single device to order a specific procedure, or lock a specific analyser that has developed a problem to prevent its use until the fault has been rectified. The management of POCT devices is improved by access to a bidirectional link to the devices, which allows the POCT coordinator to view the analyser status and patient results in 'real time', and enables remote recalibration and QC. The ability to view remotely the status of an analyser, including its 'on-board' consumables and maintenance schedules, reduces the time spent visiting POCT sites. Communication is central to the safe and effective use of pathology results, particularly in the case of a remotely generated result. The ability of a biomedical scientist to view and verify a result in real time offers the opportunity to respond quickly to abnormal results, give advice to the clinician, and reduce risk to the patient in the event of an error in analysis. Linking into the whole patient record allows for a continuous and chronologically accurate record of any given test analysis. As we move towards a true electronic record of patient data, this collation of test results will be necessary for continued patient care by those who access the records. Hospital-specific networks, the NHS central system, the Internet, or other commercial packages all provide connectivity depending on the capabilities of POCT devices. Manufacturers that currently offer connectivity for POCT devices design specific software that may allow access to devices from other manufacturers but with limited functionality. This is likely to change, as connectivity becomes more a requirement than an option. Whichever solution is available, the hospital IT department will be involved in the maintenance of the connection. It must also ensure that any connectivity package is compatible with any HIS or LIS upgrades to guard against it becoming obsolete.

The cost of connectivity for POCT devices is often significant and, indeed, may be more than the initial cost of the device. There may also be ongoing expenses for the maintenance or licence of such links and this should always be identified in the 'revenue costs' part of the business case (Section 16.3). However, the benefits of connectivity for any POCT device outweigh the costs when the total service provided is considered.

Connectivity also provides an audit (Section 16.10) trail allowing the recall of patient results, identification of the user, and verification of performance, lot numbers, and expiry dates of consumables, as well as continuing analyser status and intervention history. All of these data are requirements of accreditation and the NHS Litigation Authority (NHSLA). Audit data are required to minimize risks to the patient. In addition, such data protects the hospital trust as patient mismanagement can lead to expensive litigation cases (Section 16.11). Although failures in IT and information corruption may occur, the biggest challenge from connectivity to any POCT coordinator is ensuring that correct patient details are entered by the user. Accurate identifiers are essential to integrate POCT results with laboratory records for a given patient as the correct patient information is necessary to recall results and demonstrate a quality-assured patient management system in a legal environment. However, incorrect or incomplete information may enter the system because the patient details are simply unknown. This may happen in a major incident, on the admission of an unconscious patient, or where senility or confusion results in incorrect information being given by the patient. In can also happen when healthcare staff obtain a sample and simply forget to take the patient's details with them to the analyser. Unauthorized users may also enter made-up patient identifiers in mandatory fields to obtain results, bypassing the safeguards.

16.8 Training for point-of-care

Training and competency certification is the key to the success of any POCT service. The introduction of any point-of-care test needs to be accompanied by a SOP. This SOP must be plainly written and explain any scientific terminology and acronyms. The POCT coordinator, in conjunction with the ward sister, practice development nurse, or hospital training department must offer a comprehensive training package that covers:

- pre-analytical techniques
- analytical techniques
- post-analytical actions
- preventative actions.

Cross reference

You can read more about quality and quality assurance in Chapter 17, 'Quality assurance and management'.

Pre-analytical techniques include assessing reagent requirements, such as their storage, stability, and handling, the need to label samples, and how to obtain samples of the necessary quality. Substances that interfere with the analysis and invalidate the results should also be described. How to recognize device failure, and health and safety issues should also be included in the training package. Information about how the POCT device functions is an essential requirement of any training on analytical techniques. Quality issues should include QC and external quality assessment requirements (Section 16.9).

The operational ranges, device and analyte **reference ranges**, together with common user errors must be highlighted. It may be pertinent to include examples of serious errors or the harm caused by poor practices. Post-analytical instruction includes training in recording results, validation, and verification of results, and recognition of error codes. This is essential as corrective actions are necessary if analytical results fall outside the range of the analyser or outside the reference limits set by the laboratory manager or clinician. Descriptions of the risk reduction aspects of good record-keeping are useful. The importance of compliance, an aspect of POCT often disregarded, is apparent when an audit trail is requested during, for example, litigation.

Preventative action in the event of an analytical error is the user's responsibility, including removing the device from use and alerting the laboratory. It is essential to emphasize to trainees that the results they generate are *their* responsibility. Results from any POCT device that are inconsistent with the clinical signs and symptoms of the patient require checking by the clinical laboratory before management or treatment of the patient is initiated.

Feedback information on a training session is a constructive method of auditing quality and effectiveness. Unless trainees are observed performing POCT tests in a clinical setting, it is not possible to regard them as competent. Competency can also be lost over time if the test or technique is only performed irregularly. Competency is always the responsibility of the user. Although training update intervals may be fixed, the requirement for such training is particular to the individual, and depends on how often he or she performs the procedure and their ability to recall instructions. The use of eye-catching posters or short-form guides placed near the POCT analyser can be an effective *aide-mémoire*. The overuse of this resource, however, can result in the 'wallpaper syndrome' where users become blind to information provided in this way. When designing a training programme it is useful to be aware of the learning styles of the audience and include interactive practical demonstrations wherever possible.

Monitoring the error logs of analysers can assist in pinpointing users who may require further training. The POCT coordinator, when aware of an individual with questionable competency (as illustrated in Case study 16.5) is responsible for alerting the ward sister of the situation. Incompetent users must be prevented from using the equipment until they are retrained and are fully competent. However well trained the POCT team is in the use of their equipment, a significant part of their duties is to be out and about in wards and departments, to help maintain contact with staff, answer questions, and occasionally spot something untoward.

16.9 Quality and point-of-care

Cross reference

You can read more about QA and quality management in Chapter 17, 'Quality assurance and management'.

Quality assurance (QA) and its associated processes, QC, IQC, and EQA are all essential aspects of laboratory quality management. They are perhaps even more important when considering POCT, as testing activities may take place outside of a laboratory and be conducted by other healthcare professionals or even the patient themselves. All these quality measures are designed to help ensure the accuracy of testing and to be certain that appropriate clinical action is taken on the basis of test results.

Effective QA poses a challenge to the POCT coordinator, and is a principal aspect of the role. The implementation of a system underpinned by adequate documentation will cover:

- standardization of methods
- monitoring and maintenance of equipment

- equipment downtime
- equipment repair
- audit of consumables
- training and competency
- QC
- external quality assessment.

All POCT devices must be subjected to QC as determined by the manufacturer. This must be performed at regular device-specific intervals in order to monitor the basic functioning of the device. There may be several levels of QC material available for devices, to assess performance at low-, normal-, and high-confidence levels. QC is designed to produce a result on an instrument that can be compared with the recommended acceptable range. This involves testing a solution or sample produced by the manufacturer of the instrument, stored according to their instructions. The recommended ranges set by the manufacturers are often wide to allow for batch-to-batch variations of consumables. An individual result falling within this range may offer no indication of instrument bias or precision.

Analysers such as blood gas analysers can be programmed to perform this type of QC automatically from solutions stored in the analyser and can 'lock out' the device should it fail to achieve the expected performance. This prevents any inappropriate clinical management of the patient caused by inaccurate results and highlights the need for corrective maintenance of the analyser. For devices where automatic QC is not available, the operator must perform QC testing manually and take action in the event of its failure. These procedures must be recorded as evidence of a robust response designed to prevent use of the device prior to corrective maintenance.

16.9.1 Electronic quality control

Electronic QC involves challenging the POCT device (analyser) with an electronic stimulus designed to produce a known measurable response from the detector. This method of QC has the advantage of being integral in the performance of a measurement and is not user dependent. Where available, the POCT device should be 'quality controlled' using this procedure prior to use. The disadvantage of this method is that it may not mimic the analysis of a real sample and so may not replicate the mechanical aspects of sample flow through the testing system. Thus, the resulting QC data are qualitative rather than quantitative and give the user no guidance as to the performance of the device in clinical use.

Cross reference

Accreditation is more fully explained in Chapter 17, 'Quality assurance and management'.

16.9.2 External quality assessment

EQA involves the analyses of identical samples, usually supplied by an EQA scheme provider, to various sites that use the same POCT device. Collating the results means the performances of individual analysers may be assessed. Information regarding the accuracy and precision of a device within the distribution of results is a basis for any necessary corrective actions. An EQA report is a retrospective 'snapshot' of device performance, which can be used to manage proactively the device and highlight problems, or to indicate adverse trends. Registration of POCT devices on a suitable EQA scheme is mandatory for accreditation.

16.10 Audit

Regular audits are essential to assess the performance and clinical usefulness of all POCT devices. CPA requires horizontal and vertical audit information (see, e.g., Figure 17.3).

A vertical audit is a 'single process, all aspects' audit, such as the selection of a single test and examining a range of parameters, as listed in Table 16.2. In contrast, a horizontal audit selects one element of the vertical audit, and expands the information relevant to it. Hence, it is a one process, all aspects procedure. An example of a horizontal audit surrounding a maintenance log may contain those points given in Table 16.2.

Accreditation will require the POCT coordinator to collate analyser workload figures and cost per test, registers of training and equipment, logbooks listing any equipment replaced or periods when it was not functional (downtime logbook), and stock control information.

TABLE 16.2 Points to note when performing vertical and horizontal audits

Vertical audit	Horizontal audit
Patient identification	Device-specific training records
Date and time of test performance	Competency records
	Job description
Identity of user performing the test	Individual's contract
Analyser serial number	Evidence of continuing professional development if appropriate to the grade
Observe a test being performed	
Lot number of consumables	
Evidence of correct storage of consumables	
Training record of user performing the test	
Competency record of user performing the test	
Analyser error log	
Analyser maintenance log	
External quality control records	

SELF-CHECK 16.6

Information about a POC test, such as how to analyse for blood glucose, could be provided by which of the following?

(a) Asking a user

(b) Watch a user perform the test

(c) Examine the record book of the device

(d) Ask a patient

(e) Check the patient records against the logbook.

16.11 Problems, incidents, and litigation

A well-managed POCT service should ensure that competently trained users obtain high-quality results that benefit the patient. However, errors and incidents will inevitably occur from time to time. Such events must be investigated, remedied, documented thoroughly, and steps taken to avoid any repetition. This enables users to learn from such mistakes or malfunctions and to notify others of potential and actual risks.

16.11.1 Error logbooks

Fully completed error logbooks are used to record incidents, their seriousness, any remedial actions taken, and when such actions have been completed. Each item of POCT equipment or diagnostic kit must have its own logbook. Records must define the type of error, whether pre-analytical, analytical, or post-analytical, and if attempts at corrective action were effective. Case study 16.6 shows an example of an error and the corrective action taken to prevent it happening again. For serious errors, an incident form is required. A record must be made of notification to the POCT committee, the risk management committee, or an external body, such as the MHRA.

The error logbook of a POCT analyser can be stored with its maintenance logbook, but it may be better to keep them centrally with the POCT coordinator. This enables recurrent errors to be recognized and possible patterns of incidents or faults detected. These may identify a particular individual causing repeated errors, who requires retraining, or indicate that the siting or practical use of the equipment is causing unforeseen problems.

16.11.2 Reporting of incidents

All NHS trusts will have their own incident reporting forms, and ward staff may complete these following incidents involving patients. All such incidents must be recorded in the trust risk register.

If the incident involved POCT, as described in Case study 16.7, then a copy of the report must be forwarded to the POCT coordinator, and investigated to establish what went wrong and how it might have been prevented. If equipment malfunctions and results were found to have a clinical impact, then other users of similar POCT devices require immediate notification. The manufacturer, MHRA, and National Patient Safety Agency must also be notified. If further investigation showed a risk of harm or danger, then the MHRA may issue a Safety, Alert, or Hazard notice to warn all possible users of the problem. Each trust has a Safety Alert Broadcast System that cascades the safety notice to all possible users.

Cross reference
More information about health and safety is found in Chapter 4.

16.11.3 Litigation

The primary concern of healthcare is always the safety of the patient. A robust and effective system of error and incident reporting is necessary to minimize risk and maximize patient protection. The risk of litigation is particularly relevant with regard to POCT, where unauthorized users may not be fully trained or competent, may fail to follow instructions, or may adapt procedures without authorization. Trusts are insured against any adverse consequences arising from litigation. Any that are able to demonstrate good practices in training, documentation, and competence can obtain a reduction in their insurance premiums. This scheme is the NHSLA for Risk Management Standards. It has three levels, each requiring the trust to submit to a rigorous assessment. If the trust meets these standards, it receives a considerable reduction in its payments, with satisfying each level securing a greater reduction. The POCT service can help the trust obtain these reductions by ensuring its policies and guidelines are adhered to.

16.11.4 Human errors

User error is the major cause of incidents, which often arise from lack of training, misinterpretation of instructions, or a failure to follow the relevant SOP. Making ill-considered assumptions on the capabilities of POCT devices, a poor understanding of the chemistry underlying the test, or simply not thinking sufficiently about the procedures associated with it, can all lead to serious accidents. For example, blood gas analysers give a measurement of the pH of blood samples. Some users extrapolate this to mean that the analyser is, in fact, a pH meter and is able to determine the pH of almost any fluid. Even if the user manages to obtain a result on a non-blood fluid, they will not have any idea of reference ranges, effects of dilutions, the limitations of the analysis, or the presence of interfering substances. False assumptions about the underlying chemistry of an analytical test may lead to incorrect conclusions.

The unauthorized purchases of POCT equipment can bypass all the carefully constructed safeguards, increase the risk of harming the patient, and incur unforeseen costs. You can read about some of the problems such actions may cause in Case study 16.4. Constant watchfulness and awareness is necessary to pick up incorrect and inappropriate practices, as illustrated in Case study 16.8. QA is unlikely to identify random errors, but regular, thorough training and competency testing may eliminate most of them.

SELF-CHECK 16.7

A POCT coordinator noticed three used urine drugs of abuse screening kits in a ward sluice room. Such kits detect drug metabolites rather than the drugs themselves. It appeared that there were suspicions among the ward staff that a mother was adulterating her children's milk, orange juice, and water with her prescribed **methadone**. The screening kits had been used to test the drinks.

Suggest the most plausible of the following as to why methadone would not have been detected.

(a) There was insufficient methadone in the drinks

(b) The drinks would interfere with the test

(c) The nature of the kits.

CASE STUDY 16.1 Appointments

An audit of average turnaround times at a district general hospital showed that the time taken from a patient booking in at the outpatient haematology clinic to his or her **full blood count** result being available from the POCT analyser was 8 minutes. This included taking the blood sample from the patient and delivering the result to the consultant. Had this test been carried out in the laboratory, the turnaround time would include transport of the specimen to the laboratory, booking it in, processing the sample, testing the sample, and authorizing the result.

Prior to the implementation of POCT, the patient was required to make two visits to the outpatient department, one for the blood test, and a second follow-up appointment with the consultant for the results.

(a) What impact could this have for the patient?

(b) What are the possible benefits to hospital staff?

CASE STUDY 16.2 POCT: the patients' perspective

The use of portable blood gas analysers allows respiratory practitioner nurses to measure blood gases in a patient's home. The following comments are from two patients who experienced this form of POCT.

> I am really grateful that I don't have to have the needle in my wrist for arterial blood. The nurses take a small sample of blood from my ear, which is not at all painful, and are able to tell me the result in 2 minutes. If I am unlucky enough to need to be admitted to hospital I dread the arterial stab–it is very traumatic for me (Patient Mrs P).
>
> It would be very difficult for me to get to the hospital to the oxygen clinic. I used to attend but found that the worry about being on time made my breathlessness worse. The nurses always told me not to worry, but I did. They now visit me at home with the portable analyser and can let me know the result at once without the need to attend the hospital (Patient Mr R).

CASE STUDY 16.3 POCT and SOPs

A 27-year-old woman presented to an A&E department with severe abdominal pain. She was not taking any form of contraception. However, she stated her periods were often irregular. A urine pregnancy test was negative. She was treated for mild food poisoning and discharged. The patient was readmitted as an emergency the following day with a ruptured **ectopic pregnancy**. She underwent surgery, received massive blood transfusions, and spent weeks in intensive care. She later sued for damages.

The nurse carrying out the pregnancy test had not received training and had not followed the SOP. A false negative result was obtained because too much urine was used and had flooded the test kit.

(a) Could this situation have been avoided?
(b) Is this case an example of a pre-analytical error?

CASE STUDY 16.4 Grateful relatives

A female patient's relatives wished to thank the ward for the care she had received during a long hospital stay. The ward Sister asked them if they could raise the money to purchase a full blood count analyser, as delays in receiving results from the clinical laboratory had sometimes resulted in postponed treatments. The relatives managed to raise the full cost of the instrument. No one discussed this purchase with the POCT coordinator, clinical laboratory, or the consultant haematologists.

(a) The above scenario could result in which of the following:
 (i) An unsuitable instrument could be purchased
 (ii) Running costs could exceed budget
 (iii) Untrained staff could use the instrument
 (iv) QC and EQA could be overlooked.
(b) How could the problem of slow turnaround time have been approached more fruitfully?

CASE STUDY 16.5 Incompetency

The POCT coordinator received a phone call from a trained nurse who complained that a pregnancy testing kit was not working correctly. On investigation, it was found that the nurse had filled the sample well with urine rather than the specified amount as detailed in the SOP.

What actions should be taken to assist someone whose competence is in doubt, as is the case with this nurse?

CASE STUDY 16.6 When is total haemoglobin really total haemoglobin?

External quality assessment results from an analyser used to measure total haemoglobin were found to be increasingly inaccurate when compared with results from the clinical laboratory. An incident was reported where cross-matched blood had been requested on a patient whose total haemoglobin was determined as 79 g dm^{-3} (below critical limits) using a POCT analyser on the ward, but was found to be 141 g dm^{-3} by the laboratory (the reference range for men is 129–165 g dm^{-3} and for women 123–147 g dm^{-3}). An investigation showed that the cuvettes used with the POCT instrument were placed on a shelf in direct sunlight. The adversely high temperatures had caused them to deteriorate. Moving the instrument and cuvettes away from the window to a site with a constant room temperature cured the problem.

(a) Could the patients have been harmed by this incident?
(b) How could this problem have been detected earlier?

CASE STUDY 16.7 Whose fault?

An agency nurse was asked to monitor a diabetic patient who had been placed on an insulin infusion prior to surgery. This routine procedure is necessary to stabilize the concentration of glucose in the patient's blood. The concentration of glucose in capillary blood is measured using a glucose meter at hourly intervals. Although the nurse was unfamiliar with the meter provided, she had used a number of different types of glucose meters previously. For each test, she pressed the button to start the meter and recorded the result. All results were found to be stable and within the reference range. Unfortunately, the patient was later found in a hypoglycaemic coma and died of complications.

Suggest a plausible explanation as to why the patient died.

CASE STUDY 16.8 Blood in vomit?

Urinalysis strips are universal, simple, and easy to use. A consultant haematologist was called to a patient who was alleged to be vomiting blood. He was shown the contents of the vomit bowl and was unable to see any blood. The nurse was asked why she thought the contents contained blood. She has used a urine dipstick and it showed positive for blood. The patient was really suffering from an obstructed gastrointestinal tract and vomiting bile, which had stained the strip green. The positive colour for blood in urinalysis is also green.

(a) Can urine strips be used to test other biological fluids?
(b) Could this error have led to an incorrect diagnosis?

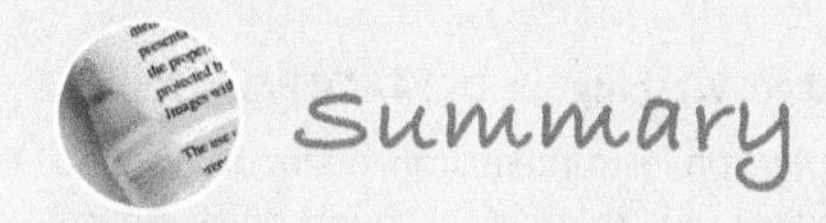

- POCT is the use of an analytical device for clinical testing outside the laboratory and close to the patient.
- POCT is governed and regulated by European legislation and national guidelines. An accreditation body conducts inspections according to these standards and its own guidelines every 4 years. An accreditation certificate is awarded to POCT services that comply fully with these standards.
- To achieve a well-managed and reliable POCT service, NHS trusts must have their own multidisciplinary POCT committee, responsible for devising its own local POCT policy. This defines the service, and how it is to be managed and organized, and how all matters related to its services are to be documented.
- Training and competence are key elements in ensuring the safety of patients by performing quality controls and external quality assessments of every IVD. Regular audit and monitoring is necessary to ensure compliance and good performance.
- POCT may be seen as a rapid process providing results almost immediately and so speeding the diagnosis and treatment; it does have potential pitfalls in terms of pre-analytical, analytical, and post-analytical errors.
- Procuring a new IVD must be based on clinical need. Care is necessary in examining cost benefits to the whole patient episode and not merely cost per test. Evaluations of new devices, their acceptance testing, maintenance, and audit are part of the role of the POCT coordinator to ensure the safety and continuing satisfactory performance of the device.
- Connectivity enables various devices to be monitored from a remote location, and corrective actions to be initiated through a computer link. Connectivity benefits the POCT service by allowing results to be downloaded to the electronic patient and laboratory computer records.
- Errors and incidents must be recorded and reported to appropriate outside bodies where necessary. Human error may result in patient harm and litigation.

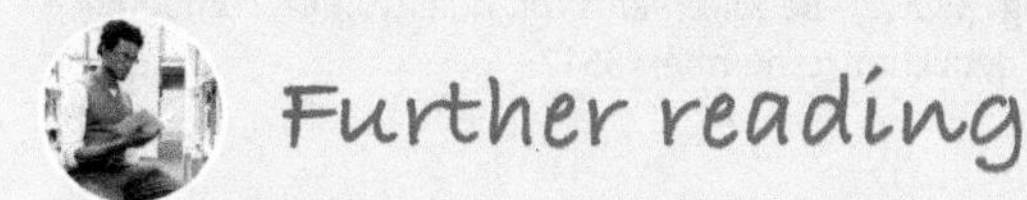

- **Becker KM, Whyte JJ. *Clinical Evaluation of Medical Devices: Principles and Case Studies*. 2nd ed. Clifton, NJ: Humana, 2005.**

  All that is necessary to know to evaluate a POCT device. The case studies are particularly useful in avoiding pitfalls and unexpected outcomes.

- **Bubner TK, Laurence CO, Gialamas A, Yelland LN, Ryan P, Willson KJ, et al. Effectiveness of point-of-care testing for therapeutic control of chronic conditions; results from the POCT in General Practice Trial. *Medical Journal of Australia* 2009; 190: 624–6.**

  This short paper summarizes the first full randomized controlled trial of the clinical effectiveness of POCT in Australian general practice. It is an encouraging endorsement of some POCT tests with an interesting approach.

- **Nichols JH, Christenson RH, Clarke W, Gronowski A, Hammett-Stabler CA, Jacobs E, et al. Executive Summary. The National Academy of Clinical Biochemistry Laboratory Medicine Practice Guidelines: evidence based practice for Point of Care Testing. *Clinica Chimica Acta* 2007; 379: 14–28.**

  Excellent discussion of what makes an effective application of POCT, some more US-based, but a good overview.

- **Price CP, Hicks JM (eds). *Point-of-Care Testing*. 2nd ed. Washington, DC: AACC Press, 2004.**

  This detailed and comprehensive work includes chapters on instrumentation, methodology, and POCT in primary care. It requires some prior knowledge but is an excellent source on all aspects for those wishing to enquire further.

- **Price CP, St John A (eds). *Point-of-Care Testing for Managers and Policymakers: From Rapid Testing to Better Outcomes*. Washington, DC: AACC Press, 2006.**

  A work intended for those in management and government which gives a useful overview from a manager's perspective.

- **St John A, Price CP. Existing and emerging technologies for point-of-care testing. *Clinical Biochemist Reviews* 2014; 35: 155–67.**

- **Westgard JO. *Basic QC Practices*. 2nd ed. Washington, DC: AACC Press, 2002.**

  Written by the foremost authority on QC and the originator of 'Westgard Rules', this work takes the reader from simple basics to more complex statistics.

Useful Websites

- www.wales.nhs.uk/documents/PFSD-August-2008-Ebook.pdf
- www.acbi.ie/downloads/guidelines-for-point-of-care-testing.pdf
- The Institute for Biomedical Science (2000) Point of Care Testing (Near-Patient Testing) Guidance on the Involvement of the Clinical Laboratory. Document Ref PR/07.04. Available at: www.ibms.org/includes/act_download.php?download=point-of-care-testing.pdf
- The Medicines and Healthcare products Regulatory Agency (2002) Management and use of IVD point of care testing devices. Available at: www.gov.uk/government/uploads/system/uploads/attachment_data/file/371800/In_vitro_diagnostic_point-of-care_test_devices.pdf
- Guidelines on Point of Care Testing (2004) Royal College of Pathologists. Available at www.rcpath.org
- ISO 15189: 2012 Medical laboratories–Requirements for quality and competence. Available at: www.iso.org/iso/catalogue_detail?csnumber=56115
- ISO 22870: 2006 Point-of-Care testing (POCT)–Requirements of quality and competence. Available at: www.iso.org/iso/catalogue_detail.htm?csnumber=35173

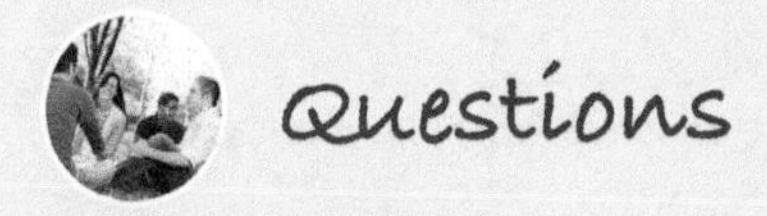

1. Which of the following records is/are necessary to prepare for an accreditation inspection of a POCT service by CPA?
   - **(a)** Staff sickness records
   - **(b)** POCT policy
   - **(c)** POCT Committee minutes
   - **(d)** Laboratory SOP
   - **(e)** POCT SOP.

2. You have just taken delivery of a new POCT analyser. Which of the following may be involved in installing the device?
   (a) Porters
   (b) Clinical engineering
   (c) Infection control
   (d) Risk management
   (e) Cleaners.

3. You find several boxes of reagents have been delivered. Although marked *Store at 2–4 °C*, they have been left in a warm room for at least a day. Which of the following is the most appropriate action?
   (a) Discard all the reagents immediately
   (b) Perform a QC check on a sample of the reagent, having refrigerated the remainder
   (c) Inform the supplier and request them to replace the reagents.

4. (a) How would you establish which POCT devices are in use in your hospital?
   (b) In which areas would you be most likely to find such devices?

5. How would you deal with any unauthorized or unexpected devices found? What actions could you consider?

6. You find that a particular user has been responsible for several instances of damage to devices, has a poor record of training, and refuses to take part in the EQA scheme. What actions do you take?

7. A consultant surgeon has asked for a POCT device to be placed in a theatre. The same test can be done in the laboratory in less than 30 minutes. How would you establish whether there was a case of clinical need and what other factors should you consider before procurement?

Answers to self-check and end-of-chapter questions are available at the end of the book.

17

Quality assurance and management

Elaine Moore

Learning objectives

After studying this chapter, you should be able to:

- Describe the features of a quality management system
- Explain the general principles underlying quality assurance and quality control
- Outline the different types of documentation found in the medical laboratory
- Outline the standards and regulatory requirements that affect the work of the medical laboratory
- Discuss how pathology affects patient safety and what is meant by a safety culture
- Outline the incident reporting process
- Describe what is meant by clinical governance and how this affects your day to day work in the medical laboratory
- Describe and understand the concept of risk management and how it is used in the medical laboratory
- Define and use tools and techniques that can be used for continual improvement
- Use knowledge and skills in everyday practice to deliver the right results to the right patient, in a timely manner, in order to diagnose and treat specific diseases

Introduction

As with many aspects of life in general, quality is at the heart of modern medicine, which, of course, includes pathology. In acknowledging this fact, it is important to define quality in a pathological sense. In this context, quality is about consistency and reproducibility, that is, getting it right the first time and every time.

However, quality is not a static event and requires continual improvement and refinement if changes occur, as new tests or equipment are introduced into a laboratory, or if reagents are withdrawn or re-classified in such a way that precludes their use.

Quality assurance, as the name implies, is about providing reassurance to users that a laboratory is addressing all aspects of quality and quality improvement in its work. However, quality cannot always be guaranteed and mistakes or errors do occur, but a quality-led laboratory will respond to such events in an entirely different way from that which may be normally expected. Aside from acknowledging an error has occurred and dealing with the immediate effects, such a laboratory will investigate the circumstances that led to any error, learn any lessons, and put into place remedial measures to help reduce the chances of similar mistakes happening again in the future.

Quality assurance, though, is not a one-way process as users must play a part in quality, too. Samples must be procured, labelled, stored, and transported correctly, and above all the sample must be appropriate for the intended investigation(s).

Quality assurance is achieved through a process of quality management, which describes the tools and techniques used to both assure and improve quality in the laboratory. Although quality management may involve significant amounts of administration and paperwork, the data collected provide objective evidence that quality services are being provided.

17.1 Why is quality important?

For any organization a quality approach is vital to maintain and grow as a business. For the medical laboratory providing a quality service is an essential part of delivering patient care. Organizations using quality techniques can expect to be safe and compliant with legislation and regulations, to deliver a service first time and every time, to be efficient and sustainable, and to be innovative, exploring and introducing new ideas and techniques. The International Organization for Standardization (ISO) defines quality as the degree to which a set of inherent characteristics fulfils specific requirements, as detailed in ISO 9000.

17.2 Quality in the laboratory

Quality has now been defined and its importance described, but quality now needs to be put into context within the laboratory setting. There are a number of processes and associated definitions that can be used to provide the framework and evidence to both users and laboratory staff alike to give confidence that quality is being addressed.

17.2.1 Quality assurance

Quality assurance (QA) is the processes and practices undertaken by a laboratory that are necessary to ensure a test on a clinical sample gives the correct result and that this finding is delivered to the appropriate clinician in a timely and efficient manner. Quality is essentially about consistency, that is, getting procedures right first time and on every subsequent occasion. QA is all the planned and systematic actions necessary to give confidence that a product or service satisfies the necessary level of quality.

There are a number of QA activities, as shown in Figure 17.1, that a laboratory can undertake to assure its service users that the level of quality it offers is of an adequate standard. These include implementing a **quality management system** (QMS), training staff, showing compliance with current regulations and standards, as well as activities such as internal and external quality measures, training, competency assessments, and audits. The medical laboratory also needs to ensure that associated clinicians, doctors,

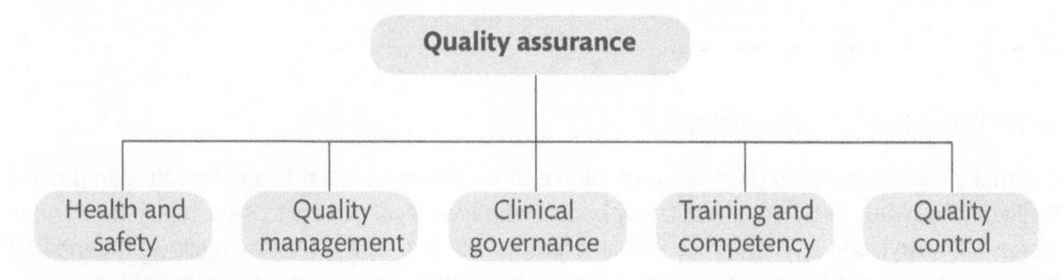

FIGURE 17.1
Diagram showing many of the facets of quality assurance.

and patients are involved in the laboratory service, through regular contact and other activities such as the production of laboratory handbooks and information sheets.

SELF-CHECK 17.1

List three of the facets of QA.

17.2.2 Quality control

In addition to QA, **quality control** (QC) is necessary to ensure accurate and consistent clinical results. QC is the operational techniques, activities, and requirements necessary to reach the desired level of consistency. In QC, the emphasis is on testing, checking, and preventing the release of incorrect results from clinical tests. Examples of QC include having a known positive control and a known negative control, which are tested alongside a patient sample to ensure the accuracy of the test results. Another example of QC is the checking of histological slides to ensure that the quality of the preparation and the staining of the tissue elements are of an acceptable standard for diagnostic purposes. QC can be classed as either internal or external.

Internal quality control

Internal quality control (IQC) is essentially performed in the medical laboratory, that is, it is 'in house'. IQC is concerned with sampling and testing to ensure that the necessary and relevant tests on a clinical sample are, indeed, carried out and that the test results are not released prior to their quality being judged as satisfactory. Thus, IQC ensures that the results supplied by the laboratory are within acceptable limits. The medical laboratory also performs the tests on control samples alongside clinical samples from patients. Given the composition and concentrations of the controls are known, they act as built-in checks on the process of analysis. Some assays require both high and low concentrations of the analyte to be determined to ensure, for example, that a chemical analyser or a procedure is operating correctly across the range of concentrations that may be encountered in the samples from patients. If the results for the controls vary from their known values, laboratory staff will immediately know that there is a problem that must be rectified before further testing of clinical samples can proceed.

IQC is also used in techniques such as special staining procedures of tissue sections and immunocytochemistry in histopathology and cytopathology laboratories. Known positive and negative controls are performed simultaneously with clinical samples to ensure that the method and reagents are working appropriately and being performed correctly by the staff. QC can also be used to determine the specificity and sensitivity of an individual test.

Laboratory assessors from bodies such as the United Kingdom Accreditation Service (UKAS) will usually examine QC records as part of the inspection or accreditation assessment.

Key points

Sensitivity is a measure of the likelihood of a test being positive in an individual who has disease. Specificity is a measure of the likelihood of a test being negative in a 'healthy' individual.

External quality assessment

External quality assessment (EQA) is currently based on two main systems: an accreditation process facilitated by UKAS, or by other regulatory bodies who have legal powers to inspect (e.g. the Human Tissue Authority); and EQA schemes that laboratories can join. The EQA system allows the medical laboratory to assess if it is in line with other organizations through a system of benchmarking.

EQA schemes assess a laboratory's performance against a defined set of criteria. Laboratory assessors will look for evidence that a laboratory participates in relevant EQA schemes and, more importantly, that the laboratory's performance is acceptable or performance issues have been addressed. All laboratories that perform tests on clinical samples are expected to participate in appropriate EQA schemes.

In EQA schemes, an outside agency, known as a scheme organizer, checks the accuracy of a laboratory's test results. This is achieved by the scheme organizer selecting file material (slides) or distributing 'known' test samples to the participating laboratories for analysis. Submitted filed material is sent to the scheme organizer and is assessed by a peer group. Distributed samples are tested by a laboratory in exactly the same way specimens from patients are treated and the results are then sent to the scheme organizer. However they are generated, the results and data from all the participating laboratories are collated and presented back to the participating laboratories as a report. These reports show the performance of a laboratory in two ways, by comparison with peer laboratories and by performance against a minimum standard. These reports will also endeavour to highlight areas of good or poor practice and indicate any trends identified when using certain reagents, kits, machines, or methods.

To ensure anonymity, each laboratory is given a unique code and scheme organizers will offer help, advice, or assistance to those laboratories identified as performing poorly by either measure. Laboratories can participate in a variety of EQA schemes, examples of which are listed in Table 17.1. In the UK, many of the schemes are organized by the UK National External Quality Assessment Service (NEQAS), which has a full list of the schemes on offer.

Following an incident at a National Health Service (NHS) trust where problems with QA were identified, an independent report into QA (chaired by Dr Ian Barnes) was commissioned by the Secretary of State for Health. This report and review was published in January 2014. These problems impacted on the care of a number of women with breast cancer. The review found that there was a need to change the current system, which had a focus on minimally acceptable standards that did not identify, incentivise, or reward those who are striving for excellence. Laboratories were unable to provide evidence to the Care Quality Commission (CQC) on the overall quality of laboratory services. The review found that the service relied heavily on the professionalism, goodwill, and commitment of staff, but the current systems of QA were no longer able to meet the needs of modern healthcare. The

TABLE 17.1 Examples of UK National External Quality Assessment Service (NEQAS) schemes

Blood Transfusion Laboratory Practice
UKNEQAS Parasitology
UKNEQAS for Blood Coagulation
UKNEQAS for General Haematology
UKNEQAS for Peptide Hormones and Related Substances
UKNEQAS for Antibiotic Assays
UKNEQAS Leucocyte Immunophenotyping
UKNEQAS for Histocompatibility and Immunogenetics
Wessex and South West England EQA Scheme
UKNEQAS for Feto-maternal Haemorrhage
UKNEQAS for Immunocytochemistry and *In Situ* Hybridization
UKNEQAS for Molecular Genetics
South East England General Histopathology EQA Scheme
UKNEQAS for Clinical Cytogenetics
UKNEQAS Haematinics Assay Scheme
UKNEQAS for Trace Elements
Welsh Assessment of Serological Proficiency Scheme (WASPS)
NHSCSP EQA Scheme for Gynaecological Cytopathology
UKNEQAS for Cellular Pathology Technique
Vitamin K EQA Scheme

review concluded that there was no national definition of an error in pathology with wide variance in error reporting within laboratory providers. To understand quality in the laboratory environment fully, the report recommended that there should be integration of the QMS with those of the host organization, such as the hospital. This should include the measurement and reporting of appropriate QA indicators. The Royal College of Pathologists has published key performance indicators (KPI) that can be used for this purpose. These KPIs cover matters such as **turnaround times** linked to patient pathways, staff training requirements, and communication of critical results to users.

A number of modest and achievable changes were suggested in the 2014 QA report to ensure that pathology services are:

- visible and transparent to patients to ensure confidence in the service
- accountable to boards and commissioners
- reliable, robust, and responsive to patients' needs
- rewarded when they make improvements in quality and patient safety
- held to account when they fail to offer the level of service that patients expect.

It is recommended in the QA review by Dr Ian Barnes that the EQA schemes chosen by the medical laboratory are also accredited. The standard that EQA schemes are accredited against is ISO 17043:2010, 'Conformity assessment–General requirements for proficiency testing'. Many areas in this EQA standard are similar to ISO 15189:2012 requirements (Section 17.4.1), with the addition of the following specific areas:

- review of requests, tenders, and contracts
- subcontracting services
- purchasing services and supplies
- service to the customer
- complaints and appeals.

17.2.3 Quality management

Quality management (QM) is any system where the management has direct control of an organization with regard to quality. Quality is not a *series* of events and therefore is *not* something with defined start and end points but is a never-ending cycle of events that requires the support and cooperation of all the people in the organization. Thus, QM is an approach, attitude, or culture within an organization. All organizations must ensure that the quality of services they provide are maintained and continually improved.

Every stage of a process, system, or project is a possible source of poor quality. The potential to provide poor quality is greatest in complicated systems such as a hospital. However, *learning organizations* have systems, mechanisms, and processes that enhance the services they provide. They are able to adapt to their environments and apply the results of their learning to achieve better results. Hence, they continually evolve and enhance their capabilities. In 2000, the Department of Health published the document 'An Organization with a Memory', which advocated that the NHS needs to learn from its mistakes and also from the mistakes of others. The recommendations that an expert group outlined in this document are designed to ensure that lessons from the past are used to reduce the risk to patients in the future. Thus, the NHS has recognized it needs to learn from **errors** and mistakes, and to refine and improve its systems and processes continually: in short, to become a learning organization. Naturally, all adverse events cannot be eliminated from complex modern healthcare but by continually learning from past errors, hospitals will be better able to achieve their stated objectives in a sustained manner, not only for the benefit of the hospital, but also for patients and the community they serve.

The ISO develops and publishes standards for international organizations to adopt. One such document, ISO 9001:2008, outlines the requirements of a QMS. A QMS is a model that may be employed to give guidance to the selection of appropriate quality activities. It is possible to apply a QMS to most areas, stages, and processes involved in the provision of a product or service. The aim of a QMS is to prevent **non-conformity** or a non-conformance to a set standard. Non-conformity is any deviation from that which is required. An effective QMS will try to satisfy customers or consumers by concentrating on the service or process. One way of achieving this is by customer surveys and monitoring their complaints. A QMS also documents processes designed to achieve consistency, providing a goal for system control, and maintaining the required level of quality at each stage in the process. (A note: as part of their processes the ISO has reviewed and reissued ISO 9001 in 2015. Although it is inevitable that changes will have been introduced, the general principles of the 2008 revision of ISO 9001 still apply. ISO 9001:2015 is described as being less prescriptive than its predecessor that embeds the 'Plan–Do–Check–Act' cycle at all levels of an organization.)

There are eight requirements of a QMS, as defined in ISO 9001:2008. These are:

1. Customer focus
2. Leadership
3. Involvement of people
4. Process approach
5. Systems approach
6. Continual improvement
7. Actual approach to decision making
8. Mutually beneficial supplier relationships.

This list forms the basis for any QMS in the medical laboratory; they all contribute to improving the quality of care for patients.

Customer-focused QM meets the needs of the clinicians who treat the patients. Medical laboratories achieve this through user surveys, monitoring complaints and compliments, and by analysing incidents.

Leadership is provided by management commitment from consultants and senior medical laboratory managers. If senior management is not involved in the quality process, then it will not work. Leadership is undertaken through the provision of quality policy and health and safety statements, and also through the actions and attitudes of senior managers. Within a pathology laboratory, the senior team may consist of a clinical director, a pathology service manager, and a lead (or head) biomedical scientist or clinical scientist for each department.

High-quality service provision only arises from the commitment of everyone in the organization. It needs everyone to be involved and aware of the systems in place. Indeed, it is often the case that the people performing specific tasks have a better understanding of problems than the manager. The organization needs a well-motivated workforce that feels valued and that takes pride in its work: excellence in the service provided is the responsibility of everybody concerned.

A process approach means individual areas must not be looked at in isolation; rather, the whole process must be examined. Often changing something in one area will have a knock-on effect in another area. For example, changing the patient request form to suit one department may mean vital information is not provided for another department. Once the process has been analysed it is necessary to consider the whole system (i.e. a systems approach). For example, a large project set out in the 2004 NHS (England) improvement plan was the 18-week pathway, which had the aim of ensuring that by 2008 all patients are seen and treated by a hospital within 18 weeks of referral by their general practitioner (GP). This approach requires the collaboration of a number of different services such as pathology, radiology, and pharmacy, and was, to a great extent, successful.

Continual efforts must be made to improve the medical laboratory service and to improve the accuracy and timeliness of results as an aid to the diagnostic process. Many tools can be used when continually improving the laboratory service; these can include audit, lean methodology, six sigma, **root cause analysis** (RCA), and process redesign. However, no single improvement tool provides all the answers to any one situation and it is essential to select the tools that best fits the task required.

Factual approach to decision making means making decisions based on facts and the analysis of data. This allows the medical laboratory to match the needs of the service users with business planning. For example, to select equipment that not only deals with the current number of tests, but also takes into account predicted future increases in workload. The use of audit, incident reporting, and statistical analysis to predict future trends all help in making the right decision.

Mutually beneficial supplier relationships improve services in a number of ways. Thus, if the laboratory and the suppliers of laboratory equipment, and its maintenance and service contracts, reagents, and consumables work together this will improve the quality of the laboratory provision.

QM must be led by senior management, which usually employs personnel to implement and maintain the QMS and the quality governance framework. These personnel are usually called a quality manager or quality governance manager. Thus, alongside the eight principles listed above, it is necessary to have a clear clinical governance framework in place in a hospital or medical laboratory. Quality governance is a form of QM that has been implemented in some organizations to improve the quality of care for patients. It is the system by which NHS organizations, such as hospital trusts, are accountable to improve continuously the quality of the services they provide and to safeguard high standards of care. The organization can achieve an environment in which clinical excellence flourishes. Quality governance requires the creation of a safety culture, as well as systems and methods of working that ensure opportunities for quality improvement are identified in all the organization's services and that over time there is an improvement in the quality of care provided throughout the NHS. It therefore provides a better experience for carers and staff.

The cost of adverse events, errors, and incidents is increasing in the NHS, and quality governance provides an opportunity to focus upon these problems. Specific types of adverse events are repeated in the NHS, which demonstrates that although the NHS is committed to becoming a learning organization, it has yet to achieve that goal fully (Section 17.5).

17.2.4 Documentation in the medical laboratory

Laboratory standards, such as ISO 15189:2012, 'Medical laboratories–Requirements for quality and competence', and ISO 22870:2006, 'Point-of-care testing–Requirements for quality and competence', outline a set of criteria that need to be fulfilled by medical laboratories (Section 17.4). These standards also include the requirement for a QMS, as outlined in ISO 9001:2008, 'Quality management systems–Requirements'. When implementing such standards in a laboratory, written, that is documented, evidence is needed to show that the laboratory is meeting the requirements specified in the standards. The minimum requirements of a QMS for a medical laboratory are a quality policy, quality manual, standard operating procedures (SOP) for the processes in place (see Figure 17.2), and process records and forms used in the sample journey. There are also requirements to have procedures to control the documents and to control changes made to any processes to ensure that changes are reviewed and approved before use.

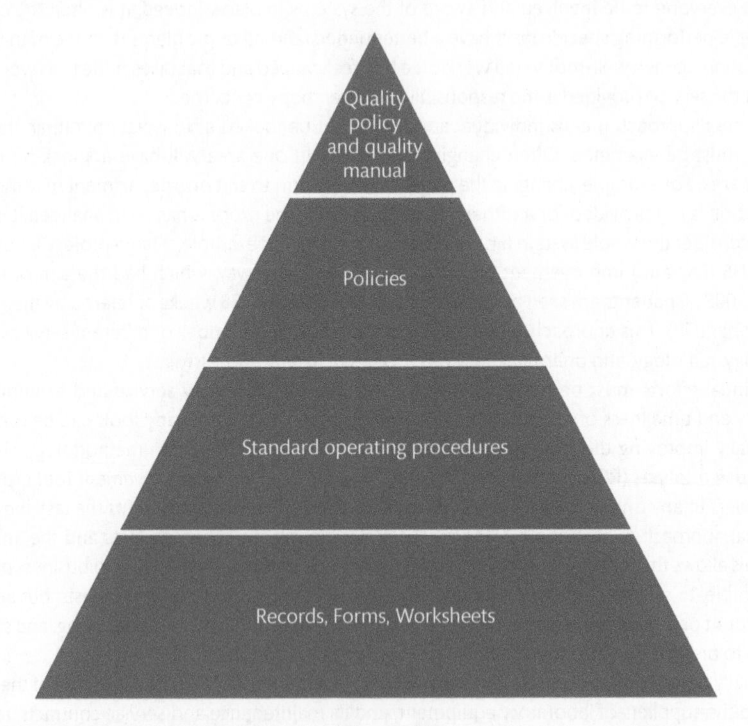

FIGURE 17.2
Triangle showing the possible hierarchy of documentation in a quality management system.

The quality policy is a short, one-page document that outlines the scope of the service the laboratory intends to provide. The quality policy contains details of the laboratory's commitment to its users and staff, and outlines the organization's **quality objectives**. It is usually signed by one of the organization's senior managers, such as its clinical director. The document is communicated and available to all staff in the laboratory, usually by displaying it on a notice board or on the organization's website and including it in the laboratory quality manual.

A quality manual is a document that outlines how the objectives stated in the quality policy are to be achieved. Documented sections within the quality manual are linked or referenced to the standard(s) with which the organization is trying to comply. A quality manual contains organizational-specific information, including its organizational structure, processes, and procedures, and the resources needed to implement quality and meet the objectives described in the quality policy.

SOPs are written instructions that describe in a clear and concise manner how procedures or methods are to be performed. They include such details as to how individual tasks are to be carried out, that is, what is to be done, by whom, and when. They accurately describe the process and methods that are in use in the laboratory and contain essential information such as health and safety instructions and the correct use of equipment and controls. It is necessary to have up-to-date SOPs that correctly describe what is done, to ensure that no step is missed and that the quality of the results is not compromised. Process records are records that are used in the sample journey and include, for example, worksheets, daily equipment maintenance, and QC records. Laboratory forms can be paper based or electronic, and are completed requesting tests or investigations.

SELF-CHECK 17.2

What is a SOP and why is it needed?

Document control is an essential part of any effective QMS. Documents associated with the QMS will change with organizational changes, changes in practices, technology, and improvements in processes. Control of these changes is necessary to ensure that the procedures correctly describe the processes, are readily available, and are up to date. For example, if the procedure to issue blood changes, these changes need to be documented so that all staff are aware of the new procedure to follow. If out-of-date procedures are followed, mistakes, such as using incorrect times to incubate microbiology samples may occur, and lead to an incorrect diagnosis.

Change control in the medical laboratory aims to ensure all changes are assessed and approved by management before they are implemented. For example, updating a computer program without considering the effect across the network may render the system unusable in some areas. Thus, changes to procedures or to equipment must be introduced in a controlled and coordinated manner to prevent time, effort, and resources being wasted.

17.3 Audit and self-assessment

ISO 9001:2008 defines an audit as 'A systematic and independent examination to determine whether quality activities and related results comply with planned arrangements and whether these arrangements are implemented effectively and are suitable to achieve objectives'. Hence, audit is an integral part of any QMS and is a requirement for accreditation by UKAS and other medical laboratory and hospital-wide standards. All audits are a means for continually improving the service provided by the laboratory.

ISO 19011:2011,'Guidelines for auditing management systems', gives essential information for managing and conducting the audit programme. It also provides guidance on how to evaluate the competence of individuals involved in the audit process. ISO 19011:2011 is applicable to all organizations that are required to conduct internal and external audits of their management system.

The concept of audit is to gather information by means of observation, interview, and sampling to identify areas where improvements can be made. This is essential if a process is to improve continually. An audit will also identify processes that are working well, and so give an opportunity to learn from good practice and to allow that knowledge to be transferred to other processes. Thus, audit enables organizations to evaluate their processes, determine deficiencies, and generate cost-effective and efficient solutions to those problems.

Key points

Audit is used to measure an organization's ability 'to do what it says it is going to do'. All audits are performed to check practices against procedures and to document thoroughly any differences. However, audits are only sampling exercises and so they cannot confirm that all aspects of a process are being complied with at all times.

17.3.1 Audit schedule

Before any audit can be undertaken, it is necessary to prepare an audit plan. The plan is called an audit schedule and contains information about what audits are to be undertaken, when they will be undertaken, and who will perform the audit. All areas of the medical laboratory should be audited every year unless there is a documented reason as to why a specific area was not audited. It is recommended that a risk-based approach be used when planning audits. For example if it is known that there has been no change in equipment in the last 12 months and no incidents or equipment breakdowns then it maybe decided that audit of equipment will not be undertaken this year. However, if, for example, a number of incidents have been reported around sample labelling, then this area will be audited twice this year.

The audit schedule is discussed and agreed by senior management at the beginning of the year. You can see an example of an audit schedule in Table 17.2.

17.3.2 Types of audit

Audits fall into three main types:

- internal audits
- external audits
- cooperative audits.

Internal audits are conducted by the medical laboratory staff itself. The purpose of an internal audit is to confirm that everything is in order and as expected, and to identify any **non-compliance** with the standards that are applicable to the laboratory. It will identify any deficiencies in the QMS and make recommendations of any actions that need to be taken to improve the laboratory processes. The type and content of the internal audit will vary according to the size and activities of the laboratory concerned.

TABLE 17.2 **Audit schedule**

Audit area	January	February	March
Audit of laboratory equipment	Horizontal audit of equipment such as analysers, centrifuges, and microscopes used in the laboratory		
Staff records		Horizontal audit of mandatory training records	
Transport of samples			Vertical audit on receipt of samples from general practitioners to the issuing of results of clinical tests

External audits are conducted by an outside person or body, such as the UKAS, which has an interest in accrediting the laboratory. They are periodic checks undertaken for a specific purpose for example to check that the laboratory meets a set of criteria or a standard.

Cooperative audits are conducted between the laboratory and another party for mutual benefit, such as clinical audits, user satisfaction surveys, patient surveys, or benchmarking activities.

17.3.3 Methods of audit

There are a number of ways to undertake audits, each of which are designed to test different processes and are often used in conjunction to provide information about the entire laboratory.

1. Vertical audit. An audit of a single sample through all medical laboratory processes.
2. Horizontal audit. An audit of many samples through a single process.
3. Witness or examination audit. An audit of both the accuracy and completeness of a SOP, performed in real time, while a person is following it. This is also used in training of staff in a new technique.

You can see an illustration of some audit methods in Figure 17.3.

A *vertical audit* examines a single item, following it from start to finish. For example, this could be the journey a single clinical sample follows once it is received by the laboratory to the result of the tests performed on it being issued to the relevant clinicians. In contrast, *a horizontal audit* examines one element in a process that is performed on more than one item. This could, for example, be an examination of 20 request forms to ensure that details of all 20 patients have been entered correctly on the computer system. A *witness audit* examines a person undertaking a task. This may be observing a person labelling samples to check that he or she has read and is following the relevant SOP. Self-assessment audit is a careful evaluation that is usually performed by the organization's own management, which results in an opinion or judgement as to its effectiveness and efficiency.

SELF-CHECK 17.3

What methods of audit do external inspectors undertake when visiting the laboratory?

17.3.4 Performing an audit

The four stages in an audit are:

1. Prepare the audit checklist
2. Undertake the audit
3. Write a summary of the audit
4. Undertake **corrective action** to improve the laboratory process and to meet the required standard.

An audit checklist is a list of the questions to be asked relating to the set standard(s). Hence, if the set standard was, 'There shall be up to date information for users', a suitable checklist question is, 'Is there up to date information available for users?' The audit checklist must also contain the names of the auditors (the person undertaking the audit), the proposed date of the audit, a unique reference number, and the area(s) being audited. The department will appoint an auditee–a person who will assist the auditor(s), accompanying them around the department to verify factual information. An audit is the process of going into a department and asking the questions in the checklist. Essentially, these questions will be of the type: WHAT ...? WHERE ...? WHY ...? WHO ...? HOW ...? SHOW ME ...? Evidence to support the answers to these questions must be examined; the auditor(s) must *see* the evidence. If

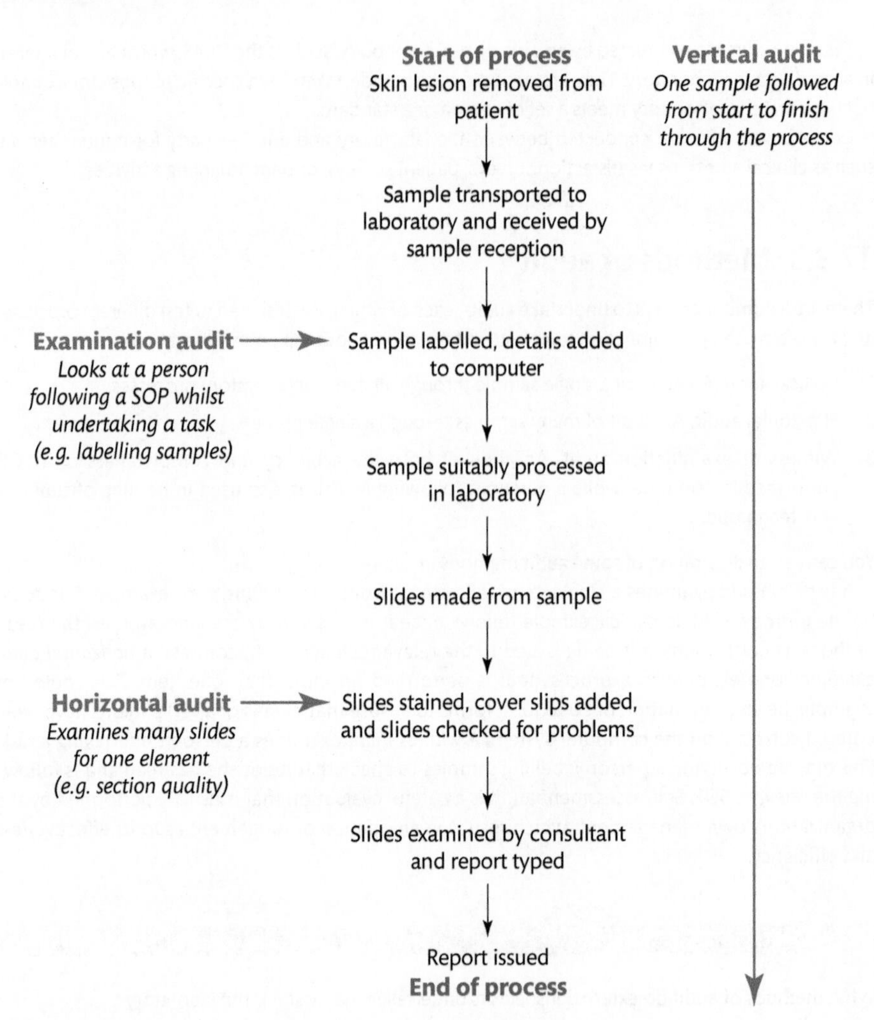
Start of process
Skin lesion removed from patient
Vertical audit
One sample followed from start to finish through the process
Sample transported to laboratory and received by sample reception
Examination audit
Looks at a person following a SOP whilst undertaking a task (e.g. labelling samples)
Sample labelled, details added to computer
Sample suitably processed in laboratory
Slides made from sample
Horizontal audit
Examines many slides for one element (e.g. section quality)
Slides stained, cover slips added, and slides checked for problems
Slides examined by consultant and report typed
Report issued
End of process

FIGURE 17.3
Diagram showing the different methods used when undertaking an audit. SOP: standard operating procedure.

it is not seen, it does not exist! Audit is all about seeing the evidence of compliance with the standard or SOP being audited.

After an audit has been completed, a summary of *all* the findings are documented. The summary outlines areas of compliance, that is, where the laboratory meets the standard, partial compliance, where the laboratory has some evidence of meeting the standard, and points of non-compliance, where evidence of meeting the standard was not provided. This is an essential stage in an audit and should not be missed out. It gives an opportunity to highlight areas of good practice and what the laboratory is doing well, in addition to identifying areas for improvement. Areas that need to be improved are issued with a non-compliance form, which outlines the area of non-compliance against the standard, what corrective actions need to be taken, by whom, and when. When all identified non-compliances against the standards are corrected, the audit can be said to be closed. However, the audit process is a never-ending cycle. Re-audit after a suitable period of time is important to see if the improvements have continued or if other issues have arisen as a result of the initial audit. This audit and re-audit cycle leads to **continual improvement** in the laboratory provision and in the service it provides to laboratory users. You can see an illustration of the cycle in Figure 17.4.

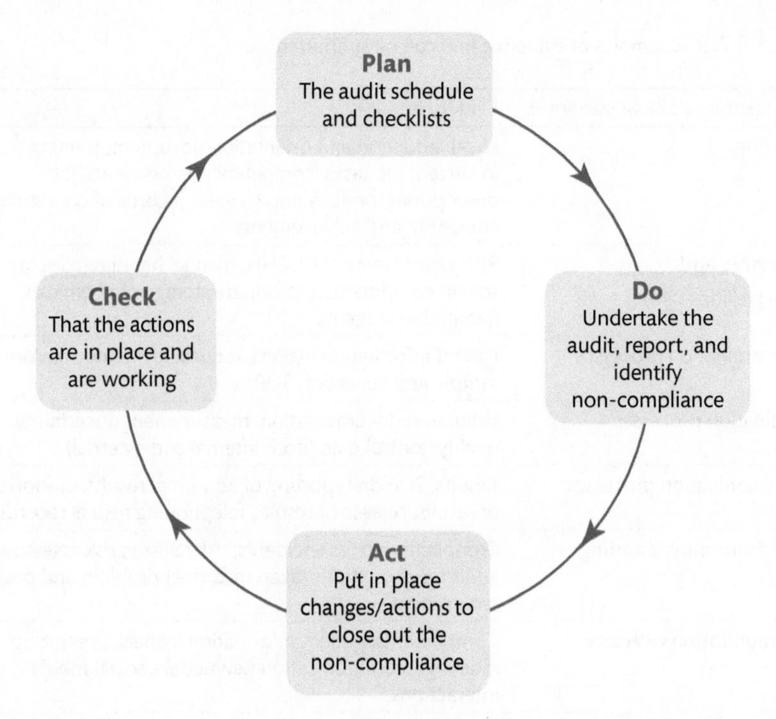

FIGURE 17.4
The audit cycle is a never-ending process of self-checking in order to implement improvements.

17.4 Regulatory requirements

Medical laboratories are subjected to rigorous inspections by a number of different assessment and accrediting bodies. Accreditation is the process by which an organization gains recognition that its activities and products have attained a certain level of compliance against a set of defined standards. In the UK, UKAS is the sole national accreditation body recognized by the government to assess against national and international standards as published by the British Standards Institute (BSI) and the ISO. The BSI develops, publishes, and markets standards in the UK. **Standards** are codes of best practice that improve safety and efficiency. The ISO is the world's largest developer and publisher of international standards. UKAS assesses medical laboratories against a number of different ISO standards, the main one being ISO 15189:2012, 'Medical laboratories–Requirements for quality and competence'. Other regulatory bodies exist, such as the Human Tissue Authority (HTA) and the Medicines and Healthcare products Regulatory Agency (MHRA), that are also able to undertake inspections of medical laboratories. These other regulatory bodies do not replace or supersede UKAS but assess defined laboratory areas covered by extra UK law, for example The Blood and Safety Quality Regulations (BSQR) (Section 17.4.3). However, both of these organizations are able to take enforcement actions and are able to close a UKAS-accredited laboratory if appropriate. This is known as a *cease and desist notice*. Medical laboratories are also subject to inspection by the CQC.

Cross reference
You can read more about the CQC in Section 17.4.

This is an independent inspection of all healthcare establishments against a set of defined criteria. As part of all regulatory inspections/assessments examples of evidence have to be either submitted before the inspection/assessment or will be looked at on the day of the inspection/assessment. Table 17.3 gives examples of evidence that could be submitted for the inspector/assessor to look at.

17.4.1 ISO 15189:2012, 'Medical laboratories–Requirements for quality and competence'

ISO 15189:2012 can be used by medical laboratories in developing their QMS. It can be used to assess the laboratories' competence. It can also be used for confirming or recognizing the quality and

TABLE 17.3 **Examples of evidence that can be submitted**

Standard/area of assessment	Evidence
Personnel	Local induction and orientation document, training in current job tasks, competency assessments, job descriptions, performance review, records of continuing education and achievements
Laboratory and office facilities	Risk assessments of COSHH, manual handling, display screen equipment, appropriate storage of chemicals, flammable reagents
Pre-examination processes	Patient information leaflets, request form information, sample and collection SOP
Examination processes	Validation documentation, measurement uncertainty, quality control data (both internal and external)
Post-examination processes	Results, SOP on reporting of abnormal results, authorizing of results, release of results, telephoning results records
Non-conformity reporting	Complaints, errors, incidents, risk register, risk assessments, audit results, actions taken to correct problem and prevent recurrences
Communication with users	Compliments, patient information leaflets, user group meetings, communication newsletters/social media interactions
Document control	Documents with correct version control in use by staff

COSHH: Control of Substances Hazardous to Health; SOP: standard operating procedures.

competence of medical laboratories by laboratory customers, regulating authorities, and accreditation bodies. Organizations that declare their adherence to a defined standard of practice and have their ability to attain these standards independently confirmed, that is, they are 'accredited' can reassure their users of the quality of service they provide to the patient. The use of ISO 15189:2012 has only come into use in the medical laboratory from October 2013. Previous to this, medical laboratories were assessed and accredited by Clinical Pathology Accreditation (CPA UK Ltd), which itself became a wholly owned subsidiary of UKAS in 2009.

17.4.2 UKAS assessment process

The assessment process involves the laboratory enrolling with UKAS, and completing an application form. When completing the application form the laboratory is required to list all of the tests they require to be assessed. This is known as the laboratory scope and is what the accreditation/assessment will be based upon. The inspection process is undertaken by trained assessors who perform an external audit of the laboratory over a number of days. The inspection team consists of a regional assessor and at least two peer assessors. The regional assessors are employed full time by UKAS and coordinate the assessment visits. Peer assessors are drawn from practising consultant pathologists, clinical scientists, and biomedical scientists. They accompany the full-time assessors and act as experts in specific subjects such as immunology or biochemistry. Inspection is undertaken using a form similar to that shown in Table 17.4.

17.4.3 Blood Safety and Quality Regulations

The BSQR 2005 subsumed two European blood directives (Directives 2002/98/EC and 2004/33/EC) into UK law, as the UK (like all European Union (EU) member states) has a duty to translate the European directives into national legislation. The BSQR require hospital blood banks and blood establishments to maintain a quality system based on the principles of good laboratory practice, which is a set of principles that provides a framework within which laboratory work is performed. The MHRA assesses the compliance of hospital blood banks and blood establishments to the BSQR.

TABLE 17.4 Outline of the ISO 15189:2012 standard

Section	Title
Forward	
Introduction	
1	Scope of the standard
2	Normative references
3	Terms and definitions
4	Management requirements 4.1 Organisation and management responsibility 4.2 Quality management system 4.3 Document control 4.4 Service agreements 4.5 Examination by referral laboratories 4.6 External services and supplies 4.7 Advisory services 4.8 Resolution of complaints 4.9 Identification and control of nonconformities 4.10 Corrective action 4.11 **Preventive action** 4.12 Continual improvement 4.13 Control of records 4.14 Evaluation and audits 4.15 Management review
5	Technical requirements 5.1 Personnel 5.2 Accommodation and environmental conditions 5.3 Laboratory equipment, reagents, and consumables 5.4 Pre-examination processes 5.5 Examination processes 5.6 Ensuring quality of examination results 5.7 Post-examination processes 5.8 Reporting of results 5.9 Release of results 5.10 Laboratory information management

The laboratory is required to submit an annual online compliance report to the MHRA, which then undertakes a risk assessment on the online report against the BSQR requirements. After the initial risk assessment the MHRA may decide to visit the laboratory and undertake an onsite assessment. This visit is usually conducted over one day by a single assessor. The assessor checks that there is accurate and complete **traceability** of blood and blood products. As part of the assessment they will examine the records kept by the medical laboratory and also visit the wards and satellite blood fridges that receive and store blood or blood products. One of the BSQR requirements is that the laboratory notifies the MHRA and **Serious Hazards of Transfusion** (SHOT) of any *serious adverse blood reactions and events* (SABRE).

This is undertaken through an online reporting system that can be accessed through the MHRA website. The reporting of SABREs allows the MHRA and SHOT teams to identify any trends in blood transfusion incidents. SHOT has recently become the formal haemovigilance organization of MHRA.

17.4.4 Human Tissue Act 2004

The Human Tissue Act covering England, Wales, and Northern Ireland came into force in 2004. The Human Tissue Act covering Scotland came into force in 2006. The Acts arose from concerns raised by events such as those at Bristol Royal Infirmary (1984–95) and the Alder Hey Children's Hospital Liverpool (prior to 1999), where it was established that organs and tissues from children had been removed, stored, and used without proper consent from next of kin. Subsequent government reports showed that the storage and use of organs and tissues from both adults and children had been widespread in the past. It became clear that current law in this area was not as clear and as consistent as it might have been for professionals or families. Therefore, the purpose of the Human Tissue Act is to provide a legal framework for issues relating to the taking, storage, and use of human tissue and organs and whole-body donation. The Act made consent the foundation for the lawful storage and use of human bodies, body parts, organs, and tissue, and the removal of tissue from the deceased, and regulates the use of tissues and cells that are to be used for transplant and stem cell research from living donors.

The Act outlined the need for a regulatory authority to be appointed: the Human Tissue Authority (HTA) is the regulatory body appointed in the UK. The HTA ensures that the activities it regulates are undertaken in a proper manner, and provides advice and guidance related to the Human Tissue Act and also to the EU Tissue and Cells Directives, which is also part of the Quality and Safety Regulations (see Chapter 4). These laws ensure that human tissue is used safely and ethically, and with proper consent.

The HTA ensures these laws are followed by setting standards that are clear and reasonable, helping people to understand the legal requirements, and providing codes of practice, advice, and support in nine different areas. These are anatomy, human application, stem cells and cord blood, public display, research, post-mortem, coroners, transplants, and DNA.

The HTA issues codes of practice that give guidance to each of the different areas. The codes differ for each but all broadly cover the need for consent, implementation of governance, and quality systems, including control of documentation and records. For example, the post-mortem standards cover consent for a hospital post-mortem, governance and quality systems, premises, facilities and equipment, and disposal of clinical material.

It is unlawful to undertake certain activities, such as post-mortems, storage of the deceased, and storage of body parts without having a licence from the HTA. A list of all licence holders is displayed on the HTA's website.

As part of its role is to ensure compliance with the Human Tissue Act, the HTA will inspect laboratories, mortuaries, and other establishments that fall under its remit. An inspection team consisting of two or more people will usually spend 1 or 2 days in the laboratory conducting an audit against the Human Tissue Act standards. The HTA requires that a named individual, known as the *designated individual*, takes responsibility for ensuring that suitable practices are used when undertaking licensed activities such as post-mortems.

Following an inspection, a report is written and any areas that need improving are identified, together with a timescale for their rectification.

The Human Tissue (Scotland) Act was introduced in Scotland in 2006. The Human Tissue Authority Codes of Practice also cover the Human Tissue (Scotland) Act in most, but not all, areas. Where the HTA refers to a coroner in England and Wales, the equivalent person in Scotland is the *Procurator Fiscal*.

17.4.5 Human Fertilisation and Embryology Authority

The Human Fertilisation and Embryology Authority (HFEA) is an independent regulator that oversees the use of gametes and embryos in fertility treatment. The HFEA licences fertility clinics and centres that undertake *in vitro* fertilization (IVF). If the medical laboratory is part of the fertility process they will need to hold a licence. The main types of licences under HFEA are as follows:

- storage
- research, for laboratories carrying out research on human embryos
- treatment, for clinics offering intrauterine insemination and other basic fertility treatments that do not involve the creation of embryos
- treatment and storage, for clinics offering IVF, intracytoplasmic sperm injection (ICSI), and gamete and embryo storage.

The HFEA publishes a code of practice. The Code of Practice is intended to help and encourage licensed centres to understand and comply with their legal requirements. It also gives guidance on how centres are expected to go about meeting those requirements. The code of practice is available on the HFEA website. HFEA will inspect organizations against the code of practice to ensure that they have robust and safe systems in place. The quality of the service clinics provide to their patients is critically dependent on the quality of the methods and equipment used in the laboratory; therefore, the laboratory plays a key part in the patient pathway.

17.4.6 Point-of-care testing

Point-of-care testing (POCT) can be defined as testing at or near the site of patient care, which increases the likelihood that the patient will receive the result of a clinical test in a timely manner. POCT is often used in accident and emergency departments, neonatal care units, GP surgeries, or even in the home of a patient (Chapter 16).

The ISO standard, ISO 22870:2006, 'Point-of-care testing (POCT)—Requirements for quality and competence', outlines the requirements for laboratory POCT. The medical laboratory must ensure that POCT is coordinated effectively and efficiently in the hospital environment. Thus, POCT policies and procedures need to be in place that outline how the equipment must be used. Staff must be trained in use of the equipment and records of this training maintained. Medico-legal considerations and the requirements of the Data Protection Act must be borne in mind when establishing these procedures. The laboratory must be involved in any purchases of new equipment for use outside of the laboratory, as well as being involved in training staff to use it correctly. Most medical laboratories have a dedicated person to look after POCT equipment, and liaise with ward and departmental staff about its correct use and cleaning. The laboratory also has a duty to ensure that the equipment is serviced in accordance with manufacturer's requirements and recommended local and/or national QA schemes (NEQAS).

POCT is not assessed or accredited in isolation but as part of the overall laboratory accreditation process.

Key points

Currently, POCT not provided as part of laboratory service that takes place in the community, such as at GP surgeries or patient homes, is neither assessed nor accredited.

17.4.7 Care Quality Commission

Medical laboratories are likely to be part of a larger organization, such as a hospital. In such cases they may be subject to pan-hospital regulatory requirements such as inspection by the CQC. The CQC regulates every healthcare activity in the UK. The regulation framework is based on the essential standards of quality and safety. The essential standards are divided into outcomes. Different outcomes apply to different providers. A full list of the outcomes can be found online (www.cqc.org.uk).

Any healthcare organization that operates in the UK must be registered with the CQC. For medical laboratories this means registering for diagnostic services and use of blood products, if applicable. The CQC inspectors use professional judgement, supported by objective measures and evidence, to assess services against five key questions, as shown in Figure 17.5.

To direct the focus of the inspection, the inspection teams use a standard set of key lines of enquiry (KLOE) that directly relate to the five key questions that are asked of all services. Having set KLOE ensures there is a consistent approach to the inspections. To enable inspection teams to reach a rating, they gather and record evidence in order to answer each KLOE. This involves an on-site inspection, as well as pre-inspection information gathering and intelligent monitoring through the use of external data, such as that from UKAS, MHRA, staff survey, and so on.

Each area inspected is given a rating. The ratings are Outstanding, Good, Requires Improvement, and Inadequate. All reports are published on the CQC website.

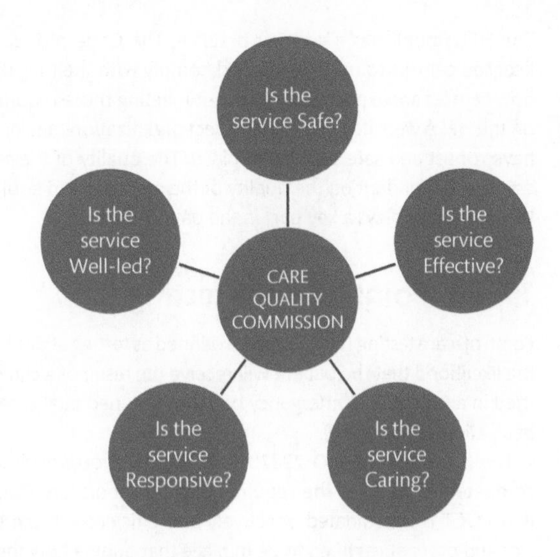

FIGURE 17.5
Care Quality Commission key questions.

Duty of candour

In April 2015, the CQC issued 'Regulation 20', which outlines what organizations are required to undertake for duty of candour. The aim of duty of candour is to ensure that organizations are open, honest, and transparent when certain incidents occur in relation to the care and treatment provided to people who use the service. This regulation is a direct response to recommendation 181 of the report into the failings at the Mid Staffordshire NHS Foundation Trust (The Francis Report) which recommended that a statutory duty of candour be imposed on all healthcare providers.

- Openness. To enable concerns and complaints to be raised freely without fear, and questions asked to be answered.
- Transparency. Allowing information about the truth in relation to performance and outcomes to be shared with staff, patients, the public, and regulators.
- Candour. Any patient harmed by the provision of a healthcare service is informed of the fact and appropriate **remedial action** offered, regardless of whether a complaint has been made or a question asked about it.

SELF-CHECK 17.4

- What are the main bodies that inspect and assess clinical and medical laboratories?
- Can you identify where these bodies overlap in terms of the areas assessed/inspected?

17.5 Quality governance and risk management

Until recently, the only duty placed upon a hospital was that of financial aspects; clinical standards while being expected were not explicitly required. This has now changed and all hospitals now have a clinical and financial responsibility to meet, consequently the role and importance of clinical governance has increased in recent years. Clinical governance naturally has an impact on laboratories, which must also play their part in the delivery of high quality healthcare.

17.5.1 Clinical governance

Clinical governance can be defined as a system through which healthcare organizations are accountable for continuously improving the quality of their services and safeguarding high standards of care by creating an environment in which excellence in clinical care is able to flourish. Key components

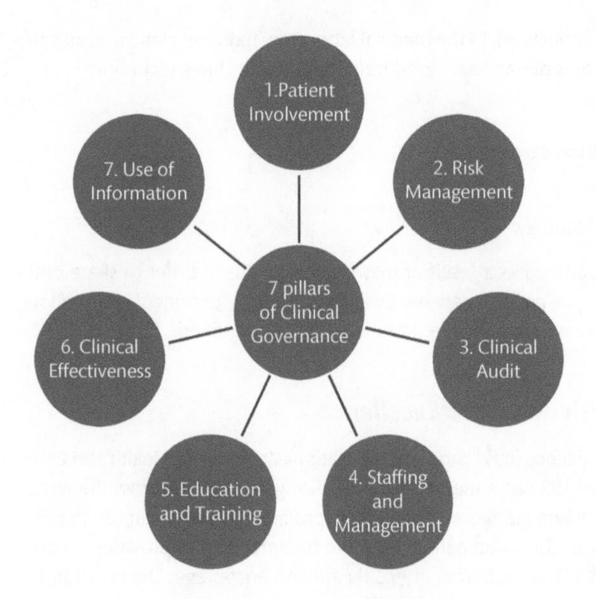

FIGURE 17.6
The pillars of clinical governance.

of quality governance in current healthcare organizations are to ensure patient safety, learning from mistakes, being open and honest, and sharing and maintaining good practice. This concept was originally based on the seven pillars of clinical governance (see Figure 17.6)

The original seven pillars could be added to, to include leadership, process and system awareness, team work, communication, and ownership. Clinical governance in practical terms means ensuring learning from mistakes and **healthcare near misses**, encouraging staff to be open and honest, sharing, developing, and maintaining good practice, and involves continual professional development and lifelong learning. The focus is a patient-centred service committed to improved patient outcomes.

17.5.2 Clinical effectiveness

Clinical effectiveness is the effectiveness of clinical practice on patient outcomes.

The Royal College of Pathologists defines clinical effectiveness as 'the extent to which specific interventions, when deployed in the field for a particular patient or population, do what they are intended to do, for example, maintain and improve health and secure the greatest possible health gain from available sources'. Clinical effectiveness involves a number of key areas as follows:

- clinical audit of patient care and outcomes
- key performance indicators
- cancer minimum datasets
- collaborative working with organizations such as the National Institute for Health and Care Excellence (NICE)
- development and standardization of pathology tests through the use of pathology catalogues.

More information can be found on the Royal College of Pathologists' website (https://www.rcpath.org/profession/clinical-effectiveness.html).

17.5.3 Clinical audit

This is a quality improvement process that seeks to improve patient care and patient outcomes through a systematic review of care. It can be used as a tool to guide best practice. Clinical audits may be at a national, a local or a regional level. In the medical laboratory national clinical audit can be seen in blood transfusion. Examples of national blood transfusion audits have included:

- the audit of the use of anti-D
- audit of use of blood in neurocritical care units
- audit of blood use in elective surgery
- audit of the use of blood in haematology.

National reports and surveys are published as a result of these clinical audits in order to share best practice and improve the standard of care given to patients. By analysing a large amount of information from a number of different healthcare organizations trends in care can be identified.

National Institute for Health and Care Excellence

The National Institute for Clinical Excellence (now known as the National Institute for Health and Care Excellence) (NICE) was introduced in 1999 as a means of standardizing healthcare across the NHS. NICE provides evidence-based advice and guidance to healthcare professionals who can use this information when taking into account decisions on patient care and treatment. NICE provides a more cost-effective rationale for use of available resources in healthcare and encourages the creation of new innovative technologies. In recent years NICE has diversified into setting standards and provides a broad range of evidence in order to improve patient outcomes. NICE guidance takes several forms, as seen in Table 17.5

NICE guidance that is applicable to medical laboratories can be found on the website and includes areas such as early and locally advanced breast cancer, interpretative tests for breast cancer, and use of molecular techniques (one-step nucleic acid amplification (OSNA)).

TABLE 17.5 National Institute of Health and Care Excellence (NICE) guidance

NICE guidance makes evidence-based recommendations on a wide range of topics, from preventing and managing specific conditions, improving health, and managing medicines in different settings to providing social care to adults and children, and planning broader services and interventions to improve the health of communities	Technology appraisals assess the clinical and cost-effectiveness of health technologies, such as new pharmaceutical and biopharmaceutical products, and also include procedures, devices, and diagnostic agents
Medical technologies and diagnostics guidance help to ensure that the NHS is able to adopt clinically and cost-effective technologies rapidly and consistently	Interventional procedures guidance recommends whether interventional procedures, such as laser treatments for eye problems or deep brain stimulation for chronic pain, are effective and safe enough for use in the NHS
Quality standards consist of a concise set of statements, with accompanying metrics, designed to drive and measure priority quality improvements within a particular area of care. They are derived from the best available evidence, particularly NICE's own guidance and, where this does not exist, from other evidence sources accredited by NICE	

NHS: National Health Service.

17.5.4 Patient safety alerts

Patient safety alerts, drug alerts, and medical devices alerts can be issued on the bases of a trend in incidents reported via the National Reporting and Learning System. The safety alert system enables issues or problems to be shared in a proactive approach to prevent harm to patients or staff. Table 17.6 shows examples of changes that have come about as a result of this system.

17.5.5 Risk management

The definition of risk management is 'the effect of uncertainty on objectives', as defined by ISO terminology guide 73, 'Risk management–Vocabulary'. ISO 15189:2012, section 4.14.6, 'Risk management', says that 'The laboratory shall evaluate the impact of work processes and potential failures on the safety of the examination results and shall modify processes to reduce or eliminate the identified risks'.

All organizations have objectives at the strategic, tactical, and operational level. Anything that makes achieving these objectives uncertain is a risk. ISO standards 31000 are a family of standards that can be used to provide principles and guidance on risk management. They provide information for the design, implementation, and maintenance of risk management processes. Effective risk management is essential in a healthcare setting as there needs to be clear identification and mitigation or resolution of risks that arise, in order to ensure the aims of patient safety are met. Risk management is a very large subject and this section will only look at providing an overview of the subject.

Risk assessment

This is a process that is undertaken to ensure that risks are properly assessed and actions are identified and acted upon to provide the safest possible outcome for patients, staff, and visitors. Risk assessments are required to estimate or measure the consequences and likelihood of a risk being realized. They can aid complex decision-making and reduce the risks in a healthcare environment to the benefit of all that use the service.

Examples of risks seen in healthcare include:

- patient at risk of a fall assessment
- patient at risk of a pressure ulcer
- staff at risk of using a noxious chemical
- environmental risks (ventilation, lighting, noise)
- risk of equipment failure.

A proactive assessment approach can be used to help prevent patient and staff safety incidents by identifying what could go wrong and what the consequences might be. Risk assessment against a requirement or set of standards can give some guidance as to what the risks to the organization or department might be. For example standard ISO 15189:2012 section 5.1.5 states that the laboratory is required to provide training for all personnel, which includes the applicable information system. If staff have not been trained in this area the risk of using the information system incorrectly is likely to be high, which could have the consequence of leading to the release of incorrect or unauthorized

TABLE 17.6 Examples of patient safety alerts

Standardize use of the '2222' number across all NHS organizations when making an emergency/crash call. Before this, organizations often had a different number making it impossible for those moving around different hospitals to remember the number, such as junior doctors
Patient safety alert on standardizing the early identification of acute kidney injury (AKI) to improve detection and diagnosis of AKI

TABLE 17.7 **Risk matrix**

Likelihood	Consequence				
	Negligible	Minor	Moderate	Major	Extreme
Almost certain	Medium	High	High	Very high	Very high
Likely	Medium	Medium	High	High	Very high
Possible	Low	Medium	Medium	High	High
Unlikely	Low	Medium	Medium	Medium	High
Rare	Low	Low	Low	Medium	Medium

results. A risk assessment of the training documentation will highlight if this risk exists. An actual risk assessment form can take many guises, but all will have a rating system in order to ascertain the level of risk. Table 17.7 is an example of what this can look like.

Risk control

Risk control, also known as risk treatment, is a step in the risk assessment process that is required to reduce or mitigate the risk that has been identified. This involves identifying the risk control options, evaluating the effectiveness of each option, and then selecting the option that gives the most benefit to reduce or eliminate the risk.

SELF-CHECK 17.5

Undertake a risk assessment of your immediate laboratory area and identify risks associated with the work that is undertaken with relation to the chemicals that are used in the laboratory. Locate the Control of Substances Hazardous to Health information and check that all risks identified have had a formal assessment and that control measures have been put into place. Discuss the results of this with your course leader, training officer or quality lead.

Risk register

The risk register is a log of risks that have been identified that could threaten an organization's success in delivering its aims and objectives. A risk register can be developed by using the risk assessment approach as described earlier. The risk register will contain key information about the risk and include a risk rating, as shown in Table 17.8. The risk register allows the organization's top management or board to have sight of the organization's risk, which helps them to understand the risk profile of the organization.

Incident reporting and root cause analysis

As has been noted by Sir Liam Donaldson, speaking at the launch of the World Alliance for Patient Safety, Washington DC, 27 October 2004, 'to err is human, to cover up is unforgivable, and to fail to learn is inexcusable'.

Incident reporting is a major tool for examining what happened when an error or problem occurred. Adequate and appropriate incident reporting systems are necessary to provide a core of sound, representative information on which to base any analysis of errors or problems and to make recommendations to prevent their repetition. Learning lessons from incidents requires timely incident reporting, which, in turn, requires an open and just culture within an organization.

Experience in other sectors, such as the airline industry, has demonstrated the value of systematic approaches to recording and reporting adverse events.

TABLE 17.8. Risk register example

Risk ID	Risk title	Date risk opened	Description	Objective	Review date	Consequence	Likelihood	Risk level	Action plan	Progress report	Resources	Risk lead
1032	Inadequate outpatient phlebotomy service	23/02/2015	Paediatric patients requiring appointments are having to wait 3–4 weeks; this is delaying diagnosis and treatment. This delay is due to staff shortages after several trained experienced staff left the phlebotomy department	Treat	23/03/2015	Moderate	Likely	HIGH	To recruit to vacant posts. To allow overtime working to fill gaps in the service	February 2015 risk highlighted at departmental meetings. Staff recruitment plan in place. Additional shifts are being covered by staff as overtime, to reduce waiting time	Staffing cost	Key manager

It is possible to identify a number of barriers that can prevent active learning, but the NHS can draw valuable lessons from particular approaches used in other sectors. It is important to have a good safety culture, where open reporting and balanced analyses are encouraged in principle and by example. This can have a positive and quantifiable impact on the performance of organizations. In contrast, 'blame cultures' may encourage people to cover up errors for fear of retribution. This acts against identifying the true causes of failure, because they focus heavily on individual actions and largely ignore the role of underlying systems. Thus, organizational culture is central to every stage of the learning process: from ensuring that incidents are identified and reported through to embedding the necessary changes deeply into practice. Detecting and accurately recording errors is a fundamental step in learning from experience. It is common sense to recognize that knowledge of what is wrong is required before steps can be taken to put it right.

Reporting of accidents and incidents must be seen as professional behaviour rather than as disloyalty to other staff or the NHS. All organizations, including the NHS, must operate a genuine two-way communication system, not just a 'top-down' system from management. Communication systems need to be in place to allow people to see what has changed as a result of an incident or the reporting of a near miss. An organization needs to recognize staff concerns. The emphasis of management should be on personal and service benefits, rather than on threats. Within the NHS, more emphasis needs to be placed on creating an informed culture, to raise the awareness of the costs of not taking risk seriously. This includes the human costs, to staff and patients, as well as financial ones.

NHS organizations have in place paper-based or electronic reporting systems, for example Datix, as a means of reporting incidents, concerns, and near misses, and these should be used by the medical laboratory to report concerns, near misses, and actual incidents that have caused harm. Incidents that result in patient harm are known as serious untoward incidents and require a full investigation to ascertain the root cause. There is a requirement for organizations to report all serious untoward incidents on the national incident database known as the Strategic Executive Information System (STEIS). Also, many clinical commissioning groups require reporting with a full investigation into any event in which a patient has been harmed.

NHS England provides information on incident reporting and publishes a list of 'never events', the reporting of which is mandatory. Never events are a subset of serious incidents. Never events are serious preventable patient safety incidents that should not occur if the available preventative measures have been implemented. For the medical laboratory the never event lists includes:

- transfusion of ABO-incompatible blood components
- transplantation of ABO-incompatible organs as a result of error.

The never events list is updated every 2 years and it can change depending on what has been reported previously.

The potential for disasters may exist but for any number of reasons those may not occur, or occur very rarely. Most accidents have the potential to produce serious injury but do not do so in practice, either because of some intervention or compensation or simply through good fortune. Confining analysis to and learning from only those events that resulted in serious harm risks skewing learning towards a small proportion of accidents and missing important lessons that could prevent future adverse events.

Adverse incidents involving medical devices, including laboratory equipment, are reported to the MHRA. Information is logged in a central database. Incidents are assigned to a specific level of investigation depending on the risks involved. The outcomes of these investigations are subject to formal reviews. Patterns or clusters of incidents can thus be identified, subjected to further risk assessment procedures, and investigated where necessary.

When an incident reveals equipment-related safety problems, the MHRA will produce a hazard or safety notice for distribution to appropriate users so that action can be taken to withdraw or modify the device.

Industry has estimated that for every major injury, 29 minor ones and 300 accidents with no injury occur (Figure 17.7). Thus, to some extent the effectiveness of a reporting system can be judged by the relative proportion of minor incidents to more serious accidents: the greater the proportion of minor incidents reported, the 'better' the reporting system is working.

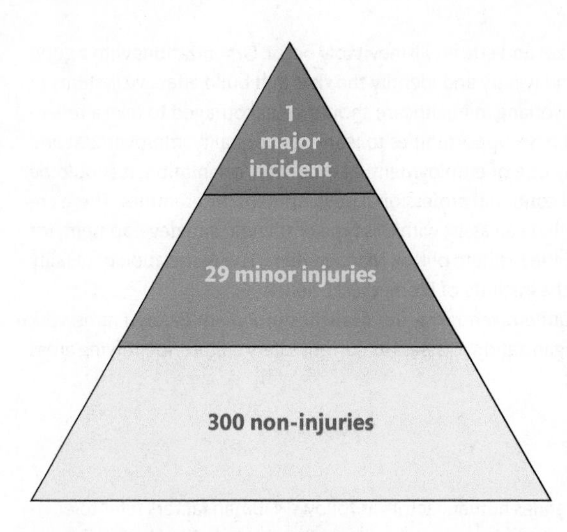

FIGURE 17.7
Triangle of incident relationships showing the proportion of near misses to harm to or death of a patient.

17.6 Patient safety

All healthcare organizations are expected to treat patients in a safe environment and protect them from avoidable harm. Harm means injury, suffering, disability, or death. The National Patient Safety Agency (NPSA), which is now part of the NHS Commissioning Board Special Health Authority, defines patient safety in seven steps, an outline of which is illustrated in Figure 17.8.

17.6.1 Safety culture

Safety culture is an attitude and behaviour exhibited by individuals and teams with a focus on the importance of patient and staff safety. It is the ability to recognize safety and risk issues and to mitigate these to improve patient outcomes.

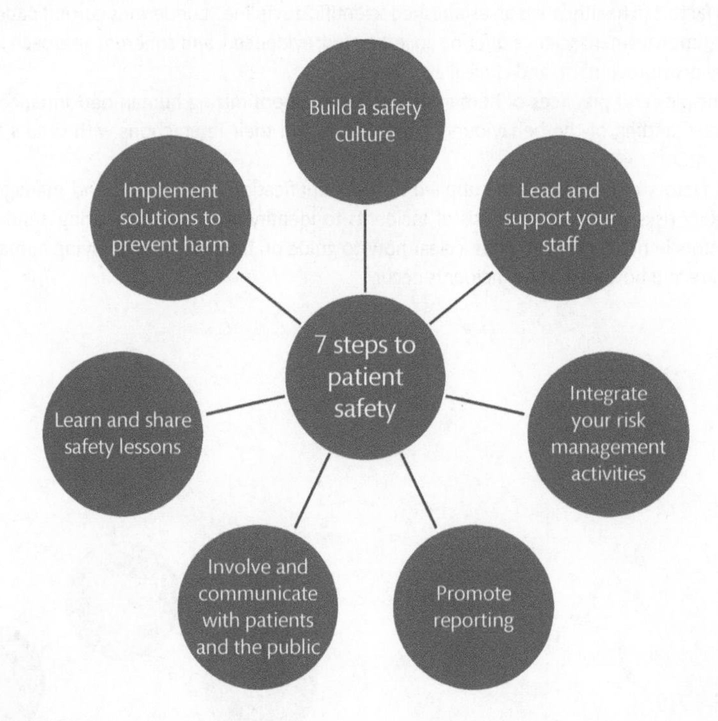

FIGURE 17.8
Seven steps to patient safety.

Humans are fallible and making mistakes and errors will inevitably occur. Organizations with a good safety culture accept that errors occur and will try and identify the risks and build effective systems to protect their patients and staff. All staff working in healthcare should be encouraged to take a reflective approach in their role. All staff must have opportunities to learn about quality improvement and patient safety. This should start from day one of employment, at induction/orientation. It should be included as part of ongoing training and continual professional development programmes. There are a number of external training providers that can assist with this type of training and development, for example the Chartered Quality Institute, the Institute of Risk Management, The Association of Quality Managers in Laboratory Medicine, and the Institute of Biomedical Science.

A safety culture assessment can be undertaken using the Manchester Patient Safety Framework (MaPSAF). The MaPSAF tool allows an organization to assess its current safety culture, identifying areas of weakness and areas for improvement.

17.6.2 Human factors

The Health and Safety Executive (HSE) defines human factors as follows: 'Human factors refer to environmental, organizational and job factors, and human and individual characteristics which influence behaviour at work in a way which can affect health and safety'. A simple way to think about human factors is to think about the three aspects shown in Figure 17.9.

Human factors have been researched in a number of safety critical settings, such as the aircraft industry, as a means of identifying how and why errors occur. This research aims to optimize human performance by understanding the interfaces between humans and human behaviour, the environment, and equipment. A clinical human factors group was established in 2005 after the death of a patient undergoing a routine operation. Human factor studies in the clinical setting look to reduce and manage error under the stresses and strains of every day work. One element to consider when looking at human factors involves examining and designing out error. For example, standardization of equipment with thorough training ensures all staff know how to use it. Working as a team where no member is more or less important than the next helps empower the team to have an approach based on avoiding errors and trapping them as they occur, thereby mitigating harm.

Human factors in healthcare is an established scientific discipline. It underpins current patient safety and quality improvement science offering an integrated, evidenced, and coherent approach to patient safety, quality improvement, and clinical excellence.

The principles and practices of human factors focus on optimizing human performance through better understanding of the behaviours of individuals and their interactions with others and their environment.

Human factors principles can be applied in the identification, assessment, and management of patient safety risks and in the analysis of incidents to identify learning and sharing. Implementing human factors in healthcare provides a clear how-to guide on the benefit of applying human factors in healthcare and how errors and incidents occur.

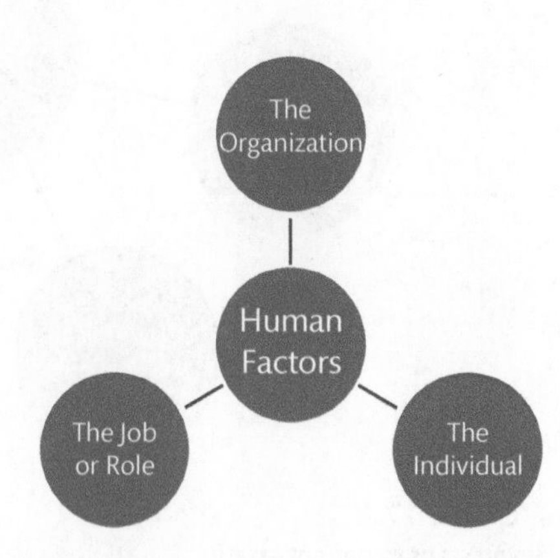

FIGURE 17.9
Human factors.

Common known human factors include:

- loss of situational awareness
- perception and cognition
- lack of teamwork and leadership
- culture.

Risks can increase when there are the following human factors involved:

- stress
- mental workload
- distractions
- the physical environment
- physical demands
- device and product design
- break down in teamwork
- miscommunication
- process and system design.

Understanding and applying human factors can:

- improve the safety culture
- enhance teamwork and improve communication between staff
- improve the design of processes and systems
- help identify what went wrong and proactively predict what could go wrong.

17.6.3 Patient experience and engagement

In the 2014 pathology QA review, Ian Barnes stated that pathology should be more accessible to its users, and this includes its patients. By informing and empowering the patient to be involved in their healthcare pathway they gain a better understanding of the value of the treatment options. This can already be seen in some areas of the medical laboratory, for example international normalized ratio testing for blood clotting diseases and blood glucose monitoring in diabetic patients.

If the patient understands their results and the consequences of not keeping to a set treatment plan or drug regime the health and well-being of the patient will be optimized. There is already a vast amount of information available for patients to access to understand their test results, such as Lab Tests Online, which is a patient-centred, peer-reviewed system to assist in understanding specific laboratory test results. Other resources are also available to allow patients to be more involved in their healthcare, for example 'patients know best' is a system that allows patients to control their own medical record. It allows clinicians to communicate with the patient and send letters and results. Keeping the patient at the centre of their treatment means that the patient will be more aware of what is required and also able to track their progress through a treatment plan. For example a patient on prostate-specific antigen (PSA) monitoring for prostate disease will be aware of when the PSA blood test is due and be proactive about ensuring this is undertaken when required. The patient will view the organization as being open and honest, and involving them in every step of their healthcare journey, thereby improving the patient experience at what can be a very frightening and difficult time.

Another aspect of patient experience is looking at putting the patient at the centre of the healthcare pathway when redesigning services. The Kings Fund has information about 'experience based on design' projects that design the services around the needs of the patient and not the needs of the organization.

A key aspect of this for the medical laboratory is the provision of point-of-care services: services delivered closer to the patient with a quick turnaround time so that treatment can begin as soon as possible.

17.7 Tools and techniques for continual improvement

As has been explained to you previously, quality is not static or a 'one-off' process, but requires continual examination and improvement as healthcare, diagnosis, and treatment evolve. Industry has understood this for many years and a number of the quality improvement tools devised for industry purposes can also be used within a healthcare setting to achieve similar outcomes.

17.7.1 Lean Six Sigma methodology

This is a collection of tools and techniques that can be used to drive up quality. It is a data-driven approach for eliminating defects and allows for standardization across processes and systems. To achieve Six Sigma, a process must not produce more than 3.4 defects per million opportunities. A defect is anything that is outside the required specification, for example in the medical laboratory the specification maybe a turnaround time for an urgent sample. One of the most commonly used tools in lean six sigma is the use of the D M A I C methodology.

D Define the problem

M Measure the current state by collecting information from the process

A Analyse the data and identify the areas that can be improved; put in place an action plan

I Improve to current state by implementing the action plan

C Control the new process; measure and re-analyse the data to ensure the desired change has been effective and sustained.

17.7.2 Root cause analysis

Root cause analysis (RCA) is a powerful tool that identifies records and visualizes the possible causes of a problem, error, or incident. It is undertaken as an investigation into the error to determine its basis, that is, its *root cause*. When an incident has occurred it is imperative that an investigation identifies its root cause, to ensure that improvements can be made and to prevent the same incident reoccurring. The basis of RCA is to break down the problem into smaller, more manageable tasks, which allows the problem to be identified and described, as shown in Figure 17.10. This approach will assist in finding a solution or corrective action that will prevent the same problem reoccurring.

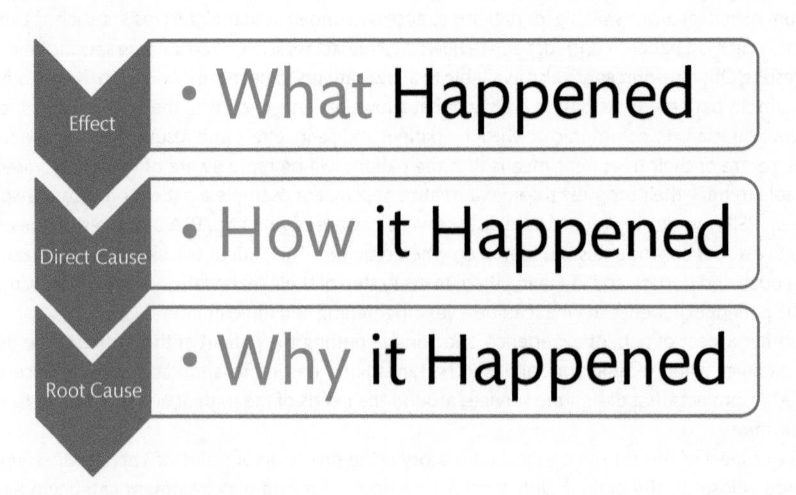

FIGURE 17.10
Getting to the root of the problem.

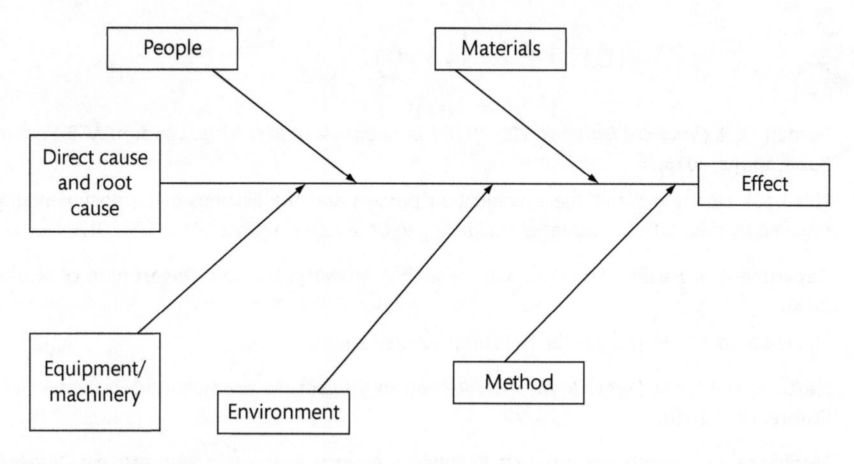

FIGURE 17.11
Cause and effect (Fishbone) diagram template.

Many methods can be used when undertaking RCA, for example a cause-and-effect diagram. Cause-and-effect diagrams provide a logical means to analyse a problem to understand its root cause. Although the cause of the failure is initially unknown, various possibilities need to be examined in detail to determine why the equipment failed. These possible causes can be mapped on the cause and effect (Fishbone) diagram, as shown in Figure 17.11, and each one analysed until the root of the problem is discovered. An action plan can then be put in place to ensure the same problem does not occur again.

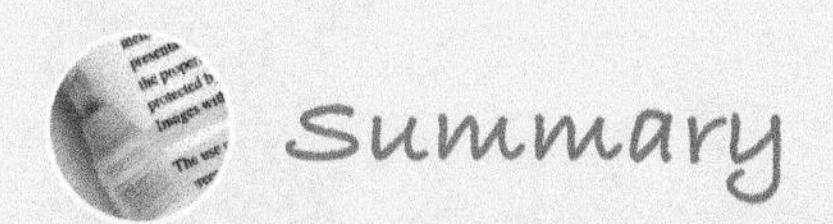

Summary

- A good pathology service can be the key to providing good patient care.
- The necessity to provide services of appropriate quality in medical laboratories, as well as the need to satisfy the users of laboratory services, has led to increasing numbers of regulations governing their operations and assessment.
- Medical laboratories are subjected to inspection and assessment by a number of different bodies, namely UKAS, the HTA, the MHRA, and the CQC.
- QA, QC, and the implementation of QMS has allowed medical laboratories to evaluate, monitor, and improve their processes and systems continually, and deliver a high-quality laboratory service to patients.
- A variety of techniques, such as audit, can be used by the laboratory to assess internally their compliance to set standards.
- Following good laboratory practices ensures the integrity of the sample audit trail in the laboratory and allows the cycle of continual improvement to be embedded into the laboratory culture.

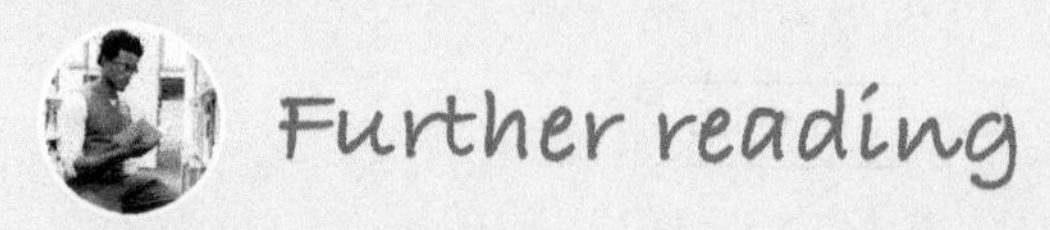

Further reading

- **Burnett D. *A Practical Guide to ISO 15189 in Laboratory Medicine*. London: ACB Venture Publications, 2013.**

  This book clearly explains the accreditation process and the accompanying documentation required by medical laboratories when looking to be assessed against ISO 15189:2012.

- **Department of Health. *An Organization with a Memory*. London: Department of Health, 2000.**

  This is a report on learning in the NHS from adverse events.

- **Haxby E, Hunter D, Jagger S. *An Introduction to Clinical Governance and Patient Safety*. Oxford: OUP, 2010.**

- **Medicines and Healthcare products Regulatory Agency. *Rules and Guidance for Pharmaceutical Manufacturers and Distributors*. London: Pharmaceutical Press, 2007.**

  Section II, 'Guidance on good manufacturing practice', is worth dipping into.

- **North East Strategic Health Authority Patient Action Team. Human factors for healthcare. Trainer's manual. Available at: http://patientsafety.health.org.uk/sites/default/files/human_factors_in_healthcare_trainer_manual_en_march_2013.pdf**

- **Risky Business. Exceptional talks sharing ideas on risk, culture, human performance, teams and leaders. Available at: www.risky-business.com**

- **The Health Foundation Inspiring improvement. Building capability to improve safety. Available at: www.health.org.uk/publication/building-capability-improve-safety**

- **The Kings Fund. www.kingsfund.org.uk/projects/ebcd/experience-based-co-design-description**

- **Vorley G, Tickle F. *Quality Management, Tools and Techniques*. 5th ed. London: Quality Management & Training, 2002.**

  Sections 3 and 4 give accounts of QMS, ISO9000, and audit.

Useful websites

- www.hta.gov.uk
- www.iso.org
- www.ukneqas.org.uk
- www.ibms.org
- www.mhra.gov.uk/index.htm
- www.npsa.nhs.uk
- www.dh.gov.uk
- www.ukas.com
- www.thecqi.org
- www.bsi-global.com
- www.hfea.gov.uk
- www.risky-business.com
- www.england.nhs.uk/ourwork/patientsafety/never-events

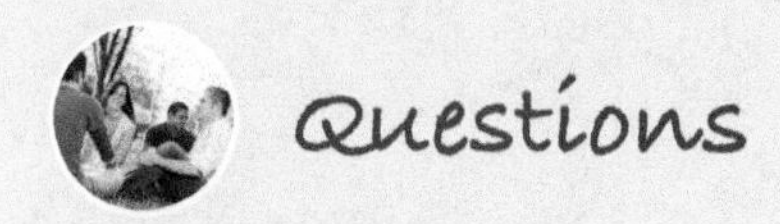

Questions

1. Which, if any, of the following has a responsibility to control quality in a medical laboratory?
   - (a) The quality manager
   - (b) Management
   - (c) The staff
   - (d) Laboratory manager
   - (e) All of the above.
2. Indicate which of the following statement(s) are **TRUE**:
   - (a) QA, QC, and the implementation of QMS has allowed medical laboratories to evaluate, monitor, and improve their processes and systems continually, thereby delivering a high-quality laboratory service to their users
   - (b) QA, QC, and the implementation of QMS has allowed the medical laboratory to eliminate all errors, thereby delivering a high-quality laboratory service to their users
   - (c) QA, QC, and the implementation of QMS has allowed medical laboratories to employ more staff to undertake audits in order to monitor their processes and systems and improve patient care
   - (d) QC means medical laboratories do not have to check their test results before releasing them to the clinicians
   - (e) EQA involves outside agencies checking the clinical test results before they are sent to the requester.
3. Outline the process of IQC as performed in the laboratory. Indicate how it differs from EQA.
4. What is meant by *document control*? State why document control is necessary.
5. What techniques can be employed to improve the laboratory service?
6. Outline the structure of a *learning organization.*
7. How do you report a concern or incident in your organization?
8. Why is it necessary to report concerns about incidents that occur in the laboratory?
9. Give an example of how a cause-and-effect diagram can be used to investigate an error.
10. Why is it necessary to have an audit trail outlining the path a clinical sample takes through a laboratory?

Answers to self-check and end-of-chapter questions are available at the end of the book.

18

Personal development

Hedley Glencross and Georgina Lavender

Learning objectives

After studying this chapter, you should be able to:

- Describe what is meant by personal development and continuing professional development
- Discuss opportunities and events that in everyday life contribute to personal development
- Discuss the types of continuing professional development needed to satisfy the requirements for registration with the Health and Care Professions Council
- Describe the opportunities provided by the Institute of Biomedical Science for professional development
- Identify opportunities within the workplace to extend personal development opportunities
- List the mandatory training that must be provided by an employer
- Select and utilize appropriate study skills for different situations
- Describe the structure of a generic performance appraisal scheme
- Explain how the performance appraisal and development structure within the National Health Service is of benefit to a biomedical scientist

Introduction

Personal development is any means by which an individual is able to expand on his or her existing knowledge, skills, and talents in all aspects of their lives. Newly qualified biomedical scientists leaving behind the academic environment and entering professional life must not imagine that this is the end of their learning (Figure 18.1). Personal development is ongoing and can be applied to anything that he or she chooses to do.

Anyone with a degree even remotely connected with science is aware that science is constantly evolving, and that today's innovations and discoveries are often old news and outdated tomorrow. Anybody listed on a professional register is required to reaffirm and review continually their knowledge base within their chosen profession; biomedical scientists are no exception. Indeed, biomedical scientists are expected to be able to demonstrate a designated level of **continuing professional development** (CPD) to remain on the Health and Care Professions Council (HCPC) register and

Cross reference

The Health and Care Professions Council continuing professional development requirements were discussed in Chapter 2, 'Fitness to practise'.

FIGURE 18.1
Undergraduate biomedical science students learning basic laboratory skills in a university laboratory.

every 2 years at the renewal of their registration must declare they have been undertaking CPD to a predefined standard and, crucially, must be able to provide evidence to support this claim.

There is a saying that *it is never too soon or too late for learning*, and this applies to both *professional* study and development and to that done for pleasure. **Lifelong learning** is often a concept that many people adopt and embrace without ever realizing that they are doing so. Lifelong learning is associated with many unexpected places, a newspaper article, a television documentary, even helping children with homework, or trying out a new recipe in the kitchen! This learning can also be carried across to biomedical science, and a keen biomedical scientist will easily be able to meet professional development targets, almost without realizing that they are doing so, as there are so many opportunities within the working environment and elsewhere for the lifelong learning that will constitute CPD.

A framework for higher education qualifications exists across Europe, the European Qualifications Framework (EQF), which is being adopted and has been referenced to the different systems within England, Wales, and Northern Ireland by the Qualifications and Curriculum Development Agency. Scotland is also undertaking a similar referencing exercise to its qualifications. These systems describe the educational level of an award or describe the expected achievement of knowledge and skills of the student at the end of a course of study (see Box 18.1).

Of the eight EQF levels, the principal ones of interest to biomedical scientists are:

- EQF Level 5 (Intermediate (I) level or levels 8/9 in Scotland: Foundation degree, HND)
- EQF Level 6 (Honours (H) level or level 10 in Scotland: Bachelors degree)

BOX 18.1 Qualification levels and biomedical science training

A biomedical science undergraduate would be described as working towards an EQF Level 6 qualification, whereas a trainee biomedical scientist (or a biomedical scientist in training) with an honours degree, but not yet HCPC registered, would be described as already being in possession of an EQF Level 6 qualification and being able to work towards an EQF level 7 qualification. These are indicators of what you are already able to achieve and what you would be able to work towards. These factors are important when choosing your study options.

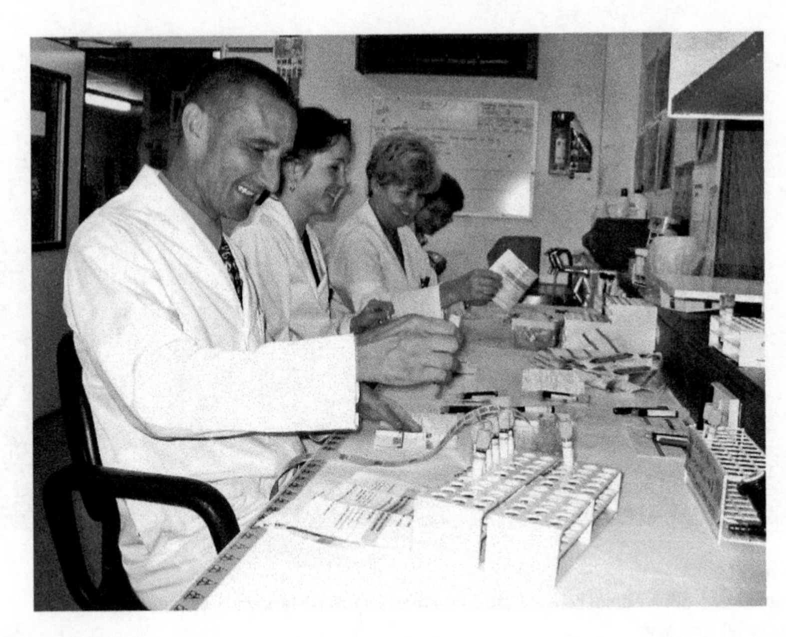

FIGURE 18.2
A group of professional biomedical scientists working in a clinical laboratory.

- EQF Level 7 (Masters (M) level or level 11 in Scotland: Masters degree)
- EQF Level 8 (Doctoral (D) level or level 12 in Scotland: Doctoral degree).

This chapter will examine the personal and professional development of biomedical scientists and, hopefully, help you to find and utilize a range of information resources that are there for the taking (Figure 18.2). All healthcare professionals have a responsibility to participate, alone or with work colleagues, to improve their knowledge and skills as a practising biomedical scientist.

18.1 Continuing professional development

CPD may be defined as the ongoing learning that is carried out throughout your professional career. It is needed in order to remain up to date in all aspects of a chosen profession. It is also necessary for promotion and is often a mandatory requirement to remain on a professional register. The two bodies with responsibilities for regulating biomedical science, the Institute of Biomedical Science (IBMS) and the HCPC, have their own definitions for CPD.

> Key points
>
> **The IBMS defines CPD as a process of lifelong learning, which enables you to expand and fulfil your personal and professional potential, as well as meet the present and future needs of patients, and deliver health outcomes and priorities. It assures that you meet the requisite knowledge and skills levels that relate to your evolving scope of professional practice.**

The definition of CPD by the HCPC is the way health professionals continue to learn and develop throughout their careers so they keep their skills and knowledge up to date and are able to work safely, legally, and effectively.

TABLE 18.1 Examples of opportunities for professional development in the workplace

Work-based learning
Learning by doing
Case studies
Reflective practice
Audit of patients
Coaching from others
Discussions with colleagues
Peer review
Gaining and learning from experience
Involvement in the wider work of your employer (e.g. being a representative on a committee)
Work shadowing
Secondments
Job rotation
Journal club
In-service training
Supervising staff or students
Visiting other departments and reporting back
Expanding your role
Significant analysis of events
Filling in self-assessment questionnaires
Project work
Evidence from learning activities undertaken

What exactly do these definitions mean to a newly qualified and registered biomedical scientist? They mean that, as a professional person, on a professional register, a biomedical scientist has a legal responsibility to meet the standards of the HCPC and also to remain abreast of current trends and developments within his or her working environment.

18.1.1 Requirements of the Health and Care Professions Council

It is a requirement of the HCPC that all registrants, including biomedical scientists, must undertake CPD to remain on the HCPC's register. Whenever he or she is asked to renew their registration, they must declare that they are complying with the HCPC standards relating to CPD. The HCPC is able to ask for evidence from registrants selected at random, for factual support that they are meeting the set standards. The standards of CPD state that a registrant on the HCPC register must:

- maintain a continuous, up-to-date, and accurate record of CPD activities
- demonstrate that CPD activities are a mixture of learning activities relevant to current or future practice
- seek to ensure that CPD has contributed to the quality of their practice and service delivery
- seek to ensure that CPD benefits the service user
- present a written profile containing evidence of CPD upon request.

The HCPC has extensively listed activities that constitute CPD (Table 18.1), together with examples of suitable supporting documents. Of the total registrants, 2.5% will be asked to submit their personal CPD records at the biennial point of renewal of registration.

SELF-CHECK 18.1

Why is CPD necessary for biomedical scientists?

18.2 Opportunities for personal development in the workplace

An induction period is normally provided with any new post or even a change in workplace. Information regarding the new employer is often given within the first few weeks of employment to all new employees to introduce them to specific organizational philosophies and policies. Induction should always be treated as an opportunity for personal development. Indeed, a formal induction programme needs to be given at the start of working at a new organization.

Induction courses are also an introduction to other employees. Presentations are ideal opportunities to find out more about your place of work, and ask seeking questions of a new employer. Frequently, presentations are given by senior members of staff and are often an excellent way to find out who does what within a large organization and to find out what they look like. It can save embarrassment at a later date if you can identify your senior managers!

Certain aspects of induction training have common themes and have to comply with legal employment and health and safety issues:

- introduction to health and safety
- safe manual handling
- fire policy and evacuation of the premises
- child protection
- cardiopulmonary resuscitation.

18.2.1 Introduction to health and safety

Any introduction to health and safety will include an outline of the Health and Safety at Work Act, which we discussed in Chapter 4. This act places legal obligations on employers and employees. Employers will outline the local policies and procedures that are in place to comply with this legislation and highlight the responsibilities of the employee to uphold these local rules. All employees should also be given the information needed to research their own health and safety concerns, such as access to health and safety publications (Figure 4.2).

18.2.2 Manual handling

Local policies will define what manual handling training is necessary for employees as part of their legal obligations. All manual handling must be carried out safely to protect the employee against serious injury. Manual handling is often subdivided into lifting people and lifting inanimate objects such as heavy boxes, awkward items of furniture, rubbish containers, or computer equipment. However, as laboratory staff are rarely required to lift patients, the manual handling in regard to direct patient care rarely applies.

18.2.3 Fire policy and evacuation of premises

It is essential that all employees are aware of the local procedure to be followed if they detect a fire and need to raise the alarm or if they have to evacuate the premises in response to a fire alarm. Among the first tasks of a new employee is to become familiar with fire exit routes, fire alarm points, and the fire-fighting equipment available in the workplace.

18.2.4 Safeguarding (children and adults)

Local policies to protect children and adults (safeguarding) working with children against physical and emotional abuse are in place. Although biomedical scientists have limited patient contact, they must be aware of and understand their professional responsibilities under safeguarding. Naturally, all such local policies, like health and safety policies, must also comply with the wider legal framework.

18.2.5 Cardiopulmonary resuscitation

Training in cardiopulmonary resuscitation techniques is often a requirement for anyone employed in healthcare professions. All employees within healthcare must be aware of their individual responsibilities should they find a patient requiring urgent medical assistance. Within the National Health Service (NHS), different levels of training apply to different groups of staff. It is likely that biomedical scientists will be encouraged to undertake a 'basic life support' training session in this respect.

18.2.6 Human resource issues

Issues related to human resources are also opportunities for employers to introduce policies that may relate to:

- sickness and absence procedures
- payroll queries
- travel or subsistence claims
- occupational health and vaccination procedures
- holidays, including annual leave
- dealing with the press
- disciplinary and grievance procedures.

Sickness and absence procedures

Sickness and absence procedures give clear instructions about the behaviour expected if employees are too ill to attend work or if there is an unforeseen emergency that prevents them from attending work. It is essential that employees follow such policies and are made aware of the consequences to the department if they fail to comply. Staff sickness is monitored by human resource and occupational health departments, and may also affect the pay of the individual concerned.

Payroll queries

Pay is always an emotive subject and new employees must understand the information given on their payslip. They must also be aware of the procedures in place to clarify pay issues, for example what should be done if there is a discrepancy between pay given and that expected. Most employers will explain taxation and national insurance procedures, and their own system for giving extra payments, for example when overtime has been worked.

Travel and subsistence claims

If any employee is required to attend meetings or conferences off site as part of their role or for educational purposes, there are local and national policies on how to claim back any out-of-pocket expenses, such as course fees, and so on.

Occupational health

Occupational health is the department that is devoted to the health and well-being of employees. Naturally, they are involved in health and safety aspects of the workplace, set policies for preventative medicine (health screening and vaccinations), and take a non-biased approach when employees may become unfit for their duties or return to work following prolonged illness.

Holidays

Paid holiday entitlement is usually written into a contract of employment, but employers and departments are likely to have their own individual policies regarding booking holidays in advance and to record any leave taken. Annual leave is the predominant reason for taking time off work, but there are other reasons that an employee may be able to take leave of absence without encroaching on their paid holiday entitlement. It is therefore necessary to be aware of your full holiday entitlement. Other discretionary leave may include jury service, carer leave, and bereavement leave.

Dealing with the press

Most employers have a policy in place that employees should follow when dealing with the press. Frequently, employees are not able to talk to the press without appropriate permission and to do so is a disciplinary matter. This also applies to all organizational human resource issues.

Disciplinary and grievance procedures

All employers have their own disciplinary and grievance procedures, which must be consistent with current employment law. Employees must be made aware of not only basic employment law, but also of their employers' individual policies, as there are certain channels to follow if they are embroiled in a difficult situation. Cases that go to court or to an employment tribunal are frequently won or lost on procedural issues, rather than on what has actually happened; it is therefore desirable to resolve disciplinary and grievance issues at the earliest possible stage.

18.3 Training and development

Most employers have a training and development department that provides access for all staff to a variety of generic training courses, appropriate to their roles within the organization. It is always worthwhile to be aware what training may be on offer from the employer outside the immediate laboratory environment. Many courses are available on request, and it may be stating the obvious, but often you will only get what you ask for. When agreeing a personal development programme, using training facilities that are readily available within the organization can reduce cost and is more likely to be agreed.

Cross reference

Chapter 3, 'Communications in laboratory medicine'.

Following the initial broad induction given to a new employee that was described in the previous section, a specific induction programme is usually provided by the individual's department. This covers all of the statutory requirements, health and safety, visitors to the department, confidentiality, dealing with telephone calls, and specific human resource issues that relate to the department. These points would include who to contact when telephoning the department to say you will be late and who to liaise with when arranging holidays. All of these more informal pieces of information are specific to a place of work and are opportunities for personal development, as not only are employers never exactly the same, but it is also unlikely that two departments within a single organization will be identical.

Leading on from what might seem an initial overload of information, it is now time to get down to applying yourself to your professional work!

Cross reference

You can read more about accreditation in Chapter 17, 'Quality assurance and management'.

To satisfy accreditation requirements and standards associated with training, it is necessary for every department to keep training records that provide evidence that an individual has been trained and has had their competence assessed. The proof of training and evidence of its success will form a large part of any biomedical scientist's personal development portfolio. Reading the background of many of the more interesting issues and finding information about many of the daily laboratory procedures and workload also gives many opportunities for **reflective learning** (Section 18.7).

It is clear that personal development is an ongoing process that arises naturally during the course of a professional person simply doing his or her job at work. This is an age of evidence collection: it is essential to document all learning and development as they arise and to keep appropriate records to satisfy statutory legislation, the needs of the employer, and those of the individual for either personal interest or professional registration (Section 18.1).

18.3.1 Additional employer-based continuing professional development activities

Many employers offer a variety of other opportunities for professional development and any available should be sought out. Many clinical laboratories provide access to presentations, lectures of interest by visiting speakers, and opportunities to visit other departments in the organization to find out more about its wider activities in a more informal setting than that presented at initial induction. Again, just because something is not immediately offered, it does not mean that it is not possible or that it does not already exist. Medical staff are usually happy to explain interesting clinical cases and biomedical scientists are often able to accompany them on visits to the wards. However, it is often up to the biomedical scientist to take the initiative and ask!

18.3.2 Tutorial programmes

If an ongoing tutorial programme is not available in the workplace, the obvious question is, *why not?* Is it simply because no one has bothered to organize one? It may be that several individuals would be prepared to speak to their peers on their favourite subject. It need not be too difficult to find a venue and put together a rotation of several lectures covering the most recent advances in the pathology disciplines. Newly qualified staff could discuss their degree projects, and those currently studying postgraduate qualifications could pass on knowledge of a particularly exciting area of research. Often a project such as tutorial organization only requires a leader, who could be any member of staff with the right amount of enthusiasm, and participation is likely to follow.

18.3.3 Journal clubs

Journal clubs involve an analysis of an appropriate article that has been published in a relevant journal. A particular journal article is selected in advance and all participants read this article and enter into a (usually) lively debate about it, or a single participant reads an article and then presents a review of it to the other members of the club for discussion. Members of a journal club usually participate equally. Again, you do not have to be a senior member of staff to instigate such activities. The skills required often have nothing to do with laboratory management and much more to do with gentle persuasion and being able to stimulate enthusiasm in colleagues.

18.4 Professional bodies and personal development

All biomedical scientists in employment will have many opportunities to develop professional contacts with others in the field. These contacts present a variety of opportunities for personal development, a number of which you can see listed in Table 18.2.

Naturally, the majority of biomedical scientists belong to their own professional body, the IBMS (Figure 1.2), which offers a variety of services to its members (Figure 18.3). The IBMS recognizes the crucial role of CPD and has many resources available to its members in all membership classes (see Box 18.2).

> **Cross reference**
> **You can read more about CPD activities in Chapter 2, 'Fitness to practise'.**

The areas where the IBMS most directly contributes to CPD are education, journal-based learning (JBL), and the organization of regional, national, and international meetings.

18.4.1 Education and the Institute of Biomedical Science

The IBMS is active in a range of educational activities, including accrediting a number of postgraduate professional qualifications (see Box 18.3). These were introduced in Chapter 2.

These qualifications are widely recognized by both employers and by higher educational institutions. Success in studying for any of them indicates that a biomedical scientist is willing to further their personal development. If you want to progress to the higher levels of the profession, such further study is increasingly necessary to demonstrate knowledge and skills at high levels of competence. Thus,

TABLE 18.2 A number of the personal development activities associated with professional contacts

Professional activities
Involvement in a professional body
Membership of a specialist interest group
Lecturing or teaching
Mentoring
Examining
Tutoring
Branch meetings
Organizing journal clubs or other specialist groups
Maintaining or developing specialist skills (e.g. musical skills)
Being an expert witness
Membership of other professional bodies or groups
Giving presentations at conferences
Organizing an accredited course
Supervising research
Acting as a national assessor
Promotion

FIGURE 18.3
Membership files of the Institute of Biomedical Science.

participation in postgraduate learning is advisable, although not yet mandatory. However, completion of a registration portfolio and the associated professional training *is* mandatory for initial registration with the HCPC. This professional training may be incorporated into an HCPC-approved degree programme or may be undertaken as a stand-alone exercise following graduation. Once registered, you will discover that there are many opportunities for professional education.

BOX 18.2 Biomedical scientists and professional bodies

The IBMS is not the only professional organization to which a biomedical scientist can belong. There are common-interest organizations that are not necessarily professional bodies. Among these are the British Blood Transfusion Society and the British Society for Immunology. All these specialist societies and groups organize series of lectures and meetings for like-minded individuals, and attendance at any such event is an opportunity for CPD.

The IBMS as a professional body has developed a comprehensive set of professional qualifications. The Specialist Diploma is awarded for the successful completion and assessment of the specialist portfolio, which is based on detailed knowledge and the presentation of evidence collected to meet a set of standards in one of the following subjects:

- cellular pathology
- cytopathology
- clinical biochemistry
- clinical immunology
- haematology and hospital-based transfusion practice
- histocompatibility and immunogenetics
- medical microbiology
- transfusion science
- virology.

Building on from the Specialist Diploma is the Higher Specialist Diploma. This is aimed at practising biomedical scientists who have skills and theoretical knowledge at EQF level 7. They are normally actively performing in a complex role, and are continuously developing a clinical, scientific, or management responsibility for a laboratory service. The Higher Specialist Diploma is open to suitable candidates who are able to pass a series of written examination papers following a period of self-directed study. The Higher Specialist Diploma is available in a series of clinical laboratory specialities and, in addition, Leadership and Management.

The IBMS also offers a series of Certificates of Expert Practice, which enable biomedical scientists with highly specific sets of skills and knowledge within their own specialist area to demonstrate their command of a subspeciality.

Moving forward from these certificates are further specific IBMS qualifications in subspecialities at differing educational levels. These include Diplomas of Expert Practice and Advanced Specialist Diplomas. These diplomas are designed to provide evidence of the highest level of professional knowledge and skills.

Differing levels of knowledge and skills may be required within a particular job description or person specification for a specific post. However, the only mandatory qualification for a biomedical scientist is the completion of an HCPC-approved degree or the registration portfolio alongside an IBMS-accredited degree to meet the threshold standards of proficiency for registration with the HCPC. All other qualifications and educational activities in the wider sense (Table 18.3), while desirable in terms of proof of personal development and as aids for promotion, are not mandatory and the decision to complete further studies after gaining a BSc degree is the personal choice of the individual.

SELF-CHECK 18.2

What does the IBMS suggest a new biomedical science graduate do next by way of professional qualifications if they are already in full-time employment?

BOX 18.3 Non-Institute of Biomedical Science providers of education in the biomedical sciences

Any opportunity to learn new skills or learning about other disciplines constitutes CPD. Of course, even taking a postgraduate qualification in the same area of practice, such as a biomedical scientist choosing to study for a Masters Degree in their own subject, constitutes professional development.

Many providers of education for biomedical scientists provide courses other than the established graduate and postgraduate degree courses. There are also specialist courses available that are appropriate for the non-science-based areas of biomedical science.

A number of universities also provide a series of short courses for special interest groups or to allow biomedical scientists to venture out of their particular disciplines. Thus, for example, haematologists may take short courses in clinical chemistry. There are also opportunities to attend courses in management topics. These may be short courses in specific areas such as audit, finance, or tutoring, or an extended postgraduate management course lasting several years.

Many colleges and universities offer not only taught courses, but also opportunities for distance and online learning in areas such as quality management, point-of-care testing, and training and development of others.

18.4.2 Journal-based learning

In JBL, the participant is expected to read a current paper published in a recognized scientific journal in a named area of biomedical science. The reference to the paper is supplied by the CPD provider to the participant. However, unless the paper is in an open-access journal or in a journal belonging to the CPD provider, as in Figures 2.4 and 18.4, the participant must obtain their own copy of the relevant paper (see Section 18.9). Any such papers made available by a CPD provider would be either an expensive process or be a breach of the copyright. The chosen paper may be either scientific or relate to a topical management issue. Typically, a CPD provider such as the IBMS then gives a list of 20 questions related to the article, which are usually printed in *The Biomedical Scientist*, *British Journal of Biomedical Science*, or on the IBMS website. Each question can be answered as either true or as false. The answers to all the questions can be found in the information given in the paper. The participant can submit his or her answers to the IBMS by post or via its website (Figure 2.5). The answers are marked, with 17 correct answers constituting the minimum pass mark.

18.4.3 Structured reading

Structured reading as an activity has now itself been removed from the CPD activities developed by the IBMS, although it remains a recommended method of study for individuals undertaking a Higher Specialist Diploma. Briefly, individuals are encouraged to select appropriate current topics for this purpose and

TABLE 18.3 Educational activities constituting personal development

Formal and educational qualifications
Additional attendance at courses
Further education
Research
Attending conferences
Writing articles or papers
Going to seminars
Distance learning
Attending courses accredited by a professional body
Planning or running a course

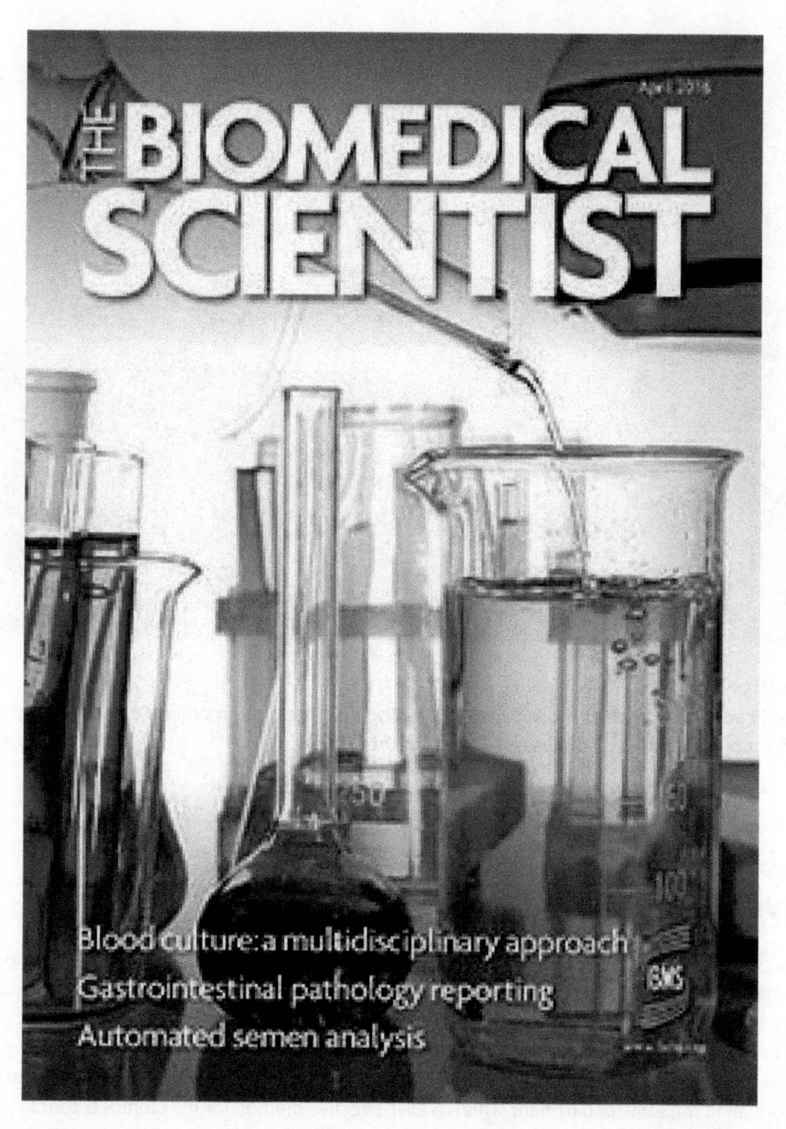

FIGURE 18.4
Front cover of a past issue of *The Biomedical Scientist*. Copyright © Institute of Biomedical Science.

write an essay following up-to-date research of current literature. The essay will be assessed at EQF level 7 and must include reasoned argument and synthesis of often conflicting theories or points of view.

18.4.4 Professional meetings

The IBMS organizes and hosts regional, national, and international meetings (Figure 18.5). These meetings may be small events for a specialist interest group in a local area or, at the other end of the spectrum, may be a full multidiscipline conference at a central venue lasting several days. The former types of events are often arranged by a willing volunteer for a small group of people who may have a common interest. Larger meetings are usually sponsored by commercial companies, and attendance gives opportunities to see the latest products and technologies and to network with fellow biomedical scientists from other hospitals, as well as listening to a formal lecture, usually by experts in their own field, or to develop expertise by talking to senior, more experienced participants.

FIGURE 18.5
Scientific meetings, such as those organized by the Institute of Biomedical Science, provide excellent opportunities to discuss all aspects of the subject.

18.5 Commercial organizations and personal development

Many of the commercial organizations that supply equipment, consumables, and services provide both short and long courses. It is possible to learn presentational and telephone skills, time management, and many other skills that may not be specific to biomedical science or even employees in broader healthcare organizations of the public or private sectors. Many companies will offer generic courses on business premises, or many even come to a workplace and give courses to several members of staff in the same department. This is a cost-effective method for the employer to train several members of staff simultaneously; so, again, if there is nothing already in place where you work, it might be worth making a suggestion to your managers.

Companies that supply scientific instruments and laboratory computer systems also offer training. Often, the courses come free of charge with the installation of newly purchased equipment, but additional courses can usually be purchased as separate items. When a new instrument is installed, the person who attends the training will usually become its 'key operator', and be the link between the laboratory and the company that supplied the instrument. This arrangement usually works well until the key operator leaves to work elsewhere. The remaining staff are then left to muddle through; their success in this depends on how much knowledge was given to them before the key operator left. This situation presents an opportunity for another member of staff to receive full training from the company and, again, this is an excellent opportunity to gain in-depth knowledge into the inner workings of laboratory equipment and thus for personal development.

Many of the major instrument and laboratory information technology (IT) suppliers also organize meetings and user groups to update the knowledge of laboratory workers. Not only is this an opportunity to network with colleagues, but it also brings together users of common equipment from different sites and so is often an excellent opportunity to assess the different ways in which one's own instrument can be utilized.

18.5.1 Trade unions

Trade unions have a commitment to personal development and lifelong learning. Many offer courses for their own members that focus on basic skills for lay persons and the development of union representatives. Funding is available through the trade unions to develop basic skills such as numeracy and literacy, and where English is not the first spoken language, opportunities to improve spoken English. All employees should be able to attain a basic standard of education to help their progress in the workplace; trade unions are able to act as vehicles to help achieve this.

Many professionals are not only members of a professional body, but also of an appropriate trade union. The professional activities undertaken by the trade unions often support and complement those undertaken by professional bodies, and are rarely in complete contradiction to one another. Trade unions offer many excellent courses in, for example, employment and health and safety law, presentation, and negotiating skills.

The trade union philosophy is that their members in the workplace should have access to the same level of information that is available to the employer. All management skills, regardless of how they were acquired, are opportunities for personal development. The government frequently asks for the support of trade unions to disseminate information, knowledge, and skills to their members. Indeed, many of the lower level educational courses run by trade unions for their membership are able to take place only because of initial government sponsorship.

18.6 Opportunities for self-guided personal development

Opportunities for personal development do not just occur in the workplace or a formal education situation. Watching television, reading a newspaper or a magazine, or surfing the Internet will present professional information, even if you are not immediately aware of it. Learning in this way is equally important and a brief paragraph of writing after reading an article in a Sunday newspaper or watching a television documentary is perfectly acceptable as a record for additional or unstructured CPD.

18.7 Reflective learning

Reflective learning is a mechanism of critically examining an event, situation, or experience after it has taken place, thinking about what has happened, the context in which it happened, and what can be learnt from the whole experience. Personal development constitutes somewhat more than listening to lectures and attending training courses. The most important aspect of personal development happens after the learning event when there is an opportunity to really use what has just been learnt. Reflective learning is an excellent mechanism to really think about the new skills and knowledge acquired and identify where they can be applied in everyday procedures.

The IBMS CPD scheme includes a reflective learning template that can be added to individual CPD activities recorded in your portfolio.

Several templates are available to help with reflective learning, but, in essence, an individual should start by asking themselves a series of questions, as shown in Table 18.4, and then try to use their new knowledge to answer them.

Clearly, each period of personal development may be equally important in terms of acquiring new knowledge and skills that may enhance the activities of the department or allow changes to the service it provides to be carried out more smoothly. However, the most beneficial development to health services occurs when the personal development of employees brings about changes that provide real benefits to the treatment and experiences of patients. Reflective practice is a means of exploring how new personal knowledge and skills can be used for the enhancement of the patient pathway. Thus, reflective practice can run alongside and enhance many of the other ways of CPD already mentioned; so, attending a conference is not only an opportunity to learn, for example, cutting edge science, but also to reflect afterwards as to how the newly acquired knowledge can be used to enhance the workplace for the benefit of all.

TABLE 18.4 Questions relevant to reflective learning

Possible questions after an incident in the laboratory	Possible questions after a period of training	Possible questions after a conference
What actually happened?	What was the purpose of the training?	What was the purpose of the conference?
Were all policies and SOPs followed?	What did I learn?	What did I learn?
If not, what was the deviation?	Was the learning useful?	Was the learning useful to me?
Was the deviation avoidable?	Can I change my current practices to incorporate what I have learned?	Will the learning be useful in my own place of work?
Was the outcome better, worse, or the same as the expected outcome?	Will my future practices benefit from these changes?	Do I need to look at any of the issues in more detail to identify where they apply in my own place of work?
What lessons have been learned from the event?		Do I need to inform anyone else of my new knowledge in order to be able to make changes?
Should the SOP be changed?		Will the changes be of benefit to the service?

SOP: standard operating procedures.

18.8 Study skills

Study skills can be put into practice informally and formally. Skills for general study and personal development are used almost all of the time and in everyday learning. For formal assessments of study, when an individual is asked to provide evidence of study such as submitting coursework or sitting a written examination, more structured skills are needed. Four basic groups of study skills are outlined in Table 18.5; each, of course, overlaps considerably.

General study skills are necessary for even the most basic academic study and are taught as part of the school curriculum. For those wanting to return to study in later life, the lack of general study skills is an enormous disadvantage over their younger peers. Younger biomedical scientists and laboratory employees often have an advantage over some of their more mature colleagues.

The basic study skills listed in Table 18.5 should be acquired easily by anyone working towards or in possession of an H-level (EQF level 6) qualification (see Introduction). Indeed, if you are working towards an honours degree in biomedical science, or are already a graduate with an approved degree for registering with the HCPC, or a graduate needing to obtain additional units for HCPC registration, you will most likely be familiar with them. The first consideration is the basic skills that should be familiar to anyone who has recently been part of any education system, regardless of the level of academic achievement. These are the skills needed for any learning, academic, or vocational qualifications, and even for leisure pursuits and hobbies. We will now expand on a number of them.

TABLE 18.5 Basic study skills

Basic study skills for everyday life	General study skills for learning	Study skills for producing coursework	Revision and exams
Reading	Literary skills	Producing coursework	Assessing objectives
Writing	Collecting data	Gathering information	Revision
Verbal communication	Essay writing	Research	Exam technique
Using a library	Case studies		
Information technology skills	Citing references		
Working as a team member	Using the web		
Numeracy	Mathematical skills		

18.8.1 Literary skills

Literary skills include the abilities to read, use language in listening and speaking, and to write. Good communication skills are essential when working in a laboratory. Appropriate levels of English language skills, written, and verbal, are needed for the safe communication of results, instructions, and information. Basic literacy is essential for biomedical scientists at every level. It is unlikely that anyone could maintain a laboratory position without literary skills equivalent to that taught at secondary school. However, a higher level of English language and grammar is needed to study for an honours degree, so if you are working towards or are in possession of a biomedical science degree, this will give others an indication of your ability to communicate in English. Graduate qualifications require more than the ability to communicate in the required standard of English; you must also be able to undertake academic work, such as taking notes, writing essays and projects, present data, and produce reports at an appropriate level.

Staff in a potentially dangerous working environment, such as a laboratory, must be able to read health and safety notices, standard operating procedures, and be able to differentiate between identification of samples, chemicals, record sheets, and manuals. Writing skills are necessary to enter data at numerous points and write safety notices, and policy documents must be read and understood.

Reading

Academic reading is a different activity from reading for pleasure: it requires an effective use of time to obtain the maximum benefit from a book, article, or webpage in the shortest possible time. The five steps to effective reading are:

1. Survey
2. Question
3. Read
4. Recall
5. Review.

Academic reading is a means of enhancing one's understanding of or gaining basic knowledge about a new topic, or finding alternative views about a particular topic and to help with further studies. However, the reading needs to be selective. Thus, skim-reading the first few paragraphs of a book, studying its index, or the summaries of chapters will all give clues as to the relevance of the contents. Something written in complicated English with indecipherable jargon is not likely to be worth the time spent trying to unravel it. It should be easy to grasp the key facts and to differentiate between these and the author's own ideas.

It is also pertinent to look at the date of publication and the identity of the author; scientific writing can quickly become outdated and authors can become discredited by their peers.

Academic publications may be expensive and are therefore frequently borrowed rather than purchased, unless it is a key book or journal. Workplaces, friends, and libraries are inexpensive sources of academic books (Figure 18.6), particularly if you are not going to use them frequently. This does mean that they have to be returned so there must be a maximum reading benefit in a fixed amount of time. While it is useful to make notes of key points while reading, it is counterproductive to copy out vast chunks of text with limited understanding of the topic.

Note taking

Notes may have to be taken from books, during demonstrations in the laboratory, and, of course, in lectures and tutorials (Figure 18.7). This is an excellent way of staying focused on what you are supposed to be learning, except when you are expected to write reams of dictated text! Notes should be concise and be an aid to self-study. There is no need to write verbatim from a lecturer or speaker.

Many ways of taking notes have been suggested; it is best to try different techniques to find which one works for you. Notes may be linear, in sentences, or expressed only as key words. They may take on more of a picture form, with a central theme in the middle of a page and ideas and topics radiating outwards. You may wish to develop your own form of shorthand. In essence, the only person who has to comprehend and use your notes is you, so it should not matter how illegible they are to others as long as they aid your understanding of a topic, help you to remember the points, and act as a revision prompt.

FIGURE 18.6
Undergraduate students reading and working in a university library.

Essay writing

Essays are not simply collections of written facts; they need structure and to be a response to a specific title or theme, and demonstrate a depth and breadth of knowledge on a specific subject. They should be written in English prose with correct spelling and grammar.

Essays should be planned from the outset and not be a collection of summary paragraphs. They require a clear introduction in the opening paragraph, followed by the main portion or body of the essay, and to end with a summary or concluding last paragraph. The body of the essay should be a collection of related paragraphs, all linked to the title of the essay. Wherever possible, use examples to illustrate the points of the essay. Care should be taken to avoid merely listing facts. A 'good' essay will demonstrate knowledge of a subject from more than one viewpoint and contain a critical appraisal of a variety of concepts or arguments.

SELF-CHECK 18.3

What features would be shown by an essay that gained high marks?

Case studies and writing reports

Case studies and report writing are one step on from essay writing. Case studies are commonplace in biomedical science and the success of a well-presented case study is demonstrated by being able to present the key facts of the topic in an interesting and understandable manner, while removing much of the peripheral and often irrelevant material. A case study will include some mention of the theory behind the case and there should be a blend of theoretical knowledge with factual data.

Report writing is also commonplace in biomedical science and many organizations ask for them to be written in a particular format or house style. Most reports require an executive summary or overview of the report presented as short, concise, bulleted points. The body of the report should include an introduction (including terms of reference if appropriate), the findings, the conclusions, and any recommendations leading on from the conclusion. References and appendices should be at the end, and are best kept separate so as not to detract from the main text. Many reports in science rely heavily on the presentation of data, and this should be presented clearly with the appropriate statistical analysis and conclusions for any non-mathematicians who may be required to read the report.

(a)

(b)

FIGURE 18.7

Photographs of undergraduate students studying in (a) a formal lecture and (b) the more informal setting of a tutorial.

18.8.2 Numeracy

Basic numeracy is also an essential skill for anyone working within an area of science, and biomedical science in no exception. The mathematics that is required for biomedical science centres on manipulation of numbers in basic calculations, and a reasonable knowledge of the collection and interpretation of statistical data. The fact that you are already in possession of, or working towards, a biomedical science degree gives others a good indication of your current mathematical abilities.

18.8.3 Using libraries

Libraries are essential to anyone studying for their CPD. It is worth getting to know the library and its librarians, particularly if you use it regularly, either for study or for pleasure. Understanding the layout of a library and the services available to you will make using the library more effective and less stressful. Contrary to the myths and stereotypes surrounding librarians, the reality is that most are knowledgeable and willing to help when approached for advice. In other words, it will pay dividends to make friends with yours!

Libraries are usually calm, quiet places and often offer a better study environment than home or the immediate workplace. They are not only a source from which books can be borrowed, but also have reference material, journals, and periodicals that can be used on the premises. If a particular book or journal is not readily available, they can usually be obtained through the inter-library services, so that if one library is unable to provide a copy of something in particular another can usually be contacted to supply the required book or article. Libraries are able to supply copies of articles from scientific journals and if you want to take part in the CPD activity of JBL (Section 18.4) you will have to source the journals needed to participate.

SELF-CHECK 18.4

What facilities are likely to be provided by local authority libraries?

18.8.4 Information technology

The ability to use IT is now as important as basic numeracy and literacy. Children are taught IT skills as part of everyday school life and it features high on the national curriculum from the age of 4 years upwards. Some students of biomedical science may find themselves out of their depth compared with others, as IT skills have only been taught as a compulsory part of education for a relatively short time. IT is an essential aid to biomedical science. Databases are part of everyday life and the use of spreadsheets and word processing for study is virtually universal. Anyone who has not received training in the use of IT to store, retrieve, and present information is extremely disadvantaged. The electronic transfer of information and the use of electronic methods of communication are central to the operations of the NHS. IT skills are frequently offered by employers as part of staff development and there are many providers of adult education in the community that offer a variety of courses for the beginner to the advanced student.

18.8.5 Team work

In the past, traditional teaching focused on the individual, but with the introduction of key skills in compulsory education, working as an active member of a team has become a skill that employers seek in their employees. Thus, further and higher educational institutions promote working with others as a key skill. The advantages of teamwork are that while everyone has clear and common objectives, an appropriate mix of people will ensure that everyone will be able to make unique contributions to the efforts of the team as a whole. Teamwork is not confined to the everyday functioning of the workplace but may extend into study and common interest groups, multidisciplinary case conferences, and leisure activities.

18.8.6 Academic study skills

Academic study skills require assessment of coursework and examination papers to monitor progress, and all students must spend some time finding out how to get the best results based on the requirements of the awarding body. Remember, academic institutions are there to pass students, rather than fail them, as a high pass rate reflects well on the teaching of the college or university. Lecturers and tutors are in the business of helping their students to achieve the highest possible standards. All help and advice given by tutors should be gratefully received and stored away for future good use.

Under- and postgraduate courses usually rely on two types of assessment: coursework and a written examination. Coursework must have defined learning criteria in the form of aims and learning outcomes, and students should be familiar with the principles by which the coursework will be assessed. The content of coursework sometimes allows the student a degree of freedom in choosing the topic and when this is the case, it is obviously better to base it on an area of personal interest wherever possible. Written examinations assess knowledge (and skills) at the end of a block of learning. Examinations should demonstrate what the student knows; they are not designed to highlight what is not known. They do, however, require careful revision and preparation. Taught courses follow a specified syllabus or curriculum; knowledge of its contents will give an indication of what material will be examined and the form the examination will take.

Assessment by portfolio is becoming increasingly popular, as this involves collecting evidence to show that you have met specific objectives or learning outcomes. Most biomedical scientists who wish to register with the HCPC should have had experience in completing the registration portfolio themselves. More experienced members of the profession have probably been involved in helping others to collect evidence and construct a portfolio in a clear and logical way. If you have not yet completed your initial portfolio for professional registration, you will become familiar with this method of assessment during your time as an undergraduate. Formative assessment is the regular and ongoing assessment during a course of study. Summative assessment is the final or end-point assessment stage.

18.9 Evidence supporting personal development

It is essential to keep records of all activities that can be offered as proof of personal development when asked. Table 18.6 lists many activities that provide evidence that a biomedical scientist has been able to learn new skills or improve him or herself in some way. It is advisable to keep copies of all documents that you have prepared or have been involved in the preparation of, and to retain attendance certificates, copies of presentation handouts, notes made during seminars, or summary documents following reading or research. The IBMS CPD scheme now includes an e-portfolio that allows the uploading and attachment of such documents to individual CPD activities, so makes retention of such documents more straightforward.

It is clear from Table 18.6 that not all types of professional development are accessible to all grades of staff, but individuals should be aware and take advantage of, every available opportunity to professionally develop themselves.

TABLE 18.6 Types of evidence showing the ability to learn new skills

Materials you may have produced	Materials obtained from others	Materials showing reflection and evaluation of your learning and work
Information leaflets	Testimonies, testimonials	Adapted documents arising from appraisals, supervision reviews
Case studies	Letters from service users, carers, students, or colleagues	Documents produced following local or national schemes related to CPD
Critical reviews	Course certificates	Evaluations of courses and conferences attended
Adaptations of student notes		Personal development plans
Policies or position statements		Approved claims for credit for previous learning or experience
Discussion, procedural, and course programme documents		
Documents about national or local processes		
Recent job applications		
Reports on project work, clinical audits, reviews		
Business plans		
Guidelines for dealing with patients		
Course assignments		
Action plans		
Presentations, journals articles, and research papers		
Questionnaires		
Funding applications		
Induction materials for new members of staff		
Learning contracts		
Contributions to the work of a professional body or of a special interest group		

18.10 Performance appraisal and personal development plans

Many employers, both in the public sector and in the commercial world, integrate the personal development of employees into their organizational culture. They ensure that employees have a framework that they can work within to develop themselves formally in ways that are of benefit to both the employer and the individual.

There is an organizational structure for targets to be set, and performance against the targets is monitored and reviewed. Usually a structured mechanism is established where an employee is able to meet with his or her manager (or a representative of the employer) on a regular basis to assess their past performance. The first of these meetings normally occurs soon after the employee begins working in their new post and then usually at regular 6-monthly or annual intervals, in accordance with employment policy at the particular place of work. These targets must be non-discriminatory and achievable, so that you (as an employee) are able to pass comment and agree to them.

Performance appraisal is not an opportunity for employers to set unreasonable targets for, or give unfair additional tasks to, an individual that are not appropriate to an existing job description for a post. Nor should it be used as part of a disciplinary process. While it is an opportunity to identify poor performance, it should be done so with the intention of offering suitable training and support to enable the individual concerned to improve and work towards meeting targets.

The NHS has introduced a programme or national strategy, called the Knowledge and Skills Framework (KSF), to standardize knowledge and skills across all the staff groups that it currently employs. This scheme does not apply to biomedical scientists working within the private health sector or commercial world, but many of these employers already have similar schemes in place. The principles of the KSF are that staff in different departments and even in different organizations, but who essentially have the same role, are measured against a series of generic standards appropriate to their grade of employment, with the aim of ensuring consistency throughout the NHS in terms of expectations and rewards. The KSF programme itself has now also been simplified into six 'core' dimensions:

- communication
- professional and people development
- health, safety, and security
- service improvement
- quality
- equality and diversity.

NHS Employers has also provided examples of the behaviours and actions that indicate whether a KSF dimension is being met. Trusts therefore now have the ability to use some or all of the original KSF outlines, or these simplified dimensions. This allows significant flexibility in the approach to appraisal discussions.

Performance appraisals are usually carried out by an immediate line manager. This should be a two-way process in which both the appraiser and 'appraisee' participate equally. Performance appraisal is a structured and formal period where the NHS employee has the opportunity to discuss how he or she is performing in line with the set KSF outlines or dimensions, and other local targets specific to the position in the laboratory. It is also the time to identify where extra help may be needed or to highlight an area that the employee may wish to develop further. This is then formally written in to a **personal development plan** (PDP). Having made reasonable requests in the PDP, the employer, both the NHS trust and the line managers, then has a duty to help the person concerned achieve the areas of personal development highlighted. Performance appraisals are also conducted to ensure that requests from employees for opportunities for personal development are being met.

The system of performance appraisal and PDP is therefore a two-way process: it should benefit both the employer and employee. It should be an opportunity for the employer to encourage and support staff to develop themselves as individuals, while giving added value to the service that the member of staff is able to provide; in short, it is part of the cycle of CPD of the individual (Figure 18.8).

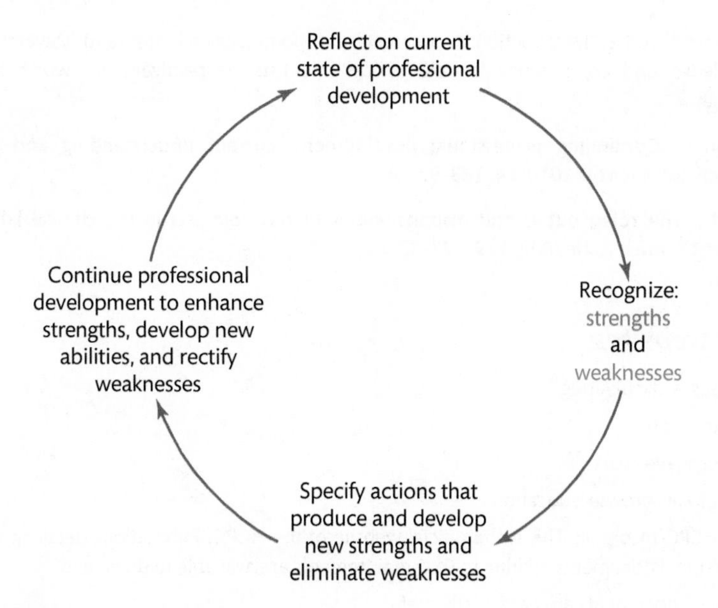

FIGURE 18.8
Overview that emphasizes the never-ending process of continuing professional development.

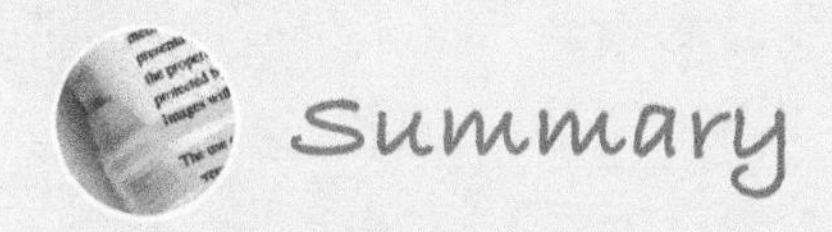

Summary

- Personal development is an ongoing process.
- The IBMS has a variety of schemes and qualifications tailored to the needs of biomedical scientists and, while not mandatory, it is advisable to take part in a variety of CPD activities.
- The HCPC has the mandatory requirement that all its registrants must be continually involved in professional development activities and be able to provide evidence to support this when asked.
- There are always opportunities for personal development wherever you choose to seek them: at work, home, and within organizations and academic institutions.
- The skills needed to develop both professionally and personally are varied and extensive, and can be adapted to suit any style of learning.

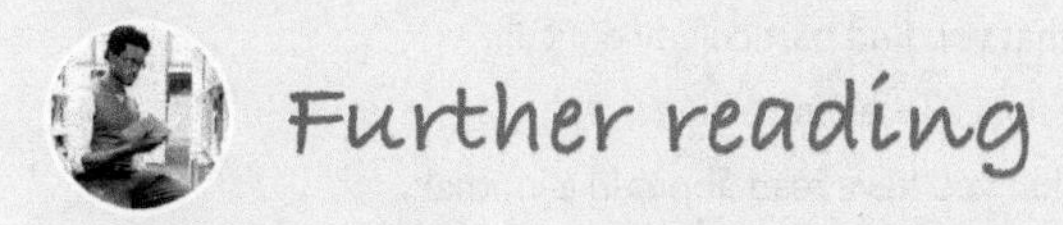

Further reading

- **Dryden G, Vos J. *The Learning Revolution*. Torrance, CA: The Learning Web, 2001.**

  An inspirational guide for anyone involved in learning, which explores various aspects to effective learning in an innovative and thought-provoking manner. Suitable for teachers and students alike.

- **Institute of Biomedical Science. *Good Professional Practice for Biomedical Scientists*. 2nd ed. London: Institute of Biomedical Science, 2005.**

- **Pitt SJ, Cunningham JM. *An Introduction to Biomedical Science in Professional and Clinical Practice*. Chichester: Wiley-Blackwell, 2009.**

Chapters 1 and 8, 'Introduction to a career as a biomedical scientist' and 'Development of knowledge and competency for biomedical scientists', respectively, are worth browsing through.

- **Wilson G. Continuing professional development: current understanding and practice. *Biomedical Scientist* 2010; 54: 168–9.**
- **Wood J. The roles, duties and responsibilities of technologists in the clinical laboratory. *Clinica Chimica Acta* 2002; 319: 127–32.**

Useful websites

- www.sussex.ac.uk/adqe
- www.ibms.org
- www.cengage.co.uk
- www.gov.uk/browse/education
- www.HCPC-uk.org.uk. This is the official website of the HCPC. Publications detailing the latest professional documents relating to CPD requirements are available to download.

The following documents are particularly useful:

- Continuing Professional Development and your Registration.
- Continuing Professional Development and your Registration: Appendix 1
- Continuing Professional Development and your Registration: Appendix 2
- Demonstrating Competence through CPD
- Standards of Conduct, Performance and Ethics.

Questions

1. Which of the following does the IBMS **NOT** offer:
   (a) Journal-based learning
   (b) Specialist M-level academic qualifications directly related to aspects of biomedical science
   (c) Access to a full-time doctoral programme
   (d) Assessment of essays based on structured reading
   (e) Information regarding suitable courses leading to accredited biomedical science degrees.
2. Performance appraisal is an opportunity to discuss one's progress with a line manager. Which of the following is **NOT** suitable for discussion at this type of meeting?
   (a) Aspects of your everyday work that you enjoy and wish to develop
   (b) Aspects of your everyday work that you find particularly difficult
   (c) Funding for a postgraduate course at a local university
   (d) Introduction of a new method that you have read about in a journal
   (e) The behaviour of another member of staff that particularly irritates you.
3. Which of the following is **NOT** true of the KSF?
   (a) It applies to professional and non professional groups of people
   (b) Is linked to targets set in a performance development plan
   (c) Applies to all employees in the NHS
   (d) Has been promoted by the Department of Education of the UK
   (e) It has been published by the Department of Health.

4. Indicate whether the following statements are **TRUE** or **FALSE**:
   - **(a)** The highest academic qualification that a biomedical scientist can achieve is at M-level
   - **(b)** Libraries that are under the control of local authorities only provide a lending service to the local community and do not routinely offer internet access
   - **(c)** The HCPC has a clear policy regarding activities for CPD that are acceptable if you wish your name to remain on the Biomedical Scientist register
   - **(d)** Only activities that are approved and regulated by an employer are considered to be professional development
   - **(e)** Biomedical scientists are not required to meet minimum standards for numeracy and literacy.
5. List the formal qualifications that are offered by the IBMS that are appropriate to a single, specialist discipline.

Answers to self-check and end-of-chapter questions are available at the end of the book.

Answers to self-check questions

Chapter 1

1.1 Experimental design, communication, team-working, negotiation, judgement, self-reliance.

1.2 Only by working in a clinical laboratory, but this may be before, during, or after academic study.

1.3 Diagnosis of illness, identification of particular illness, confirmation that an illness is not present, excision of lesions are complete, no local or distant spread.

Chapter 2

2.1 To ensure all newly registered biomedical scientists have successfully completed a minimum level of education and training that qualifies them as competent and capable of autonomous practice.

2.2 Answers to include:

- teaching students on approved degrees
- researching diseases
- novel applications of biomedical science
- professional support for registrants
- continuing professional development related to biomedical science
- patient-orientated activities.

2.3 Select from:

- criminal offences
- professional negligence
- falsifying laboratory records
- false renewal declaration
- false qualifications
- poor health affecting professional performance.

2.4 Allows career progression to be planned in a manner appropriate to laboratory roles. Professional qualifications provide evidence of differing levels of practice.

2.5 All of the statements in Table 2.5.

Chapter 3

3.1 Any reagent change may affect both the calibration and/or quality of a laboratory investigation. In case of future problems with a set of results or with performance quality of an instrument or method, it is important that all reagent associated with a particular result is traceable to a given batch.

3.2 Get the person receiving the call to write down the results and repeat them as a safety check to avoid transcription errors.

3.3 The NHS number is given to the majority of UK residents who are likely to use health services in the country. The number is unique to an individual and all electronic systems within the UK are able to accommodate the number. The number is easily transferable as a unique identifier between different areas of the NHS. The number acts as an additional check for patient identification.

Chapter 4

4.1 *Biological:* blood sample, urine, sputum, or other body fluids or unfixed tissue samples, which may contain bacteria, fungi, viruses, or prions.

Physical: equipment with moving parts, sharp objects, including scalpel blades and needles, objects that could cause trips/falls such as cables or boxes left on the floor.

Chemical: hydrochloric acid, cleaning products, dyes, any laboratory reagents in any form including powders, liquids, gases, and solids.

Electrical: any piece of electrical equipment and the cables associated with it, for example water baths, centrifuges, lamps, refrigerators, and analysers.

Fire: fuel sources, flammable materials (paper/cardboard), or chemicals (alcohols, such as industrial methylated spirits and ethanol).

Heat or ignition sources: electrical sparks, heaters, ovens, strong lamps, naked flames, and discarded cigarettes.

Oxygen-rich atmospheres: such as rooms containing or storing oxygen cylinders.

Environmental: cold rooms, noisy equipment, equipment that causes vibrations, poor ventilation, heat, humidity, lighting, and inadequate space.

4.2 Biological hazard as the blood sample may contain viruses or bacteria that could enter the body via the mouth, eye, or broken skin. Routes of entry: oral, eye, skin. Physical as some form of needle must be used to collect the sample, which presents a sharps hazard. These risks could combine if the used needle punctured the skin of the person taking the sample as any infective agent could pass directly into their bloodstream. Physical harm from the patient by being hit or scratched.

4.3 To provide guidance on how to comply with health and safety legislation and to enforce these regulations.

4.4 Examples include damaged safety equipment, such as broken safety goggles or a laboratory coat that does not fasten up correctly, any fault with equipment, such as a safety cabinet that is not reaching its safe working range or a poorly fitting centrifuge lid, a damaged fire safety sign, extinguisher, or fire door.

4.5 A DSE risk assessment should consider: the chair, ensuring it is fully adjustable and stable; the workspace or desk; the equipment (keyboard, screen) and the way it is used; the individual characteristics of the person using the equipment and the surrounding environment.

4.6 The PPE must give adequate protection against the hazard. It must be suitable for the task, the environment, and duration of use. It must not introduce any additional hazards. It must fit the individual (or all individuals) concerned. It must not cause any health problems to the wearer. All individual pieces of PPE must be compatible.

4.7 Environment, task, individual, load, and equipment.

4.8 T = Toxic, Xn = Harmful, Xi = Irritant, C = Corrosive, F = Flammable.

4.9 *Fuel sources:* flammable materials (paper, cardboard) or chemicals (alcohols such as industrial methylated spirits and ethanol).

Heat or ignition sources: electrical sparks, poorly maintained/exposed wiring heaters, ovens, strong lamps, naked flames, and discarded cigarettes.

4.10 Immediately by the quickest method possible, usually by telephone and then on the required form within 10 days.

Chapter 5

5.1 The mass would be the same as it is a measure of the matter contained in the object. As weight is influenced by gravitational force acting on the object, it would be less on the moon than on earth.

5.2 **(c)**.

5.3 Volumes of viscous solutions can be difficult to measure and deliver using a pipette (think about the problems in pouring syrup) and so weighing directly into an appropriate container can be more convenient; 18.9 g of glycerol.

5.4 Balance operation is influenced by such things as draughts, warm air, and vibrations, so they must be sited to minimize such influences. They must be calibrated using internal or external weights. This is especially necessary if the balance is moved.

5.5 The main source of potential measuring inconsistency relates to the subjective estimation of the base of the meniscus against an incremental scale on a burette or a line on a volumetric flask.

5.6 Accuracy relates to the delivery of a measured volume that is close to the true volume. An operator of a pipette will be delivering volumes with precision when repeats of the same measured volume are almost identical.

5.7 **(d)**.

5.8 Physically, a pipettor has a complex construction compared with a pipette as shown in Figure 5.5. The key difference lies in the reliability of use of the pipettor. Once the required volume is selected by the operator, the pipettor delivers that volume with both accuracy and precision as long as the correct procedure is adopted. With a pipette, the operator has to make a subjective judgement about the position of the meniscus against the scale. The use of a pipette filler can lead to further inconsistencies in volume deliveries.

5.9 The pipettor digital display reads 065. The volumes are **(a)** 6.5 μl, **(b)** 65 μl, and **(c)** 65 μl.

5.10 A mole is 'amount of substance' and contains Avogadro's number of elementary entities such as atoms, molecules, or ions. Molar mass is the mass of the elementary entities in one mole of substance. Relative molecular mass is the mass of one molecule of a substance divided by a mass equivalent to one twelfth of the mass of ^{12}C.

5.11 60.1 g.

5.12 60.0167 mol dm^{-3}; 0.835 × 10^{-3} mol (or 0.835 mmol).

5.13 When 0.1 mg of solute is added to water as solvent, there will be little change in volume of the final solution compared with the volume of solvent used. Therefore, the molal and molar concentrations will be the same. There may be some difference in volume of solution compared with solvent when the much higher amount of solute is added and so the molal and molar concentrations would be expected to be slightly different.

5.14 18 g dm^{-3}; 1.8 % w/v.

5.15 10 mmol dm^{-3}.

5.16 **(a)** Ten cycles will reach a 1024 dilution. **(b)** Three cycles reach a 1000 dilution.

Chapter 6

6.1 Check that the correct sample is obtained from the correct patient. Place the sample in appropriate container and plastic bag. Check sample and request form details match and are complete. Ensure speedy and secure transport to laboratory. Adhere to laboratory standard operating procedures for sample processing.

6.2 Surname, forename, date of birth, sex, hospital number.

6.3 **(a)** A biopsy of tissue.

(b) Tissue obtained by aspiration through a fine needle.

(c) A frozen section of tissue.

6.4 Hands must be washed and disinfected with alcohol gel. The swab sample should be obtained, before the wound is cleaned, the patient is bathed, or antibiotics commenced or changed. It should be taken directly from an infected site without contaminating undamaged skin or mucous membranes. The swab is rotated in pus and placed in transport media. Both the swab and request forms are adequately labelled and relevant information is included on the request form, particularly any antibiotic therapy and the site of the wound.

Chapter 7

7.1 Resolution is the smallest distance at which two separate points can still be distinguished.

7.2 Approximately 300 nm (more accurately 261 nm).

7.3 By using glass with different additives or (multiple) lenses of differing shapes in combination.

7.4 ¼ of a wavelength.

7.5 Ultraviolet.

7.6 In transmitted light microscopy, the specimen is illuminated from below. In incident light microscopy the specimen is illuminated from above and the operator views visible light emitted from the specimen.

7.7 By using a narrow beam of laser light that results in a light beam of constant intensity.

7.8 The oil would boil and contaminate the microscope column and specimen chamber.

Chapter 8

8.1 The required voltage would be 240 V.

8.2 Galvanic.

8.3 The voltage would be 61 mV.

8.4 Rather than pH 7, the mean pH would be pH 5.5.

Chapter 9

9.1 Absorbed.

9.2 Increase.

9.3 366.4 μg cm^{-3}.

9.4 About 530 or 580 nm.

9.5 The carboxyl group (C=O) of the carboxylic acid group.

9.6 1.8 cm^{-1}.

9.7 No.

9.8 Methyl bromide has only a single type of hydrogen atom and would show one peak (which occurs at 2.7 ppm). Methyl ethanoate has two types of hydrogens and would therefore show two peaks (at 2.1 and 3.7 ppm).

9.9 A tandem mass spectrometer is usually composed of two mass analysers in sequence. The first mass analyser separates components of a sample, and selects a specific ion or component for further analysis. This ion or component is transferred into the collision induced dissociation (CID) cell where it collides with an inert gas and fragments. The second mass analyser then provides a mass spectrum (known as an MSMS spectrum) that shows all the fragment ions obtained and which can be used to deduce the structure of the ion or component selected.

9.10 Gas–liquid chromatography (GLC) and high-performance liquid chromatography (HPLC).

Chapter 10

10.1 24,500 rev min^{-1}.

10.2 The binding of the inhibitor to the enzyme results in a conformational change in the complex, which has a larger frictional coefficient (f) than the free enzyme.

10.3 **(a)** 2040 tonnes; **(b)** 2004 tons.

10.4 On no account should the rotor be used. It should be tested for safety by the manufacturer and passed as safe to use or condemned. You should endeavour to find an alternative rotor and centrifuge suitable for isolating the virus.

Chapter 11

11.1 All are unitless, because the concentrations in both phases are in the same units and therefore cancel each other out.

11.2 No, it is not possible as the sample molecules cannot move faster than the solvent front. As values for Rf are relative to the solvent front, they should remain constant, despite changes in room temperature. However, an increase in room temperature may reduce the time required to do the chromatography and vice versa.

11.3 Morphine 0.16, cocaine 0.91, nicotine 0.46, methadone 0.31, and heroin 0.74. The analyte of interest is most likely to be methadone.

11.4 Sephadex G100.

11.5 Approximately 4500 μm.

11.6 Glutamate, alanine, lysine.

11.7 The peak height ratio for drug X is 1, as both it and the internal standard have the same peak heights. Using the calibration graph gives a corresponding concentration of 37 mg dm^{-3} for X in the sample.

11.8 Decanoic acid, followed by tetradecanoic, hexadecanoic, and octadecanoic acids, and, finally, eicosanoic acid.

Chapter 12

12.1 0.01 V cm^{-1}.

12.2 Albumin.

12.3 Prepare gel in a fume cupboard. Wear safety gloves and glasses throughout the procedure.

12.4 10%.

12.5 Coomassie G-250 has extra methyl (CH_3) groups.

12.6 Approximately 1 ng.

12.7 SDS denatures proteins and so eliminates most enzymatic activities.

12.8 Ethidium bromide stains by intercalating between the base pairs of double-stranded nucleic acid molecules.

Chapter 13

13.1 **(a)** Bone marrow, thymus, the spleen, lymph nodes, mucosa-associated (MALT), and gut-associated lymphoid tissue (GALT). **(b)** That which is present at birth; is non-specific and does not develop immunological memory. **(c)** To destroy microorganisms and parasites.

13.2 **(a)** Immunoglobulin molecules are Y-shaped and consist of two heavy and two light chains joined by disulfide bonds.

(b) Plasma cells produce antibodies. **(c)** Genetic predisposition and viral infections.

13.3 Adjuvants are thought to slow down the release of the antigen and create a series of small 'hits' to the immune system.

13.4 **(a)** Affinity is an indication of the strength of antigen-antibody binding. Avidity is a measure of the number of binding sites. **(b)** Equivalence is reached just before maximum precipitation occurs.

13.5 **(a)** An immune complex forms that shows up as a line of white precipitate. **(b)** Prozone phenomenon.

13.6 **(a)** Labelled or unlabelled; type of label, for example, radioisotope, enzyme, or fluorochrome; method of signal detection; competitive or non-competitive assay; whether the assay requires a separation step. **(b)** Health and safety regulations. **(c)** Adsorption with activated charcoal; precipitation with polyethylene glycol; use of a secondary binding antibody; attachment to a solid phase.

13.7 **(a)** Washing steps are necessary because any residual substance that remains in the well may interfere with subsequent steps. **(b)** Antibodies or cytokines.

13.8 **(a)** Cluster of differentiation (CD) numbers. **(b)** Many antibodies or antigens can be measured simultaneously.

13.9 **(a)** Direct immunofluorescence detects the presence of antigens in sections of tissue; indirect immunofluorescence detects antibodies by demonstrating their attachment to appropriate sections of tissue. **(b)** Any two of fluorescein, rhodamine, or phycoerythrin.

Chapter 14

14.1 TACTCCTAACGGCAT.

14.2 Approximately 2 m.

14.3 2.75 mg cm^{-3}.

14.4 Only a very weak one, arising from the unfolding of intrachain double helical regions.

14.5 BamHI, *Hind*III, *Kpn*I, *Msp*I, *Pst*I, and *Taq*I.

14.6 ... ATCATGTCATAGCTGTTTCCTG.

14.7 Western blotting is used to identify a specific protein or group of proteins by immunoblotting (detection by antibodies). Southern blotting is used to identify DNA molecules/fragments with specific sequence of bases. Northern blotting is used to identify RNA molecules/fragments with specific sequence of bases.

14.8 Isolate the relevant sections of DNA and sequence it or use PCR across the potentially amplified region and determine the length (size) of the amplified fragment and thus estimate the extent of any amplification.

Chapter 15

15.1 A reduced test turnaround time will mean a result will be available to a clinician so that decisions may be made faster. For example, if the clinician suspects that a patient has acute renal failure and has the results of renal markers, including creatinine and urea, in 30 minutes rather than 1 hour, treatment may be initiated more quickly.

15.2 The universal preservation of blood sample constituents cannot be achieved with a single type of sample. A number of different types of additives are added to collection tubes, including lithium heparin, sodium EDTA, sodium citrate, and potassium oxalate:sodium fluoride. The last, for example, is the preferred additive for the analysis of glucose as it inhibits the process of glycolysis. Consequently, when a number of investigations are required a corresponding set of blood specimen tubes will need collecting to be suitable for those tests requested.

15.3 False. Specimen barcodes may be applied at a number of points on the specimen journey including at the point of request in the clinical area or the laboratory's specimen reception.

15.4 The progress of the specimen tube can be followed by either re-reading the barcode at barcode reading stations or by using a radiofrequency tag.

15.5 Automated decappers remove the requirement for manual removal of caps from specimens, a task that if repeated many times in the course of a day can cause repetitive strain injury.

15.6 All of these laboratories can benefit from automation.

Chapter 16

16.1 **(a)** True. **(b)** False. **(c)** True.

16.2 Diagnostic testing is physically closer to the patient, giving a rapid turnaround time for the availability of results.

16.3 **(b)**, **(c)**, and **(e)**.

16.4 **(a)** and **(c)**.

16.5 Designating a person to take responsibility for deliveries ensures that all deliveries are correctly handled.

16.6 **(a)**, **(b)**, **(c)**, **(d)**, and **(e)**.

16.7 **(c)**.

Chapter 17

17.1 Any three of the following: health and safety; quality management; clinical governance; training and competence; quality control.

17.2 SOP are written instructions that lay out a procedure or method in a clear and concise manner. They detail how individual tasks are to be carried out, that is, what is to be done, by whom, and when. They accurately describe the process and methods that are in use in the laboratory and contain vital information such as health and safety instructions, correct use of equipment, and controls.

17.3 Four methods are used by external inspectors when undertaking an audit: vertical audit; horizontal audit; witness; examination audit.

17.4 The main bodies that inspect medical laboratories are: UKAS/CPA; HTA; MHRA.The overlap between the three bodies is limited but more with MHRA as it looks at the collection, storage, supply, and disposal of blood and blood products. This will inevitably involve the laboratory to a large degree. The HTA, however, looks more at the use of tissue, ensuring that appropriate consent has been obtained and that it is only used for scheduled (or consented) purposes, and that it has been disposed of correctly and respectfully. The major difference is that the MHRA and HTA have legal powers attached to their activities and can initiate a prosecution or shut down a blood bank, whereas UKAS/CPA only inspect against an ISO standard.

17.5 Discussion points should include some/all of the following:

- Identification of the appropriate chemical hazards-explosive, flammable, oxidising, corrosive, skin irritants, respiratory irritants, environmental hazards
- Relate these hazards to the chemical state-solids, liquids, vapours, gases, dusts
- Ensure that no less hazardous chemical is available or a less hazardous procedure can be utilized
- If not, ensure the working environment is appropriate for the hazards encountered
- If not, ensure the appropriate PPE is both available and functional
- Calculate the relative risk on the 5 x 5 table
- Following the calculation, as HSAW is about protection of others too, ensure those who are at more significant risk of harm (pregnant, chronically ill-asthma, eczema, etc.) have any extra correct PPE or are taken off such activities.

Chapter 18

18.1 Biomedical science is a rapidly developing subject and its practitioners need to be continually abreast of new developments. Furthermore, biomedical scientists need to develop continually as individuals. In these ways, biomedical scientists can improve the service offered to clinicians.

18.2 For a biomedical scientist employed in a training laboratory, completion and assessment of the Specialist Portfolio in the discipline relevant to the workplace. For a biomedical scientist employed elsewhere, a part-time MSc in a relevant subject. There is no reason why the Specialist Portfolio and MSc cannot run concurrently, if an employer agrees.

18.3 It will show the following:

- an introductory paragraph on the topic
- content matter that is applicable to the title
- knowledge of appropriate depth and breadth
- contain material that has not been plagiarized from unacknowledged sources
- a summary or concluding paragraph
- show appropriate spelling and grammar
- be appropriately referenced.

18.4 All libraries offer facilities to borrow books available at the premises, facilities to order books from other libraries to then be available for loan, newspapers, and magazines that may be read on site or borrowed, reference books that cannot be borrowed, scientific and specialist documents to order, internet facilities, study area facilities, videos and DVDs for hire, and also many reading groups and leisure activities.

Answers to case studies

Chapter 5

5.1 *Comments on scenario A.* In A, you need to dilute the stock electrophoresis buffer 20-fold. As the final volume is 200 cm^3, one-twentieth volume part (10 cm^3) of the stock buffer solution will be needed, which must be diluted to 200 cm^3 by the addition of 190 cm^3 of distilled water. What is the best way to measure these volumes? Think about the accuracy of the dilution that is required. A buffer is being diluted for electrophoresis and not, for instance, a solution of a specific concentration being prepared for an important assay. If the buffer dilution was 19.9- or 20.1-fold, rather than 20.0-fold, the discrepancy would have a negligible, if any, impact on the electrophoresis. Of course, good laboratory practice would demand a consistent approach to how the dilution was achieved. You could proceed in one of various ways. Appropriate measuring cylinders could be used to deliver 10 cm^3 and 190 cm^3, mixing the two volumes in a large beaker. You will, however, see later in this section that for consistent volume measurements, the use of measuring cylinders is not ideal. They provide a lower level of accuracy than pipettes and volumetric flasks. A more accurate approach would be to use a 10 cm^3 pipette to deliver the stock buffer into a 200 cm^3 volumetric flask, completing the dilution by adding deionized water to the calibration mark on the flask. With either approach, the end result would probably be acceptable for the required purpose of diluting the buffer.

Comments on scenario B. B requires a greater attention to accuracy and precision, convenience, and reliability because the results of such assays could have importance in informing a diagnosis and treatment. The assay protocol will have identified minimum levels of accuracy and precision needed, and the procedures for measuring volumes must be consistent with these requirements. It would be necessary to use pipettes in the 0–1 cm^3 volume range for all volume measurements to make these dilutions consistent with the accuracy and precision requirements.

These two scenarios illustrate the need for an awareness of the limitations, as well as the advantages, of different pieces of laboratory equipment and taking these into account where alternatives are available.

5.2 The haemoglobin concentrations for tubes 1–7 are 0.00, 0.17, 0.33, 0.67, 1.00, 1.50, and 2.00 mg cm^{-3}.

Chapter 6

6.1 The hypocalcaemia is artefactual. On investigation, it was learnt that the blood was collected into a tube containing EDTA, which is an anticoagulent and chelates calcium ions. Such tubes are often used for haematological investigations. After collecting blood into this tube, the nursing staff had transferred the sample into a correct clinical chemistry tube but by that time any blood calcium had been chelated.

Chapter 7

7.2 The most probable cause is infection with an adenovirus. The lack of evidence for bacterial contamination and the presence of adenoviral particles in all faecal samples examined are strong evidence for this conclusion. The ages of the patients and the relatively mild symptoms are supportive of this conclusion. Adenovirus-induced gastroenteritis in infants and young children is usually associated with types 40 and 41.

These types of infections are impossible to prevent, given the presence of the virus in such a large proportion of the population and the ease with which it can spread from person to person and from surfaces to person. The risks of infections can, however, be minimized by isolating patients as soon as they exhibit symptoms, thorough cleaning of communal areas, and imposing a regime of strict personal hygiene.

Chapter 8

8.1

(a) Before training, Philippa's lactate threshold (LT) was achieved when she was running at only 10km/h; above that her body produced lactate faster than she could metabolize it and blood lactate levels rose sharply. This would occur when the body was becoming increasingly hypoxic. After training, likely improvements to her pulmonary cardiovascular system enable her to run at higher speeds without becoming hypoxic. The training regime is therefore likely to improve her ability to run the marathon.

(b) Studies have shown a blood lactate level of 4.0 mmol dm^{-3} is associated with a 25-day in-hospital mortality rate approaching 30% and a 3-day mortality rate of over 20%; hence, this condition is very serious and must be diagnosed and treated promptly.

Possible causes to be considered are those resulting from:

- inadequate oxygen delivery, especially septic shock, trauma, or profound dehydration
- excessive oxygen demand, including hyperthermia or seizure
- inadequate oxygen utilization, including diabetes and HIV infection.

(c) Matt has not undertaken significant recent exercise and his lactate level of 4.0 mmol dm^{-3} is suggesting the possibility of a serious physical condition. Philippa's normal level of <2 mmol dm^{-3} is typical of a healthy adult and only rises to 10 mmol dm^{-3} during exercise. Once exercise ceases this should return to a pre-exercise level within 20–30 minutes.

Chapter 9

9.1 The increases in the patient's serum creatinine and urea concentrations following fluoroquinolone administration indicates a hypersensitive reaction to fluoroquinolone causing nephrotoxicity and acute renal failure. The lack of evidence for liver malfunction, where urea is synthesized; the return of renal function with the decreases in serum creatinine and urea nitrogen concentrations to reference range values; and the disappearance of purpura and erythematous skin lesions when fluoroquinolone treatment was discontinued. The treatment with intravenous fluids, furosemide, and prednisone initiated support this initial diagnosis.

Chapter 11

11.1 **(a)** Joe's urine most likely contains galactose. **(b)** He is probably suffering from galactosaemia.

Chapter 13

13.1 The presence of rubella-specific IgM antibodies indicates this was the woman's first exposure to rubella virus and she is therefore probably infected. The absence of IgG antibodies indicates she is not immune, which is in accordance with this being her first exposure to the rubella virus, and her unborn baby is therefore at risk. The woman requires advice regarding these points, and counselling about the possibility of a therapeutic abortion.

Chapter 14

14.1 Although the woman's *BRCA1* sequence is normal, the mutation in *BRCA2* indicates that the woman has a higher-than-average risk of developing breast or ovarian cancer. She should be counselled as to the possible needs for preventative mastectomies and removal of her ovaries

Chapter 16

16.1 **(a)** Reduces patient anxiety, although they will not necessarily shorten the appointment time and may not be cost-effective. **(b)** Testing can be carried out by trained, but unqualified, staff, releasing laboratory staff to concentrate on other work.

16.3 **(a)** Yes by the clinical laboratory performing the pregnancy test. It was performed in the A&E department because a fast turnaround time was necessary. **(b)** No, it was an analytical error. There is no evidence to suggest that the sample used was inappropriate.

16.4 **(a)** All of (i)–(iv). **(b)** Discussing the situation with the laboratory staff to achieve a quicker turnaround time for the tests would have been a more effective action.

16.5 Check that they have read and signed the relevant SOP and/or prevent them from using the device until retrained.

16.6 **(a)** Yes. **(b)** Selecting a suitable storage site. Carrying out quality control checks on a daily basis. Making staff aware that the cuvettes are temperature sensitive.

16.7 Investigations showed that the nurse had pressed the memory button and recorded the previous quality control result each time, rather than that of the patient. As an agency nurse, she was not included in the mandatory training scheme for permanent employees.

16.8 **(a)** No. **(b)** Yes, urine test strips are only licensed for use with urine because they are liable to react unpredictably with other fluids.

Answers to end-of-chapter questions

Chapter 1

1. **(b)** and **(c)**.

2. Answers to include:
   - signs or symptoms often require tests from more than one discipline
   - knowledge and skill in all disciplines allows a registrant to make more informed judgements
   - all individuals once registered are able to work in any discipline.

3. Answers to include:
   - case-mix and sample profile of the training organization
   - prior knowledge and/or experience
   - the point at which training is delivered, that is, before, during, or after academic study.

4. **(a)**, **(d)**, and **(e)**.

Chapter 2

1. **(a)**, **(c)**, **(d)**, and **(e)**.

2. **(a)**, **(b)**, **(c)**, **(d)**, **(e)**, and **(f)**.

3. Answers to include:
   - protecting the public
   - legal right to practise using a protected title
   - minimum standards of proficiency/competence established
   - self-declaration about health and CPD
   - legal right to investigate allegations
   - legal right to sanction practise up to and including removal from register.

4. Answers to include:
   - professional code of conduct, help, and advice
   - source of information
   - professional qualifications
   - CPD scheme
   - networking
   - ability to represent the profession.

5. Answers to include:
   - formal requirement of registrants
   - professional body schemes can be used as evidence
   - shows commitment to continuing education
   - allows personal and professional reflection on knowledge and skills
   - basis of a professional portfolio.

Chapter 3

1. **(b)**.

2. **(d)**.

3. **(b)**.

4. **(a)**, **(b)**, and **(e)**.

5. Results may be given to a patient when there is a consultant-led service and a biomedical scientist is acting under the instructions of a medically qualified person. Examples may include pregnancy tests, therapeutic drugs for compliance, blood groups for the purpose of blood donation, or MRSA screening to eliminate the negative individuals from a population.

6. The consequences of a loss of stored medical electronic data can be subdivided into two categories. First, if data are lost, a patient may have an incomplete medical record or history. Essential information may be missing that could lead to a misdiagnosis or a failure to treat an existing condition properly. Cumulative records may be incomplete and a trend may not be noticeable. The presence of a pre-existing erythrocyte antibody may be overlooked. Incomplete blood transfusion records would make tracing the fate of a particular unit of blood practically impossible. Second, confidential data relating to the health of an individual may find its way into the public domain. This could lead to personal distress for the person concerned as friends and family may gain access to private information that the person would prefer them not to know. Potential employers and insurers may be able to access medical histories. The information could prejudice future employment or insurance for the individual concerned even if the data are not current or valid.

Chapter 4

1. **(c)**.

2. **(a)**.

3.

ACDP	Advisory Committee on Dangerous Pathogens
ACOP	Approved Codes of Practice
DSE	Display screen equipment
HSE	Health and Safety Executive
WRULD	Work related upper limb disorder
DSEAR	Dangerous Substances and Explosive Atmosphere Regulations
HASAWA	Health and Safety at Work Act
HSC	Health and Safety Commission
IOSH	Institute of Occupational Safety and Health
LOLER	Lifting Operations and Lifting Equipment Regulations

MHO	Manual Handling Operations
WEL	Workplace exposure limits
PUWER	Provision and Use of Work Equipment Regulations
RSI	Repetitive strain injury
RIDDOR	Reporting of Injuries, Diseases and Dangerous Occurrences Regulations
PPE	Personal protective equipment

4.

- Identify hazards.
- Identify who is at risk and how.
- Evaluate the risks. Decide on control measures.
- Record and implement findings.
- Review and/or update the assessment when changes occur or as part of a regular review process.

5. **(a)** Mouth, nose, eyes, and skin. **(b)** If Inhaled through the mouth or nose, control is by a mask or respirator; absorption through the eyes by safety goggles; or absorption through the skin controlled by a laboratory coat or protective suit and gloves suitable for the chemical in question. The optimal control would be to contain the mist within a safety cabinet if possible or to have LEV at the source of the mist in combination with PPE.

6. To take reasonable care for your own health and safety and for that of others who may be affected by what you do.

7. Cooperate with your employer on health and safety issues. Correctly use work items provided, including personal protective clothing, in accordance with training or instructions. Do not interfere with or misuse anything provided for your health, safety, and welfare.

8. **(a)** Oxygen, fuel, and heat/ignition. **(b)** Removal of oxygen.

9. **(a)** Mechanical controls include fume hoods and classes 1–3 safety cabinets and LEV. **(b)** Such equipment must have a thorough examination at least every 14 months; the records of which must be kept for a minimum of 5 years.

10. The staff in a laboratory using universal precautions will wear PPE, such as gloves, masks, safety spectacles, or goggles at all times for all procedures. The staff in a laboratory using risk assessment-based control will wear the same sorts of PPE but will use ones that are specific for individual tasks and possibly wear only a laboratory coat for those of low risk.

Chapter 5

1. **(a)** False. You should always maximize the accuracy and precision of a volume measurement by choosing a pipette or pipettor with a maximum volume as close as possible to the volume to be measured.

(b) False because it depends upon the specific requirements of the laboratory.

(c) True. A standard curve is used to measure the concentration of an unknown.

2. **(a)** 0.104 mol dm^{-3}.

(b) 0.104 mol dm^{-3}.

(c) 0.052 mol dm^{-3}

(d) 0.094 mol dm^{-3}.

(e) 2.1 mmol dm^{-3}.

3. **(a)** (ii) 1×10^{-3} g per 100 g is the same as 10 g per 10^6 g. **(b)** (i) 0.056 mol dm^{-3}; (ii) 1 % (w/v).

4. You will need to review the information on the location of balances (e.g. to exclude draughts and avoid uneven surfaces), balance calibration, and correct mode of operation for that model (including an awareness of the meaning of any error codes). You would have to be aware of any other activities in that area of the laboratory that might affect what you were doing. If you were handling dangerous material, other workers would need to be informed accordingly.

5. Your table will probably show scoring values similar to those in Table 5.6. Cost will vary with the size of a particular item and so, for example one pipette may be much cheaper than another. Your level of confidence in reading a meniscus level may influence your scoring of convenience of use for some items. You will probably conclude that pipettors are best overall, but the desirable levels of accuracy and convenience of use are bought at a high cost. By comparison, Pasteur pipettes are bargain-basement items with considerable limitations, but they are very simple and good enough for certain applications.

TABLE 5.6 Grading of features of volume measuring equipment. Each feature is given a low (--) to a high (++) rating

	Cost	Simplicity of construction	Convenience of use	Level of accuracy
Pasteur pipette	–	++	+	– –
Beaker	–	++	+	– –
Measuring cylinder	–	+	– –	–/+
Pipette	–	+	–	+
Pipettor	++	– –	++	++

6. Molar concentrations take into account molecular size because a given weighed amount gives more moles of smaller molecules and fewer moles of larger molecules. Both a balance and volume measurement device will be required for producing molar concentration solutions. By comparison, % w/w concentrations only involve the use of a balance. Molecular size is not a consideration because moles are not involved.

7. In dilution series of this type, the first dilution is made and then the second dilution is achieved by using some of the first dilution. This approach is then repeated for each successive dilution. If a mistake is made at any point in this dilution series, all dilutions including and beyond the incorrect dilution will be wrong. Care has to be taken to use correct and accurate volume measurements at every step.

Chapter 6

1. **(b)**.

2. All are true.

3.

Additive	Mode of action	Tests
Ethylenediamine tetraacetic acid	Chelation	Full blood count; sickle cell tests; glandular fever tests; blood group and compatibility test; antibody screening
Trisodium citrate	Chelation	Coagulation (clotting) tests, e.g. prothrombin time, activated partial thromboplastin time
Trisodium citrate	Chelation	Erythrocyte sedimentation rate
Lithium heparin, heparin	Chelation	Blood glucose, calcium
Fluoride oxalate	Inhibitor of glycolysis	Blood glucose
Gel clot activator	Shortens clotting times	Tests requiring serum

4. Centrifugation must be performed at the appropriate speed and time to separate the serum or plasma from the blood cellular components. It is essential that cellular components do not leak into the serum or plasma because concentrations of some analytes will be altered. For similar reasons, samples must be transported to the clinical laboratory in a reasonably short time so that they can be analysed, or refrigerated or frozen until tested.

5.

- Abstain from the following foods for 48 hours before, and for 24 hours during, the collection of a urine sample: bananas, fruit, coffee, chocolate, vanilla, and flavourings.
- Check with the laboratory or medical office about any medication you are taking.
- Begin the collection on arising in the morning.
- Empty the bladder and discard the urine, noting the exact time.
- After that, all urine voided should be put into the plastic container provided.
- Do not empty out any preservative fluid in the container before beginning collection.
- On the next day, at exactly the same time as the collection started on the previous day, empty the bladder and put this urine into the container. The urine collection is then complete.
- Clearly label the container with your name and the dates and time of collection.

Chapter 7

1. **(c)**.

2. **(a)**, **(b)**, and **(e)**.

3. **(a)** Chromatic aberration is the inability of a lens to bring light of different wavelengths to a common focal point. **(b)** It is corrected by adding fluorite to the lens glass and by using a combination of lenses of different curvatures and shapes.

4. First, adjust the condenser to as high a position as possible and then focus on a sample using a low power objective lens (normally ×10 or ×4). Second, close the field diaphragm in the condenser and sharply focus the diaphragm onto the image of the sample using the condenser focus controls. The field diaphragm can be considered to represent the illumination axis and should be adjusted using the x and y condenser adjustment screws. The condenser lens system focuses as much light onto the sample and the lamp is focused onto the plane of the field diaphragm of the condenser. Now open the field diaphragm until it is just outside the field of view.

5. Resolution is affected by the wavelength of the radiation and the numerical aperture of the lens.

6. Electrons are extremely energetic and react violently with gas molecules in air and so would travel less than 1 mm rather than the desired two metres or so.

7. Your diagram should resemble Figure 7.12, with a description of the main components. The major uses in biomedical science include, for example:

- immunofluorescence techniques
- fluorescence *in situ* hybridization (FISH)
- fluorescence dye staining such as those for amyloid and tuberculosis bacilli.

8. ×800.

Chapter 8

1. **(b)**.

2. **(c)**.

3. **(a)** False; **(b)** true.

4. **(a)** The pH electrode would produce a 61.5-mV change in potential for every tenfold change in the concentration of H^+, whereas the Na^+ SE would only produce a 30.8-mV change. **(b)** The steeper slope of response would, in theory, give the pH electrode double the accuracy of the Na^+ ISE.

5. Electrochemical techniques are used in many situations where a single analyte must be measured in a complex mixture, good examples being environmental, pollution, food, and beverage analyses. The portability of these techniques is often of particular importance as *in situ* measurements are often required for analytes that are unstable and would degrade during the return to a centralized analytical laboratory.

6. **(a)** Most pH electrodes have their isopotential point set at pH 7 and rely on this as the first calibration value, before using buffers of pH 4 or 10. Some manufacturers might require more

or, indeed, fewer calibration points. **(b)** Cleaning procedures should be fairly common but might differ depending on the other components of the system. For example, if the glass bulb is set in a plastic casing this might have an influence on some of the cleaning materials that may be used. Similarly, storage methods might differ depending on the nature of the electrode itself but also, to an extent, simply on the manufacturer's own preferences and the storage solutions that they might supply.

7. The rate of reaction could be followed using an oxygen electrode to measure the rate of disappearance of the co-substrate. It could also be measured using a pH electrode, as the product, gluconic acid, dissociates to gluconate and H^+, acidifying the sample. Hydrogen peroxide is not naturally found in blood; therefore, its presence and concentration can be evaluated using a peroxide electrode. Any measurement of oxygen concentration would be more difficult as the 'background' oxygen concentration in blood samples is variable depending on the physiological condition of the donor and possibly the location in the body from which the sample was taken. Similarly, measurement of pH would be problematic owing to the variability in the initial pH of the blood samples, and also owing to them having differing buffering capacities, both of which are known to be affected by the physiological condition of the patient.

Chapter 9

1. **(e)**.

2. **(a)** FALSE
   (b) FALSE
   (c) TRUE
   (d) FALSE
   (e) FALSE

3. **(a)** ε is 600 M^{-1} cm^{-1}; **(b)** a is 60 m^2 mol^{-1}.

4. The single value may be outside the concentration limit set by the Beer–Lambert law. A standard curve avoids this, as the concentration limit can be seen (see Figure 9.5).

5. 31.6%.

6. Some of the light is absorbed, reflected, or scattered by the particles. The remainder is transmitted and detected by the colorimeter.

7. Hydrogens of the CH_3 for the 1.5 ppm and the CH for the 3.8 ppm.

8. A mass spectrometer comprises an ionizer, an analyser, and a detector, for example MALDI, ESI, and TOF or quadrupole ion trap.

9. Any three from: proteomics, metabolomics, identifying protein modifications, biomarker discovery, protein/peptide quantitation, sequencing (proteins/peptides/sugars), mass spectrometric imaging, identifying microorganisms.

Chapter 10

1. 1100 *g*.

2.
   (a) Bacterial cells–bench centrifuge
   (b) DNA molecules–microfuge
   (c) Mitochondria–high-speed refrigerated centrifuge
   (d) Viral particles–ultracentrifuge.

3. **(b)**

4. 2.5 minutes.

5. Enzymes are globular proteins with roughly spherical molecules, but those of DNA are rod-like. Asymmetric molecules, like DNA, have larger frictional coefficients (f) than the spherical enzymes; therefore, the $\mathbf{s}$ for the enzymes would be larger.

6. 2.72 cm.

7. M_r 70,150.

8. No. Wash the rotor with an appropriate detergent, rinse it thoroughly.

Chapter 11

1. Chloramphenicol, paracetamol, and fluorouracil.

2. **(b)**.

3. **(e)**.

4. Catalase and fibrinogen will coelute (in the void volume), followed by haemoglobin, carbonic anhydrase, chymotrypsinogen, and insulin.

5. Plot a graph of V_e on the y-axis and logarithm M_r on x-axis for proteins of known M_r, which gives a straight line. As the V_e is known for catalase, its M_r can be determined from the corresponding antilogarithm as approximately 245,000.

6. Convert the amino acids to volatile esters. One of the most common types of derivatives formed are trifluoroacetylated amino acid methyl esters.

7. **(d)**.

8. Pass a buffer containing EDTA through the column. The EDTA will chelate and remove the Mn^{2+}, releasing the enzyme.

9. **(a)** Selectivity (α) = (ty – t0)/(tx – t0) = (330 – 90)/(210 – 90) = 2.
   (b) Resolution (R) = 2(ty – tx)/(Wy + Wx) = 2(330 – 210)/(38 + 42) = 3. A resolution of 3 suggests a good separation of the two peaks was achieved.

Chapter 12

1. 133.3 mV.

2. Glycerol is a dense, viscous alcohol that allows the sample to sink into the wells during the application of samples to the gel. Bromophenol blue acts as a marker because it migrates more rapidly than most of the molecules investigated by electrophoresis.

3. Yes, they are compatible. The protein is most probably a tetramer, with each of the four polypeptide chains having an M_r of 16,000, giving the complete protein an M_r of 64,000.

4. **(b)**.

5. Proteins: **(b)**, **(c)**, and **(e)**. Nucleic acids: **(a)** and **(d)**.

6. A calibration graph of logarithm of M_r on the *y*-axis and relative mobilities on the *x*-axis, respectively, can be used to give an M_r of 21,400 for the unknown protein.

7. A calibration graph of pH on the *y*-axis and distance from anode on *x*-axis, respectively, is required. The *p*I values for A and B can be determined from this graph as 7.20 and 4.95, respectively.

Chapter 13

1. **(b)**, **(c)**, and **(e)**.

2. **(b)**.

3. Answer should include:

- antibodies are produced in response to antigens
- the polyclonal response is the natural response to infection
- monoclonal antibodies are not normally found in healthy individuals
- both are produced *in vitro* for research therapeutic and commercial purposes
- pros and cons of polyclonal and monoclonal antibodies
- diagrams of antibody production.

4. Drawbacks of polyclonal antibodies include limited reproducibility, their cross-reactivity, low quantity, and the high cost of maintaining laboratory animals.

5. Answer should include:

- radioisotopes–safety restrictions, importance of half-life and background count, detection equipment, examples of radioisotopes; for example, competitive binding assay
- enzymes–main advantage of use, examples of enzymes and substrates, detection equipment; for example, ELISA
- fluorescent–principle of fluorescence, detection equipment, examples of fluorochromes; for example, flow cytometry.

Chapter 14

1. **(b)**, **(d)**, and **(e)**.

2. **(a)**, **(c)**, and **(e)**.

3. SDS, like many detergents, dissociates protein–lipid complexes of cell membranes.

4. Add a ribonuclease to digest the RNA or add an alcohol, and recover the precipitated DNA by centrifugation.

5. Ethanol precipitates DNA but most of the RNA remains soluble. Thus, ethanol precipitation is effective in separating DNA from contaminating RNA.

6. Uridine 3′-monophosphate and cytidine 3′-monophosphate.

7. **(c)** and **(e)** because they are not palindromic.

8. Natural DNA molecules, such as plasmids and their restriction fragments are too large to penetrate polyacrylamide gels.

9. **(a)** Nylon and **(b)** nitrocellulose membranes.

10. Infection with the AIDS virus will cause the release of specific antibodies against the virus into the patient's blood. The serum antibodies can be separated from other proteins by gel electrophoresis and the proteins transferred onto a suitable membrane. The presence of the viral-specific antibodies on the membrane could then be detected using an appropriate probe, such as a labelled antibody able to recognize and bind to epitopes of the antiviral antibodies.

Chapter 15

1. Centrifugation isolates the serum and plasma, the portion of the blood that is required for most laboratory investigations. Automation of the step enables consistent and efficient delivery of serum and plasma samples to the analysers.

2. **(d)** is true.

3. 8.85.

4. Automation units reduce the risk of infection to laboratory staff by reducing the numbers of times they handle open uncapped clinical samples.

5. True.

6. True.

7. **(a)** and **(b)** are true.

Chapter 16

1. **(b)**, **(c)**, and **(e)**.

2. **(b)**, **(c)**, and **(d)**.

3. **(b)**.

4. **(a)** Conduct a horizontal audit of all wards and departments on the site. You would need to note the type of device, its serial number and asset mark if appropriate. **(b)** The areas most likely to have POCT devices are A&E, intensive therapy units, and all critical care areas. Labour wards and neonatal care wards are also likely, but urinalysis strips and blood glucose meters may be found in almost any clinical setting.

5. You need to ask the following questions:

- Is this device in current use? Is it asset marked?
- Is the laboratory and relevant pathology consultant aware if it?
- When was it purchased and who ordered it? Ask to see invoices.
- Was the purchase approved by the POCT committee, and is there paperwork to show this?
- Has it been acceptance tested and has an evaluation been carried out?
- Have staff received training from the company and where are the training records?
- Where is the SOP for the device?

If the answers to these questions are unsatisfactory and the device appears to be unapproved with no current training records or SOP, inform the ward manager of this, and immediately notify the relevant pathology consultant and chair of the POCT committee and seek their advice. If there is any question of risk, you should also inform the lead for clinical risk. You may remove the device from the ward if you are advised to do so, or shut it down and unplug it. If the device is obsolete, you can remove it for storage. The safety of the patient is always paramount.

6. In the first instance, approach the user and offer training immediately. Ask if there is a particular feature that they have trouble with or do not understand. Always try to help the user become safe and competent. Explain the necessity for EQA; how it safeguards the user and the patient from harm and litigation. Remind them of their obligations under the trust's POCT policy and show them a copy. If the user is uncooperative or obstructive, immediately inform the ward manager, pointing out that the user is not deemed competent and must be prevented from using the device. Inform the user that they are not complying with policy and remove their password from the device or take their name from the authorized users list. Inform the lead for clinical risk and the POCT committee of the situation and the actions you have taken. Ensure that the user understands the reason for this and that they may be subject to disciplinary action.

7. Approach the consultant and ask if he or she is aware that the laboratory can offer this test result in less than 30 minutes. Try to establish whether the time to results is the real issue. If so, can the laboratory produce a result in a suitably short turnaround time by changing its practices or by altering transport arrangements? The consultant is responsible for establishing his or her clinical need and providing a business case for consideration by the POCT committee. You may advise on purchase costs and possible alternative devices. Ensure that total running costs, including QC and EQA have been fully considered. Who will use, maintain, and service the device? How will results be recorded and interpreted? Is there a safe, suitable site with the required facilities? How will staff be trained, checked for competence, and monitored? A service level agreement will be required. Dialogue and communication are key here, in ensuring that the appropriate service is provided.

Chapter 17

1. **(e)**.

2. **(a)**.

3. Internal quality control is concerned with sampling and testing to ensure that the necessary and relevant tests are actually carried out and those results are not released prior to their quality being judged to be satisfactory. EQA schemes are those that assess a laboratory's performance against a set of criteria. They can be retrospective and give information on how your performance rates amongst your peers. All laboratories that perform tests on clinical samples must participate in EQA schemes.

4. Document control is an essential part of an effective QMS. Documents associated with the QMS will change with organizational changes, changes in practices, technology, and improvements in processes. Control of these changes is necessary to ensure that the procedures correctly describe the processes, are readily available, and are up to date.

5. Techniques to improve the laboratory service include, but are not limited to:

(a) audit

(b) incident reporting

(c) root cause analysis

(d) use of cause-and-effect diagrams.

6. A learning organization will have systems, mechanisms, and processes that are used to enhance the service to the patient. This will allow the achievement of sustainable objectives, not only for themselves, but also for their patients and the community in which they operate. Learning organizations are able to adapt to their environment, continually evolve and enhance their capabilities, and apply the results of learning to the achievement of better results.

7. Dependent on their nature, concerns may be reported to your line manage or through a formal reporting mechanism such as DATIX. A formal reporting system is used to identify trends or highlight significant issues that may affect more than the area within which it occurred.

8. Reporting of concerns is necessary in order to learn from near misses, mistakes, and errors.

9. Cause-and-effect diagrams can be used to obtain the root cause of a problem, such as when equipment fails.

10. If an error or problem is uncovered, an audit trail allows the sample path to be recreated.

Chapter 18

1. **(c)** and **(d)**.

2. **(e)**.

3. **(d)**.

4. **(a)** False

(b) False

(c) True

(d) False

(e) False

5. The following specialist subjects are available for completion of a specialist portfolio and are complementary to an appropriate MSc. Assessment is by portfolio:

- cellular pathology
- cytopathology
- clinical biochemistry
- clinical immunology
- haematology and transfusion science

- histocompatibility and immunogenetics
- medical microbiology
- transfusion science
- virology.

The following specialist subjects are available for completion of a higher specialist diploma and an applicant is required to hold an appropriate MSc. Assessment is by written exam.

- cellular pathology
- clinical chemistry
- cytopathology
- haematology
- immunology
- histocompatibility and immunogenetics
- management and leadership
- medical microbiology
- transfusion science
- virology.

Glossary

Absorption spectrum graph illustrating the fractions of incident electromagnetic radiation absorbed by a material over a range of different wavelengths.

Acceleration is the rate of speed change.

Acceleration voltage the potential difference in volts necessary to produce electrons from the tungsten filament.

Accreditation an inspection and examination procedure to ensure participants comply with published practice and documentation standards.

Accuracy is how close a measurement is to the true value of the substance being measured.

Achromatic a lens corrected for two colours only.

Acquired immunity or adaptive response one that produces antibodies specific to antigens on pathogens and develops an immunological memory.

Activity a measure of an ion's ability to affect chemical equilibria and reaction rates.

Adhesion molecules mediate the binding of cells to each other or to other biological surfaces.

Adjuvant substance that enhances the immune response to an antigen.

Adsorption accumulation of molecules from a solution on to the surface of an adjoining solid without penetrating its interior.

Adverse incident any event that may affect the reliability of a result issued by the laboratory or any delay in the diagnosis and treatment of a patient following laboratory investigations.

Affinity measure of the strength of antibody–antigen binding.

Affinity chromatography form of column chromatography that relies on the specific binding of the analyte of interest to an immobilized ligand.

Agglutination clumping together of cells or particles by antibody binding to antigens on their surfaces.

Amperometry a form of voltammetry that involves the reduction or oxidation of an electroactive molecular species at a constant applied potential.

Ampholytes synthetic polyamino-polycarboxylic or polysulfonic acids of low *Mr* that differ from one another in the values of their isoelectric points (*p*I).

Analyser polarizing filter placed in the light path fixed within the body of the microscope. It is used together with a substage polarizer.

Analyte specific substance in a sample under investigation to be identified and/or quantified.

Anions negatively charged ions that are attracted to the **anode**.

Anode the negatively charged electrode.

Anode plate disc in the electron gun of an electron microscope that is maintained at earth potential and causes the cloud of electrons produced by the filament of the gun to be accelerated into the microscope column.

Anthropometry study of human body measurement related to capabilities.

Antibody immunoglobulin with the ability to bind to a specific antigen.

Antigen molecule that can elicit the production of and bind to an antibody.

Antigenic determinant or epitope site on an antigen that is recognized by an antibody.

Antihuman globulin polyclonal antibodies produced against an epitope on a human immunoglobulin molecule.

Antinuclear antibodies antibodies to nuclear proteins.

Antiserum one that contains antibodies and may be used as an analytical reagent.

Apochromatic a lens corrected for three colours and which is the best quality type of lens available.

Approved Codes of Practice explain legal regulations and the detailed requirements necessary to comply with them.

Astigmatism the inability of a lens to bring light or electrons passing through different parts of the lens to a common focus.

Autoimmune disease condition in which the immune system of an individual reacts against self antigens.

Autoradiography sensitive method of producing a photographic image of a specimen containing radioactively labelled material(s).

Avidin-biotin system Avidin is found in egg white and has a high affinity for biotin, which is a co-enzyme. Together they can form complexes with antibodies, enzymes, radioisotopes and fluorochromes.

Avidity ability of an antibody to bind multiple antigens.

Avogadro's number is a constant and refers to the number of particles, i.e. 6.022×10^{23} in one mole of a substance.

B lymphocyte type of white blood cell that can form a plasma cell.

Balancing of a rotor consists of distributing the centrifuge tubes in the rotor, i.e. the weight, evenly around the centre of rotation.

Barrier filter a filter used in the fluorescence microscope, which is only capable of transmitting light in the visible spectrum and which prevents the transmission of ultraviolet light.

Bile green-coloured product of the liver stored in the gall bladder.

Binding site variable part of an antibody molecule that can bind to an epitope on an antigen.

Biomedical science the application of the natural sciences, to medicine, in the study of the causes, consequences, diagnosis, and treatment of human diseases.

Biomedical scientist an individual who practises biomedical science and is entitled to use the legally protected job title. Only those individuals who have met the threshold requirements of the Health and Care Professions Council may use this term to describe themselves within the UK.

Biosensor an analytical device that incorporates a biological component, such as an enzyme, and an electrochemical transducer.

Birefringence the ability of a substance to split a ray of light into two components, the ordinary and extraordinary ray.

Blood sciences laboratory a commonly used term for an automated laboratory in which the predominant analyses are blood investigations and which encompasses the traditional disciplines of biochemistry, haematology, coagulation, immunology, and virology. Sometimes called a core automated laboratory.

Blotting transfer of the molecule of interest from an electrophoretic agarose or polyacrylamide gel to a piece of nitrocellulose, paper, or a nylon membrane.

Calibration a process using calibrants of known concentration or value to set the performance parameters of a device.

Cannula flexible plastic tube carried on a needle that is inserted into the body to obtain a sample of fluid or introduce medication.

Capacity is the weighing range of a balance, e.g. up to 200 g, 1000 g, or more, respectively.

Cathode the positively charged electrode.

Cations positively charged ions that are attracted to the **cathode**.

cDNA library collection of clones (transformed cells) each containing a recombinant DNA prepared using mRNA molecules extracted from a single source (organism).

CE mark a mark placed on an *in vitro* diagnostic device to show that it is compliant with European Directives, and is fit for the purpose stated by the manufacturer.

Centrifugal field an acceleration directed outwards from the centre of rotation.

Centrifugation term applied to the mechanical process of separating mixtures by applying centrifugal forces.

Centrifugation tubes containers for samples used in the rotor during centrifugation.

Centrifuge mechanical device used to separate or isolate substances suspended in a fluid by spinning them about a central axis to produce a suitable centrifugal force.

Certificate of Competence an award made to individuals who have completed an Institute of Biomedical Science-accredited honours degree and their registration portfolio. This award allows an individual to apply to the Health and Care Professions Council to be admitted to the biomedical scientist register.

Chemokine protein that stimulates the migration and activation of cells.

Chromatic aberration the inability of a lens to bring light or electrons of different wavelengths to a common focal point.

Chromatography collective term for a family of analytical techniques used to separate the components of mixtures of molecules for their identification and possible estimation of their concentrations in the original mixture based on differences in their partitioning between two immiscible phases.

Chromophore atom, molecule, or part of a molecule that becomes excited by absorption of electromagnetic radiation.

Chronic ill health a long-term illness or condition that may not be curable but can be managed using laboratory tests and their results.

Clinical governance a method by which a procedure can be controlled, documented, and audited.

Clinical risk a measure of anything that harms, or potentially harms, a patient.

Clone population of cells derived from a single stem cell.

Coherent light rays of the same amplitude and wavelength, which are in phase with one another.

Combination electrode a pH electrode in which the external reference electrode is built into the glass electrode in a concentric double-barrel arrangement.

Commensals a term used to describe all the natural bacteria that live on and in a healthy person. The main areas for commensal organisms are the skin, mouth, upper respiratory, gastrointestinal, and urogenital systems.

Competent ability of an individual to perform a task repeatedly to an agreed level of quality.

Competitive tender where companies bid for a specified business contract.

Complement system group of proteins that act directly or with antibodies to kill pathogens.

Continual improvement ethos allowing an organization to build constantly on lessons learnt.

Continuing professional development any learning associated with your chosen profession, rather than your personal development. The two are not exclusive.

Continuous polyacrylamide gel electrophoresis uses the same type of buffer to dissolve the sample as that present in the gel and in the compartments of the electrophoresis equipment. Compare **discontinuous polyacrylamide gel electrophoresis**.

Control of Substances Hazardous to Health set of laws that require employer controls materials that are hazardous.

Core automated laboratory an alternative name for the blood sciences laboratory.

Corrective action that taken to eliminate the cause of a detected non-conformity or other undesirable situation.

Corrosive substance that will destroy or irreversibly damage another that it comes into contact with.

Coulomb is a unit of electrical charge equal to the charge transferred by a current of 1 ampere s^{-1}.

Council for Professions Supplementary to Medicine predecessor of the Health and Care Professions Council.

Critical illumination achieved when a substage condenser is adjusted so that the light it emits comes to a **focal point** at the level of the specimen.

Curettings scrapings of tissues.

Cuvettes specialized transparent container in which the absorbance of a solution is measured.

Cytocentrifuge or slide centrifuge a low-speed centrifuge with a modified rotor used to separate cells and deposit them as a monolayer onto the slides for cytological examination.

Cytokine protein that directs the behaviour of other cells.

Dalton a unit equal to one-twelfth of that for ^{12}C, which is sometimes used to express the **mass** of an atom or molecule.

Decalcifiying agents solutions of, e.g., ethylenediaminetetraacetic acid, formic, or hydrochloric acid that reduce the mineral content of samples of tissues.

Delta check a facility that allows a current result to be compared with previous results from an individual.

Destaining the removal of excess stain following staining to give a gel of clear background.

Differential centrifugation technique that separates particles during sedimentation mainly on differences in their sizes.

Discontinuous polyacrylamide gel electrophoresis uses a buffer that differs in composition and pH to dissolve the sample to that present in the gel and in the compartments of the electrophoresis equipment. Compare **continuous polyacrylamide gel electrophoresis**.

Dissertation involves a literature review and a written presentation of the findings of that review. It is a component of an honours degree programme.

DNA cloning the preparation of many identical copies of a DNA molecule.

DNA-dependent DNA polymerase enzyme that replicates DNA. Usually abbreviated to DNA polymerase or DNA pol.

Double junction reference electrode an external reference electrode that has an additional chamber between the reference electrode and the external solution.

Dress code a clear set of written instructions that specifies what can and cannot be worn in a place of work.

Dusts fine particles of solid substances or pellets that may be harmful or flammable.

Ectopic pregnancy implantation of the fetus outside the uterus, often within the Fallopian tubes, and a life-threatening condition.

Electric field the potential difference or voltage applied between two electrodes. In electrophoresis it is normally quoted in units of V cm^{-1}.

Electroactive a molecular species that can undergo oxidation and reduction.

Electrode potential a spontaneous potential difference that is established across an electrode/electrolyte interface due to oxidation or reduction events occurring within the half-cell.

Electrolytic cell an electrochemical cell in which reactions are not spontaneous, only occurring if an external voltage is applied across the two electrodes.

Electromagnetic radiation spectrum the full range of electromagnetic radiation.

Electro-osmosis occurs in support media that possess negatively charged groups on their surface and so attract hydrated **cations** present in the buffer. When an **electric field** is applied the movement of cations and associated water molecules is towards the cathode. Electro-osmosis may be sufficiently strong that weakly charged anions can be carried towards the cathode.

Electrophoretic mobility velocity of an ion per unit of **electric field**.

Eosinophils granulocytes that are associated with combating parasitic infections and causing allergic reactions.

Ergonomics study of designing a job, equipment, and workplace to fit the worker.

Error failure to complete a planned action as intended or the use of an inappropriate plan of action to achieve a given aim.

Exciter filter one used in a fluorescence microscope which transmits ultraviolet light of the required wavelength and prevents transmission of unwanted visible light.

External reference electrode a separate electrode that generates a stable electrochemical potential in potentiometric analysis. Requires electrical continuity with the working electrode.

Fine-needle aspiration is a form of biopsy in which a thin needle is inserted in to a region of tissue of abnormal appearance to collect a sample of fluid or cells for clinical testing.

First aid provision of initial care for an illness or injury.

First voided urine the initial urination of any given day.

Flammable substance that will easily ignite causing fire.

Fluorescence virtually instantaneous emission of radiant energy by a chromophore following the absorption of radiation of a higher energy.

Fluorochrome fluorescent dye used to label or stain a biological substance.

Focal length the distance between the centre of the lens and its **focal point**.

Focal point the point at which a lens brings all of the light passing through it to a common focus.

Frictional coefficient property of a particle sedimenting through a solvent and equal to the frictional resistance opposing its movement divided by its sedimentation velocity.

Fronting production of an asymmetrical peak during column chromatography by a slow rise at the beginning of the peak and a sharp fall after the peak.

Full blood count number of haematological measurements of blood indicating the numbers and sizes of various types of blood cell.

Fume mist or vapour containing very small metallic particles.

Galvanic cell an electrochemical cell in which reactions occur spontaneously at the electrodes when they are connected externally by a conductor.

Gel electrode a pH or ion-selective electrode in which the electrolyte of the reference electrode is retained within a gel layer, enhancing portability, but meaning that the electrolyte cannot be refilled as in conventional electrodes.

Gel filtration is a form of column chromatography that separates molecules according to differences in their sizes and to some extent shapes.

Gene is a unit of heredity consisting of DNA and on a specific location on the chromosome and determines a particular characteristic of the individual.

Gradient elution use of a buffer of gradually changing pH or one whose concentration of salt progressively increases to wash materials out of a chromatography column. Compare **isocratic elution**.

Granulocyte type of leucocyte that has granules in its cytoplasm containing enzymes that can digest microorganisms.

Group III pathogens one of the four categories of pathogenic microorganisms recognized by the Advisory Committee for Dangerous Pathogens.

Haematoma a localized swelling consisting of blood that has leaked from veins and capillaries into the tissues.

Haemolysis breakdown of erythrocytes leading to release of free haemoglobin into the plasma.

Half-cell an electrochemical system comprising an electrode surrounded by an electrolyte.

Harmful any chemical that causes or is likely to cause harm.

Hazard substance, activity, or process that may cause harm.

HCPC register list of all those individuals who have met the threshold requirements of the **Health and Care Professions Council** and have been admitted to the register.

Health state of well being in body and mind.

Health and Care Professions Council (HCPC) one of the statutory regulatory bodies operating within the UK health services. Its principal function is to protect the public by regulating a number of healthcare professionals.

Health and Safety Executive a non-departmental public body responsible for the encouragement, regulation, and enforcement of workplace health, safety, and welfare.

Health and safety guidance documents written materials concerned with legal and best practices relating to health and safety produced by the **Health and Safety Executive**.

Healthcare near misses situations in which an event or omission, or a sequence of events or omissions arising during clinical care fail to develop further and so injuries to a patient do not occur.

Hep2 cells human epithelial cells derived from tumours and which are cultured for use in research and diagnostic tests.

High-performance liquid chromatography any of a variety of column chromatography methods that use the relatively high pressures produced by pumps to force the mobile phase through the column to increase the resolution and gain faster separation times.

Hybrid DNA molecules comprised of portions from two or more different sources. Also called **recombinant DNA**.

Hybridoma hybrid cell formed by fusing two different types of cells, e.g. a specific antibody-producing B lymphocyte with a myeloma cell to produce monoclonal antibodies.

Hydrophobic effect the tendency of molecules in an aqueous environment to fold so that their hydrophobic portions are buried in the interior of the molecule.

Hydrophobic interaction chromatography form of column chromatography that relies on the specific binding of the analyte of interest to an immobilized hydrophobic ligand.

Hyperkalaemia refers to a K^+ concentration that is higher than the quoted laboratory reference range.

Hypoglycaemic coma a state of unconsciousness caused when blood glucose falls to a dangerously low level.

Immiscible phases ones that do not mix.

Immobilized enzyme an enzyme that is either bound to a surface or entrapped within a structure.

Immune complex consists of an antibody and antigen.

Immune surveillance capacity of the immune system to continually monitor for and detect foreign antigens.

Immune tolerance failure of the immune system to respond to an antigen.

Immunoassay any analytical method that uses an antibody to detect and/or quantify an analyte in a sample.

Immunodeficiency situation in which the immune response is reduced or absent.

Immunofluorescence methods employing the fluorescence microscope to identify the presence of specific antigens within tissue samples using fluorescently labelled antibodies.

Immunoglobulins group of large protein molecules that includes all types of antibodies and the membrane-bound receptors on B lymphocytes.

Immunophenotyping technique used to categorize cells by the proteins expressed on the surfaces of their plasma membranes.

Improvement notice legal order to improve practice issued by the **Health and Safety Executive**.

In vitro literally 'in glass': occurring in the laboratory. Compare ***in vivo***.

In vivo occurring within a living organism.

Incident individual occurrence or event.

Innate immunity that which is present at birth.

Institute of Biomedical Science a professional body whose principal aims are to promote biomedical science and its practitioners.

Internal reference electrode an electrode found within the glass bulb of a pH electrode.

Ion-exchange chromatography use of a charged resin to separate ions on the basis of differences in their charge densities.

Ion-selective field effect transistor a field effect transistor with the gate modified to respond to specific ions.

Ionic strength adjustor a solution of known high ionic concentration. Proprietary adjustors may also contain pH adjustors and decomplexing agents or agents to remove species that interfere with the measurement of the ion of interest.

Ions atoms or molecules that possess an electrical charge.

Irritant substance that causes irritation.

Isocratic elution use of a buffer of constant composition and pH to wash materials out of a chromatography column progressively. Compare **gradient elution**.

Isopotential point the pH at which a pH electrode, or the ion concentration at which an ion-selective electrode, generates zero potential, and where temperature has no effect on potential.

Isopycnic density gradient centrifugation technique to separate particles based mainly on differences in densities during their sedimentation through a column of liquid whose density increases from the top of the centrifuge tube towards its bottom.

Journal club an arrangement between a group of like-minded individuals who agree that one member should read a particular journal paper and then present the contents to the other members of the group.

Köhler illumination achieved where the substage condenser is used to focus the maximum amount of light onto the specimen.

Laboratory information management systems computer databases that hold complete patient records of incoming requests and outgoing results and allow for the electronic exchange of laboratory data between computers.

Lens piece of glass ground into a spherical or ovoid shape that produces a magnified image. Two or more lenses may be used in combination to produce higher magnifications.

Lifelong learning term used to encompass any new skill or knowledge that you may acquire at any time during your life.

Lipaemia high levels of fats such as cholesterol and triacylglycrerols, especially found after eating a meal with a high fat content, resulting in cloudy or cream-coloured plasma.

Litigation legal action against an individual or organization.

Lymphocyte leucocyte that mediates the activities of antibodies and cellular elements of the immune response.

Macromolecules those with Mr greater than 5,000.

Magnification the process of enlarging something only in appearance, not in physical size.

Mass is a measure of the amount of matter present in an entity.

Mast cells cells found in connective tissue that play a major role in allergic reactions.

Mediator a soluble electroactive molecule that can accept electrons from an enzyme and then diffuse to an electrode surface and donate the electrons to that electrode.

Melting separation of the two strands of DNA by breaking the intrachain hydrogen bonds.

Metal chelation chromatography form of column chromatography that relies on the specific binding of the analyte of interest to an immobilized metal ion.

Metal oxide semiconductor field effect transistor the electrical transistor found in nearly all electrical devices.

Methadone a heroin substitute used to wean addicts off the drug.

Michaelis–Menten equation one description of the kinetics of an enzyme-catalysed reaction.

Microscope an instrument that forms an enlarged image of an object that would often be too small to be seen with the naked eye.

Microscopy the use of a microscope to examine and analyse objects that would often be too small to be seen with the naked eye.

Mobile phase the fluid phase used in chromatography; can be a gas or liquid.

Modified glass electrode a type of ion-selective electrode that relies on a glass bulb to generate a potential.

Molality is way of expressing concentration where the **mass** of the **solvent** is used rather than the total volume of **solution**. A 1.0 molal solution would contain 1.0 **mole** of **solute** per kilogram of **solvent**.

Molar mass is the **mass** of all the elementary entities in one **mole** of a substance.

Molarity a way of expressing concentration as the number of **moles** of **solute** in a total volume of **solution** of 1 dm^3 or litre.

Mole the amount of a substance that contains as many elementary entities, such as photons, atoms, or molecules, as there are atoms in 0.012 kg (12 g) of ^{12}C. This value is called Avogadro's number and is equal to 6.022×10^{23}.

Molecular biology study of biology in terms of the structure and functions of the atoms and molecules concerned.

Molecular sieving the separation of ions during electrophoresis by porous support media that restrict the flow of ions. Smaller ions experience a lower frictional resistance and so migrate through the gel faster and travel further than larger ones for a given time.

Monochromatic refers to light of the same wavelength and colour.

Monochromatic radiation electromagnetic radiation of a single wavelength.

Monochromator a device for separating a beam of electromagnetic radiation into its component parts on the basis of their wavelengths.

Monoclonal antibody single type of antibody that binds to a single epitope and is produced by a single clone of B lymphocytes.

Natural killer cell lymphocytes that are part of the innate immune response and have a role in the destruction of some tumour cells.

Near miss unplanned event that did not result in injury, illness, or damage but had the potential to do so.

Nephelometry analytical technique in which the intensity of light scattered by a suspension is measured at an angle to the incident beam.

Nernst equation an equation that describes the potential developed by a potentiometric electrode.

Neutrophils phagocytic granulocytes that engulf and destroy pathogenic organisms.

Nomograms arrangements of three scales, such that a line connecting a value on one scale to that on one of the others gives the required value on the third scale. Nomograms relating the radius of rotation, speed of the rotor, and relative centrifugal field are useful in centrifugation.

Non-compliance non-fulfilment of a requirement or standard.

Non-conformity non-fulfilment of a requirement, including incidents, problems, errors, risks, and complaints.

Non-verbal communication any form of communication that does not involve the spoken or written word.

Nucleoside chemical combination of a base and sugar.

Nucleotide chemical combination of a phosphate(s) and a nucleoside.

Numerical aperture quantitative measure of the light gathering capacity of a lens. The greater the numerical aperture, the more efficient the lens.

O rings gaskets formed of a flat ring of rubber or plastic used to make joints air and water tight.

Occult blood small amounts of blood in faeces that are indications of possible disease, but while invisible to the naked eye may be detected using a biochemical test.

Ohm's law a description of the relationship between the current passing between any two points of a conductor, the potential difference between the two points, and the resistance of the conductor.

Oxidation a reaction in which there is a loss of electrons.

Palindrome word, phrase, or sequence that reads the same in both directions.

Paraprotein identical immunoglobulin or monoclonal antibody molecules produced by plasma cells in the bone marrow in excessive amounts in the condition multiple myeloma.

Partial specific volume (of a particle) change in volume that occurs when the particle is added to a large volume, i.e. excess of solvent.

Partition or distribution coefficient numerical descriptor of the way in which a substance distributes at equilibrium between two immiscible phases.

Pathobiology is the study of disease processes and their effects on humans. It is a principal subject component of an honours degree programme in biomedical science.

Pathogen a disease-causing organism.

Pathology the study of disease (pathological, relating to disease).

Pellet sedimented material found at the bottom of centrifuge tubes or bottles following centrifugation.

Performance appraisal a formal interview held between an employee and their line manager to assess the workplace performance of the individual, both retrospectively and looking to the future.

Personal development a strategy by which an individual is able to expand on their existing knowledge and skills, in their personal life and in their professional life.

Personal development plan a strategy agreed at a performance appraisal that includes the future training and educational needs of the employee.

Personal protective equipment a series of protective clothing or devices designed to minimize or completely remove the potential for an individual to be exposed to harm during work activities.

Personalized medicine is using an individual's molecular biological, genetic, and epigenetic information to provide that patient with the right treatment in the best way and at the right time.

Phagocytic cells see monocyte and neutrophil.

Phase plate device within the substage condenser of a phase contrast microscope that causes light to be retarded by one-quarter of its wavelength.

Phlebotomist a person trained to collect blood samples in a safe and appropriate manner.

Phosphor coating on the viewing screen of an electron microscope that emits visible light when excited by electrons to produce an image of the sample on the viewing screen.

Photons 'particles' or discrete packages of electromagnetic radiation.

Plasma the bodily fluid obtained following centrifugation of a blood sample that has been collected and preserved using an anticoagulant such as lithium heparin.

Plasma cells dedicated antibody-producing cells derived from B lymphocytes.

Plasmid extrachromosomal circular and double-stranded DNA molecule found in bacteria and some yeasts.

Point-of-care testing is the provision of a diagnostic pathology testing service outside the traditional clinical laboratory and physically closer to the patient.

Polarized light is light where the waves vibrate in a single plane.

Polarizer filter placed in the light path of the microscope beneath the substage condenser, which only allows the transmission of plane polarized light.

Polarography a form of voltammetry that employs a dropping mercury electrode as the working electrode.

Polyclonal antibodies mixture of different types of antibodies produced when a number of B-lymphocyte clones respond to different epitopes on the same antigen or a mixture of antigens.

Polymer membrane electrode a type of ion-selective electrode that incorporates an ion exchange material in an inert matrix such as polyvinylchloride, polyethylene, or silicone rubber to generate a potential.

Polymorphisms occurrence of two or more distinct forms of the same kind of structure, be it organism or molecule.

Polynucleotide polymers of nucleotides, therefore alternative name for nucleic acid.

Potential gradient the electric field found by dividing the potential difference between two electrodes by the distance separating them.

Potentiometry electrochemical techniques that employ galvanic cells and measure the spontaneous potential of the electrode with minimal current flow.

Precision a measure of the closeness of a series of measurements of the same substance.

Preventive action action to eliminate cause of a potential non-conformity or other potentially undesirable situation.

Primary fluorescence substances or tissue elements that naturally exhibit fluorescence under ultraviolet light (also called autofluorescence).

Primary immune response response of the immune system on first encountering an antigen; usually this results in the production of an IgM antibody.

Primary sampling the analysis of a sample using aliquots from the initial sampling tube.

Probe labelled single-stranded polynucleotide that is complementary to, and used to detect and characterize, specific nucleotide sequences in RNA and DNA samples by forming DNA–DNA, RNA–RNA homoduplexes, or RNA–DNA heteroduplexes.

Professional body organization of like-minded individuals who have specialized knowledge which sets standards of practice and examinations on that knowledge.

Prohibition notice legal order to cease activity issued by the Health and Safety Executive.

Project a piece of individual work involving a literature review, research, and the written and oral presentation of the findings. It is a principal component of an honours degree programme.

Proteomics the study of the complete complement of proteins encoded by a genome.

Prozone phenomenon occurs when high concentrations of antigen or antibody produce a false-positive result due to saturation of binding sites.

Pyogenic bacteria bacteria that in the process of infection form pus, which is composed of dead neutrophils.

Quality Assurance Agency for Higher Education body in the UK that helps ensure HE qualifications are of a suitable standard.

Quality control the techniques and activities used to reach the desired level of quality.

Quality control charts graphical representations that clearly show how laboratory analytes or instruments are performing.

Quality control procedures internal laboratory checks carried out to ensure that a scientific method or analytical instrument is working properly and producing results that are reliable and valid.

Quality management system establishment of relevant quality objectives and the policy necessary to achieve those objectives.

Quality objective anything sought or aimed for related to quality.

Quenching refers to a process which reduces the fluorescence being given off by a substance.

Radiation is energy that travels in the form of waves or very small particles.

Radius of rotation distance of a particle from the centre of rotation.

Rate-zonal density gradient centrifugation technique to separate particles based mainly on differences in their sedimentation coefficients during sedimentation through a column of liquid whose density increases from the top of the centrifuge tube towards its bottom.

Readability the number of decimal places to which a balance weighs accurately.

Reagent books a means to record the reagent changes and lot numbers that are associated with a clinical test or instrument.

Real image a representation of an object (source) in which the perceived location is actually a point of convergence of the rays of light that make up the image. If a screen is placed in the plane of a real image the image will generally become visible on the screen.

Reanneal re-association of two complementary DNA strands to form the original double helical structure.

Recombinant DNA one whose molecules are comprised of strands from two different sources.

Reduction a reaction in which there is a gain of electrons.

Reference electrode within potentiometric devices, an electrode that generates a stable electrochemical potential and thereby enables measurement of the potential difference with the working electrode.

Reference range a range of values found within the 95th percentile of the population for a given substance. Sometimes incorrectly referred to as a 'normal' range.

Reflective learning assessing an event, its strengths and weaknesses, what has been learned from it, and how this could be put to future use.

Refractive index a numerical measure of the ability of a substance to deviate light rays.

Relative centrifugal field one expressed as multiples of the earth's gravitational field (*g*), which is an acceleration of 981 cm s^{-2}.

Remedial action action taken to mitigate the immediate effects of a non-conformity.

Reports issued by a laboratory are the permanent record of the results and conclusions for a single or series of laboratory investigations.

Requests the investigations that a laboratory is asked to carry out on patient samples.

Resolution ability to separate (or resolve) completely one analyte or one point from another.

Resonance absorption of radiofrequency radiation by nuclei with low-energy spins that changes their orientation to that of a higher-energy spin state.

Restriction endonuclease endonuclease that catalyses the hydrolysis of phosphodiester bonds at identical positions within a palindromic site of DNA.

R_f (relative to front value) distance travelled by any component relative to that moved by the solvent during chromatographic analyses.

Rhesus blood group blood group system defined by the presence of the D antigen on erythrocytes.

Risk likelihood, high or low, that somebody or something will be harmed by a hazard.

Root cause analysis a systematic, quantified, and documented approach to the identification, understanding, and resolution of underlying causes of non-conformities.

Rotor specialized container that holds liquid samples in centrifuge tubes and which is rotated during centrifugation.

Safety prevention of physical injury.

Salt bridge a solution of ions, often saturated potassium chloride, which enables electrical continuity between two half cells.

Scope of practice range of tasks that a registrant has been trained to undertake and been assessed as competent to perform.

Secondary fluorescence the binding of fluorochromes to substances or tissue elements in order to make them visible under ultraviolet light.

Secondary immune response response of memory plasma cells to produce rapidly specific antibodies at a higher intensity against an antigen the immune system has previously encountered.

Sedimentation coefficient (of a particle) ratio of its velocity to the applied centrifugal field ($v/\omega^2 r$) when sedimenting during centrifugation.

Selectivity ability of a chromatography column to distinguish and separate two similar analytes.

Sensing electrode the electrode that generates a variable potential at its surface with the analyte, dependent on analyte concentration. Also called a working electrode.

Serious Hazards of Transfusion the UK's independent, professionally led haemovigilance scheme. Where risks and problems are identified, SHOT produces recommendations to improve patient safety. The recommendations are put into its annual report, which is then circulated to all the relevant organizations, including the four UK Blood Services, the Departments of Health in England, Wales, Scotland, and Northern Ireland, and all the relevant professional bodies, as well as circulating it to all of the reporting hospitals.

Serum the fluid derived following centrifugation of a blood sample that has been allowed to clot.

Sodium dodecyl sulfate polyacrylamide gel electrophoresis perhaps the most widely used form of polyacrylamide gel electrophoresis. It is particularly useful in assessing the purity of proteins, to determine their M_r, and was used extensively in sequencing DNA molecules.

Solid-state electrode a type of ion-selective electrode that incorporates one or more salt crystals to generate a potential.

Solute the substance dissolved in a **solvent** to form a **solution**.

Solution a liquid consisting of a **solute** dissolved in a **solvent**.

Solvent the liquid used to dissolve a **solute** when making a **solution**.

Spectroscopy the study of the interactions between electromagnetic radiation and matter.

Staining the detection of separated components following electrophoresis by converting them to visible, coloured complexes. Compare **destaining**.

Standard curve method of plotting the measurements from a known series of concentrations to determine that of an unknown sample.

Standard operating procedure clear set of written instructions to ensure that all staff act in a consistent manner when carrying out all aspects of laboratory work.

Standards codes of best practice that improve safety and efficiency.

Stationary phase the non-mobile phase used in chromatography; can be a solid or an immobilized liquid.

Supernatant liquid found above the pellet that contains unsedimented material following centrifugation.

Svedberg units unit of sedimentation coefficient (*s*) equal to 1×10^{-13} seconds.

System set of interdependent elements that interact to achieve a common aim. These elements may be human, equipment, and technologies.

T lymphocyte type of lymphocyte that develops in the thymus and has a variety of functions in the immune response.

Tailing production of an asymmetric peak during column chromatography characterized by a normal rise before the peak but a slow fall after the peak.

Tare the operation used to set the readout of a balance to 0 g (zero).

Teamwork a means by which a group of individuals all work together to provide a common outcome.

Terms of reference rules governing how a committee is appointed, how it acts, and how this is documented.

Thermocycler automated instrument for performing polymerase chain reaction.

Titre a measure of antibody concentration based on serial dilutions to a visible end point, e.g. colour intensity, agglutination, or fluorescence.

Toxic substance capable of causing injury or damage to an organism.

Traceability value or measurement of a standard in relation to stated references, which allows comparisons through an unbroken chain.

Tracking dye small coloured substances that pass through the pores in an electrophoresis gel and so act as a marker for the electrophoretic front. Bromophenol blue is often used.

Transcription synthesis of a complementary RNA molecule using a single strand of DNA as template.

Transcription transfer of results from a print-out or direct from a device into the patient's notes.

Transducer within a biosensor, the electrochemical device that converts a biological or biochemical signal or response into a quantifiable electrical signal.

Transformation process of a cell taking up recombinant DNA.

Transition change between any two energy levels that occurs when an atom or molecule absorbs an amount of energy exactly equal to that between the two relevant states.

Transport emergence card written instructions carried by drivers when transporting dangerous goods that give details of the substance, its category name, United Nations number and class, what to do and who to contact in an emergency, and first aid information in case of spillage.

Troubleshooting sheets a means to record problems experienced with clinical tests or instruments and to record how particular problems are solved.

Turbidity degree of opacity or cloudiness of a fluid caused by suspended particles.

Turnaround time that taken for a test result to be available to a clinician after the sample has been obtained from the patient.

Ultracentrifuge one capable of producing an RCF in excess of 100,000 *g*.

Vacutainer container used for blood sampling via venepuncture in which a vacuum has been created prior to sealing. Thus, blood enters the tube from the vein by gentle suction.

Vapour phase between the liquid and gaseous states of matter.

Vectors carrier of the DNA into a recipient cell for cloning. Vectors are usually plasmids or bacteriophage DNA.

Vernier scale a device used on a microscope to accurately identify the position of specific points within a specimen.

VIR triangle Ohm's law relates voltage (V) to current (I) and resistance (R). The equation that describes Ohm's law may be written in three distinct ways and this is best remembered using the VIR triangle, so that when any two of the values are known, the third unknown may be calculated using this triangle.

Virtual image is an image in which the outgoing rays from a point on the object always intersect at a point. A simple example is a flat mirror where the image of oneself is perceived at twice the distance from oneself to the mirror. That is, if one is half a metre in front of the mirror, one's image will appear to be at a distance of 1 metre away (or half a metre inside or behind the mirror).

Voltammetry electrochemical techniques that employ electrolytic cells and measure the current passing through an electrode when an external potential is applied.

Wavelength the distance (measured in the direction of propagation) between two points in the same phase of any two consecutive cycles of a wave.

Weight the force (measured, e.g., in Newtons, grams, or pounds) reflecting the effect of gravity at a particular location upon a **mass**.

Welfare condition of well being, happiness, and comfort.

Wired enzyme electrode an electrode that incorporates an immobilized enzyme within a network of fixed electroactive centres that are able to shuttle electrons from the enzyme to the electrode surface.

Working electrode the electrode that generates a variable potential at its surface with the analyte, dependent on analyte concentration. Also called a sensing electrode.

Workplace exposure limit extent to which a person may be safely exposed to a hazardous substance without endangering health.

Workplace inspection audit of a workplace in terms of its health and safety.

Index

J

K

L

M

N

T

U